ALLEN COUNTY PUBLIC LIBRARY

ACPL ITEM
DISCARDED

Y 574

BIOLOGY

DO NOT REMOVE
CARDS FROM POCKET

ALLEN COUNTY PUBLIC LIBRARY

FORT WAYNE, INDIANA 46802

You may return this book to any agency, branch,
or bookmobile of the Allen County Public Library

DEMCO

BIOLOGY

A Personalized Approach

BIOLOGY

A Personalized Approach

James L. Kelly
Alan R. Orr
Barbara W. Saigo
Roy H. Saigo

KENDALL/HUNT PUBLISHING COMPANY
2460 Kerper Boulevard P.O. Box 539 Dubuque, Iowa 52004-0539

Allen County Public Library
Ft. Wayne, Indiana

Cover photo: Ivy covered wall, Normandie. Copyright © Francisco Hidalgo.

Copyright © 1989 by Kendall/Hunt Publishing Company

Library of Congress Catalog Card Number: 88–83135

ISBN 0–8403–4332–9

All rights reserved. No part of this publication may be reproduced, stored in a retrieval system, or transmitted, in any form or by any means, electronic, mechanical, photocopying, recording, or otherwise, without the prior written permission of the copyright owner.

Printed in the United States of America
10 9 8 7 6 5 4 3 2 1

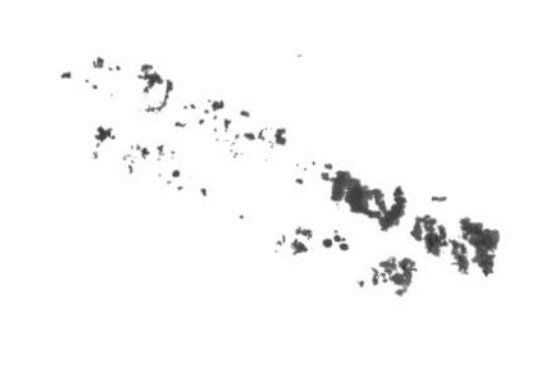

Contents

Preface

There are many biology textbooks. Why write another one? Indeed, what makes this one different from the rest? Both are questions that should be asked, and both are questions that are highlighted in this preface.

Most textbooks are written as a reading book with questions placed at the end of the chapter. Most textbooks, while associated with a laboratory manual, have little connection to hands on experiences, especially if the teacher elects not to use the laboratory manual. *Biology: A Personalized Approach* was written to offer an alternative to the model offered by most textbooks.

Biology: A Personalized Approach was written to tie up the loose ends often found in other books. This program of study follows a different teaching model, one that tries to capture the best of several teaching models. In so doing, the learner is at the focal point of learning and the benefactor of our labors.

Biology: A Personalized Approach uses structured-pacing as its associated teaching model. This model allows teachers to structure whole class learning, by using the beginning days of a chapter in large group discussion, demonstration, or whatever means he or she finds effective with his or her students.

Once the basics of the chapter, motivational schemes or advanced organizers have been placed before the students, the next step is to allow them to pace through the chapter. Pacing allows each student to work through the chapter at a pace conducive to their abilities. The total time allotted for pacing, however, is controlled by the teacher. The structure is still in the hands of the teacher. This simply means that if nine days have been assigned to chapter two for completion during the pacing mode, the students must complete the chapter in this amount of time.

For some students, the nine day time frame may be just right. For other students, this amount of time may mean a few extra hours after school are needed to complete the assigned chapter. Still for others, the chapter can be completed in seven days. There are many individual differences and they should be accommodated as best as can be done under the circumstances.

We've found this model matches up very well with nearly all students. The time frame meets their needs to be successful. The students that must put in a few extra hours, do it. They come into our lab during the noon hour, as well as after school. They take the responsibility to do this. The more capable students use the extra time to study the material for mastery or they will work on extra credit laboratory activities that are provided from the long list of other worksheets that have been accumulated through years of teaching. In the end, all students seem to have a great deal of success matching up to the given deadline. Once the chapter has been completed, the class comes back to a large group mode to wrap up the concepts learned, to show closure to the chapter. This part is, once again, led by the teacher. In the end, the directed structure by the teacher at both ends of a chapter, the pacing of each student through the chapter and the subsequent interaction of the teacher with the individual students during the pacing mode, leads to successful learning.

Successful learning comes from active involvement. A program like *Biology: A Personalized Approach* promotes active learning. Students are always doing something. *Biology: A Personalized Approach* has questions built into the text where they need to be asked. It also has the laboratory activities placed where they are needed. Such placement enhances a concept or supports the development of a concept. None of these are simply at the end of the chapter or in an ancillary manual. The development is sequential, with each part important to the whole. Each chapter lays the groundwork for future chapters.

We believe the student is the reason for developing a new and different type of biology book. The format is different from other biology books on the market. The intent of how to use the book is different from other biology books. In order to really feel the differences, you have to use this book and try structured-pacing to experience, first hand, the thrill of teaching by this method. *Biology: A Personalized Approach* lets all of this happen.

Your success as a teacher is important to us. The success of your students is equally important. We wish you much success as you begin to discover the exhilaration of teaching and watching that same excitement happen in your students as they learn biology.

To the Student

You are about to embark on a new and different biology program of study. The program you are going to use this year is called *Biology: A Personalized Approach.* It is different in that it uses a teaching method called structured-pacing. Structured-pacing allows your teacher to set the tempo for your study by offering large group instruction (that's the structured part). At the same time, this program, along with your teacher, is able to provide a flexible program that matches your individual differences (that's the pacing portion).

Biology: A Personalized Approach was designed to meet the needs of high school students like you. While using this biology program, you will assume a lot of the responsibility for your success. This is due, in part, to the design of the program. This program walks you through the content, exposing you to concepts enhanced with activities and laboratory experiences. Your role is to respond to pertinent questions, complete laboratory experiences, and compile and interpret appropriate data. The combination of these parts lends itself to learning. A lot of the success you have, depends on how well you apply yourself to the individual parts.

Look through this book. Notice the content. Pay attention to the illustrations. Like the content, they are an integral part of your studies. The illustrations and photographs enhance the written content. Don't overlook or fail to take their value into your studies.

As you read and work through this text, you will be writing your answers in your log book. The log book is a place where you write your answers to questions and/or supply data. In some cases, your answers will be one or two word answers. In other situations, however, you will need to give considerable explanation. The use of complete sentences and detailed responses will enable you to effectively express yourself. The construction of complete answers will help you become a successful learner. If you do a thorough job of writing, you should be better able to comprehend the content. And, comprehension means success.

Laboratory experiences require careful and accurate observations, as well as careful measurements. Careful and thorough collection of data reflects a commitment for excellence.

Excellence doesn't always come easy, but the pacing part of the structured-pacing design gives you the time and opportunity to meet the challenge of reaching excellence in your work. This biology program allows you to apply the old adage, "You get from your work what you put into it." In this case, your success is measured by your involvement, so don't short change yourself. You will find this biology program offers an effective and fun way to learn.

Good luck with your studies. Between you, your teacher, and *Biology: A Personalized Approach,* you have a winning combination.

James L. Kelly
Alan R. Orr
Barbara W. Saigo
Roy H. Saigo

How to Use the Logbook for *Biology: A Personalized Approach*

As you work through *Biology: A Personalized Approach,* you will need to write answers and data, draw illustrations, construct graphs, and just keep some basic notes. The place to do all of this is in your log book.

As you progress through each chapter of this biology program, you will note that each section is numbered. The number for each section has a corresponding number, with an appropriate space for your answer, in your log book. For example, in the first chapter, you need to write a short answer on the first page. The section in the book is numbered 1.1. The corresponding 1.1 in your log book is where you need to respond to the question. Look for 1.1 in your log book. Do you see where you will write your answer? Good!

Note that you are asked to write a rather extensive answer in 1.1 following the prior short line answer. Lines have been provided for an extensive answer for you in your log book. Do you see these lines? Can you distinguish where you begin one answer from another? If you cannot, ask your teacher for help right now. It is important that you are able to follow the text and log book with ease.

On occasion, you will need to write a report or something that is quite a bit longer than a typical answer. In such cases, your text book will suggest that you use a separate sheet of paper. This separate sheet will need to be attached to your log book entries for evaluation.

Your log book is sequentially designed to assist you as you work through this biology program. We believe you will have no difficulty using your log book.

Keep notes in your log book. Write observations. Make your graphs here too. But keep one thing in mind as you write your answers, use complete sentences where necessary and whenever possible. The more complete your answers, the better service your log book will give you as you study biology. Your log book could become a very good friend to you.

When you need to hand in a chapter for evaluation, simply tear out the necessary sheets, attach any extra sheets that you might have used to complete the chapter, staple the pages together by placing a single staple in the upper left-hand corner, and hand the pages in for evaluation. After your evaluated work is returned to you, keep it in a 3-ring binder for future use.

Good luck with your studies. We hope you find your log book to be a neat part of your biology studies. We wish you much success with your biology studies.

James L. Kelly
Alan R. Orr
Barbara W. Saigo
Roy H. Saigo

BIOLOGY

A Personalized Approach

Introduction: The Power of Observation

1

Why is scientific observation important and what kinds of observations can you make?

1.1

At what point our early Neanderthalic or Cro-Magnon ancestors began to understand the significance of the interrelationships that exist in life is not clearly known. However, the drawings they left on cave walls suggest they were aware that plants and animals were necessary for their well-being.

Today we, too, are also aware of the importance of plants and animals. We all seem to be conscious of our natural surroundings. We recognize we are dependent on many things so that we can participate in the so-called good life. We certainly know about our need for shelter, food, and clothing, the same as our early ancestors. But today these essentials have changed so much that our early ancestors would probably not recognize the relationship that exists between all living things.

High technology, the advancement of medicine, the refinement of our eating habits, the changes in land management and conservation practices, to name a few, have entered into this interrelationship scheme. Consider the broad categories of crime, medicine, environmental pollution, substance abuse, conservation, engineering and high technology, food research, and population growth. Which category do you feel has the greatest impact on you at this time? ____________ Now consider your choice. Why have you chosen this category? Before you write your answer, keep in mind that one of the objectives of this biology program is to help you become a better writer. Writing effectively and correctly requires practice and time. When you write the reason for your choice above, use complete sentences, proper punctuation, and correct spelling.

A tough goal for a biology program? Yes. But, you will do a lot of writing this year, and by practicing, you should improve, even if you are a good writer now. Now back to the original question. Why have you chosen the category that you did? How does this category affect you?

Write your answer in your log book. Your log book will be your record as you progress through this program of study.

1.2

This course is about the science of life. By its very nature, biology is an exciting science course to study. The science of life has been studied, or at least observed, longer than any other science.

Bio means "life," and *ology* means "study of." These root words come from the Greek language. Does the science of biology have any relationship with the broad category you chose in section 1.1? (yes/no) ______ If you

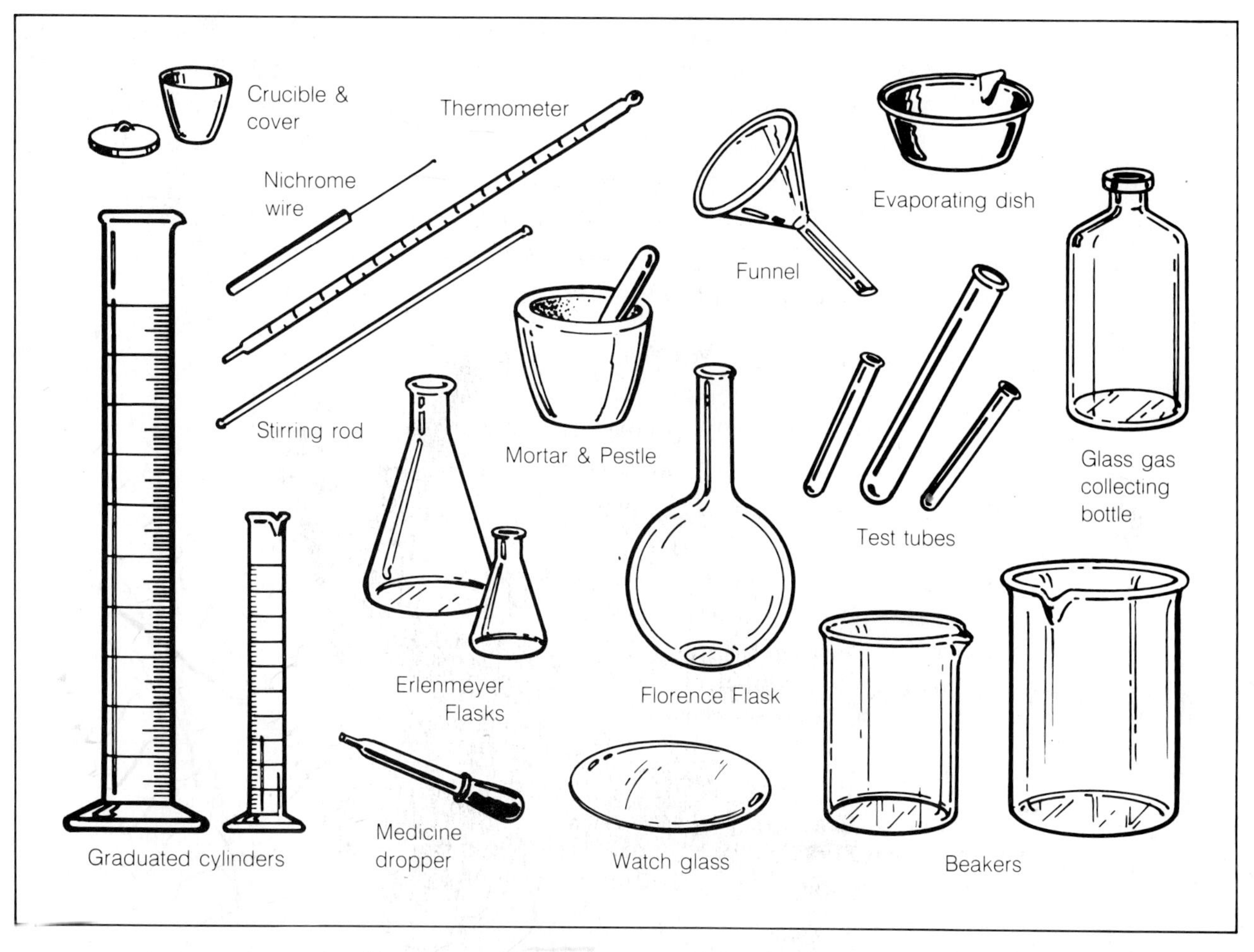

Figure 1.1 Common laboratory ware you will use with this course of study.

answered yes, write a brief paragraph explaining why you think the two have a connection. If you answered no, discuss the reasons for your answer. Three to five sentences should be sufficient for your answer. As before, write your answer and the corresponding explanation in your log book.

1.3 This biology program will be somewhat different than other science courses you may have taken. This course is written and published in such a way that you literally consume the log book. You will write your answers in the space provided or attach a separate sheet of paper. You will illustrate your observations in the spaces provided and fill in the data tables with your data. You will create graphs from data. You may need to attach additional reports to your chapter, when you submit your chapter for evaluation.

We alluded to your writing complete answers in section 1.1. But let us expand on this. By writing accurate, complete, and detailed answers, you will be able to refer back to your work for study and know what you intended to say.

To analyze this further, a hypothetical question is asked: "How high did the temperature rise when an acid and a base were put together?" You can answer this question the following two ways.

First

It went up 9°.
My question to you is, "What is meant by *it?*"
Also, you are lacking units of accurate measurement.

Second

The temperature went up 9° Celsius (9° C) when the acid and base were put together.

Which answer is better? (first/second) ______________

Which answer will make more sense to you in a couple of months?

(first/second) ______________

Use a noun in your sentences that tells you, or anyone else reading your work, what is actually happening. Strive for detail and leave shortcuts alone. Many students look and react too quickly without thinking about what they are going to write. In the end, they render shallow written answers, which often are not satisfactory.

In the space provided in your log book, write a couple of sentences that best describes your long-range goals for this course.

1.4

Observation is a skill that also needs to be developed and refined. From observation, you are able to describe what is happening around you. A major part of science is observing and describing. It is important to be able to make accurate observations in science. One reason for this chapter is to help you develop and refine your skills of observation.

To illustrate that not everyone sees the same thing, describe what you see in figure 1.2. This drawing was done for this chapter by a high school student in a modern art class. Use the space provided in your log book to write your description.

1.5 Your written description of figure 1.2 tells the reader how you perceived this sketch. You may have based your description on facts, or you may have tried to describe the sketch from the feelings you got from your observations. Is your description based on feelings, facts, or both? ________________ It is usually necessary to separate factual observations from interpretive (feelings) observations. Why?

1.6 Reexamine figure 1.2. Based on your observations of the diagram, write four one- or two-word observations that can be considered as factual and four one- or two-word observations that are interpretive.

Facts	*Feelings*
1. ________________	1. ________________
2. ________________	2. ________________
3. ________________	3. ________________
4. ________________	4. ________________

Are your factual statements similar to the statements of others in your class? (yes/no) ________ Are your interpretive statements similar to the statements of others in your class? (yes/no) ________ Explain why your factual and interpretive statements are or are not the same as other class members.

1.2 Use this illustration to answer questions in 1.4, 1.5 and 1.6.

1.7 Many things that are observed are not always observed by others, or at least not from the same perspective. You have identified observations as either factual or interpretive. Now let us become more scientific. Observations can also be classified as *qualitative* or *quantitative.* A qualitative observation is one that describes the object. It lacks the measurements that would be found in a quantitative observation. Quantitative observations require a measurement expressed in a unit, such as inches, feet, millimeters, grams, and so forth. For the next exercise, obtain a plant from your teacher and list fifteen qualitative and fifteen quantitative observations. All quantitative measurements should be done in metric.

Qualitative	*Quantitative*
1. ______________	1. ______________
2. ______________	2. ______________
3. ______________	3. ______________
4. ______________	4. ______________
5. ______________	5. ______________
6. ______________	6. ______________
7. ______________	7. ______________
8. ______________	8. ______________
9. ______________	9. ______________
10. ______________	10. ______________
11. ______________	11. ______________
12. ______________	12. ______________
13. ______________	13. ______________
14. ______________	14. ______________
15. ______________	15. ______________

1.8 After each observation below, identifiy it as being either qualitative or quantitative.

1. The color of the water is red. ______________
2. The temperature of the water is 78° C. ______________
3. The tree has green leaves. ______________
4. A glass container has 10 mL increments. ______________
5. A white solid forms when an egg is cooked.

   ____
6. It required 4 minutes to complete the observations of the chemical reaction. ______________

1.9 Quantitative measurements are important measurements. Accurate measurements, however, are often not done because an individual is impatient and rushes his/her observations. This results in human error. Patience and concentration reduces error and increases the accuracy of the data.

This next exercise will help you to collect and record data with another person. You will need three 600 mL beakers or three large containers of equal size and three Celsius thermometers. Note the increments on the thermometers.

1. What is the numerical value of one increment? ____________
2. Can you estimate to the nearest one-half of a degree? ______
3. If you do estimate to the nearest one-half of a degree, is this more accurate than estimating to the nearest whole degree? ______
4. Would you be more accurate if you could estimate to the nearest one-tenth of a degree? ______

1.10 You will need at least 25 to 30 minutes to complete this investigation. Add 200 mL of tap water to each beaker. Set the water temperature of each beaker at 10° C. Check this very closely with your thermometer. Add ice, if necessary. Once 10° C is reached, remove the ice. However, maintain the temperature of the water at 10° C in each container until it is used. Place your hand in the first container and leave it there for 5 minutes. Do not move your hand at any time during this investigation. Leave the thermometer in the beaker where you can easily read the degree markings and record the temperature every 30 seconds. Estimate the temperature to the nearest one-half of a degree. Record your data in data table 1.1 for container 1.

Data Table 1.1 Temperature Readings for Container 1

Time		Temperature °C
Seconds	Minutes	
0	0	
30	½	
60	1	
90	1½	
120	2	
150	2½	

Time		Temperature °C
Seconds	Minutes	
180	3	
210	3½	
240	4	
270	4½	
300	5	

1.11 Remove your hand from the first container and *immediately* immerse the same amount of the same hand into the second container, which also has a starting temperature of 10° C. As you did in section 1.10, take a temperature reading every 30 seconds for 5 minutes. Record this data in data table 1.2 for container 2. Estimate the temperature to the nearest one-half of a degree.

Data Table 1.2 Temperature Readings for Container 2

Time		Temperature °C
Seconds	Minutes	
0	0	
30	½	
60	1	
90	1½	
120	2	
150	2½	

Time		Temperature °C
Seconds	Minutes	
180	3	
210	3½	
240	4	
270	4½	
300	5	

1.12 To the third container, which also has a starting temperature of 10° C, record the temperature every 30 seconds for 5 minutes, as you did earlier. *Do not* insert your hand. Allow the water to warm up by room temperature only. Record your data to the nearest one-half of a degree in data table 1.3 below.

Data Table 1.3 Temperature Readings for Container 3

Time		Temperature °C
Seconds	Minutes	
0	0	
30	½	
60	1	
90	1½	
120	2	
150	2½	

Time		Temperature °C
Seconds	Minutes	
180	3	
210	3½	
240	4	
270	4½	
300	5	

1.13 The data from data tables 1.1, 1.2, and 1.3 must now be illustrated for further analysis. A graph gives us the best means for showing data. However, a graph must have certain things present if it is to be useful.

First, a graph should have a descriptive title. The title should tell the person looking at the graph what the graph is all about. The title should be placed with the graph.

Second, each axis should be properly and completely labeled. In addition to the proper labels, appropriate scales should be used. There is a rule to follow when labeling and placing a scale on each axis. Place the dependent variable on the Y axis (vertical) and the independent variable on the X axis (horizontal) of a graph. The scale will usually begin at the junction of the X and Y axis and read zero (0). From zero, the scale will proceed at equal intervals along each axis.

Third and last, place each data point on the grid in its appropriate location. The points are usually connected. A line that strikes an *average of all* the points may be drawn if desired. This line is called the *best fit line.*

To distinguish different sets of data on the same graph, use symbols for data points. Symbols that can be used are ○, ●, △, □, ■, and so forth. Symbols act as a key to the specific data and should be stated as such. Colored pencils may also be used to separate data. A key should be used to distinguish between the colors and their representative data.

A sample graph, figure 1.3, follows. It shows the proper way to use the symbols.

Referring to the graph, let us go through the various steps for making a good graph.

1. A graph should have a descriptive title. What is the title for the given graph? ______________________________

   ____

   Does this title tell you what the graph is about? __________
2. What is the dependent variable? ____________________

   What is the independent variable? ____________________
3. What increment is used for the Y axis? ______________

   What increment is used for the X axis? ______________
4. Is the key visible so the data can be recognized with its appropriate line? ______________________________

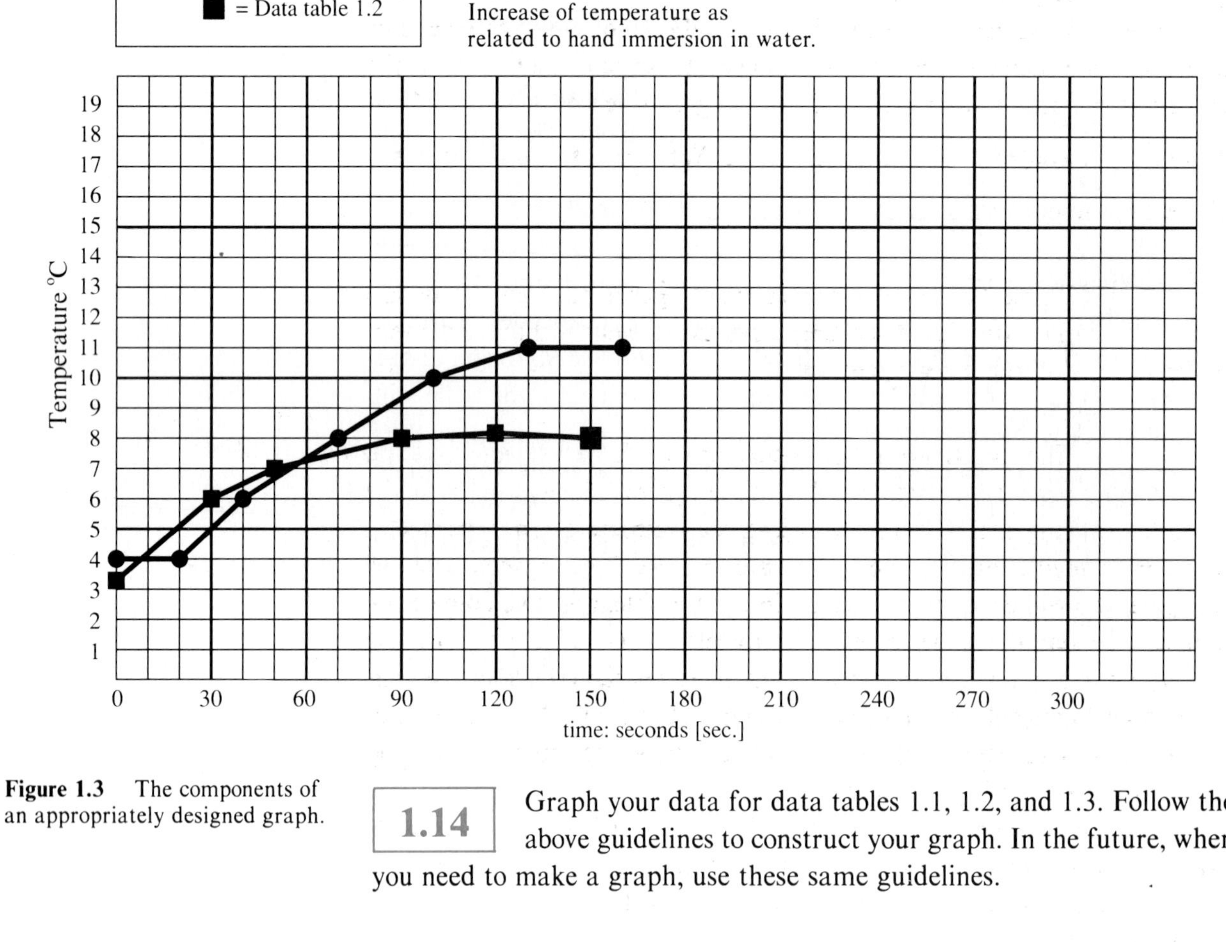

Figure 1.3 The components of an appropriately designed graph.

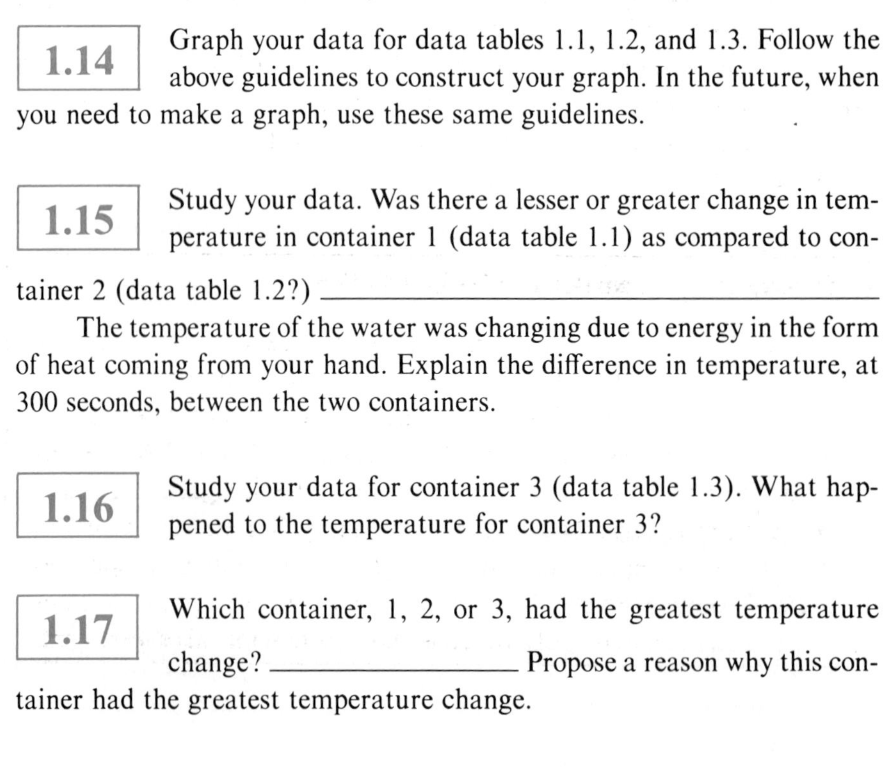

1.14 Graph your data for data tables 1.1, 1.2, and 1.3. Follow the above guidelines to construct your graph. In the future, when you need to make a graph, use these same guidelines.

1.15 Study your data. Was there a lesser or greater change in temperature in container 1 (data table 1.1) as compared to container 2 (data table 1.2?) ______________________________

The temperature of the water was changing due to energy in the form of heat coming from your hand. Explain the difference in temperature, at 300 seconds, between the two containers.

1.16 Study your data for container 3 (data table 1.3). What happened to the temperature for container 3?

1.17 Which container, 1, 2, or 3, had the greatest temperature change? ______________ Propose a reason why this container had the greatest temperature change.

1.18 Whenever an experiment is undertaken, a *control* is necessary. A control is a standard of comparison maintained without a change in the variables. This makes it possible to observe the results or to change one factor (variable) at a time and compare the results to the control to determine the magnitude of change. Which container (1, 2, or 3) acted as a control? ________________ What was the variable in the above experiment? ____________________

1.19 Using the idea of variables in an experiment, why did you use the same hand for the above experiment rather than use a different hand in the second container?

1.20 The hypothesis is an important tool for the scientist. While often described as an "educated guess," the hypothesis is the basis for making a prediction regarding the outcome of an experiment. Sometimes predictions will be inaccurate or false, but, regardless of the outcome, you will have greater insight into the problem than you did before you carried out the experiment. The hypothesis (prediction) gives you something to aim for, and the experiment gives you the vehicle to test your hypothesis. Being able to make good predictions takes time. You may reduce the amount of inaccuracy in your hypothesis by doing more reading and thinking before writing your prediction.

This next exercise requires you to make a serial dilution, that is, to start with a known concentration of solution and dilute it to a lesser concentration. The diagram and description below will help you complete the serial dilution.

Description for Completing a Serial Dilution

Follow figure 1.4 as you read through these instructions. In test tube 1, you have 10 mL of the original colored solution. From test tube 1, extract 1 mL of solution and put it into test tube 2. Add 9 mL of tap water to test tube 2 and stir thoroughly. Extract 1 mL of solution from test tube 2 and put it into test tube 3. Again, add 9 mL of tap water to test tube 3 and stir thoroughly. Repeat this procedure for the remaining four test tubes. If you use a medicine dropper to transfer solutions to the test tubes, you will find that 20 drops is approximately equal to 1 mL, which is close enough for this experiment.

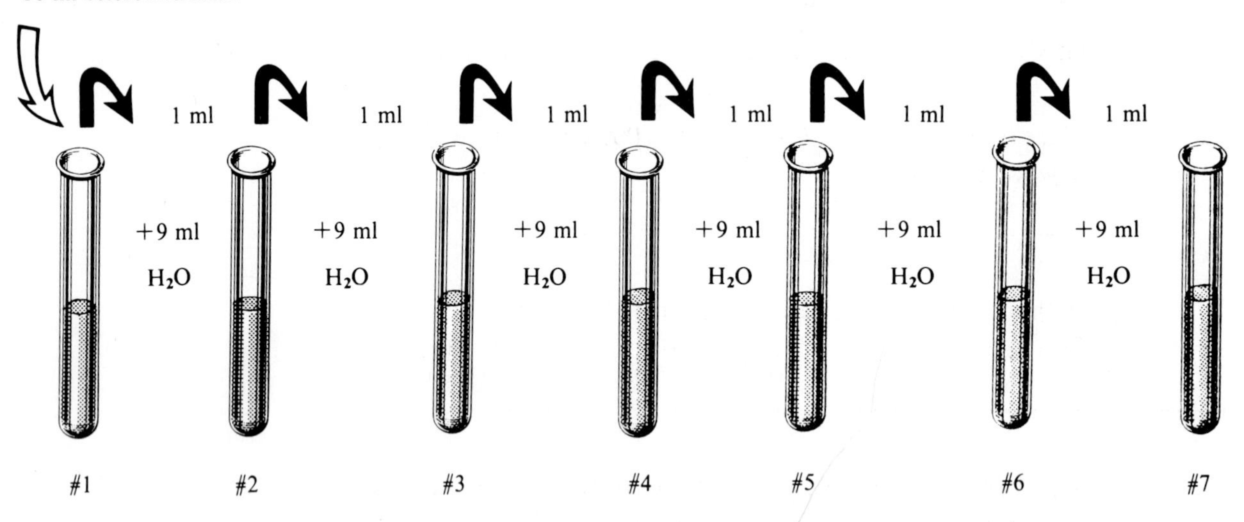

Figure 1.4 A visual procedure for completing a serial dilution.

Serial dilution shows the ratio to which concentrated solutions can be diluted. This is important today since a lot of the pollution levels are expressed in units that are related to serial dilutions, as you will see later.

Now that you are aware of the steps needed to complete a serial dilution, obtain seven 13 × 100 mm test tubes and complete a serial dilution using the stock colored solution provided. First, however, make a prediction of what is going to happen to the color intensity as you proceed with each dilution step.

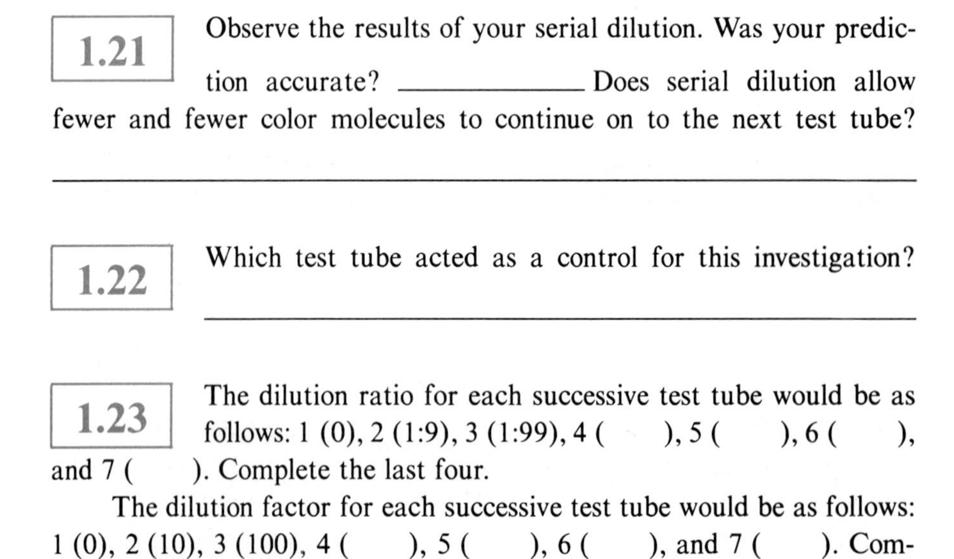

1.21 Observe the results of your serial dilution. Was your prediction accurate? ____________ Does serial dilution allow fewer and fewer color molecules to continue on to the next test tube?

____

1.22 Which test tube acted as a control for this investigation?

____

1.23 The dilution ratio for each successive test tube would be as follows: 1 (0), 2 (1:9), 3 (1:99), 4 (   ), 5 (   ), 6 (   ), and 7 (   ). Complete the last four.

The dilution factor for each successive test tube would be as follows: 1 (0), 2 (10), 3 (100), 4 (   ), 5 (   ), 6 (   ), and 7 (   ). Complete the last four.

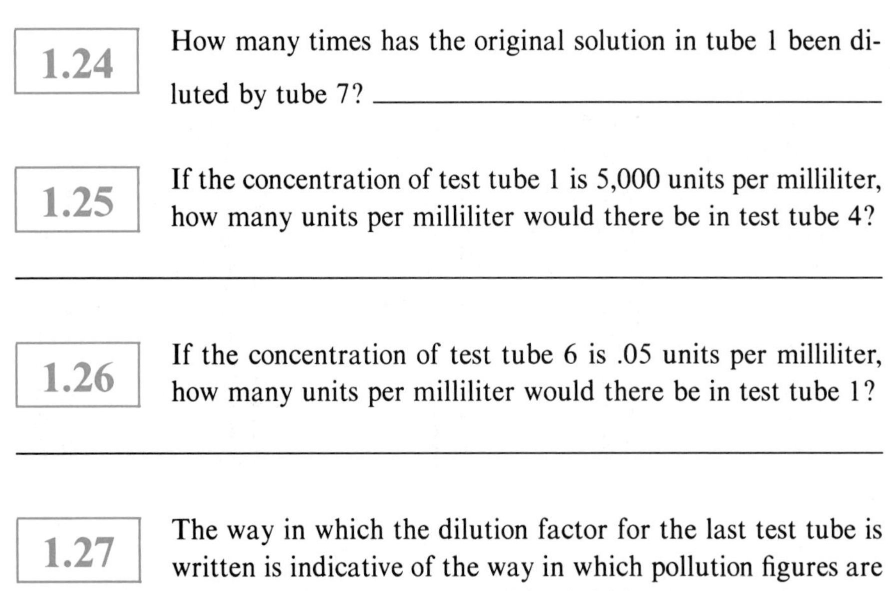

1.24 How many times has the original solution in tube 1 been diluted by tube 7? ____________________

1.25 If the concentration of test tube 1 is 5,000 units per milliliter, how many units per milliliter would there be in test tube 4?

____

1.26 If the concentration of test tube 6 is .05 units per milliliter, how many units per milliliter would there be in test tube 1?

____

1.27 The way in which the dilution factor for the last test tube is written is indicative of the way in which pollution figures are expressed. They are read in "parts per million." This means you have one part of solute from test tube 6 in 1 million parts of water, as found in test tube 7. You have heard or read about government figures pointing out that a particular food is unsafe for human consumption because it contains five parts per million of mercury or something just as toxic. This does not seem like much, since your last tube looks quite clear. But, would you be willing to drink or eat something that contained a toxic chemical? What if the initial control solution was raw sewage? Would you drink the water in test tube 7 then? Supposedly, the water you drink from the tap has been purified so that it is safe for your consumption.

Relating now back to section 1.20, which test tube (2–7) has the greatest dilution factor? ____________ Which test tube (2–7) has the least dilution factor? ____________ In terms of color, then, make a general statement comparing the color of the solution to the amount of dilution.

1.28 Observations are fun and informative. Within individual plants and animals there are various structures that operate as a part of the whole organism. You can observe these. Most students enjoy dissecting, and this next laboratory activity is to dissect the leopard frog. The purpose of the dissection is to develop a technique for dissection and to refine your observation skills. You will be making many observations of the external and internal organs associated with each of the body systems.

The eleven body systems of the frog (as well as human) anatomy are: (a) the integumentary system, (b) the muscular system, (c) the skeletal system, (d) the respiratory system, (e) the circulatory system, (f) the digestive system, (g) the endocrine system, (h) the urinary system, (i) the nervous system, (j) the reproductive system, and (k) the lymphatic system.

The instructions for a good, complete dissection must be followed. The accompanying guide sheets will assist you in recognizing frog anatomy. Use them as guidelines while you are dissecting. Your teacher may want you to use additional guide sheets as well.

Your frog may be a preserved one or a fresh one. In either case, check with your teacher for any instuctions before continuing with the dissection.

CAUTION: dissecting equipment should be handled with care due to sharp edges and points.

1.29 You will need a dissecting tray and dissecting materials. These can be found on the supply table. Obtain a frog from your teacher.

Observe the external anatomy of your frog. Note particularly the position of the eyes. Also, note the **tympanic membrane** below and back from the eyes. The tympanic membrane is the ear for the frog. The position of the eyes enables the frog to remain mostly submerged in water while looking because its eyes are above the surface of the water.

Observe and examine the webbing of the hind feet carefully. You should be able to see many small blood vessels. Because these vessels are only a single cell's distance from the surface of the skin, oxygen is able to pass from the water environment to the blood vessels. As a result of this phenomenon, frogs can carry on respiration while submerged for great lengths of time.

Also, examine the front feet. The sex of the frog can be determined by the size of the thumb pad. A large thumb indicates the frog is a male (♂), and the small thumb, a female (♀). Are you able to determine the sex of your frog? ______________________ You will know for certain when you check to see if your frog has testes or ovaries.

1.30 Once you have finished studying the external anatomy, study the mouth. Figure 1.6 illustrates the anatomy of the mouth. Locate each anatomical part illustrated here on your frog.

- The internal **nares** are air passages that connect with the external nares.
- The maxillary, or upper jaw, teeth are used for gripping.
- Vomerine teeth are used to prevent the prey from escaping as they point inward toward the esophagus.
- The **esophagus** is the entry point to the digestive system.

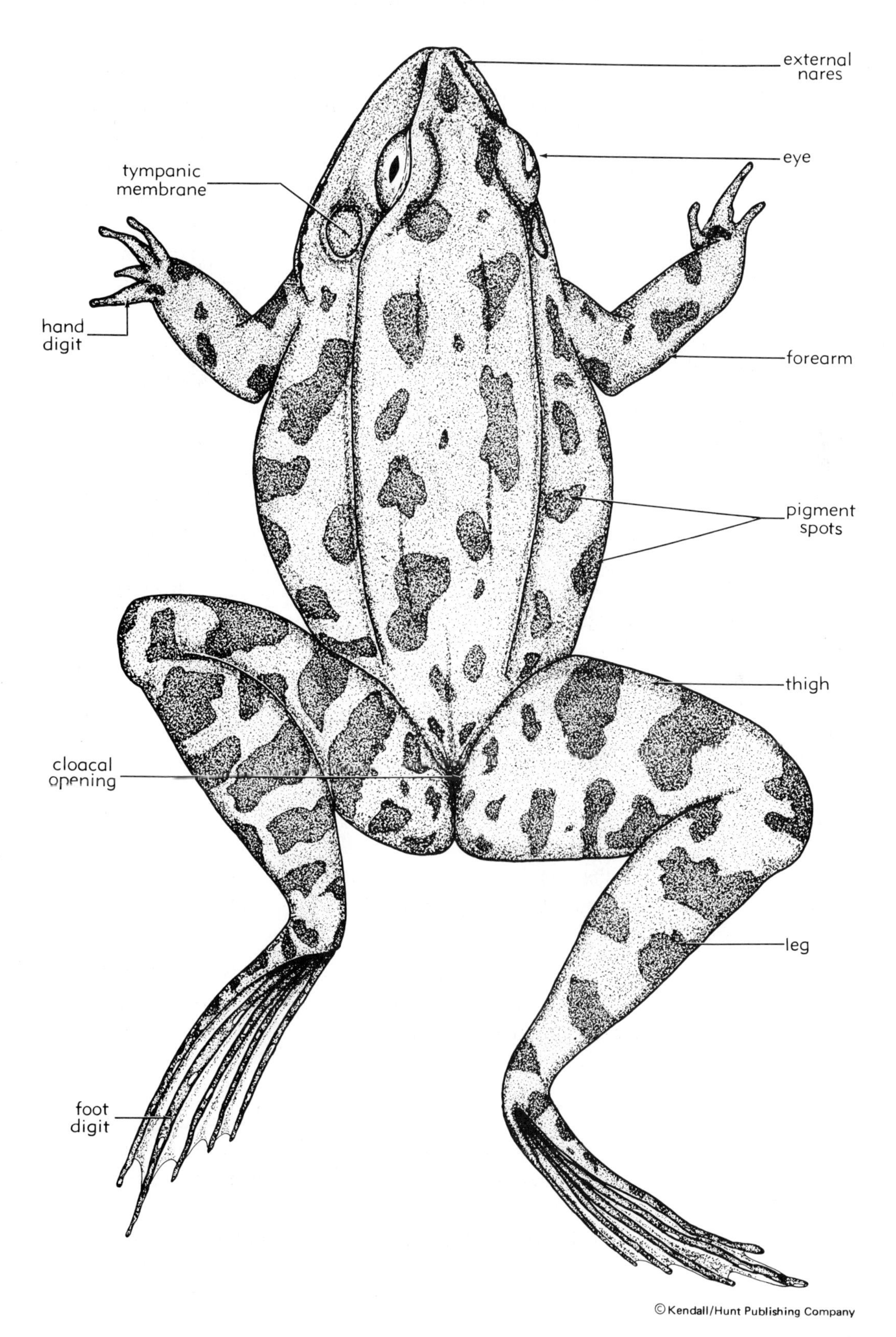

Figure 1.5 External anatomy of the frog.

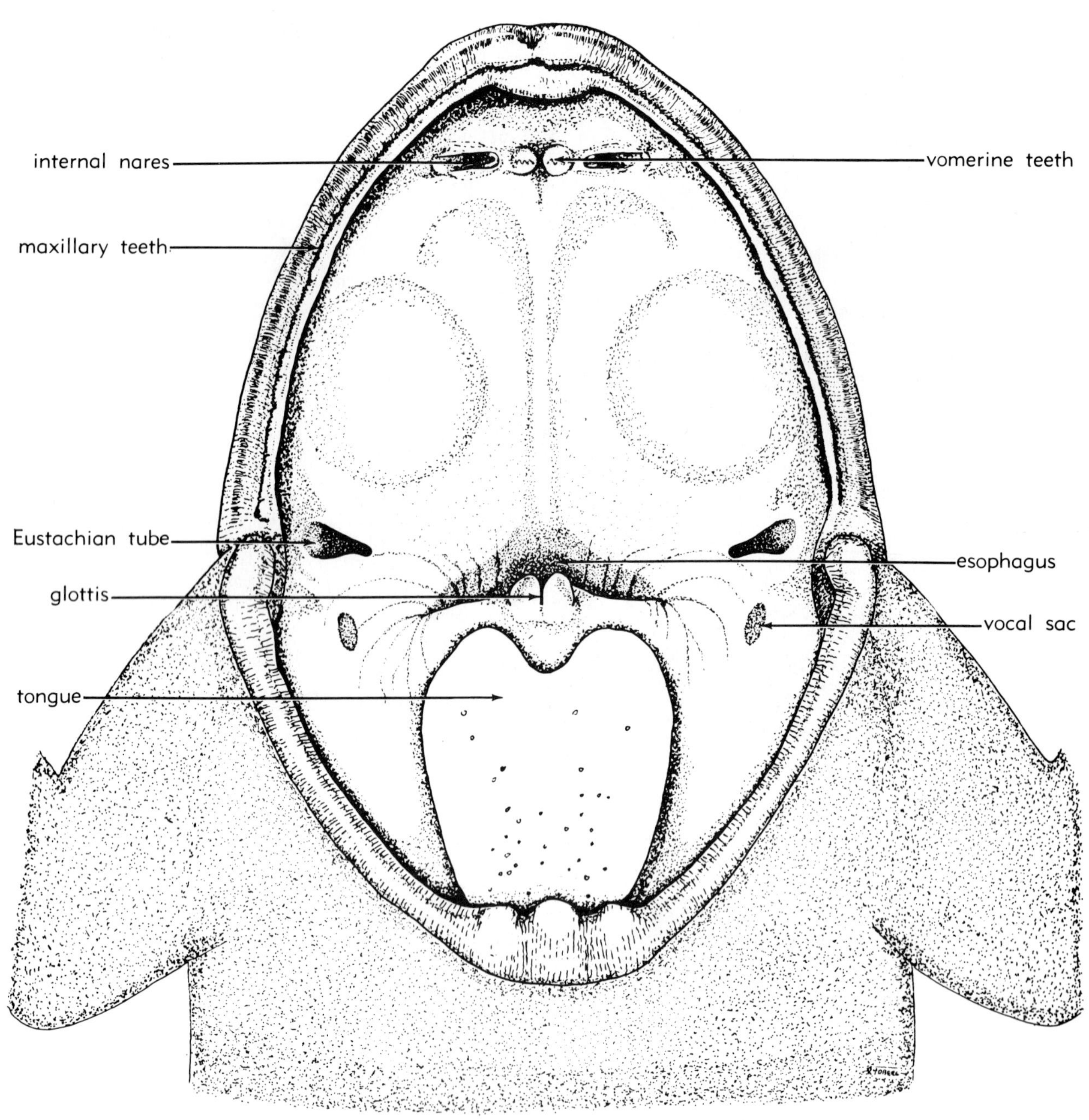

© Kendall/Hunt Publishing Company

Figure 1.6 Anatomy of the mouth.

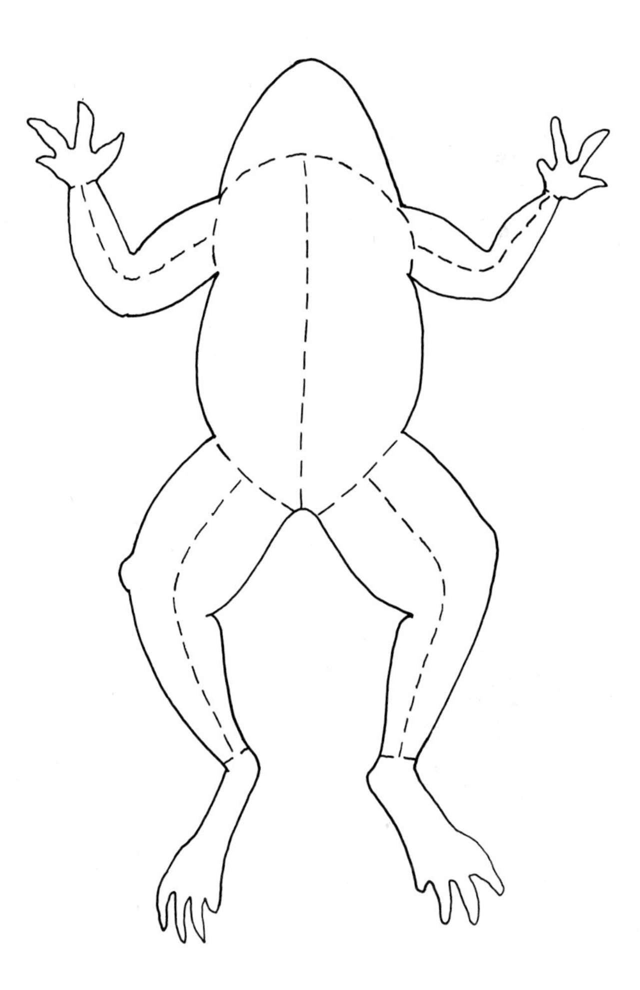

Figure 1.7 Cutting lines to remove skin from the frog.

- The **glottis** guards the passage of air from the mouth to and from the lungs. As food is taken in and swallowed, the glottis is closed. During normal breathing, the glottis is open.
- The tongue is attached to the front of the mouth and is used to catch insects, flipping caught ones back into the mouth.
- The openings to the vocal sac allow the male frog to produce its sound.
- The two **eustachian tubes** establish an equilibrium between the pressure within the ear and external air pressure.

To begin the dissection process, complete the following steps:

a. Pin the frog down on a dissecting tray.
b. Cut the frog along the dashed lines as indicated in figure 1.7.
c. Pull back the skin to expose the muscle structure. Study it according to figures 1.8 and 1.9. Be able to identify the muscles indicated in these two figures.

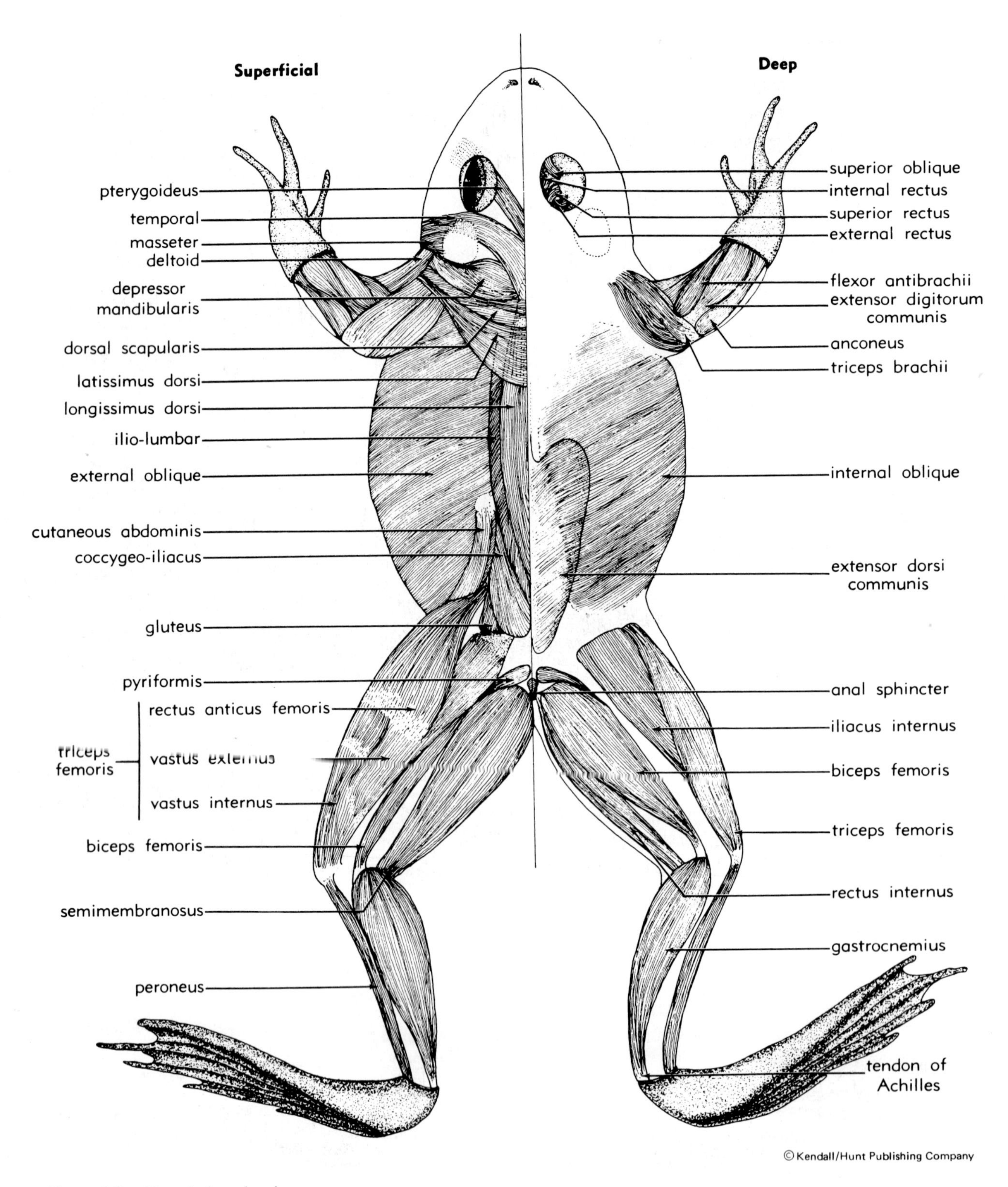

© Kendall/Hunt Publishing Company

Figure 1.8 Dorsal view showing muscles of the frog.

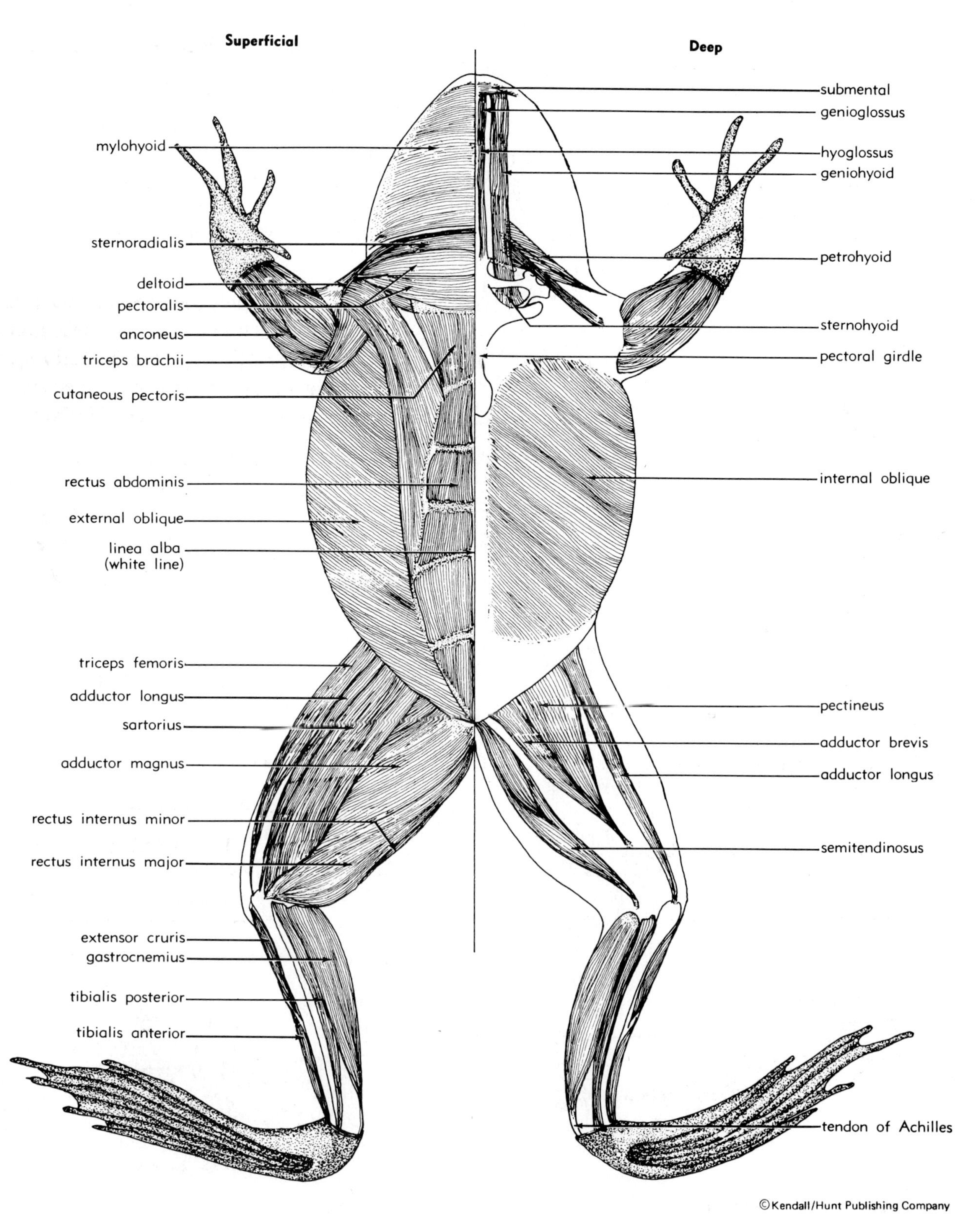

©Kendall/Hunt Publishing Company

Figure 1.9 Ventral view showing muscles of the frog.

d. Carefully cut through the muscle tissue of the abdomen and the ribs to expose the internal viscera, heart, and lungs. Be careful. Find each of the organs illustrated in Figures 1.10, 1.11, 1.12, and 1.13.

The organs illustrated in figures 1.10–1.13 are listed below along with their respective body systems.

Organ	**System**
heart	circulatory
lungs	respiratory
liver	digestive
gall bladder	digestive
spleen	lymphatic
stomach	digestive
small intestine	digestive
large intestine	digestive
oviducts	reproductive
anus	digestive
pancreas	endocrine
urinary bladder	urinary
postcava	circulatory
cloaca	digestive
kidney	urinary
fat body	digestive
ovary	reproductive
testis	reproductive
Wolffian duct	reproductive

Remove each organ separately. Carefully examine each organ; then dissect it. Pay particular attention to its texture and physical structure. Compare each organ with the others as you remove and examine them.

Food is passed from the mouth to the stomach through the esophagus. There the food is fragmented and passed on to the **small intestines** where digestion is completed. Bile, which is stored in the **gallbladder,** is secreted by the liver. Bile helps to break down fats as well as neutralize stomach acid.

The tan-colored **pancreas** secretes digestive enzymes. While some of the nutrients taken in are stored in the liver, many are converted to fat and stored in the yellowish fingers called **fat bodies.**

Unlike the human heart, the frog heart has only three chambers, two thin-walled **auricles** and one thick-walled **ventricle.** Oxygen-deficient blood enters the right auricle, while oxygenated blood from the *lungs* enters the left auricle. The ventricle does the pumping.

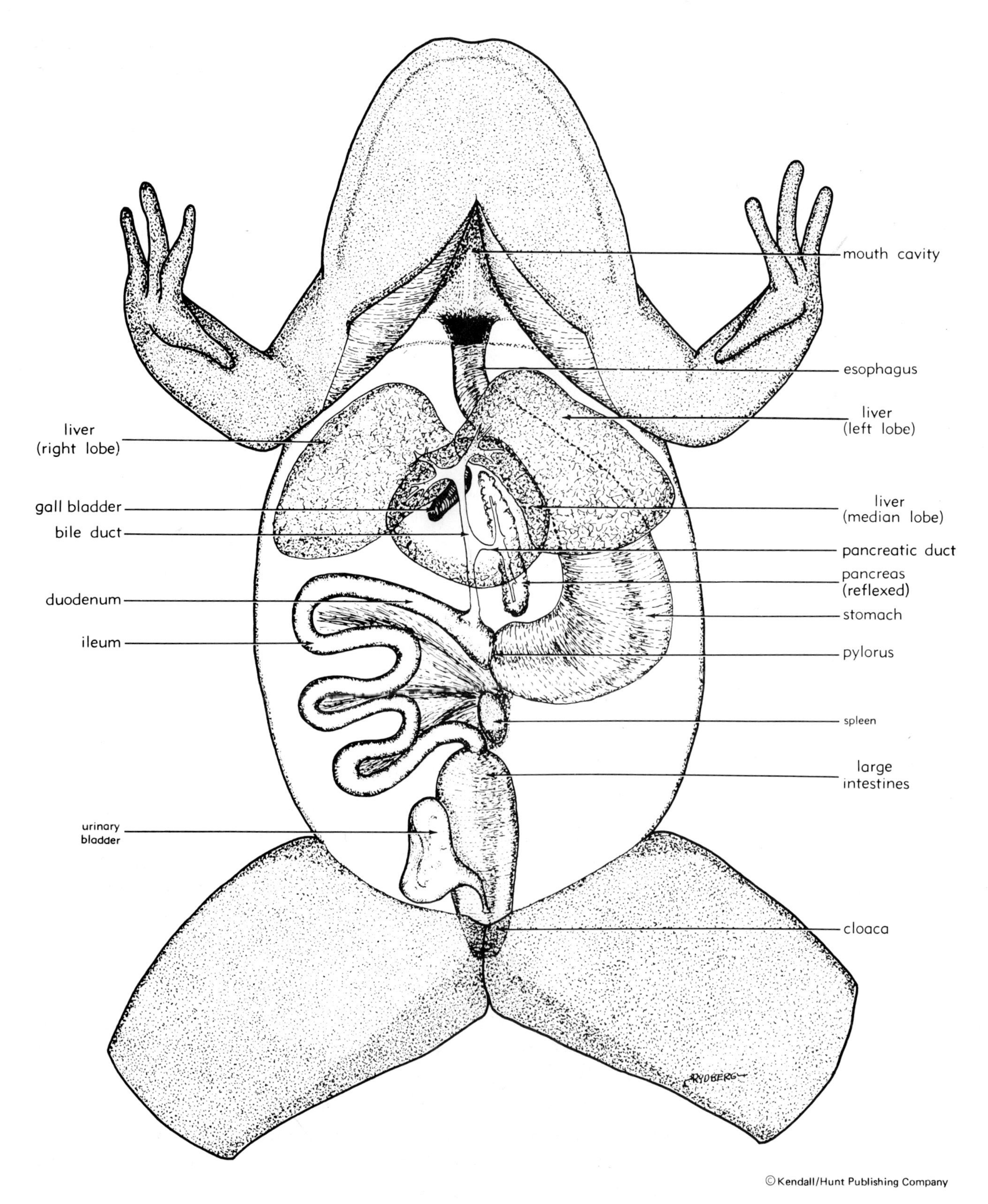

©Kendall/Hunt Publishing Company

Figure 1.10 Digestive system of the frog.

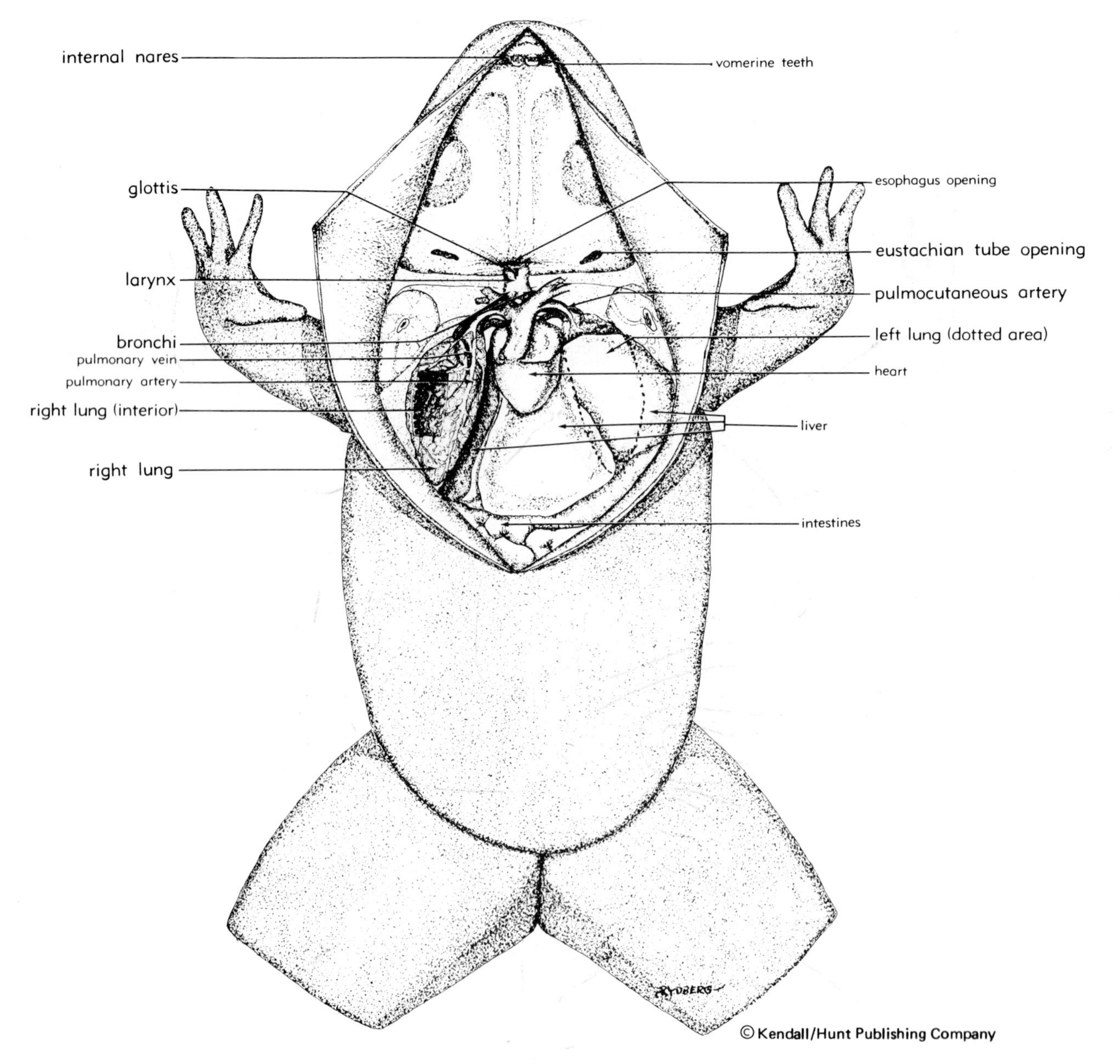

Figure 1.11 Respiratory system of the frog.

Blood cells are formed in the bone marrow and the **spleen.** The spleen also takes care of old blood cells.

The two dark **kidneys** lie on the dorsal side. Small **ureters** lead from the kidneys to the **cloaca** and then to the **urinary bladder.** Urine is stored in the bladder, later to be expelled with solid waste through the **anus.**

The cloaca and anus also serve as a depository for expelling sperm from the testes of the male frog. The testes are small, white oval-shaped organs anterior to the kidneys.

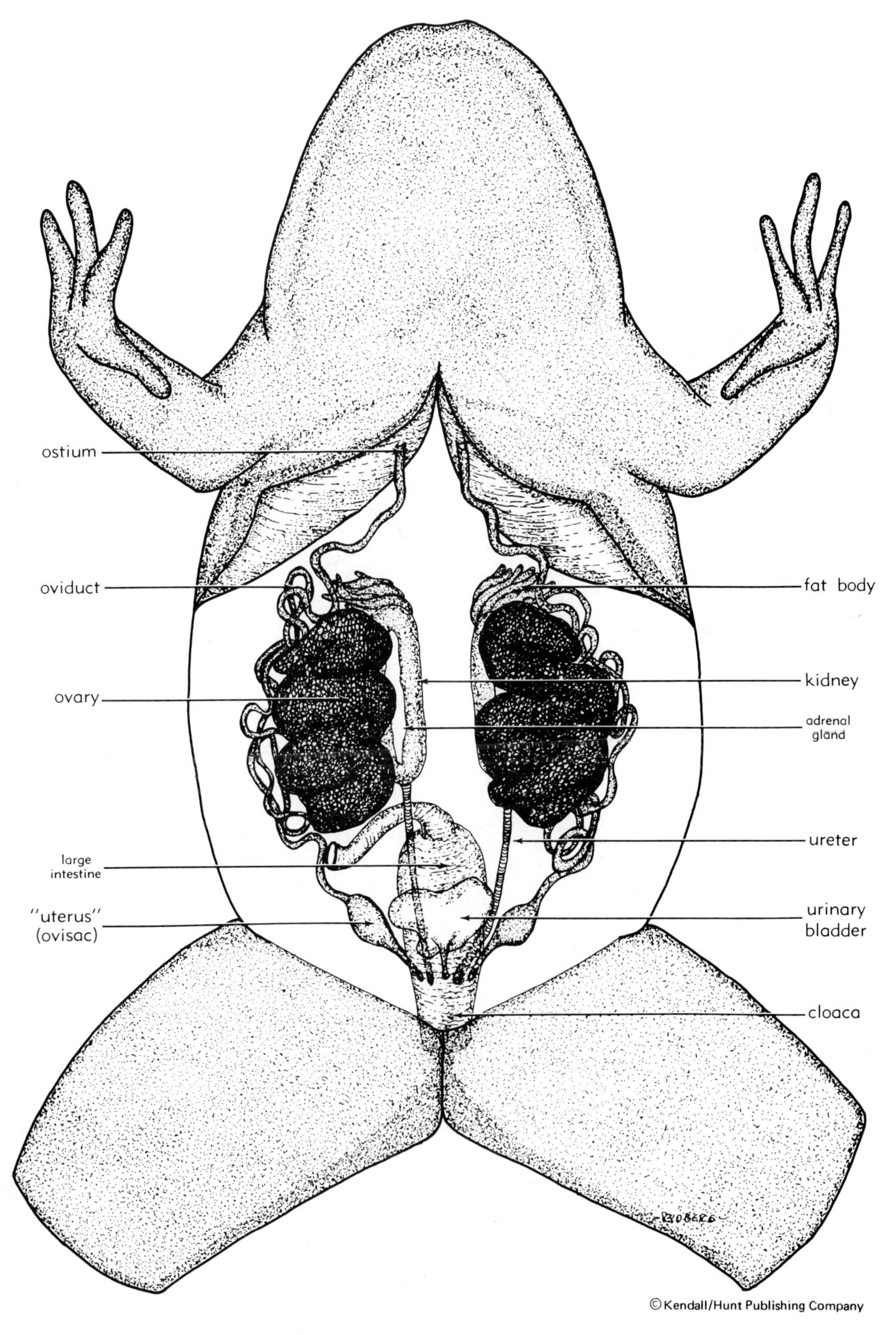

© Kendall/Hunt Publishing Company

Figure 1.12 Female frog urogenital system.

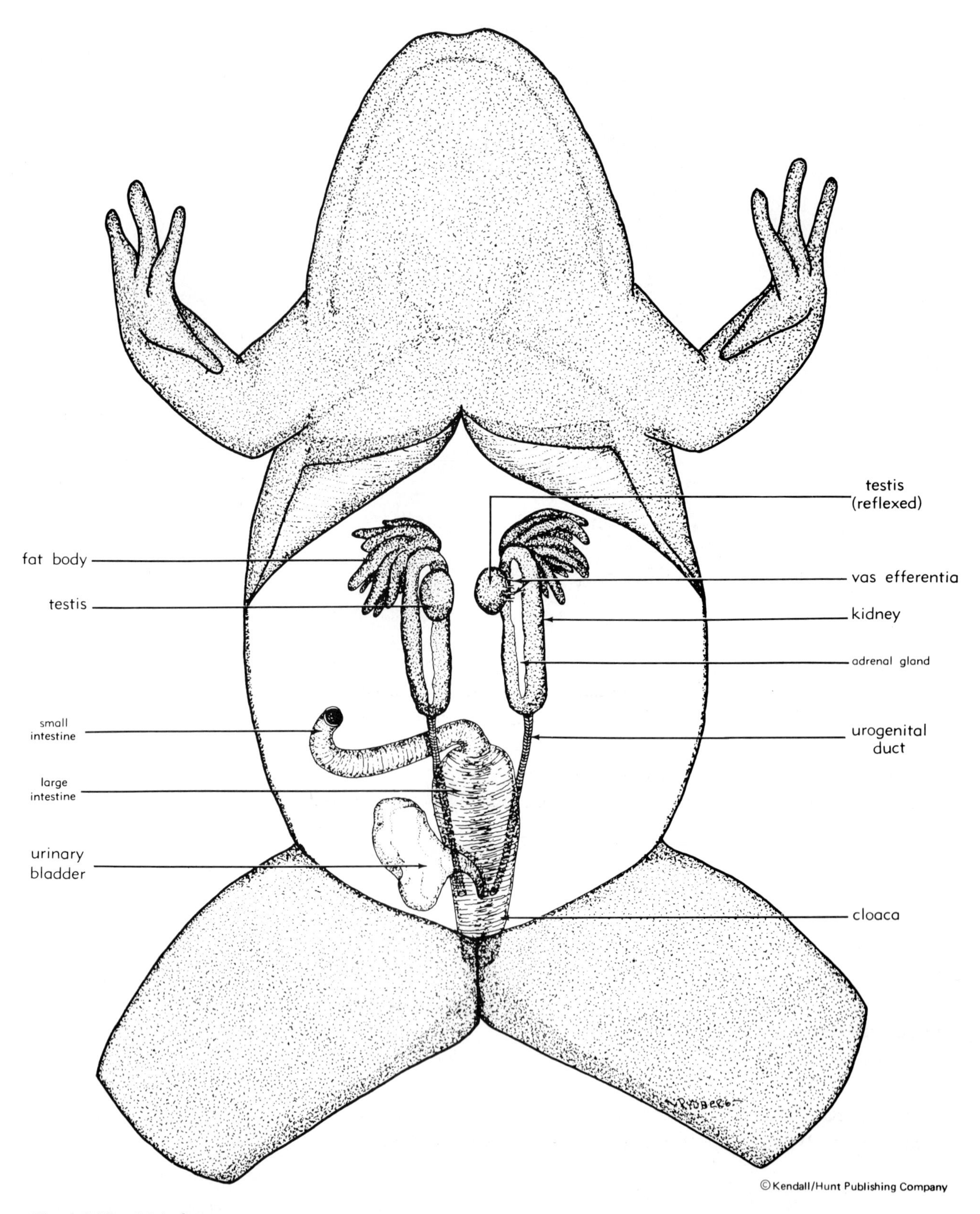

Figure 1.13 Male frog urogenital system.

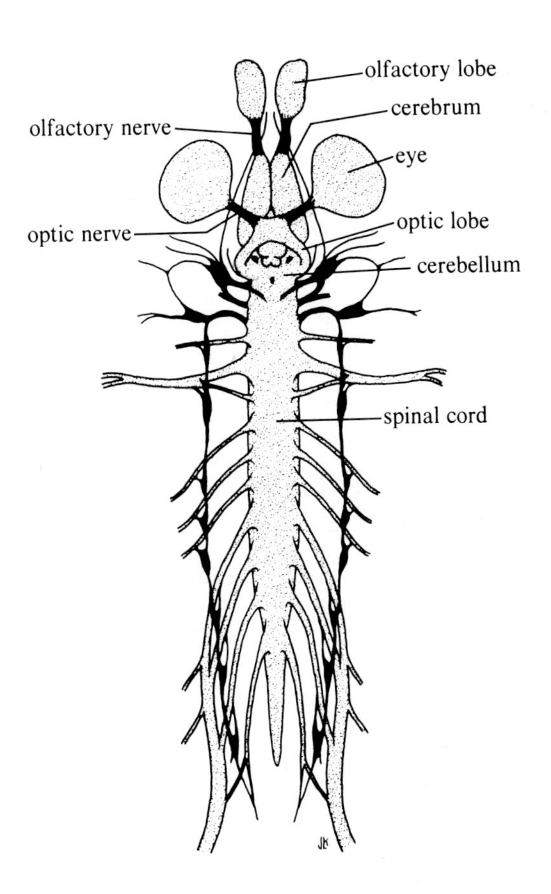

Figure 1.14 Central nervous system of the frog. Courtesy of James Koevenig.

The ovaries and stringy oviducts allow eggs to pass to the cloaca and anus of the female frog. The eggs are initially stored in the **uterus.** The eggs have a gelatinous covering, so when they are expelled they will become attached to surrounding aquatic vegetation.

Fertilization takes place when the sperm cells combine in the external environment with the egg cells. The embryos metamorph to tadpoles and ultimately adult frogs.

1.31

The last part of this dissection is to expose the brain. This is particularly tricky, because the brain is easily destroyed while removing the skeletal covering. Carefully make a slit in the center of the skull by using your scalpel. Peel away the skeletal covering with a pair of tweezers. Be careful! The brain lies just below the skeleton. Once you have exposed it, examine each part.

Figure 1.14 will assist you in identifying the various lobes of the brain.

You have completed the dissection. You should be able to identify the various organs. You may be asked to look up and study further the function of each organ. If this happens, use a reference book to assist you.

1.32 To become aware of the world that surrounds you, go outside some evening before dark and sit down in your yard, on a creek bank, by a pond, in a wooded area, or wherever you think is a good place. Listen and observe. Jot down on a piece of paper what you see and what you hear. You will be surprised at the different kinds of things that go on around you. You may need to look closely, but you will find many types of living things surround you. Write your observations in your log book.

1.33 To summarize this chapter, write a brief statement answering the questions listed.

1. What is the difference between qualitative observations and quantitative observations?
2. Why is it important to obtain a lot of data rather than only one or two pieces of data?
3. When a graph is made from established data, it is important to label each axis completely, put a title on the graph, and use symbols to represent the data. Why?
4. What is the function of a hypothesis?

1.34 Because of the nature of this chapter, let's do one more activity before we go on to chapter two. Cut out any newspaper article and underline the qualitative statements with a pencil. Then underline the quantitative statements in any other color. Attach the article to this chapter before you hand it in for evaluation.

1.35 Make certain you have completed the entire chapter before you hand it in to be evaluated. Be satisfied that you have done the best you can do. If you can do better with some additional editing of your answers, do it! Now is the time to establish good habits.

The study points for this chapter are listed below. To assist you in studying for the quiz for this chapter, read over the study points carefully. If you are able to answer a specific study point, place a check (✓) in the box next to that objective. However, if you cannot relate to the study point, then study that part of the chapter that enables you to have a mastery of the objective.

While many of the activities in this chapter can only be done in the classroom, studying the concepts can be done at home or in the library. You should get into the habit of reviewing your work and each part of this and all future chapters from 15 to 30 minutes each day. Such a habit should yield success, and that is what you want. Right?

On completion of this chapter, you should be able to do the following:

☐ 1. Differentiate between qualitative observations and quantitative observations.

☐ 2. Use appropriate laboratory skills to acquire data that will enable qualitative and quantitative observations.

☐ 3. Accurately record data that is obtained from observations of laboratory experiences. Also make conclusions from the attained data, as well as predictions for future results.

☐ 4. Make an appropriate and accurate graph of collected data.

☐ 5. Dissect a frog, using good technique.

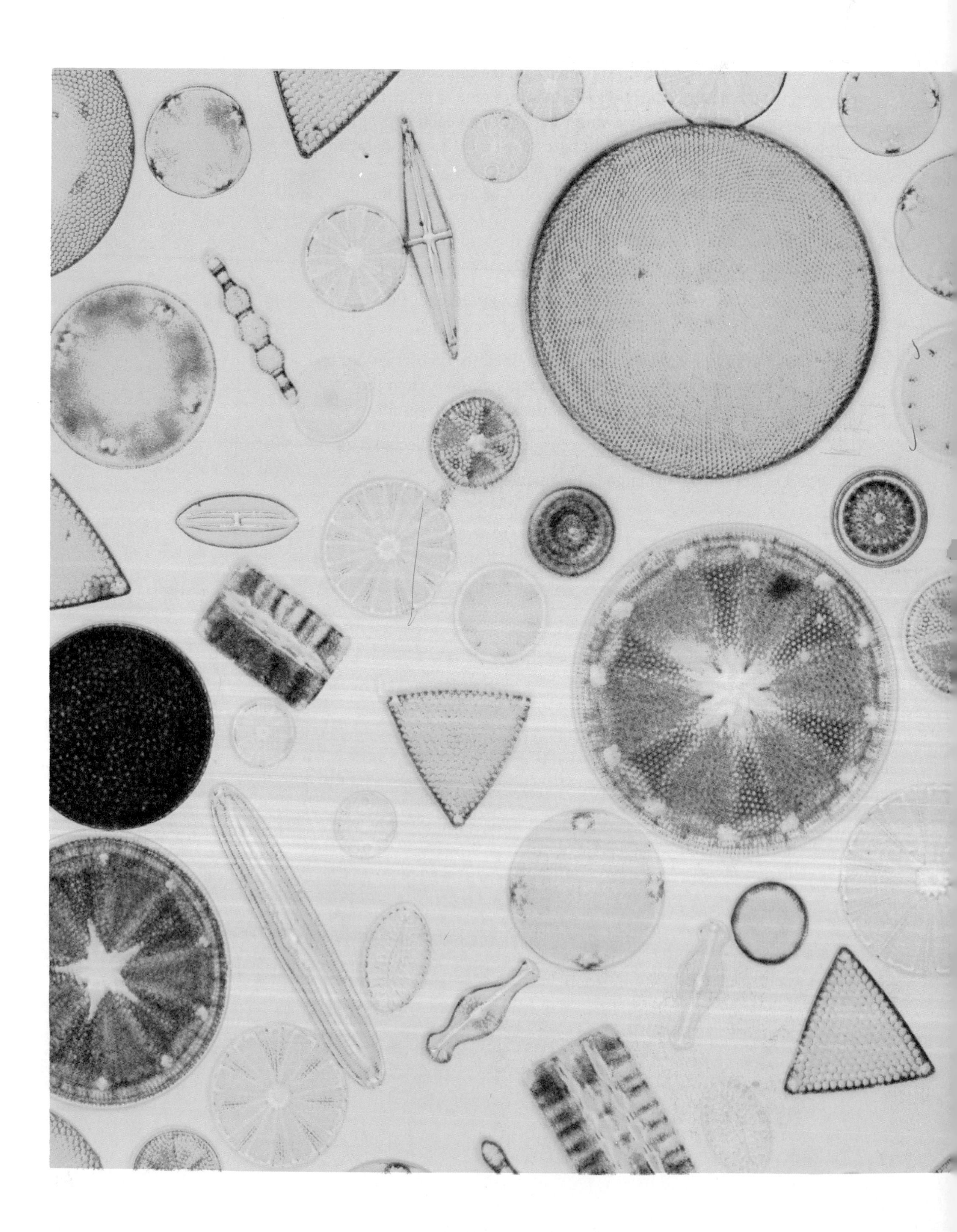

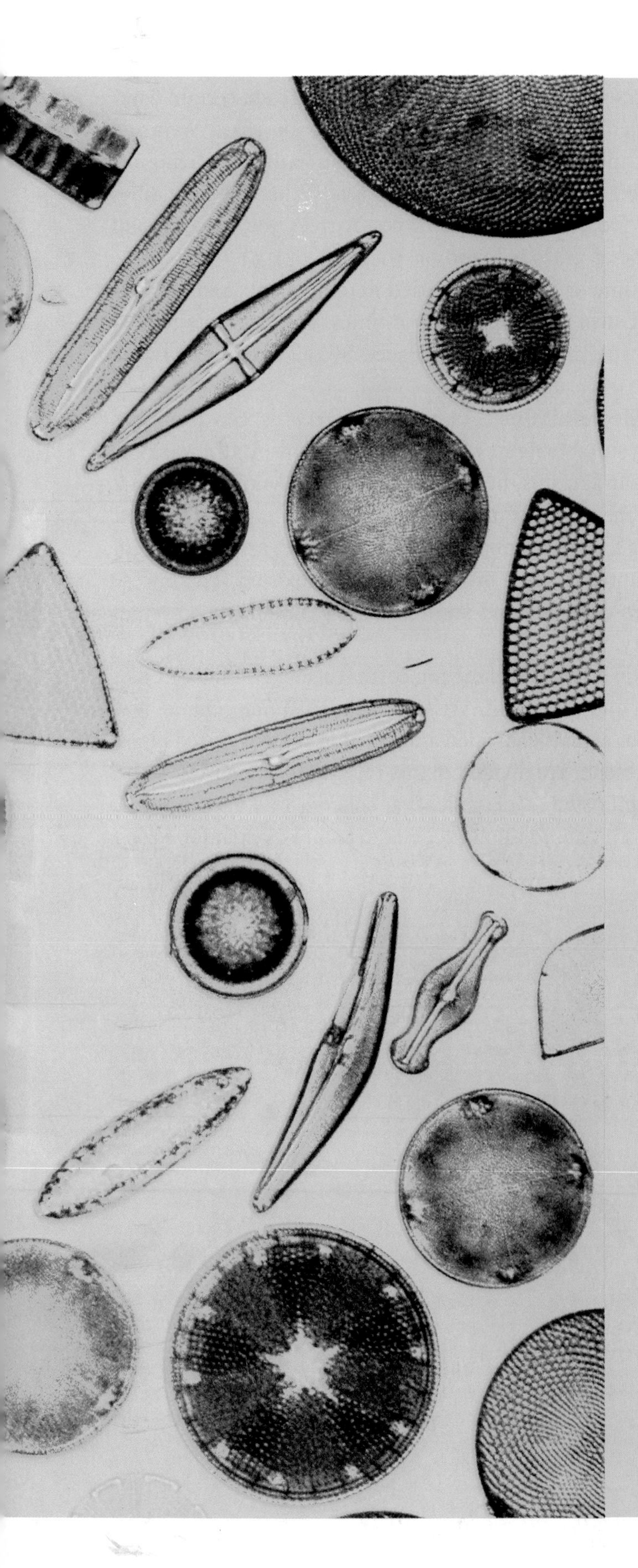

2

Taxonomy

How do you know what to call living things and how do you make a taxonomic key?

2.1 You often see new or different objects and are not certain what to call them. For some physical objects you ask, "What do you call that?" Sometimes you are not certain of the name of a living organism. Do you recognize all of the animals you see? What are the names of all of those plants? Do you know what types of trees you have around your house? What type of grass is found on your school lawn?

Everyone at one time or another has tried naming plants and animals with common names. During early prehistoric times, our ancestors named or described *creatures* in some way, as evidenced by the drawings on cave walls.

It was during Medieval times that history records the problems scholars had in finding suitable names for living organisms. Latin was normally used for naming organisms, because it was the language of scholars at that time. However, the choice of names was not entirely limited to Latin.

During the 1700s, Carolus Linnaeus (1707–1778) began renaming all living things. He attempted to place order where chaos existed. There were so many different types of plants and animals that a classification system had to become a reality.

Linnaeus, a Swedish botanist, recognized that using common names was causing problems, and the use of Aristotle's classification scheme was intolerable. Aristotle, for example, divided animals into those with red blood and those without red blood. He divided plants into three categories: trees, shrubs, and herbs. While such a classification scheme was used prior to Linnaeus, the problems it caused allowed a chaotic state to exist for a long time.

Figure 2.1 Carolus Linnaeus developed the two-name taxonomic system.

Linnaeus' system was based on the similarities and differences of living organisms using four classification groups—**order, class, genus,** and **species.** He is recognized most of all for giving each living organism a binomial technical name. His system of giving organisms two names is called **binomial nomenclature.** (*Binomial* means "two," and *nomenclature* means "giving a name to that being identified.") With this system, the first part is the genus name, and the second part is the species name. If the binomial or scientific name of an organism is used, biologists, regardless of their nationality, will recognize the organism.

Today, modern classification still uses much of Linnaeus's scheme. Yet, some changes have occurred. For one, in addition to the two-name system of binomial nomenclature, botanists often use a third name for varieties within a species. This practice is especially common in the naming of marketable new varieties of cultivated plants. You can recognize varietal names because they are separated from the specific name by the abbreviation for the word variety (var.).

Taxonomy is a science for classifying all living systems and placing such living systems into groups based on similarities. **Classifying** is the process of placing organisms in categories according to a taxonomic system. Aristotle's system was a simple classification system. While placement of an organism is determined by similarities, the biggest problem in taxonomy is the placement of an organism in a proper category.

To assist you with categorizing, this next activity should be of help. The first chapter helped you to refine your observation skills. In a science like taxonomy, observation skills are essential. You need to observe carefully the characteristics of the object or organism being considered, so you can place it in a proper category.

Get kit 2.1 from the supply table. To help sort the objects, literally put one object in each box of data table 2.1 and sketch it. Next, carefully describe each object, noting as many qualitative and quantitative characteristics as possible.

Data Table 2.1 Characteristics of Screws and Nails

1.	6.
2.	7.
3.	8.
4.	9.
5.	10.

2.2 From your recorded observations of the objects in kit 2.1, place them into two major categories using common names.

1. ____________ 2. ____________

Which numbers from data table 2.1 fit your first category? ____________

Which numbers fit your second category? ____________

2.3 What characteristic prompted you to place the objects in these respective categories? ____________

____

____

2.4 Could the objects in category 1 be subdivided into smaller categories? ____________ On what basis would you make the division? ____________

____

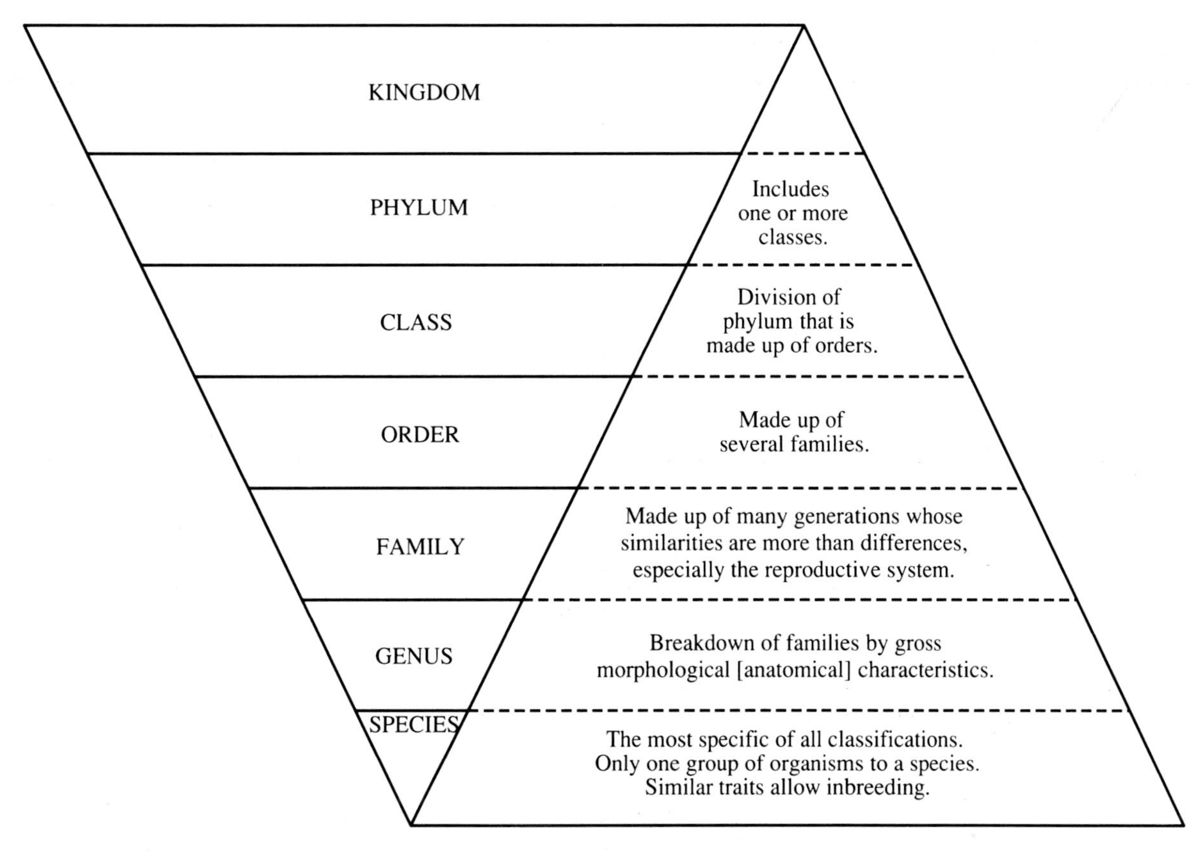

Figure 2.2 The classification pyramid. Note how each category becomes more specific as you move from top to bottom.

2.5 Could the objects in category 2 be subdivided into smaller categories? ________________ On what basis would you make the division? ____

____

2.6 This exercise has allowed you to observe objects, relate to specific characteristics, and place them into categories based on similarities. This is the process taxonomers go through as they try to place newly found living organisms into a specific category.

The systematic means for classifying a living organism is illustrated in figure 2.2. Study this closely.

The kingdom is the largest category. It contains all of the organisms classified to belong in this category. From the kingdom, each category becomes more specific, based on the similarities of the organisms that make up that category. Which category is the second largest classification category? ________________ Which category would you consider to be the most specific? ________________

Thinking of Linnaeus's binomial system, which two categories are used to scientifically name living organisms? ________________ and ________________ Each kingdom has several phyla (plural of phylum), and each phylum has several classes. Would you expect each order to have several families? __________ Would you expect each family to have more than one genera (plural of genus)? __________

2.7 To further illustrate a systematic taxonomic scheme, refer to data table 2.2. The example compares humans to the gorilla. (The arrows to the left of the table suggest each category becomes more specific as you progress from kingdom to species.)

At what point is there a difference in the classification scheme between humans and gorillas? ________________ Using data table 2.2 as your guide, complete data tables 2.3 and 2.4. Data table 2.3 compares the dog and the wolf. Data table 2.4 compares the pea plant with the corn plant. All of the hints for your success with this activity reside in your careful study of data table 2.2. The names for each category are present, but scrambled. You will need to unscramble them, and then write them in their proper place.

1. Where in the taxonomic scheme does the wolf differ from the dog? ________________
2. Where in the taxonomic scheme does the corn plant differ from the pea plant? ________________
3. What trait of the binomial name will always tell you which part is the species name?

   ____

   ____

2.8 All living organisms are currently placed in one of the five existing kingdoms. The five-kingdom system was established by R. H. Whittaker (1924–1980). The fifth kingdom, named Monera, is made up of organisms that are only *prokaryotic* (from Greek *pro,* meaning "before," and *karyon,* meaning "nucleus"). This kingdom is essentially made up of bacteria, and blue-green algae, now more formally called cyanobacteria.

Data Table 2.2 The Classification System

Classification	Man	Ape
1. Kingdom	Animalia	Animalia
2. Phylum	Chordata	Chordata
3. Class	Mammalia	Mammalia
4. Order	Primate	Primate
5. Family	Hominidae	Pongidae
6. Genus	Homo	Gorilla
7. Species	sapien	gorilla

Data Table 2.3 Classifying the Dog and Wolf

Common Name Is Dog		Common Name Is Wolf
Carnivora, Canis, Animalia, familiaris, Mammalia, Chordata, Canidae		**Animalia, Canis, Mammalia, lupus, Chordata, Canidae, Carnivora**
1. Kingdom		1. Kingdom
2. Phylum		2. Phylum
3. Class		3. Class
4. Order	Carnivora	4. Order
5. Family		5. Family
6. Genus		6. Genus
7. Species		7. Species

Data Table 2.4 Classifying Corn

Common Name Is Peas		Common Name is Corn
		mays, Plantae, Graminales, Zea, Tracheophyta, Graminacae, Angiospermae
1. Kingdom	Plantae	1. Kingdom
2. Phylum	Tracheophyta	2. Phylum
3. Class	Angiospermae	3. Class
4. Order	Rosales	4. Order
5. Family	Leguminosae	5. Family
6. Genus	Pisum	6. Genus
7. Species	sativum	7. Species

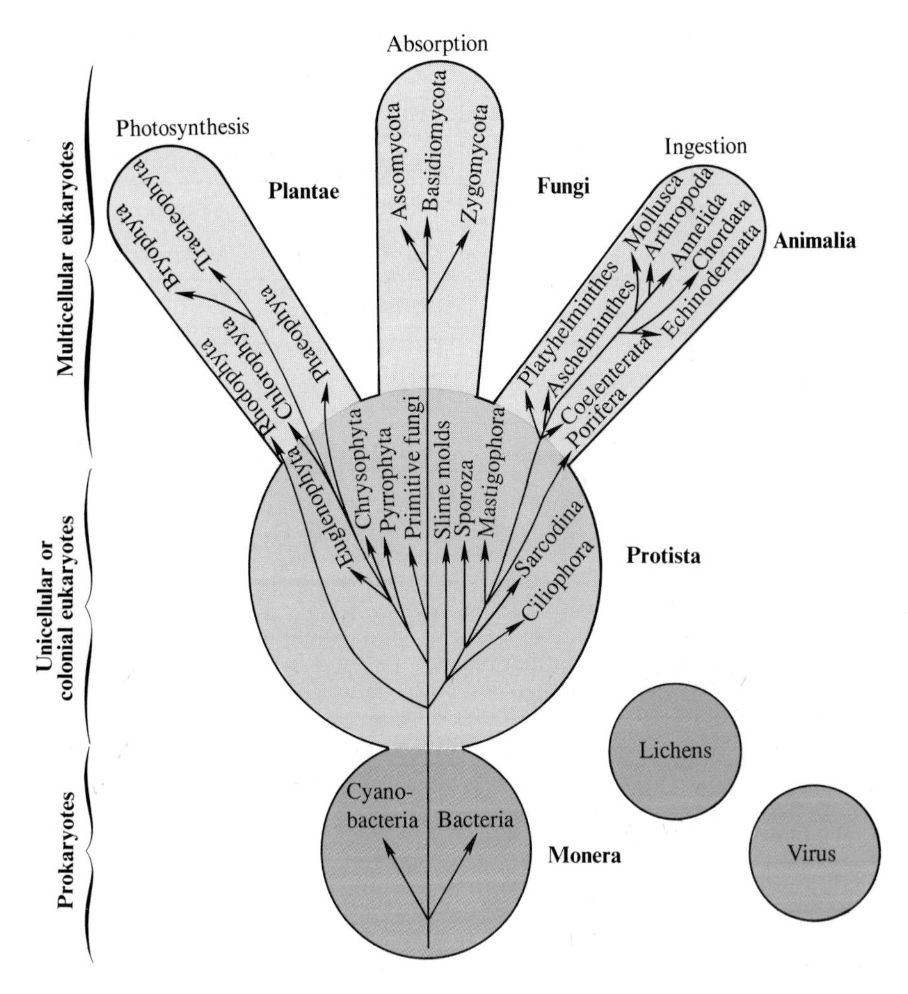

Figure 2.3 Whittaker's model displaying the five kingdoms. At present, the placement of the lichens and viruses is undetermined.

The other four kingdoms are considered to be *eukaryotic*. These organisms all have a membrane-bounded nucleus containing chromosomal material. The remaining four kingdoms are: the plant kingdom, animal kingdom, protist kingdom, and fungi kingdom.

Of the four kingdoms, the protist kingdom is unusually diverse. This statement is based on diverse form and biochemistry of the organisms found in the kingdom. For instance, three phyla, the red, brown, and green algae, have been previously classified as plants and placed in the plant kingdom. Some biologists still classify these algae as plants. However, more classification schemes and taxonomic keys now reflect these algae as a part of the protist kingdom, even though they carry out the photosynthetic process for making food and are multicellular.

2.9 The use of a taxonomic key becomes an important tool for a biologist. No one remembers all of the different names. Consequently, it becomes necessary to look up names from time to time or use a key to figure out a name in a systematic fashion. A picture key is handy,

Johanssen, John, 1681 W. 1st	839-4621
Johnson, A. G., 428 Clark	839-5522
Johnson, Mabel, 821 2nd Ave. N.	839-1362
Johnson, Rbt. H., 1404 1st Ave.	839-2003
Johnson, Tom, 409 W. 7th St.	839-5679
Johnson, Walter L., R. R. 1	839-6681
Jones, Andrew, Fairview Dr.	839-1026
Jones, Jack, 1801 2nd Ave. N.	839-7532
Jones, X. P., 921 Flower Circle	839-8220
Joplin, Mary S., 101 1st Ave. N.	839-9108
Jotten, Mark, 926-C St. E.	839-4553
Justen, Wm. P., R. R. 4	839-0089

Figure 2.4 Using a taxonomic key is like using a telephone directory. Section 2.9 gives you a further explanation.

but usually does not contain many of the species. It is generally limited to the more common species. It is not important to memorize a lot of organisms by their scientific names, but it is important to be able to key an organism if the situation should ever occur. Keep in mind that a key is an artificial tool for taxonomy that illustrates the *diversity* as well as the *similarities* of organisms.

This activity is designed to acquaint you with the use of a taxonomic key. A taxonomic key is used in a fashion similar to the way you would use a telephone directory. For instance, to find Jack Jone's telephone number, you would first turn to the "J" section of the directory. Second, you would look for "Jones." Once you found Jones, you would look for "Jones, Jack." If there were more than one "Jones, Jack," you would then need to know the address to find the specific Jack Jones with which you wished to speak.

In this analogy, Jones represents the genus name, and Jack, the species name. Many times a key to the birds of Eastern North America or a key to the wild flowers of the Midwest will give a short description of the habitat (home or location) of the plant or animal. In the case of the telephone directory, this would be represented by the street address found there.

Let us continue now by using an actual taxonomic key. Taxonomic keys begin at the beginning and, though they are not always complete, through a series of steps they end up with the name of a species. Vocabulary becomes a major part of a key because keys mainly refer to the anatomy of the organisms. For this reason, figure 2.5 will aid you in keying the fish used in this activity.

The following identification characteristics are used to key common fish. They will be keyed to common names rather than to scientific names, but the procedure for keying any organism is handled in a similar fashion, and this is the main purpose for this exercise. Please note that each step gives you two choices per step and, depending on which choice you make, will name the organism or direct you by number to another set of choices.

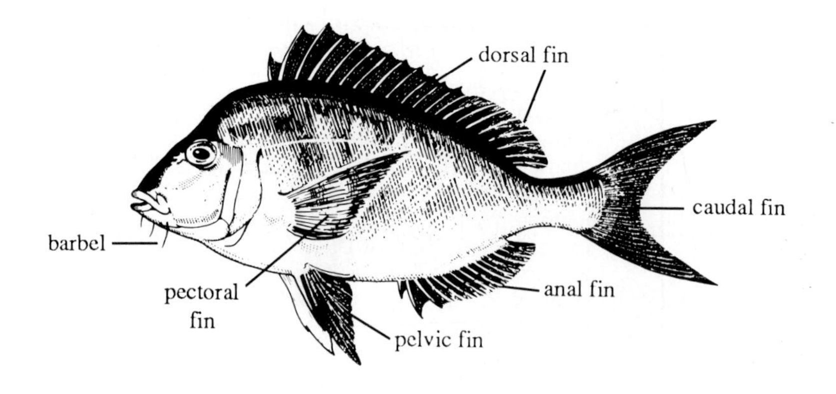

Figure 2.5 Fin identification for fish. Use this figure to assist you as you key the fish illustrated in figure 2.6.

1a.	Body cover is predominantly scales.	2
1b.	No or few scales are present; body covering is that of skin.	8
2a.	Single dorsal fin only.	3
2b.	Two dorsal fins are present. The fins may be joined or separated or spines.	6
3a.	Body is elongated and slender. Normally one-fourth as deep as long.	4
3b.	Body is short and deep.	5
4a.	Fins spotted, scales on cheeks.	northern pike
4b.	No spots, long slender snout.	gar
5a.	Mouth positioned ventrally; no barbels.	sucker
5b.	Mouth positioned ventrally; barbels.	carp
6a.	Dorsal fins and spines present.	7
6b.	Dorsal fins joined.	sheepshead
7a.	Dorsal fin preceded by four to six spines.	five-spined stickleback
7b.	Dorsal fin preceded by eight to eleven spines.	nine-spined stickleback
8a.	Snakelike body; continuous dorsal fin from dorsal midpoint to ventral midpoint.	eel
8b.	Body not snakelike.	9
9a.	Barbels or whiskers present around mouth.	10
9b.	Barbels or whiskers not present around mouth.	trout
10a.	Deeply forked caudal fin.	11
10b.	Caudal fin is rounded, not deeply forked.	bullhead
11a.	Dorsal fin rounded.	channel catfish
11b.	Dorsal fin pointed.	blue catfish

Key the fish illustrated below, listing each step you take to get its name. Do so in the same manner as indicated by the example.

Example: Number sample 0

1b – 8, 8b – 9, 9a – 10, 10a – 11, 11a
Name: Channel catfish

Number 1

Name: ____________

Number 2

Name: ____________

Number 3

Name: ____________

Number 4

Name: ____________

Number 5

Name: ____________

Number 6

Name: ____________

Teacher's initials

Number 1

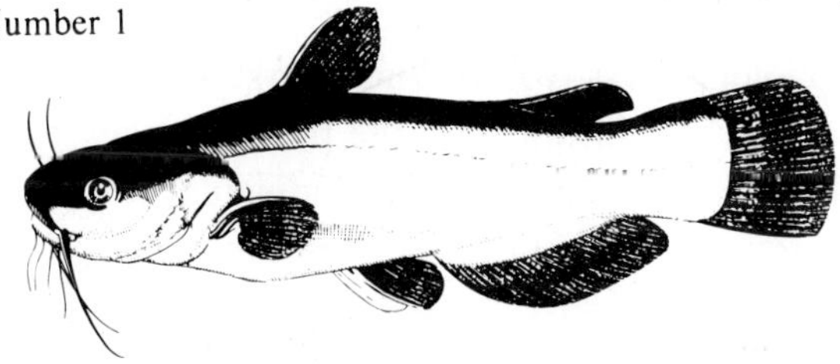

Number 2

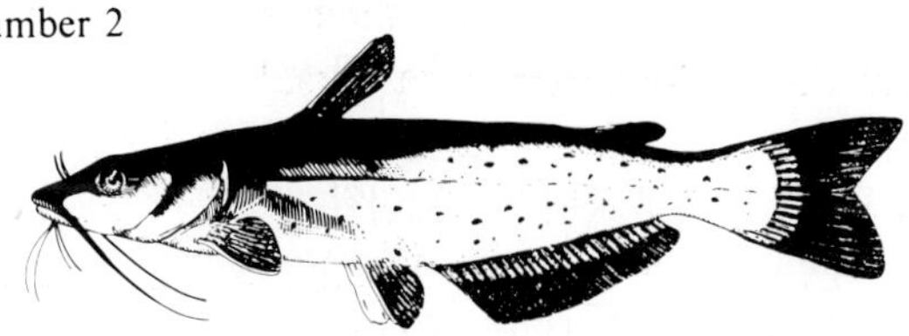

Number 3

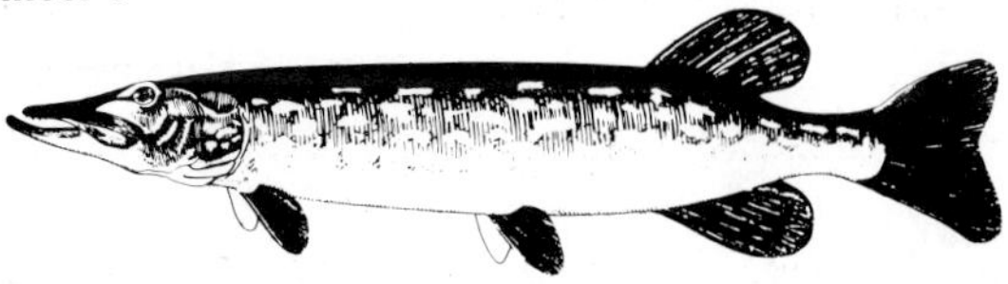

Number 4

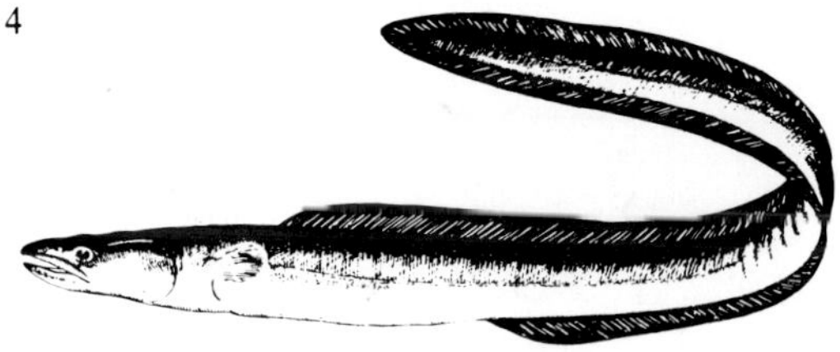

Number 5

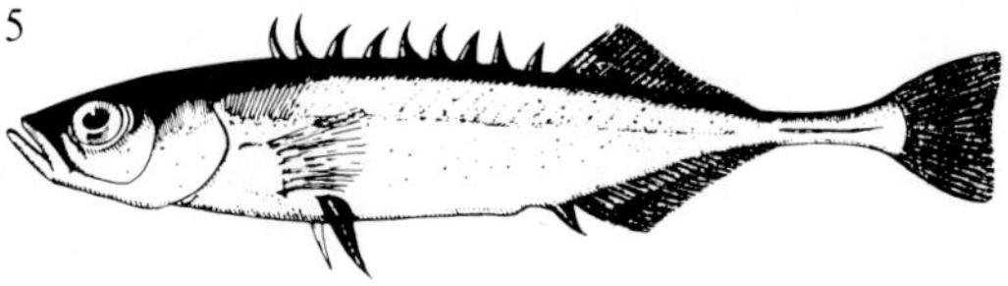

Number 6

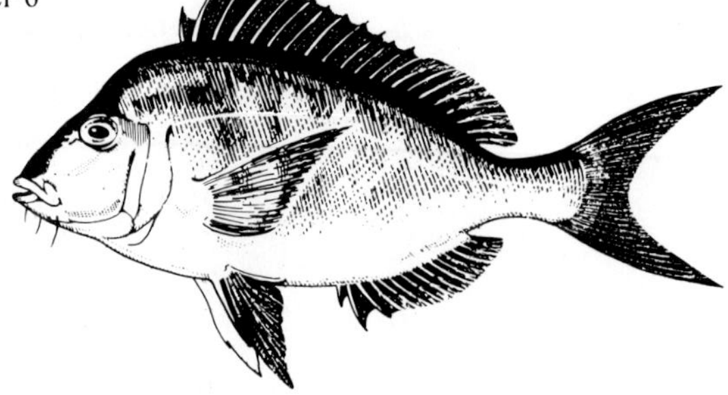

Figure 2.6 Six fish to key out. Numbers 3, 5 and 6 have scales. Numbers 1, 2 and 4 do not have scales.

2.10 Using the key below, key the type of duckweed found in ponds, streams, or in your classroom aquarium. You will need to use a dissecting microscope for this investigation. Your teacher may wish to give you a few instructions on the proper way to use a dissecting microscope.

Lemnaceae: Duckweed Family

1a.	Each plant contains several roots.	2
1b.	Each plant has only one root.	3
1c.	Plants do not have roots.	4
2a.	Diameter of plant is 3 to 7 mm.	Spirodela polyrhiza
3a.	Diameter of plant is 2 to 4 mm.	Lemna minor
3b.	Plant has a slender stalk present.	Lemna triscula
4a.	Diameter of plants less than 1 mm.	Wolffia punctata
4b.	Top of plant concaved.	Wolffia columbiana

Name of plant keyed is ____ .

2.11 Biological keys are usually based on pairs of choices. This should be obvious to you, having just used two keys. To understand better how to use biologic keys, you will now make your own key. In so doing, you should gain further insight into the process of keying an individual. Figure 2.7 contains shapes that represent individual organisms. These same shapes are also found in your log book. Remove this page, and cut out each separate shape.

2.12 Spread the shapes, or individual organisms, on the table before you with their letters (names) up. Do not use the individuals named U-1 through U-5 at this time. You will note the shapes are different in several ways. These differences can be called characteristics. Please note that the letter-name is not a characteristic. List *all* characteristics of the shapes that you can identify in the appropriate box in data table 2.5.

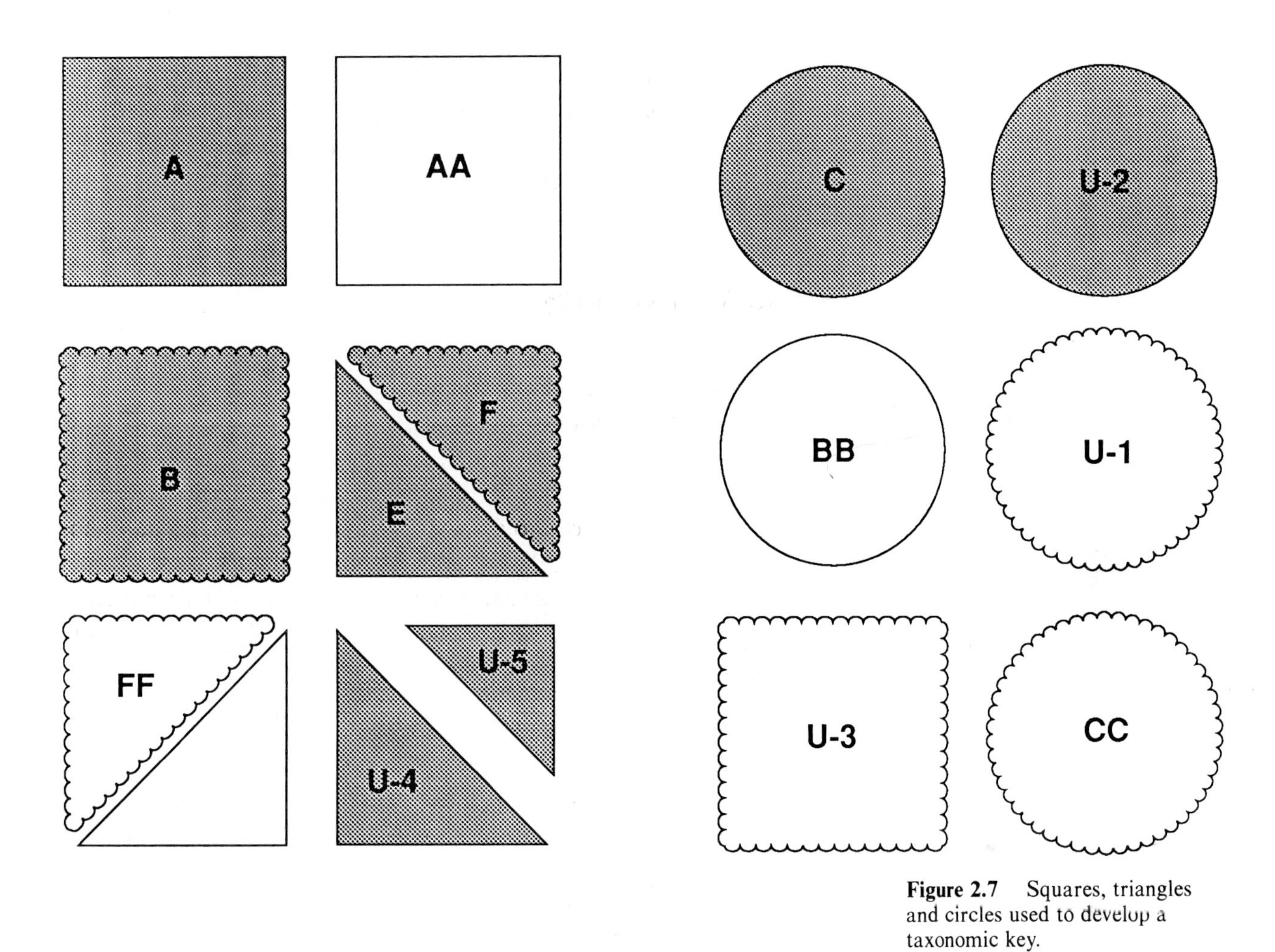

Figure 2.7 Squares, triangles and circles used to develop a taxonomic key.

Data Table 2.5 Making a Taxonomic Key

Name	Characteristic
A	
AA	
B	
BB	
C	
CC	
E	
EE	
F	
FF	

2.13 To make a biologic key, you begin by grouping the shapes, or individual organisms, into two groups. You did this with the nails and screws. You can also do this with these shapes. For example, you can separate the shapes into two groups on the basis of being shaded or unshaded. You may also divide the individuals into other groups using other characteristics. Three options are given to you. Write options 4 and 5 on the lines below.

1. Triangles or not triangles
2. Shaded or unshaded
3. Smooth edge or rough edge
4. ______________________________
5. ______________________________

2.14 The key has been started for you in data table 2.6. The decision was made to start with *triangular* or *not triangular.* These two choices have been placed in 1a and 1b. Do you see this? _____ If not, check with your teacher.

2.15 Place the group of individuals that are triangular in shape in front of you. Note the key directs you to choices 2a and 2b by placing the number 2 opposite 1a. Next separate the triangles into two groups on the basis of *shaded* and *unshaded.* The choices are placed in the key, and the number 3 is placed opposite 2a—individual unshaded—and directs you to choices 3a and 3b. You are now left with only two individuals that may be separated and named. Notice that 3a has been named EE, and 3b has been named FF. Do you confirm this? ___________

2.16 Now use the shaded triangles in front of you. Write the number 4 opposite 2b—individuals shaded. Since there are only two individuals in this group, they can be named. Turn to the outline. What name has been assigned to the shape identified with 4a? ____________

4b? ____________________

2.17 Place the remaining individuals, not triangular, in front of you, and place the number 5 opposite lb—individual not triangular. You are now ready to complete this key with fewer written directions. Remember, if you have only one individual, you must name it by letter as you did in the previous example.

Data Table 2.6 Outline for Making a Biologic Key

Number Group	Characteristics	Continuation Number or Name
1a	Individual triangular in shape	2
1b	Individual not triangular in shape	
2a	Individual unshaded	3
2b	Individual shaded	
3a	Individual smooth edge	EE
3b	Individual rough edge	FF
4a	Individual smooth edge	E
4b	Individual rough edge	F
5a		
5b		
6a		
6b		
7a		
7b		
8a		
8b		B
9a		
9b		

2.18 The key you just made may now be used to identify unknown individuals. Place the figures labeled U-1, U-2, U-3, U-4, and U-5 in front of you. Proceed with each unknown individual through the key you have just completed. Write the name identifying each unknown in data table 2.6. The choice of names you have are EE, FF, E, F, CC, C, B, BB, A, and AA. U-1 has been named for you. Work through your key with U-1. Do you agree with the name? ___________

2.19 After you have named, by letter(s), each unknown individual, compare it to the individual used to make the key. Do the unknowns compare? Record a *yes* or *no* in data table 2.7.

Data Table 2.7 Classifying Unknowns

Unknown	Name	Individual Compares to Data Table 2.6
U-1	CC	Yes
U-2		
U-3		
U-4		
U-5		

2.20 You may have trouble identifying one of the unknown individuals. If so, try keying it again. If giving the unknown individual a name remains a problem for you, it probably means that a specific characteristic of the unknown was not used to make the key. Therefore, this key does not work for this individual. This is a possibility for any taxonomic key you might use. Which unknown appears to have no name?

_______________ What causes this key to fail in naming this unknown? _______________

____

____

2.21 The fact that individuals will miskey or be misnamed is important for you to remember. It is usually a good idea to compare an unknown organism to an already identified specimen. Taxonomists and museums maintain collections of plants and animals, and their facilities are available for comparisons. If you do not have these resources, a good idea is to compare the unknown to a photograph. Many popular keys have a written description and a picture of the specimen for comparison.

Variations between species, as well as within a species, have contributed to the difficulty in taxonomic studies. Linnaeus brought uniformity to chaos, but thousands of new species of plants, insects, and animals are discovered each year, and the problem of where they fit in the taxonomic structure repeats itself time and time again. This is especially true when the variation in the species is so slight that it verges on the line of being a new species or an already-identified species.

The taxonomic key that appears below is a good key to use when keying out evergreen trees, the gymnosperms. The next exercise will require you to spend some time outdoors keying evergreen trees. Your teacher will give you instructions for using this key and the expectations of this exercise.

Data Table 2.8 A Taxonomic Key for Evergreen Trees

CLASS. GYMNOSPERMAE

Order 1. Ginkgoales

Family 1. Ginkgoaceae *Ginkgo family*

1. *Ginkgo* *Maidenhair tree*

1. Lvs. broad and notched at apex, wedge-shaped at base, 5–8 cm across; introduced from China. .... G. biloba.

Order 2. Coniferales **Conifers**

1. Low shrubs with seeds in a red fleshy cup; lvs. continued down stem in a narrow ridge. .... 3. Taxaceae
1. Trees or shrubs, mostly evergreen and not as above. .... 2. Pinaceae

Family 2. Pinaceae *Pine family*

1. Deciduous (shedding leaves in winter). 2
1. Evergreen. 3
2. Cone-bearing; lvs. clustered on short shoots. .... 5. Larix
2. No cones evident; lvs. one in a place. 4
3. Lvs. 1 in a place. 5
3. Lvs. in bunches of 2, 3, or 5 needle-shaped. .... 6. Pinus
3. Lvs. scalelike. 9
4. Lvs. needle-shaped, slender. 5
4. Lvs. scalelike, opposite, pressed close to stem. 9
5. Lvs. spirally arranged. 6
5. Lvs. opposite or whorled in 3s. .... 9. Juniperus
6. Lvs. 4-sided in cross section. .... 4. Picea
6. Lvs. flat, with distinct upper and lower sides. 7
7. Lvs. 8–12 mm long, narrowed to a distinct slender stalk. .... 2. Tsuga
8. Lvs. 20–60 mm long, without distinct stalk or petiole. 8
8. Buds very sharp, chocolate brown, lf. scars slightly raised. .... 3. Pseudotsuga
8. Buds rounded, yellow-brown; lf. scars not at all raised. .... 1. Abies
9. Twigs 4-sided or nearly cylindrical. .... 9. Juniperus
9. Twigs flat, in flat branching fronds. .... 8. Thuja

1. *Abies* *Fir*

1. Lvs. 2 cm long or less. .... A. balsamea
1. Lvs. 3–6 cm long; cult. (White Fir). .... A. concolor

2. *Tsuga* *Hemlock*

1. Lvs. 8–12 mm long; cones soft, 1–2.5 cm long. .... T. canadensis

3. *Pseudotsuga* *Douglas fir*

1. Lvs. about 2 cm long; cones with a 3-lobed bracht between the scales; from northwestern U.S. .... P. taxifolia

*Adapted with permission of the family from Henry S. Conrad, *Plants of Iowa* (Grinnell, Iowa: Published by the Author, 1958), pp. 1–4.

Data Table 2.8 A Taxonomic Key for Evergreen Trees—*Continued*

CLASS. GYMNOSPERMAE

4. *Picea* *Spruce*

1. Twigs orange-yellow; lvs. green; cones 10–15 cm long. .... P. abies
1. Twigs ashy-yellow; lvs. ashy; cones 2.5–5 cm long (White Sprue). .... P. glauca
1. Twigs pale; lvs. bluish-green, very sharp (Blue Spruce). .... P. pungens

5. *Larix* *Larch, tamarack*

1. Cones 2.5–3 cm long; tree symmetrical (European Larch). .... L. decidua
1. Cones 1–2 cm long; branches and trunk crooked (Tamarack). .... L laricina

6. *Pinus* *Pines*

1. Lvs. strictly in 2s. 2
1. Lvs. in 2s and 3s on the same tree (Ponderosa Pine). .... P. ponderosa
1. Lvs. in groups of 5. 5
2. Low and shrubby. .... P. montana mughus
2. Tree. 3
3. Branches covered with old cones; lvs. short (Jack Pine). .... P. banksiana
3. Cones few, not remaining long on tree. 4
4. Lvs. 6–14 cm long; twigs dull-orange color (Red Pine). .... P. resinosa
4. Lvs. 6–14 cm long; twigs dull gray (Austrian Pine). .... P. nigra austriaca
4. Lvs. 4–6 cm long; bark papery, yellow (Scotch Pine). .... P. sylvestris
5. Lvs. slender, soft, dark-green; common (White Pine). .... P. strobus
5. Lvs. stiff, curved, 4–7 cm long; rare, cult. .... P. flexilis

7. *Taxodium* *Southern cypress*

1. Lvs. 1–2 cm long; twigs deciduous. .... T. distichum

8. *Thuja* *Arbor vitae*

1. Twigs horizontal with distinct upper and lower sides. .... T. occidentalis
1. Twigs standing vertically, with both sides alike. .... T. orientalis

9. *Juniperus* *Juniper, red cedar*

1. Lvs. whorled in 3s, nearly at right angles to stem, white lined above. 2
1. Lvs. opposite and scalelike, or needlelike and at about 45 degrees to stem. 3
2. Erect, with many shoots; cult. .... J. communis
2. Sprawling and spreading; native n.e. .... J. c. depressa
3. Trees; heartwood red (Red Cedar). .... J. virginiana
3. Shrubs. 4
4. Scalelike lvs. sharp pointed; creeping widely; rare n.e. .... J. horizontalis
4. Scalelike lvs. blunt; branches straight, obliquely ascending; cult. (Pfitzer). .... J. chinensis pfitzeriana

Family 3. Taxaceae *Yew family*

1. *Taxus* *Yew*

1. Sprawling shrub; 1.5–2 mm wide, with slender midrib; native n.e. .... T. canadensis
1. Erect or spreading; lvs. 2–3 mm wide; midrib prominent; cult. .... (Japanese Yew) T. cuspidata

2.22 Chapter 2 should have some long-lasting effects for you, because regardless of where you go, there will always be an organism that has a name. You may not know the name of the plant or animal, but you will have an idea of how to look up a name or use a taxonomic key.

A unique feature of taxonomy is that none of the sources of data ever become totally obsolete. Although taxonomy was one of the earliest of the biological sciences, the process of naming living organisms is not yet finished. Thousands of new organisms are discovered each year.

The study points for chapter 2 are listed below. As you review them, place a check (✓) in the appropriate box when you have mastered that particular objective.

On completion of this chapter, you should be able to do the following:

☐ 1. Use a taxonomic key effectively for the purpose of keying an organism.

☐ 2. Relate to the major groups of taxonomy; namely, the kingdom, phylum, class, order, family, genus, and species.

☐ 3. Identify the organisms in the five-kingdom scheme of R. H. Whittaker.

☐ 4. Distinguish between a prokaryote and a eukaryote.

☐ 5. Make a division of objects, whether living or nonliving, based on similar characteristics of the objects, into subgroups.

☐ 6. Construct a taxonomic key.

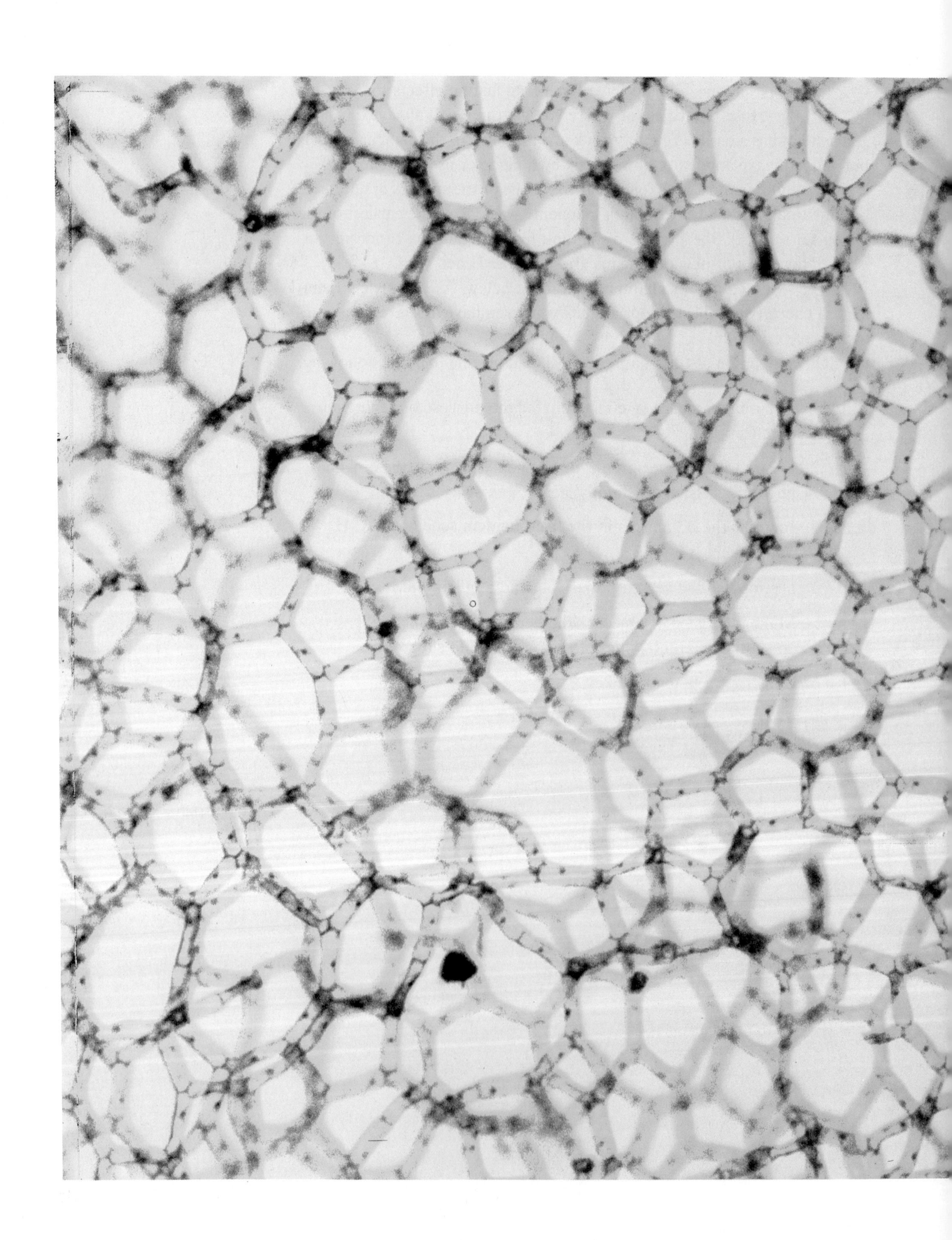

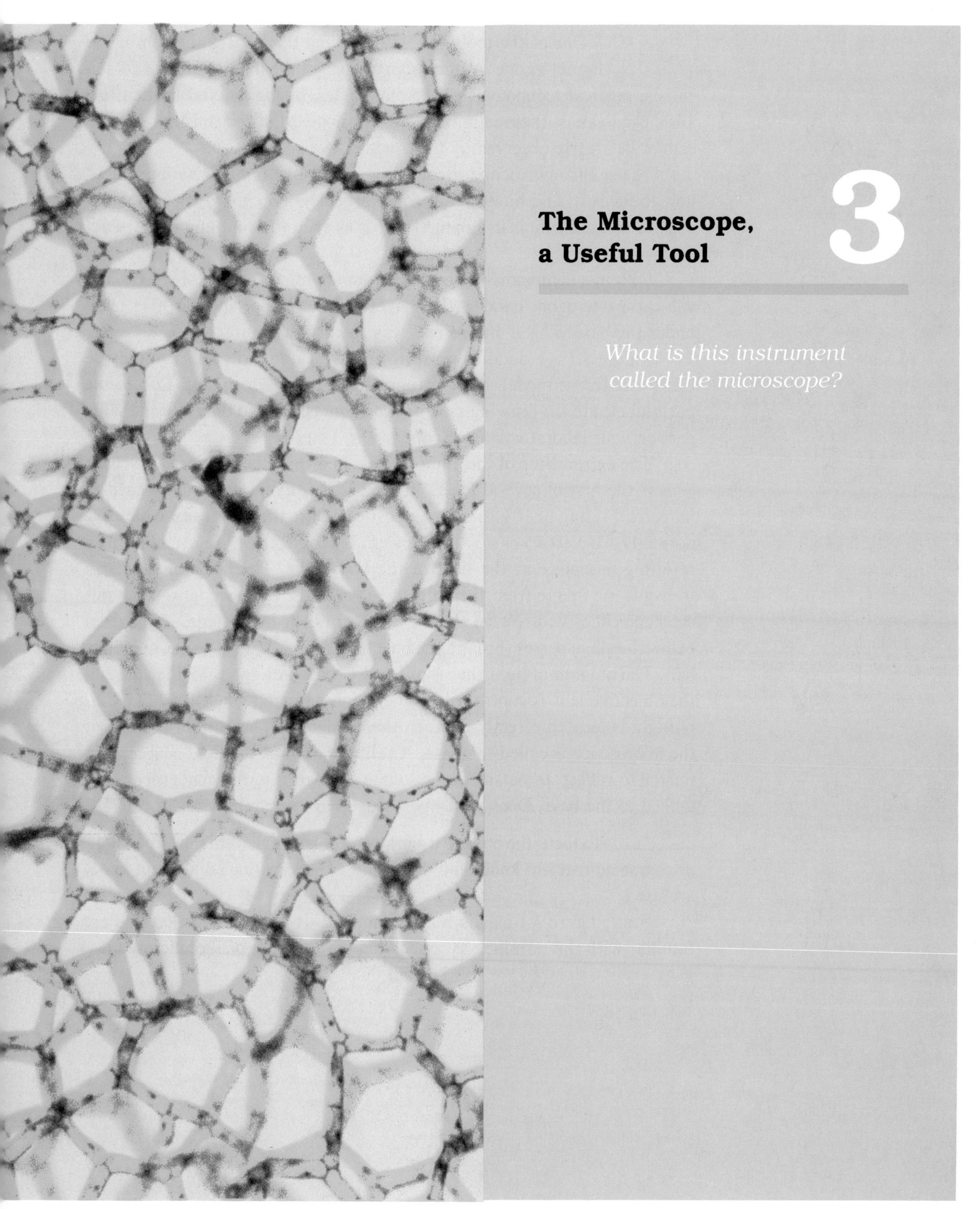

The Microscope, a Useful Tool

3

What is this instrument called the microscope?

3.1 One of the tools you will be using this year is the compound microscope. A microscope is one of the many instruments scientists, medical technologists, doctors, nurses, biologists, and horticulturalists use to assist them in making observations. Many organisms and cells cannot be clearly observed by the unaided eye.

To use the microscope effectively, you must be familiar with its parts, how to focus it, and how to care for it properly. The microscope is an expensive and delicate instrument. Treat it as such to avoid unnecessary accidents.

Obtain a compound microscope from the microscope cabinet. Be sure you carry it with one hand on the arm of the microscope and the other hand under the base. This is the correct way to carry a microscope.

Figure 3.1 is a detailed drawing of a typical microscope. You may be using a different model, but the functioning parts are similar. Locate the parts labeled in the drawing as you read about them, and then find the same part on your laboratory microscope.

The extreme top of the microscope is known as the **eyepiece,** or **ocular.** This is the first of several lenses in the system. Eyepieces usually carry a numerical rating of 10×. At the opposite end are the **objectives,** which are normally 4×, 10×, or 43× in power. The objectives are attached to a **revolving nosepiece** at the base of the **body tube.** The body tube acts to transmit the image from the objective lens to the eyepiece lens. Your microscope slides will rest on the flat platform called the **stage.** Light from below the stage passes through the hole in the stage and enters the objective lens. The amount of light that is allowed to pass through the slide and into the objective lens is controlled by the **diaphragm** and **condenser.** The diaphragm is located directly beneath the condenser. The bottom support of the microscope is called the **base.** An illuminator, as the light source, or a mirror to reflect an outside light source upward into the condenser is attached to the base. Does your microscope have an illuminator or mirror?

________ To focus the objective on the object being viewed, the large knob, or **coarse adjustment knob,** and the smaller knob, or **fine adjustment knob,** are used.

Study figure 3.1 as you read section 3.2. It is important that you are familiar with the terminology associated with the use of the microscope.

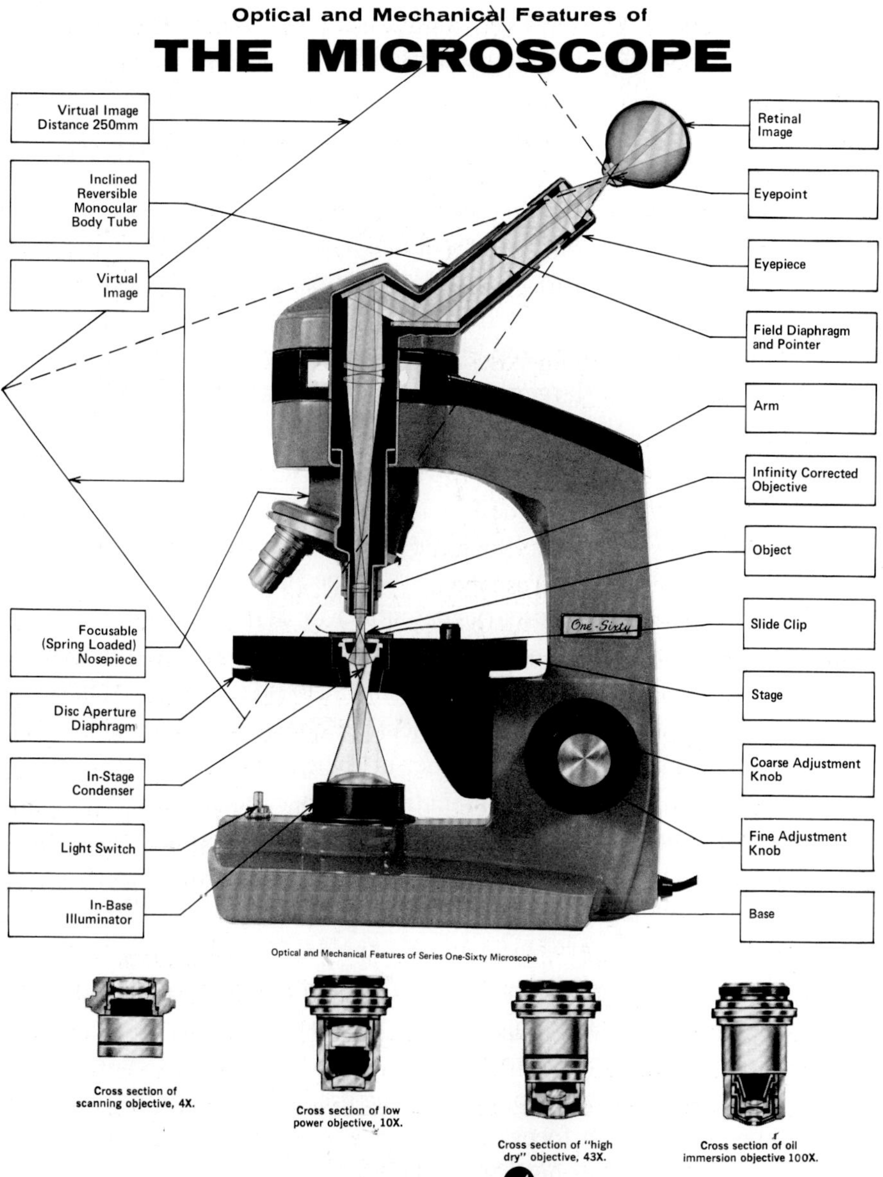

Cambridge Instruments
Optical Systems Division
P.O. Box 123, Buffalo, New York 14240-0123
Tel: (716) 891-3000

Figure 3.1 Optical and mechanical features of the microscope. Courtesy of Cambridge Instruments, Inc., Optical Systems Division.

1. The terms with modified explanation courtesy of Cambridge Instruments, Inc., Optical Systems Division.

3.2

The terms below will familarize you with microscopes, their parts, and how they are used.[1]

Common Optical and Mechanical Microscope Terms

Aperture diaphragm. A rotatable disc diaphragm, or iris diaphragm, is located beneath the microscope condenser. It is used to control the light directed to a specimen on the slide and entering the objective lens. Resolution and contrast (the ability to see) of the specimen significantly depends on the proper setting of the diaphragm opening.

Coarse adjustment. The larger knobs actuate the nosepiece assembly. They are used for rapid, or coarse, focusing onto specimens mounted on very thick or well-type slides. Use only when the low-dry objective is in position.

Compound microscope. This optical instrument is used to *magnify* and *resolve* fine detail within a transparent specimen. It differs from the simple microscope (ordinary magnifier) in that it has two separate lens systems: an objective, located near the specimen, that magnifies the specimen in a different amount; and an eyepiece that further magnifies the image formed by the objective. The *magnification* observed by the eye is equal to the product of the magnification of both lens systems. The primary magnification number of objectives and eyepieces, followed by an *X,* are engraved on each objective and the ocular.

Condenser. These lenses collect light rays and converge them to focus on the specimen. The condenser is located directly beneath the microscope stage.

Cover glasses. These square, rectangular, or circular coverslips of thin, flat glass are used to cover microscope slide specimens.

Depth of focus. This phrase describes the thickness of the specimen that may be seen in focus at one time. The greater the magnification, the thinner the layer in focus at one time. The lesser the magnification, the thicker the layer in focus at one time.

Diaphragm. This assembly of thin-metal leaves is controllable by a lever to produce a variable-sized opening. The diaphragm is generally associated with microscope condensers and illuminators of intermediate and advanced types.

Dry objective. Microscope objectives are designed to be used dry, that is, without immersion oil. The 40×, 43×, or 45× objective is quite frequently referred to as the *high-dry* objective; the 5× or 10×, *low-dry* objective.

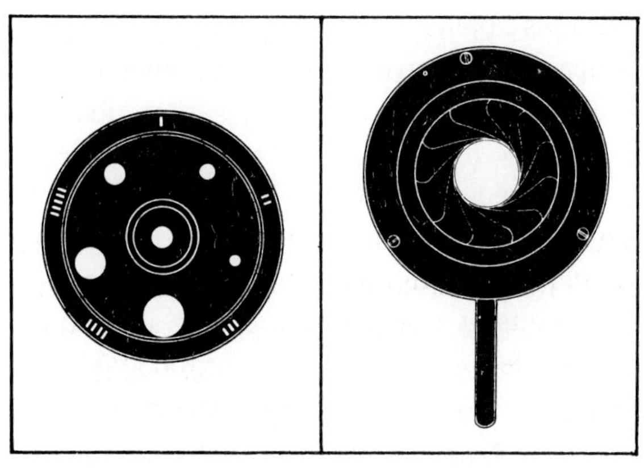

Figure 3.2 Typical diaphragms associated with microscopes. These are used to control the amount of light that reaches the object being viewed. Courtesy of Cambridge Instruments, Inc., Optical Systems Division.

Empty magnification. Large magnification increases size without enhancing resolution of the details of the specimen.

Eyepiece. The topmost optical lens of a microscope, sometimes referred to as the *ocular,* is used by the observer to further magnify the primary image transmitted and amplified by the objective.

Field of view. This phrase refers to the visible area seen through the microscope when a specimen is in focus. The field of view varies directly with the magnification: the greater the magnification, the smaller the field of view.

Fine adjustment. The smaller set of focusing knobs controls the precise, fine movement of the objectives. *Always* use fine adjustment when high-dry (40×, 43×, or 45×) objective is in position.

Lens. The term refers to a convex or concave transparent glass used to change the direction of the rays of light and, thus, magnify the apparent size of objects.

Light. The human eye can see only with radiation of 400 to 700 millimicrons and is most sensitive at 555 millimicrons, or .555 microns, which is yellow-green light. (1 micron is equal to one-thousandth of a millimeter.) Light is radiant energy of the wavelengths described above that, upon reaching the retina of the eye, stimulate nerve impulses to produce the sensation of vision. White light is composed of a mixture of colored light of different wavelengths. When specimens are too transparent to be seen effectively, they may be stained, which enables the specimen to be seen by the color image formed as the dye absorbs certain wavelengths of light and transmits the others to the eye.

Objective. The objective is the complex lens system located directly above the object or specimen.

Parfocal. This term is applied when practically no change in focus has to be made when one objective is changed for another. The objectives on the revolving nosepiece of a microscope are parfocalized so that only a slight turn of the fine adjustment is required when changing from a low to a higher power objective.

Resolving power (resolution). This phrase refers to the ability of a microscope to reveal fine detail. It is stated as the least distance between two lines or points at which they are seen as two, rather than as a single blurred object. Increased magnification without increased resolution is called empty magnification. Resolving power is a function of the wavelength of light used and the greatest cone of light that can enter the objective.

Working distance. This phrase refers to the distance between the objective and the top of the cover glass when the microscope is focused on a specimen preparation. The greater the magnification of the objective, the smaller the working distance.

3.3

The *stereoscopic dissecting microscope* is used for a specific purpose, to magnify objects only. This microscope is capable of magnifying a rock; whereas, a compound microscope is only capable of magnifying a thin, transparent organism or object that allows light to pass through it.

The stereoscopic dissecting microscope has two eyepieces and, usually, one objective. Figure 3.3 illustrates this microscope. Get a stereoscopic dissecting microscope from the storage cabinet, and compare the photograph to your microscope. Be able to identify the parts as you use this microscope in your work.

3.4

You must become proficient at caring for the microscope you use. There are four major areas where the microscope can get dirty and hamper your viewing. The first is the lens of the eyepiece. Oil from your eyelashes will smear the lens and, as a result, you will have a blurred image. Second, the lenses of the objectives also can become smudged if you let them touch solutions. This also will blur the image. The only cleaning aid you should use on the two surfaces is *lens paper.* This is a special tissue designed for the purpose of cleaning the lenses. Other materials may scratch a lens because it is a precision, soft, ground glass. Use only the lens paper provided.

The third way a dirty microscope can impair your performance is to place a dirty slide onto a dirty stage. Both should be kept clean and dry. Nothing special is needed to do this. Fourth, the mirror or light source should be kept clean. Again, use soft material such as lens paper or facial tissue to clean it.

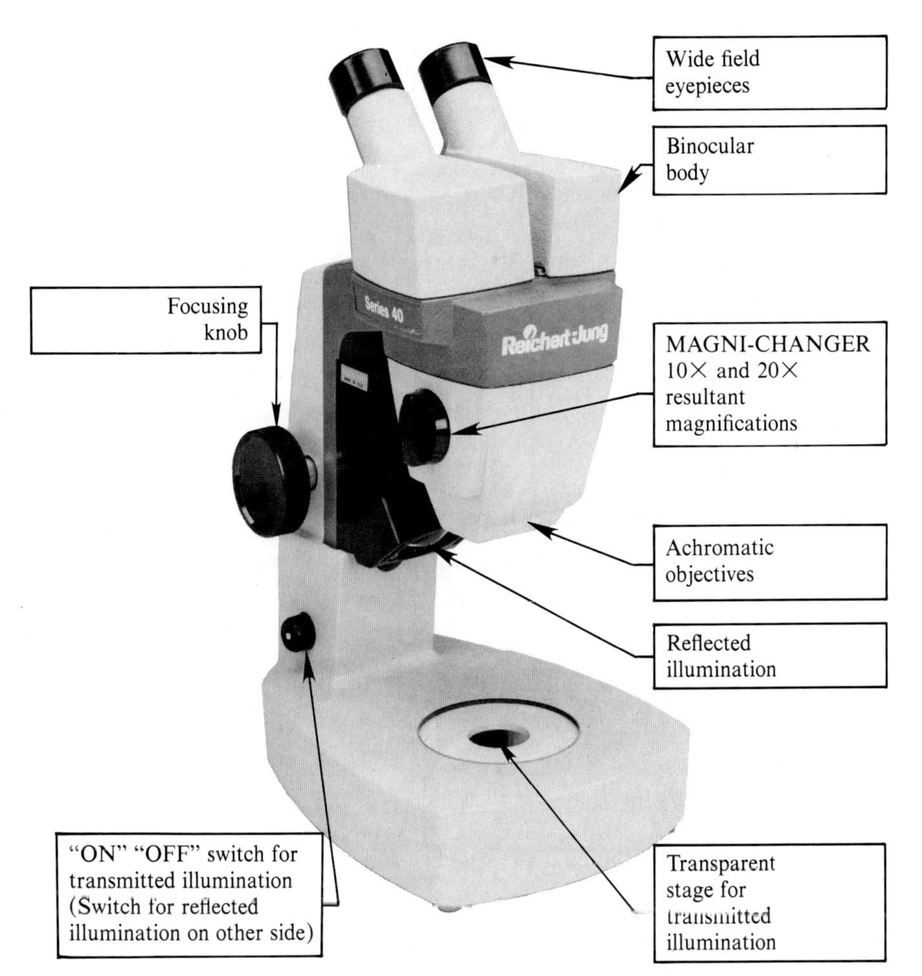

Figure 3.3 Mechanical features of a stereoscopic dissecting microscope. Courtesy of Cambridge Instruments, Inc., Optical Systems Division.

Many times dust covers of some type are available. These help keep dust away from the mechanical parts and lenses. If your microscope has one of these, be sure to put it on before you return it to the storage cabinet.

Using lens paper, clean your microscope now.

3.5 There are three images involved with microscopy. The lens system of your microscope forms an image of the object you are viewing. The primary image is called the *real image*. The real image may be larger or smaller than the object. It is formed by the rays of light that pass through the lens. This image could be focused on a screen within the ocular. Study figure 3.4. Locate the real image. How does the real image position itself with regards to the object? ______________________

____

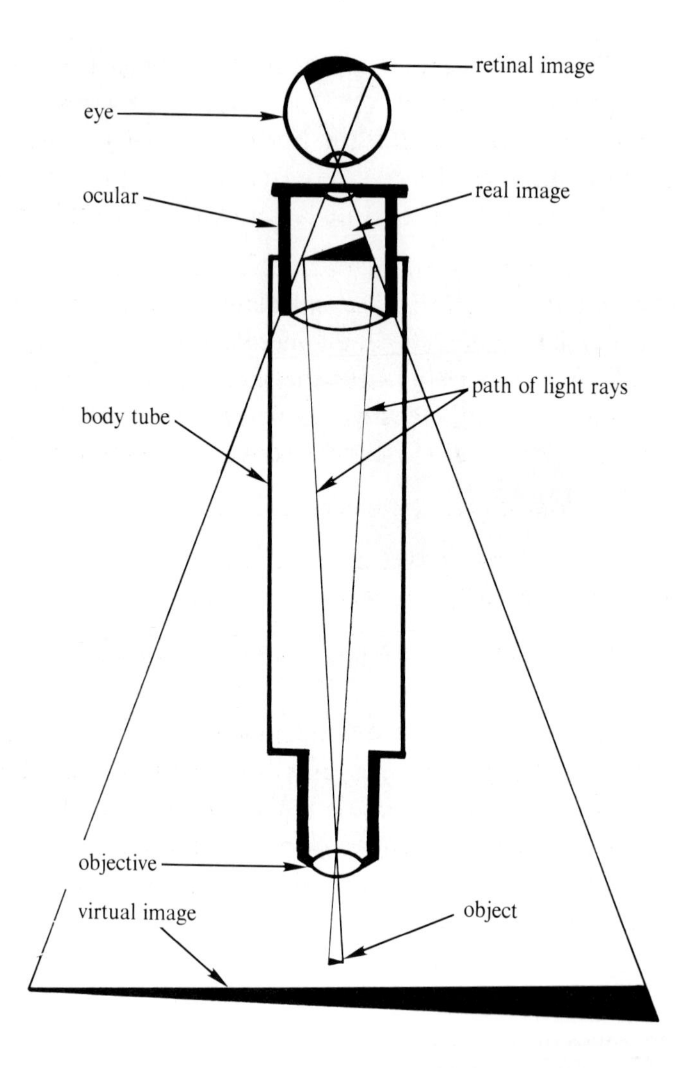

Figure 3.4 The formation of the retinal image and virtual image with a compound microscope.

3.6

The two other images formed with microscopy are the *retinal image* and the *virtual image*. The retinal image, as the name implies, is formed on the retina. The retina is a lining on the back of the eye containing special sensitive cells called rods and cones. The rods are sensitive to varying degrees of light but not color. The cones are sensitive to color. The retinal image appears to the eye to be in the position of the virtual image. And because of the magnification properties of a microscope, the retinal image gives the impression of magnification. The virtual image is produced by your sense of sight, which includes the eye and the brain.

Again in figure 3.4, locate the position of the retinal image and the virtual image. How do they compare to the relative position of the object and the real image? ______________________________

3.7 Total magnification results in the combined product of the power of the ocular, or eyepiece, and the objective. The power of each is normally stamped on them. For example, if you are observing an object with a 10× objective and the eyepiece is 5×, then the total magnification would be 5 × 10 or 50 times larger than the object being viewed.

What is the power of your eyepiece? ________________ Using the different objectives of your microscope, calculate the total magnification for each objective. *Please note* that not all microscopes have the same number of objectives, nor have all objectives the same power, so make sure you check the power of the objective to know the total magnification. The parenthetical term is also used to describe low power. On your microscope, identify low power (scanner) ______________, middle power ______________, and high power ______________.

If your scope has a fourth objective, usually an oil immersion lens, indicate the total magnification for this objective: ________________

3.8 Before you can use your microscope, you must first be able to *focus* on an object properly. Focusing is not easy for the beginner. Many times you will focus on a piece of lint or a particle of dust rather than on the object. Focusing can be accomplished easily each time if you get into the proper procedural habit. Get a prepared slide from the supply table. It does not matter what slide you use. Place it on the stage, center the object you wish to see directly over the hole in the stage, and fasten it down with the clips. *Now follow the steps listed below* to correctly focus an object.

a. Always begin with the lowest-power objective.
b. Viewing from the side, not through the eyepiece, lower the objective as low as it will go. This method prevents scratching the objective lens and breaking the slide.
c. Looking through the eyepiece, slowly raise the objective with the coarse adjustment knob until you can make out an image. It is best if you can become adjusted to viewing with either eye and not close the opposite eye.
d. Bring the image into sharp focus with the fine adjustment knob.

Practice focusing until you become proficient at it with all of the objective lenses. If your microscope is parfocal, you should be able to change the low objective to the next with only minor focusing corrections. *Never* use the coarse adjustment knob with a high-dry objective.

3.9 There are three distinct characteristics of a functioning microscope. They are: (1) the working distance, (2) depth of focus, and (3) the viewing field. The viewing field is the area you are able to observe with any given objective.

To relate to these characteristics and enhance your ability to use the microscope, get a metric slide. Observe the slide with each of the objectives, excluding oil immersion, and record the number of millimeters (mm) you can observe with each objective.

1. low-power ______
2. middle-power ______
3. high-power ______

Which objective has the largest viewing field? ______

Which objective has the smallest viewing field? ______
Make a general statement relating power of objective to viewing field.

3.10 To illustrate depth of focus, you will need a prepared colored thread microscope slide. View this slide using low power.

1. Which colored thread comes into focus first? ______
2. Which colored thread comes into focus last? ______
3. Which colored thread has the greatest depth of focus? ______
4. Which colored thread has the shallowest depth of focus? ______

Always return slides to a predesignated place.

3.11 When you are scanning a slide, you will find the microscope does not always respond to your hand adjustments. Place any prepared slide on the microscope stage. View it with low power. Move the slide to your right. What direction does the slide appear to be moving while viewing it with the microscope? ______ Move the slide to your left. Does the same kind of phenomenon occur? ______ Predict what will happen if you pulled the slide towards you and away from you while viewing it.

3.12 Using the same prepared slide, adjust the diaphragm to see the effect that light has on the object you are viewing. Record your observations in the space provided.

When finished, return the microscope slide and the microscope to its proper place.

3.13 The use of the microscope is essential when you view the microscopic world. The word *micro* refers to the small, not-able-to-be-seen-with-the-naked-eye world. The existence of the *macro* world is dependent on the microscopic world. This places the micro world at the bottom of the food pyramid, making it extremely important. Humans are found at the top of the food pyramid, which is basically the least significant. The higher an organism is in the food pyramid, the more dependent it is on the organisms below.

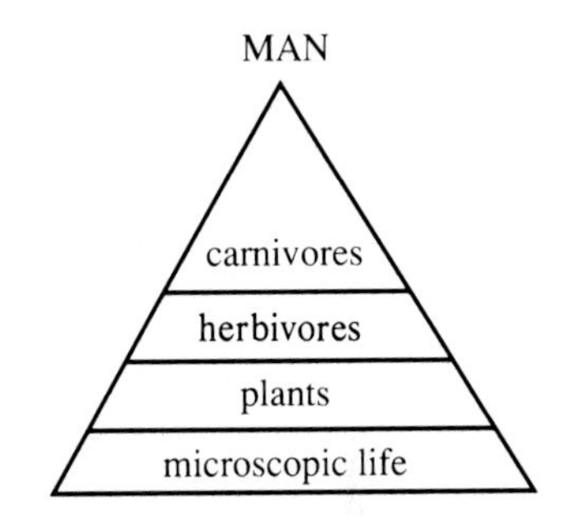

Figure 3.5 A simplified model of a food pyramid.

The microscopic world was first observed by Anton van Leeuwenhoek, a Dutch naturalist. He was able to see small creatures, suspected of being present, but whose existence was doubted by most of his contemporaries because of their superstitious beliefs. Van Leeuwenhoek called the small creatures he observed "animalcules." His drawings were quite accurate, though they lacked detail because his microscopes were not capable of much magnification. As the microscope improved, scientists discovered multitudes of microscopic species.

Figure 3.6 Anton van Leeuwenhoek was credited with much of the discovering of the microscopic world.

From van Leeuwenhoek's time, the microscope has evolved into a major research instrument. With some modifications, the basic compound microscope became a *medical microscope*. This, in turn, became a *binocular research microscope*. Finally, when optical microscopes were unable to do a proper job, scientists developed the latest and most powerful magnifying instruments, the *transmission electron microscope* (TEM) and the *scanning electron microscope* (SEM).

Figure 3.7 illustrates what each of the above microscopes accomplishes. Look at each photograph carefully. If you wanted to study the basic anatomy of a housefly quickly, which microscope would you most likely use? ________________ What microscope would you use if you wanted to look at red blood cells at a highly magnified level, giving three-dimensional detail? ________________

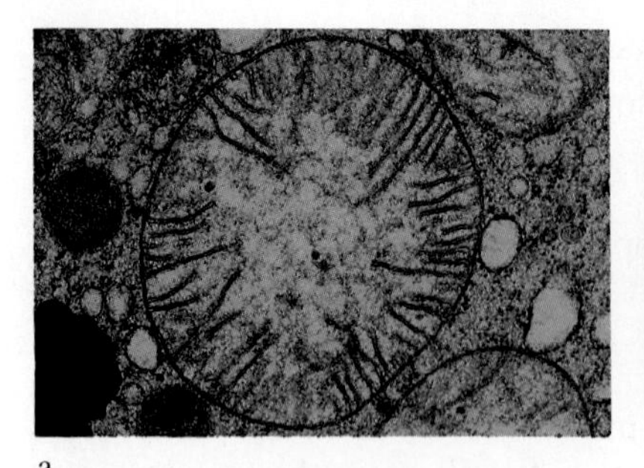
a.

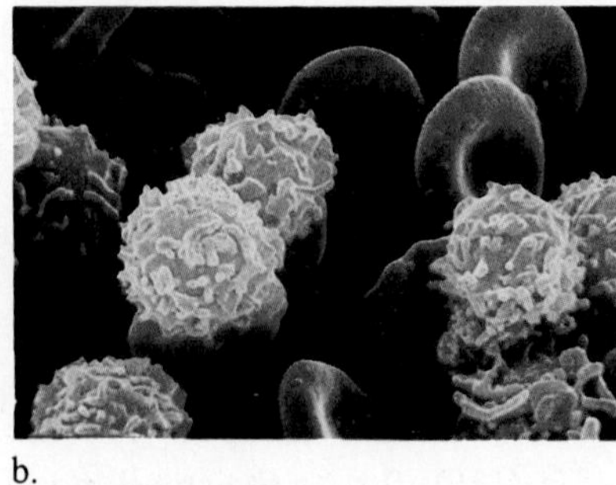
b.

c.

Figure 3.7 Observations made with different types of microscopes: (a) mitochondria magnified by a transmission electron microscope, (b) blood cells as seen with a scanning electron microscope, and (c) paramecia seen with a compound microscope.

3.14

There are few experiences in science that match the excitement of seeing the living, microscopic world. Can you imagine how van Leeuwenhoek would have reacted if he had had the microscopic technology of today? Even the most experienced microscopist looks on in awe at the hidden world that resides in a drop of pond water.

With the microscope it is natural to want to look at the smaller, living world, so let us do that. You will need to collect some pond water samples and transfer a part of that water to a specially prepared culture water.

To make the cultured water solution, follow the procedures below:

a. Add several pieces of timothy hay or other clover-type hay to 400 mL of boiling water. Boil the hay for 3 to 5 minutes.
b. In a separate container, boil 10 grams of rice for 5 minutes.
c. Transfer your pond water samples to assigned culture dishes. Add to your culture dish, ten to twelve 2-inch pieces of boiled hay and a couple of grains of rice. The rice and hay infusion will ensure a high bacteria count to provide a food source for many protozoans.

Initially you should have a good number of organisms available. However, within a couple of days, your culture should bloom.

The microscopic world you will be exploring will probably have many different forms of life in it. The two kingdoms under which these life forms are classified are protista and animalia. The following phylums represent the major classification of the two kingdoms.

The organisms listed in data table 3.1 are a few of the more common ones you will find in most freshwater pond samples. Figure 3.8 illustrates some of these organisms. As you scan pond water samples, locate, draw, and name twelve of the above microscopic organisms.

When you draw microscopic organisms, spend enough time to render them as accurately as you are able. Here is your chance to make careful observations and communicate these to your instructor and fellow students through good scientific illustrations.

Data Table 3.1 Information Concerning Specific Microorganisms

I. Protist Kingdom

Phylum	*Comments*
1. Rhizopoda	Single-celled amoebas, either naked or having a shell.
2. Chrysophyta	A group of algae that have yellow-pigmented plastids and two undulipodia.
3. Euglenophyta	Capable of locomotion by means of a whip-like tail called flagellum. They also contain green chloroplasts, allowing for photosynthesis when holozoic or animal-like ingestion is not possible.
4. Cryptophyta	Like euglenophytes, crytophytes are capable of photosynthesis or holozoic feeding habits.
5. Bacillariophyta	As many as ten thousand different species exist. Both marine and freshwater diatoms, as they are commonly called, are usually brownish to yellowish in color.
6. Gamophyta	Green algae, with the most common forms being Closterium, Desmidium, and Spirogyra.
7. Chlorophyta	Green algae that differ from gamophyta due to sexual forms. Common members of the chlorophyta include Chlamydomonas, Chlorella, Nitella, Oedogonium, Ulothrix, Ulva, and Volvox.
8. Ciliophora	Unicellular, heterotrophic organisms that are most familiar to the novice microscopist include Blespharisma, Didinium, Euplotes, Paramecium, Spirostomum, Stentor, Tetrahymena, and Vorticella.

II. Animal Kingdom

Phylum	*Comments*
1. Platyhelminthes	A flatworm that is adaptable to many habitats. Planaria are the most familiar organisms to the novice biologist.
2. Gastrotricha	Freshwater gastrotrichs prefer plant-choked environments.
3. Rotifera	Named for the crown of cilia at their heads, these aquatic animals can also be found on lichenous material, moss, or hay.
4. Nemotoda	Also called roundworms, nematodes are found where rotifers are found. Basically parasitic nematodes may lay one-half million eggs per day.
5. Tardigrada	Also called water bears, this aquatic animal is also found in habitats similar to rotifers and nematodes.
6. Arthropoda	A large class of arthropods is the class Crustaceae. In this class are commonly found the Cladocera (water fleas), Ostracoda (ostracods), Anostraca (fairy shrimp), and Copepoda (copepods).

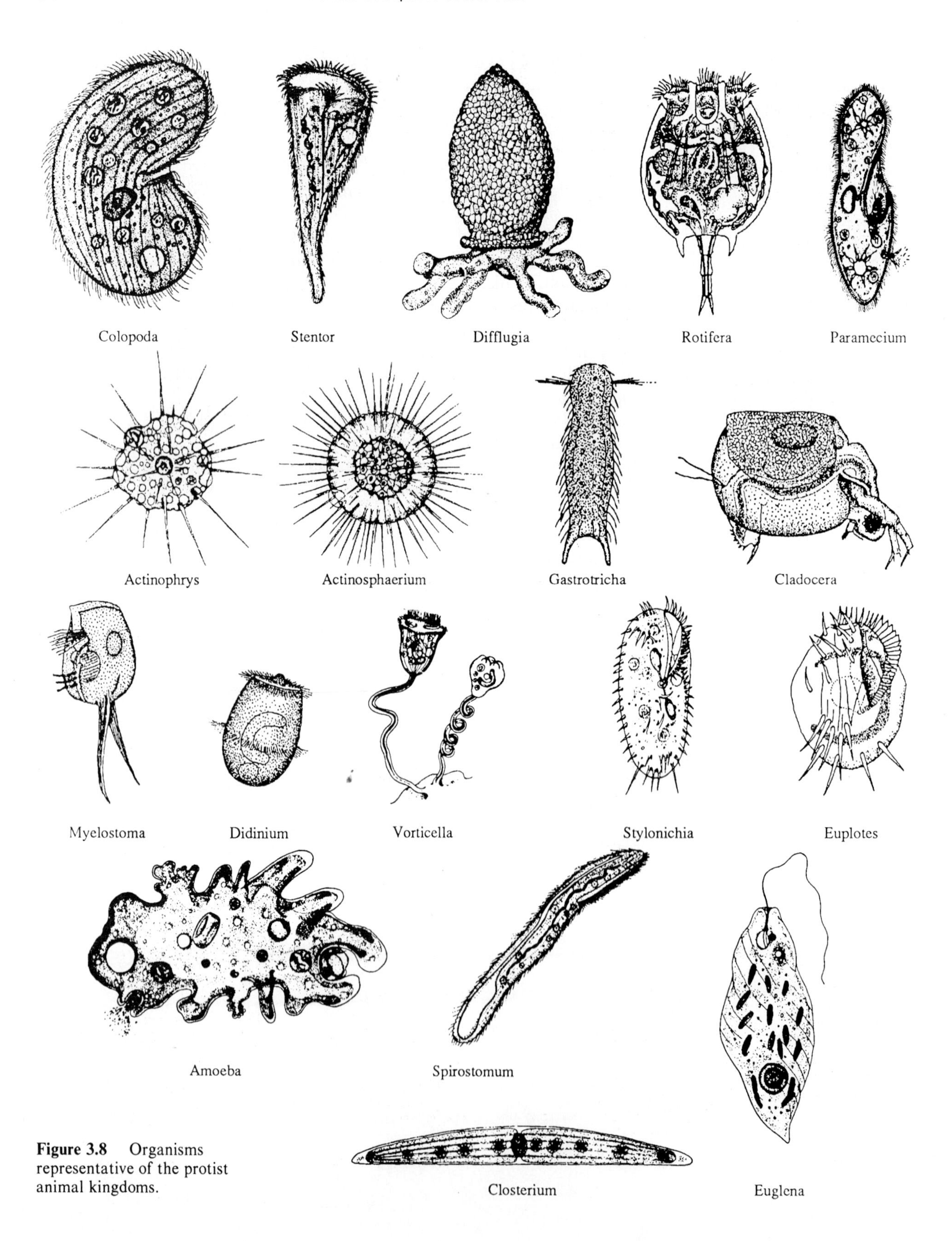

Figure 3.8 Organisms representative of the protist animal kingdoms.

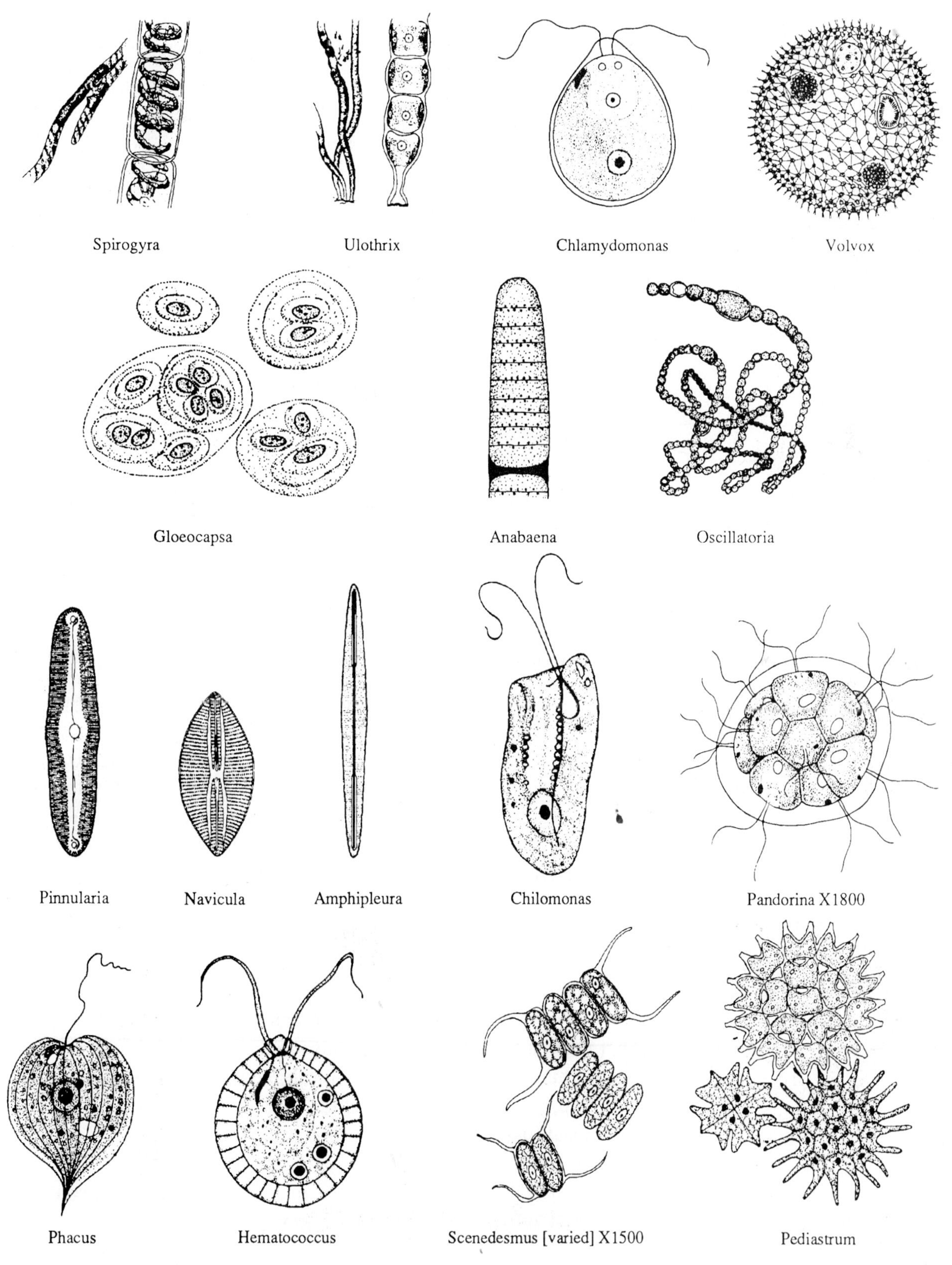

Figure 3.8 *Continued*

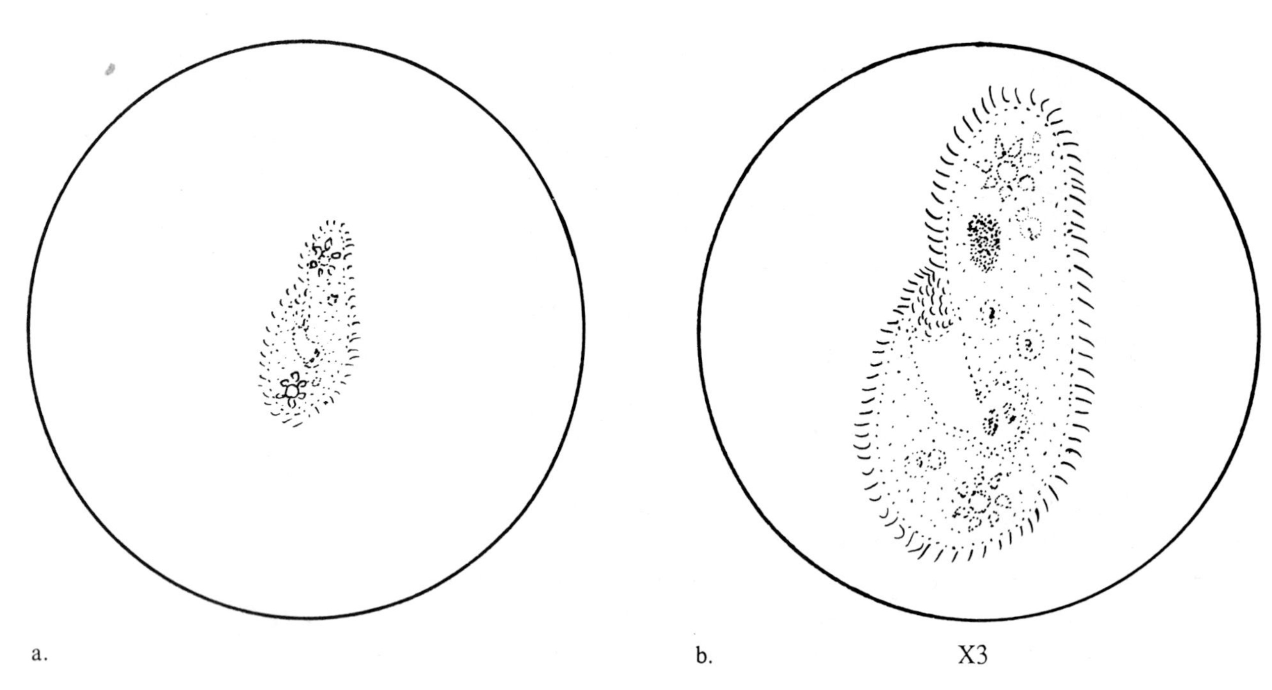

Figure 3.9 Proper illustration for your work should be increased as shown in diagram b. This allows you to enhance the detail of the organism.

The sketches of the microorganisms can be increased to enhance the detail. Do not draw an organism like drawing A when you mean for it to look like drawing B in figure 3.9. Viewing fields (figure 3.10) have been provided.

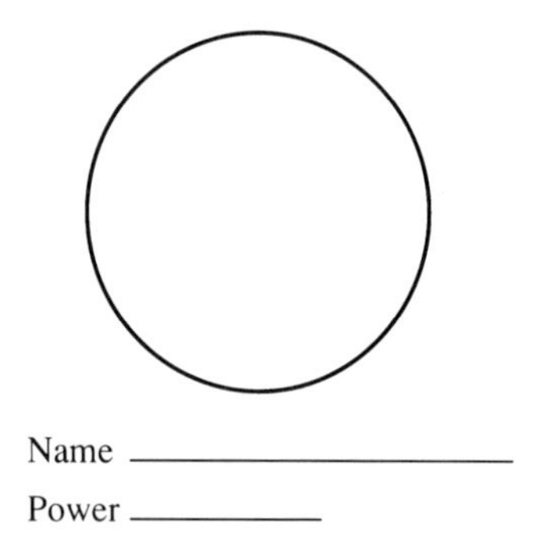

Figure 3.10 Viewing fields provide a format for illustrating microscopic organisms. Notice the information required. If you increase the size of your illustration, identify the approximate reference, such as 2X or 3X, etc.

3.15 Pond water samples offer a chance to view and marvel at the diversity of living organisms. Microscopic organisms can also be found in other places, far away from a water source. The classic example is called the hay infusion. You simply put some pieces of hay in tap water that has been allowed to stand 24 hours to rid the water of excess additives. Within 2 to 3 days you should be able to check the water and find a variety of living organisms.

Moss samples and lichen samples collected in forests reveal an abundance of living organisms when prepared and studied. In fact, an unusual tale evolves from lichens.

Lichens can be found clinging to the bark of trees. A part of the fungi kingdom, lichens support a large microscopic population of several species. When it rains, a thin film of water clings to the lichen. This film is enough to produce many hidden life forms (tardigrades, nematodes, rotifers, and others). And, as long as this film exists, there is an environment for them to move about, carrying out normal life processes.

When conditions become less favorable and the thin film of water evaporates, most of the microscopic organisms will carry out a **cryptobiotic** process called **anhydrobiosis.** Refer to figure 3.11.

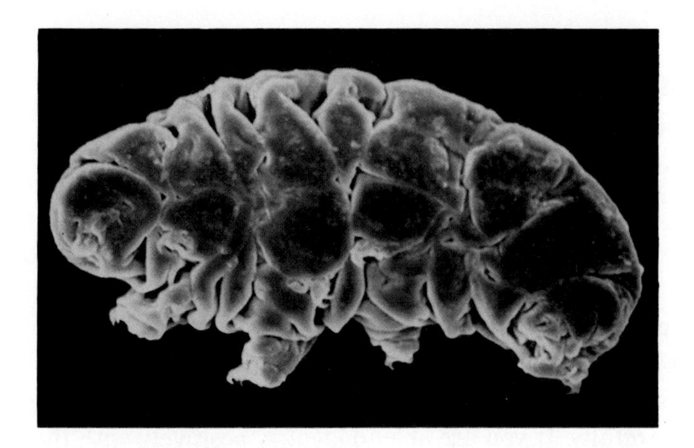

Figure 3.11 A microscopic organism called a tardigrade. Tardigrades can withstand a temperature range of less than 200° C to greater than 100° C for short periods of time.

A tardigrade is approximately 85 percent water, but during anhydrobiosis that amount drops to 3 percent and the organism takes on a contracted form called a *tun.* This form apparently keeps any further water from desiccating from the tardigrade's body. As conditions improve, the body of the dehydrated tardigrade will swell, similar to a dry sponge soaking up water.

The purpose of this next activity is first to find the main types of life that exist on lichens, and, second, to simulate an anhydrobiotic situation that will enable you to see visually animals become active after having previously been in this state.

To prepare lichen material, soak it in an inch of water in a petri dish for 24 hours. At the end of this time, remove all lichens and/or bark and scan the bottom of the petri dish with a stereoscopic dissecting or replace dissecting for stereoscopic microscope. The tardigrades will be viewed walking on the bottom of the petri dish. Rotifers and nematodes should also be visible, but we will work with the tardigrade only. If the others are transferred with the tardigrade, that is okay. They perform the same anhydrobiotic process as tardigrades do. However, tardigrades are the chosen organism for this activity.

Once you have found a tardigrade, carefully draw it into a medicine dropper and transfer it to a depression slide. There you can view it further with the aid of a compound microscope.

By allowing the water to evaporate, you should cause the tardigrade to move into its anhydrobiotic state. View the slide to determine that this is true.

Next, you want to watch the tun rehydrate. Add a couple of drops of distilled water to the depression slide. Rehydration will occur within 10 to 20 minutes. You may not need to watch constantly, but you will want to keep close attention to your work.

3.16 Were you successful at invoking an anhydrobiotic state? _______ Were you successful with rehydration? _______

3.17 The microscopic world has so many wonders to see and study. There simply is not enough time in a general biology class to completely cover all aspects of microscopic protist and animal kingdoms.

One thing is for certain, though. The first microscopes were not very effective, but they led the way for modern development. Science and medicine have certainly advanced because of this one instrument.

The study points below are based on the context of this chapter. Read each one carefully. If you are able to identify positively with the objective, place a check (✓) in the appropriate box. If you have difficulty with a specific study point, review the corresponding material from the chapter.

On completion of this chapter, you should be able to do the following:

☐ 1. Have command of the various parts of a compound microscope and a stereoscopic dissecting microscope.

☐ 2. Make a temporary mount of a microscope slide, and correctly use the compound microscope to observe an object on a microscope slide.

☐ 3. Examine pond water for protists and invertebrate animals, and identify them using various types of taxonomic keys.

☐ 4. Locate tardigrades and possibly rotifers and nemotodes after soaking lichen material. Having found them, effectively bring about an anhydrobiotic condition followed by a reanimated one.

☐ 5. Identify the major characteristics of a good microscope.

☐ 6. Focus a parfocal microscope properly.

☐ 7. Differentiate between the real image, retinal image, and virtual image.

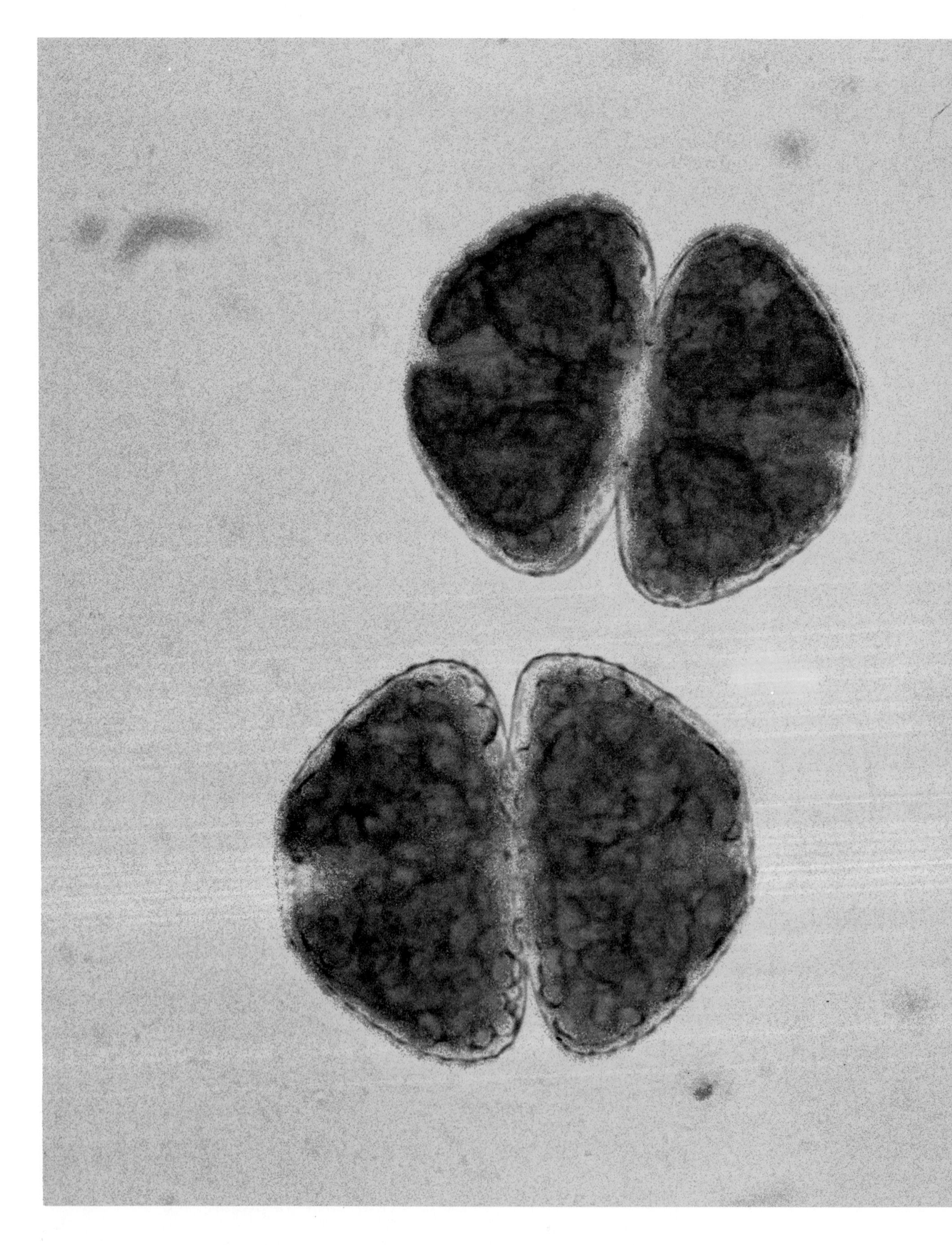

4 The World of the Cell

What is the structure of the cell?

4.1 There exists within each living system a common pattern of structure—a unit that unifies the living world. Every living thing is made of cells, and everything a living thing does is done by cells. Cells are the basic structural and functional units of life in creatures as diverse as rose plants, birds, humans, bacteria, fish, protozoa, and fungi. All of an organism's activities are the result of cellular activities. Life exists and reproduces itself through cells.

Understanding cellular activities contributes to solving problems in medicine, industry, and agriculture. The new frontier of medical science lies inside the microscopic world of the cell. Understanding the basic workings of cells is the key to controlling the major health problems of today: cancer, genetic diseases, mental illness, and deficiencies in the immune system. To control these and other disorders in living systems, researchers need to know much more about what goes on inside the cell. Cell biologists continue their efforts to understand how cells orchestrate the activities of living things.

What Are Cells?

To learn this, you will figuratively have to enter the world of the cell to make a trip like Alice made through the looking glass, "where everything is smaller and smaller, and curiouser and curiouser."

Since comparisons of size, mass and volume utilize the metric system, it is important to be familiar with metric language to enjoy fully your trip through the cell. The use of the metric system makes it easy to change from one level of measurement to another by moving a *decimal point* and changing the *prefix* of the unit of measure. Three units of the metric system are *meter* (linear measure), *gram* (mass measure), and *liter* (capacity measure). Data table 4.1A shows the chief prefixes of the metric system and the relationship of these units to each other. Data table 4.1B shows the prefixes used with each unit of measure and the symbol used for each measurement. Record the answers to the following questions in your log book.

1. If the diameter of an object is 2 centimeters, how big is the object in millimeters? ______________________
2. If you have a cell whose diameter is 20 micrometers, how big is the object in cm? ______________________
3. One-fourth (1/4) of a liter equals how many milliliters? ______

Let us begin your journey into the cell by starting with a familiar object such as a tree (figure 4.1). Far away, a tree is an indistinct mass of color. Coming closer, your eyes resolve the mass of color into individual leaves (figure 4.2). Closer still, each leaf has detail such as hairs and veins (figure 4.3).

At this level you have reached the limit of seeing leaf structure, smaller than 100 micrometers, with your unaided eyes. Yet leaves are made of tiny units called cells. Most plant and animal cells have a linear measurement between 20 and 30 micrometers. The average human cell is approximately 20 micrometers in diameter. Other structural parts of the human cell range from 5 m in length to 75 Å in width. Data table 4.2 shows the metric range of eukaryotic cells, organelles, molecules, and atoms.

Data Table 4.1 Metric Units

A. Chief Prefixes of the Metric System

Prefix

kilo = $1000X = 10^3x$
deci = $1/10X = 10^{-1}x$ = 10 deci
centi = $1/100X = 10^{-2}x$ = 100 centi
milli = $1/1000X = 10^{-3}x$ = 1000 milli
micro = $1/1{,}000{,}000X = 10^{-6}x$ = 1,000,000 micro
angstrom = $1/10{,}000{,}000{,}000X = 10^{-10}$ = 10,000,000,000 angstroms
X = meter, gram, or liter

B. Chief Units, Prefixes, and Symbols of the Metric System

Linear Measure (meter)	**Symbol**
1 kilometer = 1000 meters	km
1 meter = 100 centimeters	m
1 centimeter = 10^{-2} meters	cm
1 millimeter = 10^{-3} meters	mm
1 micrometer = 10^{-6} meters	μm
1 nanometer = 10^{-9} meters	nm
1 angstrom = 10^{-10} meters	Å
Mass Measure (gram)	
1 kilogram = 1000 grams	kg
1 gram = 1000 milligrams	g
1 milligram = 10^{-3} grams	mg
Capacity Measure (liter)	
1 kiloliter = 1000 liters	kl
1 liter = 1000 milliliters	l
1 milliliter = 10^{-3} liters	ml

Figure 4.1 View to a forest of trees.

Figure 4.2 Individual leaves of a single tree in the forest.

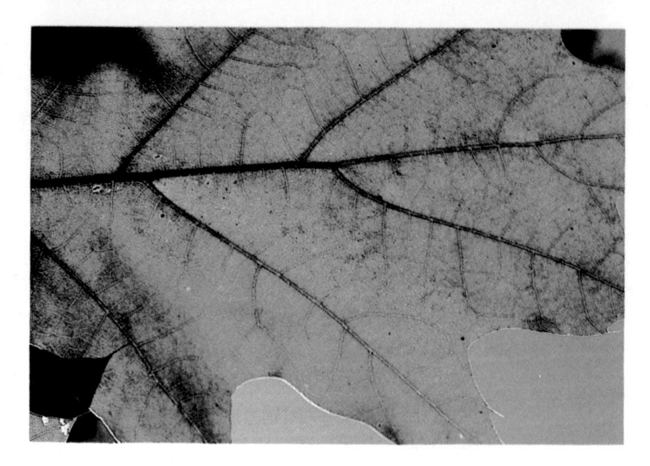

Figure 4.3 Close-up view of a single leaf with veins.

Data Table 4.2 Approximate Range of Metric Dimensions of Cells, Organelles, Molecules, and Atoms

	Metric Units
Cells	μm to cm
Organelles	nm to μm
Macromolecules (proteins)	5 to 500 nm
Small molecules (amino acids)	1 nm
Atoms	Å

To visit the world of a cell requires the use of a microscope. With a microscope, you can see that the leaf has a complex architecture based on units called cells (figure 4.4). Each cell, too, has an internal structure consisting of many individual parts functioning together (figure 4.5). At higher magnification with the electron microscope, you can see that these individual parts, or organelles, also have an internal organization and are quite different from one another (figure 4.6).

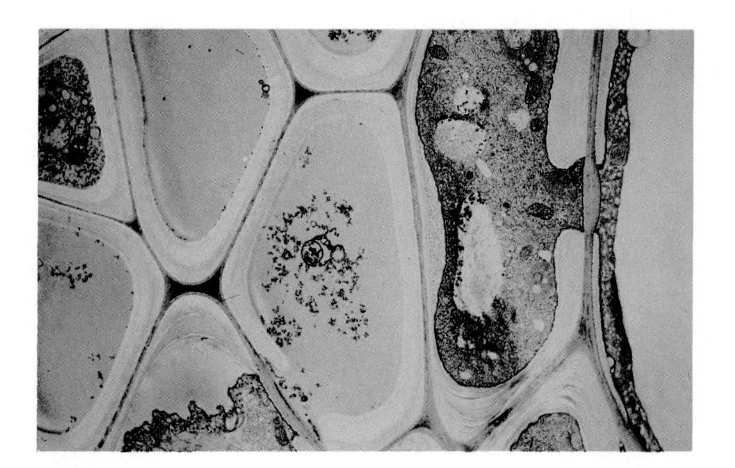

Figure 4.4 Light microscopic photograph of several leaf cells. Only one complete cell is visible.

4.2

Now let us take a look at an animal cell, with a light microscope. The *animal cells* you will observe are called **epithelial cells.** These cells form the outermost layer of your skin and the inner lining of your intestines, stomach, and mouth.

With the aid of a wooden applicator stick, gently scrape the skin on the inside of your cheek. This will remove cells from the epithelial lining of the inside of your mouth. Place a single drop of water on a clean slide, and stir the tip of the applicator stick containing the epithelial cells in the drop of water. Next, add a drop of methylene blue stain to the drop of water. This will help you see the cells.

View the cells with the aid of a compound light microscope at high-power magnification. Make a diagram in your log book of two cells, and label the drawing with a letter to represent the different cell parts that you see. It is not necessary to name each cell part at this time. However, if you want to make a guess, go ahead. How many *different* cell parts did you find? ________. Record in your log book, and be careful not to count air bubbles. These represent air trapped under the coverslip when you prepared the material for microscopic examination.

Save your cheek cell slide until you complete sections 4.3 and 4.4.

4.3

Let us continue to look at other living cells. First, obtain an *Elodea* leaf. Place the plant leaf on a clean microscope slide, add water, and cover with a coverslip. Observe a leaf at different focal points with a compound microscope. The leaf will appear to be several cells thick so you will note some overlapping of the plant's cells. This is similar to the overlapping of bricks in a brick wall (figure 4.5).

Elodea leaves have spike cells along the margin of the leaf. Also, examine a spike cell with your microscope under high-power magnification.

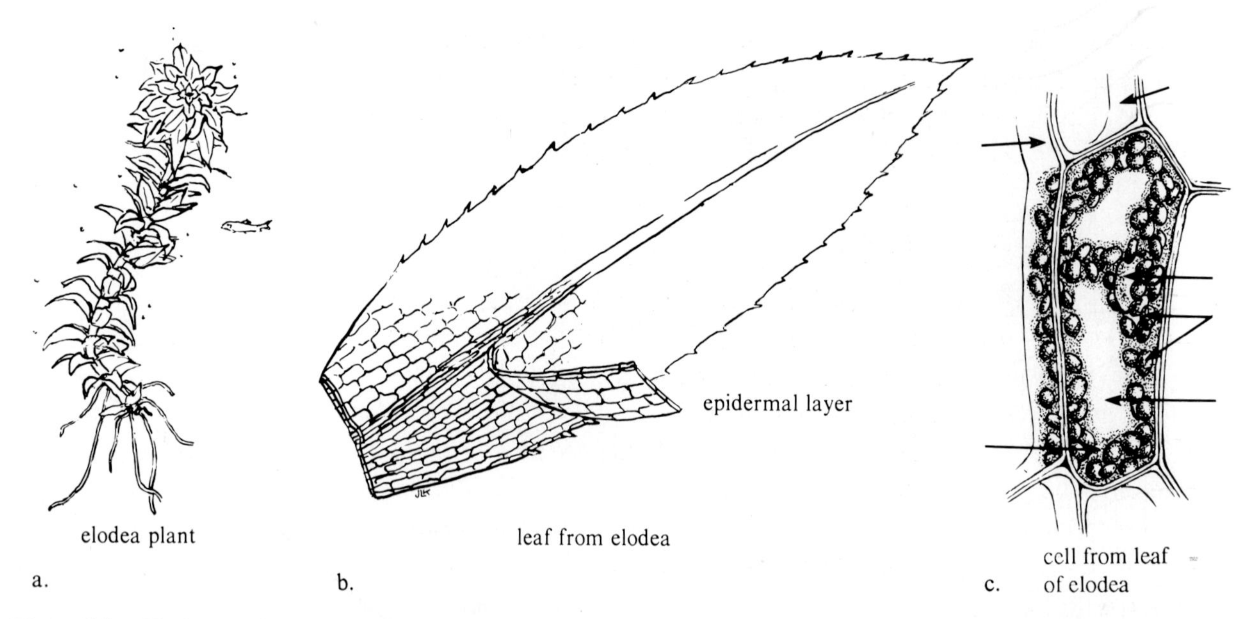

Figure 4.5 Elodea. a. plant. b. leaf. c. cell. Courtesy James Koevenig.

Label in the log book all arrows in figure 4.5c with different letters (*a–e*). Find the cell parts in your elodea leaf that are marked with an arrow in figure 4.5c. It is not necessary to name each part at this time. How many did you find? ________ . Place an asterisk (*) on log book figure 4.5c beside any part you cannot find.

1. Do plant cells and animal cells appear to have similar parts?

   ______

2. Do plant cells and animal cells appear to have different parts?

   ______

Save your slide preparation until you complete section 4.4.

4.4 Using methods similar to these and the light microscope, biologists discovered that cells had certain components in common. Today cell biologists, as the explorers of the cell are called, have produced a generalized map of the cell (figure 4.6). The bodies within the cell, the **organelles,** are named and defined. Now, let us follow the map that these cellular explorers have provided us through their research studies, and enter into and journey through the world of the cell.

A **cell wall** encloses most plant cells, but it is not found with animal cells. The wall that surrounds a growing plant cell is called the *primary wall.* Many plant cells deposit additional *secondary wall* layers inside the primary wall, but outside the plasma membrane (figure 4.7).

Plasma Membrane—The Cell's Shelter

Just inside the cell wall of plant cells is the **plasma membrane** (figures 4.6, 4.7). Cell biologists using a light microscope cannot see the plasma membrane directly, but when cells are placed in a solution that causes water to leave the cell by osmosis, the plasma membrane can be hypothesized as the outer boundary of cytoplasm, or cell contents. Experiments have demonstrated that the plasma membrane exhibits selective permeability and, thus, controls the movement of substances into and out of the cell. Animal cells are not enclosed by cell walls. The plasma membrane is the outer covering of animal cells.

Find and label in your log book the cell wall and plasma membrane in your cheek cell drawing and figure 4.5c.

Another structure that is characteristic of a mature plant cell is a large **vacuole,** a membranous sac filled with water and salts. The vacuole pushes the nucleus and other organelles against the periphery of the cell. Some plants store a red pigment called anthocyanin in their vacuoles. This pigment gives leaves and some fruits and vegetables their red color. Succulant plants will temporarily store the CO_2 fixed during photosynthesis in their vacuoles for later sugar production. Find and label in your log book a vacuole in figure 4.5c. Young plant cells (figure 4.6*a*) and some animal cells usually have smaller vacuoles.

Nucleus—The Cell's Command Center

The **nucleus,** bounded by a double membrane unit called a **nuclear envelope,** is found in almost all plant and animal cells. Inside the nucleus another structure called the **nucleolus** can be seen. The **chromatin,** the major material contained in the nucleus, is primarily deoxyribonucleic acid, DNA. The DNA molecules within the nucleus store the genetic information of the cell. During certain times in the life of all cells, the chromatin contracts to form a mass of tangled threads, or short cylindrical objects, called **chro-**

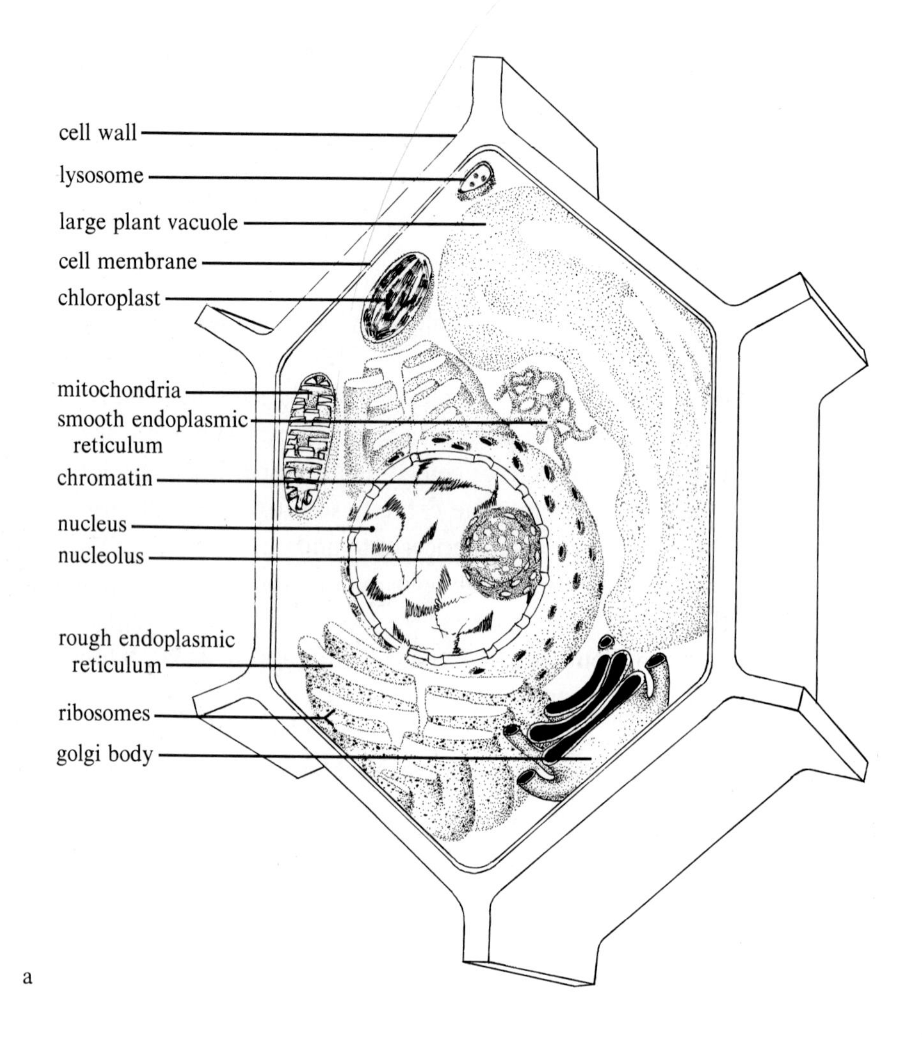

Figure 4.6 A typical cell.
a. A drawing of a plant cell to show the various parts of the cell.
b. (opposite page) A drawing of an animal cell to show the various parts. Compare to the plant cell sketched in 4.6a.

mosomes and the nuclear envelope disappears. These are seen only in dividing cells. Find and label in your log book the nucleus in your cheek cell drawing and figure 4.5. Can you find the nucleolus in your cheek cells? Label it on your drawing if you see it using the microscope.

The Chloroplast—The Cell's Light Transducer

Also easily observed in many plant cells, using the light microscope, are small green bodies called **chloroplasts** (figure 4.6). Chloroplasts produce the food for most plants, and indirectly for animals by using the energy of light to transform carbon dioxide and water into carbohydrates. This process is called **photosynthesis.** Storage **plastids,** bodies similar to chloroplasts but not green, contain various chemical substances, such as starch.

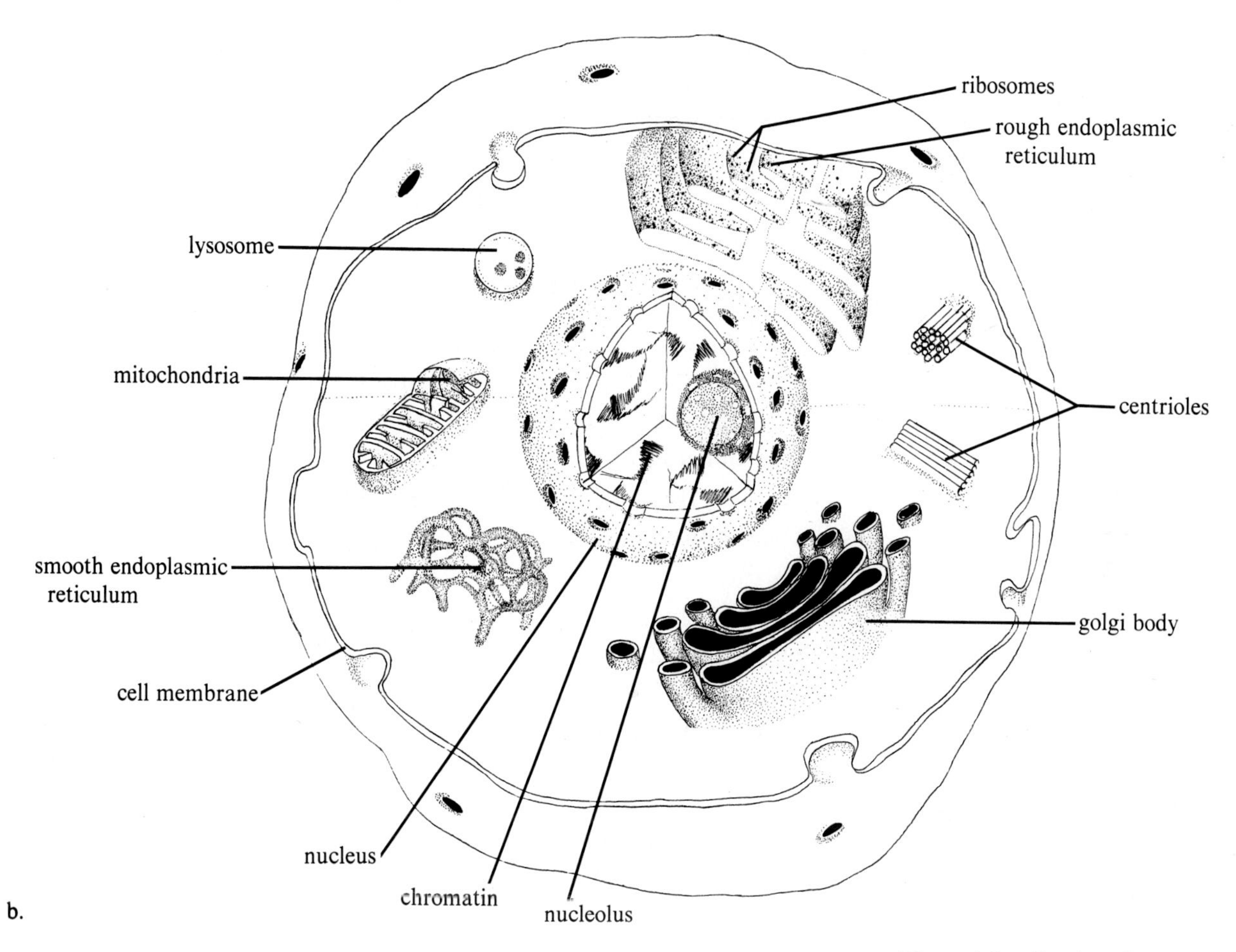

Figure 4.6 *Continued*

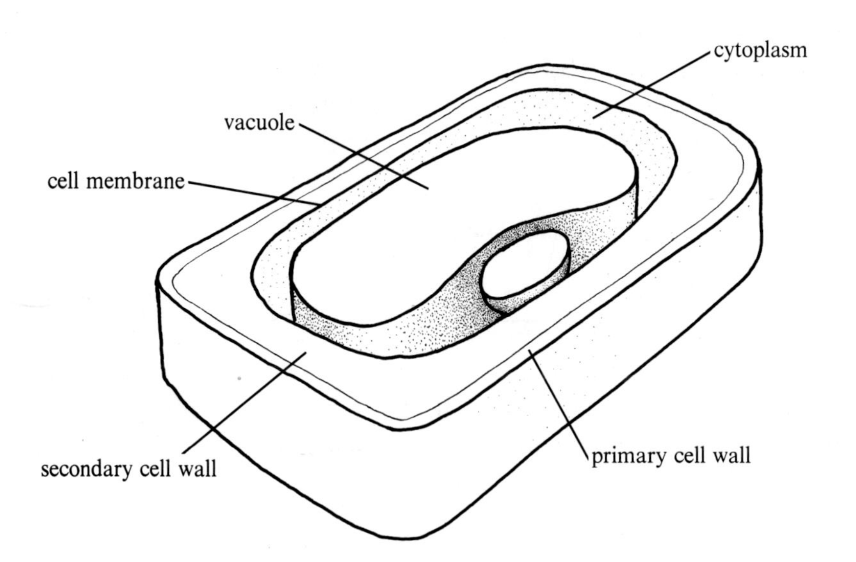

Figure 4.7 Sketch of a plant cell. Note the primary and secondary cell walls.

Find and label in your log book a chloroplast in figure 4.5c. During certain physiological processes, such as fruit ripening and seasonal changes in leaf coloration, chloroplasts may be transformed into chromoplasts. In these plastids the chlorophyll is largely replaced by red or yellow pigments known as carotenoids.

Mitochondria—The Cell's Power Plant

After the glass lenses of the light microscope were improved, biologists discovered small particles called **mitochondria** in cells of plants and animals (figure 4.6). Mitochondria contain enzymes that allow it to generate large amounts of chemical energy by cellular respiration. Mitochondria together with chloroplasts provide most of the energy for the work of the living world.

4.5 Cellular biologists can see inside cells by cutting them open with a knife (figure 4.8) into thin layers (sections). These thin sections are placed on a glass slide, stained with dyes that color the cell parts, and a coverslip added. Obtain a prepared microscope slide of a section of liver tissue and place the slide on the stage of a light microscope. Examine the tissue under high-power magnification. Make a drawing in your log book of two or three liver cells. Find and label the plasma membrane, nucleus, nucleolus, mitochondria (if visible), and the cytoplasm.

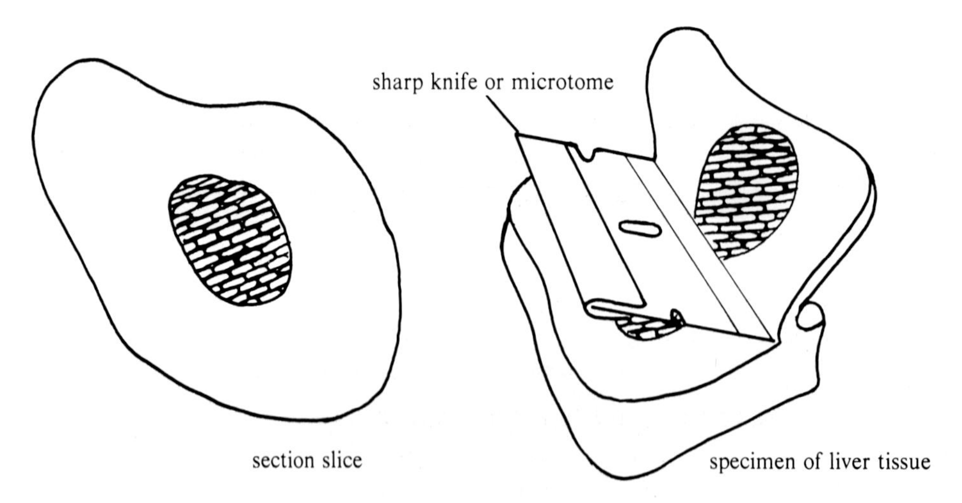

Figure 4.8 Method to obtain hand-made slices of fresh tissue. The liver tissue is carefully sectioned with a *sharp* razor blade to get a slice of the tissue.

4.6 Prepare a wet mount of a *small, fresh* piece of liver tissue.

a. The piece of liver tissue is obtained by carefully *slicing* with a razor blade a small (1 to 2 mm^3) piece from the liver sample pro vided in the laboratory.
b. Place this piece on a clean glass slide, and add 2 drops of 6 percent sucrose solution. Place a clean glass coverslip on top. Place a paper towel over the top, and wrap the end of your thumb with a large band-aid.
c. Place your wrapped thumb on top of the paper towel over the piece of liver. *Squash* the piece of liver by applying a direct downward pressure on the paper towel. Press very, very hard.
d. Place the liver cell squash on the stage of the microscope, locate a single liver cell, and then identify the parts of the cell under high-power magnification. Draw a cell in your log book, and label the parts you observed in the prepared liver slide.

4.7 For the next part of your journey through the cell, it is necessary to change microscopes in order to see the smaller parts of the cell. With a light microscope you cannot see objects smaller than one-half the wavelength of light. Because white light has an average wavelength of 5,500 angstroms (Å), the smallest object can be no smaller than 0.27 micrometers (μm). This limit for the light microscope is the approximate size of a mitochondrion. Also any two objects that are closer than 0.27 μm apart from each other will appear as a single object, provided the two objects together are larger than 0.27 μm. Therefore, to resolve cell parts smaller or closer together than 0.27 μm, it is necessary to use an electron microscope.

Sections of plant or animal tissue prepared for the electron microscope must be very thin and stained with heavy metals that scatter electrons. Reexamine figure 4.6—the drawings of plant and animal cells—and then place an asterisk (*) beside those cell parts in log book figure 4.6 you did not see with a light microscope.

The cell wall consists of numerous fibers and layers that completely surround each individual cell (figure 4.25*c*, levels 2 and 3). Just inside the cell wall, the plasma membrane is visible. This structure, which bounds all cells, was hypothesized from experimental observations with the light microscope. But now, with the greater resolution and magnification of the electron microscope, you can see its structure. In cross section, the plasma membrane consists of two dense layers separated by a less dense space (figure 4.9).

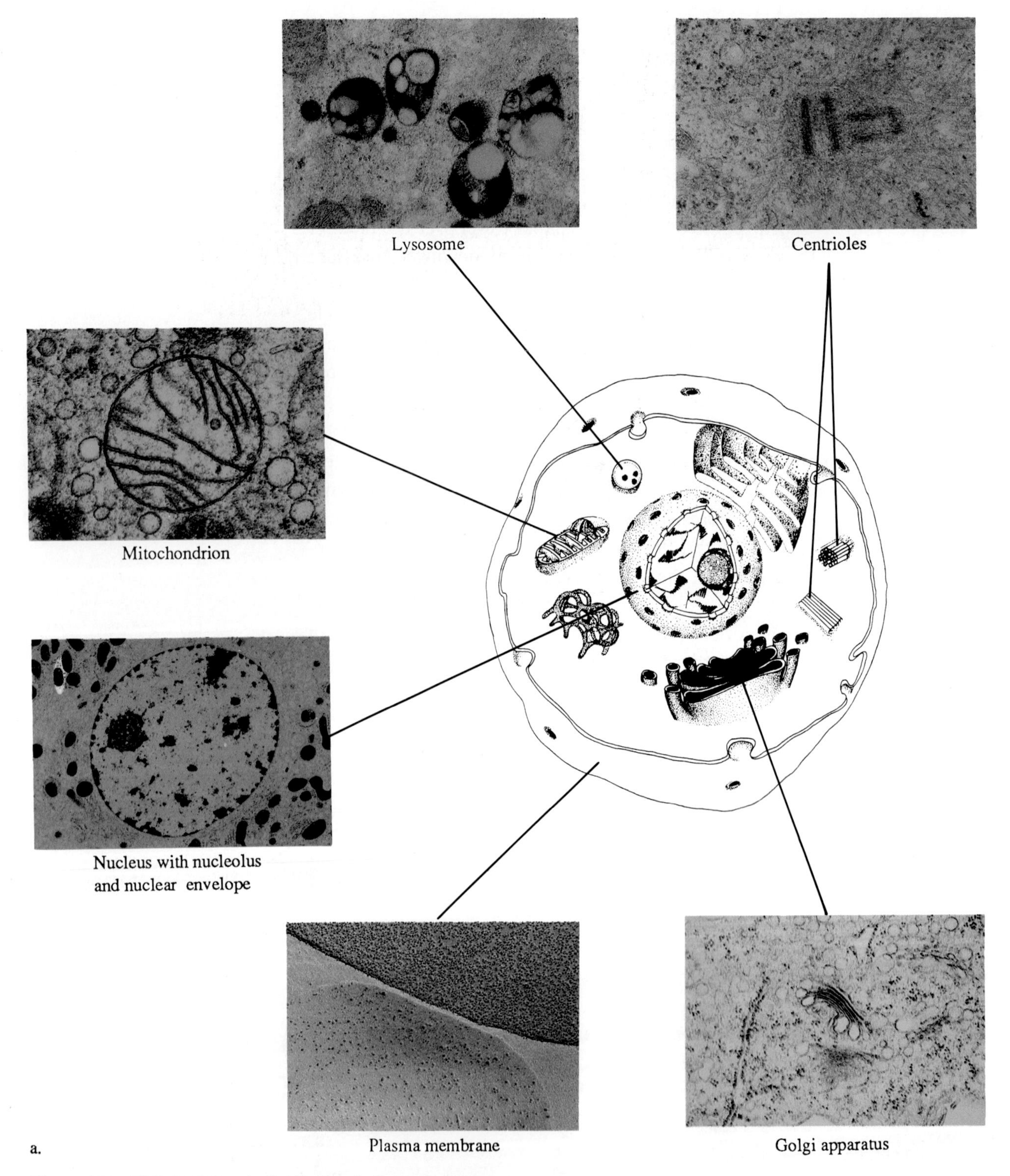

Figure 4.9 Sketch of a typical animal and plant cell with electron micrographs to show internal parts and structure. a. animal cell. b. plant cell.

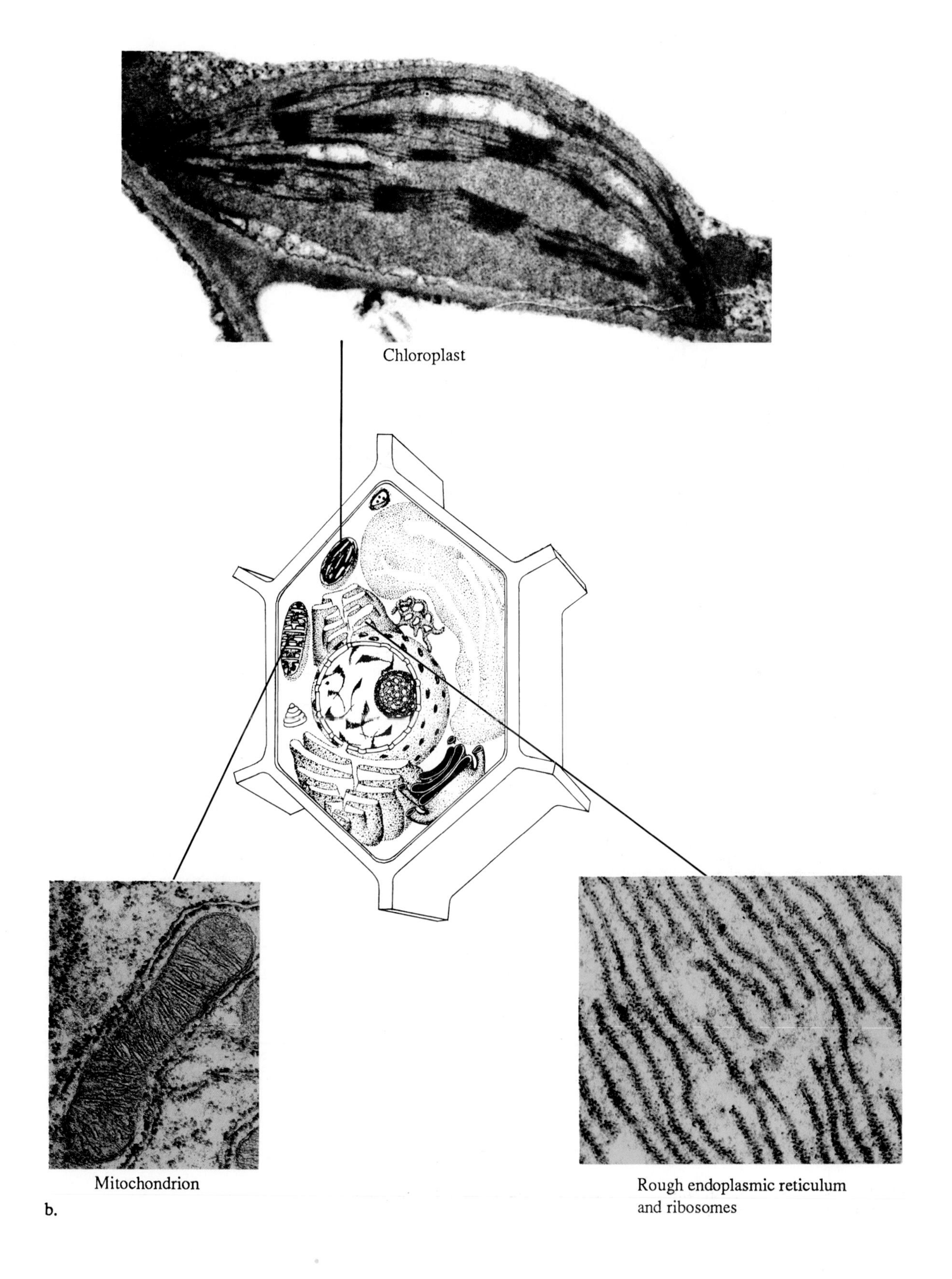
Chloroplast
Mitochondrion
Rough endoplasmic reticulum
and ribosomes

b.

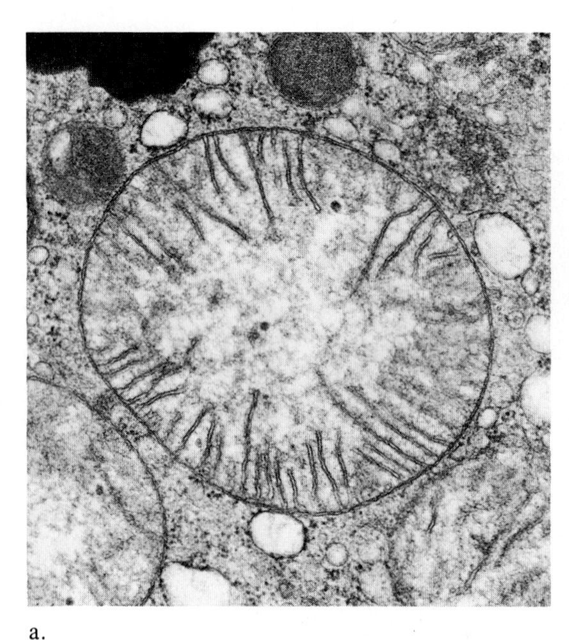

a.

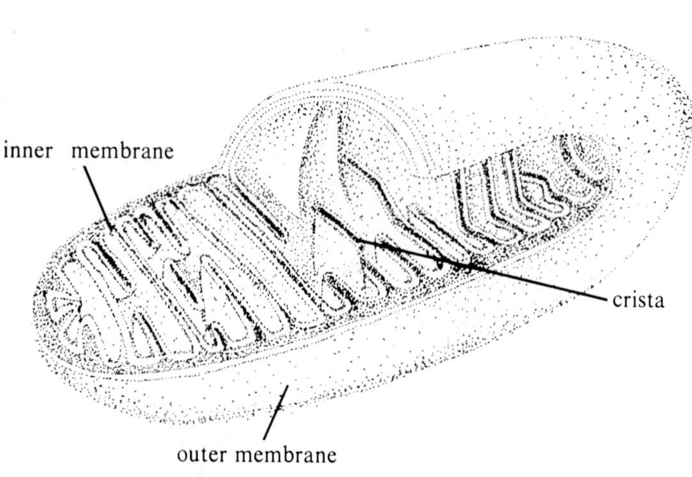

b.

Figure 4.10 Mitochondria. a. An electron micrograph of a mitochondrion cut across its long axis. b. Sketch of a mitochondrion showing inner (crista) and outer membranes.

An examination of the organelles with the electron microscope reveals new detail. The mitochondria are enclosed by two membranes (figure 4.10). The inner membrane, called a **cristum,** folds into a series of tubes or sheets (cristea) and contains many of the enzymes necessary for aerobic respiration. Label the inner and outer membrane in figure 4.10*a*. Chloroplasts are constructed with three membranes: two outer membranes and an inner membrane, called a **thylakoid,** that contains green-colored chlorophyll pigments and many of the enzymes necessary for photosynthesis (figure 4.11). The inner membranes may stack into saclike units called *grana*. Label the inner membrane, grana, thylakoid sac, stroma, and outer two membranes in log book figure 4.11.

The nucleus also is surrounded by two membranes (figure 4.9 and 4.12). However, in contrast to mitochondria and chloroplasts, the membranes surrounding the nucleus are penetrated by a number of pores. These pores form channels for the transfer of molecules between the nucleus and the cytoplasm. Inside the nucleus, the chromatin can be seen as a collection of fibrillar material. The nucleolus is not bounded by a membrane. As might be expected, the nucleus is constantly chemically active. Before cell reproduction, new DNA must be made. The genes in the chromatin direct the synthesis of thousands of enzymes and other proteins.

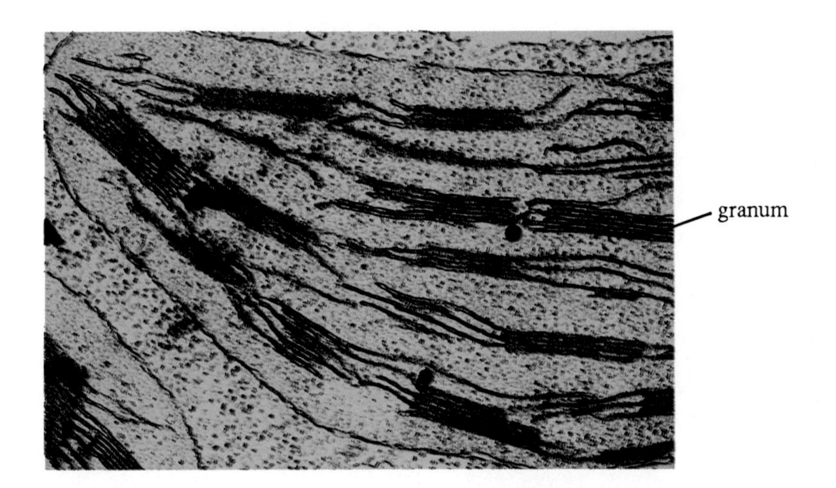

Figure 4.11 An electron micrograph of a small portion of a chloroplast.

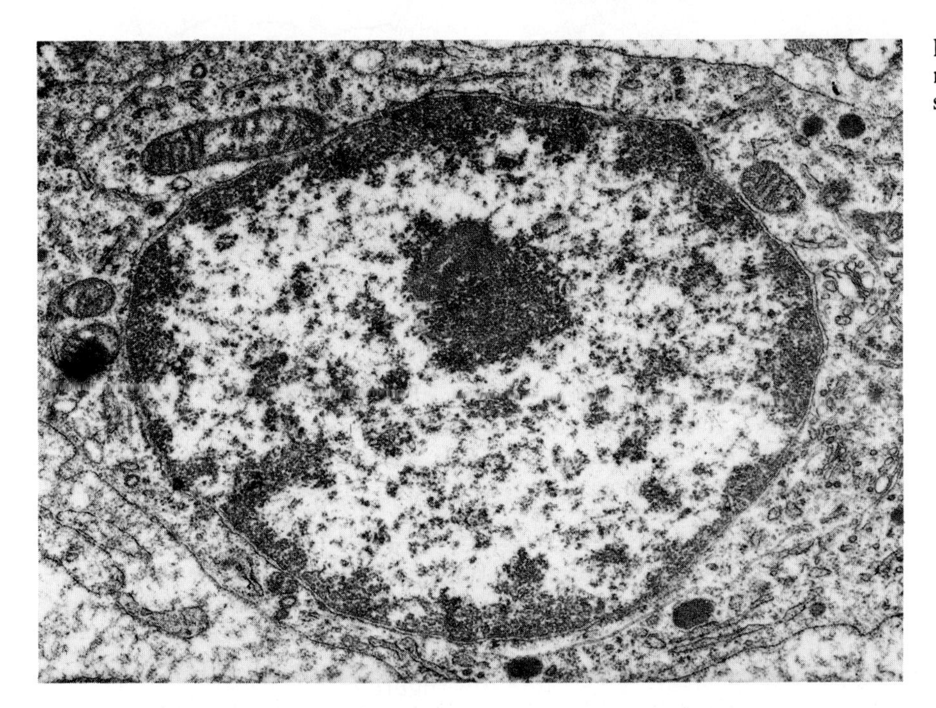

Figure 4.12 An electron micrograph of a nucleus and surrounding cytosol.

4.8

In addition to new knowledge of existing cell structures, cell biologists also discovered new cell parts. Whenever the cell needs a protein, the blueprint for making the protein is transmitted from the nucleus to structures in the cytoplasm of the cell called **ribosomes** (figure 4.9). Ribosomes are very tiny: approximately 20.0 Å (0.02 μm) or less than one millionth of an inch. Most cells contain thousands and sometimes millions of tiny ribosomes that are part of a protein factory assembly line. Ribosomes are found free in the cytoplasm or attached to a branched network of membranous sacs called **endoplasmic reticulum** (ER) (figures 4.9 and 4.13*a*).

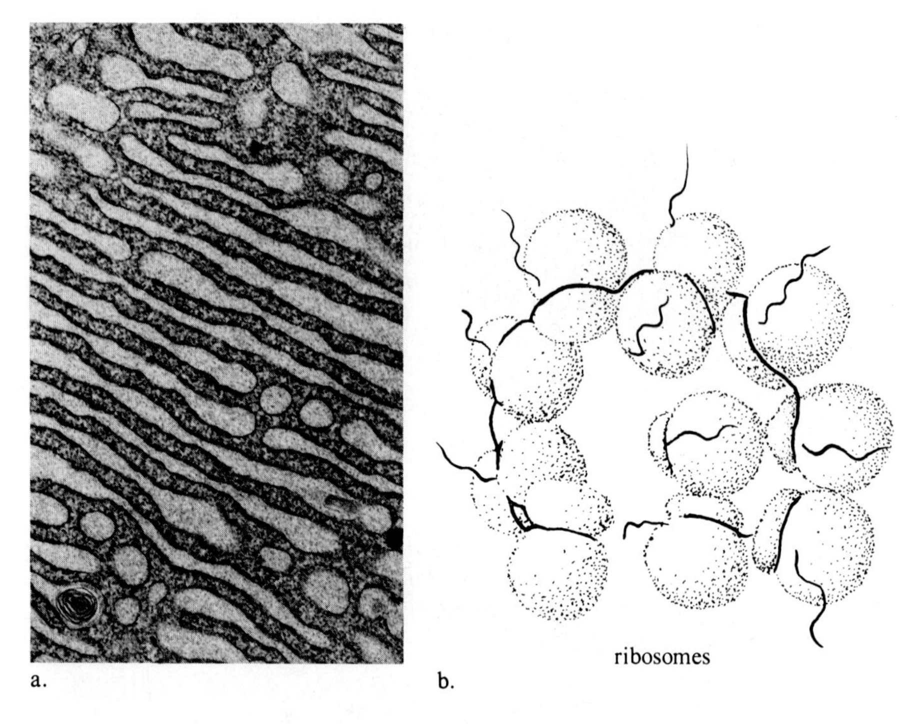

Figure 4.13 Endoplasmic reticulum and associated ribosomes. a. An electron micrograph of rough endoplasmic reticulum. b. Sketch of a small cluster of ribosomes bound together by ribonucleic acid. This polysomal cluster occurs for the synthesis of some types of protein.

Free ribosomes may occur singly or be grouped in small clusters called *polysomes* (figure 4.6 and figure 4.13*b*). Both polysomes and ER ribosomes are sites of protein synthesis. ER dotted with ribosomes is called rough. Label in your log book ER-bound ribosomes in figure 4.13*a* and figure 4.10*a* and ER in figure 4.9. Label rough ER and the ER canals in figure 4.13*a*. ER without ribosomes is called smooth ER. The chief function of the ER is storage, segregation, and transportation of chemical substances to various parts of the cell or to the outside of the cell.

Electron microscopists confirmed the presence of **Golgi** bodies, stacks of membrane-bound discs with a number of vesicles at the periphery (figure 4.9). Detailed study of Golgi bodies showed that small vesicles pinch off from the central discs (figure 4.14) and fuse with the plasma membrane to release their contents outside the cell (figure 4.15). Experiments with radioactive tracers have confirmed that the Golgi apparatus plays a major role in the packaging and secretion of substances from the cell. For example, the digestive enzymes produced in the pancreas and carbohydrates for cell walls are packaged, delivered to the plasma membrane, and secreted outside the cell. Label a Golgi body and a secretory vesicle in log book figure 4.9*a*. Golgi bodies also package proteins marked for delivery to internal parts of the cell, such as lysosomes.

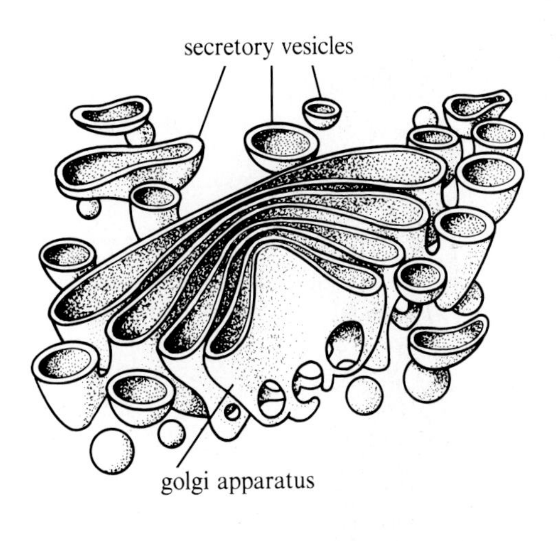

Figure 4.14 Sketch of a golgi body.

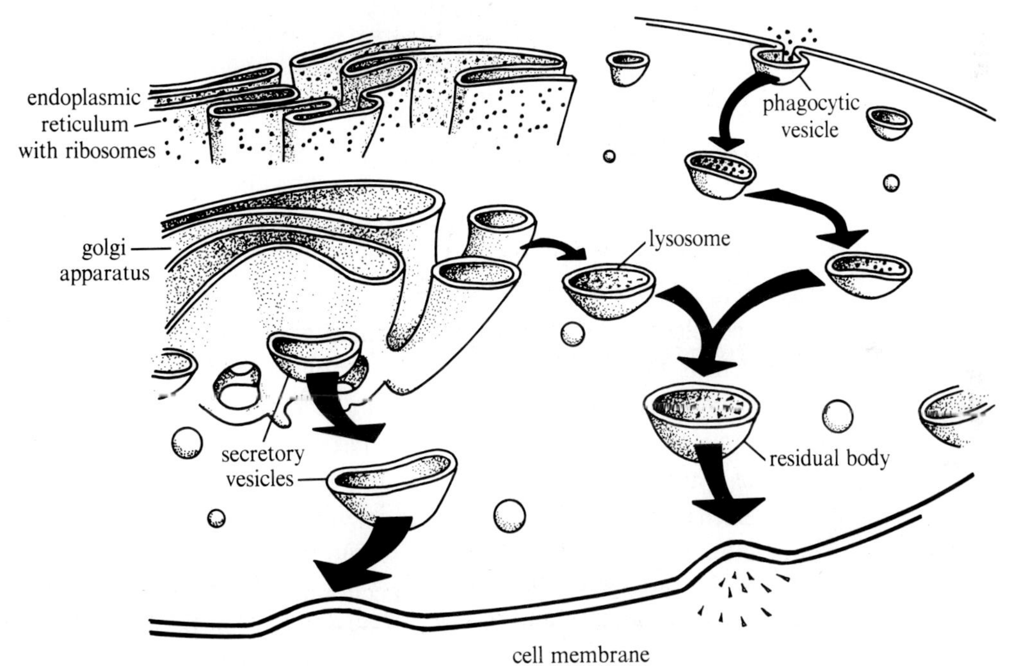

Figure 4.15 Sketch of a white blood cell to show a function of lysosomes. Note the infolding of the plasma membrane to form a phagocytic vesicle which fuses with a lysosome.

Lysosomes are spherical in shape, bound by a single membrane (figure 4.9*a*) and are formed from Golgi (figure 4.14 and 4.15). They vary in size, with an average diameter of 0.5 μm. Each lysosome contains digestive enzymes that function to break down materials injested by the cell. For example a human's white blood cells defend the body against bacterial infection by injesting bacteria into the cell. The white blood cell's plasma membrane folds around the bacterium, forming a phagocytic vesicle (figure 4.15). The lysosome then fuses with the vesicle, and the lysosomal enzymes destroy the bacterium by enzymatically digesting its chemical structure. The useful products of this digestion move into the cytoplasm, and the undigestible material accumulates in a residual body (figure 4.15). In some

Figure 4.16 Electron micrograph of 'healthy' red blood cells.

cases, the residual body moves to fuse with the plasma membrane in order for the undigested material to leave the cell.

Red blood cells (figure 4.16) sometime become damaged as they move through the capillaries of the circulatory system. These damaged cells are removed from the blood stream by macrophage cells that phagocytize the red blood cells by engulfing them (figure 4.17). Digesting the red blood cells is similar to that described for phagocytized bacteria.

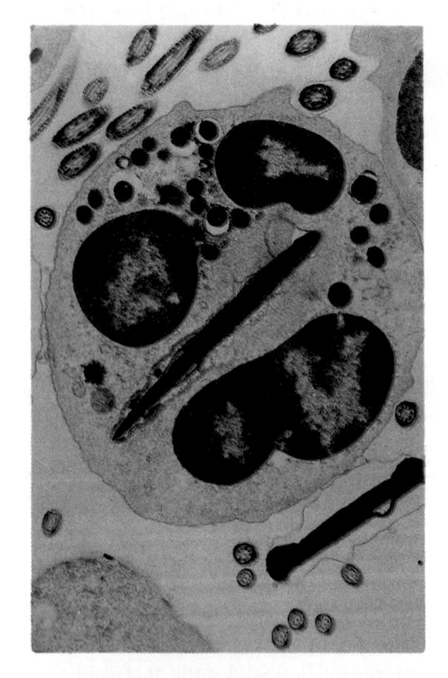

Figure 4.17 Electron micrograph of damaged red blood cells being phagocytize by white blood cells called macrophages.

In the human population, more than thirty hereditary diseases are linked to missing enzymes in the lysosomes. In each case, an accumulation of waste material leads to cellular pollution that may cause death. Tay-Sachs disease is caused by an enzyme missing from lysosomes of certain ethnic persons of European descent. This single gene deficiency results in brain-damaged cells from the accumulation of a toxic product.

Lysosomes can also digest parts of a cell in which they are housed. Because of this autodigestion, lysosomes are called "suicide bags." This process is useful in removing damaged cells or those programmed to die. When cells are programmed to die, for example, in the metamorphosis of insects or when tadpoles lose their tails to become frogs, the lysosomes release their enzymes into the cells' cytoplasm.

4.9 As cell biologists continued their journey into the world of the cell, they soon discovered in the cytoplasm a system of physical support for the cell, as well as a contractile mechanism for cellular movement. **Microfilaments** and **microtubules** (figure 4.18) are part of the **cytoskeleton** (or framework) of the cell that helps maintain cell shape, anchors organelles (figure 4.18), and helps cells move.

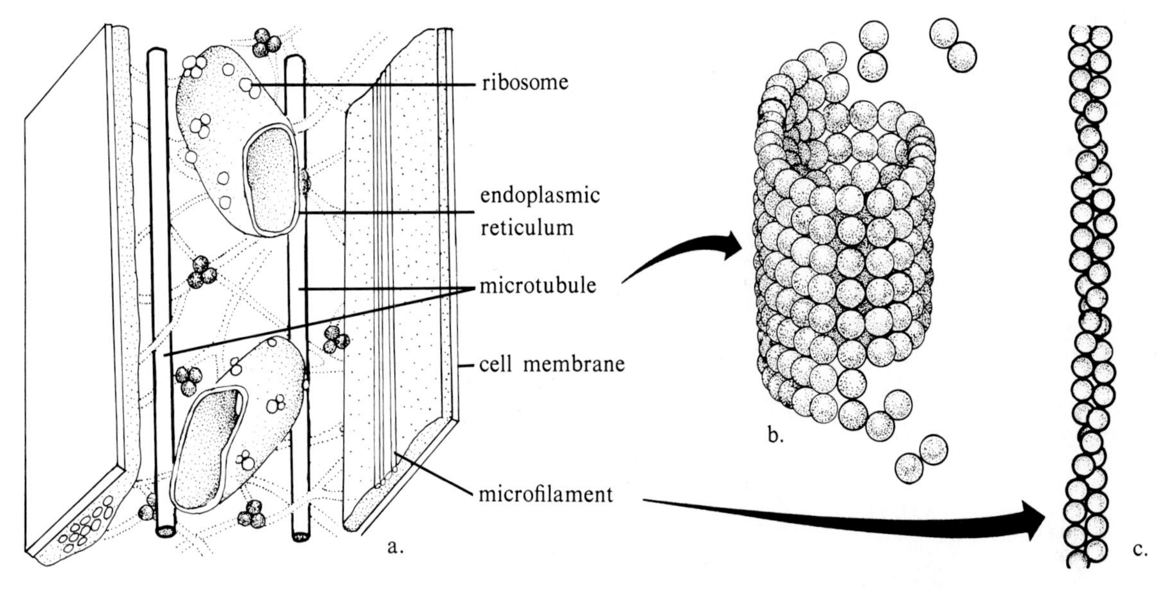

Figure 4.18 Microtubules and microfilaments. a. Sketch of microtubules and microfilaments within the cytosol. b. Diagram of a microtubule showing how the tubulin subunits (sphere) are arranged to form the cylindrical wall. c. Diagram of a microfilament showing the arrangement of actin molecules (sphere) to form a single long strand. Note two strands intertwine to form the filament.

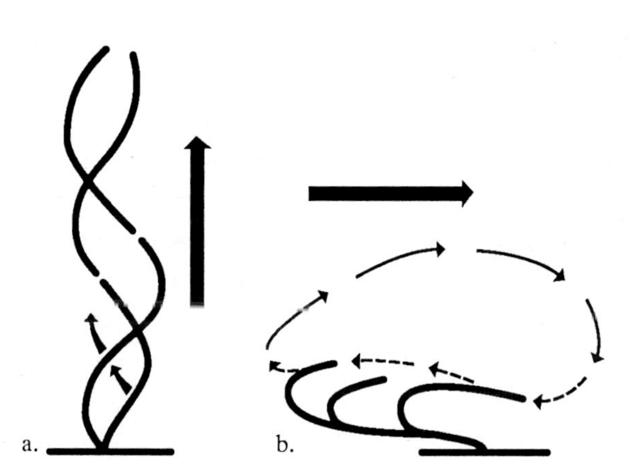

Figure 4.19 Wave motions of cilia and flagella used to propel the cell or to move material past the cell surface. a. undulation motion. b. stroke motion. Large arrows indicate direction of water movement.

Microtubules are thin, hollow cylinders (.025 μm) of variable length and are constructed of spherical protein molecules called *tubulin* (figure 4.18*b* and *c*). During the process of cell reproduction, called mitosis, microtubules aid in the movement of chromosomes by forming a bundle called spindle fibers. Microtubules also maintain the shape of red blood cells and stabilize the shape of nerve cells.

Cilia and **flagella,** whiplike structures that project from the surfaces of cells and move rhythmically (figure 4.19), are made of microtubules bounded by a plasma membrane. Large numbers of cilia are found on cells of the epithelial layer that line the respiratory system—for example, cells lining the trachea and lungs—where they sweep out dust and debris. Both cilia and flagella play an important role in human reproduction: the beating

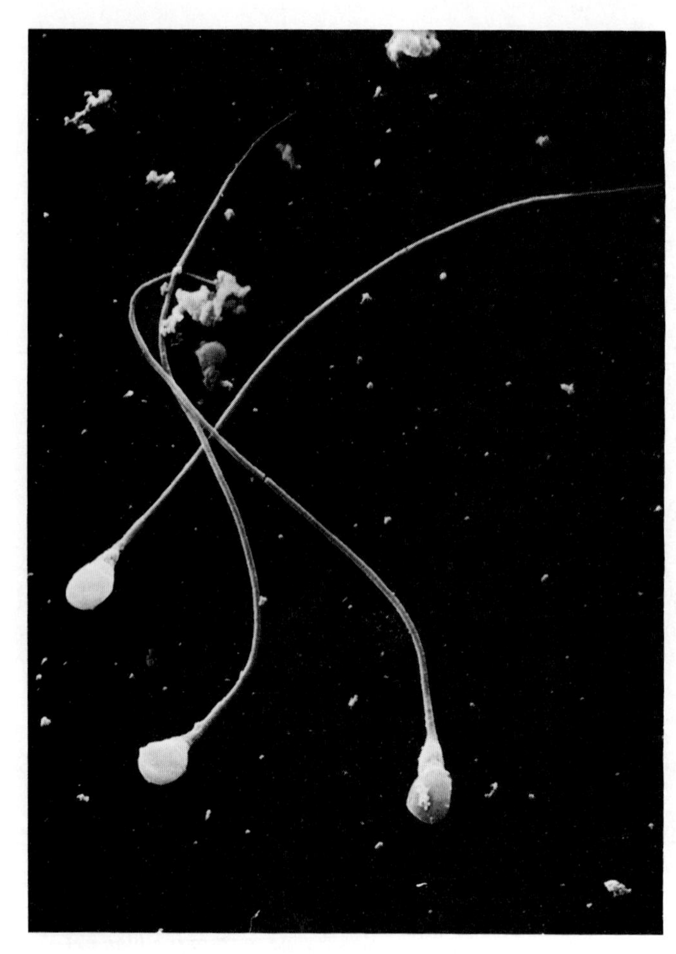

Figure 4.20 Sperm cell. Wave motion of these flagella is undulant.

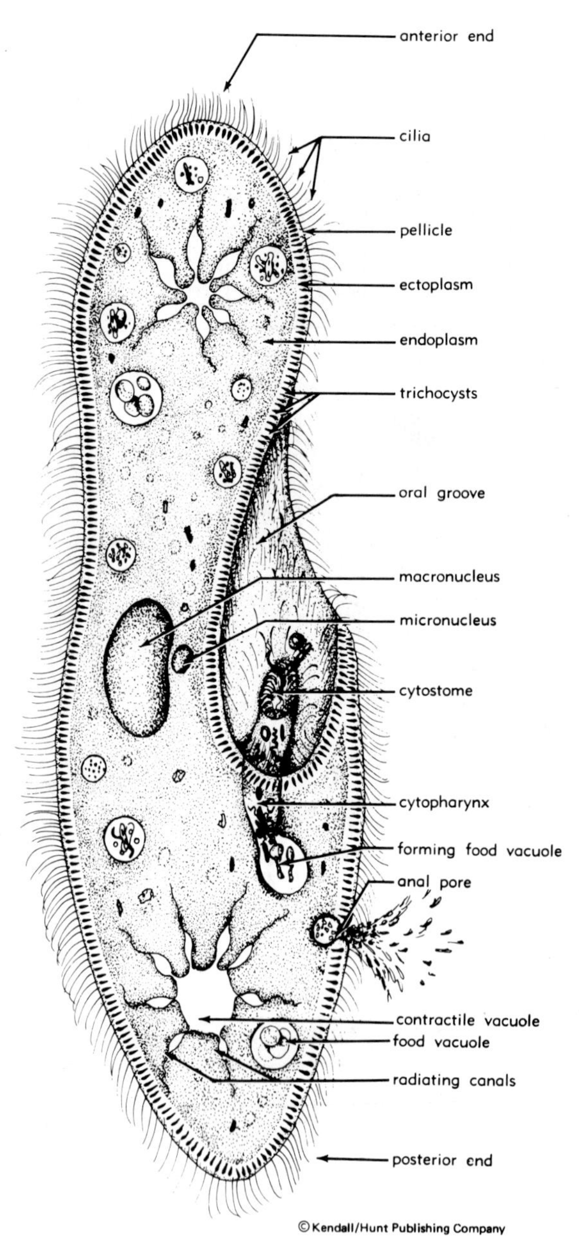

Figure 4.21 Sketch of a paramecium cell showing cilia. Wave motion of these cilia is a stroke.

of cilia on cells lining the female oviduct produces a current that draws the female egg into the uterus, while the rapidly moving tail of the male sperm is a flagellum (figure 4.20). Some single-celled organisms, such as a paramecium, use cilia and flagella to move (figure 4.21).

Microfilaments (figure 4.18) are thin strands made of a spherical protein called **actin.** Microfilaments also allow cells to crawl along surfaces. Amoebae (figure 4.22), white blood cells, and macrophages move by the action of microfilaments forming extensions called **pseudopods.**

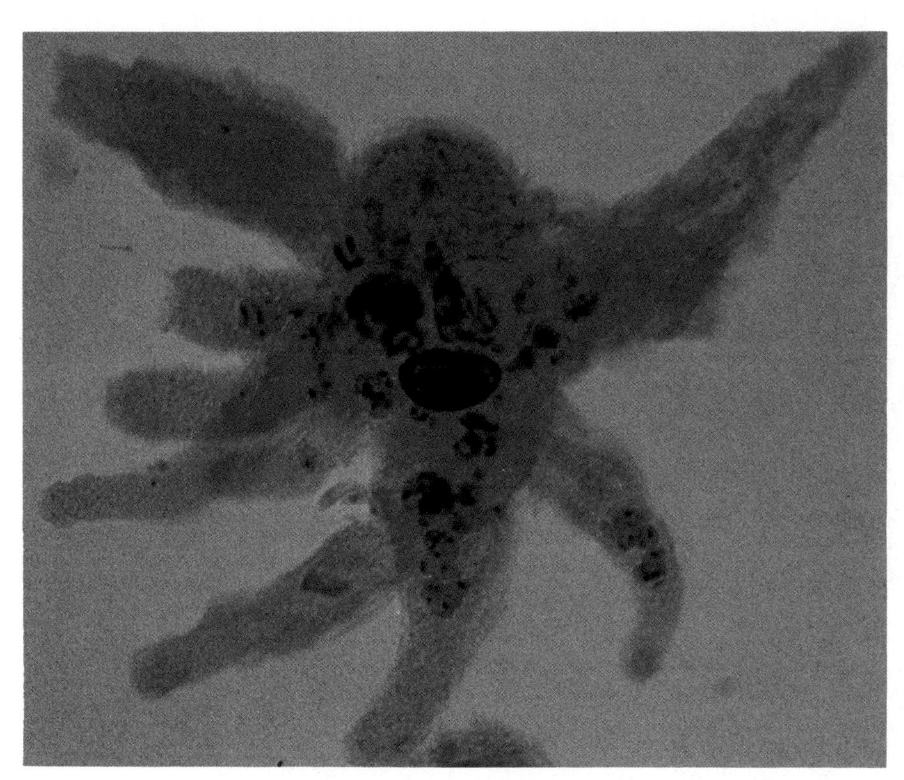

a.

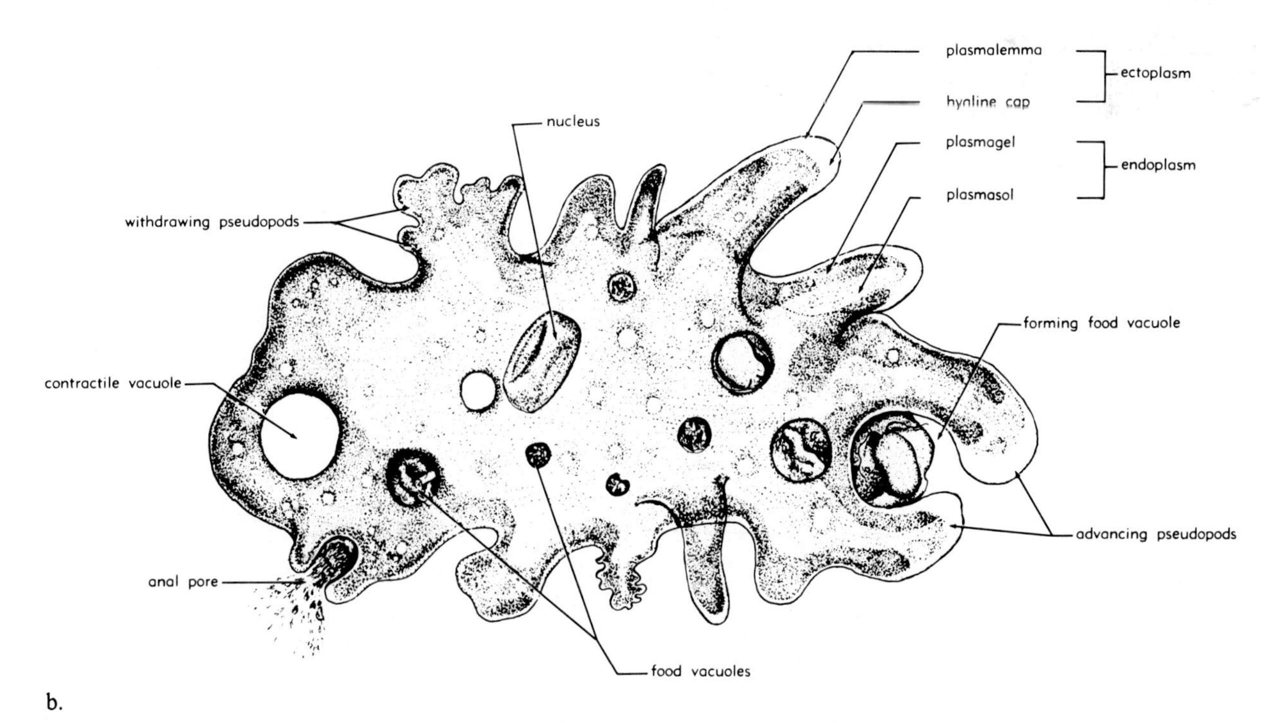

b.

Figure 4.22 Amoeba.
a. Photograph of amoeba.
b. Sketch of amoeba. Note the advancing pseudopods by which the animal moves and the engulfment of food material by the pseudopods.

Place a drop of amoeba or paramecium culture on a clean slide. Place a coverslip on top. Observe with a light microscope. Locate the following: plasma membrane, nucleus, cytoplasm, nucleolus, food vacuoles, and contractile vacuole. A contractile vacuole is an organelle that removes excess fluid from a protozoan. Also find the amoeba's pseudopodium and the paramecium's cilium. The contraction of skeletal muscles is produced by the motion of microfilaments sliding by each other (figure 4.23).

4.10 Complete data table 4.3 in your log book by writing in the *function* of each cell organelle listed.

Data Table 4.3 Eukaryotic Organelles

Name	Function
Plasma membrane	
Vacuole	
Nucleus	
Nucleolus	
Chloroplast	
Mitochondrion	
Endoplasmic reticulum	
Ribosome	
Golgi body	
Lysosome	
Microfilament	
Microtubule	

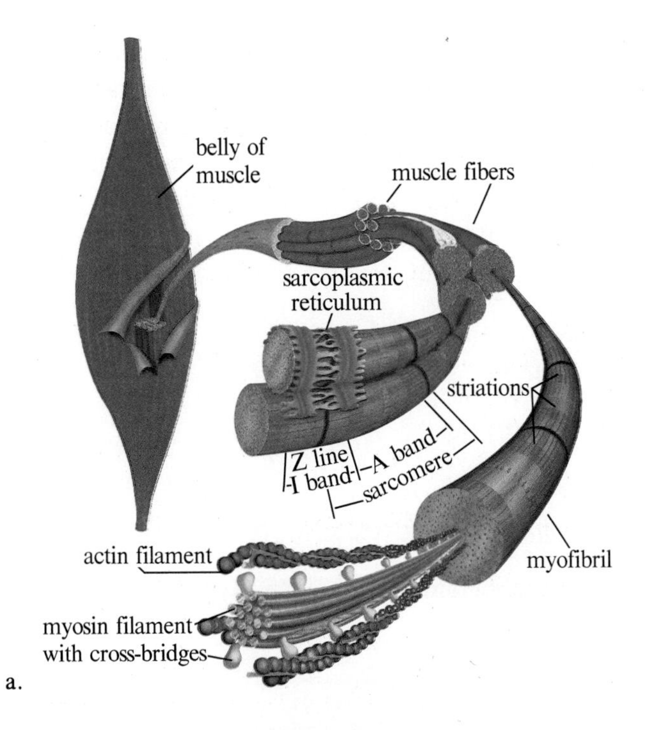

Figure 4.23 Levels of skeletal muscle. a. Exploded sketch of muscle. The muscle tissue contains bundles of muscle fibers (muscle cells). Each cell contains many myofibrils. A single myofibril contains bundles of microfilaments (actin and myosin) aligned with the long axis of the muscle cell. b. Micrograph of several myofibrils showing striations produced by microfilament alignment. c. Electron micrograph of striation pattern of a fiber. Striations are the result of actin and myosin overlapped and linked with cross bridges. Muscle contraction involves sliding of the actin microfilament past the myosin microfilament with the assistance of the cross bridges.

4.11 Observe the electron photomicrograph of a plant cell in figure 4.24*a*. Identify and label the numbered cell structures on the line drawing in log book figure 4.24*b*. You should be able to find the following structures. Place a check (✓) in the log book box when you have found the structure.

- ☐ chloroplasts
- ☐ chromatin
- ☐ cytoplasm
- ☐ cell wall
- ☐ nucleolus
- ☐ nuclear envelope
- ☐ endoplasmic reticulum
- ☐ mitochondria
- ☐ ribosomes

4.12 Life may be studied at many different levels of structural organization that ranges from the **atom** to the entire world itself, called the **biosphere.** All levels are interacting and are important to each other.

Atoms combine to make organic **molecules** that, in turn, are organized into larger *macromolecules*. Macromolecules assemble into *organelles* that make up a **cell** (figure 4.25*a, c*). The cell is the first and smallest unit of organization that biologists recognize as life. An organism may be composed of one cell, such as protozoans or bacteria, or combined into tissues and organs making a multicellular *organism*. You are a multicellular organism made up of atoms, molecules, and cells, and you are a part of a community, an ecosystem, and a biosphere.

In our journey into the world of a cell, we missed seeing several components of the cell because they cannot be clearly resolved with the microscope. At level 2 (figure 4.25*a, b*), macromolecules are only observed as a fuzzy mass in the electron microscope. Physical and chemical techniques used by molecular biologists can give us a view of these parts at level 1. In chapter 5, we will look at levels 1 and 2, the world of the molecular biologist.

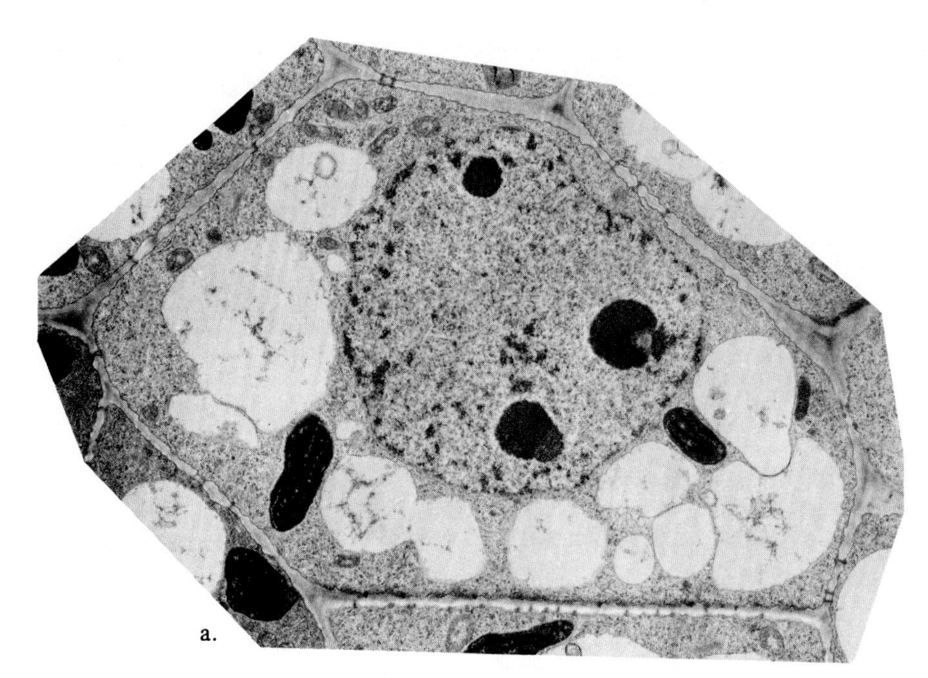

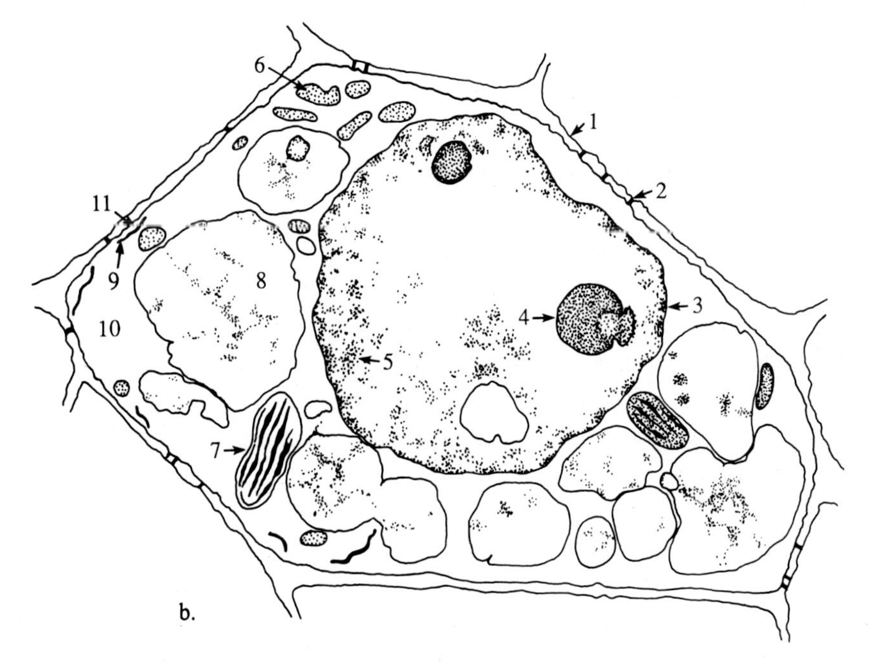

Figure 4.24 Plant cell. a. Electron micrograph. b. Sketch of cell in 4.24a.

Figure 4.25 Assemblage of a cell from organic macromolecules. a. Whole cell at level 4. b. Cellulose at level 2. c. Small organic molecules (level 1) are made from simple inorganic substances and are joined together to form macromolecules (level 2). Macromolecules assemble into supra-molecular structures (level 3) of which the cell and its organelles are constructed (level 4).

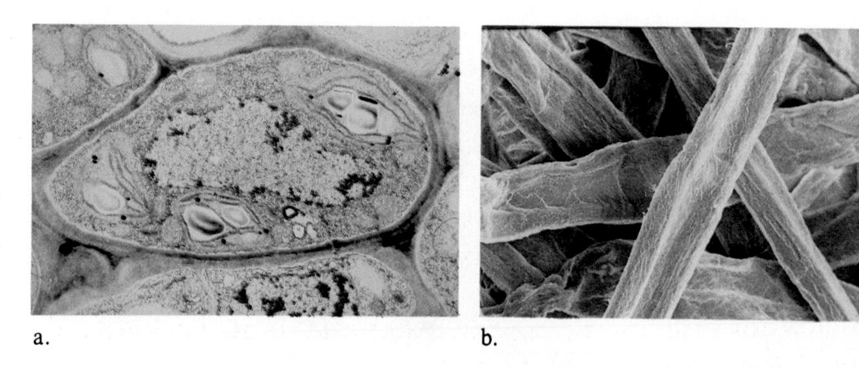

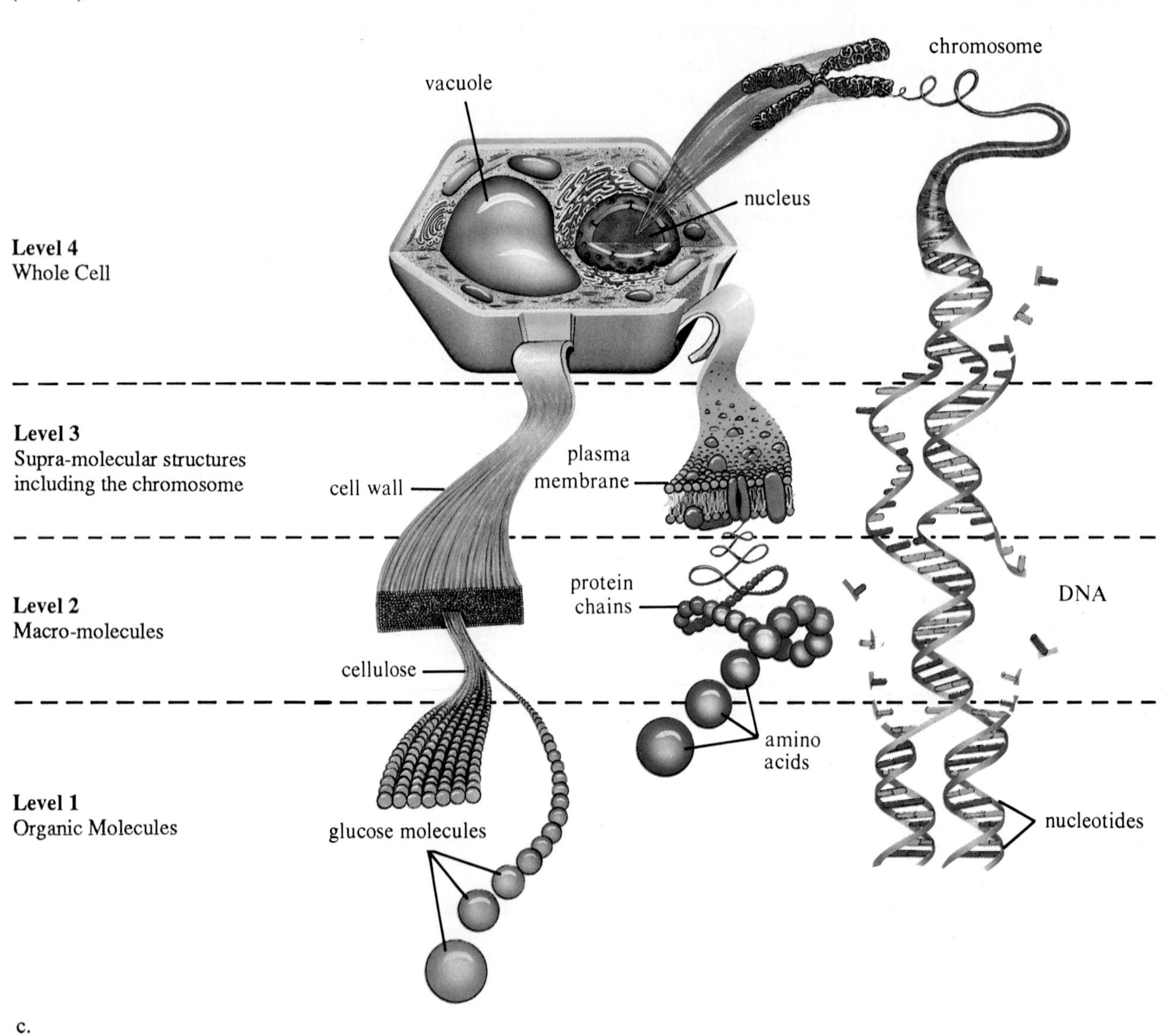

4.13 The cell is the structural and functional unit of prokaryotes, plants and animals. The picture of the cell presented thus far has been predominantly one of a relatively uniform, undifferentiated plant or animal cell. Organisms of the living world come in an endless array of forms and functions, and so do the cells of which they are composed. Cells make up multicellular organisms in a manner similar to apartments that make up apartment buildings. Just as apartments in the same building are built from the same basic plan, they still may be decorated differently to serve different functions. So are cells in the body of a multicellular organism different or varied from a basic design to perform specific functions.

Each cell type has its own pattern or arrangement of organelles. Your teacher may have a display on a bulletin board of the many different types of specialized plant and animal cells. Write a brief summary of the specialized function of each cell type and attach it to your log book.

4.14 There are two major cell groups. One group, the group you have been studying, includes plant and animal cells (except the blue-green algae). These are called **eukaryotic** cells. These cells have a membrane surrounding the nucleus. The second major group includes bacteria and cyanobacteria (blue-green algae). These are called **prokaryotic** cells and lack a membrane around the nucleus. These two cellular groups represent a taxonomic classification based on the anatomy of cells. The Kingdom Monera contains the prokaryotic cells.

Bacteria are so small they can barely be seen with a light microscope. Usually it takes special stains to show the outline and size of bacteria. To compare the average size of bacteria with some animal cells, see figure 4.26. There are three major shapes of bacteria: **bacilli, cocci,** and **spirilla.** Bacilli are cylindrical, or rodlike; cocci are spherical; and spirilla are corkscrew, or helical.

Examine the prepared microscope slides of various bacteria. Draw in your log book the bacteria you see in the viewing fields found in figure 4.27. Label each different bacterial type using the singular terms bacillus, coccus, and spirillum.

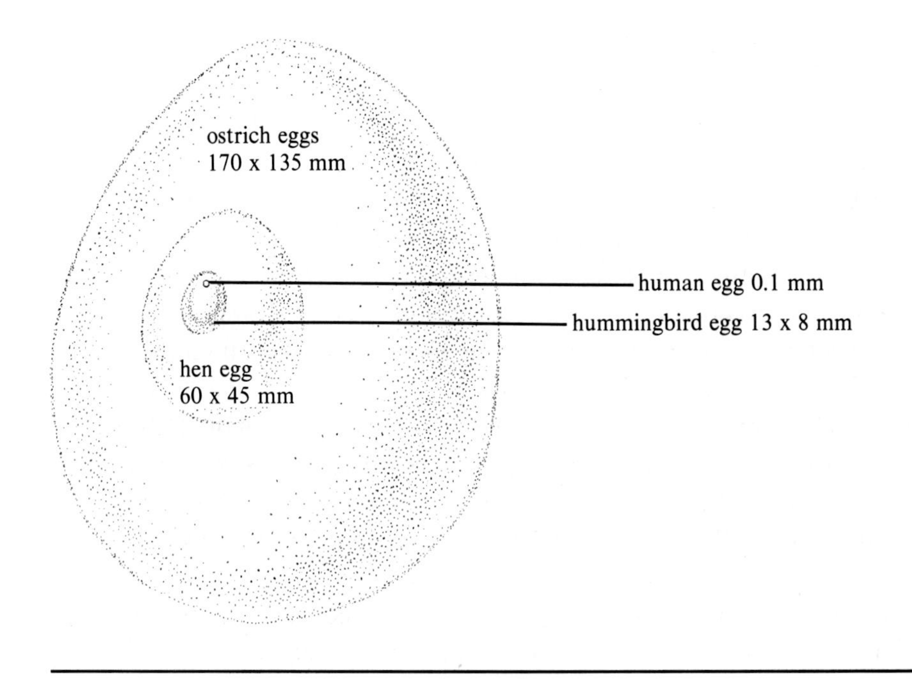

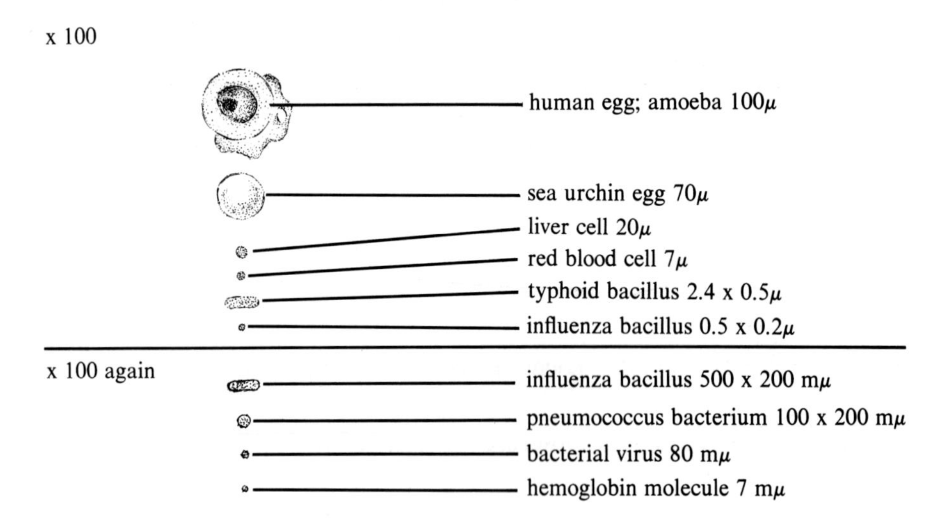

Figure 4.26 Comparative size relationship of eukaryotic cells, prokaryotic cells (bacteria) and viruses. The average (2.5μm) bacterium is 10× smaller than the average (25μm) plant or animal cell.

4.15 When an electron microscope is used to study bacteria, an internal cellular anatomy becomes visible. The diagram of a photograph in figure 4.28 shows the anatomy of a typical bacterium. Bacteria have DNA, but no nuclear membrane; instead, the DNA is localized in a region called a *nucleoid*. Many bacteria often have a cell wall outside the plasma membrane.

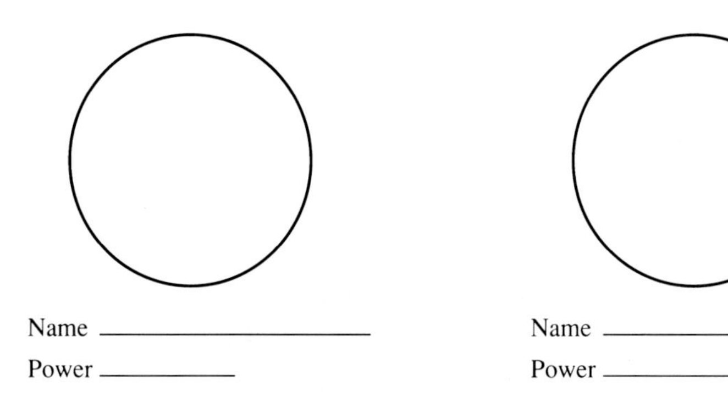

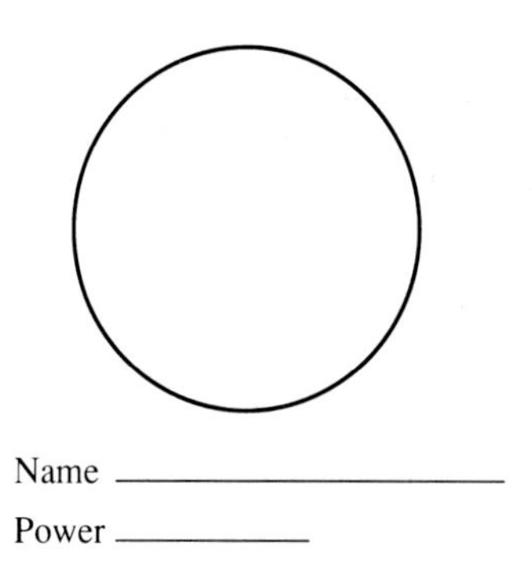

Figure 4.27 Viewing fields.

The chemistry of bacterial cell walls is associated with differences in sensitivity to antibiotic action. For example, penicillin interferes with proper cell wall formation. The result is the bacteria will die without a cell wall. Outside the cell wall of some bacteria, a sheath or gelatinous capsule is found. This sheath of carbohydrates or proteins protects infectious bacteria against phagocytosis by white blood cells. In such a case, other body defense systems then take over the destruction of bacteria.

Data Table 4.4 Major Cell Structures of Prokaryotes and Eukaryotes

	Prokaryotes	Eukaryotes	
Structure		**Plants**	**Animals**
Plasma		+	+
Nucleus		+	+
Nucleolus		+	+
Nuclear envelope		+	+
Mitochondrion		+	+
Chloroplast		+	−
Ribosome		+	+
Endoplasmic reticulum		+	+
Golgi complex		+	+
Microtubule		+	+
Microfilament		+	+
Vacuole		+	+
Cell wall		+	−

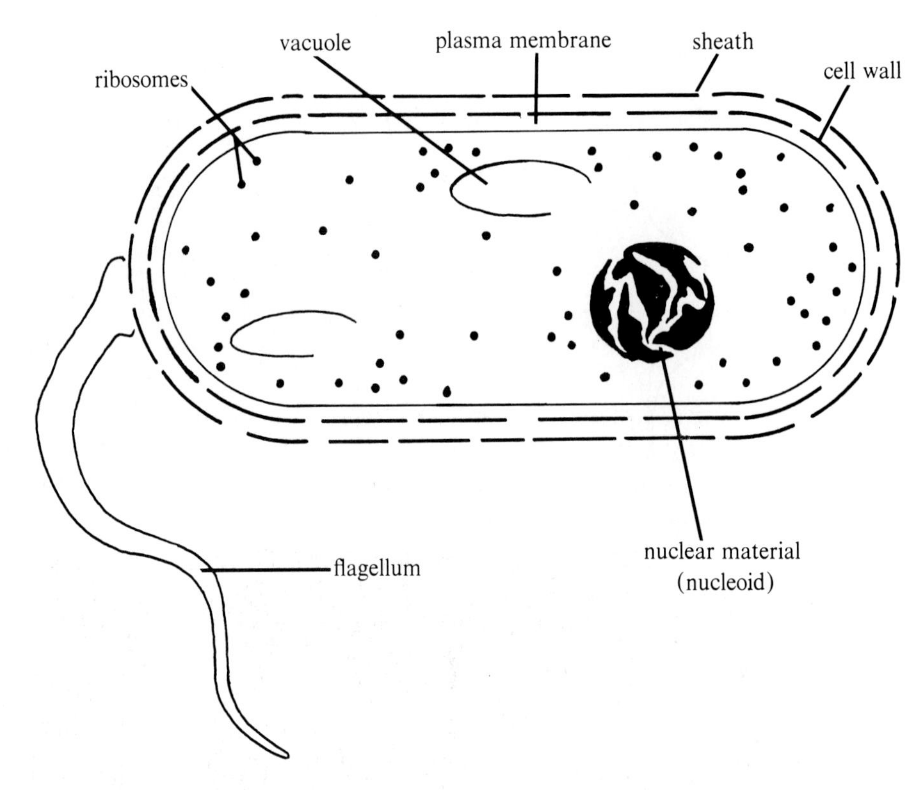

Figure 4.28 Sketch of a bacterium to show internal cell parts. Compare internal cell parts to a plant and animal cell.

Complete data table 4.4 in your log book by marking a plus (+) beside those anatomical features that are associated with bacteria, and mark a minus (−) beside those that are not found associated with bacteria.

4.16 Studies of bacteria are often centered around harmful bacteria (data table 4.5). Louis Pasteur, who developed the idea that bacteria can cause disease, is considered the "father of bacteriology" (study of bacteria). Write a report on Louis Pasteur's contributions to the improvement of human health and attach it to your log book.

Data Table 4.5 Infectious Diseases Caused by Bacteria

Respiratory Tract	**Nervous System**
Strep throat (sometimes causing rheumatic and scarlet fever)	*Tetanus
Pneumonia	Botulism
*Whooping cough	Meningitis
*Diphtheria	**Digestive Tract**
*Tuberculosis	Food poisoning (salmonella, botulism, and staph)
Skin Reactions	*Typhoid fever
Staph (pimples and boils)	*Cholera
*Gas gangrene (wound infections)	**Venereal Diseases**
	Gonorrhea
	Syphilis

*Vaccines are available. Tuberculosis vaccine is not used in this country. Typhoid fever, cholera, and gas gangrene vaccines are given if the situation requires it. Others are routinely given.

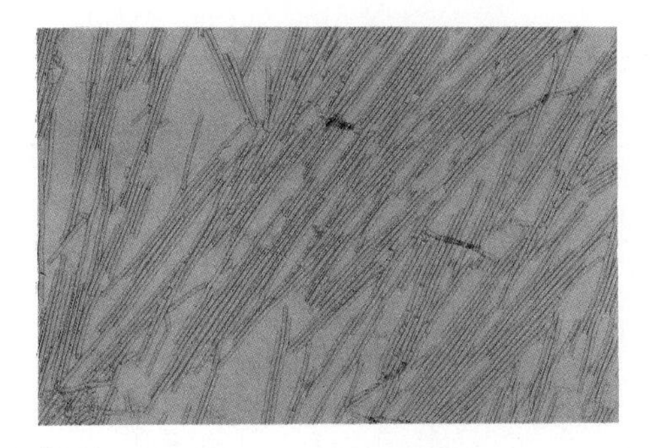
a.

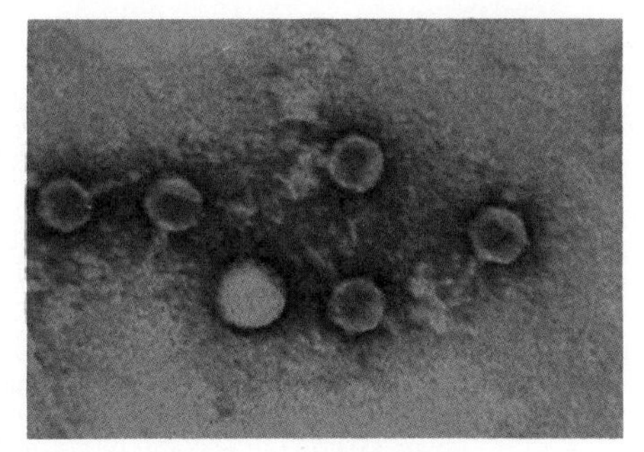
b.

Figure 4.29 Electron micrograph of viruses. a. Rod shape. b. T shape.

4.17

Not all bacteria produce harmful effects. Bacteria play an important role in the production of food stuffs (vinegar, cheese, and yogurt) and in the manufacture of vitamin K required for blood clotting. Bacterial action is an important part of modern-day sewage treatment and in the natural recycling chemical of elements from decaying plants and animals to the soil. Select one of the above roles, write a report on the importance of bacteria to human affairs, and attach it to your log book.

4.18

Now let us take a look at the structure of viruses. **Viruses** are a unique biologic system, neither living nor nonliving, but sort of a borderline bridge between these two categories. Viruses are not cells.

Viruses are particles of matter smaller than the cell (figure 4.26). They vary in size from .5 micrometers (largest) to .01 micrometers (smallest). The electron microscope is required to study viruses. It has been calculated that about twenty thousand viruses could fit inside a small bacterial cell.

Most viruses are polyhedral-like; some are helical or rod-shaped; others are shaped like a T (figure 4.29). Molecular biologists sometimes speak of viruses as *nucleoprotein* molecules because a typical virus has a molecular coat of protein molecules that surrounds and protects an inner coiled molecule of nucleic acid (the genetic material). A virus has no organelles and no membranes.

Viruses are found in eukaryotic and prokaryotic cells. There are more than three hundred different viruses that can infect humans. It is estimated that 60 percent of human infectious diseases are caused by viruses. Some of the more common infectious viruses are influenza, measles, mumps, smallpox, colds, warts, polio, rabies, some venereal diseases, and AIDS.

Influenza (flu) was once described as the last of the great plagues. Almost every 10 years a new strain of the influenza virus appears and spreads around the world, causing illness and death. In 1918, influenza struck the world, causing an estimated 20 million deaths. Pneumonia, a secondary infection (caused by bacteria), is the chief cause of death. Quite often people with the flu will obtain an antibiotic (pill or shot). This helps fight the pneumonia-causing bacteria, not the viruses.

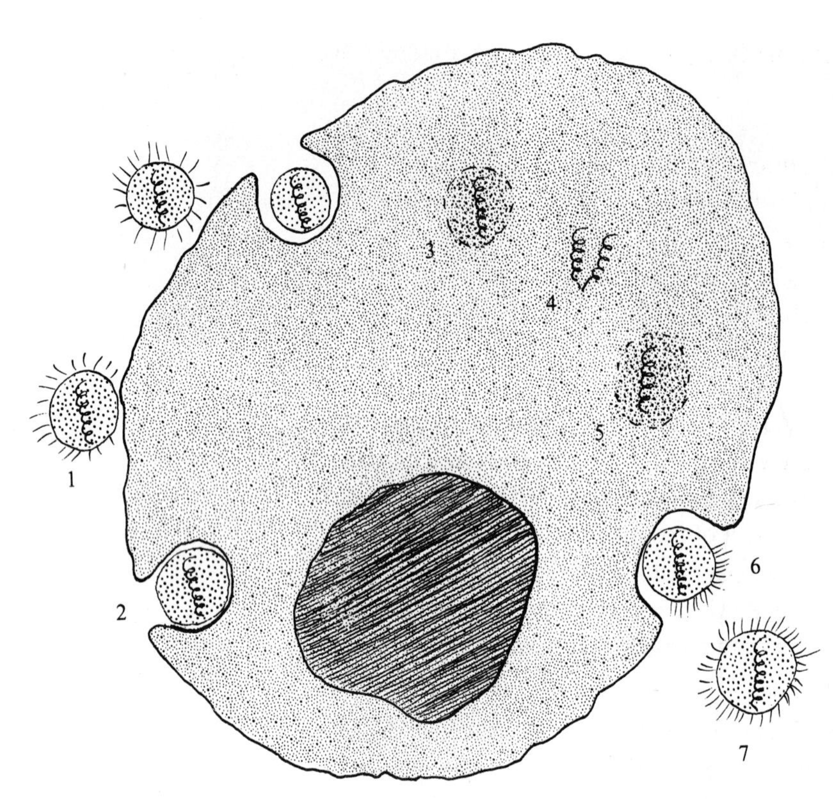

Figure 4.30 Sketch of an influenza virus invading a host cell (ciliated epithelial cell).

Infection of a human cell by the influenza virus begins by inhalation of air containing droplets with viruses, such as those expelled during a sneeze or a cough from an infected person. The virus particles attach to ciliated epithelial cells lining the nose or throat (figure 4.30). The virus particles fuse to the plasma membrane of the ciliated epithelial cell and are engulfed into the cell by phagocytosis. Once inside the cell, the influenza particle breaks up by shedding its coat. The nucleic acid (genetic material) is replicated to form many informational molecules, and many new protein coats are produced. These are assembled together to make new viral particles and then released from the cell. The newly released particles may then infect other cells.

Why do you think you get a sore throat with the flu?

Vaccines can be made against the influenza virus. Prepare a report describing a vaccine and how it prepares the body against future viral infections. You will want to consider the immunity system in your study. Attach your report to your log book.

The defeat of polio was made possible by a vaccine developed in the laboratory of Jonas Salk. Write a report about this significant discovery and attach it to your log book.

Since people discovered that viruses can cause diseases, medical research has been in a constant battle with these tiny invaders. Our most recent research battle is to find out if viruses cause cancer in man, and the solution to the AIDS disease.

Plant diseases caused by viruses amount to millions of dollars lost each year in crop production. Viruses affect apples, potatoes, barley, and many other crops. Viruses can cause gall formations on plants, damage the photosynthetic machinery in leaves, and produce dwarf plants. One of the hopes of controlling infectious viral diseases is to breed or produce through genetic engineering virus-resistant crop plants.

This chapter has helped you understand the structural world of the cell and has given you some clues to what cells do during their life processes. Before you began, did you know that an understanding of the world of the cell holds a key to solving the major health problems of today? Did you know that the cells of your body have their own energy-producing power plant, their own digestive system, and their own communication system? Were you aware that you started life as a single cell with all the directions (information) for making a "you" found in a special chemical molecule called DNA?

Life begins at the cellular level, the point of origin of living things. Today biologists believe that new cells come from preexisting cells, that is, there must be a cell before another new one can be produced. But, that is another story.

4.19 Have you ever wondered where eukaryotic cells of plants and animals originated? It is generally agreed that eukaryotic cells evolved from ancestoral prokaryotic cells. Prokaryotic cells are considered ancestral to eukaryotic cells because the fossil record shows "look-alike" fossil microorganisms in 3.5-billion-year-old Precambrian rock, while eukaryotic cells first appeared in the fossil record approximately 1.5 billion years ago (figure 4.31). In more recent geologic time, oxygen-producing, photosynthetic, blue-green, algaelike prokaryotes are found in rocks formed 2.7 billion years ago. In fact, filamentous blue-green algae that appear identical to the present-day form of algae occurs in the 2-billion-year-old Gunflint Iron formation in Canada.

Because of structural differences between prokaryotes and eukaryotes, there is much speculation and differences of opinion regarding the evolutionary path from prokaryote cells to the eukaryote cells of plants and animals. One group of biologists suggests that the evolution of eukaryotic cells was by a gradual, step-by-step process of accumulating mutationally-derived, inheritable genetic information that led to the structural and functional characteristics found in eukaryotic cells, but not found in prokaryotic cells.

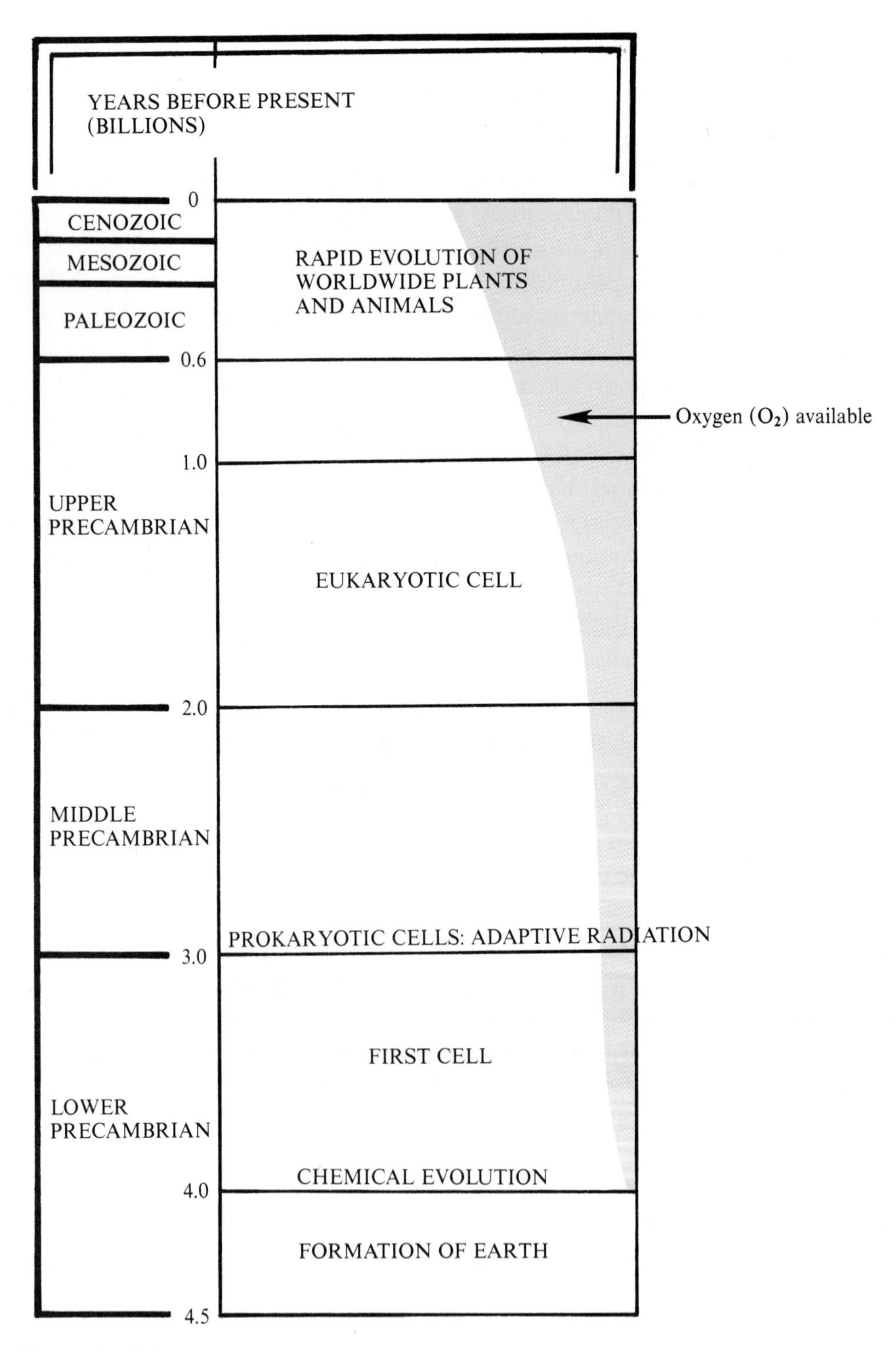

Figure 4.31 Time scale of the geological periods showing the evolution of cells. Note the estimated origin of the first cells, the prokaryotic cells and the eukaryotic cells.

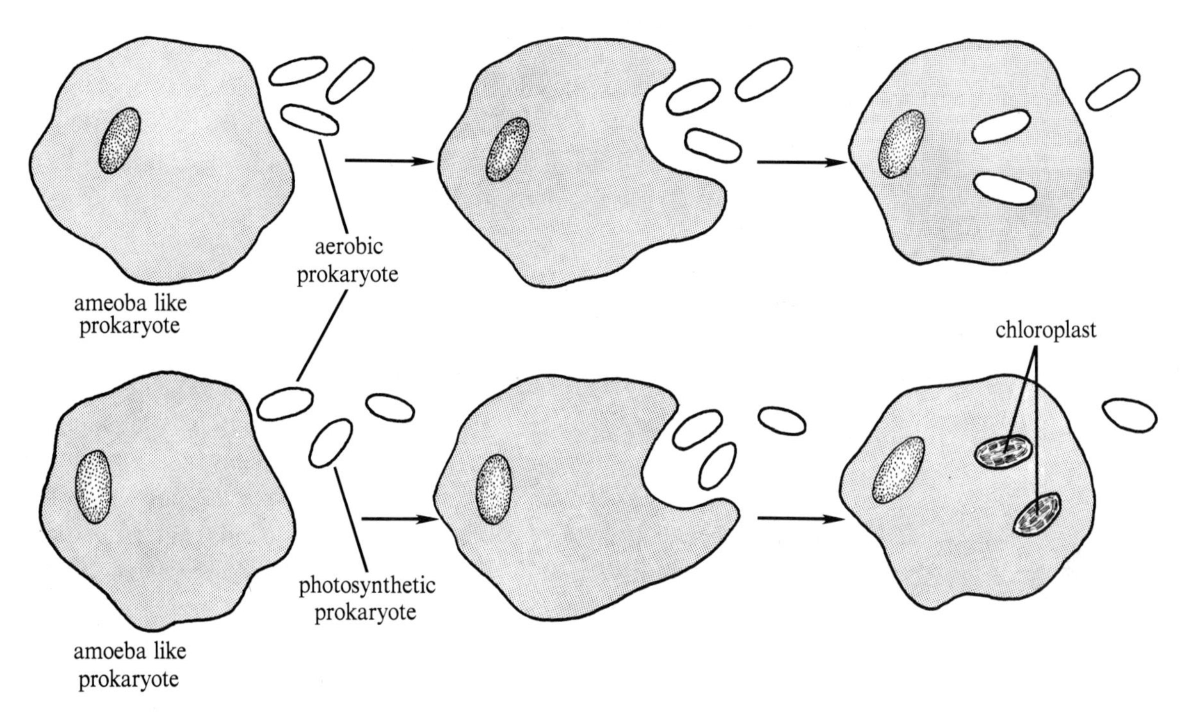

Figure 4.32 Sketch of the endosymbiotic hypothesis for the origin of eukaryotic cells from prokaryotic cells.

The most popular hypothesis, called *endosymbiotic,* is supported by a group of biologists who believe eukaryotic structure and function come from the descendants of large prokaryotic cells that engulfed smaller prokaryotic cells (figure 4.32). That is, small, free-living prokaryotic cells were brought into prototype eukaryotic cells by infoldings of the plasma membrane but were not digested. This association was for the mutual benefit of both cells, i.e., symbiosis. The prokaryotes continued to function, ultimately becoming specialized structures. By this route of evolution our ancestoral cells obtained important organelles, such as mitochondria and chloroplasts. In other words, genetic information was gained not through a step-by-step accumulation of mutational changes, but by a series of symbiotic events.

We may never know how eukaryotic cells evolved or which hypothesis is the correct one, but we do know that the evolution of many internal membranes permitted both an increase in cell size and a way for concentrating cellular chemicals that allowed for the next major evolutionary event, the evolution of **multicellular** organisms.

4.20

All living things share one common property. That property is the ability of self-reproduction. At one time, the beginning of life was believed to be one of **spontaneous generation,** meaning that living things grew out of dead material. The work of L. Spallanzani and Pasteur, however, ended such notions and showed that only life was capable of giving new life. This means that every living thing is a descendant of a living lineal past. That is, cells came from preexisting cells.

The question of life's beginning has been sought by many persons. If you examine the data that has been collected on the formation of stars, several gases always seem to be present. They are hydrogen (H_2), methane (CH_4), and ammonia (NH_3). Based on this information, it has been concluded by scientists that these gases were also present in the primitive atmosphere of Earth. Harold Urey was honored with a Nobel Prize for his work in the evaluation of possible chemical compounds that existed during the time of primitive Earth. It was suggested by A. I. Oparin, and supported by Stanley Miller's research, that the right amount of energy could cause a spontaneous chemical reaction to occur and, thus, form proteinlike compounds. In his experiment, Miller used an energy source similar to lightning, an electrical-spark-discharge apparatus. The medium for the formation of proteinlike compounds could have been the ancient oceans. Miller also experimentally simulated this part. Figure 4.33 illustrates the apparatus that Miller used for his research on the chemical evolution of the first cells. Miller was successfully able to create amino acids, the building blocks of proteins, with his spark-discharge apparatus.

It was Oparin and Miller's viewpoint that the primitive gases, H_2, CH_4, H_2O, and NH_3, combined to form proteinlike molecules in the presence of an energy source. Water molecules then attached to the protein particles. The result was the formation of protein clusters called *coacervates.* Figure 4.34 illustrates this principle. The coacervate resembles a very primitive single cell from which today's cells may have evolved. The idea is that coacervates developed into more complex forms of cell structure. In contrast, S. W. Fox argues that the first prototype cells may have been proteinod microspheres (figure 4.35). Heated amino acids that form long chains of amino acids are called proteinods. When these structures are placed in water they form microspheres.

Whichever hypotheses is correct, one of the critical steps in any evolutionary sequence of living systems is the development of mechanisms for growth and reproduction. In order to perpetuate life, there must be cell division, and it must occur in such a way to ensure distribution of essential genetic messages into new *daughter cells.*

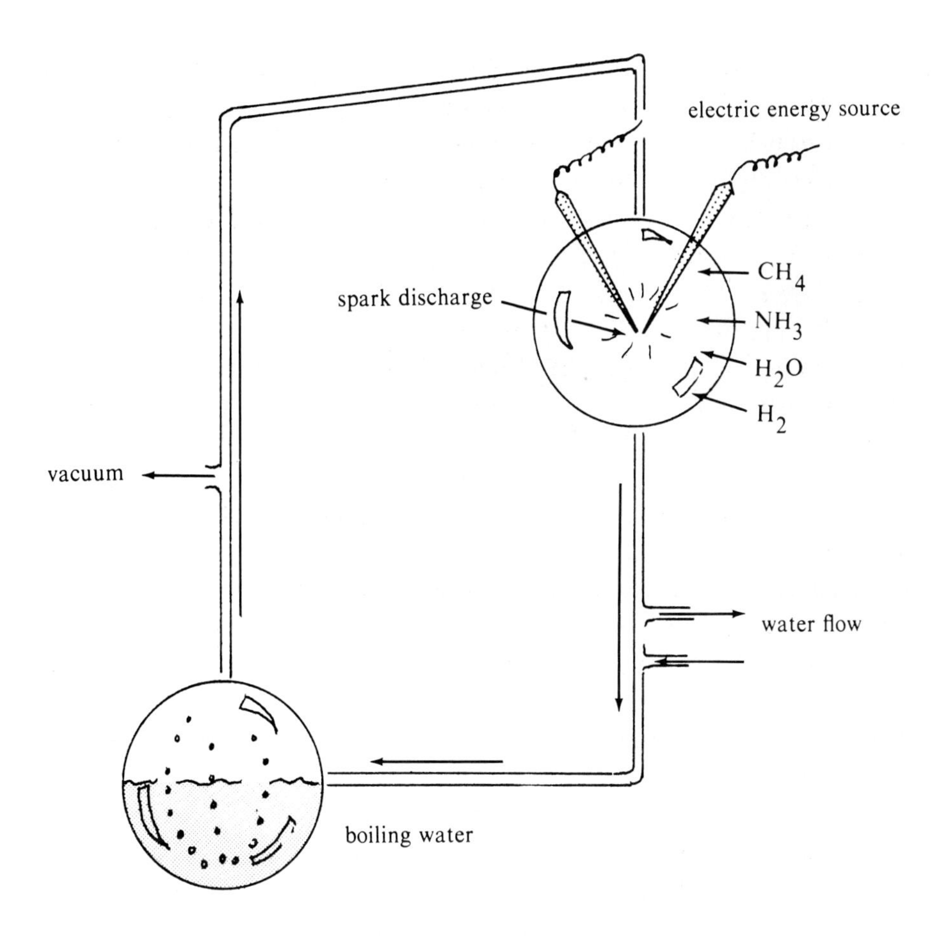

Figure 4.33 Diagram of S. Miller's spark-discharge apparatus. Note the simple inorganic chemicals which gave rise to the organic amino acids.

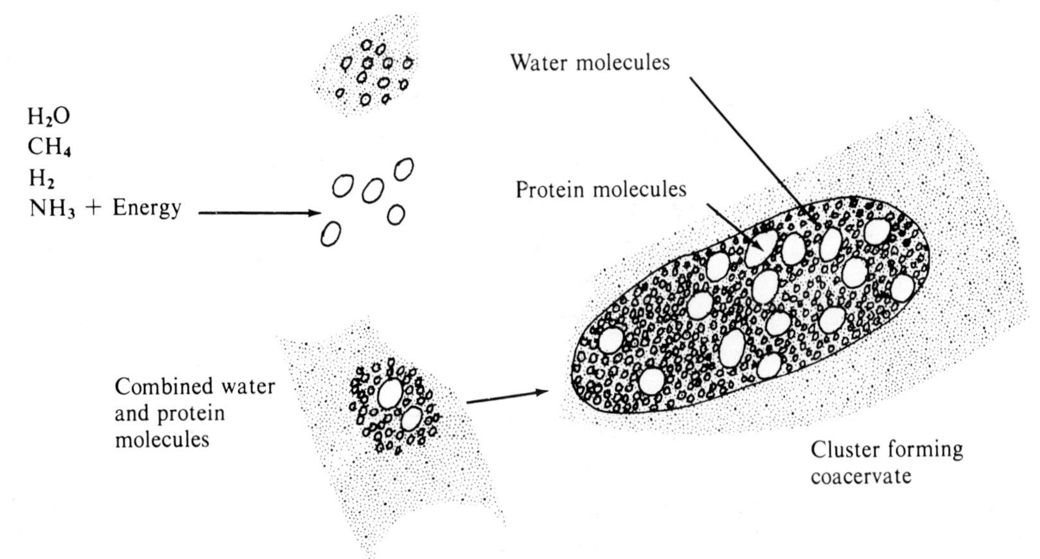

Figure 4.34 Formation of coacervates.

a.

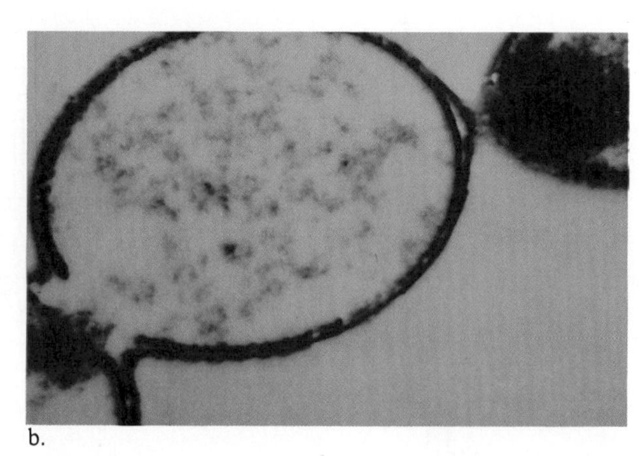
b.

Figure 4.35 Electron micrograph of proteinoid microspheres. a. Surface view. b. Interior view.

4.21 This chapter has given you a look at the cell. If you finish this course understanding the cell, you have a basis for understanding a lot more about life science. Review the study points below. Place a check (✓) in the log book box when you understand the concept.

On completion of this chapter, you should be able to do the following:

☐ 1. Relate the comparable sizes of cells and organelles contained within the cell.

☐ 2. Identify the basic organelles found in an animal cell and a plant cell.

☐ 3. Compare the differences and similarities between the plant and animal cells.

☐ 4. Illustrate the sequential organization of life from atom to biosphere.

☐ 5. Distinguish between eukaryotic and prokaryotic cells.

☐ 6. Discuss viruses and the role they play in life.

☐ 7. Relate the origin of eukaryotic plant and animal cells.

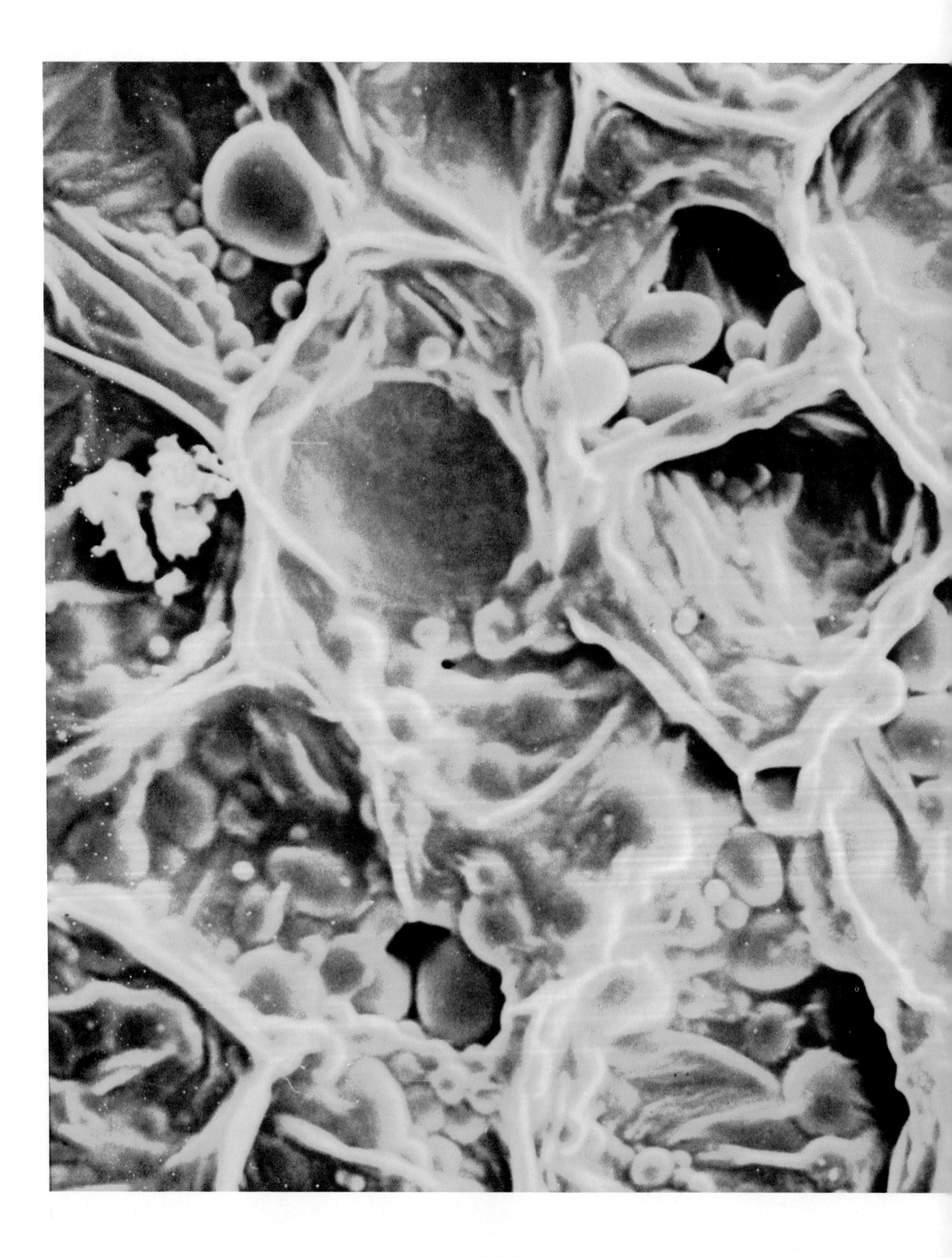

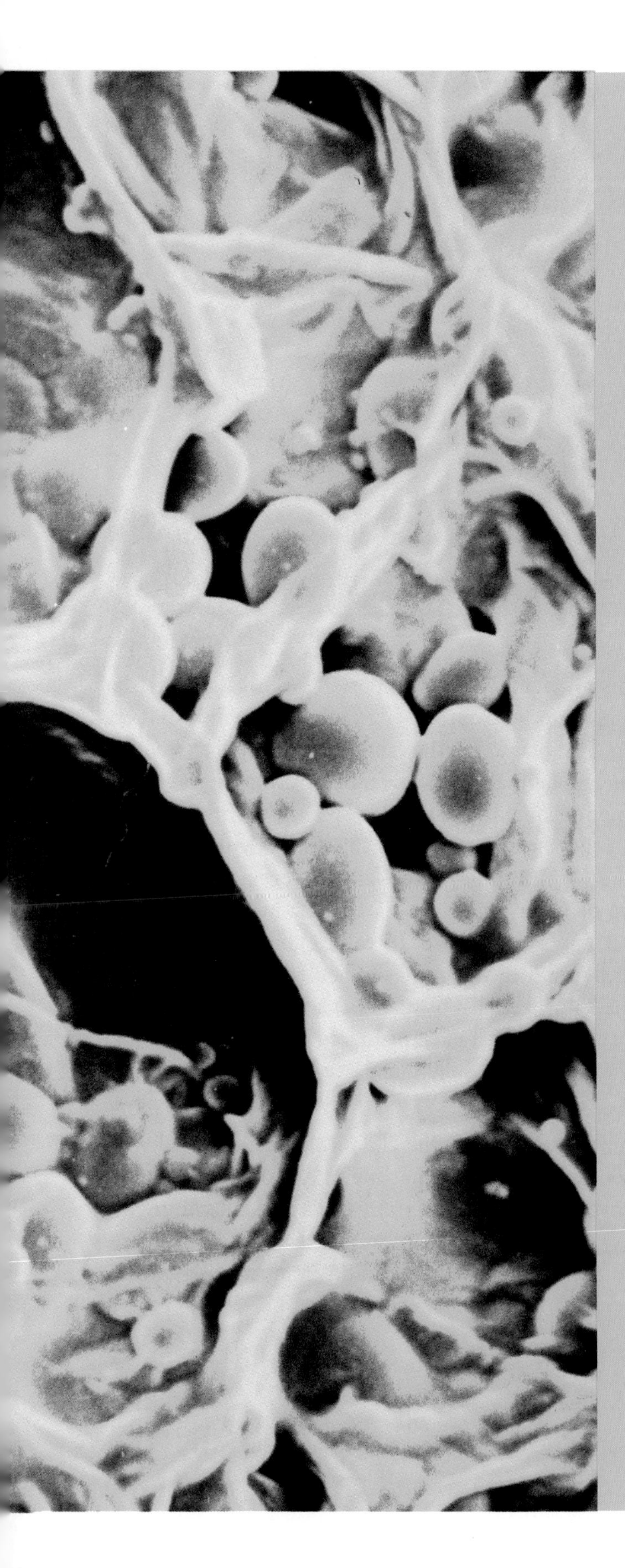

Chemistry for Biology

5

What is the chemistry of life?

Figure 5.1 Photograph of the different environmental spheres.

5.1 The chemistry of life is very important. Basically it is the chemistry of water and carbon-containing molecules. But let us start at the beginning. The planet Earth is the only known place to contain living things. Plants and animals are surrounded by and interact with nonliving matter of the environment. This includes: (*a*) the **lithosphere**—soil and rocks closest to the surface of the Earth; (*b*) the **hydrosphere**—fresh and salt waters; and (*c*) the **atmosphere**—gaseous envelope. The **biosphere** is a narrow zone of living things on Earth that interact with the lithosphere, hydrosphere, and atomosphere. In your logbook label each of the nonliving spheres in figure 5.1. Which of the nonliving spheres might contain plants and animals? Make a list of ways (a) a turtle interacts with the atmosphere, (b) you interact with each of the environmental spheres, and (c) a tree interacts with the atmosphere and the lithosphere.

5.2 What is matter? Matter is the chemical substance of living things. It occupies space and has mass. An **atom** is the basic unit of all chemical substances in living matter. Cells and, therefore, all living things are built with atoms. Atoms are joined together in different ways to form the different chemicals, structures, and forms of an organism or cell. In chemistry, atoms are called **elements.** The many elements found in the lithosphere, hydrosphere, atmosphere that make up the biosphere can be seen by looking at a chemical periodic chart of the elements. Determine the number of recorded elements in the periodic chart and enter the number in your logbook. ______

Data table 5.1 shows the common elements in living things. Observe that only a few of the many known elements are found in the living members of the biosphere. However, scientists remain uncertain about the exact number of different elements in the living world. Data table 5.2 shows the elemental composition of the lithosphere and the human body. Each number represents the percent of the total elements present. For example, nine and one-half out of every one hundred atoms found in a sample of the human body are elements of carbon, while there are only nineteen atoms of carbon in every ten thousand atoms of the lithosphere. Determine the comparative number for the hydrogen element and enter it in the logbook. ______

Data Table 5.1 Common Elements in Living Things

Element	Symbol	Atomic Number	
Hydrogen	H	1	
Carbon	C	6	
Nitrogen	N	7	These elements make up most biological molecules.
Oxygen	O	8	
Phosphorus	P	15	
Sulfur	S	16	
Sodium	Na	11	
Magnesium	Mg	12	These elements occur mainly as dissolved salts.
Chlorine	Cl	17	
Potassium	K	19	
Calcium	Ca	20	
Iron	Fe	26	These elements also play vital roles.
Copper	Cu	29	
Zinc	Zn	30	

One of the strategies of life is to take up certain elements that are scarce in the nonliving world and concentrate them within living cells. Based on information in data table 5.2, enter in your logbook those elements found in the highest amount in living things?

Data Table 5.2 Elements and the Human Body

Composition of Lithosphere		Composition of Human Body	
Oxygen	47	Hydrogen	63
Silicon	28	Oxygen	25.5
Aluminum	7.9	Carbon	9.5
Iron	4.5	Nitrogen	1.4
Calcium	3.5	Calcium	0.31
Sodium	2.5	Phosphorus	0.22
Potassium	2.5	Chlorine	0.03
Magnesium	2.2	Potassium	0.06
Titanium	0.46	Sulfur	0.05
Hydrogen	0.22	Sodium	0.03
Carbon	0.19	Magnesium	0.01
All others	<0.1	All others	<0.01

5.3

Atoms are made up of three basic parts. They are the **electron** (indicated by e^-), the **proton** (p^+), and the **neutron** (n). The simplified diagrams (figure 5.2*a, b, c*) of an atom indicate an atom is constructed of protons and neutrons in the center of a nucleus while electrons orbit around a nucleus. Electrons are arranged (or orbit) at different distances (energy levels) from the nucleus. In figure 5.2*b*, negative-charged electrons are represented as a cloud surrounding a positive-charged nucleus. Note the atom does not have a sharp boundary. Figure 5.2*c* shows the atomic structures of hydrogen, carbon, and oxygen. Study this figure.

The number of electrons in an atom is equal to the number of protons. Therefore, if oxygen contains eight electrons, it would also have eight protons. The number of neutrons in an atom is normally not equal to the number of electrons or protons. However, there are exceptions. The number of protons in an atom is referred to as the **atomic number.** Record in your logbook the number of protons found in an atom of oxygen. ______ Record in your logbook the atomic number of oxygen. ________ All parts of an atom determine its atomic weight, but the majority of an atom's atomic weight is determined by the number of neutrons and protons. Electrons have very little weight and are usually not considered in calculating the atomic weight.

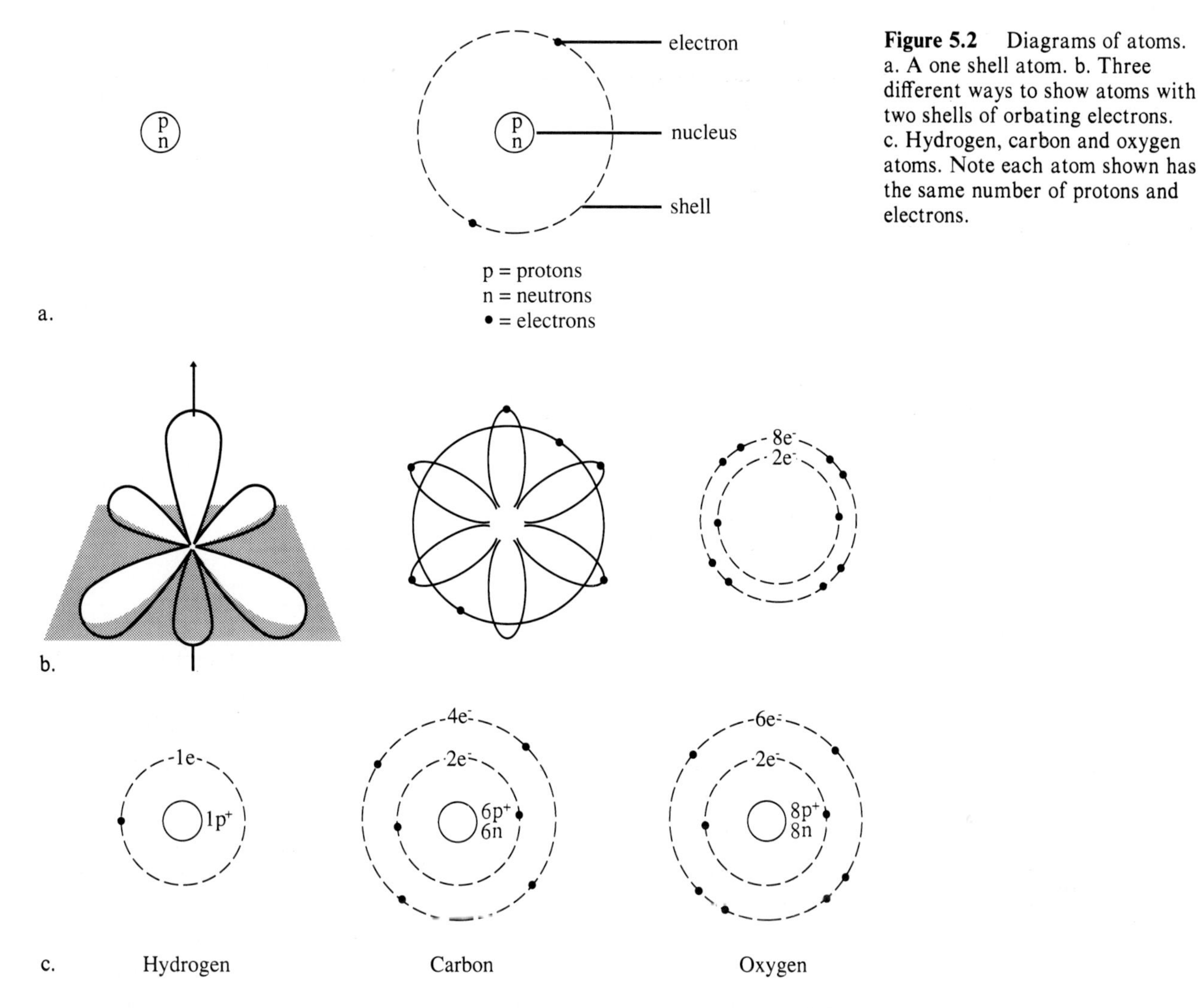

Figure 5.2 Diagrams of atoms. a. A one shell atom. b. Three different ways to show atoms with two shells of orbating electrons. c. Hydrogen, carbon and oxygen atoms. Note each atom shown has the same number of protons and electrons.

Record in data table 5.3 of your logbook the elemental symbol and atomic number of the atoms listed (see data table 5.1). Let us look at oxygen as an example to complete the data table. Since the number of electrons is equal to the atomic number, the number of electrons is eight. If the number of protons is equal to the number of electrons, then the number of protons is also eight. The number of neutrons is found by subtracting the number of protons from the atomic weight ($16 - 8 = 8$). Now complete data table 5.3 in your logbook.

Data Table 5.3 Atoms and Their Basic Parts

Element Name	**Symbol**	**Atomic Number**	**Atomic Weight**	**No. of e^-'s**	**No. of p's**	**No. of n's**
Oxygen	0	8	16	8	8	8
Copper						
Nitrogen						
Sulfur						
Phosphorus						
Sodium						
Nitrogen						
Carbon						
Hydrogen						

5.4

Two or more atoms can combine to form a **molecule.** Two examples are illustrated below.

atom + atom = molecule
O + O = O_2
Na + Cl = NaCl

Answer all of the questions in your logbook.

Molecules may be **compounds** (figure 5.4*a*). Compounds are made of two or more kinds of atoms in a fixed proportion. Is sodium chloride a compound? ________ Why?

Matter (that which occupies space and has mass) made of the same kind of atoms is an *element* (Figure 5.4*b*). Thus, elements are made of atoms with the same atomic number. Is O_2 an element? ____________ *Please note* that a single atom is also called an element. Is NaCl an element? ________ Figure 5.4*c* shows a **mixture.** Define a mixture.

If a compound (solute) mixes (dissolves) with water (solvent) or some other liquid (solvent), a *solution* is formed. Some compounds will not form solutions, but instead become suspended in the liquid. This is called a *suspension.* A suspension is a mixture.

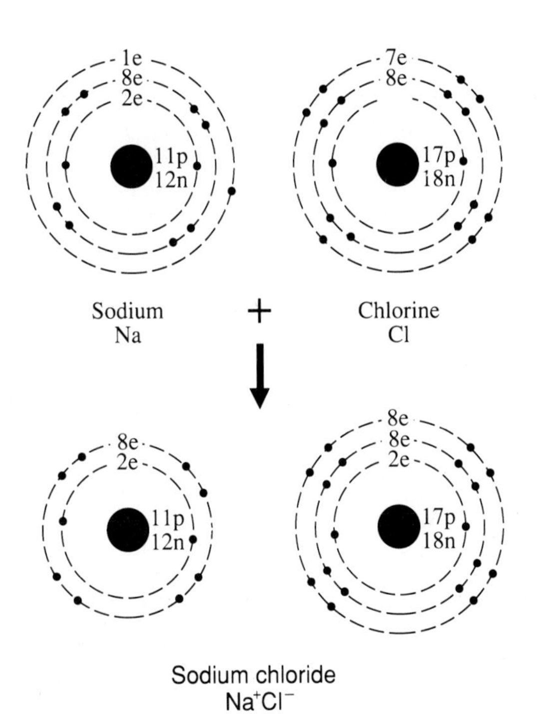

Figure 5.3 Ionic reaction showing sodium donating an electron to chlorine. Note the products of the reaction each have eight electrons in the outer shell.

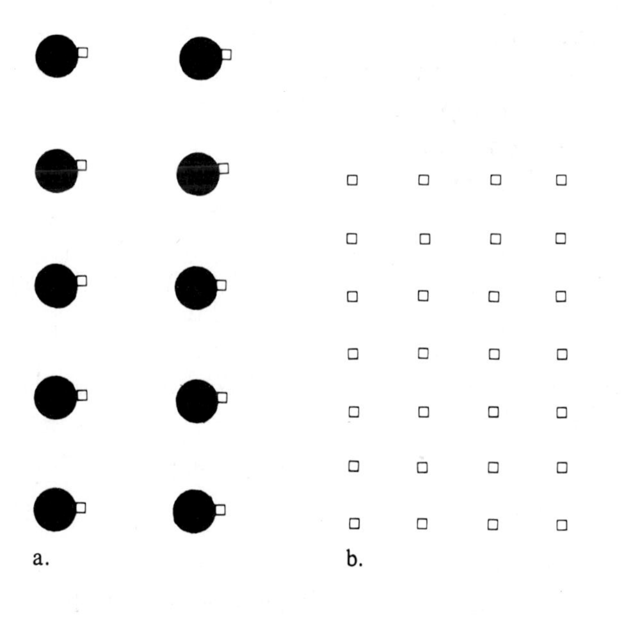

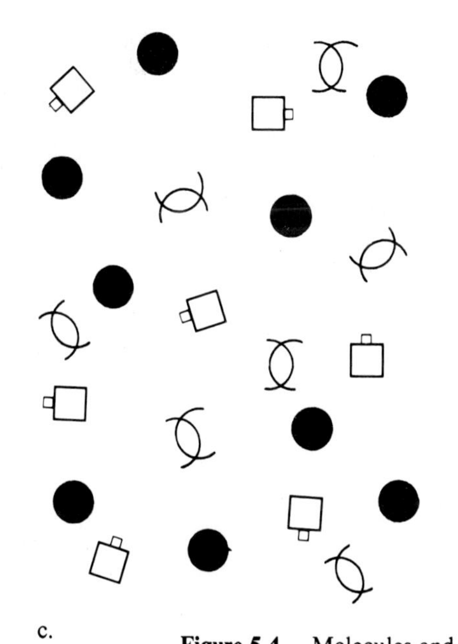

Figure 5.4 Molecules and compounds. a. Compounds. b. Elements. c. Mixture.

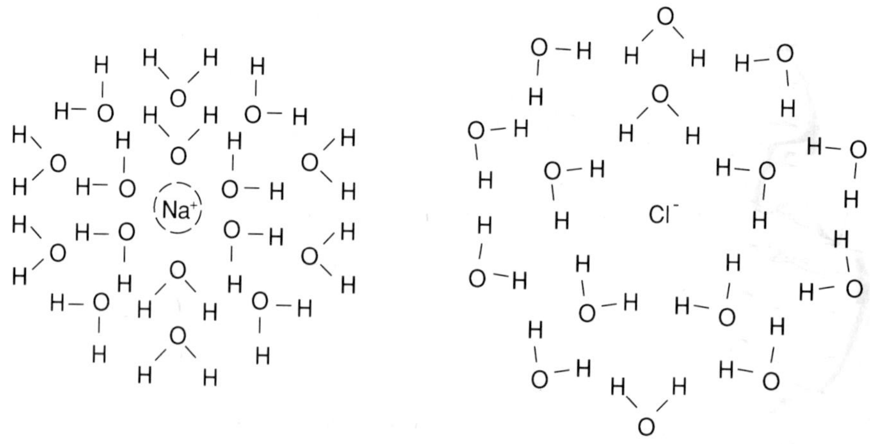

Figure 5.5 Diagram of sodium chloride dissolved in water. Note the polar nature of the ions.

5.5

The energy force that holds atoms together is a **chemical bond.** Atoms may join together by losing or gaining electrons to form an *ionic* bond. Sodium chloride (NaCl) is a compound formed by Na losing an electron from its outer shell and Cl gaining an electron in its outer shell (figure 5.3). Now both atoms have 8 electrons in their outer shells. *Please note* that each atom of the compound has a charge imbalance because of the exchange of electrons. Na has eleven p^+ and ten e^-. Cl has seventeen p^+ and eighteen e^-. Such charged atoms are called *ions,* and opposite-charged ions attract each other. When the compound NaCl is placed in water, the ions separate from each other and are surrounded by water molecules (figure 5.5).

Water is a universal solvent. It enables chemical activities to occur in living things. In fact, living organisms have high percentages of water. Oxygen and hydrogen are bonded together to make water by sharing electrons (figure 5.6). This type of bond is a *covalent* bond. Usually the atoms of each compound have a charge balance because the sharing of electrons between atoms is approximately equal. The charge balance is a chemical mark of covalent bonds and gives the bond a *nonpolar* characteristic. But sometimes one atom of a covalent bond possesses a greater negative charge than its atom partner; a slight charge imbalance occurs. The unequal sharing of electrons in a covalent bond is called a *polar covalent bond* or *hydrogen bond.*

Part of a water molecule has a slight negative end (that is, electronegative), and the other part has a slight positive charge (that is, electropositive) (figure 5.7). *Please note* that the covalent bonds in figures 5.6 and 5.7 are shown by a line joining two atoms together. If the oxygen atom has a slight negative charge, then is the electron of the hydrogen atom closer

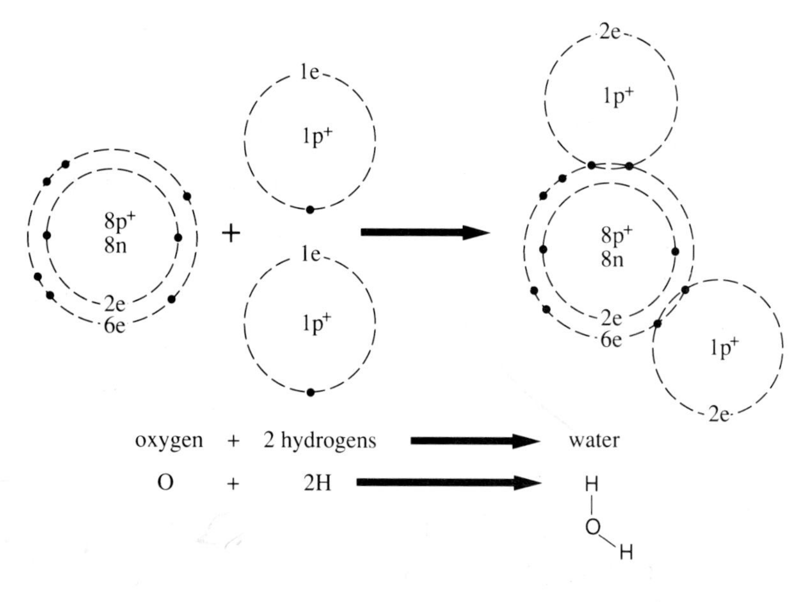

Figure 5.6 Formation of water. Oxygen is sharing a pair of its electrons with each of two hydrogen atoms. Note a covalent is represented by a short line joining two atoms.

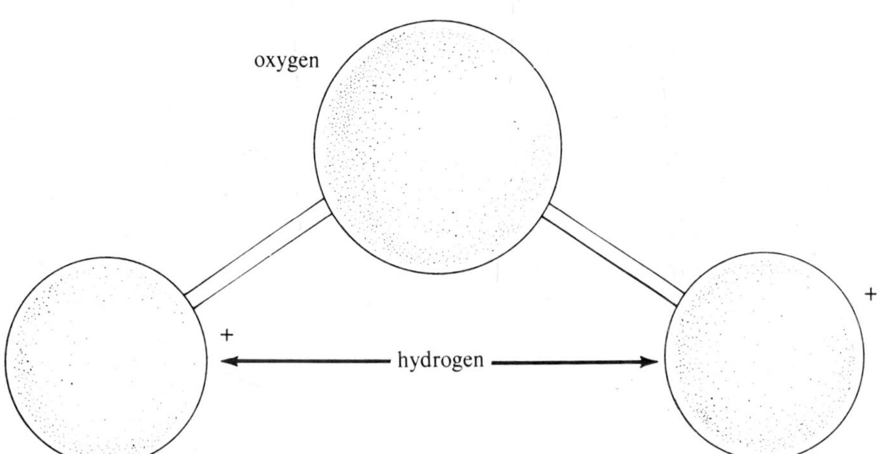

Figure 5.7 Diagram of a single polar water molecule.

to the hydrogen nucleus or the oxygen nucleus? ________ It is this slight charge imbalance that allows water to be a universal solvent because the electronegative or electropositive ends attract other molecules with either a negative or positive charge. Examine figure 5.5 and label on the diagram in your logbook the positive charge and negative charge ends of the water molecules. Water molecules are cohesive: water flows but does not break apart. Explain in your logbook why water molecules stay together.

One of the reasons water is abundant in living systems is because it can dissolve (disperse or separate the molecules from each other) small or large molecules that are either polar or ionic. Therefore, *if* a compound dissolves in water, *then* it or a portion of it must be either ____________

or ____________ .

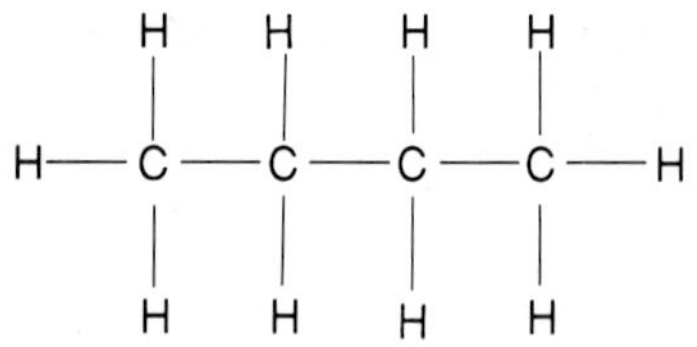

a. Straight chain

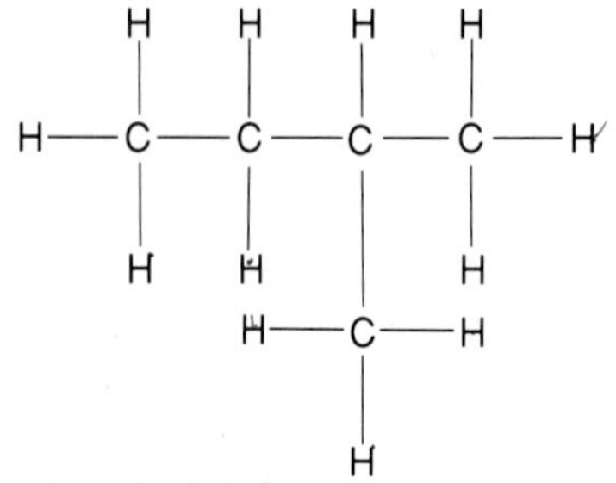

b. Branched chain

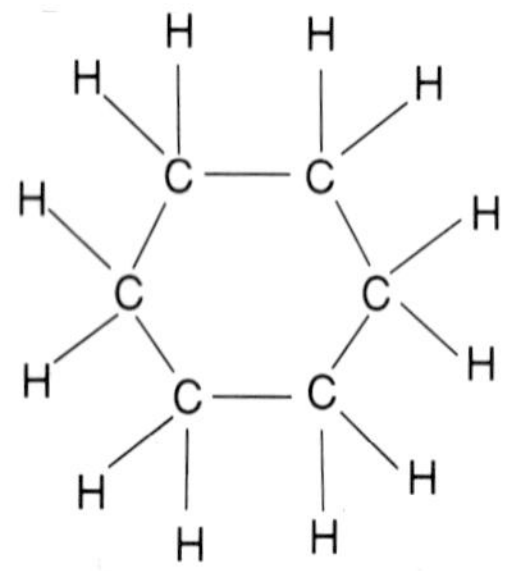

c. Carbon ring

Figure 5.8 Different carbon-carbon backbones illustrate diversity of carbon compounds. a. Straight chain. b. Branched chain. c. Ring chain.

5.6 The matter of living systems is found in the form of molecules and ions instead of elements. In fact, living systems are aggregations of molecules. Cellular structures such as membranes, mitochondria, and ribosomes are aggregates of molecules. The molecules of living matter may be classified into two groups: *inorganic* and *organic.* Water is the most common inorganic compound in living organisms. Other inorganic compounds include minerals and metals.

Organic chemistry is the chemistry of *carbon compounds.* Carbon compounds are built with a backbone that provides a strong carbon-carbon covalent bond. Carbon compounds can form either a straight chain, branched chain, or ring (Figure 5.8*a*, *b*, *c*). *Please note* that the short line (bond) represents two electrons being shared between two atoms, that is, a covalent bond. Because a carbon atom has four electrons in its outer shell (figure 5.2*c*), it may share with up to four other atoms to form a covalent bond. Observe figure 5.8 and count the number of covalent bonds in each chain or ring. How many covalent bonds can be formed between each carbon atom? ________ How many different atoms could be linked to a single carbon atom? ________ How many covalent bonds can be formed between each hydrogen atom? ________

Carbon may share more than a single pair of electrons with another atom and form one *double* covalent bond (Figure 5.8*a*). *Please note* that two lines between two atoms means two covalent bonds, or one double bond. In your logbook draw a straight chain organic compound containing four carbon atoms, six hydrogen atoms, and only two double bonds.

You have just drawn the *structural* formula for an organic compound; the *empirical* formula of your compound is C_4H_6. Record in your logbook the empirical formula for the compound in figure 5.8*b*? ________

Organic compounds can be written as empirical formulas or as structural formulas. The empirical formula is the simplest *numerical relationship* of atoms in a compound. The structural formula is both the numerical relationship and a diagram of the *arrangement* of the atoms in a compound.

5.7 Data table 5.4 shows some common organic compounds. These are the major *macromolecules* of living things. Macromolecules are built from the aggregation of smaller *unit molecules* called *monomers.* The process of aggregating or joining monomers together is a *condensation,* or *synthesis,* reaction (figure 5.9*a*). The breakdown, or deaggregation, of a macromolecule is a *digestive* reaction (figure 5.9*b*). Both synthesis and breakdown occur within plant and animal cells. Sometimes digestion occurs outside the cell by enzymes that are moved across the

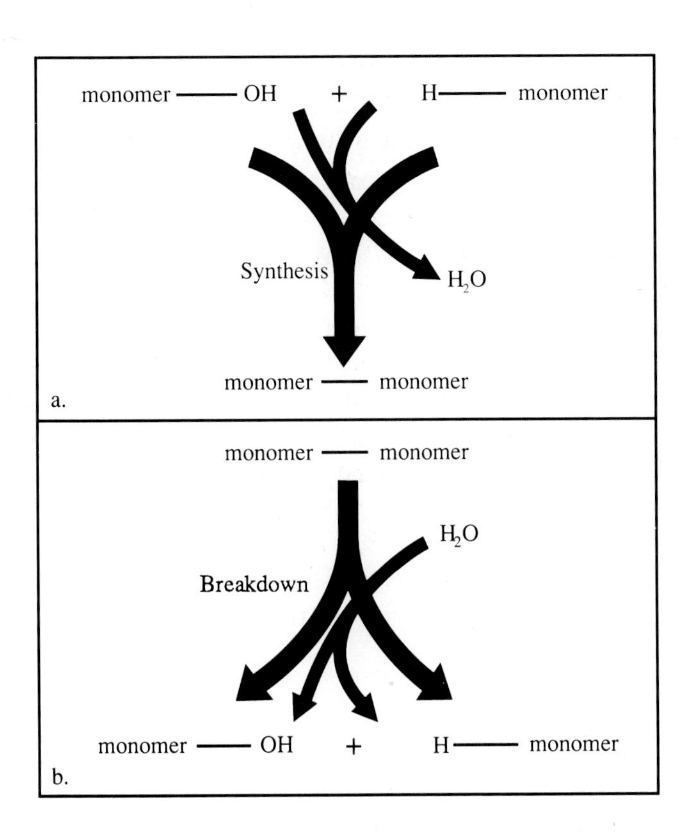

Figure 5.9 Typical synthesis and breakdown of cellular macromolecules. a. Synthesis. b. Breakdown.

Data Table 5.4 Major Macromolecules of Living Things

Selected Macromolecules	Unit Molecules	Common Atoms	An Important Function in Cells
Polysaccharides			
Starch and glycogen	Glucose	C,H,O	Short-term storage of energy
Cellulose	Glucose	C,H,O	Plant cell walls
Lipids			
Neutral fats	Glycerol + 3 fatty acids	C,H,O	Long-term storage of energy
Phospholipids	Glycerol + fatty acids + phosphate	C,H,O,P	Cell membrane structure
Steroids	Ring compounds + hydrocarbon chain		Various
Proteins	Amino acids	C,H,O,N	Enzymes are globular proteins.
Nucleic acids	Nucleotides	C,H,O,N,P	Genes are nucleic acids.

plasma membrane. Several digestive enzymes found in the human body are listed in data tables 5.5 and 5.6. These are synthesized in the cells of your body and moved across the plasma membrane to the outside. The digestive action of these enzymes is the breakdown of macromolecules into their monomers. These monomers are moved across the plasma membrane into the cell and used for energy, or used as unit monomers to build new macromolecules.
Answer the following questions in your logbook.

1. What macromolecules did you eat today or yesterday in your diet?
2a. What is the enzyme responsible for breaking down the starch molecules you ate? ______ 7
3a. In what part of your body is starch digested? ______ 8
4a. What monomer is formed by the breakdown of a starch macromolecule? ______ 9

Ask and *answer* (2b, 3b, 4b) the same questions about the protein molecules you ate.

Data Table 5.5 Some Digestive Enzymes of the Human Body

Reaction	Enzyme	Gland	Site of Occurrence
starch + $H_2O \longrightarrow$ maltose	*a.* Salivary amylase *b.* Pancreatic amylase	*a.* Salivary *b.* Pancreas	*a.* Mouth *b.* Small intestine
maltose + $H_2O \longrightarrow$ glucose*	Maltase	Intestinal	Small intestine
protein + $H_2O \longrightarrow$ peptides	*a.* Pepsin *b.* Trypsin	*a.* Gastric *b.* Pancreas	*a.* Stomach *b.* Small intestine
peptides + $H_2O \longrightarrow$ amino acids*	Peptidases	Intestinal	Small intestine
fat + $H_2O \longrightarrow$ glycerol + fatty acids*	Lipase	Pancreas	Small intestine

*Absorbed by villi—very small cellular extensions.

Food is largely made up of carbohydrate (starch), protein, and fat. These very large macromolecules are broken down by digestive enzymes to small molecules that can be absorbed by intestinal villi. The table indicates the steps needed for carbohydrate digestion (starch and maltose), protein digestion (protein and peptides), and fat digestion (fat) and shows that they are all hydrolytic reactions.

Data Table 5.6 Enzymic Digestion of Molecules

Enzyme	Source	Optimum pH	Type of Molecule Digested	Product
Salivary amylase	Saliva	Neutral	Starch	Maltose
Pepsin	Stomach	Acid	Protein	Peptides
Pancreatic amylase	Pancreas	Alkaline	Starch	Maltose
Lipase	Pancreas	Alkaline	Fat	Glycerol; fatty acids
Trypsin	Pancreas	Alkaline	Protein	Peptides
Nucleases	Pancreas	Alkaline	RNA, DNA	Nucleotides
Peptidases	Intestine	Alkaline	Peptides	Amino acids
Maltase	Intestine	Alkaline	Maltose	Glucose

5.8 **Carbohydrates** are compounds that contain carbon, hydrogen, and oxygen atoms. The term *carbohydrate* is used because most have atoms of hydrogen and oxygen in the same numerical ratio found in water. Carbohydrates are classified as (a) **monosaccharides,** (*b*) **oligosaccharides,** and (*c*) **polysaccharides.**

Monosaccharides may exist in living things as monomers, or they may be assembled together to form oligosaccharides or polysaccharides. A general empirical formula for a monosaccharide is $C_NH_{2N}O_N$, where N refers to the number of atoms. The ratio of a monosaccharide is 1:2:1 ($C_NH_{2N}O_N$).

Determine the ratio of hydrogen to oxygen for the structural formula shown in figure 5.10 and record it in your logbook. ______ Is this ratio the same as water? ______

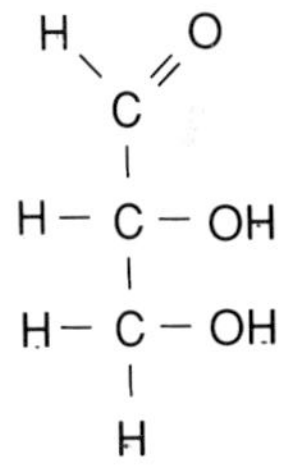

Figure 5.10 Glycerol.

5.9 Monosaccharide sugars can be illustrated by their structural formula in two ways: (*a*) an *open linear* chain and (*b*) a *closed ring* formation. Figure 5.11 represents the relationship between an open chain (*a*) and a closed ring (*b*) for a six-carbon monosaccharide. This structure shows the numerical ratio and the structural arrangement for glucose. The generic name for a six-carbon sugar is hexose. Determine the generic name for a five-carbon sugar and record it in your logbook. ______

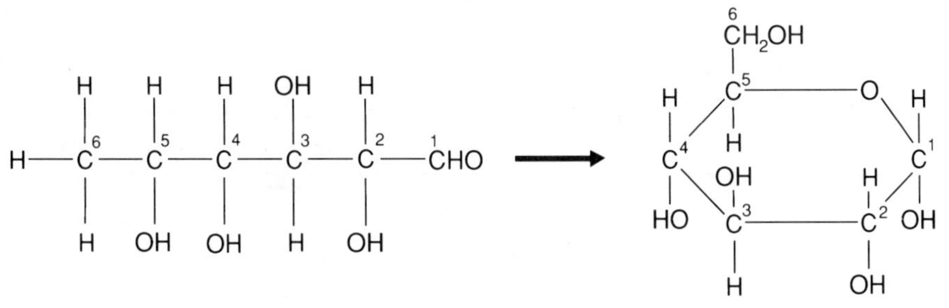

a. b.

Figure 5.11 Glucose. a. Straight chain. b. Ring chain.

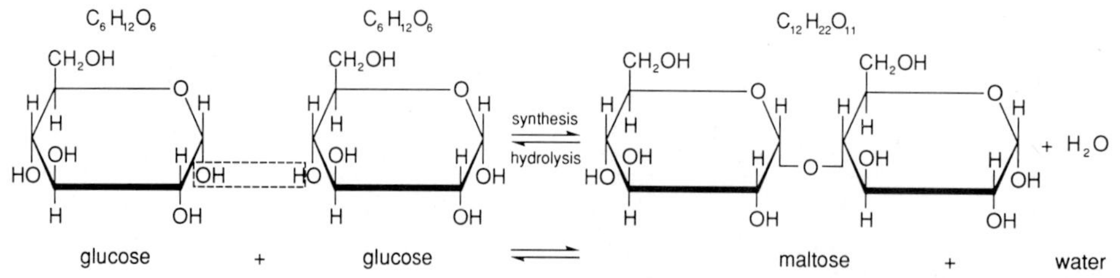

$C_6H_{12}O_6$ $C_6H_{12}O_6$ $C_{12}H_{22}O_{11}$

CH_2OH CH_2OH synthesis / hydrolysis CH_2OH CH_2OH + H_2O

glucose + glucose ⇌ maltose + water

Figure 5.12 Synthesis of the disaccharide maltose from two glucose molecules.

Figure 5.13 Cellulose fibers.

5.10 *Oligosaccharides* are made by joining a few monomers of monosaccharides. The simplest oligosaccharide is a **disaccharide** formed by uniting two monosaccharides with a covalent bond. Examples of disaccharides are sucrose, maltose (malt sugar), and lactose (milk sugar). Sucrose is formed in plants like sugar cane and sugar beets by joining the two hexose molecules, glucose and fructose. Maltose, a common sugar of cereal grains, is formed by joining two glucose molecules (figure 5.12).

5.11 *Polysaccharides* are large chains of many monosaccharides joined together. Starch (figure 5.13), cellulose, and glycogen are common polysaccharides. Cell walls of plant cells contain cellulose (figure 5.13). Label a cellulose fiber (many cellulose molecules) on figure 5.13. Paper is made from cellulose.

Starch and glycogen also are long chains of the monomer glucose (figure 5.14). Glycogen is similar to starch, but has more side branches than starch.

Glycogen is a storage form of carbohydrates in animal cells. After you eat, your liver cells may store excess glucose as glycogen. Label the glycogen granules (black dots) in figure 5.15*a*. Plants may store carbohydrates as starch. Observe the starch grains in figure 5.15*b*.

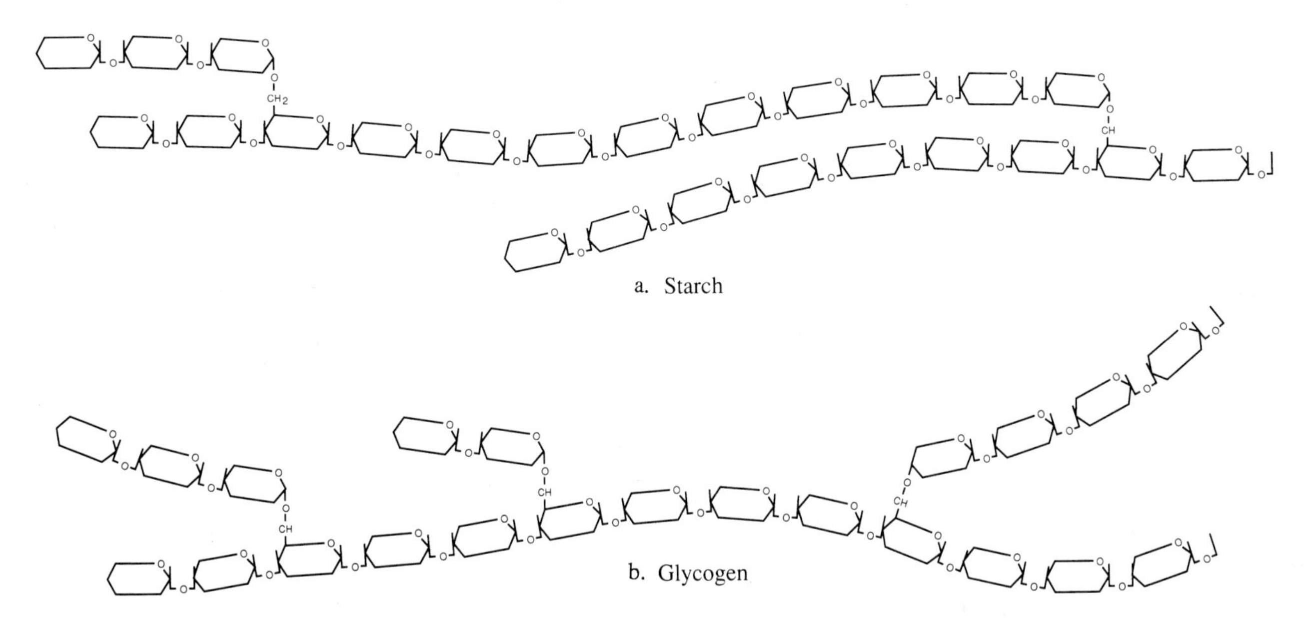

Figure 5.14 Polysaccarides. a. Starch. b. Glycogen. Note each is made of glucose molecules in the ring configuration.

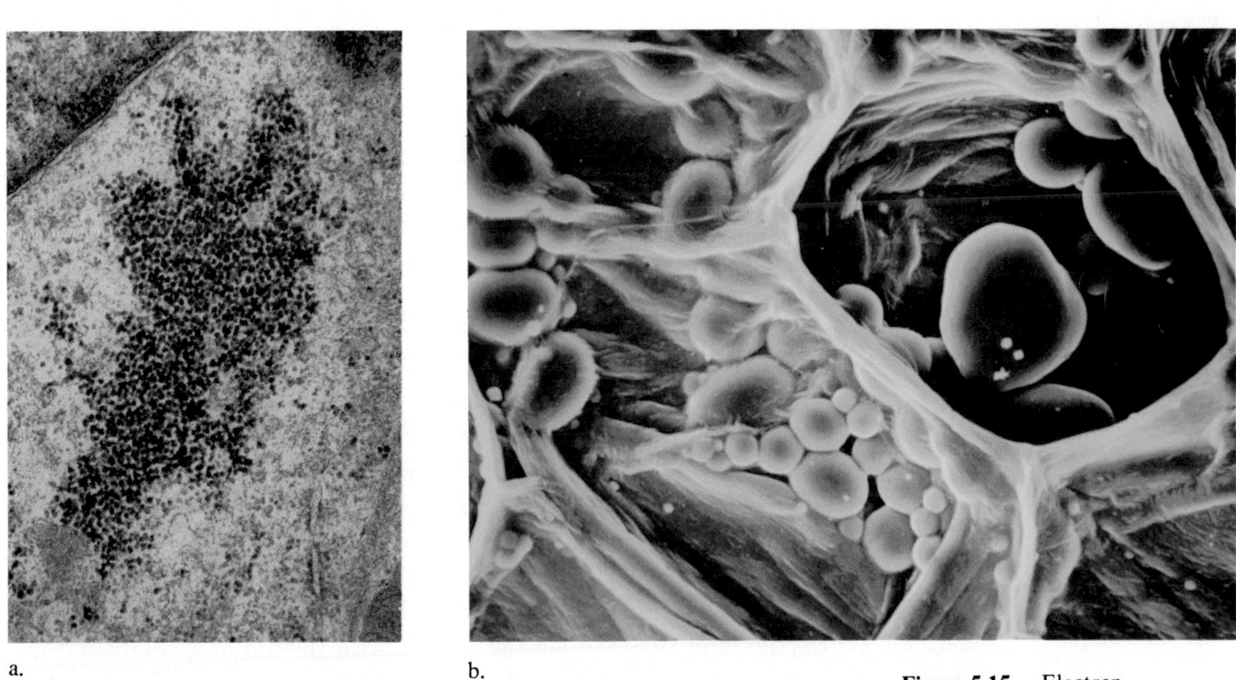

Figure 5.15 Electron micrograph. a. Animal cell with glycogen granules. b. Plant cells with starch grains.

5.12 Disaccharides and polysaccharides are part of a human's diet, but before the human body can use these carbohydrates, it must break them down within the alimentary tract into simple monomer sugars. Macromolecules are too large to pass through the plasma membrane of the intestinal villi cells, therefore, they are broken down (digested). Biochemically speaking, another name for digestion is hydrolysis. Hydrolysis of disaccharides and polysaccharides occurs enzymatically with the addition of water to the compound (figures 5.9*b* and 5.16).

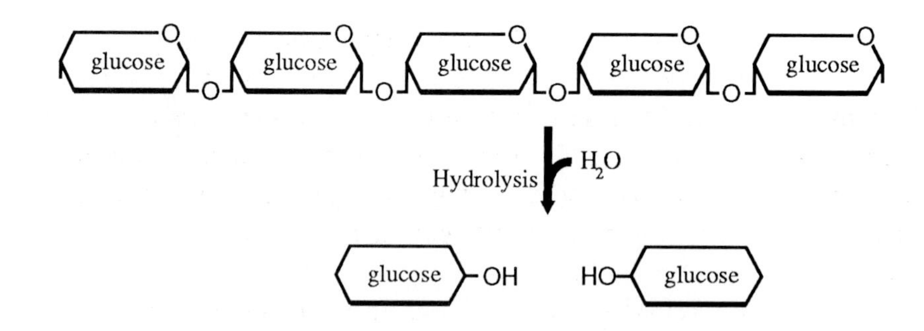

Figure 5.16 Breakdown (hydrolysis) of a polysaccharide to the monomer level.

5.13

Digestion is an organism's way of acquiring energy to do the *work* of living. When you digest a complex macromolecule such as starch, the salivary enzyme, amylase (ptyalin), present in your mouth breaks down the starch molecules to disaccharides. Additional enzyme reactions with the addition of water molecules will break down the disaccharides into glucose molecules (figure 5.16).

Cells transform glucose into useful chemical energy called ATP (adenosine triphosphate). ATP is a compound that cells use to do work. This concept will be covered in other chapters. It is important, however, to remember the relationship between the carbohydrates you eat and the energy they provide through cellular processes.

Although **glycogen** can be digested into glucose by your own liver cells, you and most animals cannot digest cellulose because the cells lack the proper enzymes. However, grass-eating animals such as cows and sheep can digest cellulose. These animals have a special stomach chamber (rumen) where cellulose is digested to glucose molecules and used for food. Other organisms, such as termites (wood eaters) and snails, also digest cellulose into glucose for their food.

5.14

Some food material manufactured by plants is stored as a complex carbohydrate. For example, the starch grains in living potato cells are synthesized from glucose produced by photosynthesis. To find starch grains in living potato cells: (*a*) with a razor blade, slice a *very thin* section from a piece of potato; (*b*) place the section on a clean slide and add a drop of Lugol's solution on the section; (*c*) cover the section with a coverslip; and (*d*) examine the section with the aid of a compound microscope.

Sketch and color in your logbook a rendering of a potato cell. Label your illustration in your logbook by identifying the cell wall, cytoplasm, and starch grains. You can recognize the starch grains because they color blue-black or purple with Lugol's solution. Lugol's solution is a test indicator solution for starch.

5.15 Place 1/2 g of starch and 2 mL of water into a large test tube. Heat the contents by placing the test tube into a hot water bath until the starch thickens. With a test tube holder, remove the tube from the hot water bath. Add another 2 mL of water, mix, and divide the contents of the tube into approximately two equal parts by transferring one-half the contents to a clean test tube. Label one test tube *L* and the other *B*.

To test tube L, add 1 or 2 drops of Lugol's solution. Record the color of the solution in your logbook. ____________ Does this color indicate the presence of starch? ____________

Save test tube B for section 5.17.

5.16 *If* polysaccharide starch molecules are made of glucose monomers (see figure 5.14), *then* the breakdown of starch molecules (see figure 5.9*b*) should produce monomers of glucose. To test this prediction, a method is needed to break down starch and to observe the molecules formed.

Hydrochloric acid (HCl) is a chemical compound that will break down starch into its building block monomer units of glucose. Benedict's test is a way to observe the presence of glucose. An orange or red color is a positive test for glucose when glucose and Benedict's solution are heated together.

5.17 To test tube B from section 5.15 add 10 drops of concentrated HCl. *Be careful*—this acid burns human skin! *Wear goggles!* Place the test tube into a boiling water bath for 3 minutes, remove it, and cool the tube for several minutes. Next, check the solution with a piece of *blue* litmus paper by inserting the litmus paper part way into the solution. Remove the paper and observe the color. If the paper is red, add 10 drops of 3 M NaOH and mix. Check the solution again with *red* litmus paper. When the paper looks slightly blue, the solution is close to a neutral acid-base condition. Add additional drops of 3 M NaOH until the solution is approximately neutral.

To check for the presence of glucose, add 1 or 2 drops of Benedict's solution to test tube B. Reheat the test tube in a boiling water bath for 5 minutes. A reddish orange color will indicate the presence of glucose. A pale green color also may be observed if only a few glucose molecules were broken away (hydrolyzed) from the starch molecules. In your logbook describe your results and prepare a written statement to explain your results.

Do you accept or reject the prediction made in section 5.16? __________
Explain your answer.

5.18 Many seeds and fruits are storage containers for starch molecules. These carbohydrates are used for the germination processes or the nourishment of animals. Do corn grains (both a seed and fruit) contain starch? Complete the following statement in your logbook by making a *prediction:*

IF corn grains contain starch, *then* ______________ should be observed when the corn is tested with Lugol's solution.

Grind a corn grain in a mortar dish with a pestle. Add water to make a thin paste. Add 1 or 2 drops of Lugol's solution. Do you accept or reject your prediction? _________ . Explain.

All living things must have fuel (food) to obtain energy to stay alive, that is, to do the work of living. Carbohydrates are the main source of energy for living organisms and cells. Also, carbohydrates protect and support cells. For example, cell walls in plants are made of carbohydrates; in invertebrate animals, a carbohydrate complex forms an outer body skeleton (exoskeleton). Carbohydrates are essential parts of nucleic acid (DNA and RNA) molecules and, thus, play an important chemical role in the reproduction and transmission of hereditary traits.

5.19 **Lipids** are macromolecular compounds that are similar to carbohydrates because lipids contain carbon, hydrogen, and oxygen atoms. However, the hydrogen-oxygen ratio of lipids is different from the 2:1 H:0 ratio of carbohydrates.

Simple (neutral) lipids are made in cells by joining three fatty acids (figure 5.17*a*) and one alcohol molecule called glycerol (figure 5.17*b*). This lipid production occurs at the endoplasmic reticulum membrane to form the structure seen in figure 5.18. Fatty acids vary in chemical composition and contain a chain of many carbon atoms. Two basic types of fatty acid molecules found in lipids are *saturated* (figure 5.17*a*) and *unsaturated* (figure 5.19).

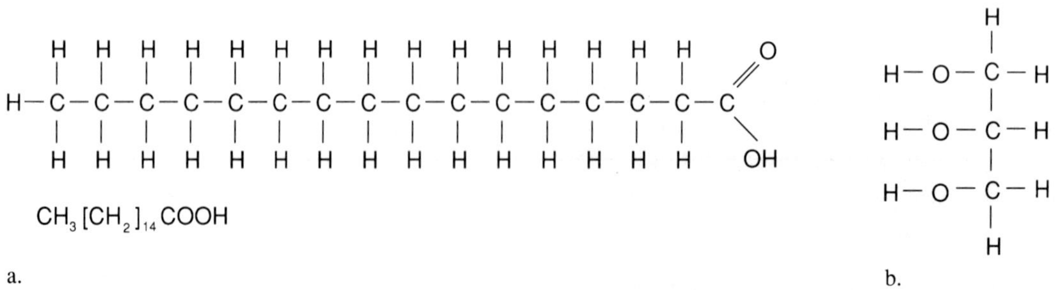

```
   H H H H H H H H H H H H H H H    O
   | | | | | | | | | | | | | | |   //
 H-C-C-C-C-C-C-C-C-C-C-C-C-C-C-C-C
   | | | | | | | | | | | | | | |   \
   H H H H H H H H H H H H H H H    OH
```

$CH_3[CH_2]_{14}COOH$

a.

```
          H
          |
   H-O-C-H
          |
   H-O-C-H
          |
   H-O-C-H
          |
          H
```

b.

Figure 5.17 Monomers of lipid molecules. a. Fatty acid. b. Glycerol.

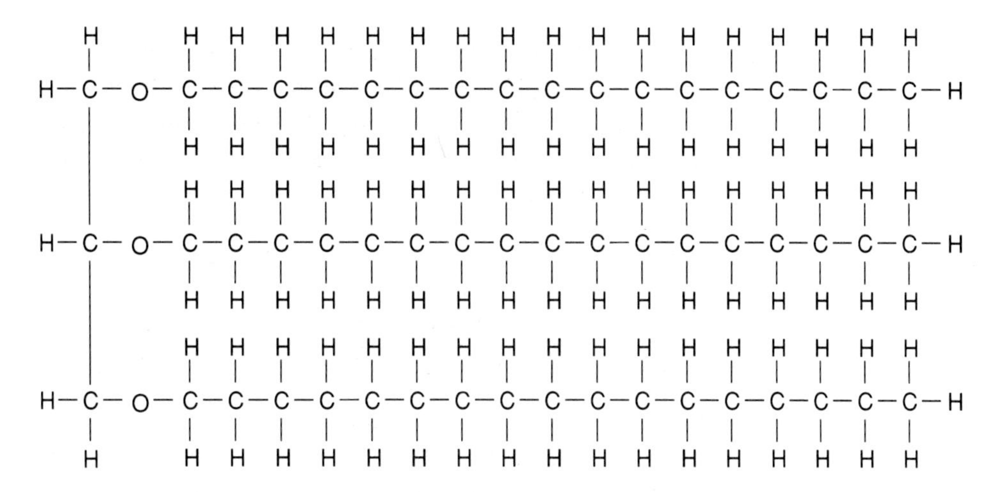

Figure 5.18 Structure of a neutral lipid with a glycerol backbone and three saturated fatty acids attached.

Please note that a saturated fat contains more hydrogen atoms than an unsaturated fat. Saturated fats are solid and hard compared to unsaturated fats. Lard and butter are lipids. Do they contain saturated or unsaturated fatty acids? ________ (Record answer in your logbook.)

5.20 Classify the foods in data table 5.7 as either high saturated or high unsaturated in fats by placing a check (✓) in the appropriate box in your logbook.

Data Table 5.7 Fat Classification

Food Fats	High Saturated	High Unsaturated
Soybean oil		
Egg yolk		
Seafood		
Corn oil		
Red meats		
Chicken		
Turkey		

5.21 Lipids are found in all living cells, especially in human body cells called **adipose cells.** These cells store simple fats. Interestingly, approximately 50 percent of the total simple fats stored in adipose cells of the human body is located in subcutaneous skin tissue. These stored lipids serve as heat insulators by slowing down the exchange of heat between the body and its environment.

5.22 Certain fatty acids are essential in human diets. A chemical molecule is essential if its absence will cause a disease to occur in the human body. *If* the human body cells cannot manufacture the fatty acid, they must obtain it from the diet. Lipid molecules function in the human body to (*a*) produce energy, (*b*) protect vital organs by providing a general padding, (*c*) provide the architectural framework for cellular membranes, and (*d*) reduce heat flow to and from the environment.

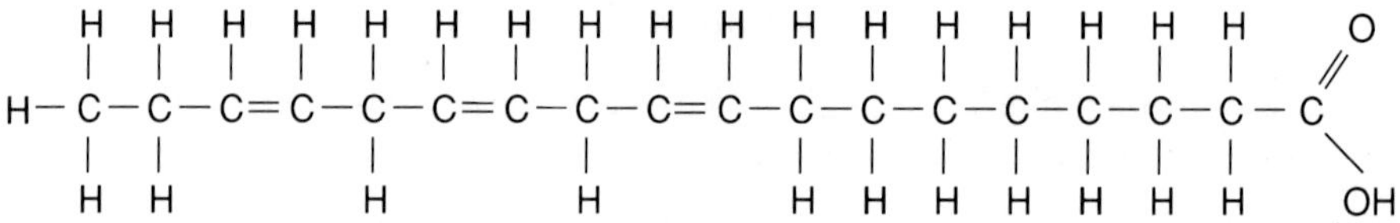

Figure 5.19 Unsaturated fatty acid. Note the reduction of hydrogen atoms compared to a saturated fatty acid shown in figure 5.17a and 5.18.

```
     H   H   H   H   H   H   H   H   H   H   H   H   H   H   H   H   H      O
     |   |   |   |   |   |   |   |   |   |   |   |   |   |   |   |   |     //
H— C— C— C= C— C— C= C— C— C= C— C— C— C— C— C— C— C
     |   |           |           |           |   |   |   |   |   |   |     \
     H   H           H           H           H   H   H   H   H   H   H      OH
```

$CH_3CH_2[CH{=}CHCH_2]\ [CH_3]_6COOH$

Gram for gram, the use of lipid molecules for food will supply more energy to the body than carbohydrates. A pat of butter, for example, will contain more calories than an average size potato. The high caloric content of neutral fats is due to the large number of hydrogen atoms. Assume that two different lipid molecules have an equal number of carbon atoms. Would a lipid containing unsaturated fatty acids be higher or lower in calorie value than a lipid constructed with saturated fatty acids? ________

When fats are digested by the human body, their component monomer molecules are produced (figure 5.20). Use data tables 5.5 and 5.6 in your logbook to answer the following questions. In what part of the human digestive system are fats broken down? ________ What is the general name of the enzyme responsible for this hydrolysis? ________

5.23 **Proteins** are essential to the structure and function of all living things. Proteins (*a*) manage the chemical reactions of cells (enzymes); (*b*) repair tissue damage; (*c*) combat infections (antibodies); (*d*) regulate chemical activities (hormones); and (*e*) affect digestion, respiration, circulation, and secretion.

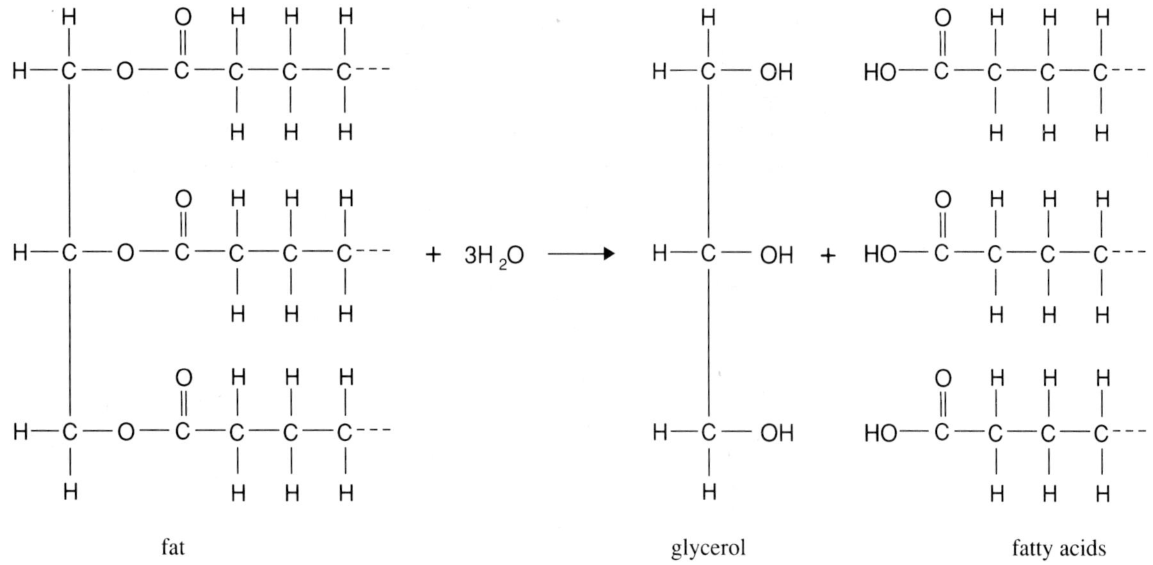

fat + $3H_2O$ → glycerol + fatty acids

Figure 5.20 Breakdown (hydrolysis) of a lipid molecule.

Proteins, like carbohydrates, contain carbon, hydrogen, and oxygen, but unlike carbohydrates, they also have nitrogen atoms in their chemical makeup. Sulfur and phosphorous are also important elements of many proteins.

Proteins are large macromolecules made at ribosomes, and they are the indirect products of hereditary information stored in the nucleus (that is, the nucleic acid called DNA—deoxyribonucleic acid).

The production of proteins at ribosomes occurs by joining monomer unit molecules together (figure 5.9*a*; data table 5.4). What are the unit monomers of proteins? ______________ The general structural formula for a monomer is seen in figure 5.21*a*. These monomers are joined together by a covalent bond to make a long chain of monomers (figure 5.21*b*). The covalent bond linking two amino acid monomers is called a **peptide bond,** and the long chain of amino acids is called a **polypeptide.**

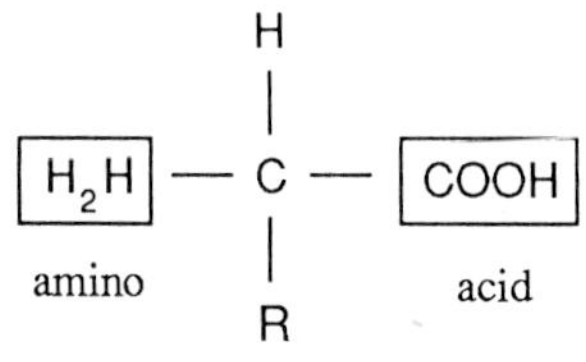

H_2H (amino) — C(H)(R) — COOH (acid)

a.

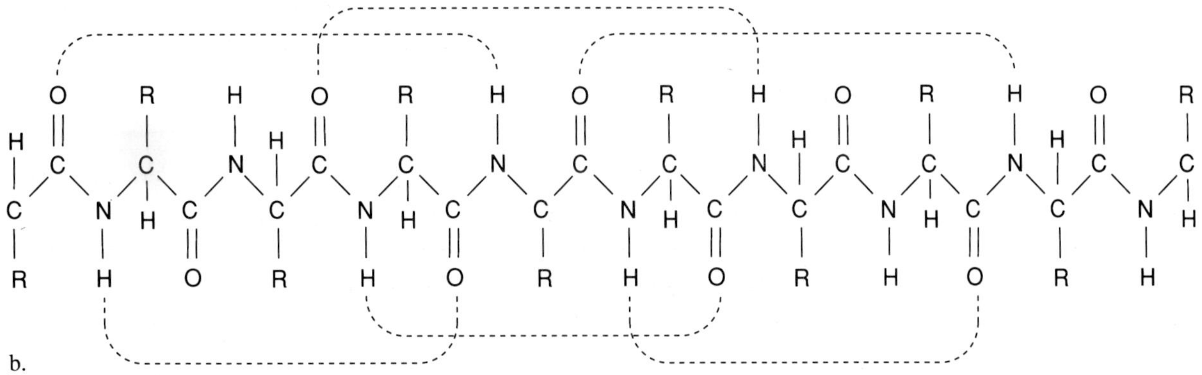

b.

Figure 5.21 Structure of a. an amino acid and b. a peptide made of several amino acids joined together.

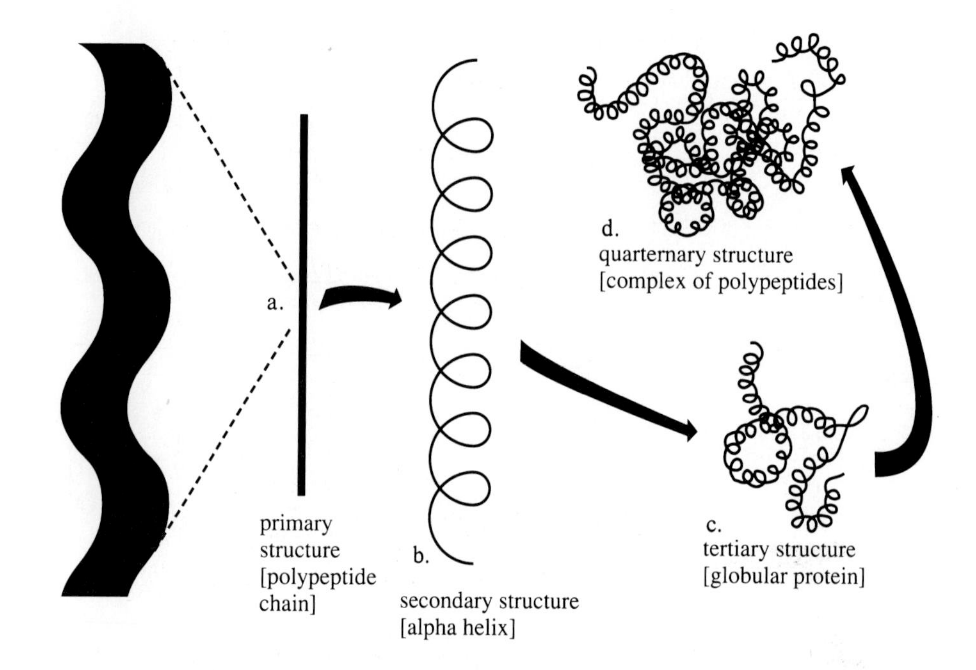

Figure 5.22 Four possible structural configurations of protein molecules. a. Primary. b. Secondary. c. Tertiary. d. Quarternary.

Polypeptide molecules in living cells exist in different structural arrangements: primary, secondary, tertiary, and quaternary (figure 5.22). Note that a quaternary arrangement includes two or more polypeptides. Thus, a protein molecule may be a single polypeptide, or the protein molecule may be two or more polypeptides. In cells, protein molecules function as *structural materials, enzymes, contractile elements, carrier molecules,* and *recognition molecules,* to name a few examples.

Proteins are very important in the diet of many types of animals. Cells use amino acids of dietary proteins to build protein molecules for use by the cell. Animals, including humans, require essential amino acids in the protein portion of their diet because they cannot synthesize certain amino acids necessary in building polypeptides. Proteins in the human diet are digested (hydrolyzed) into amino acids (figure 5.9*b*) before passing into body cells. Look at data tables 5.5 and 5.6. Where are proteins digested?

_____________ Where are the enzymes produced that managed this digestion? _____________

5.24 Chapter 5 contains a lot of background information. You should reread this chapter completely so that you can assure yourself of mastery. Look at the study points below. Review them carefully. Place a check (✓) in the box before each study point when you are certain you have mastered it.

On completion of this chapter, you should be able to do the following:

☐ 1. Identify the nonliving spheres of the environment and demonstrate a general understanding of how living things interact with them.

☐ 2. List the common elements found in living things.

☐ 3. Recognize the basic structure of the atom, and determine the numerical amount of each component in the atom.

☐ 4. Recognize that molecules are made from the combination of atoms and that compounds are made from combinations of molecules.

☐ 5. Distinguish between a covalent bond and an ionic bond.

☐ 6. Distinguish between polar covalent bonds and nonpolar covalent bonds.

☐ 7. Recognize the differences between inorganic and organic chemistry.

☐ 8. Recognize that macromolecules are made up of monomers.

☐ 9. Distinguish between empirical and structural formulas and write examples of each.

☐ 10. Relate the components and forms of carbohydrates.

☐ 11. Discuss lab procedures for proving the presence of carbohydrates.

☐ 12. Relate the components and forms of lipids.

☐ 13. Relate the components of proteins.

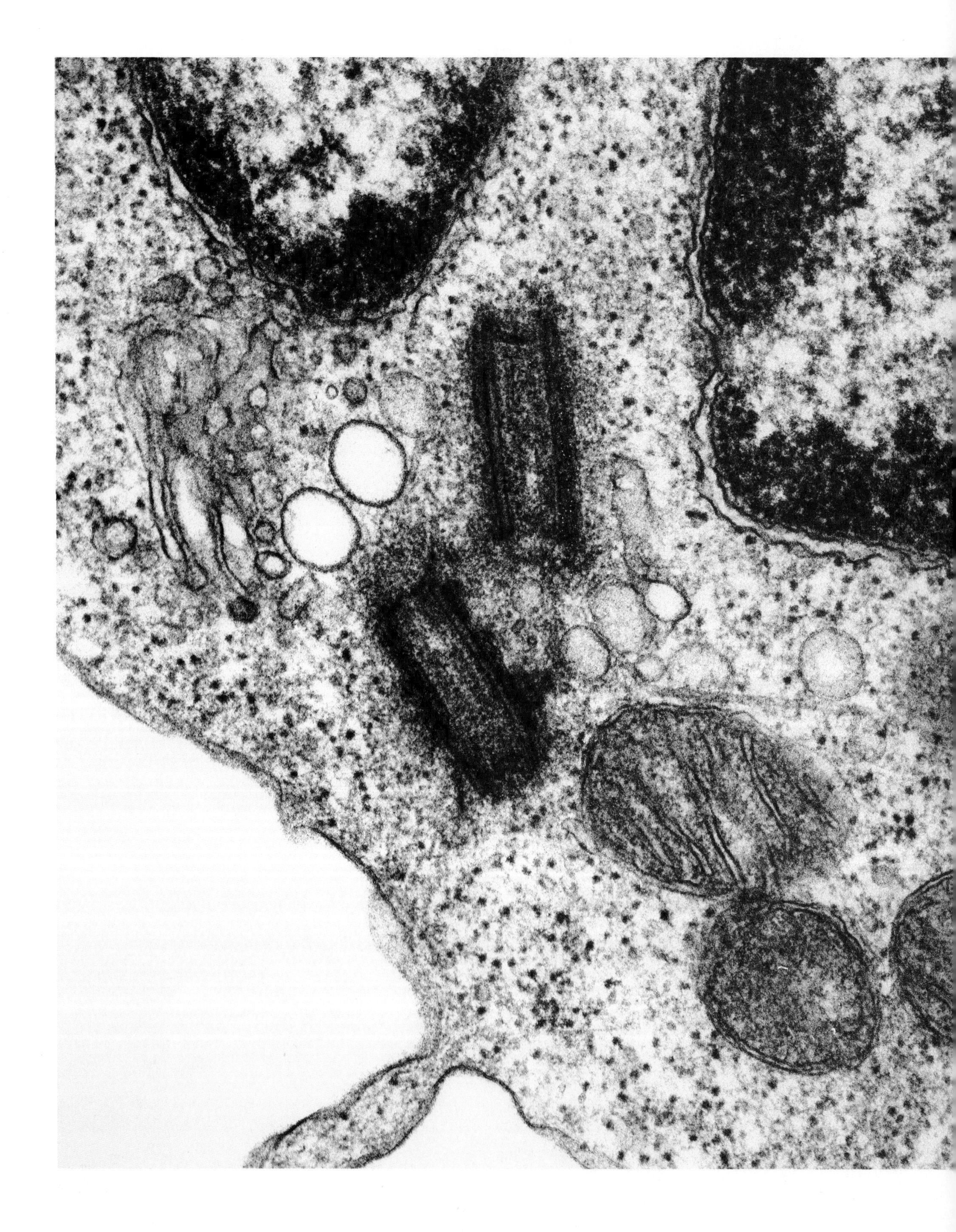

Membranes and Their Cellular Functions

6

Why are membranes so important to cells?

6.1 In the early beginning of organic evolution, protocells (see chapter 4) had a boundary layer of molecules that separated the chemicals inside the protocells from those outside the cells. The chemical activity of all cells in existence today depends on the presence of this chemical wrapping, called a membrane, to enclose chemical reactions. In eukaryotic cells the primary function of all membranes (outer membrane and many internal membranes) is to separate chemical reactions from each other. In figure 6.1 several kinds of internal membranes are illustrated, and the functional chemical activities they enclose are suggested.

In your logbook label the following membranes on figure 6.1: nuclear, mitochondrial (two kinds), Golgi, endoplasmic reticulum (rough), and lysosomal. Also write beside the label the major functional activity occurring within the membrane-bound organelle compartment. You may need to use information found in Chapter 4.

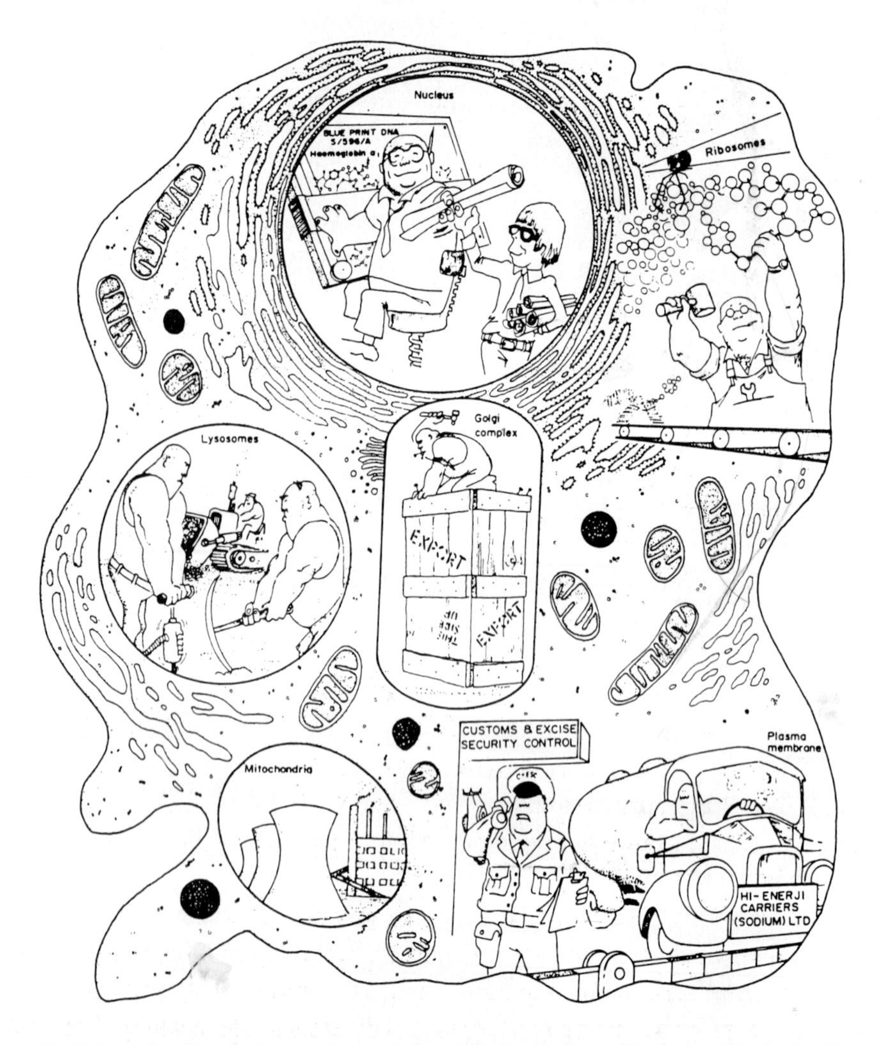

Figure 6.1 Some internal membranes and their functions.

W. M. Becker, *The World of the Cell*, copyright © 1986 by Benjamin/Cummings Publishing Company. Reprinted by permission.

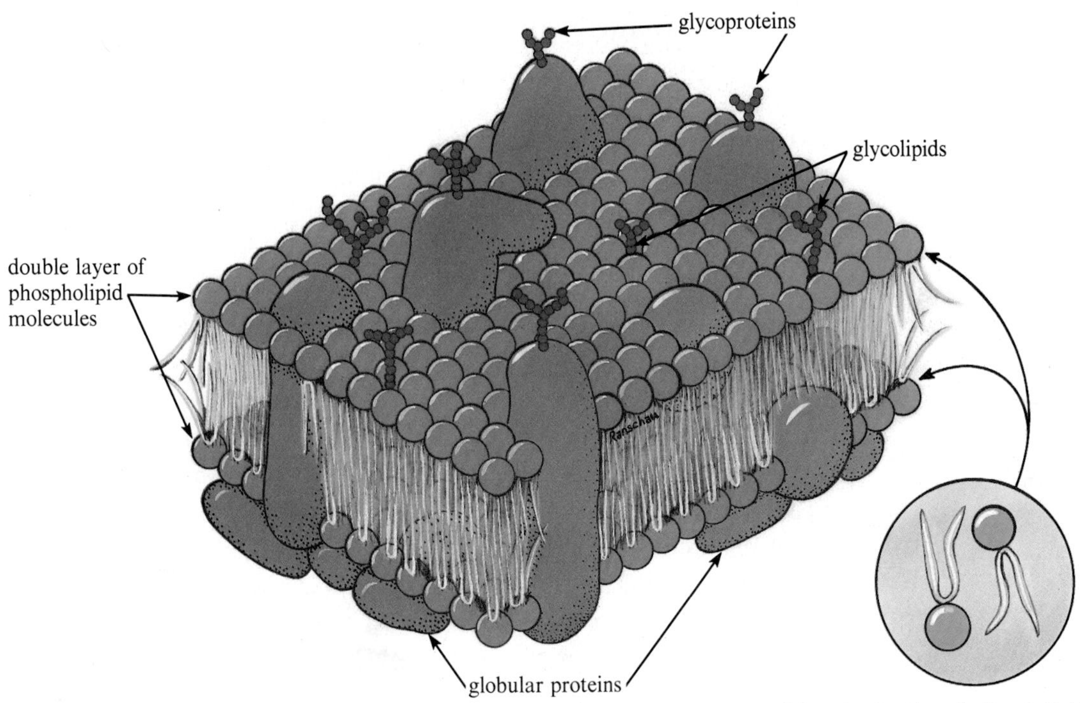

Figure 6.2 Model of membrane structure. A bilayer of phospholipid molecules with membrane proteins and associated carbohydrates.

6.2 All cells are bounded by an aggregate of macromolecules called a plasma membrane. In your logbook label the plasma membrane in figure 6.1. As you may have guessed, the plasma membrane controls the entry and exit of molecules to and from the cell. The secret to both preventing and allowing substances to cross the plasma membrane lies in the chemical composition and structural arrangement of the macromolecules that form the plasma membrane boundary layer.

The basic structural framework of all cellular membranes is a bilayer (a structure consisting of two layers) of lipid molecules and a mixture of protein molecules embedded or attached to the surfaces of the lipid bilayers (figure 6.2). All membranes are built on the same pattern, but each kind may contain its own specific chemical composition (amount and type) of lipids and proteins.

The most common lipids found in membranes are called phospholipids. These are similar to those described in chapter 5, but, in addition, they have a phosphate group attached. When phospholipid molecules are placed in a watery environment, the molecules will form into two layers with their water-"loving" portion facing the water and with their oily portion in the middle of the bilayer. Because of this oily middle portion, most cellular molecules, such as proteins, sugars, amino acids, ions, and hormones, cannot cross the bilayer because they are soluble in water. The lipid bilayer is an effective barrier that stops almost all movement of biological molecules across it, except for water and dissolved gases (CO_2, O_2). However, the oily portion is a very flexible barrier; lipid molecules wiggle and move about in a chaotic motion. This flexibility permits very small gaps to occur in the lipid bilayer. Water molecules may enter and leave a cell or an organelle by squeezing through the gaps that open up in the bilayer.

Two-way traffic across the cellular membrane is essential to the maintenance of cellular life. Since the lipid bilayer stops the movement of most molecules across the membrane, how do nutrient supplies enter the cell or how do manufactured products leave the cell? Traffic across the boundary layer is controlled, in part, by globular proteins that are located within the lipid bilayer (figure 6.2). These molecules function as security control gates by permitting only those molecules with a proper chemical identification to pass across the border.

A characteristic of plasma membranes and other membranes (Golgi, mitochondria, chloroplasts, and lysosomes) is the presence of carbohydrates. In the case of plasma membranes, these carbohydrates (glycos) are attached to lipids and proteins (figure 6.2). Many of these glyco molecules function as receptors for the binding of specific molecules to the surface of the plasma membrane. There may be a large forest of receptors at the surface of plasma membranes to catch, and sometimes engulf, molecules with the correct chemical identification. Bacteria and viruses may be chemically identified as foreign to an individual's own tissue because their glyco molecules function as **antigens.** Antigens are molecules that stimulate antibody production by the immune system of higher animals. Antibodies are produced by a subpopulation of white blood cells called **lymphocytes.**

Cell-to-cell adhesion in the formation of tissues is presumably a function of glycolipids and glycoproteins. Plasma membrane carbohydrates are responsible for the different human blood types (A, B, AB, O, MN).

6.3

What kinds of molecular traffic move substances across membrane barriers? Cellular membranes form compartments (that is, Golgi bodies, mitochondria, lysosomes, and others) whose boundary layer selectively restricts the access of chemical substances. These boundary layers, therefore, are selectively permeable. There are four kinds of traffic between a cell and its environment as well as between the cytoplasm and organelles: **diffusion, facilitated transport, active transport,** and **bulk transport.**

Before you begin your study of these traffic patterns, it is necessary to briefly consider energy, because molecules that pass through a cellular membrane must be in motion. Energy is said to exist in basically two forms: potential energy and kinetic energy. In a nutshell, potential energy is energy at rest, while kinetic energy is energy in motion. All things exist as a form of matter, namely, a gas, liquid, or solid. Water is one substance that can exist in all three forms. Which form of water (ice, liquid, or water vapor) would you expect to possess the greatest amount of kinetic energy if molecules in a gaseous state have greater kinetic energy than molecules in a solid state? _______________ Which form would possess the least amount of kinetic energy? _______________

Brownian (chaotic) movement is an exhibit of kinetic energy where small particles of solid matter in a liquid move in an erratic random course.

Would you expect to find Brownian movement in a cell? __________ Explain your answer in the logbook.

A reasonable way to observe kinetic energy in matter is to view your surroundings. Motion is all around you. But what about the "invisible particles"? In chapter 4 you observed that cheek cells take on the color of an applied stain. Your cheek epithelial cells were colored by the methylene blue. Explain in your logbook how this was possible. __________

To further investigate the movement of cellular substances, your teacher has set up four test tubes. Two test tubes are labeled *beginning,* and the other two test tubes are labeled *24 hours.* Locate these test tubes, examine the two systems carefully, and record your observations of the *beginning* test tubes and the *24-hour* test tubes in data table 6.1.

Data Table 6.1 Comparing the Rate of Diffusion of Food Coloring in Two Liquids

Initial Observations for Water and Food Coloring	Initial Observations for Karo Syrup and Food Coloring
Final Observations for Water and Food Coloring	**Final Observations for Karo Syrup and Food Coloring**

Which solution showed the most rapid dispersion of food coloring?

_______________ Propose a reason for this phenomenon.

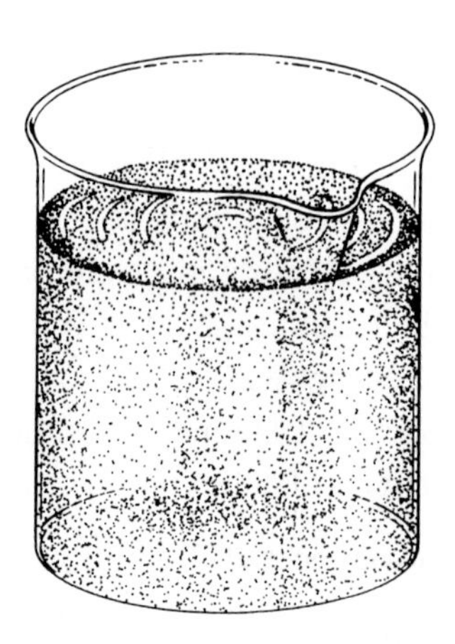

Figure 6.3 Diffusion of solute molecules (dots) in solvent molecules (water).

6.4

Diffusion occurred when the food-coloring molecules, or **solute** (the substance being dissolved), moved through the water, or **solvent** (the substance used to dissolve the solute). Given enough time, the food-coloring molecules would become equally dispersed as a result of the random movements and kinetic energy of the solute and solvent molecules. Figure 6.3 will help you visualize diffusion of solute molecules from a high-concentrated region to a lower one. Diffusion is the molecular traffic of atoms and molecules intermingling because of their kinetic energy and their solubility. It results in the net flow of substance particles from a place of higher concentration to a place of lower concentration. A molecule or atom that is found in a cell can be considered a particle. Opening a bottle of fine perfume in a room is another example of diffusion. If a bottle of fragrant perfume is opened in a room that is free of other, stronger odors, the room will soon have the fragrance of the perfume. Again, this is due to diffusion of perfume molecules from a region of high concentration to a region of lower concentration.

6.5

The net passage of particles through a cellular membrane by diffusion is called *permeation.* This type of molecular traffic requires no other energy beside the thermal kinetic energy of the moving molecules. If the numbers of identical solute molecules are unequal on the two sides of the membrane, a non-equilibrium state, the overall (net) movement of the solute molecules will be from the side of the higher concentration (more molecules) toward the side of lower concentration (fewer molecules) until an *equilibrium state* is reached. In figure 6.4 a semipermeable membrane is illustrated with "thermal gaps" in the lipid bilayer. These thermal gaps (transient holes) are created by the motion of the lipid molecules. *A* molecules are small enough to squeeze through the gaps, but *B* molecules are too large to move through.

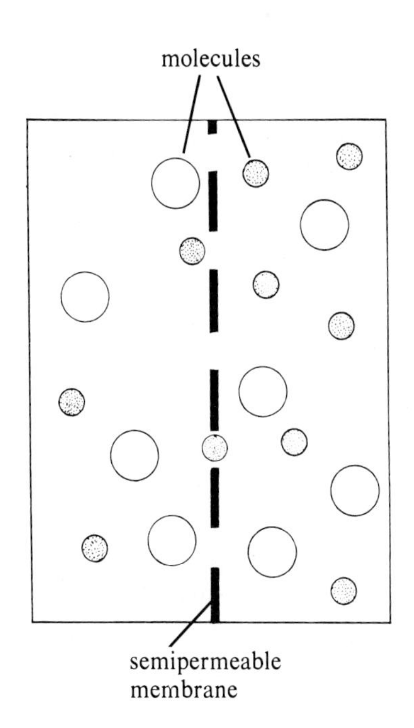

Figure 6.4 Diffusion of molecules through a semipermeable membrane. Small circles, A molecules; large circles, B molecules.

Predict which direction the *A* molecules will move across the membrane (right to left, left to right, or in both directions). ________ *Please remember* that all molecules are in random motion.

Draw an arrow on figure 6.4 in your logbook to show the direction that the majority of the molecules will move before equilibrium is reached.

When *A* molecules are at a state of equilibrium, how many will be on the right side of the membrane? ________ How many will be on the left side of the membrane? ________

Diffusion of molecules is very important for your own health. Oxygen molecules you breathe must diffuse into the lung cells, then to the blood, and finally into each living body cell, where they ultimately diffuse to the mitochondria. Carbon dioxide (CO_2), a waste product of cell respiration, must diffuse a reverse route, out of the body cells, before it is exhaled.

6.6 Let us apply the idea of diffusion to the movement of molecules across a cell. Could a stain diffuse through a cell membrane to color the cell? ________ Would there be a time when the stain might not diffuse into a cell? ____________

You may not be able to predict an answer to these two questions at this time, but the following activity should help.

Place 1 mL of yeast *suspension* in each of two small test tubes (10 × 75 mm). Label them *1* and *2*. Heat the contents of number 1 to boiling. *Be careful!* You have only 1 mL of suspended solution, and you could boil it away.) Allow the solution to cool to room temperature and then add 3 drops of Congo Red to each of the two test tubes. Mix. Place 1 drop of the mixture of each test tube on a clean microscope slide, cover with a coverslip, and observe with the aid of a microscope. What noticeable difference in yeast cells are observed? ____________ Record in logbook.

Did the heated yeast culture or the unheated yeast culture have the majority of stained cells? ________ Since you allowed the heated culture tube to cool to room temperature, can temperature be considered a cause for staining? ____________

The lipid and protein molecules of a cell membrane have kinetic energy and are in motion. Would heating increase the movement of membrane molecules? ________ Could these molecules gain enough kinetic energy to fly apart from each other? ________

So far we have directed the questions mainly towards the stained cells. Offer some explanation or hypothesize why some cells did not stain. This may be a guess. ________

Atoms, ions, and molecules move from a high-concentration area toward a lower-concentration area until equal distribution occurs. Is it possible that diffusion is the reason for staining? ________ Did the heated yeast cells stain because they were heated? ________ Did the *cell membrane* keep the stain molecules out of certain yeast cells? ________ Is heat the answer? ________ Can the molecular size of the stain molecules be a reason for staining differences? ________

6.7 Does the size of the solute molecule affect its movement across a cell membrane? Let us perform an experiment to obtain an answer.

You will need a 250 mL Erlenmeyer flask with stopper (see figure 6.5). Put 200 mL of iodine-water solution in the flask. Make a small sack with a 5-inch piece of pre-soaked dialysis tubing. Pre-soaking makes the dialysis tubing easier to open into a tube. If you place the tubing between your thumb and forefinger and move the two sides of the tubing back and forth, you will be able to separate the layers. If you have difficulty, check with your teacher. Once you have the dialysis tubing to the point that it will form a tube, fold over one end and tie it shut with a piece of string or fish line. Figure 6.5 illustrates this. If you do not use this method, the dialysis sack will leak.

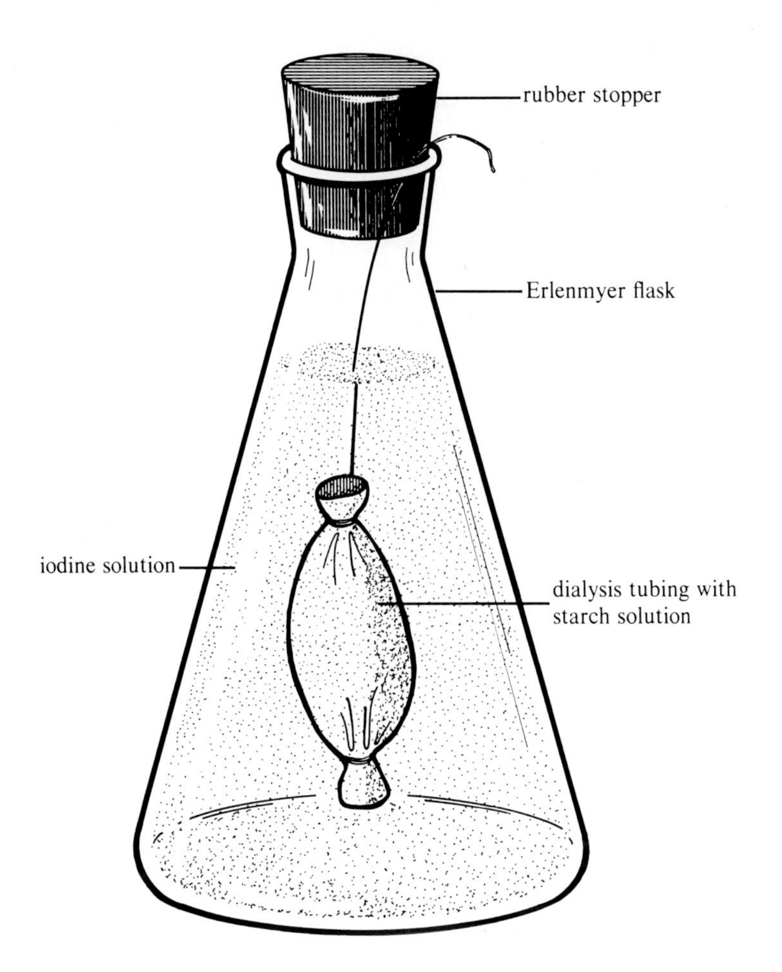

Figure 6.5 Components and their arrangement for experiment 6.7.

Fill the sack approximately 3/4 full with starch-water solution. Fold over the open end and tie it shut. Use an extra long piece of fish line because this will allow you to suspend the dialysis sack in the Erlenmeyer flask (refer to figure 6.5).

1. What are the two kinds of solute molecules in this experiment?

   ________________ and ________________ .

2. What is the solvent? ____________
3. What is the initial color of the solution in the flask?

   ________________ in the sack? ________________

Allow the flask to stand. Make observations by the end of the class period. Record those observations in data table 6.2 of the logbook. Also, record your observations towards the end of the school day and after 24 hours have passed.

Data Table 6.2 Osmosis Observations: Color and Size

	Observations	
	Color	*Size of Sack*
End of class:		
End of school day:		
24 hours:		

After you set-up the flask, place a drop of starch-water solution and the iodine-water solution together on a slide. Record the color in your log-book. ____________

Dialysis tubing is made of cellophane. It resembles but is not identical in structure and function to a cellular membrane. Both membranes are considered **semipermeable** because they allow certain size particles to pass through them. From your observations, which solute molecule is too big to pass through the dialysis tubing? ________________ Does the dialysis membrane stop the flow of solvent molecules? ________

6.8 A special kind of diffusion is called **osmosis.** In biological systems, osmosis is *the movement of water* through a semipermeable membrane from a high concentration of solvent (water) molecules to a lower concentration. Both diffusion and osmosis result from the kinetic energy of molecules.

To illustrate osmosis, obtain four pieces of pre-soaked dialysis tubing from the refrigerator. Immerse them in distilled water and form a sack with each one.

By folding over one end and tying it with a piece of string, you now have a cellophane sack. Fill each dialysis sack three-quarters full with only one of the following solutions:

No. 1: colored distilled water
No. 2: 20 percent sucrose solution
No. 3: 40 percent sucrose solution
No. 4: 60 percent sucrose solution

Fold over the remaining open end and tie secure as you did the first experiment. Quickly weigh each sack to the nearest hundredth (0.01) of a gram, and place the sack in a container of distilled water. There should be just enough water to cover the entire sack. Record the initial weight of each sack in the time slot marked *0* in data table 6.3. Thereafter, record the *change in weight* at 5-minute intervals, and record these figures in data table 6.3. (You should carefully, but quickly, blot dry each sack before weighing.)

Assuming each sack was filled with an equal volume, answer the following questions in your logbook before the experiment is completed.

1. Which molecules represent the solvent? ____________
2. Which sack has the greatest number of water molecules? ________
3. Which sack has the fewest number of water molecules? ________
4. *IF* osmosis occurred, *then* would some sacks change weight? ________ Explain.
5. Which sack do you predict will gain the most weight? ______

Allow the sacks to sit for 24 hours in the distilled water. Weigh them at the end of this period, the same as you did earlier, and record your data in data table 6.3

Plot your data from data table 6.3 on a sheet of graph paper from the logbook. Plot *weight* (in grams) vs. *time*. Plot time on the *X* axis and weight change on the *Y* axis. Which axis is the dependent variable? ________ Which axis is the independent variable? ________ Which dialysis sack gained the greatest amount of weight? ________ Which sack gained the least or lost the most weight? ____________ Molecules tend to disperse from regions of higher concentration to regions of lower concentration. Does this statement tend to support your observations? Explain.

Data Table 6.3 Osmosis Observations: Weight

Time in Minutes	Change in Weight of the Sacks in Grams			
	No. 1	No. 2	No. 3	No. 4
0				
5				
10				
15				
20				
25				
24 Hours				

The graph you plotted indicates that the sack in the highest percent sucrose solution (lowest in water molecules) gained the most weight, or at least it should have. You are aware that some particles or molecules were too large to pass through the pores of the membrane. Are sucrose molecules too large? ________ Because an equilibrium for total dispersion of water molecules is being strived for, and because the sucrose molecules are unable to penetrate through the membrane (they are too big), the likely conclusion is the water molecules moved inside of/outside of the sack and dispersed the sucrose molecules. This should lead to an equilibrium, provided enough space is available in the sack to hold a greater number of water molecules.

6.9 The concentration of water molecules in a cell is lower than in "pure" water. When the space inside a cell is limited in volume, the force of the solvent (water) may exert a pressure (**osmotic pressure**) on the plasma membrane as the water molecules enter the cell.

Plant cells have a rigid cell wall outside the plasma membrane and, if placed in water, can withstand high osmotic pressure without bursting. This is similar to sack 1 at the end of the experiment. This swollen or rigid condition is called **turgor,** and a plant cell is said to be *turgid* (data table 6.4). Thus, the cell wall sustains a **turgor pressure** due to the high osmotic pressure.

Data Table 6.4 Effect of Osmosis on Cells

Tonicity of Solution	Description	Results	
		Animal	*Plant*
Isotonic*	Equal concentration of water on both sides of cell membrane	No change	
Hypertonic*	Lesser water and greater solute concentration outside the cell than inside	Shrink	Plasmolysis
Hypotonic*	Greater water and lesser solute concentration outside the cell than inside	Swell to bursting	Turgor pressure

Iso means "same as," *hyper* means "more than," and *hypo* means "less than."

Most animal cells placed in water burst because of the high *osmotic pressure* on the inside surface of the plasma membrane (data table 6.4). Osmotic pressure is the amount of force that must be exerted to prevent a net flow of water across a selectively permeable membrane. There is no cell wall to protect the animal cell. Your body fluids, such as blood plasma, must be in osmotic equilibrium with the cells they bathe. This simply means there is a pressure balance between the two sides of the plasma membrane and equal numbers of water molecules *moving* in both directions across the membrane. Such a condition is called **isosmotic,** or **isotonic** (data table 6.4). The diagram in figure 6.6 will show this condition.

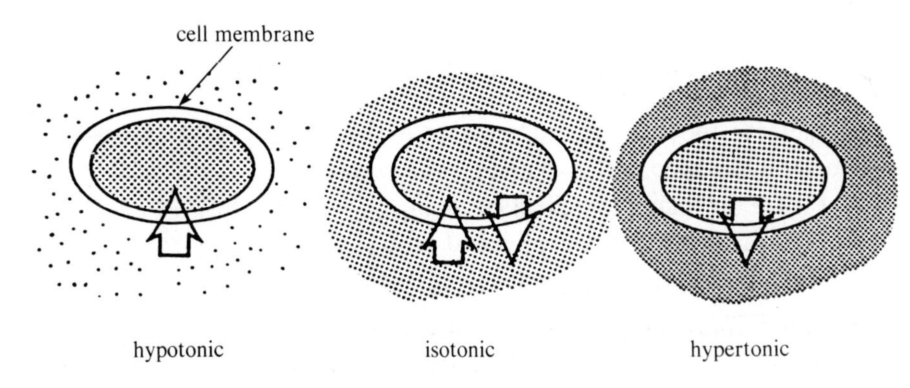

Figure 6.6 Osmosis. Net directional flow of water across a cellular membrane is affected by the tonicity of solutions on both sides of the membrane.

Solutions that have unequal osmotic pressure are said to be either **hypotonic** or **hypertonic** (data table 6.4). Describe in the logbook the hypotonic and the hypertonic conditions illustrated in figure 6.6. ________

At the start of the experiment in section 6.8, sack 1 would be in which of the three osmotic conditions? ____________________

6.10 Up to this point, you have studied *simulated* models of what happens to the movement of molecules across a plasma membrane. You will now test the model, using living red blood cells.

Obtain three microscope slides and three glass coverslips. Be sure they are very clean. Mark each slide near the end as follows: H_2O, 0.85 percent NaCl, and 2.0 percent NaCl.

Wash the end of one of your fingers with a cotton ball dipped in alcohol; after the alcohol dries, stick your finger tip with a *sterile lancet* and produce a drop of blood by squeezing your finger. *Do not use another person's lancet! Do not touch another person's blood with your blood!*

Place 1 drop of blood on each slide and immediately add 1 drop of H_2O, 0.85 percent NaCl, and 2.0 percent NaCl respectively to slides 1, 2, and 3 (figure 6.7).

Place a coverslip on each solution and observe with a compound microscope. Record your observations in your logbook for each slide.

No. 1 (H_2O):

No. 2 (0.85% NaCl):

No. 3 (2.0% NaCl):

Indicate opposite the slide number in data table 6.5 which condition (isotonic, hypertonic, and hypotonic) was observed.

Figure 6.8 shows the effects of tonicity on red blood cells.

Medical technologists are called on to *break open* or *hemolyze* red blood cells for patient blood analysis. Record in the logbook which slide contained *hemolyzed* red blood cells. ____________________

Data Table 6.5 Effect of Osmosis on Blood Cells

Slide	Osmotic Condition
No. 1	
No. 2	
No. 3	

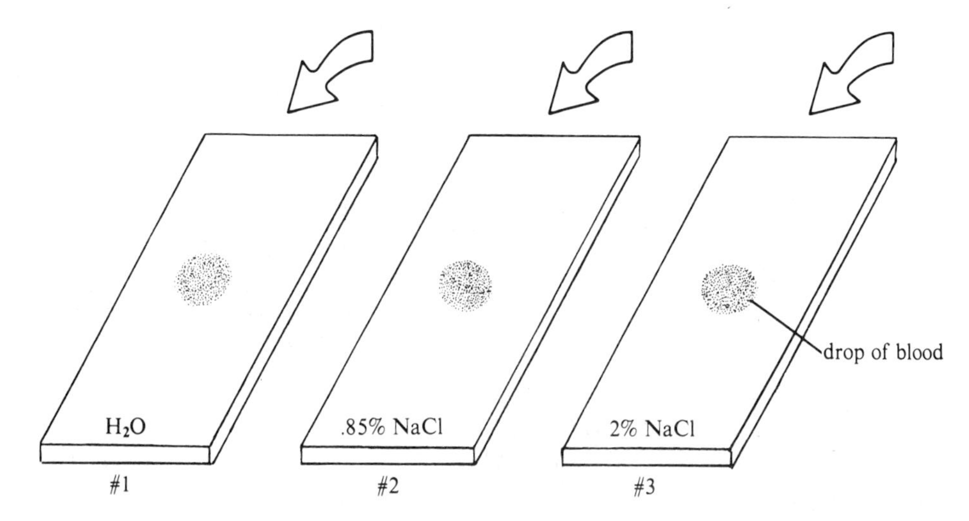

Figure 6.7 Osmotic pressure application (arrow) of solutions to drop of blood on slide.

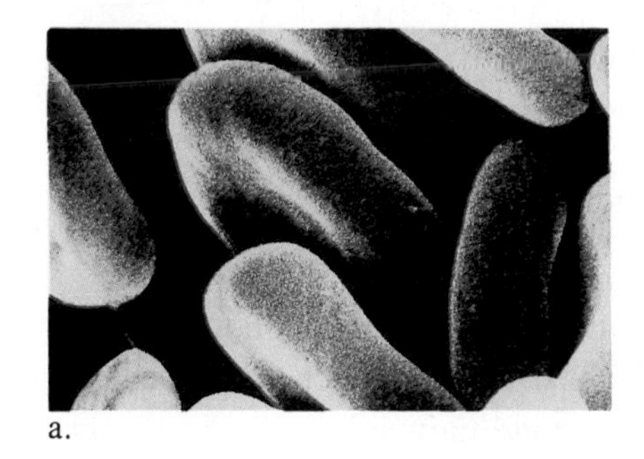
a.

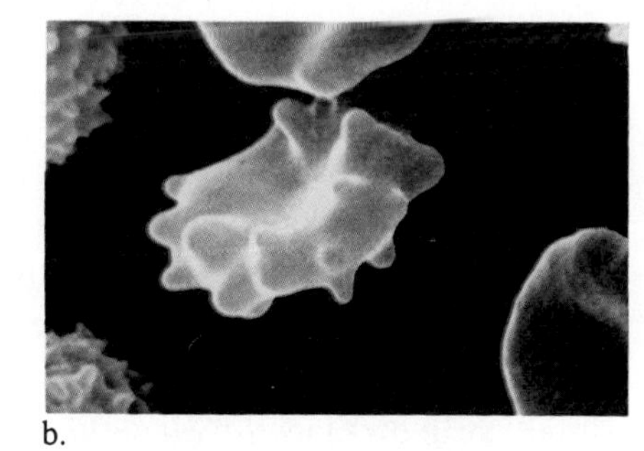
b.

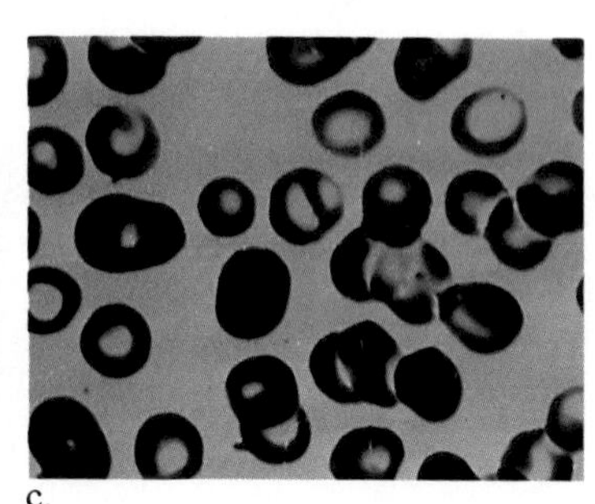
c.

Figure 6.8 Effects of tonicity on red blood cells. a. isotonic (SEM). b. hypertonic (SEM). c. hypotonic (light microscope photograph).

6.11 Another example of molecular movement across a membrane can be illustrated with pea seeds. Place fifteen pea seeds in a 50 mL beaker. Cover the seeds with water and allow them to stand for 24 hours. In your logbook describe the general condition of the seeds before and after the 24-hour period.

1. Observations at the beginning:

2. Observations after 24 hours:

3. Has there been a change in the pea seeds? ________________

4. What was the major change according to your observations?

6.12 **Imbibition** is the process of water adsorption (adhesion) onto the surface of certain particles (molecules). This forces the particles further apart and causes swelling. Imbibition may occur in wood because of cellulose molecules present in dead cell walls. During imbibition the volume of the *imbibant* will increase. In this case, the imbibant is the wood, such as a jammed door, or as above, the pea seed or, more specifically, the seed coat. Once water has saturated the cellulose (H_2O molecules moved in between the cellulose molecules) of the pea seed cell walls, the water moves into the cell through the cell membrane by osmosis and becomes adsorbed onto the surface of the internal cellular molecules.

Imbibition, like diffusion and osmosis, is affected by temperature. As the temperature increases, so does the rate of imbibition.

6.13 Shrinkage, on the other hand, of cytoplasm from the cell wall of plant cells is called **plasmolysis** (figure 6.9).

To illustrate plasmolysis, place 10 mL of each of the following solutions in a 15 × 150 mm or medium size test tube. To each test tube, add a small sprig of Elodea. Allow the Elodea to remain submersed in the solutions for approximately 20 minutes. After that time, remove each sprig and examine one leaf of each with the aid of a compound microscope. Record in your logbook the observations for each solution in the data table 6.6. Note the distribution of chloroplasts.

Write a description of your observations recorded in data table 6.6. Use cell structure, plasmolysis, and toxicity terms.

Data Table 6.6 Effects of Sucrose on Cell Size

% Sucrose	Elodea Observations
0	
20	
40	
60	

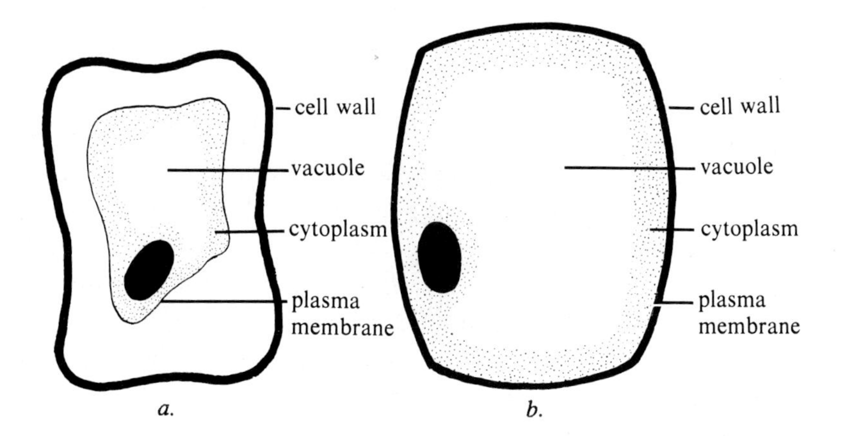

Figure 6.9 Schematic of plant cells. a. plasmolyzed cell. b. turgid cell.

6.14

The simplest traffic across the plasma membrane, diffusion and osmosis, is passive transport, which requires no additional cellular energy. The flow of this molecular traffic is slow. Most molecules (other than water) traffic across the plasma membrane with the assistance of protein *carriers* in the lipid bilayer. If molecules are carried across the membrane from a high concentration toward a lower concentration of its own kind, the process is called **facilitated transport** (figure 6.10). The assistance by carrier molecules makes this a fast rate of traffic flow. Because the traffic moves from a high to low concentration, no energy is required.

Building block (monomers) molecules, such as amino acids, sugars, and nucleotides, are sometimes moved in and out of cells by facilitated transport. There are a number of human diseases that are related to a defective facilitated transport mechanism. In a human an increase in the level of blood sugar triggers the release of insulin into the bloodstream by pancreatic cells. Insulin molecules are moved by the blood to glucose receptor-carrier protein molecules in the plasma membrane of specific cells, such as

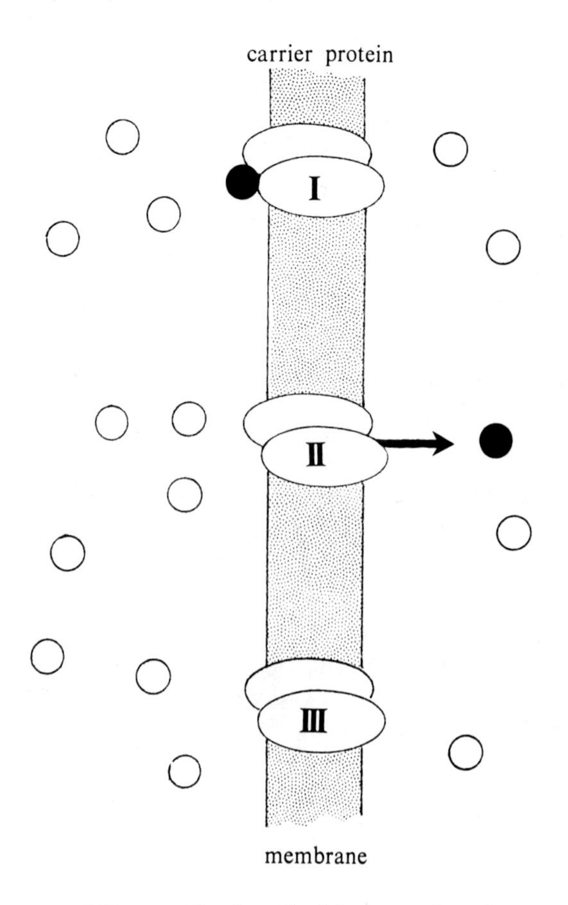

Figure 6.10 Model of facilitated transport of molecules across a cell membrane. (I) Molecule binds to carrier. (II) Molecule released from carrier on the other side of membrane. (III) Available carrier for transport.

muscle, adipose, and liver cells. Insulin binds to the glucose carrier molecules and, thus, increases the rate of glucose uptake by facilitated transport. In a person with diabetes mellitus, a reduced secretion of insulin slows down the uptake of glucose; thus, many symptoms of diabetes are the result of a defective facilitated transport mechanism.

6.15 A third type of traffic across the plasma membrane is called *active transport*. The rate of traffic flow by this process is fast and requires cellular energy. Energy is required because the transport usually occurs *against* a concentration gradient. Solute molecules are transported with the assistance of carrier protein molecules in the lipid bilayer of the plasma membrane (figure 6.11). These carriers are generally called pumps. Pumps use the chemical energy of the cell to move the solute particles from one solvent mass across the lipid barrier to another solvent mass.

Active transport is an important process that helps feed cells. As the contents of a human meal traverse the alimentary canal, enzymes digest the foodstuffs into monomer-size molecules. Starches are broken into glucose monomers, and proteins are broken into amino acid monomers. As the monomers of food molecules move through the small intestine, they are moved by active transport across the plasma membrane into cells lining the alimentary canal. These cells then move the monomers across their plasma membrane by facilitated transport into the circulating blood.

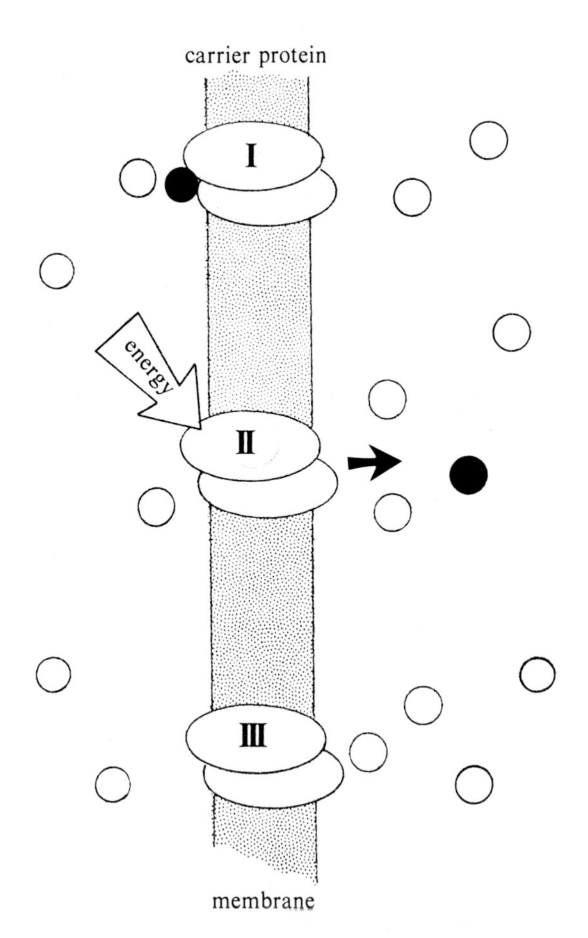

Figure 6.11 Model of active transport of molecules across a cell membrane. (I) Molecule binds to carrier. (II) Energy used to assist carrier to move molecule across membrane. (III) Available carrier for transport.

Most mammalian cells possess an electrical potential (voltage) difference across their plasma membrane. This electrical potential is produced by actively transporting positively charged ions, especially of hydrogen (H^+), calcium (Ca^{2+}), sodium (Na^+), and potassium (K^+), across the plasma membrane. Conservation of energy in chloroplasts (photosynthesis) and in mitochondria (cellular respiration) depend on the electrical potential across their internal membranes. The active transport of ions across animal plasma membranes is responsible for nerve transmission, brain function, and muscle excitation.

6.16 There is a *fourth* type of traffic pattern across the plasma membrane that permits uptake or release of bulky substances too large to cross the membrane either by diffusion or transport (facilitated and active). Instead, these large substances travel in membrane packages, called **vesicles,** formed by Golgi bodies or formed by the plasma membrane itself. If the large substances are packaged-up in a Golgi body membrane and dumped outside the cell, it is called *exocytosis* (figure 6.12). The role of exocytosis is to *secrete* manufactured products to the outside of cells. Examples include: (*a*) the secretion of proteins from the pancreatic and salivary gland cells into the alimentary digestive tract; (*b*) neurotransmitters (e.g., noradrenalin) from nerve cells; (*c*) hormones from pituitary cells (e.g.,

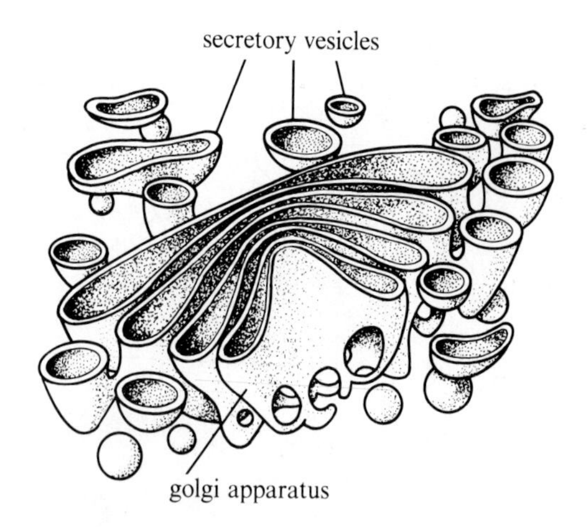

Figure 6.12 A form of vesicular transport used for exocytosis.

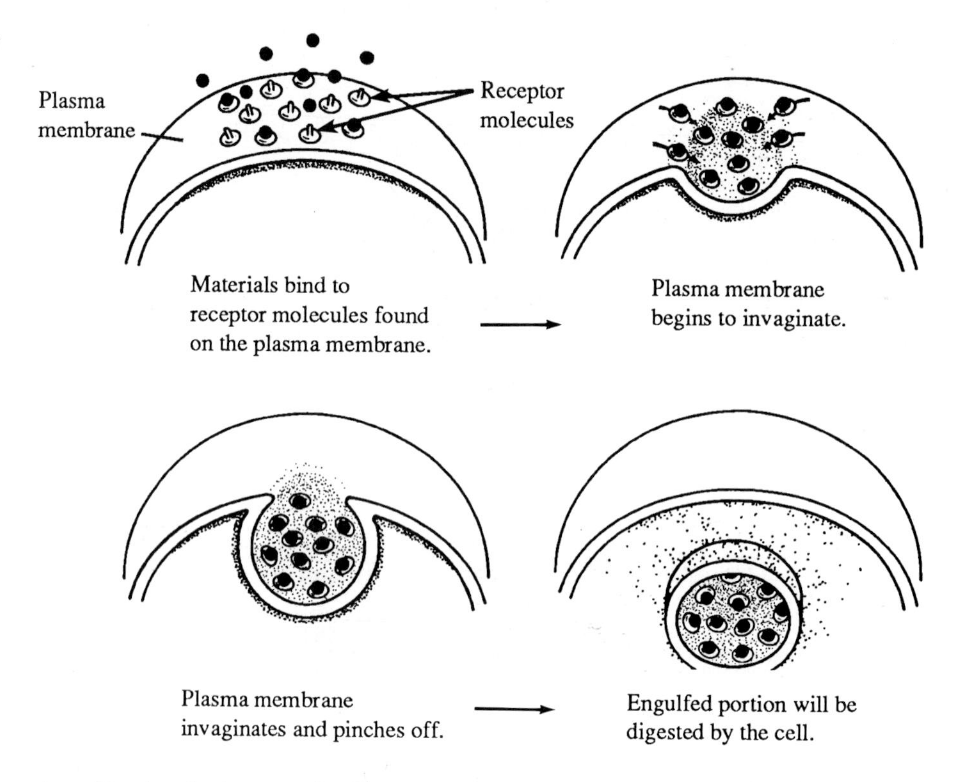

Figure 6.13 A form of vesicular transport used for endocytosis.

thyroid-stimulating, growth, and follicle-stimulating); (*d*) insulin from pancreatic cells; and (*e*) histamine from mast cells. Plant cell walls are produced by the secretion of pectic materials and carbohydrates from cells by exocytosis.

In a reverse process, *endocytosis,* large substances enter the cell by first binding to receptor molecules of the plasma membrane (figure 6.13). The plasma membrane then folds into the cytoplasm to envelop the substances. This forms a membrane-bound vesicle which pinches off the plasma membrane into the cell. When the vesicles are formed by engulfing large particles, the process is called **phagocytosis.** When protozoans, such as an

Data Table 6.7 Cell Function and Type of Cell Membrane

Cell Function	Membrane
Diffusion of small ions and molecules	All membranes
Active transport of molecules	All membranes
Phagocytosis, pinocytosis	Plasma membrane
Secretion of products out of cell	Golgi
Convert light energy into electrical impulse	Retina (eye) rod membrane
Photosynthesis	
Respiration	
Contains digestive enzymes	
Protein synthesis	
House's genetic information (DNA)	

ameoba, feed, they phagocytize their food. If the vesicles engulf mainly fluids that contain small dissolved particles, the process is called **pinocytosis.** Pinocytosis is the major way in which fluids are moved from inside the blood capillaries to surrounding cells.

Thus, phagocytosis and pinocytosis are specific types of endocytosis. In either endocytosis or exocytosis the process is done without damaging the plasma membrane—it remains intact. A role of endocytosis is to feed cells. For single-celled protozoans it is the major way these cells obtain their food and, thus, their energy for growth and reproduction. Eukaryotic cells of multicellular organisms are "fed" by both molecular transport (facilitated and active) and endocytosis. After a substance has been endocytized, it is digested. This is done by a fusion of the endocytic vesicle with a lysosome. Multicellular mammalian organisms defend themselves against unwanted bacteria by phagocytizing them with white blood cells.

6.17 All cellular membranes function to regulate the traffic of molecules between compartments (data table 6.7). However, all cellular membranes have at least one additional functional role. For example, the plasma membrane and Golgi membranes function in endocytotic and exocytotic traffic control. The ability to see this printed page is made possible by the retina membrane converting light energy into electrical impulses. This membrane and those involved in photosynthesis and cellular respiration function to convert and conserve energy for the work of living cells. Complete in your logbook the column marked *membrane* in data table 6.7 with the name of the specific organelle membrane that is associated with the listed function. These membrane functions will be discussed in the next few chapters.

The cell membrane has an active role. You can see from this chapter how the membrane has multiple functions. Review this chapter carefully. Understanding this chapter is important for future studies. Look over the study points. Check (✓) the box of each in your logbook if you have mastered the content associated with the study point.

On completion of this chapter, you should be able to do the following:

☐ 1. Recognize and label specific cell organelles.

☐ 2. Discuss the makeup of the cell membrane.

☐ 3. Discuss completely the various ways materials move through the cell membrane.

☐ 4. Distinguish between the similarities and differences of membrane traffic.

☐ 5. Identify the conditions of equilibrium.

☐ 6. Distinguish the *conditions* of a cell that put them out of equilibrium with their surrounding environment.

7 Energy in Life

How do cells meet their energy needs?

7.1 What is energy? The biological world is made of matter organized and maintained through its interaction with energy. This *energy* may be observed when one of its forms interacts with nonliving or living matter. There are several basic forms of energy. They include mechanical, heat, radiant, sound, chemical, nuclear, and electrical energy. So that you have an understanding of each form of energy, define each with a working definition in your logbook. You may need to use a reference book to assist you.

1. Mechanical energy:
2. Heat energy:
3. Radiant energy:
4. Sound energy:
5. Chemical energy:
6. Nuclear energy:
7. Electrical energy:

7.2 The equation below shows how energy may be *transformed* from one form into another form. *If* radiant energy, for example, is transformed when it interacts with matter, *then* you should be able to observe the change in the form of energy following the transformation (T).

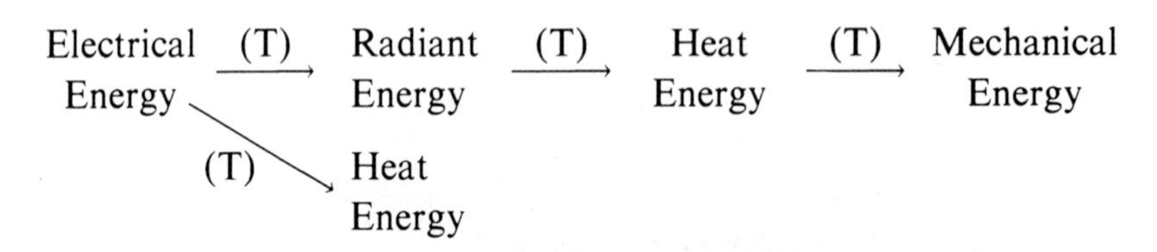

To illustrate this, shine an electric light source on the blades of a radiometer. List some of the forms of energy you observe in data table 7.1. The first observation and conclusion is already listed. How do you know that electrical energy is converted to radiant energy? Write your answer in the logbook.

____

____

Did you observe that mechanical energy was produced? ________ Explain in your logbook how light energy moves the blades of the radiometer.

Data Table 7.1 Data for Radiometer

Form of Energy Observed	Suggested Energy Transformation
Light	Electric → Radiant

7.3 Another example of energy transformation is the burning of gasoline in an automobile engine. Make a flow diagram in your logbook to show the different forms of energy and the direction of transformations. Use the equation in section 7.2 as a guide.

List in your logbook several everyday examples of energy transformations in nonliving systems in data table 7.2, and list the form of energy associated with each. Draw an arrow between each form to show the direction of transfer. Use data table 7.1 as a guide.

Data Table 7.2 Data for Everyday Energy Transformations

Form of Energy Observed	Suggested Energy Transformation

7.4 Study figure 7.1 and determine what forms of energy are involved in the burning of wood. Describe in your logbook the direction of energy transformation: name the form of energy being transformed and the forms of energy generated by the transformation. Although the chemical composition of the wood is transformed into other forms, the total amount of energy remains the same.

7.5 Turning on an electric light bulb transforms electrical energy into radiant and heat energy, but energy is neither lost nor gained in this transformation, the total remains the same—it is conserved. Let us examine the transfer (flow) of heat energy in water when hot water is added to cooler water. Do you predict the hot water would cool down?

________ Would the temperature of the cooler water change? ________
IF heat energy is conserved, *then* you might predict the quantity of heat

Figure 7.1 Burning wood.

Wood → CO_2 + H_2O + ash (minerals)

energy lost by the hot water to be equal to the quantity of heat energy gained by the cooler water. Do the following experiment to test this hypothesis and record your observations in data table 7.3 of your log book.

a. Pour 50 mL (50 g) of water into a small beaker and begin heating it on the burner stand over the burner.
b. Pour 100 mL (100 g) of the cool water marked for this experiment into a thermos bottle. Insert a rubber stopper, fitted with a thermometer, into the mouth of the bottle. Read the stabilized temperature to the nearest 0.5° C and record it in the data table. This represents the initial temperature (T_1) of the cool water.
c. When the heated water is *near* boiling, remove it from the stand, and *quickly* read the temperature to the nearest 0.5° C. This is the initial temperature (T_1) of the hot water. *Immediately* pour all heated water into the thermos bottle. Immediately reinsert the rubber stopper fitted with the thermometer. Record the initial temperature in the data table.
d. Carefully shake (swirl) the bottle. Wait for the temperature of the water to stabilize, then read it to the nearest 0.5° C. This is the final temperature (T_F) for both the cool and the hot water. Record it in the data table.
e. Calculate the change in temperature (ΔT) for both the hot and cool water.

f. Calculate the heat lost ($M \times \Delta T$) by the hot water and the heat gained ($M \times \Delta T$) by the cool water. The quantity of heat energy is measured in units called **calories.**
g. Prepare a written statement in your log book summarizing: (*a*) your observations recorded in data table 7.3; and (*b*) your conclusions. Do you accept or reject your prediction that heat energy may be transferred from one place to another place without a gain or loss in energy? Why?

Data Table 7.3 Data for Heat Transfer in Water

	Cool Water	Hot Water
Mass (M)	100 g	50 g
Initial temperature (T_1)	°C	°C
Final temperature (T_F)	°C	°C
Change in temperature (ΔT)	°C	°C
$M \times \Delta T$	Calories	Calories

7.6 Energy can also be transformed by *living systems*. The form of energy may be altered, but the amount of energy remains the same. Living systems can transform the radiant energy of the sun into chemical energy by photosynthesis and store the chemical energy in the form of polysaccharides that can be used by animals or plants (figure 7.2). Notice in figure 7.2 that the sun is the ultimate source of energy for the plants and animals of your biosphere, and it is the ultimate source of chemical energy stored in fossil fuels.

All living biological systems must do work to remain alive. To do work requires a source of energy. The sun is an energy source; food is an energy source. Some organisms can manufacture their own food for the work of living by using the radiant energy of the sun (figure 7.3). These are called **autotrophic** organisms. Are autotrophs plants or animals? ____________
Other organisms, called **heterotrophs,** depend upon the autotrophs for their energy. Another name for these is **herbivores.** Are heterotrophs plants or

*Δ is the symbol for the Greek letter, delta, which is used in mathematics and science to designate a change.
**Calorie is a unit of quantity of heat energy: one calorie is the quantity of heat required to raise the temperature of 1 gram of water 1 degree Celsius.

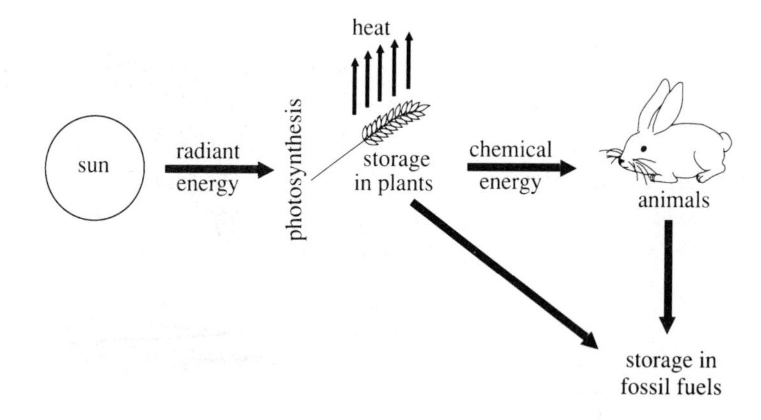

Figure 7.2 Energy transformation by living plants and animals.

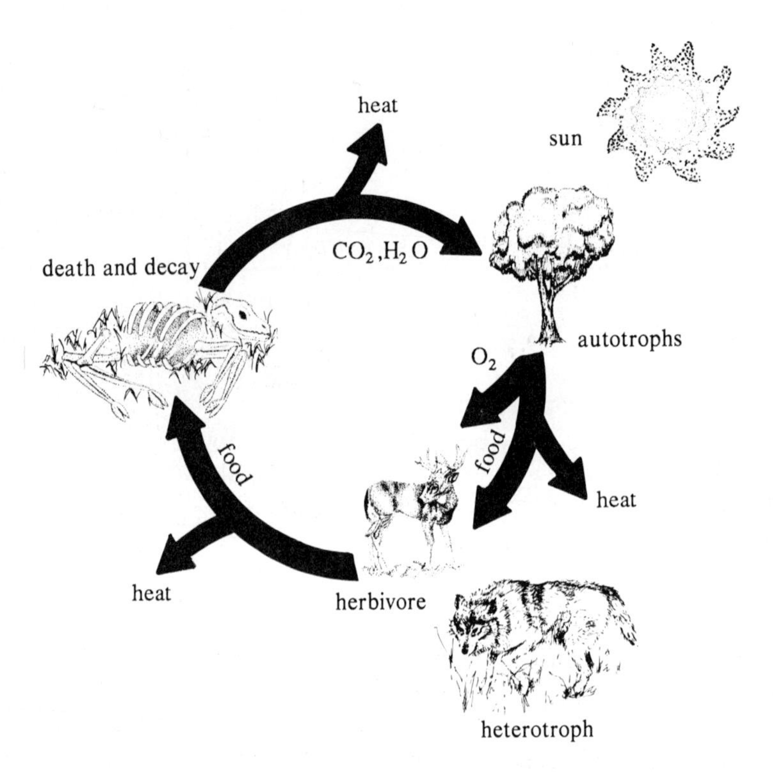

Figure 7.3 Flow of energy through the biosphere. The flow of energy is accompanied by a cyclic flow of matter between autotrophs and heterotrophs.

animals? _____________ Organisms that eat herbivores for energy are called **carnivores.** Heterotrophs that eat either plants or animals are called **omnivores.** What kind of heterotroph are you? _____________ Examine figure 7.4 and be aware that energy transformations must occur for organisms to utilize the sun's nuclear energy. Life on earth continues only because of the constant flow of energy from the sun. Plants transform radiant energy into chemical energy. These are the producers of food. Animals eat the plants for energy, but lesser amounts of energy are available compared to what the plants were able to acquire directly from the sun. These animals are consumers.

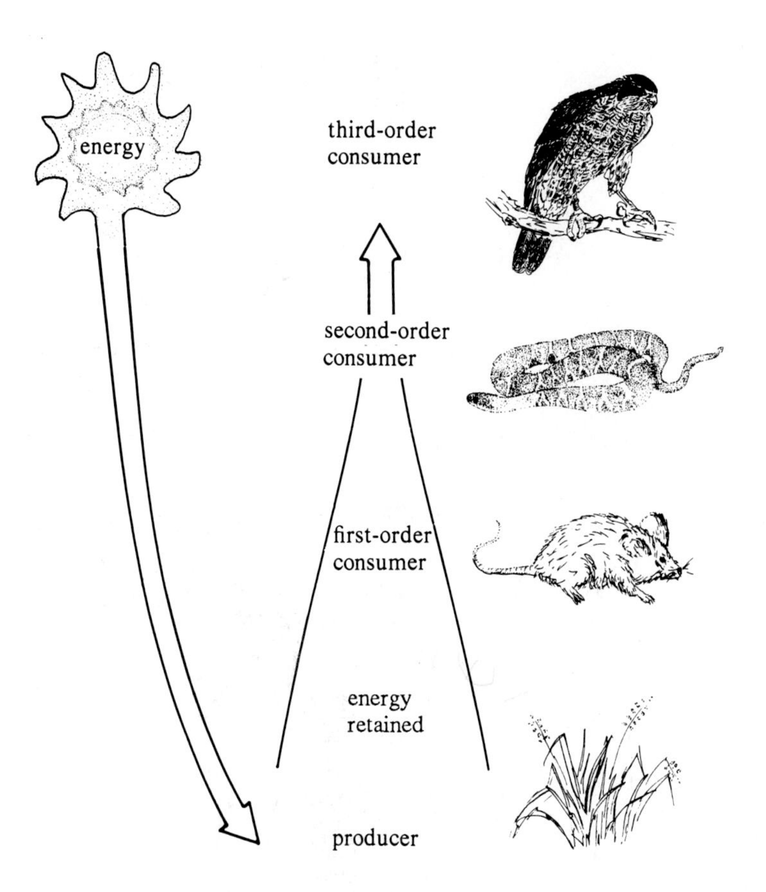

Figure 7.4 Flow and transformation of energy from the producers through the consumers. At each higher order step less of the energy originally captured by the producer plants is available to do the work of living.

When a first-order consumer such as a cow eats grass, about 9 percent of the sun's energy that was stored in the grass leaves is lost duc to natural body work done by the cow. Does a second-order consumer have *more* or *less* energy available for work than a first-order consumer? ___________

Would it be more energy efficient for you to eat steak or wheat? _______
It is important to know that when chemical energy is used by consumers for food, a part of the energy is transformed into heat, which is lost to the atmosphere. Therefore, energy does not cycle, and living systems must continually receive a supply of nuclear energy from the sun.

7.7 How do living things do work? Energy may be defined as the potential to do *work*. Gasoline has the potential to do the work of moving a car, but first a car must transform this chemical energy by burning it.

Living cells require a constant supply of energy from their ecosystem because they need to do the biological work of living things. In living systems, work is done by the transfer of chemical energy. This idea was introduced in the equation in section 7.2. Unlike the radiometer, however,

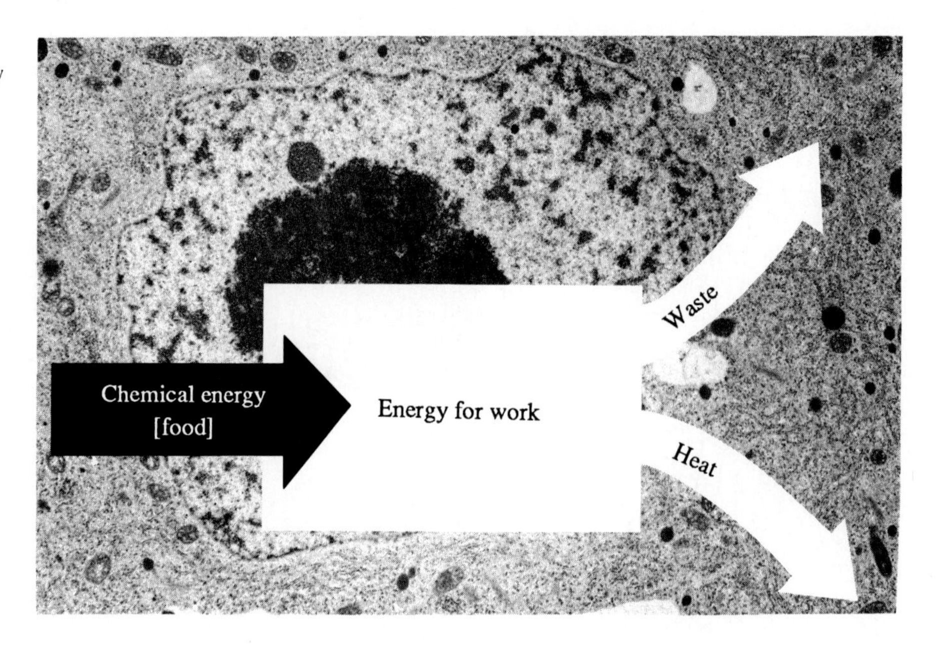

Figure 7.5 Cellular transformation of chemical energy for the work of living.

work cannot be done by living systems with heat flow. In all cellular transformations, the total amount of energy a cell receives is equal to the sum of energy used for work plus the amount of heat energy produced and the energy of the waste products (figure 7.5). Energy is neither created nor destroyed by the cellular transformation shown in figure 7.5.

Cells do work by *linking* (*coupling*) energy yielding chemical transformations to energy requiring work tasks. You can compare the work of a cell to the work of lifting a block to the top of a building (figure 7.6). Work is done (lifting block B upward) by joining a heavier block A to a lighter weight block B and placing a pulley (coupler) in between. The energy that is available from the downhill drop of the heavier block is coupled by the pulley to the uphill work of lifting the lighter block.

Cells obtain their energy by coupling downhill energy-yielding reactions to uphill energy-requiring work tasks (figure 7.7). In living cells, the pulley or coupler is a chemical molecule called **ATP** (adenosine triphosphate). ATP couples downhill energy-yielding reactions to uphill energy-requiring biological work tasks.

Where do the cells of plants and animals get their ATP energy to drive the energy-requiring work tasks? Cells and, thus, most life in the biosphere obtain ATP by two main energy transformations: **photosynthesis** and **cellular respiration.** Notice in figures 7.3 and 7.8 that a human's source of energy depends on the radiant energy of the sun being transformed by the photosynthetic process of autotrophic plants into chemical, energy-rich organic compounds. During photosynthesis, ATP couples the radiant energy

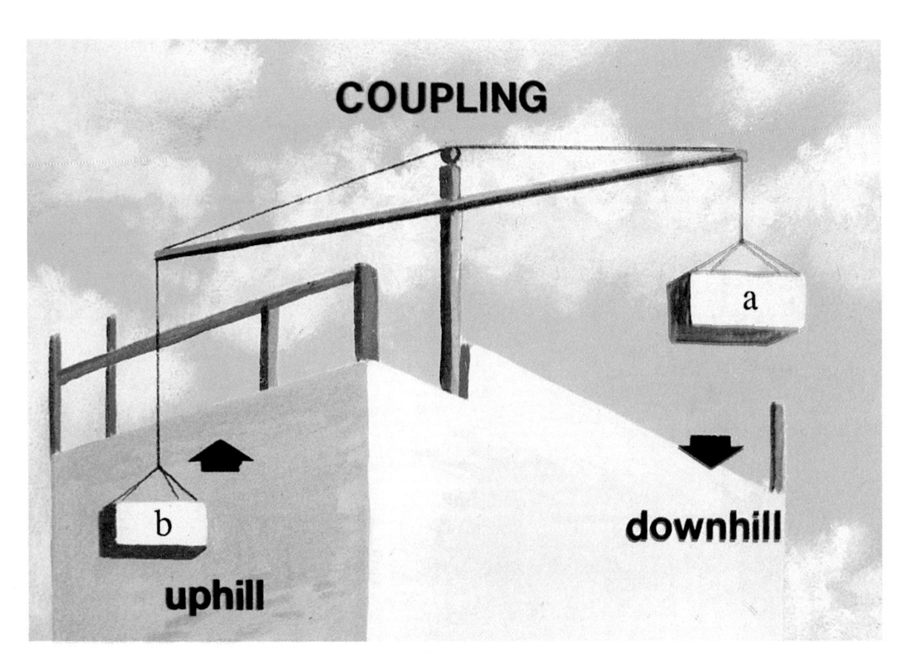

Figure 7.6 Coupling the movement of a heavy (A) mass (representing energy yielding) to the movement of a lighter (B) mass (representing energy requiring).

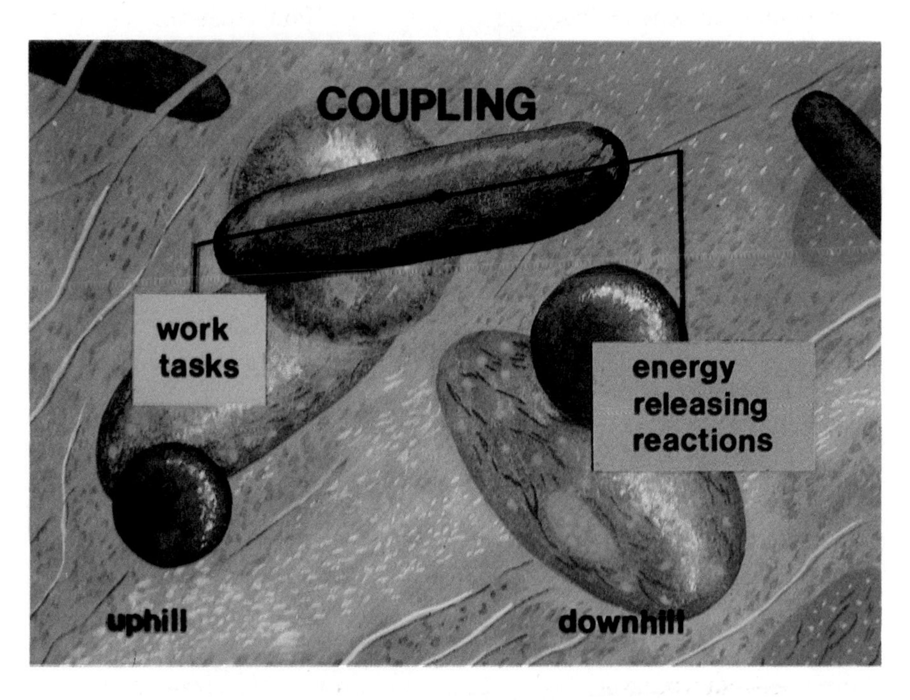

Figure 7.7 Coupling the energy requiring work activities of cells to the energy yielding reactions of cells. A major cell coupler is ATP.

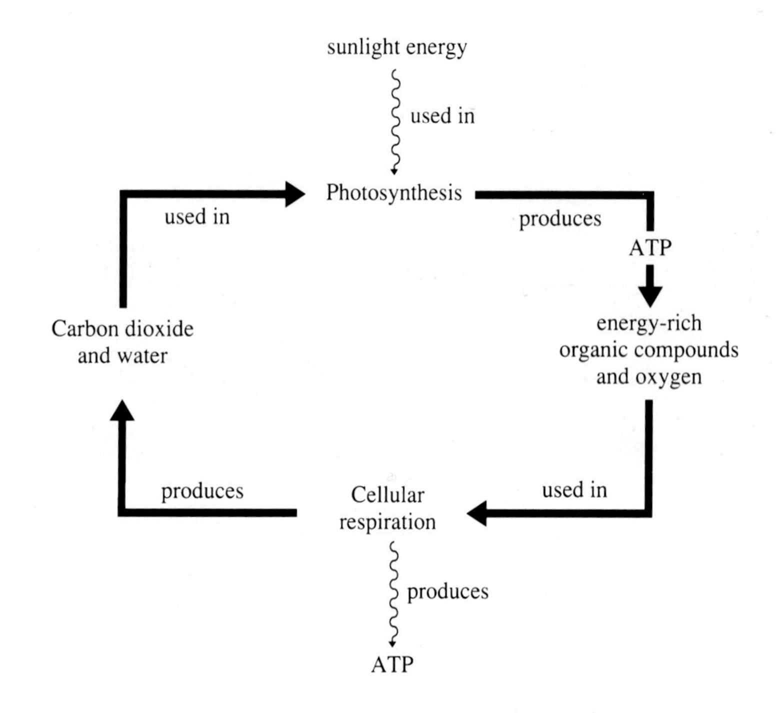

Figure 7.8 Transformation routes for the production of the chemical coupler, ATP.

of the sun to the manufacture of the energy-rich organic foods. Heterotrophic animals, like yourself, obtain chemical energy for work by eating autotrophs and/or herbivores. The cells of heterotrophic animals transform the chemical energy into ATP. Then ATP couples the energy-yielding chemical transformation to biological work.

Energy transformations that occur in living cells are called cellular **metabolism.** Metabolism is divided into two processes: **anabolism** and **catabolism.** Anabolism is the synthesis of chemical molecules, and catabolism is the breakdown of chemical molecules. Is the joining of amino acids together to make proteins an anabolic process or a catabolic process? _____ Is the separation of starch molecules into glucose molecules an anabolic or catabolic process? ________________ Catabolism is a downhill process that yields energy, and anabolism is an uphill work task that requires energy.

Both photosynthesis and cellular respiration supply energy for the synthesis of ATP (figure 7.9). The breakdown of ATP supplies energy for biological work. ATP is like the pulley in figure 7.6; it is the couple that links the energy-yielding activities of photosynthesis and cellular respiration to the energy-requiring biological work activities.

7.8 Photosynthesis makes chemical food energy available to both plants and animals to do the work of living. This food energy must first be broken down by cellular respiration in order to make ATP (figure 7.10).

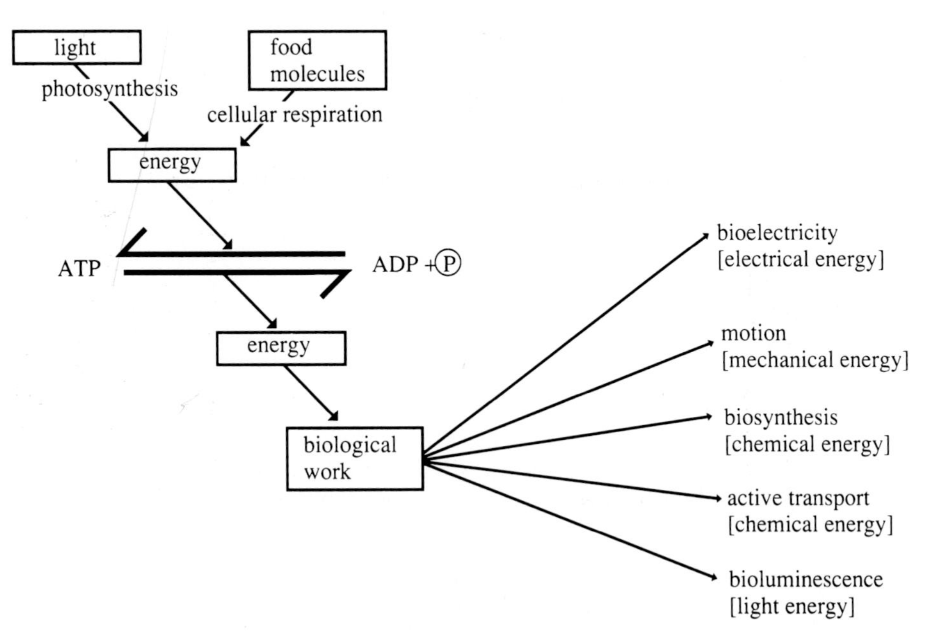

Figure 7.9 ATP couples photosynthesis and respiration to biological work. Each work event listed requires energy obtained by cellular transformation.

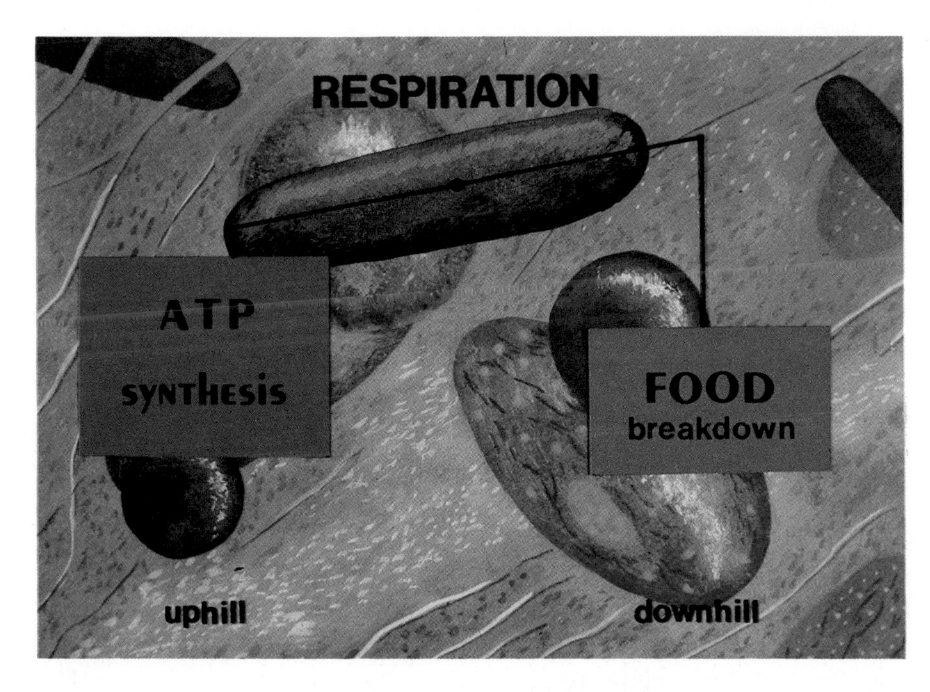

Figure 7.10 Cellular respiration couples the breakdown of foods (downhill = energy yielding) to the synthesis (uphill = energy requiring) of ATP.

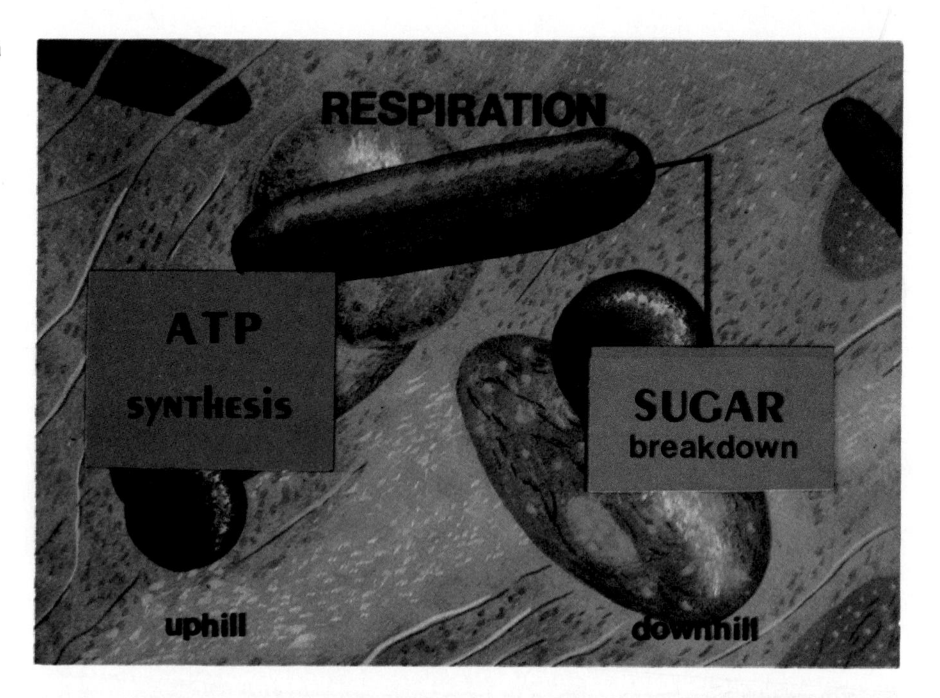

Figure 7.11 Cellular respiration couples the breakdown of glucose (food) to the synthesis of ATP.

Carbohydrates, fats, and proteins are used as foods by plants and animals to make ATP. However, the most common food source is carbohydrates such as glucose or starch. These sugars are broken down by cellular respiration and much of the energy in the sugars is used to synthesize ATP (figure 7.11). Compare figures 7.6, 7.7, and 7.11. In what way is the cellular respiration of plants and animals similar to lifting a heavy object to the top of a building?

____

____

7.9 Food molecules, such as the sugar glucose, contain great amounts of energy stored in their chemical bonds. This chemical energy is released by cellular respiration and used to synthesize ATP and to produce building block monomers (figure 7.12). The release or breakdown is called *catabolism.* There are two kinds of catabolic reactions that are a part of cellular respiration: **aerobic** respiration and **anaerobic** respiration. Aerobic respiration requires oxygen and conserves approximately 18 to 19 times more energy from the sugar molecules for ATP synthesis than is conserved by anaerobic respiration. Use figure 7.12 and other

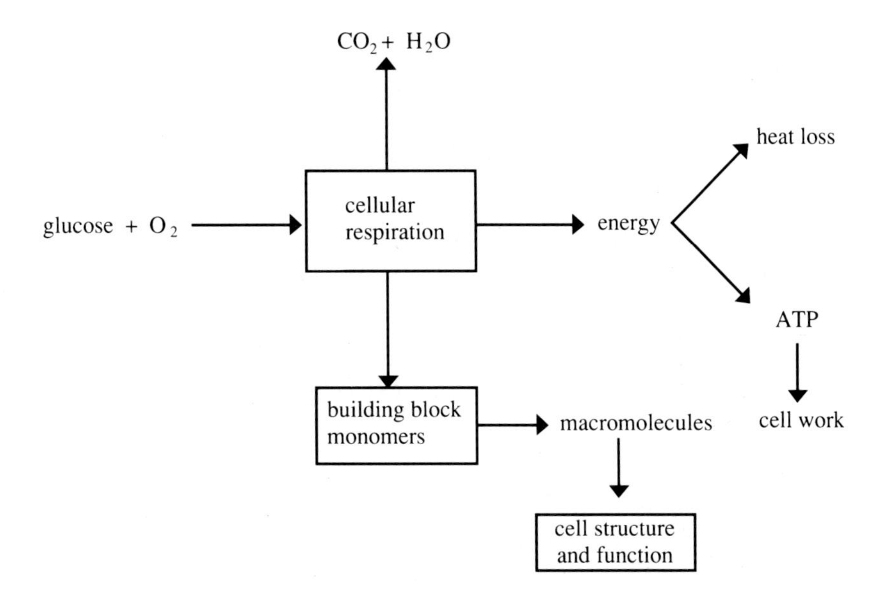

Figure 7.12 Cellular respiration of sugars is coupled to the production of both ATP and macromolecules.

information in this chapter and chapter 5 to answer the following questions in your logbook:

1. If glucose and oxygen are metabolized by cells, what are the products?
2. What kinds of biological work can be done with the ATP?
3. How do cells use the building block monomers?
4. Do cells use the heat energy to do work?
5. What happens to the carbon dioxide?

7.10 How does aerobic respiration conserve the energy of food molecules? Aerobic cellular respiration includes three catabolic phases: **glycolysis, Kreb's cycle** and the **electron transport system** (figure 7.13). An input-output chemical equation below summarizes the three phases.

$$C_6H_{12}O_6 + 6\ O_2 \rightarrow 6\ CO_2 + 6\ H_2O + 36/38 \text{ units of ATP}$$

[input] [output]

Each phase includes a series of chemical reactions that may be studied by examining a reference textbook. Whether 36 or 38 units of ATP are produced depends on the type of cells respiring the glucose molecules.

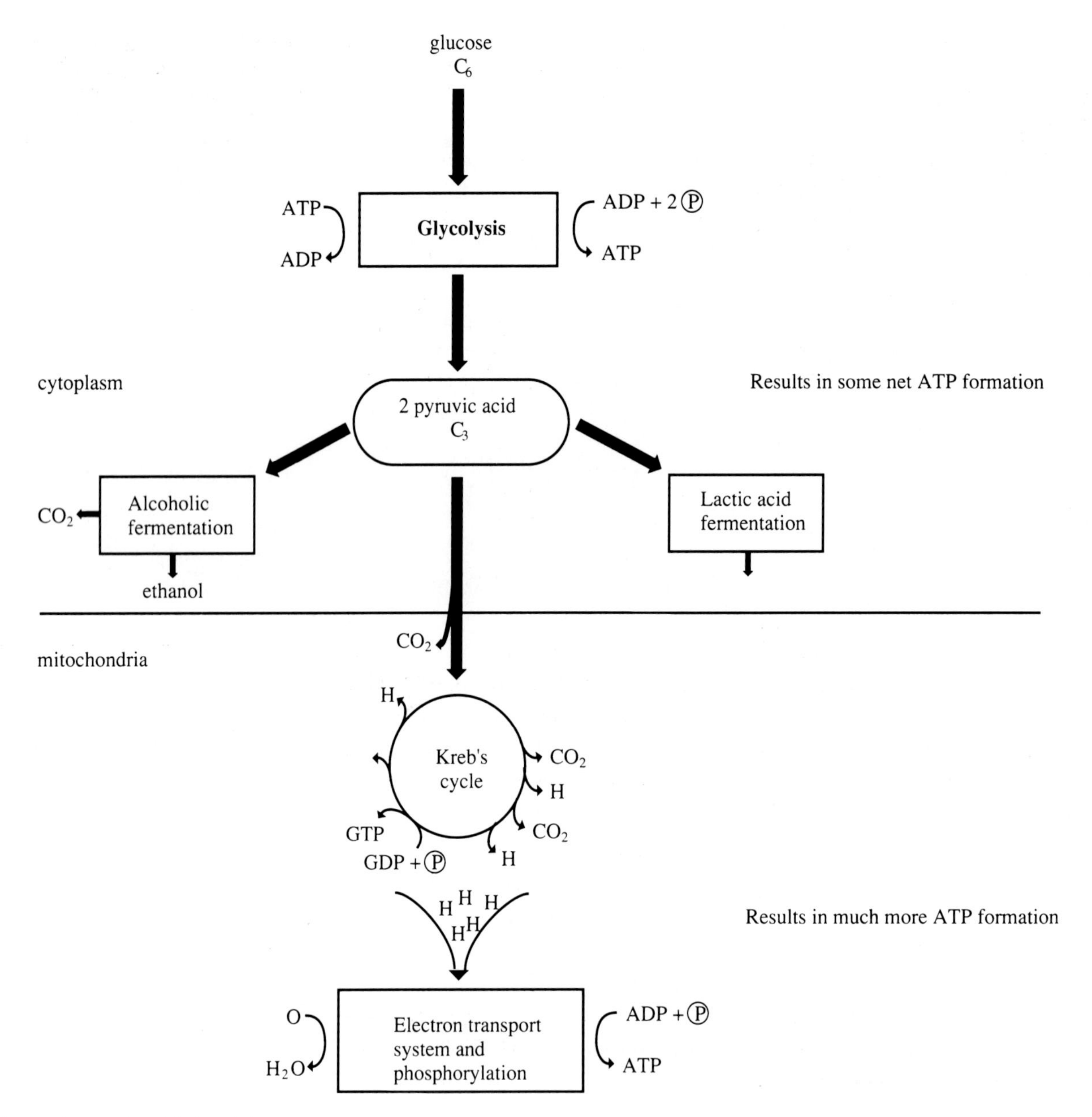

Figure 7.13 Aerobic breakdown of organic foods (e.g., glucose) by three metabolic routes: glycolysis, Kreb's cycle, and the combination of electron transport and phosphorylation.

Glycolysis is the breakdown of glucose to pyruvic acid. This occurs in the cytoplasm of cells. Pyruvic acid is transported into the mitochondria where it is broken down to hydrogen atoms (H) and carbon dioxide by the Kreb's cycle (often called the citric acid cycle). The carbon dioxide will diffuse out of the mitochondria, exit the cells, and ultimately leave the organism.

How do you get rid of the carbon dioxide produced by aerobic cellular respiration?

Figure 7.14 Cascade of water flow (energy yielding) coupled to the production of electricity (energy requiring). Energy transformation occurs at the dam sites.

7.11

The Kreb's cycle is a big supplier of hydrogen atoms to the electron transport system phase of aerobic cellular respiration. Cells have chemical "trucks" in their mitochondria to carry the hydrogen atoms from the Kreb's cycle to the third and final phase of aerobic respiration, the electron transport system. The hydrogen atom, you remember, carries a single electron. It is these electrons and the protons of hydrogen atoms that are delivered to electron transport system. Oxygen ultimately accepts these electrons after they pass through the electron transport system. After this electron addition oxygen adds two protons to form water (H_2O).

The electron transport system is a catabolic hill. The catabolic hill is like a river flowing downhill through a series of power dams (figure 7.14). As the river flows downhill, energy is available to do work. This available energy turns the turbines at the power dams, resulting in the generation of electricity. Similarly, as a river of electrons flow through the electron transport system, energy is available to synthesize ATP.

7.12

If the oxygen supply to the cells is stopped, the flow of molecules through the Kreb's cycle and the flow of electrons through the electron transport system will be stopped. This is much like stopping the flow of water to the power dam, which will eliminate the energy available to generate electricity. When the operation of the Kreb's cycle and the electron transport system are stopped, the cells will not be able to produce enough ATP energy to keep the whole organism alive.

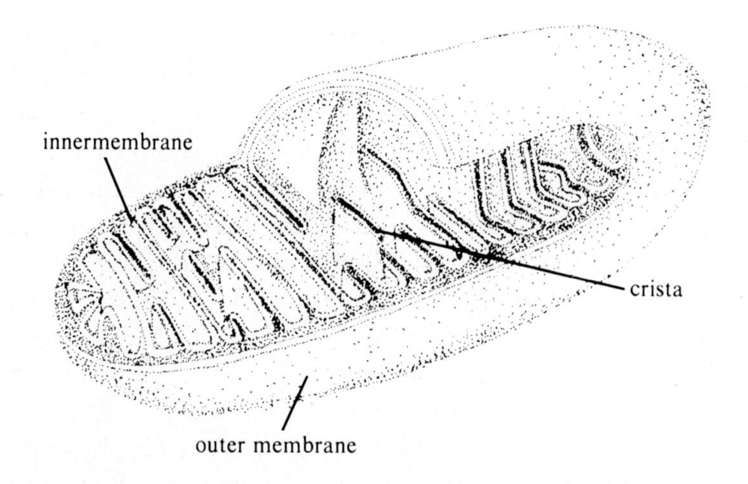

Figure 7.15 Drawing of a mitochodrion.

If we assume that by cellular respiration 38 ATP molecules are produced from the breakdown of one glucose molecule and, *if* we assume that only 4 ATP molecules are produced during glycolysis and the Kreb's cycle phase, *then* 34 ATP molecules must be produced during the electron transport system phase. Therefore, approximately 90 percent of the ATP is produced in the mitochondria (figure 7.15), which are thus nicknamed the powerhouses of the cell. This means that most of the energy available in the glucose molecules becomes available to make ATP only if oxygen is supplied to the cells and to the mitochondria. Because the electron transport system and the synthesis of ATP is located at the cristae membrane, the structure of this membrane is very important to the production of ATP. Lead poisoning often occurs to small children when they eat chips of old paint containing lead. Today we know that lead (Pb) destroys the inner membrane of mitochondria, which slows down the making of ATP.

Biologists also know that alcohol affects membranes. Human liver cells of chronic alcoholics will show fewer "healthy" mitochondria per cell and reduced numbers of cristae membrane per mitochondrion. Lower numbers of mitochondria means these damaged cells cannot make as much ATP as normal healthy cells.

7.13 Can some living organisms survive without oxygen? Anaerobic respiration also is a way for cells and, thus, organisms to extract energy from food molecules. The most familiar are the *fermentation* processes that microorganisms, such as bacteria and fungi, use to obtain their energy. Two input-output summary equations of fermentation are shown below.

A. $C_6H_{12}O_6 \xrightarrow{\text{bacteria}} 2 \text{ lactic acid} + 2 \text{ ATP}$

B. $C_6H_{12}O_6 \xrightarrow{\text{fungi}} 2 \text{ ethanol} + 2\ CO_2 + 2 \text{ ATP}$

One example of bacterial fermentation is the lactic acid production during the souring of milk. Interestingly, lactic acid is also produced in the skeletal muscle cells of animals when oxygen is scarce, but not by bacteria.

When a person exercises so hard that the ATP supply cannot keep up with the work demand, additional amounts of ATP energy are made available by the fermentation of more glucose molecules through glycolysis to pyruvic acid, and then to lactic acid. In this way, extra ATP is provided for the biologic work of muscle contraction. However, this extra ATP production can occur for only a short period of time because the lactic acid buildup will eventually inhibit muscle contraction, and it may be responsible for some of the painful side effects of extensive exercise.

Another example is the production of alcohol and CO_2 by unicellular fungi, such as yeast cells. There are many different end products produced by microorganisms fermenting all kinds of compounds. The purpose of the fermentation reactions is for the cell to obtain chemical energy for work. Compare the amount of usable ATP energy produced aerobically and anaerobically and answer the following questions in your logbook. Which cellular respiratory process yields the most energy from the breakdown of glucose? ________________ Which metabolic phase of aerobic respiration also is common to anaerobic respiration? ________________

7.14 There are several ways to determine if an organism is engaged in cellular respiration: measure the amount of (*a*) glucose used, (*b*) oxygen consumed, (*c*) carbon dioxide released, and (*d*) heat released.

IF your body cells are producing carbon dioxide during aerobic cellular respiration, *then* you should be able to detect carbon dioxide in the exhaled portion of your breathing. When CO_2 is added to water, carbonic acid is formed.

$$CO_2 + H_2O \longrightarrow H_2CO_3 \xrightarrow{d} HCO_3^- + H_3O^+$$

$$\text{Carbonic dioxide} + \text{water} \longrightarrow \text{carbonic acid}$$

Note the dissociation (d) of carbonic acid produces two ions, the negative *anion* (HCO_3^-) and the positive *cation* (H_3O^+). In this case the cation is called a *hydronium* ion and is formed when carbonic acid gives up a proton (H^+) to H_2O. It is the activity of the hydronium ion that allows the carbonic acid to be detected. As the amount of acid increases, the amount of hydronium ion increases and, thus, the greater the acidity of the water solution. Draw a line on figure 7.16 in your logbook to represent the change in acidity with increasing amounts of carbon dioxide (CO_2).

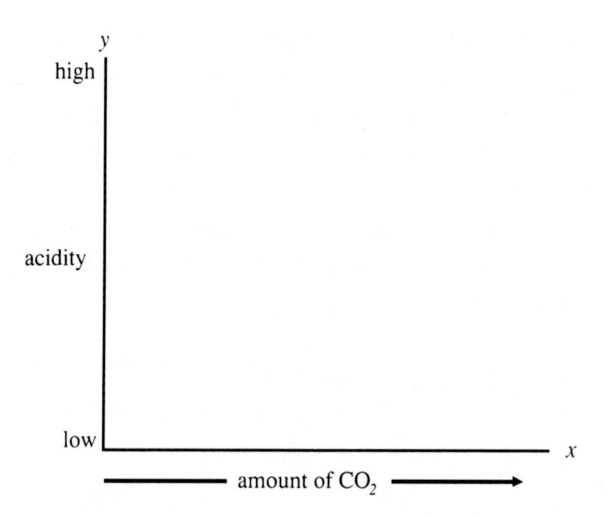

Figure 7.16 Graph correlating acidity and amount of CO_2.

Certain chemical compounds are available to detect changes in the amount of acidity; and some even change color with differing amounts of H_3O^+. The dye bromthymol blue is a good chemical detector of acidity. A yellow color indicates an acid condition of the water.

Does your breath contain CO_2? ________ Let us test your prediction.

a. Add boiled, deionized (contains no CO_2) water to a clean glass test tube until the tube is approximately two-thirds full.

b. Add bromthymol blue (BTB) solution, drop by drop, to the water. Mix each drop of BTB immediately after adding it to the water. Add only enough drops of BTB to turn the color of the water a light green or light blue. A green or blue color indicates a basic condition (few H_3O^+). A yellow color indicates an acidic condition (many H_3O^+).

c. Use a soda straw to *gently* blow your breath into the water. *Caution:* Blowing too hard will cause the water to exit the test tube.

1. Is there CO_2 in your exhaled breath? ________________
2. Where in your body was the CO_2 produced? ________________
3. How was the CO_2 produced? ________________
4. If CO_2 were removed from the solution in the test tube, what would happen to the acidity of the solution? ________________

7.15

Carbon dioxide and oxygen are gases. Many studies are designed to detect cellular respiration by measuring changes in the volume or the pressure of a gas. Any changes in volume or pressure in a closed system where an organism breaks down glucose would indicate cellular respiration.

If cellular respiration occurs in cells, *then* it should be observable in yeast cells. You can test this prediction by determining if a gas is produced in a system containing yeast cells.

a. Obtain eight clean, small glass test tubes (5 × 45 mm or 6 × 50 mm) and eight small vials. Label four test tubes *A* and four test tubes *B*. Mark, with a wax pencil, horizontal lines 1 cm apart on all test tubes (see figure 7.17).

b. Place all tubes in a vertical position. With an eyedropper, add yeast cell suspension (provided by your teacher) to all tubes until each tube is approximately one-third full. Be sure to add the same number of drops of yeast to each tube. *Note:* Yeast cells will settle to the bottom of the container. Shake to evenly distribute the cells in suspension before adding them to the tubes.

c. To tubes marked *A*, add enough water to fill the tubes completely. To tubes marked *B*, add enough 10 percent glucose solution to fill the tubes completely.

d. Examine figure 7.17. You are to place a glass vial over *each* test tube according to the following directions:

- Hold filled test tube upright.
- Place a vial over the tube.
- Hold the test tube tightly against the bottom of the vial.
- Invert the vial and tube quickly to the position shown in figure 7.17.
- No bubble (or only a *small* bubble) should be present in the upper end of each test tube.

e. Estimate the length of the bubble in each tube to the nearest 0.5 cm and record in data table 7.4 of your logbook.

f. Place all eight vials into an incubation chamber set at 37° C.

g. Examine and estimate the length of the bubble in each tube every 30 minutes for 3 hours. Make a final observation at 24 hours.

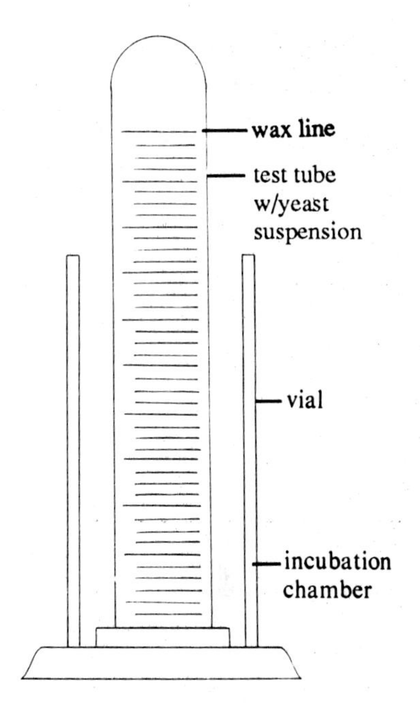

Figure 7.17 Cellular respiration apparatus.

Data Table 7.4 Data for Cellular Respiration in Yeast Cells

	Length of Bubble (cm) Time/minutes							
Tube A	0	30	60	90	120	150	180	24
1								
2								
3								
4								
Average								
	Length of Bubble (cm) Time/minutes							
Tube B	0	30	60	90	120	150	180	24
1								
2								
3								
4								
Average								

7.16 Plot your average data values on a sheet of graph paper from your logbook. Plot the length of the bubble (*Y* axis) against time (*X* axis). Use ▲ to represent yeast cells and water, and use *O* to represent yeast cells and 10 percent glucose. Ask your teacher to help you add the following two lines to the graph: one line that best fits data points for tube A and one line that best fits data points for tube B.

1. Which test tubes function in the experiment as the controls?

   ____

2. What is the gas that formed the bubble? ____________________
3. Is cellular respiration occurring in the yeast cells?

   ____________________ How do you know?

4. Which metabolic phase of 7.13 best explains how the yeast cells formed the gas by the cellular respiration of glucose?

7.17 *If* yeast cells used the glucose to make the gas, *then* is it logical to predict that the amount of glucose would decline? Let us test your prediction.

a. Obtain four plastic clinical centrifuge tubes. Label two tubes *A* and 2 tubes *B*.
b. Add 3 mL (60 drops) of a yeast cell suspension to each tube marked *A* and to each tube marked *B*.
c. To each tube A, add 0.5 mL (10 drops) of water.
d. To each tube B, add 0.5 mL (10 drops) of 10 percent glucose solution.
e. Cover *one* tube marked *A* and cover *one* tube marked *B* with aluminum foil and place both tubes into an incubator set at 37° C for 24 hours.
f. Place the second tube marked *A* and the second tube marked *B* into a clinical centrifuge. Centrifuge the second set of tubes at maximum speed for 20 minutes to pack the yeast cells at the bottom of the tubes.
g. Obtain two clean glass test tubes (15 mm × 60 mm) and label one *A* and the other *B*.
h. Remove the second set of test tubes from the centrifuge and carefully pour off the upper liquid in each tube into clean glass test tubes and save it (pour A into A and B into B). Discard the yeast cells packed at the bottom of the centrifuge tubes.
i. Add .1 mL (2 drops) Benedict's solution to each tube of the second set and heat the tubes in a boiling water bath for 5 minutes. Record in your logbook the colors which indicate a positive Benedict's test?

______ or ______

j. Record your observation in data table 7.5 in your logbook as either a positive (+) or negative (−) Benedict's test.
k. After 24 hours, remove the first set of tubes from the incubator, centrifuge as directed in step *f*, and repeat steps *g* through *j* for the first set of test tubes marked *A* and *B*.

Data Table 7.5 Benedict's Test

	A	B
Initial		
Final		

1. What is the experimental purpose of tubes A?
2. Was the Benedict's test for tube B (final) about the same, more positive, or less positive than tube B (initial)? ______

3. Do you think there is a change in the amount of glucose in tube B final? ______ Is cellular respiration occurring in yeast cells? ______

7.18 Prepare a written statement to explain your observation.

Prepare a written statement describing the relationship between the two previous experiments with yeast cells: gas production and glucose detection.

Do you think some living cells can survive without oxygen? Explain.

7.19 Do living organisms require energy for biologic work? One type of work done by cells is biosynthesis. Figure 7.18 illustrates the major stages where ATP energy is needed for the biosynthetic production of macromolecules, cellular parts, and new cells.

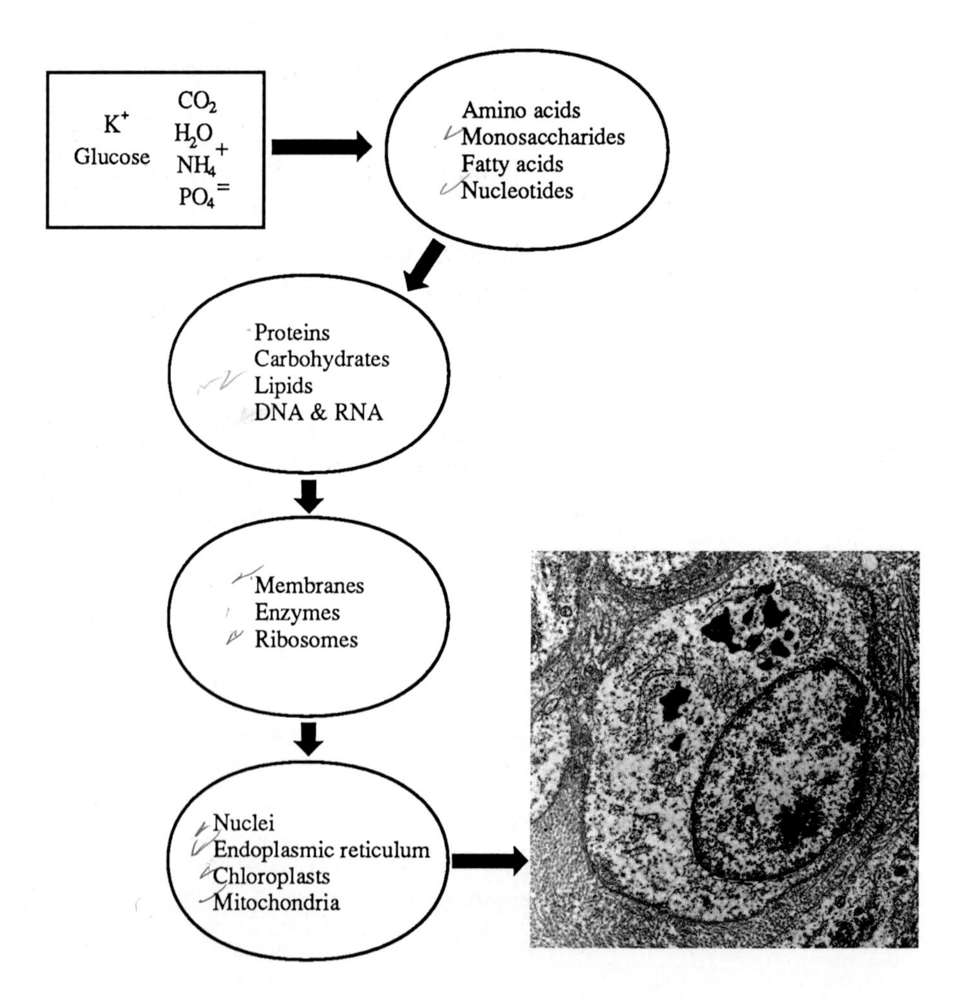

Figure 7.18 Major assemble stages of cell synthesis.

Most organisms consist of a population of cells. The adult human consists of approximately 60 thousand billion (6×10^{13}) cells. Some cells of the human body are continuously dying and are replaced through cellular reproduction. Approximately 1 to 2 percent of an adult human's cells die each day and, therefore, several million cells are synthesized each day.

Use figure 7.18 to explain in your logbook how the cells of your body use the food they receive daily to replace lost cells.

7.20

By this time, you may believe that all the energy in food molecules is either lost as a heat or used to make ATP (see figure 7.12). However, a careful study of figure 7.12 will show that some of the energy of food is used to make building block monomers. Examine figure 7.18 and mark with an asterisk (*) those building block monomers. Label with a check (✓) the macromolecules listed in figure 7.18.

The energy of food molecules like glucose is used to produce both ATP energy for cellular work and building block monomers for macromolecular synthesis.

7.21

So far you have observed cellular respiration by measuring the production of carbon dioxide and measuring the use of a carbohydrate (glucose) as a source of energy. Another way to observe cellular respiration in living organisms is to measure the amount of oxygen being used.

Seeds of plants contain within them a young plant called an embryo. When seeds germinate, the embryo grows, and a young seedling emerges from the seed coat. Do germinating seeds use oxygen? Figure 7.19 shows an experimental apparatus used by biologists to determine if pea seeds use oxygen during germination. *If* pea seeds use oxygen during germination, *then* would this indicate that the cells of the pea seed embryo carry out aerobic cellular respiration? The results of such a study are already entered in data table 7.6.

The experimental setup consists of two glass chambers that can be closed to the outside (figure 7.19*d*), after a dye is added to the capillary tubing (figure 7.19*d*), by clamping the rubber tubing at the top of each glass chamber. Because the cells of the plant embryos produce CO_2 during cellular respiration, and because the biologist-researcher wants to observe changes in gas pressure only by oxygen usage, potassium hydroxide (KOH) is added to each chamber (figure 7.19*a*) above the seeds in order to remove the CO_2. The equation below explains the chemical reaction taking place.

$$KOH + CO_2 \longrightarrow K^+ + HCO_3^-$$

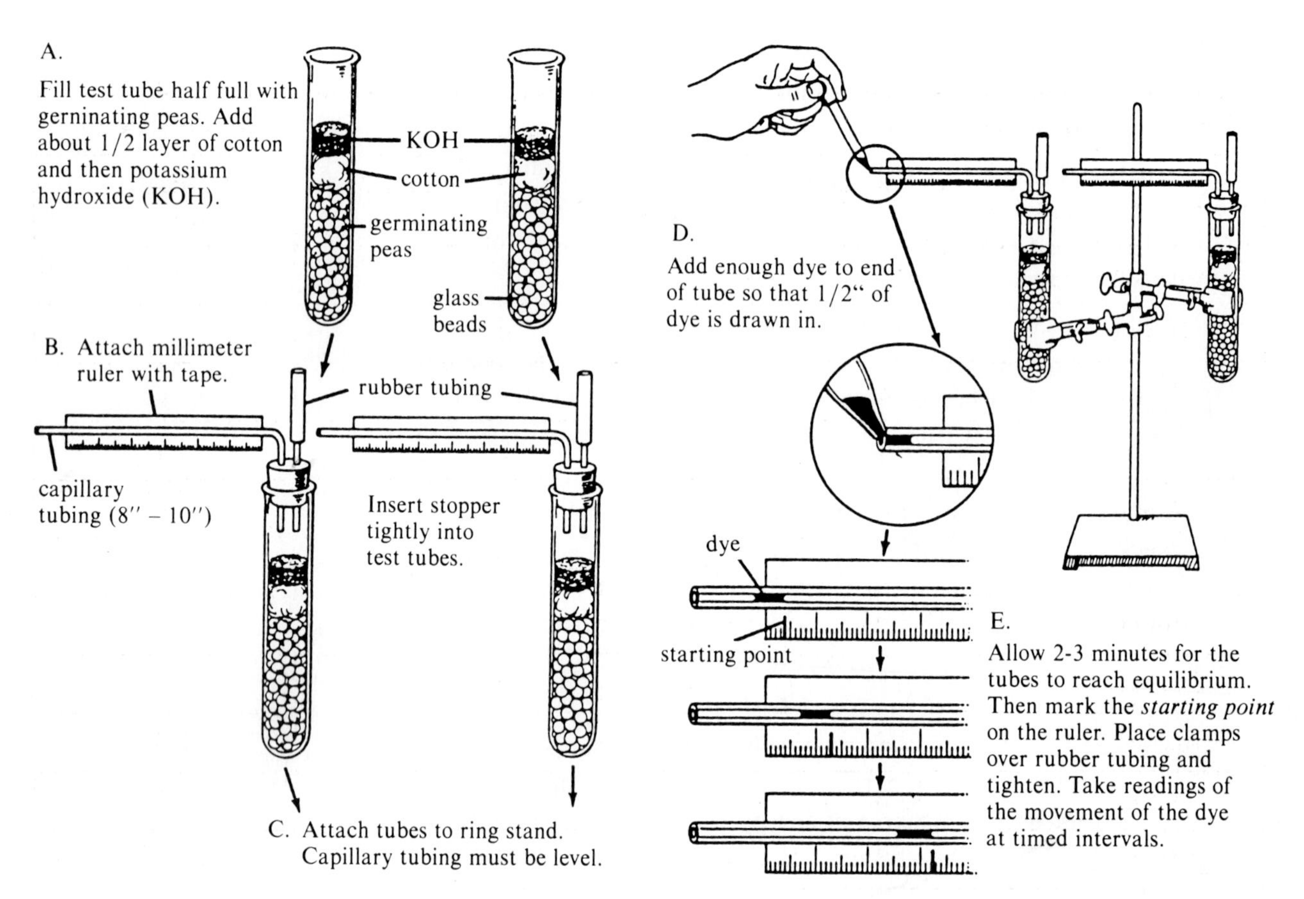

Figure 7.19 Apparatus for measuring cellular respiration of seeds.

The cotton plug over the seeds prevents the KOH from damaging the seeds. However, it is loosely packed to permit a flow of air (source of oxygen) through the cotton. The chamber containing the glass beads is a control chamber. Observations from the control chamber are used to correct the observation from the chamber containing germinating pea seeds. The correction is necessary because of possible changes in temperature and atmospheric pressure during the study.

If the oxygen (O_2) in the test tube is utilized by the germinating pea seeds, *then* the dye will move toward the chamber. (This is due to the outside air pressure attempting to equalize the pressure inside the test tube because a vacuum will be established when the oxygen in the test tube is used up.)

After the dye was placed into the capillary tube, the researcher began obtaining data based on dye movement every 20 seconds for a total of 5 minutes. The data is recorded in data table 7.6.

Data Table 7.6 Cellular Respiration of Pea Seeds

Time	Germinating Peas	Beads	Peas Minus Beads
	Total Distance Bubble Moved (mm)	Total Distance Bubble Moved (mm)	Total Distance Corrected
Beginning	0	0	0
20 sec.	0.5	0	0.5
40 sec.	0.8	0	0.8
1 min.	1.1	0	1.1
80 sec.	1.4	0	1.4
100 sec.	1.5	0	1.5
2 min.	1.9	0	1.9
140 sec.	2.1	0	2.1
160 sec.	2.3	0	2.3
3 min.	2.6	0	2.6
200 sec.	2.8	0.5	2.3
240 sec.	3.2	0.5	2.7
4 min.	3.3	0.5	2.8
280 sec.	3.6	0.5	3.1
300 sec.	4.0	0.6	3.4
5 min.	4.2	0.6	3.6

7.22 Plot a graph on a sheet of graph paper from your logbook, showing the total distance (corrected for control) the bubble moved against the time. Plot the total distance moved on the vertical (*Y* axis) and the time in seconds on the horizontal (*X* axis). Label each axis.

1. Was respiration greatest in the germinating seeds or the tube containing the nongerminating seeds and glass beads?
2. Why did the dye move toward the chamber containing the germinating pea seeds?

3. Was oxygen used by the germinating pea seeds?
4. Why do the cells of the plant embryos use oxygen?

7.23

Does photosynthesis supply energy for biological work? Plants play a unique role in transforming light energy into usable chemical, energy-rich molecules (figure 7.20). These energy-rich molecules serve as building block monomers for synthesizing new cellular materials and chemical energy fuels for cellular respiration. **Photosynthesis** is the process by which all green plants and some microscopic prokaryotes transform the radiant energy of the sun into chemical energy. Animals do not have the chemical equipment to perform this energy transformation and, therefore, must depend on plants for their energy-rich organic fuel molecules.

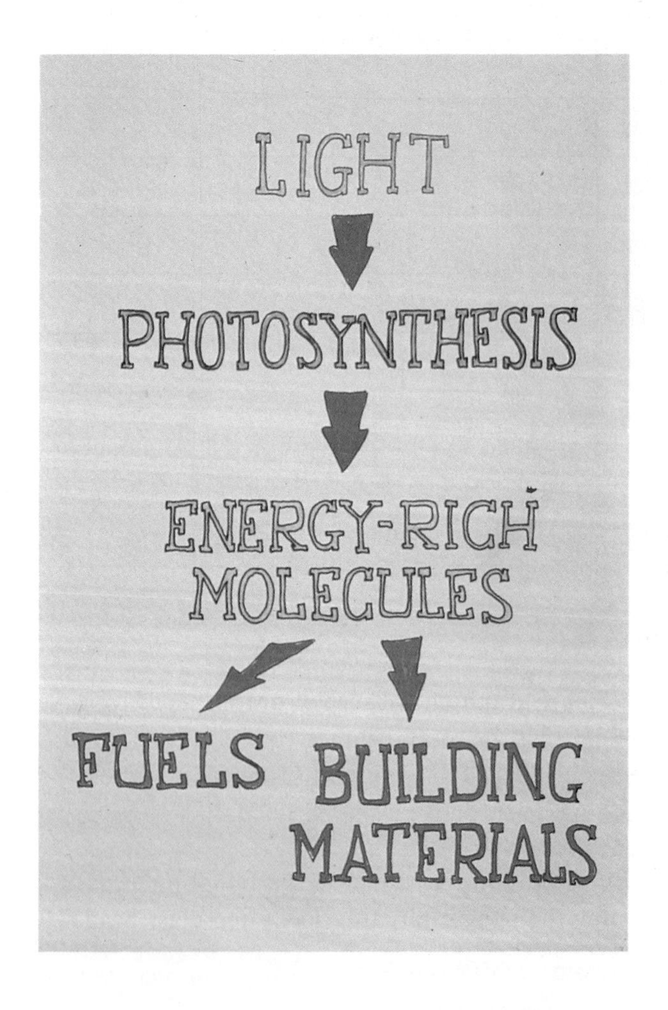

Figure 7.20 Transformation of light energy into chemical energy by plants.

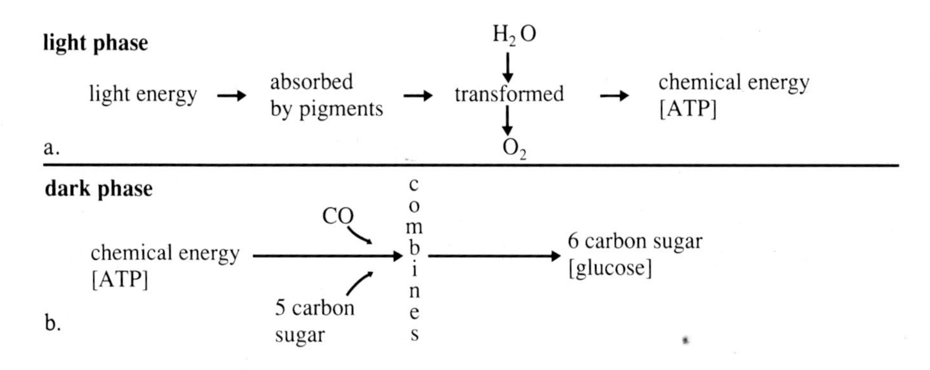

Figure 7.21 Two phases of photosynthesis. a. Light phase. b. Dark phase.

The input-output equation below summarizes the photosynthetic process of transforming light energy into chemical energy.

$$6\ CO_2 + 12H_2O \xrightarrow[\text{Chlorophyll}]{\text{Light}} \underset{\substack{\text{Sugar} \\ \text{(Glucose)}}}{C_6H_{12}O_6} + 6\ O_2 + 6\ H_2O$$

The synthesis of energy-rich sugar is an anabolic reaction that requires light energy to perform this work task. The work is done by the plant absorbing and transforming the radiant (light) energy of the sun into the chemical energy of ATP and another energy-rich molecule (figure 7.21*a*). ATP is used to combine H_2O and CO_2 into sugars (figure 7.21*b*).

Therefore, ATP *couples* the energy-yielding transformation of light energy into chemical energy to the energy-requiring (work) synthesis of sugar (figure 7.22).

Photosynthesis is completed in two phases (figure 7.21). The first phase (*a*), the *light cycle,* involves light energy being absorbed by chlorophyll pigments and transformed. The second phase (*b*), the *dark cycle* (which only occurs in the light), of photosynthesis joins CO_2 to a five-carbon sugar in the stroma of the chloroplast. This joining reaction that produces glucose requires the ATP made in the light phase. Therefore, is light energy absorbed by chlorophyll? ________ The energy of light is eventually used to bond a P (phosphate) atom to ADP (adenosine diphosphate) molecule to form ATP (adenosine triphosphate), and H_2O is split, releasing oxygen to the air. ATP is used to do work: to join CO_2 with a five-carbon sugar during the *dark* (*cycle*) to form a sugar ($C_6H_{12}O_2$).

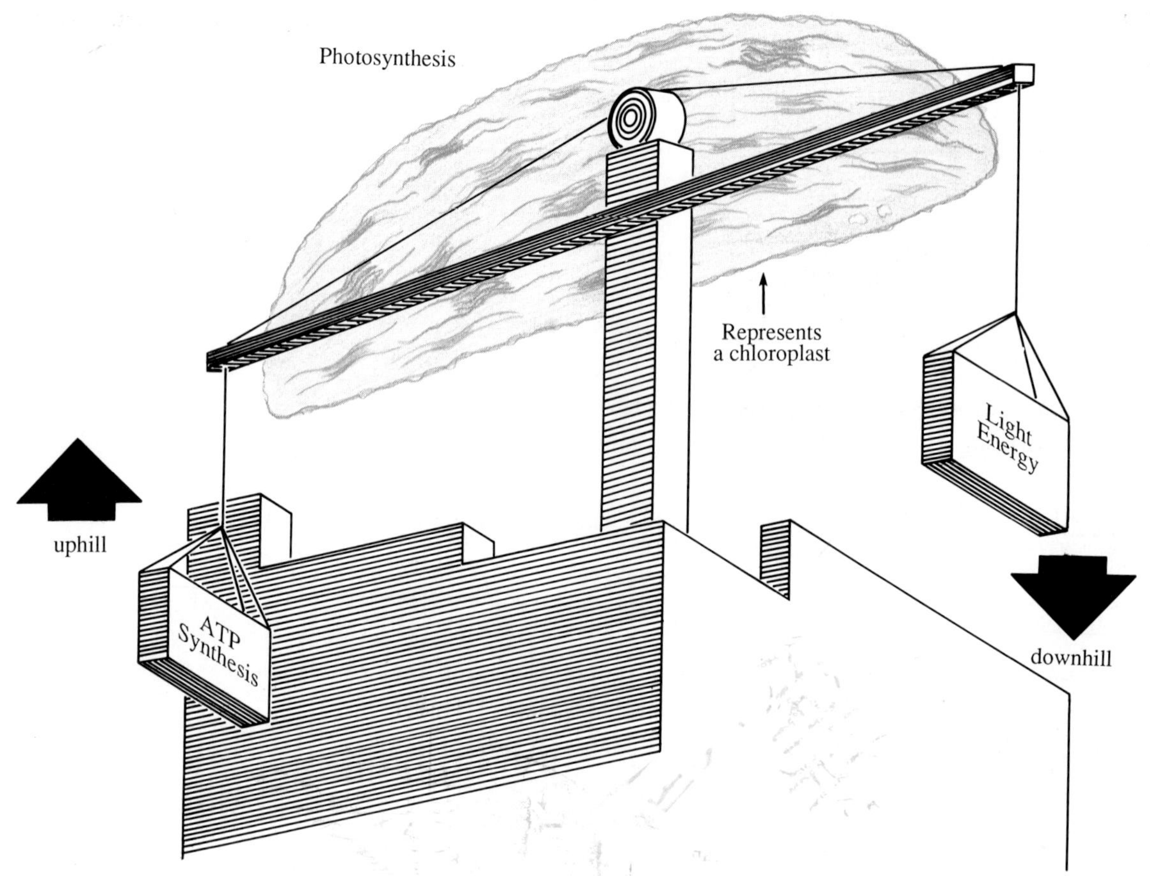

Figure 7.22 Photosynthesis couples light energy to ATP production. Downhill = energy yielding; Uphill = energy requring.

7.24 Where does photosynthesis occur? During photosynthesis, carbon dioxide is taken up and oxygen is released. Therefore, the plant structure where photosynthesis occurs must permit an exchange of CO_2 and O_2. All plants (except fungi) photosynthesize, so let us study the leaf-bearing angiosperms. There are tiny openings called **stomata** in plant leaves, through which CO_2 and O_2 move.

a. Obtain a leaf of wandering Jew or tradescantia from a laboratory plant. Peel off a piece of the lower surface and mount in a drop of water (outer side up). Please note that the lower layer can be removed by a twisting and tearing of the leaf in one motion.
b. Add a cover glass and examine under a microscope with low-power magnification.
c. Locate a stoma (the opening) and examine under high power.

Can you see the adjacent *guard* cells? They control the opening and closing of the stoma. During the day most plants have their stomata open. At night most stomata are closed. Figure 7.23 shows an open stoma and two adjacent guard cells.

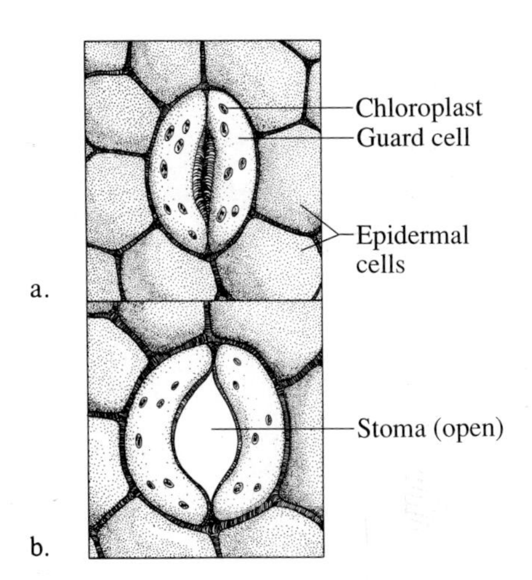

Figure 7.23 Illustration of a guard cell with a) closed stoma and b) open stoma.

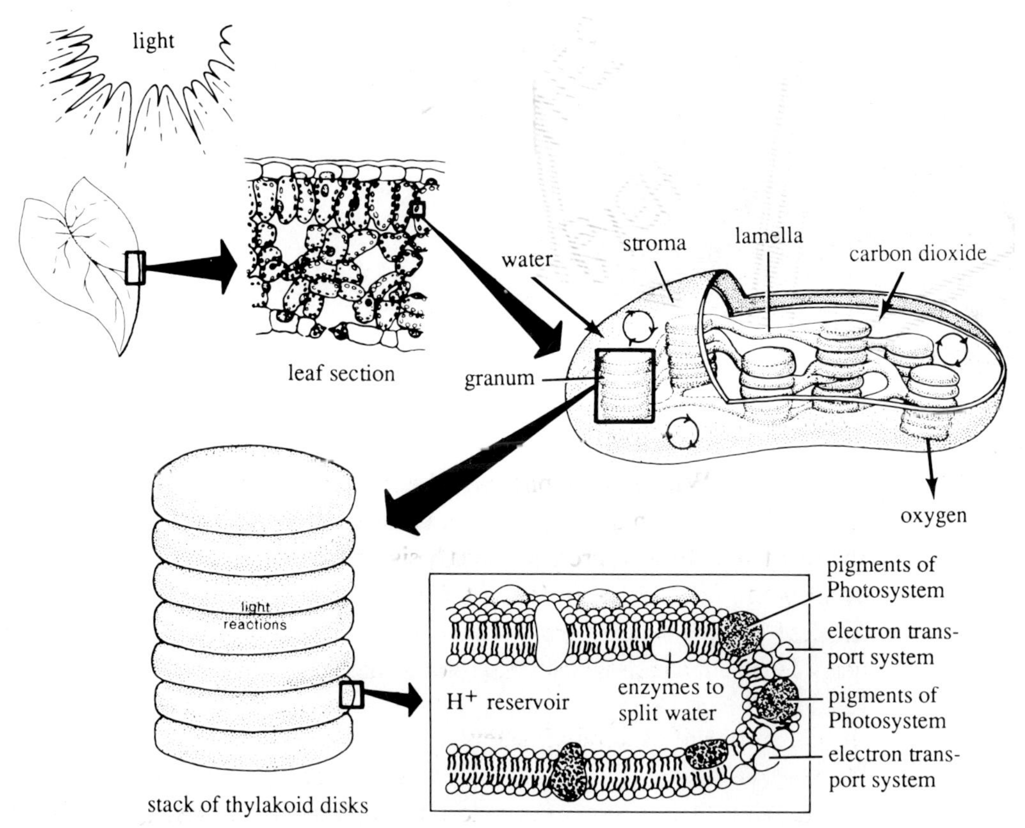

Figure 7.24 Sketch of grana membrane-cellular and leaf location.

7.25 Figure 7.24 is a series of diagrams showing the relationship between a leaf, a leaf cross section, a chloroplast of a single leaf cell, and the grana membranes. Use this to help visualize the sites of photosynthesis and to help locate the structures shown in figures 7.25–7.27.

Examine a *prepared* slide of a cross section of a leaf. Locate and identify the tissues and structures shown in figure 7.25.

Figure 7.25 A leaf. a. Cutaway view. b. Light microscopy photograph of a cross section slice.

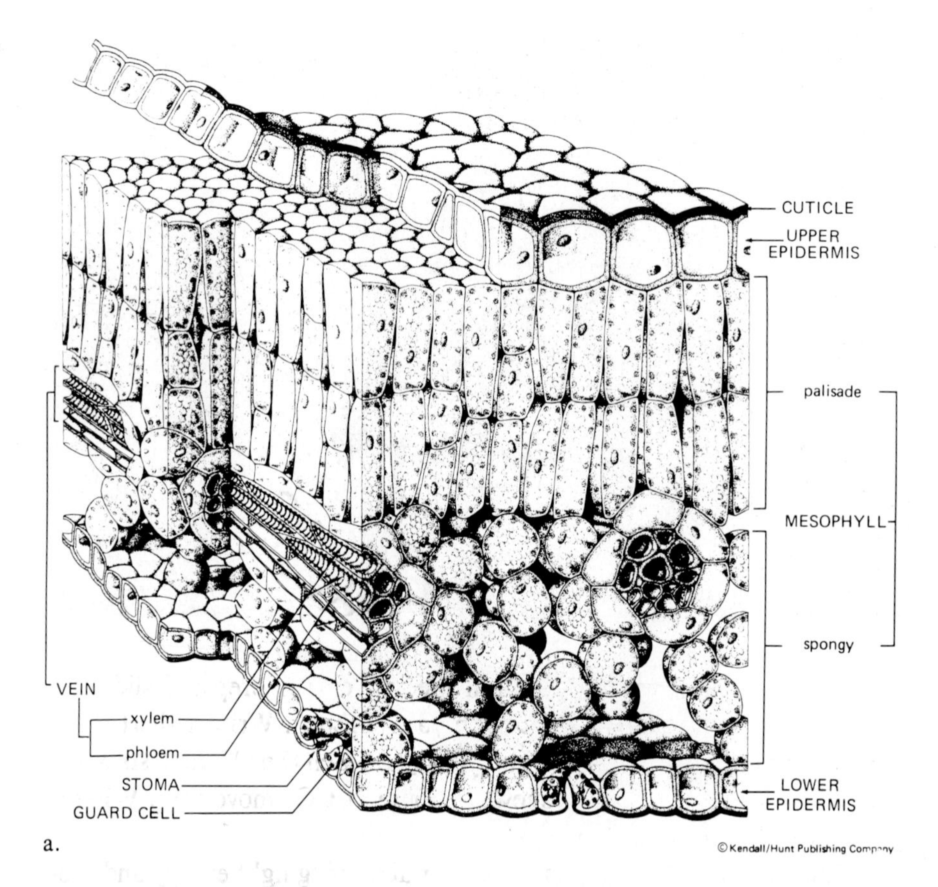

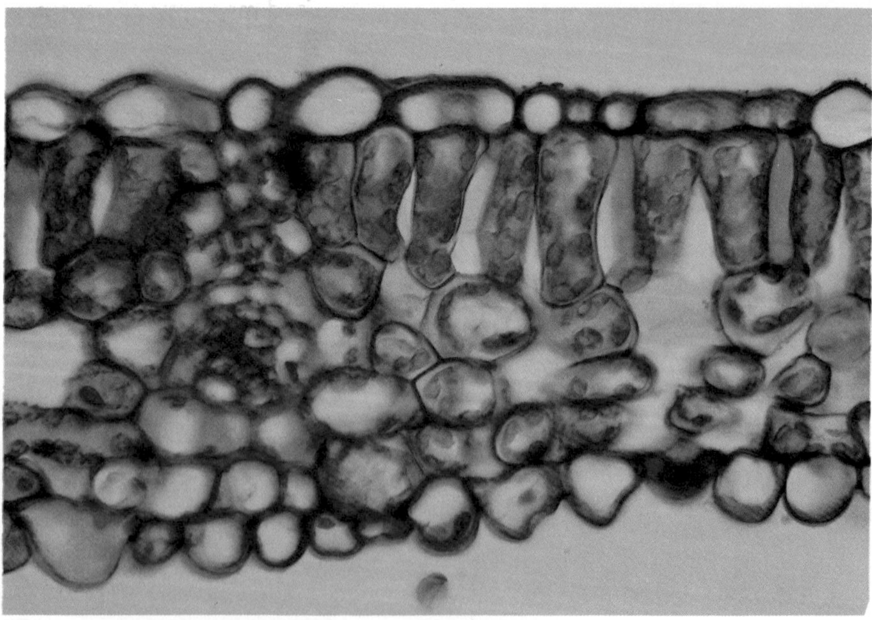

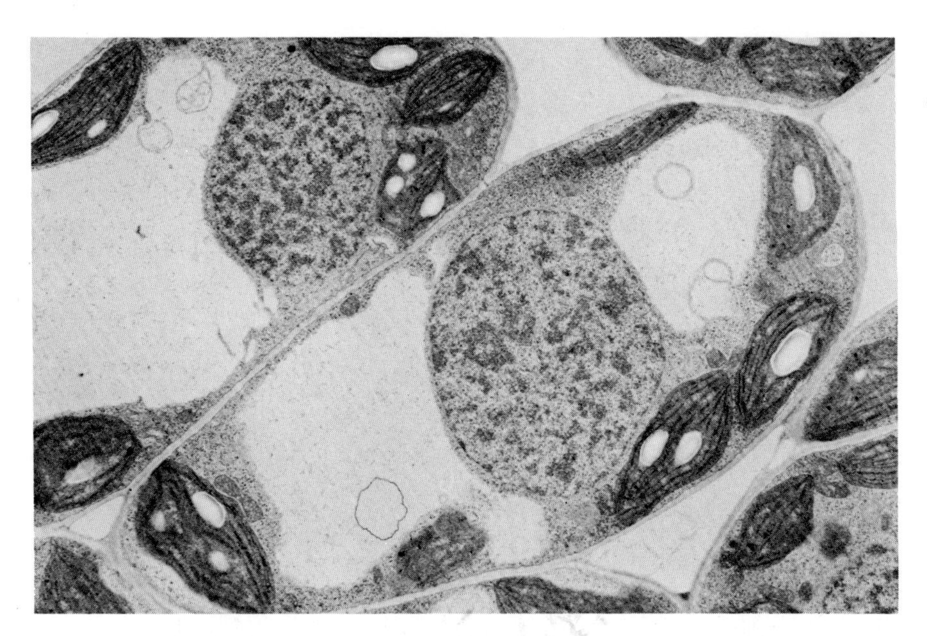

Figure 7.26 Electron micrograph of a plant cell with chloroplasts.

Find the cellular chloroplasts in the leaf on the prepared slide and label a chloroplast in figure 7.25*a, b* in your logbook. Write a short statement explaining the pathway taken by CO_2 to reach the chloroplasts of the leaf cells. Make a diagram to show the route of CO_2 movement. What is the route of O_2 movement?

The blade of a leaf is constructed for absorbing light energy and producing energy-rich sugar molecules. Most of the food made by angiosperms is produced in the leaves. Where does photosynthesis occur in gymnosperms? ________________ .

A leaf blade is usually thin and contains many cells with numerous chloroplasts (figure 7.25). The TEM photomicrograph (figure 7.26) shows a thin slice through a leaf cell containing chloroplasts. Label figure 7.26 in your logbook with the names of the cell parts identified. You should identify the cell wall, the nucleus with chromatin material, vacuoles, and chloroplasts. Note the layers of membrane within the chloroplasts. These form the grana.

Figure 7.27 is a TEM photograph (*a*) and a drawing (*b*) of a single chloroplast from a corn leaf. Note the grana is a stack of membranes. The stroma is the "cytoplasm" of the chloroplast. Photosynthesis occurs in the grana (light reaction) and the stroma (dark reaction).

Light, H_2O, and CO_2 enter the chloroplast. What are the products of photosynthesis that exit the chloroplasts? __________ , __________ , and __________

Reexamine figure 5.15 in chapter 5 and answer the following question.

How are starch grains made in chloroplasts?

Figure 7.27 Chloroplast.
a. Electron micrograph. b. Sketch.

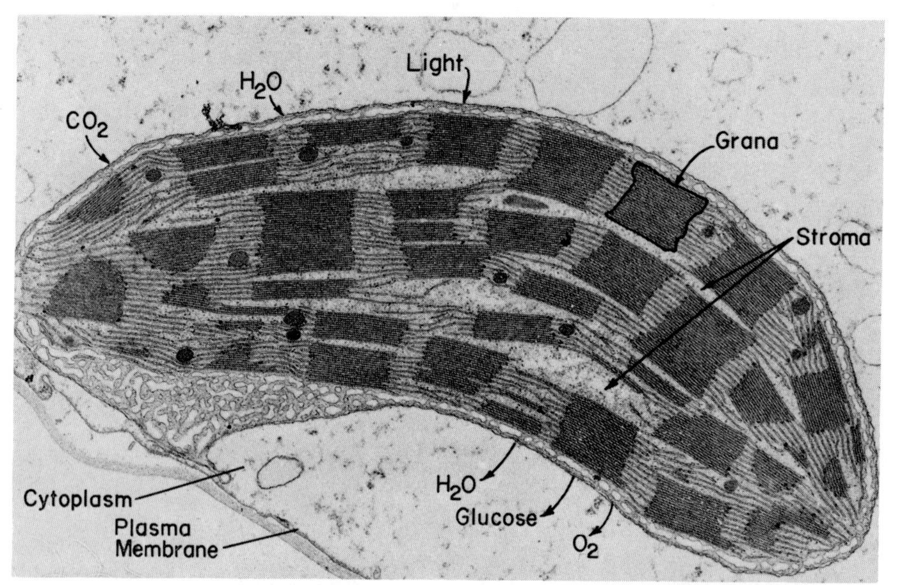

a.

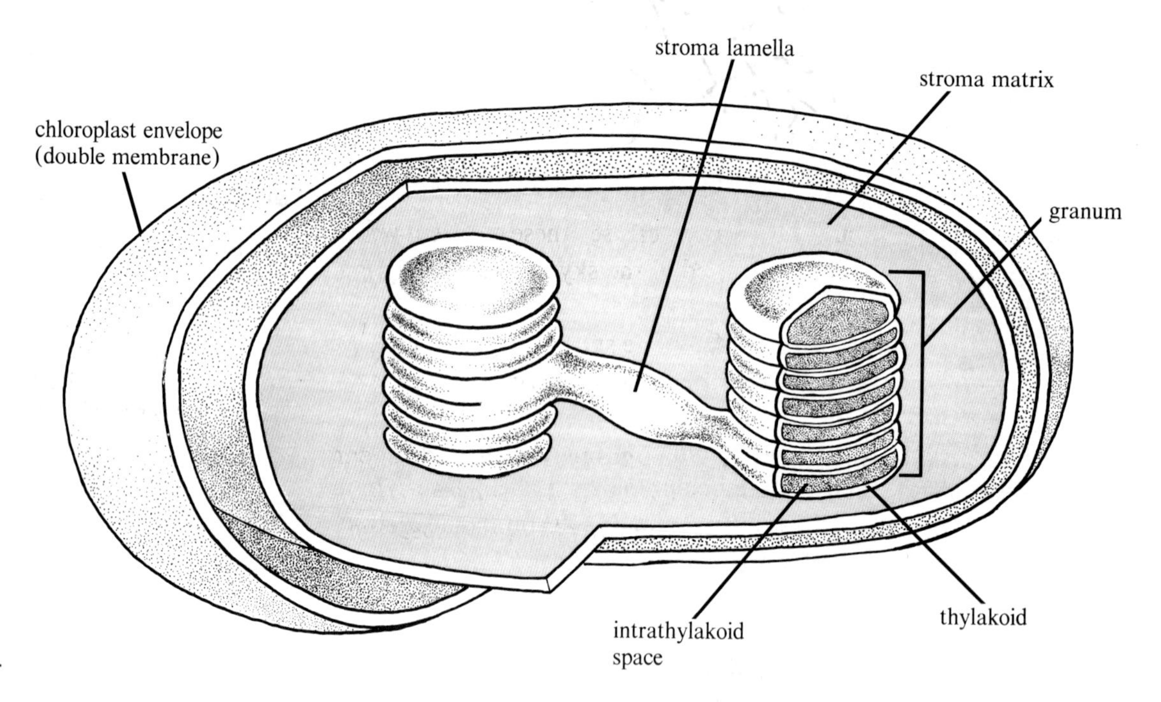

b.

7.26 What radiant energy is used for photosynthesis? The radiant energy from the sun is categorized into different radiation wavelengths ranging from short gamma rays and X rays to the long microwaves and radiowaves (figure 7.28). Record in your logbook the wave type which has the most energy. ________________

All of the categories together are called the *electromagnetic spectrum.* Photosynthesis uses the visible portion of the electromagnetic spectrum. When visible light passes through a prism, it is separated by the prism

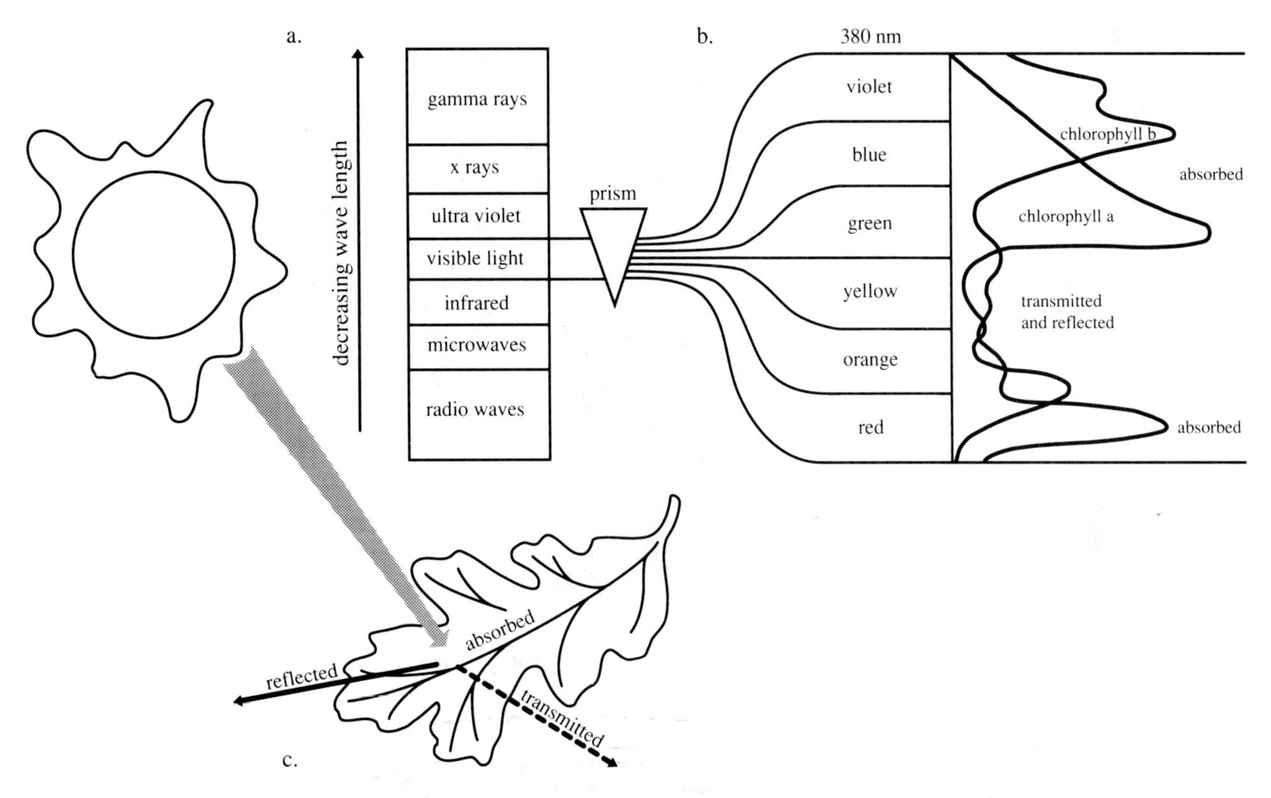

Figure 7.28 Radiant energy. a. Wavelengths of the electromagnetic spectrum. b. Color spectrum of visible light, some wavelengths are absorbed by chlorophyll. c. Leaf color is due to reflection and absorption of specific wavelengths of light by plant pigments, e.g., chlorophyll.

into different wavelengths (figure 7.28*b*). You can see these different wavelengths of energy when a rainbow is formed in the sky by the water prism effect.

Radiant energy comes to the earth from the sun packaged in units of energy called *quanta,* or *photons.* These photons of energy move along a wave pattern.

Plot the data in data table 7.7 on a piece of graph paper found in your logbook. Label the *Y* axis *energy content* and the *X* axis *wavelength.* A nanometer is 1×10^{-9} meters. The graph is called an *absorption spectrum.*

Data Table 7.7 Wavelengths and Their Energy Content

Wavelength (NM)	**Energy (CAL)**
400	71,000
500	57,000
600	48,000
700	41,000

Which color of visible light contains the least amount of energy?

7.27 What is the relationship between wavelength and energy? Figure 7.28*c* shows that only a portion of the visible light is absorbed by plant leaves. Most of the light that strikes the surface of the leaf is *reflected* or *transmitted* through the thin leaf.

The wavelengths of light absorbed and reflected by the leaf is primarily dependent on the pigments in the leaf. **Chlorophyll** is the most abundant pigment in plant leaves. Figure 7.28*b* shows that chlorophyll pigments (type *a* and *b*) absorb certain wavelengths of light and transmit other wavelengths of light. This is called an *absorption pattern.*

Record in your logbook the color (wavelength) least absorbed by the chlorophyll? ________________ Why does a leaf appear green to the human eye?

7.28 *If* light energy for photosynthesis is absorbed by chlorophyll pigments, *then* which wavelength of visible light do you predict would yield the highest photosynthetic activity? ________________.

The lowest photosynthetic activity? ________________.

The amount of photosynthetic activity can be measured relative to the different wavelengths of visible light. The researcher can determine the amount of oxygen given off or the amount of CO_2 used as a measure of photosynthetic activity. A biologist exposed plant leaves to various wavelengths of visible light and measured the amount of CO_2 used. The data from the experiment is recorded in data table 7.8. Plot the data from data table 7.8 on the graph paper found in your logbook. Label the *Y* and *X* axis.

The graph you made is called an *action spectrum.* Compare the action spectrum graph to the absorption spectrum graph in figure 7.28*b,* and answer the following questions in your logbook.

1. Do you accept or reject the prediction you made? __________.
2. In what colors of light should plants be grown to obtain maximal photosynthesis and therefore, maximal chemical energy to do the work of living? ____________ and ____________.

Growing plants under artificial light requires the use of special lamps or a combination of incandescent and fluorescent lamps. An incandescent lamp emits light in the far-orange and red region, and a fluorescent lamp emits light in the blue region of visible light.

Data Table 7.8 CO_2 Used as a Function of Wavelength

Wavelength (NM)	CO_2 Used
380	—
400	73
420	75
440	74
460	62
480	50
500	37
520	31
540	28
560	27
580	30
600	31
620	34
640	40
660	63
680	41
700	11

7.29 What do plant pigments do? When light energy is absorbed by chlorophyll pigments, negatively charged electrons are moved into a different orbital of a higher energy value around the nuclei (figure 7.29*a*). The electrons are said to be "excited." This activity can be observed with isolated chlorophyll molecules.

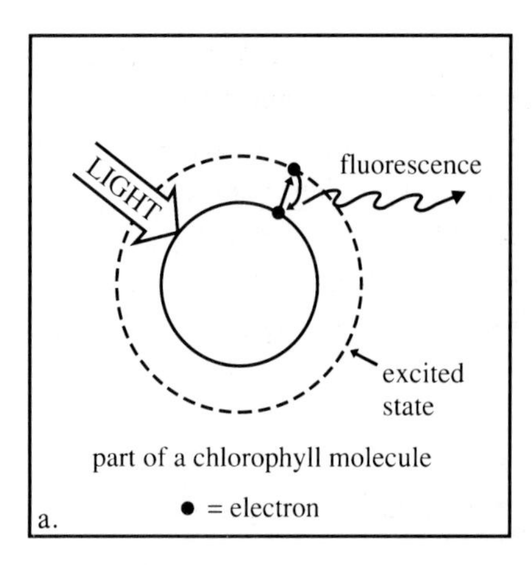

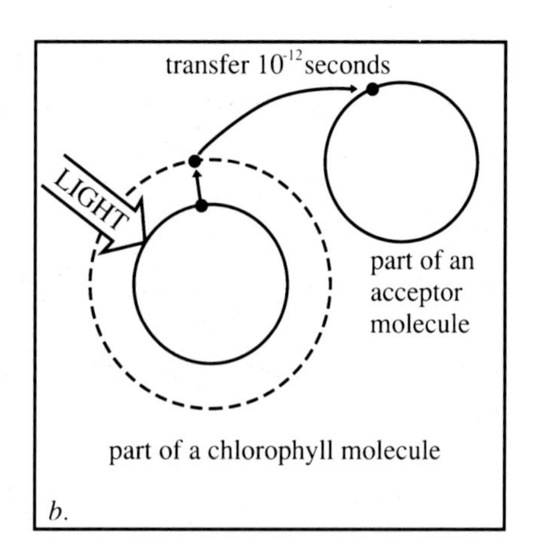

Figure 7.29 Change of a) electron orbital path by the absorption of light energy by chlorophyll and b) electron transfer from chlorophyll to electron acceptor molecule.

Use the following technique to obtain isolated chlorophyll molecules. Keep your chlorophyll solution for use in figure 7.30.

a. Grind some spinach or other available leaves in a mortar with a pestle.
b. Add 80 percent acetone to just cover the leaves. *Keep acetone away from heat or flame.*
c. Wait at least 10 minutes or until the solution becomes dark green.
d. Filter the solution through a paper filter and collect the green solution in a glass container. *Close* this container, label, and keep for the next investigation. The green solution contains chlorophyll pigments dissolved in a *flammable* solvent. *Keep this solution away from heat or flame!*
e. Place a sample of the chlorophyll solution into a small glass test tube and position the tube in the beam of light from a high-intensity incandescent lamp of a binocular microscope or some other source.

1. What colors do you see? ______________________
2. What color is reflected? __________ transmitted? __________

The red color you saw is called *fluorescence.* Fluorescence happens when the excited electrons from chlorophyll pigment molecules return to the nonexcited energy level and give off absorbed light energy as red light (figure 7.29*a*). This is evidence that light is absorbed by chlorophyll pigment.

7.30 The chlorophylls that exist in the plants are partially responsible for photosynthesis. Chlorophyll *a,* a blue-green pigment, and chlorophyll *b,* a yellow-green pigment, make up the bulk of the pigments found in plants. Normally, carotene, a yellow pigment, and xanthophyll, a greenish-yellow pigment, are associated with the chlorophyll *a* and *b* pigments in photosynthesis. The other pigments can be placed under a general heading of anthocyanins. Anthocyanin pigments are particularly common in flowers, fruits and vegetables and give them a red-to-purple coloration.

If these different pigments are present in plant leaves, *then* it should be possible to separate different plant pigments by using paper chromatography.

a. Use the chlorophyll solution from the experiment in section 7.29 or grind some more spinach or other available leaves in a mortar with a pestle.
b. Obtain a strip of chromatography paper. Avoid touching the paper except along the outer edges.
c. Draw a light pencil line 2 cm from the bottom of the paper as shown in figure 7.30*b*.
d. Place a small drop of pigment solution on the line using a capillary tube. Place the dot along the line in the center of the paper.
e. Repeat step *b* after the spot has dried by placing a second drop on the first. This will concentrate the pigment spot. Repeat this process 10 times.
f. Place in 2 mL of solvent (acetone and petroleum, 8 percent: 92 percent) into the chromatography chamber (figure 7.30*a*). *Keep solvent away from heat or flame.*
g. Carefully place the spotted paper into the chamber. The bottom edge of the paper must be inserted into the solvent, but the pigment spot must not be inserted into the solvent.

Be careful that the paper does not touch the sides of the glass tube. Set the tube in a vertical position and record your observations as the pigments begin to separate.

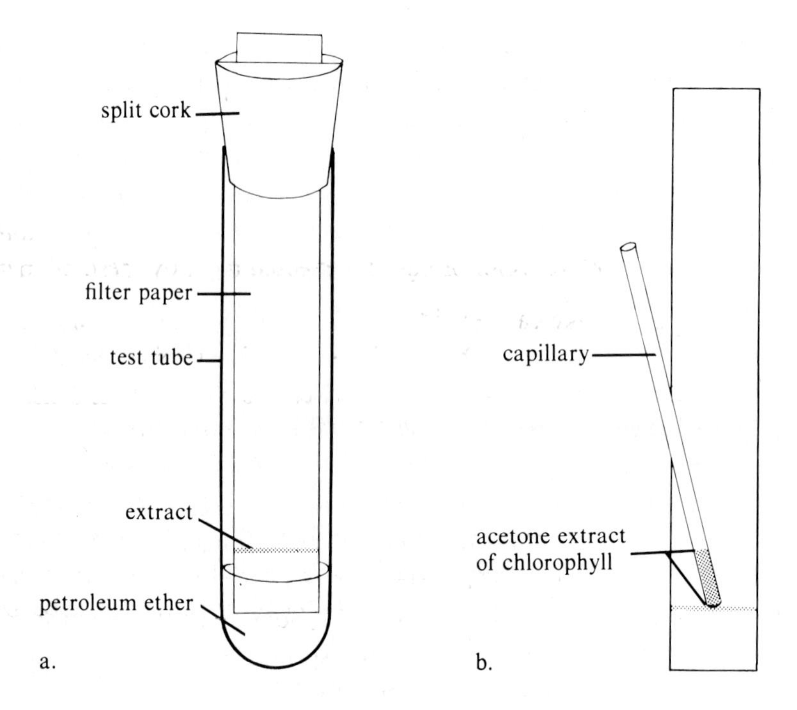

Figure 7.30 Chromatography a) apparatus and b) application technique.

1. How many color bands do you see? ________
2. How many different pigments are evident? ________

Note: (yellow-orange = carotenes, yellow = xanthrophylls, blue-green = chlorophyll *a,* and yellow-green = chlorophyll *b*).

When the chromatogram is complete, label the pigments with a pencil and attach to your logbook.

7.31 Where do you find leaf pigments? The green pigments of the plants are located in the grana membranes of the chloroplasts. While chlorophyll is the most common pigment in plants, the chloroplast contains other pigments. These additional pigments help in photosynthesis and are the yellow carotenes, red phycoerythrins, blue phycocyanins, and brown fucoxanthins. A plant that contains one of these particular pigments would show color dominance over the chlorophylls: (*a*) red kelp (marine algae) contains phycoerythrins and chlorophyll; (*b*) brown fucus (marine algae) contains fucoxanthins and chlorophylls. In leaves, other pigments may be present, but are usually masked (hidden) by the large amounts of chlorophyll pigments.

Chlorophyll pigment molecules break down during the growing season, but are continually being replaced in the leaves. During autumn in the northern hemisphere, the breakdown of chlorophyll exceeds its resynthesis, which results in the unmasking of other pigments present in the leaf. The *carotenoids* will show through as the chlorophyll disappears and the leaves display a yellow or orange coloration. Carotenoids are in the chloroplasts within the grana membranes and will participate in photosynthesis. These same pigments give corn grains, daffodils, egg yolk, and bananas their yellow coloration.

The reds and purples that decorate autumn foliage are due to *anthocyanin* pigments. Which color of light is reflected most by these pigments?

____________ Unlike carotenoids, anthocyanin pigments are not present in most leaves during most of the growing season. These pigments are made in the late summer and are found in the vacuoles of leaf plant cells.

Do you think anthocyanins participate in photosynthesis? ________ Explain in your logbook.

The formation of anthocyanins is dependent on the breakdown of sugars during periods of *bright* light. This process maybe triggered by the loss of phosphate from leaf cells. Anthocyanin formation is enhanced by cool days and chilly, but not freezing nights. Sometimes the edges of young leaves just emerging from the bud in early spring are tinted by anthocyanin pigments in the outer epidermal layers of the leaf.

In your logbook list some fruits that are colored red.

7.32 Have you noticed that the pattern of gas exchange in photosynthesis is the opposite from that observed in respiration? During the day, with sufficient light, photosynthesis exceeds plant respiration. Therefore, the O_2 production of photosynthesis exceeds the O_2 consumption of respiration in plants.

During the day would CO_2 consumption be *less than* or *more than* CO_2 production? ____________

In figure 7.31, note that A and D indicate points where no photosynthesis occurs, but respiration is taking place. Points B and C are places where photosynthesis and respiration are occurring at equal rates, that is, the amount of CO_2 produced by photosynthesis is equal to the amount of CO_2 consumed by respiration. Between points B and C, above the horizontal dashed line, photosynthesis exceeds respiration.

1. Between points A and B and between points C and D, is respiration more or less than photosynthesis? ____________
2. At which point on the CO_2 curve, where photosynthesis is occurring, would the plant eventually starve to death because of low organic fuel food? ____________

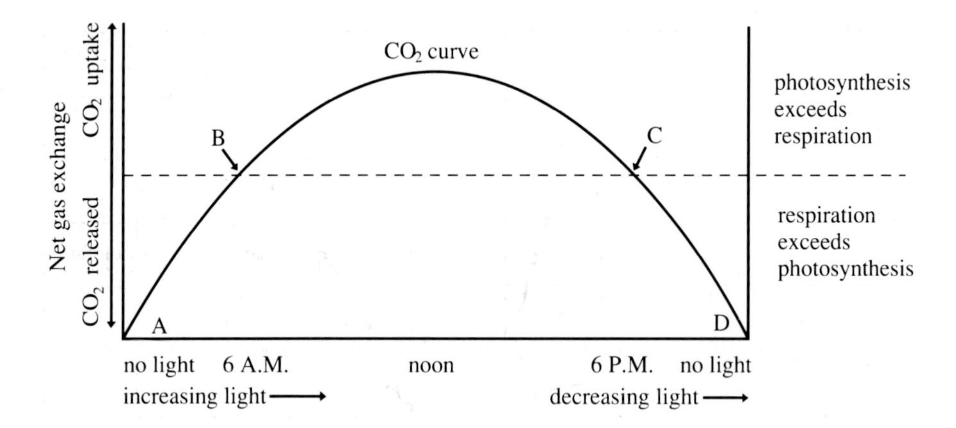

Figure 7.31 CO_2 gas exchange between the leaf and the atmosphere during changes in light intensity.

If you analyze Figure 7.31, it should be evident that during the daylight hours there is excess oxygen produced by photosynthesis. This excess O_2 usually exits the plant through the stomata. While the plant is in the *dark,* no oxygen is produced, but at the same time oxygen is being utilized by respiration.

Respiration and photosynthesis are two key processes used by plants. Let's look closer at photosynthesis.

7.33 The data in figure 7.32 indicates how various factors can affect the rate of photosynthesis in plants. The vertical axis on all graphs represent the relative rate of photosynthesis. You can think of the relative rate as either O_2 production or CO_2 consumption.

1. In general, what effect does light have on the relative rate of photosynthesis? ________________
2. Would there be a greater or lesser rate of photosynthesis during a cloudy day? ________________
3. Would there be a greater photosynthetic rate during sunny days? ________________
4. Does the relative rate of photosynthesis and thus, the relative note of energy production, change with light intensity and the rate of carbon dioxide consumption? ________________
   Explain.
5. What affect does temperature have on the relative rate of photosynthesis? ________________

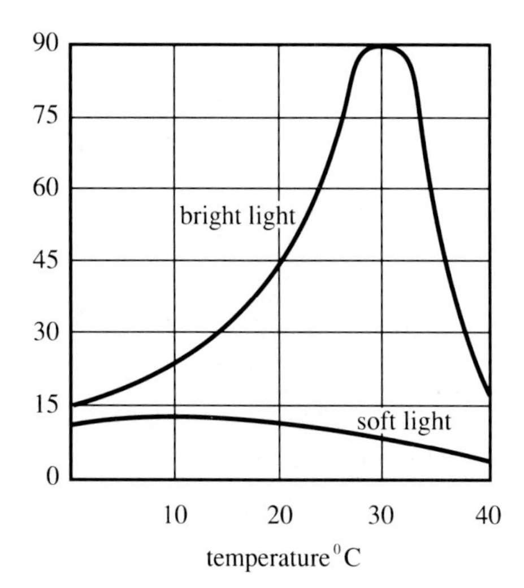

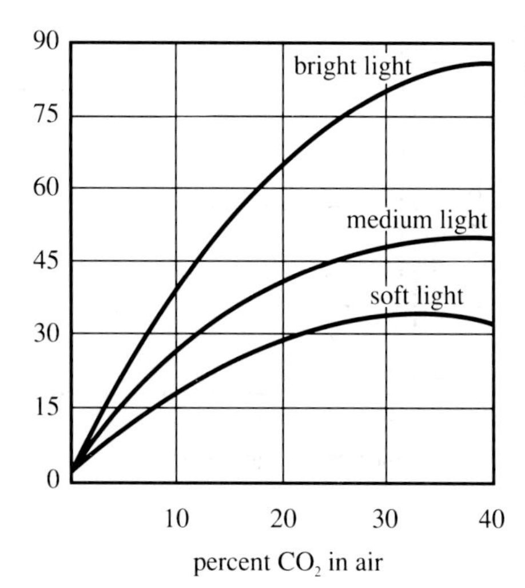

Figure 7.32 Dependence of photosynthetic rate on light, temperature and CO_2.

7.34 Why does photosynthesis decline above 30° C? Inside the chloroplasts the absorbed light energy is transformed into the excited state energy of chlorophyll electrons (figure 7.29). This absorbed energy is transferred to an electron acceptor molecule located next to the chlorophyll molecule in the grana membranes (figure 7.29*b*). This completes the first step in trapping the sun's energy. The second step in the light phase of photosynthesis uses the trapped light energy to make ATP. ATP is made in the chloroplast very much like it is made in mitochondria (figure 7.14). However, oxygen is *not* used in the making of ATP by photosynthesis. In fact, oxygen is produced.

Many herbicides used by farmers to control weed growth contain molecules that interrupt the flow of electrons and stop the production of ATP. Atrazine and diuron are two types of herbicides used to stop ATP synthesis. If these herbicides stop photosynthesis by blocking ATP production, and this stops plants from growing, what does this tell you about the use or role of ATP produced during the light reaction of photosynthesis? ________

____

7.35 An examination of figure 7.21*b* shows that ATP is used to make sugar during the dark phase of photosynthesis. This phase, however, actually occurs in the light.

If CO_2 is being used during photosynthesis, *then* you should be able to measure its disappearance from the environment. Do submerged aquatic plants use CO_2? You recall that CO_2 dissolves in H_2O and forms carbonic acid, which then ionizes to a hydronium ion (H_3O^+) and a carboxylate ion (CO_2^-).

The equation below summarizes this process.

$$CO_2 + H_2O \rightleftharpoons H_2CO_3 \longrightarrow H_2O^+ + CO_3^-$$

Does CO_2 added to water make the solution acidic? ________ Bromthymol blue is an indicator of acidic and basic solutions. Bromthymol blue is a yellow color in a basic solution and a green or blue color in an acidic solution see 7.14.

If you blow CO_2 into water and add bromthymol blue, what will be the color of the solution? ________ If the carbohydrate ion dissolves in water, the carbonic acid that is formed will turn the water a ________ color.

Let's look at CO_2 and the dark phase. If we can show the CO_3^- ion is present in the water, we can suggest it was produced during the dark phase of the plant's respiratory cycle. To test this, follow the steps below.

a. Obtain four large test tubes and fill each one-half full with *boiled distilled* water removes CO_2. Add 3 drops of bromthymol blue until the water just turns green or blue in each test tube, indicating a basic condition. It is absolutely necessary that you use a clean test tube, free from any contaminants, including soap.
Number two of the test tubes #1 and #2. Using a soda straw, gently blow your breath into the two test tubes. When you blow into these test tubes, you are adding a gas from exhalation. Make each tube approximately the same light yellow color. Number the other two tubes #3 and #4. Do not blow your breath into tubes #3 and #4. What gas did you add to the solutions in test tubes #1 and #2? ________________

b. Place a 3-inch sprig of Elodea in all four test tubes. Place tubes #1 and #3 in bright light, and tubes #2 and #4 in complete darkness. *Note:* It may be necessary to leave your experiment set up for 24 hours. At the end of the experimental time, pull the Elodea out of the test tubes and analyze the color of the solutions. When you check your results, hold a white piece of paper behind the test tubes. This will allow you to determine the color of the solution.
Color of solution
Test tube 1 (with sprig in light): ________________________

Test tube 2 (with sprig in dark): ________________________

Test tube 3 (with sprig in light): ________________________

Test tube 4 (with sprig in dark): ________________________

1. Which solution(s) change(s) color? ________________________
2. Which solution changed from a light green or light blue color to a yellow color? ______ (A loss of any color is significant and acceptable.)

3. Is CO_2 being used in test tube 1? ______ tube 2? ______ tube 3? ______ tube 4? ______
4. How is CO_2 being removed from the solution? ____________
5. Is light required to remove CO_2? ________
6. What is the purpose of the third and fourth test tubes? ______

____

Write in your logbook an experiment to answer the following question: Is the Elodea plant required to remove the CO_2 from the water?

7.36 Where is the sugar (starch) produced and stored during photosynthesis? Write a short statement predicting the answer to the above question in your logbook.

7.37 You can show the production of sugar in plants. Obtain a 250 mL beaker and boil 175 mL of water. While the water is heating, number three medium-sized test tubes and fill each tube one-half full of water. To test tube 1, add 5 grams of sugar. To test tube 2, add 5 grams of starch. Tube 3 will only have water in it. Shake well to mix well.

Add 3 drops of IKI solution (Lugol's solution) to each of the three test tubes. If starch is present, then the solution will turn a dark-blue to purple color. Record your results in your logbook.

1. Color of sugar mixture: ____________
2. Color of starch mixture: ____________
3. Color of water mixture: ____________
4. Which of the three samples has the greatest color change?

____

5. What tube contains starch? ____________
6. Does your observation indicate a positive color for starch?

____

Figure 7.33 Sketches of unboiled leaves; a) light grown, b) dark grown.

Figure 7.34 Sketches of boiled leaves; a) light grown, b) dark grown.

Now to the boiling water. Complete the experiment below by following steps *a* through *d*. Safety glasses must be worn while completing the heating portion of this experiment.

a. Take a leaf from each of two coleus plants. One has been kept in the dark and the other has been exposed to light. Mark each leaf some way so you will know which is which. Coleus leaves have visible pigment coloration. Color the pattern of each leaf (figure 7.33) in your logbook and label the one kept in the dark and the leaf kept in the light.
b. Put both leaves in boiling water for 2 minutes.
c. Remove the leaves from the boiling water bath to a petri dish. Cover them with ethyl alcohol and heat carefully on a hot plate for two minutes. *Do not heat alcohol over an open flame!* The chlorophyll pigment will be extracted from the leaves into the alcohol.
d. Now place the leaves in a clean petri dish after you have blotted them dry with a paper towel. Be careful. Do not tear the leaves. You want them to remain flat. Cover the leaves with IKI. Record your results by showing the color pattern of each leaf in figure 7.34.

1. Which leaf contained the most starch? ______________
2. Where is the starch found in the leaf? ______________
3. Explain why is there a difference in the two leaves?

7.38 Photosynthesis is a complex chemical reaction that results in the formation of oxygen. There was a time without photosynthesis on primitive earth and, thus, the Earth's atmosphere lacked O_2. All living things respired anaerobically during this time. The evolution of living things with a photosynthetic capability permitted the accumulation of oxygen in the atmosphere. Life then evolved to the aerobic-cellular-respiration level. Hence, a greater efficiency of energy conservation by cells in an aerobic (O_2) atmosphere.

On the planet earth, there is a ratio of carbon dioxide (CO_2) and oxygen (O_2) that is established by two processes—cellular aerobic respiration and photosynthesis. Both processes are essential for life on Earth. Without photosynthesis, the oxygen and organic compounds on this planet would be depleted by respiration. Therefore, photosynthesis is the essential source of almost all the food energy used by organisms, both autotrophic and heterotrophic organisms. On this planet autotrophic and heterotrophic organisms are found associated together in food chains. The primary source of energy in these food chains is photosynthesis. It has been suggested that without photosynthesis, the food on this planet would be gone in less than 4 years.

At first, long ago on our planet, most of the ultraviolet (UV) radiation light in the solar radiation reached the Earth's surface. Ultraviolet radiation damages the genetic (hereditary) material, nucleic acids, and thus a high level of genetic mutations will occur. The release of oxygen by photosynthesis resulted in the formation of the ozone layer (O_3) in the atmosphere. Then, as the amount of readily available oxygen increased to form the ozone level, the amount of ultraviolet radiation decreased. The ozone layer, thus, provides a protective shield to the living organisms on this planet.

Today, the amount of ozone in the Earth's atmosphere is being reduced. List some of the ways ozone is being reduced.

7.39 This has been a long, involved chapter. A lot of material has been covered. Therefore, careful review is necessary. Look at the study points below. Check (✓) the box in the logbook if you have mastered the contents that reflect on the main idea of each study point.

On completion of this chapter, you should be able to do the following:

☐ 1. Have a working definition of the various types of energy.

☐ 2. Distinguish between autotrophs and heterotrophs.

☐ 3. Distinguish between herbivores, carnivores, and omnivores.

☐ 4. Identify the two main ways cells obtain ATP.

☒ 5. Distinguish between anabolism and catabolism.

☐ 6. Distinguish between aerobic and anaerobic respiration.

☒ 7. Show understanding of how aerobic respiration conserves the energy of food molecules via glycolysis, Kreb's cycle, and the electron transport system.

☐ 8. Discuss the fermentation process.

☐ 9. Measure processes that show organisms are engaged in cellular respiration.

☐ 10. Determine the role of photosynthesis as a supplier of energy.

☐ 11. Contrast and compare the light cycle and the dark cycle.

☐ 12. Demonstrate the procedures needed to show where photosynthesis occurs.

☐ 13. Discuss the components of radiant energy and how they apply to photosynthesis.

☒ 14. Identify the various pigments found in chlorophyll.

☐ 15. Relate the effect of environmental factors on photosynthesis.

☐ 16. Identify where sugar is produced in plants.

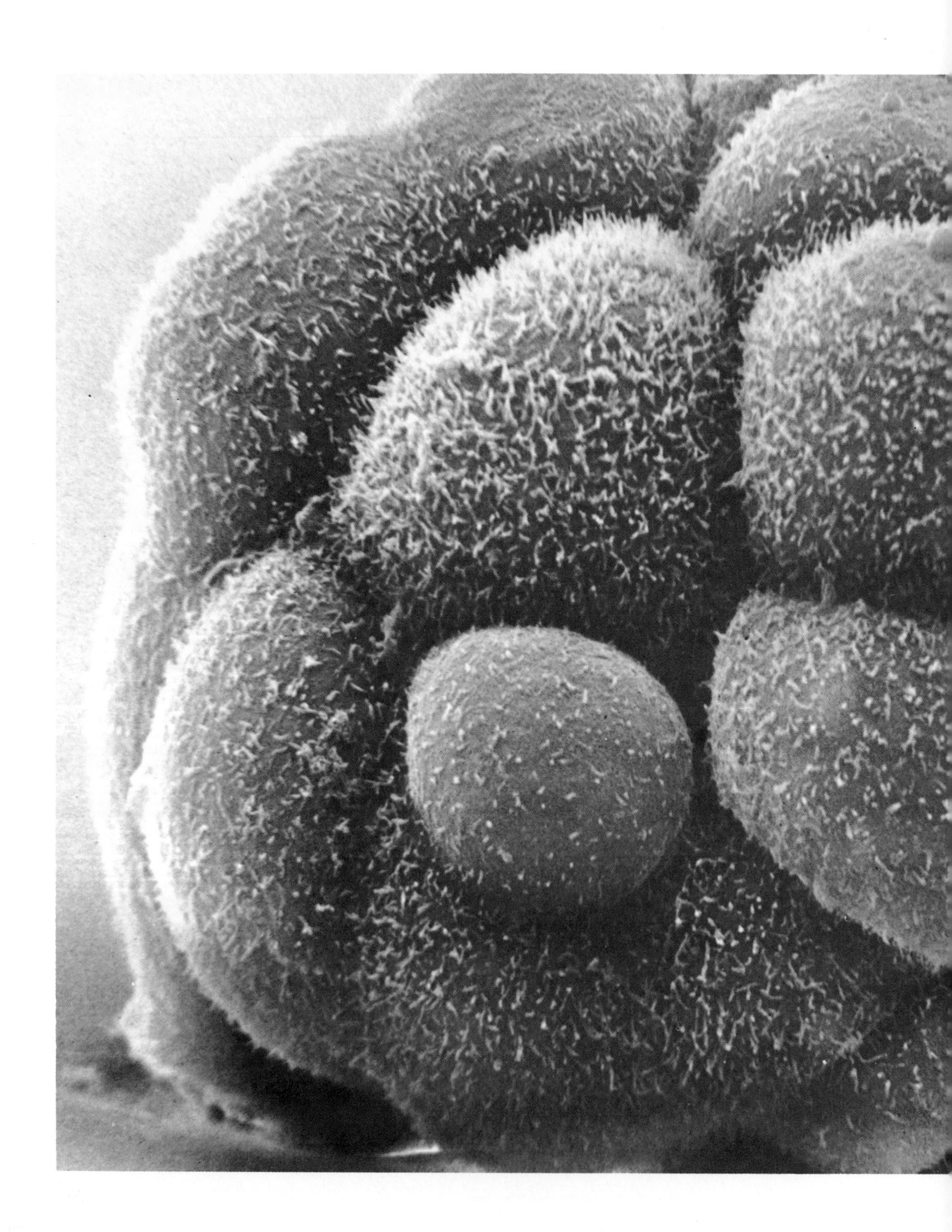

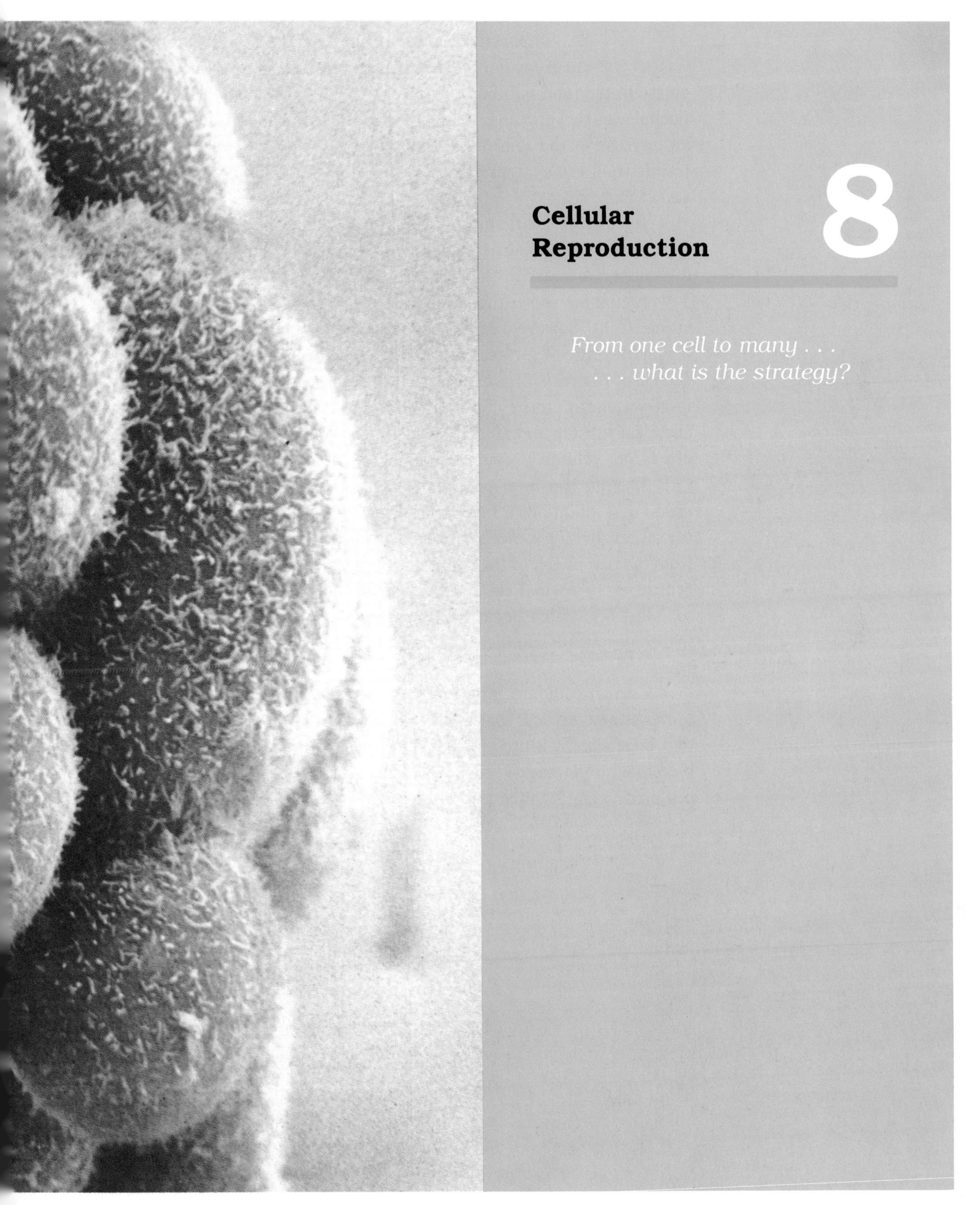

8 Cellular Reproduction

From one cell to many . . .
. . . what is the strategy?

8.1

An important characteristic of living organisms is the ability to grow. For a single-celled organism, growth produces an increase in the total number of individual cells in the population. In a multicellular eukaryotic animal or plant, growth is achieved, in part, through an increase in the total number of cells. Growth is a cellular event that results from the synthesis and assembly of nucleic acids, lipids, proteins, and carbohydrates.

In eukaryotic organisms, there are two types of cellular growth. One type produces an enlargement of individual cells and the other type, cell division, an increase in the number of cells.

Figure 8.1 shows the difference between cell enlargement and cell division. Cells, however, cannot enlarge forever because as the cell grows bigger, the ratio of surface area (plasma membrane) to volume (cytoplasm) decreases (figure 8.2). Less plasma membrane surface area relative to the total volume of the cytoplasm would eventually reduce the amount of chemical substances entering the cell relative to the total metabolic needs of a larger volume of cytoplasm.

Therefore, a brief period of cell growth is often followed by *cell division* that maintains a favorable surface-area-to-volume ratio. A key feature of cell division is the faithful transmission of genetic information (genes) from the parent cell to the two daughter cells. In other words, the daughter cells of every normal cellular division are genetically identical, or almost so. Most organisms begin life as a single cell that may give rise to 100 trillion genetically identical cells, the number estimated for an adult human being.

Correct cellular reproduction ensures that new cells will be identical to those they replace: that muscle cells will produce new muscle cells, and that bone cells will produce new bone cells. If this were not true, the replacement of damaged tissue would be impossible. The strategem for new cell production is called the cell cycle.

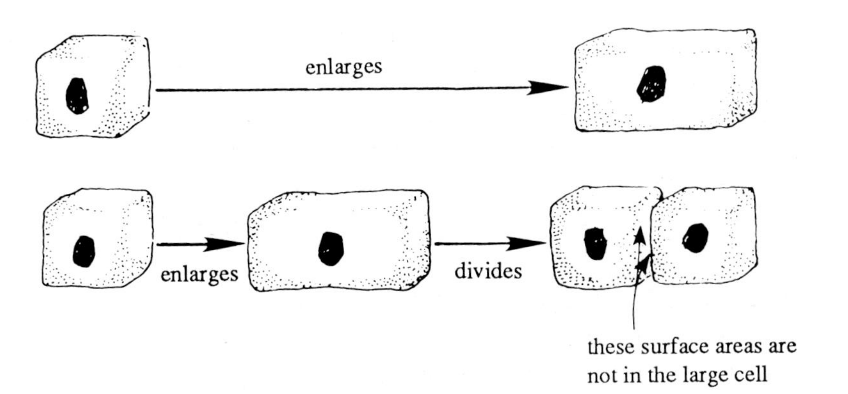

Figure 8.1 Cell enlargement and cell division. Courtesy of W. B. Saunders Company and Biological Sciences Curriculum Study, Inc.

	radius	surface area [A]	volume [V]	$\frac{A}{V}$
	4 cm	201.06 cm^2	268.08 cm^3	0.75
	2 cm	50.26 cm^2	33.51 cm^3	1.50
	1 cm	12.57 cm^2	4.19 cm^3	3.0

Figure 8.2 Relationship of surface area to volume.

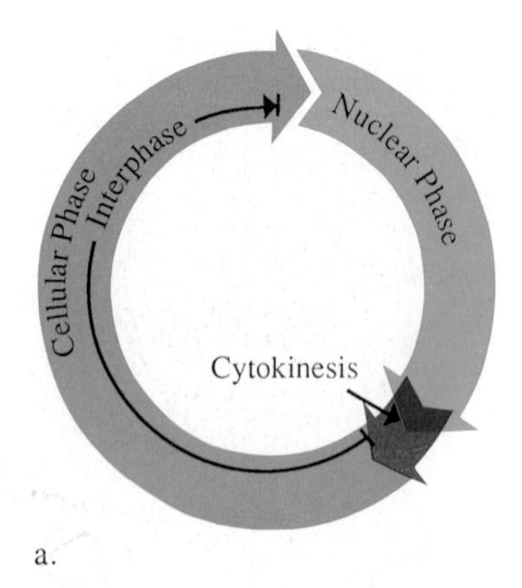

a.

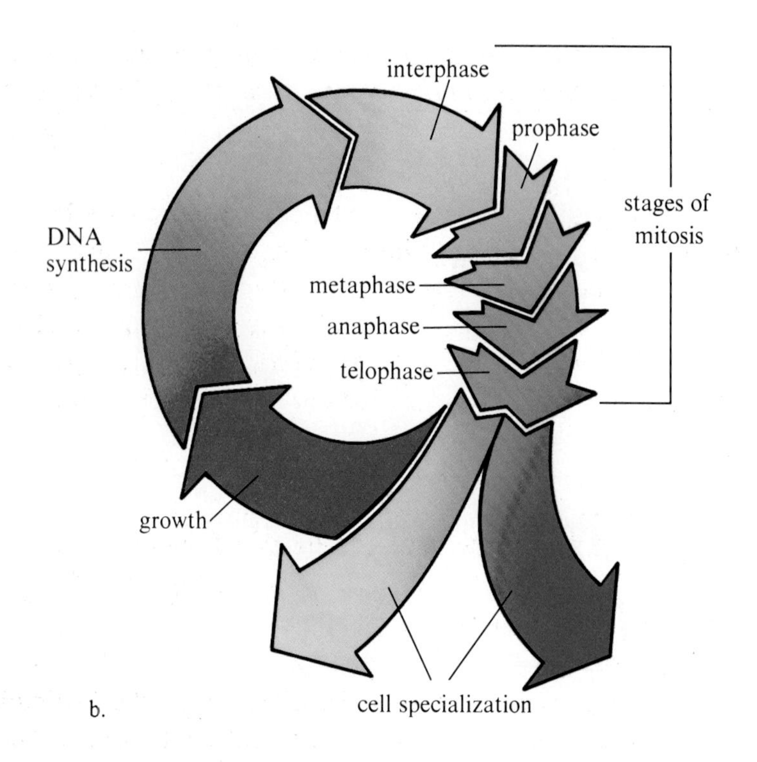

b.

Figure 8.3 The eukaryotic cell cycle. a. The cycle consists of a nuclear phase and a cellular phase. Note the cellular phase includes cytokinesis. b. Events that are part of the cellular phase (growth and DNA synthesis) and the nuclear phase (interphase and mitotic stages). Note that cytokinesis results in two new daughter cells which may develop specialized activities.

8.2

The cell cycle has two main events or phases—a *nuclear* phase and a *cellular* phase. The cellular phase includes the enlargement of cells and the division (**cytokinesis**) of the cytoplasm. The nuclear phase is called **mitosis.**

The cell cycle involves a doubling of cellular parts, especially DNA, which is followed by mitosis (nuclear phase) and cytokinesis (cellular phase). Mitosis and cytokinesis are actually two overlapping processes which results in two new cells (figure 8.3*a, b*).

Mitosis and cytokinesis are reproductive processes that (*a*) add new cells to preexisting ones during the growth of a multicellular plant or animal from the embryonic stage to the adult stage, and (*b*) maintain the total number of cells making up the adult multicellular organism. Cellular reproduction is occurring in your body at many locations at this very instance to replace cells that are being destroyed. It is a continual process that ensures enough cells will be around to perform the functions of life.

In higher plant cells (eukaryotic), the formation of new cells is limited to special regions called **meristems.** Two of the major meristematic regions are the shoot tip and root tip. In addition, some higher plants have lateral meristems that increase girth through their cell reproduction activities.

A cell spends most of its time in the interphase: two growth phases (G_1 and G_2) and a DNA synthesis period (figure 8.3). The **interphase** designates the time before and after mitosis and cytokinesis, that is, the interphase defines the state of the nucleus between mitotic divisions.

8.3 The interphase is characterized by a nuclear envelope, nucleolus, chromatin, and chromosomal duplication. In figure 8.4 of your logbook, locate and label these nuclear interphase features.

8.4 Observe an interphase cell with the microscope by first obtaining a prepared slide of onion root tips. Use the low-power objective of your microscope for your initial study to determine the outline features of the root tip. With a root tip in focus at low power, change to the high-power objective and look for an individual cell that appears to be in interphase. In your logbook describe the features that characterize the onion root interphase cell? ______________________________

____

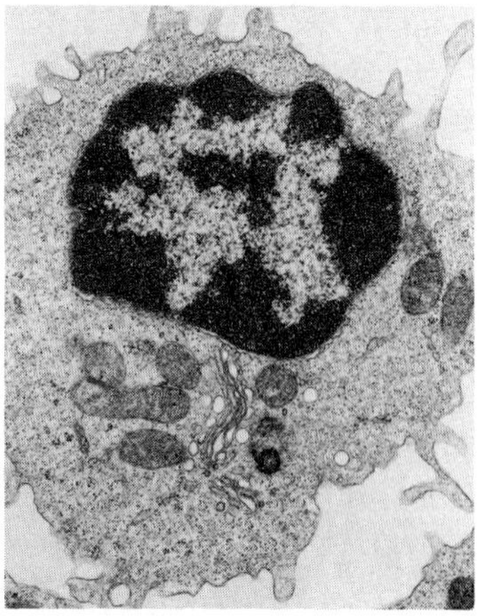

Figure 8.4 Interphase nucleus.

In summary, the nucleus spends most of its time in the interphase genetically managing the activities of the cell and preparing for the mitotic phase by synthesizing (duplicating) DNA.

8.5 The process of mitosis is the strategem for the distribution of genetic information from the parent cell to the two daughter cells. The duplicated DNA of the duplicated chromosomes is dispersed by the mitotic process (figure 8.5). Observe that the first large cell is the parent cell and the two smaller cells are the daughter cells of nuclear and cellular

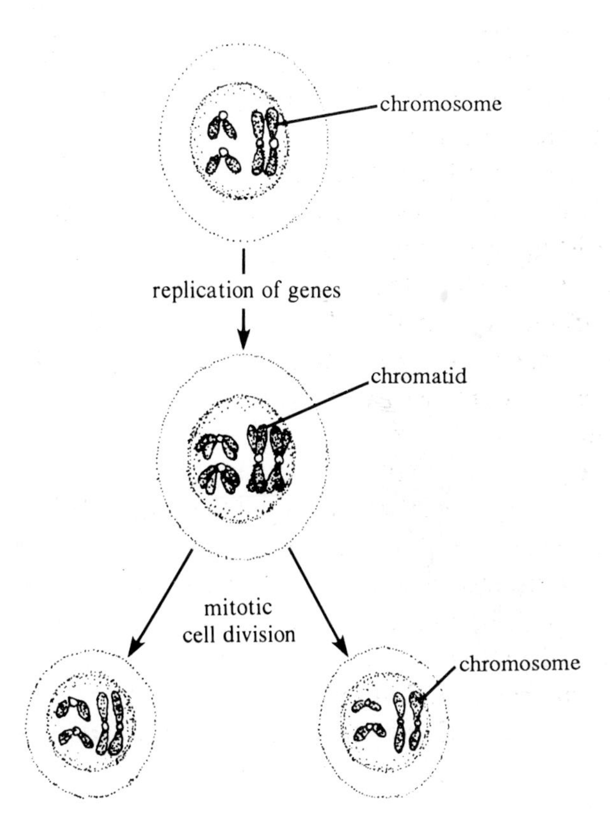

Figure 8.5 Chromosome duplication and mitotic cell division.

division. Each daughter cell in figure 8.5 contains a nucleus with four chromosomes, the same as the parent cell.

It is during the interphase that each parent cell doubles its DNA and chromosome number. This is marked in figure 8.5 as replication (doubling) of genes. The larger middle cell in figure 8.5 has twice (2×) the DNA, and each duplicate chromosome is now called a **chromatid.**

8.6

Mitosis consists of a sequence of four continuous but different phases. The four phases, in the order of occurrence are **prophase, metaphase, anaphase, and telophase** (figure 8.6).

When a cell enters prophase, the nucleus changes its appearance. The nuclear envelope begins to break down, the chomatin starts to condense into threadlike structures called chromatids, and the nucleolus slowly disappears. At the end of prophase, the nucleolus and nuclear envelope are all gone. Place a prepared slide of an onion root tip on the microscope stage and using figures 8.6 and 8.7, locate one or two onion root tip cells in prophase.

8.7

Metaphase marks the time when the chromatid pairs (duplicate chromosomes) are aligned at the approximate equator of the cell (really the equator of the mitotic spindle, which will be described later). At this stage of mitosis, the chromosomes are clearly seen as a pair of attached chromatids. The place where the two sister chromatids are attached is called a **centromere** (figure 8.6). This structure plays a role in chromatid movement. Each chromatid contains a copy of parent DNA. Locate one or two metaphase cells in the onion root tip.

8.8

The next phase, **anaphase,** is recognized by the separation of the sister chromatids from each other. Each chromatid becomes an independent structure during anaphase and, therefore, we begin naming them chromosomes again. Locate an onion root tip cell in anaphase. Record in your logbook how you identify an anaphase cell from a metaphase cell and an interphase cell?

8.9

Telophase is the final phase of mitosis and marks the final distribution activity of the chromosomes. In telophase, the chromosomes change from a condensed threadlike appearance to a dispersed non-threadlike appearance. Note there are two chromosomal masses

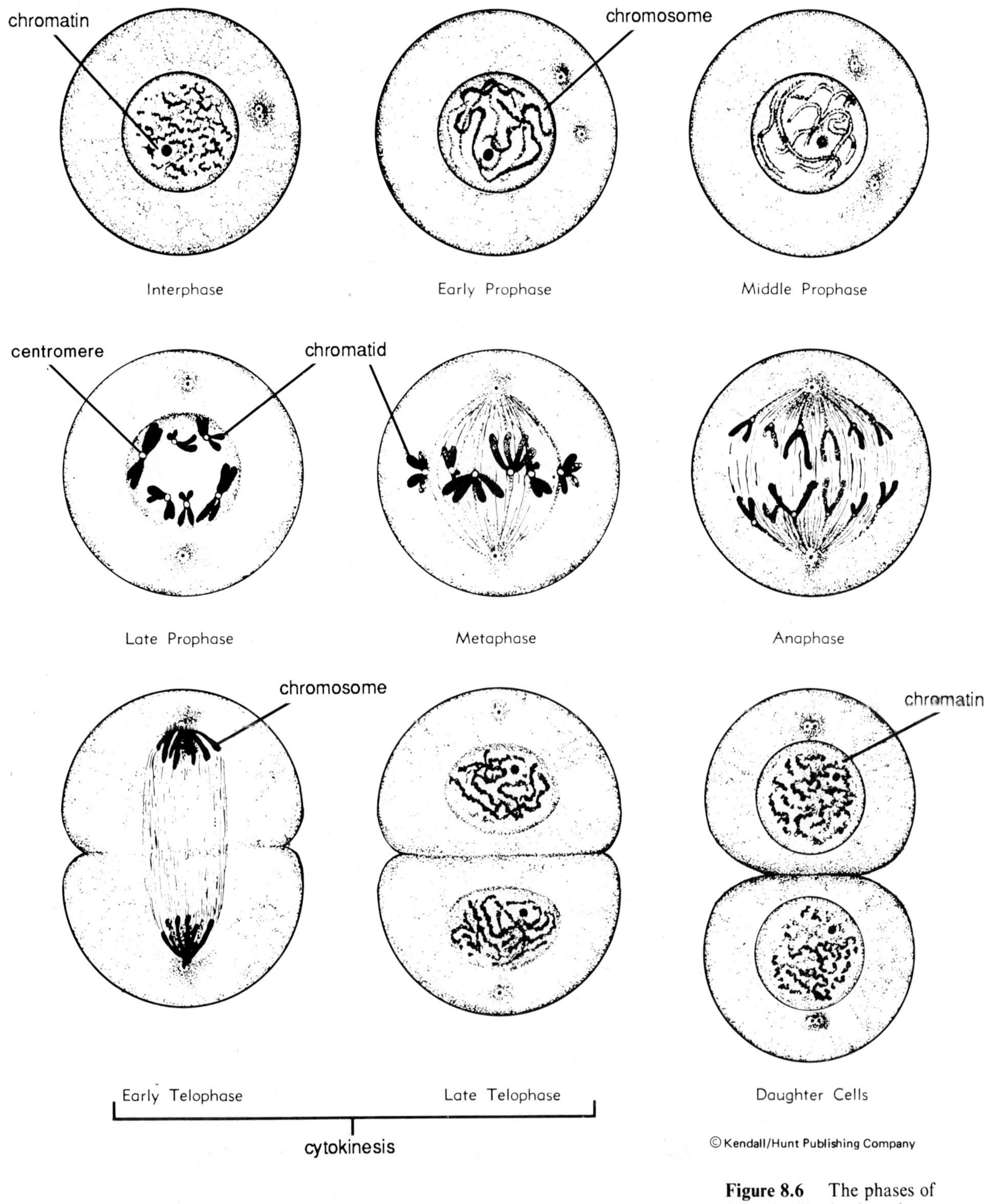

© Kendall/Hunt Publishing Company

Figure 8.6 The phases of animal cell mitosis and cytokinesis. (o = centromere)

Figure 8.7 The phases of plant cell mitosis and cytokinesis.

in the cell. It is near the end of telophase that the nuclear envelope is resynthesized around each of the chromosomal masses. During this restoration of the nuclear envelope, the chromosome mass is reorganized into an arrangement called *chromatin.* Have you noticed that we have used several names for the same genetic information of the cell—DNA, chromosome, chromatin, and chromatid? Be sure you understand when to use the proper word.

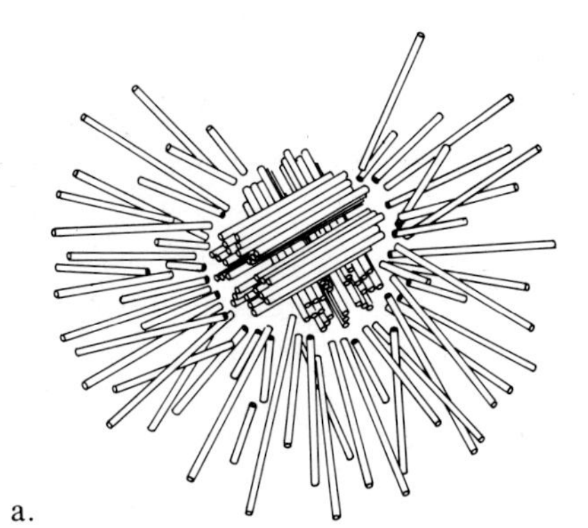

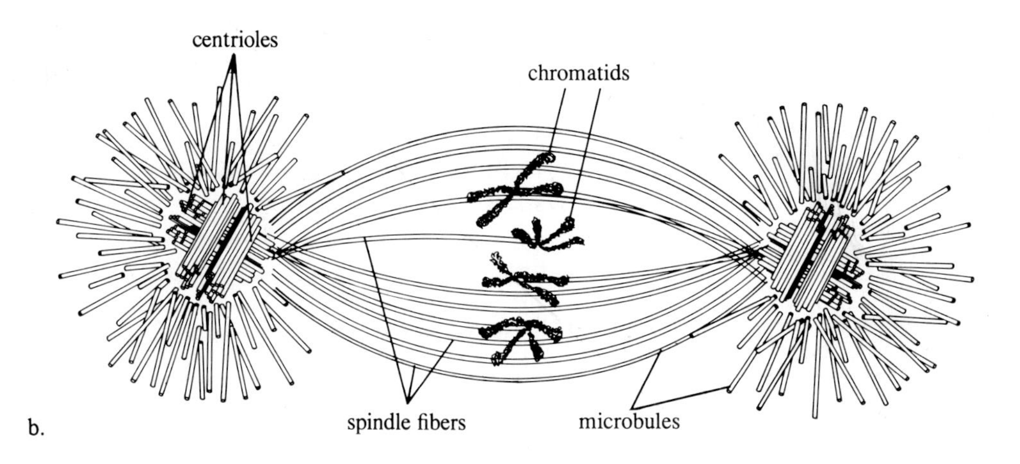

Figure 8.8 Animal cell mitosis. a. Aster. b. Spindle apparatus.

Also during telophase, *cytokinesis* usually starts and is marked in animal cells by a furrow (figure 8.6) and is marked in plant cells by a cell plate (figure 8.7). The furrow in animal cells pinches the parent cytoplasm into two daughter cells. In plants, the cell plate is formed in the approximate equatorial plane of the mitotic spindle, and this marks the synthesis of the new cell wall for each daughter cell.

Locate in the onion root tip a telophase cell and a cell in cytokinesis.

8.10 Be prepared to identify the stages of mitosis in an onion root tip using prepared slides, drawing or photographs.

8.11 Let us take a closer look at how chromosomes move. A structural rigging is used to (*a*) align the chromosomes (a pair of chromatids) at the equatorial plane during metaphase, and (*b*) separate the chromatids during anaphase and telophase. The rigging is made of (*a*) **microtubules** (some of which are called spindle fibers), and (*b*) **centrioles** (only in animal cells and some mobile plant cells). This rigging plus the chromosomes (or chromatids) is called the spindle apparatus (figure 8.8*b*).

The two centrioles are duplicated during the interphase portion of the life cycle (figure 8.6). Therefore, at the beginning of mitosis, there are two sets of centrioles (two per set). The two sets separate from each other during prophase, at the time the spindle microtubules begin to form. Finally, the spindle fibers extend pole-to-pole in the parent cell and a set of centrioles surrounded by microtubules is located at each pole. The centrioles and these associated microtubules mark the **aster** (figure 8.8*a*).

At the start of metaphase (figures 8.6 and 8.7), some of the spindle fibers attach to the centromere of each chromatid, and the pairs of chromatids are moved to a position along the equatorial plane of the spindle. The spindle elongates slightly, and sister chromatids separate from each other in anaphase. The spindle fibers are also part of the mechanics that distribute the chromosomes toward the asters during telophase.

8.12 Use the information from figures 8.6, 8.7, and 8.8 and from your study of onion root tip cells to number and name the events in data table 8.1 of your logbook. Number the events in mitosis in the order in which they occur.

Data Table 8.1 Activities During Life Cycle of a Cell

Event	Number		Name
Chromosomes not visible as distinct structures in the nucleus of the cell.	1	*a.*	interphase
Cell plate forms in center of cell, beginning the division of the cell that follows mitosis.	____	*b.*	____
Centromeres lined up on metaphase plate and attached to spindle fibers.	____	*c.*	____
Centromeres divide, allowing chromatids (now called chromosomes) to separate and move in opposite directions.	____	*d.*	____
Chromosomes become visible in the nucleus, in a shorter, thicker (replicated) state.	____	*e.*	____

8.13 Use figure 8.9 to complete data table 8.2 in your logbook. This activity focuses you on a comparison of the amount of DNA and the number of chromosomes during the life cycle of a cell. Note that G_1 and G_2 are two periods of cell growth during interphase (see figure 8.3).

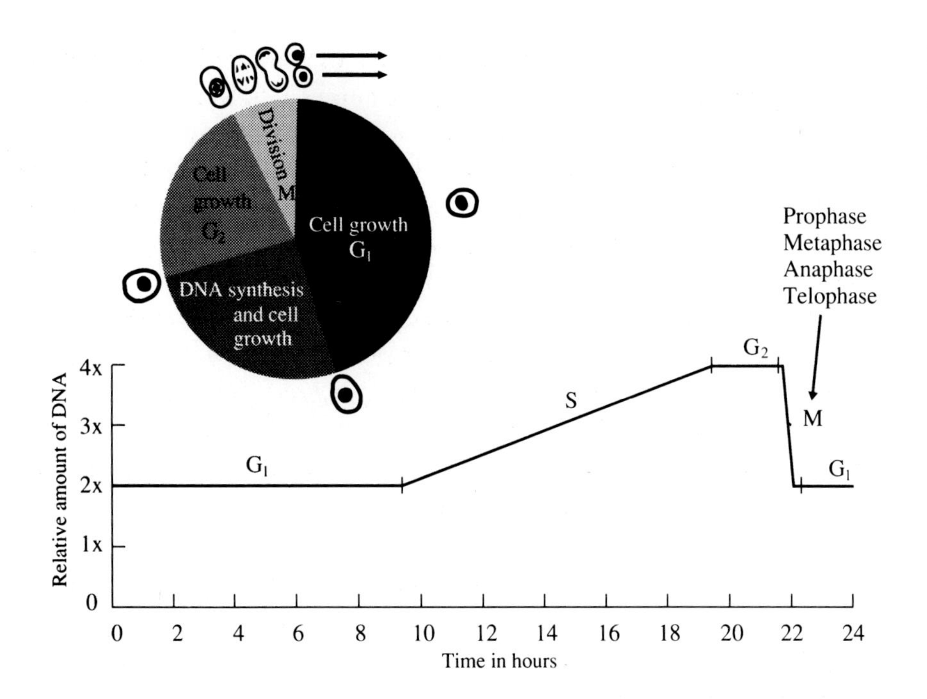

Figure 8.9 Comparison of total DNA and chromosome number during the life cycle of a cell. Courtesy of W. B. Saunders Company and Biological Sciences Curriculum Study, Inc.

Data Table 8.2 Interphase and Mitotic Events

		Relative Amount of DNA/ Cell	Number of Chromatids/ Chromosome	Relative Number of Chromosomes/Cell
Interphase	G_1	2×	1	2n
	S			
	G_2			
Prophase				
Metaphase				
Anaphase				
Telophase				

In summary, mitosis is the strategem used to distribute identical genetic information (genes) from a parent cell to two daughter cells. Thus, each new cell produced by nuclear and cellular division receives a copy of the parents DNA.

8.14 How do identical twins happen? Study figure 8.10 and prepare a written answer in your logbook to this question. How do identical human twins develop from a single fertilized egg?

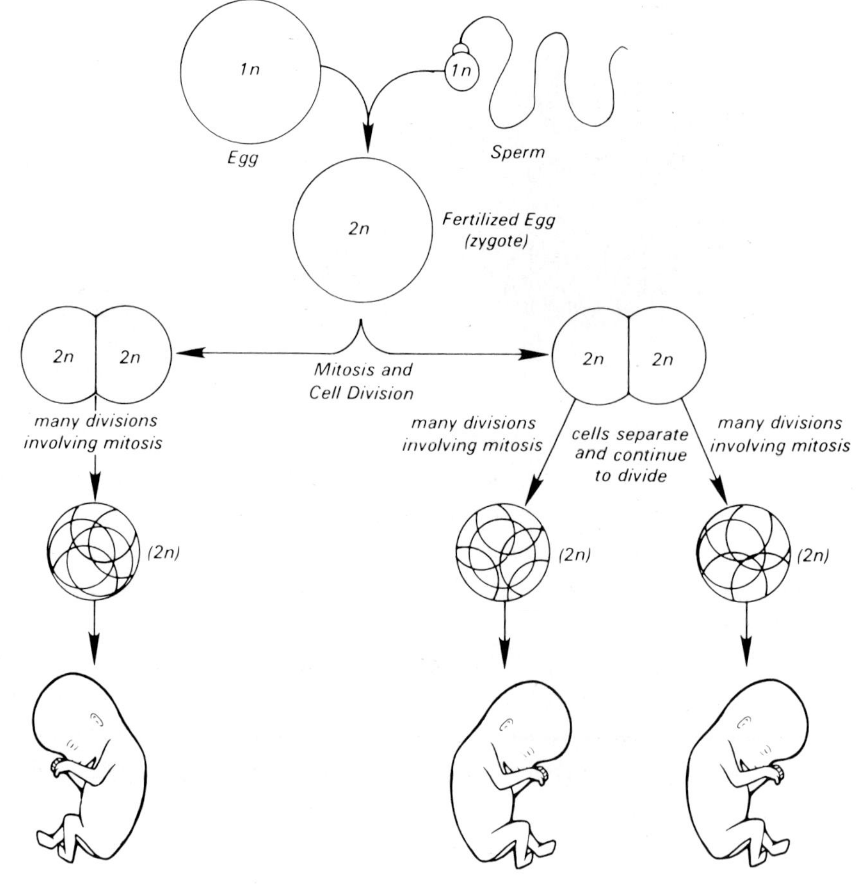

Figure 8.10 Mitotic process for the development in humans of single infants and identical twins. Courtesy of W. B. Saunders Company and Biological Sciences Curriculum Study, Inc.

Asexual and Sexual Reproduction

One of the critical steps in any evolutionary scheme (model) of living systems is the development of strategies for growth and reproduction. In order to perpetuate life, cell division must occur, and it must occur in such a way as to ensure the distribution of essential genetic messages onto new *daughter cells.*

There are two basic types of cellular reproduction that ensure the formation of new cells. They are asexual and sexual.

a. **Asexual.** Maintenance of genetic information; each cell is genetically like its original. The process is called **mitosis.** Living things use mitosis for asexual reproduction.

b. **Sexual.** Production of a new combination of genetic information in which there exists the potential to adapt to the environment and survive. The process is called **meiosis.**

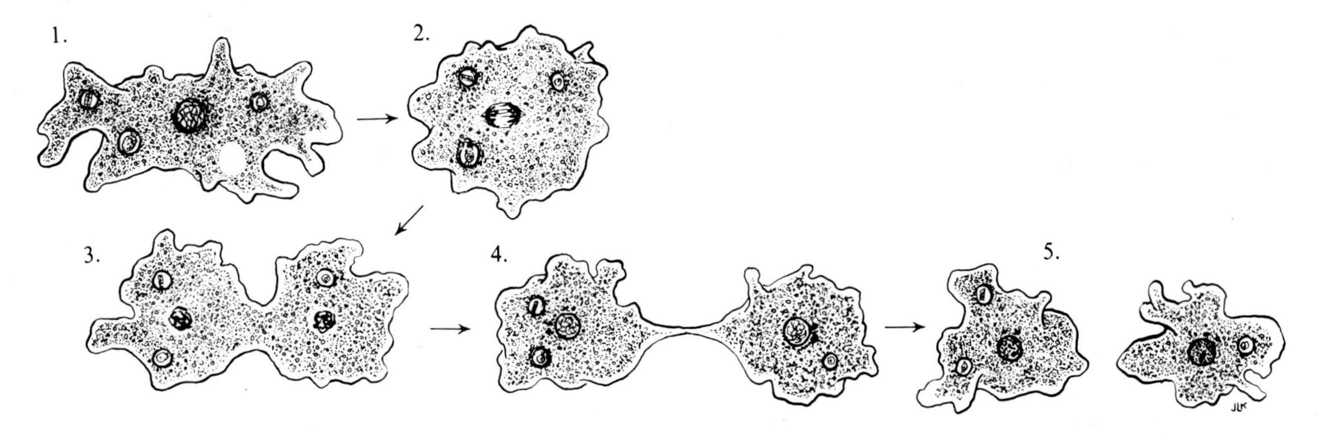

Figure 8.11 Asexual reproduction of an amoeba by fission. Courtesy of James Koevenig.

8.15

An example of *asexual,* or *vegetative,* reproduction is when the parent cell divides into two or more parts, each of which becomes a new individual or cell. This form of reproduction is common to all plants and lower-order animals. Division of the organism or cell is preceded by mitosis.

Asexual reproduction can occur in single-celled organisms or multicelled organisms. In single-celled organisms, such as the amoeba, the most common form of asexual reproduction is called **fission.** Through fission, the organism merely divides into two parts. Figure 8.11 represents an amoeba reproducing by fission.

A second form of asexual reproduction in single-celled and also in multicellular organisms is called **budding.** Budding results in the formation of a similar organism. After the new organism has developed, it will pinch off from the parent organism to become an independent organism. An example of budding can be seen in figure 8.12*a.* This is a hydra that is budding. The hydra is a multicellular organism. A second example of budding is illustrated in yeast cells (figure 8.12*b*) and Kalanchoe plants (figure 8.12*c*). Yeast cells are single-celled. The Kalanchoe is multicellular.

A third form of asexual reproduction is found in plants only and can be referred to as being strictly **vegetative.** Many plants store food underground in the form of tubers. An example of a tuber is a potato. Each tuber on a potato plant is capable of producing several new plants.

Other forms of vegetative reproduction include **stolons, rhizomes,** and **corms** (figure 8.13). Runners (stemlike branches) found above ground that are capable of forming new plants are stolons. Underground, runners are called rhizomes, and many of the garden flowers come from corms. These forms of reproduction are asexual reproduction.

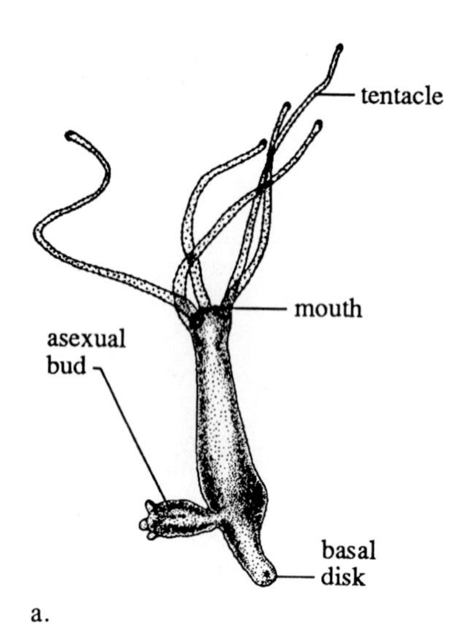

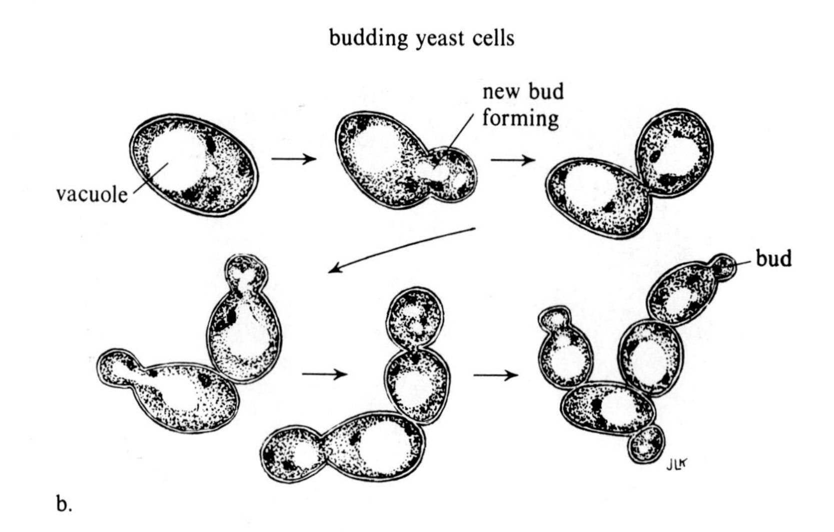

Figure 8.12 Asexual reproduction by budding in a. Hydra and b. Yeast. c. Vegetative asexual reproduction in kalanchoe plants. Courtesy of James Koevenig.

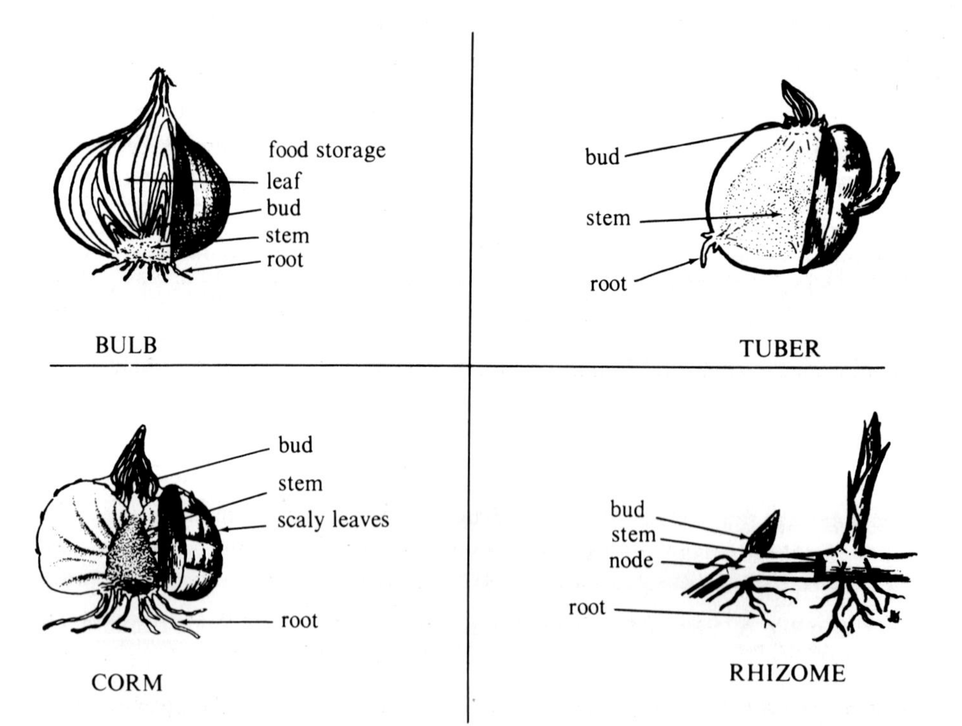

Figure 8.13 Various kinds of asexual reproduction in plants. Courtesy of James Koevenig.

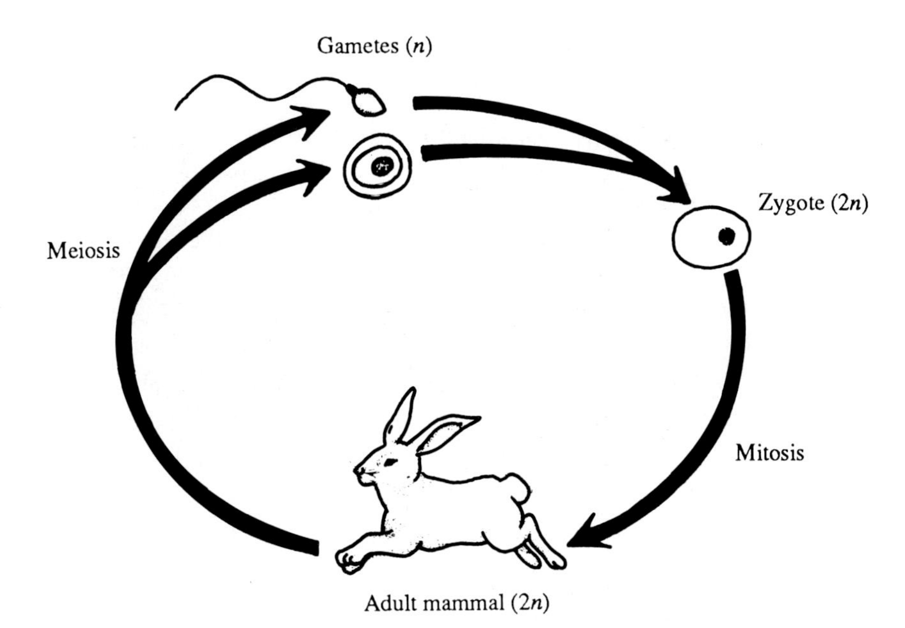

Figure 8.14 Sexual reproduction cycle in mammals.

8.16

In **meiosis,** or sexual reproduction, the production of new cells or some organelles comes from the bringing together of 2 sets of chromosomes. The number of chromosomes per set is different in different organisms. Human beings have 2 sets of chromosomes with each set containing twenty-three chromosomes. How many total chromosomes do you have in each of your cells, gametes excluded? ______ If the fruit fly, *Drosophila melanogaster,* has a total of eight chromosomes per cell, how many chromosomes are there per set? ______ Each cell that contains 2 sets of chromosomes is called **diploid** and has $2n$ number of chromosome sets (N or n = a set of chromosomes). If a cell has 1 set of chromosomes, it is called a **haploid** ($1n$) cell. How many human chromosomes are there in a human haploid cell? ______ Where are these chromosome sets located in a cell? ______

8.17

A haploid cell also is called a **gamete.** Both plants and animals produce gamete (haploid) cells that contain 1 set (1n) of chromosomes. There are two types of gametes produced by plants and animals: sperm and egg. These are shown in the life cycle of a mammal (figure 8.14. How many chromosomes are found in the nucleus of a human sperm gamete? ______ How many chromosome cells in an egg gamete?

______

8.18 When the two types of gametes (egg and sperm) are joined together, a different cell is produced, called a **zygote.** How many chromosome sets are contained in a zygote? ______ Is the zygote diploid or haploid? ______

Chromosomes, as you already know, are chemical aggregates found in the nucleus of eukaryotic cells and carry the genetic information (genes) for what a cell looks like (**phenotype**) and what a cell does (functions). In a multicellular organism, like yourself or a tree, all the cells together determine what the organism does and looks like. Within each diploid cell, there are 2 sets of "look-alike" chromosomes that carry genetic information for the same trait (that is, for eye color, leaf color, and so forth). These look-alike chromosomes are genetic **homologues** of each other. Diploid cells, therefore, contain homologous pairs of chromosomes. If you were asked, "How many homologous pairs of chromosomes are found in a diploid human cell?" you might say "twenty-three." You would be almost correct in your answer except for 1 pair that is not homologous. This is the so-called sex pair of chromosomes (*xx* in females; *xy* in males). The male pair do not quite look-alike. A zygote then contains 2 sets of homologues (chromosomes): 1 set of homologues is *inherited* from the sperm, and the other set of homologues, or look-alikes, is *inherited* from the egg.

8.19 Let us take a closer look at meiosis and why it is an important strategem in the living world. Figure 8.15 shows the effects on chromosome number if (*a*) mitosis is the mechanism for chromosome distribution or if (*b*) meiosis is the mechanism for chromosome distribution in the sexual part of the life cycle of mammals. Calculate the effect on chromosome number using both model *a* and *b* of the chromosome distribution pattern. Write the proper number by each question mark in figure 8.15 in your logbook.

Which model is correct? ____________ Write a short statement in your logbook explaining your selection.

8.20 An important event in meiosis is the decrease in the number of chromosome sets from 2 sets to 1 set ($2n$ to $1n$) per cell. In human beings, this results in (*a*) forty-six chromosomes reduced to twenty-three chromosomes and (*b*) 2 chromosome sets reduced to 1 chromosome set per cell.

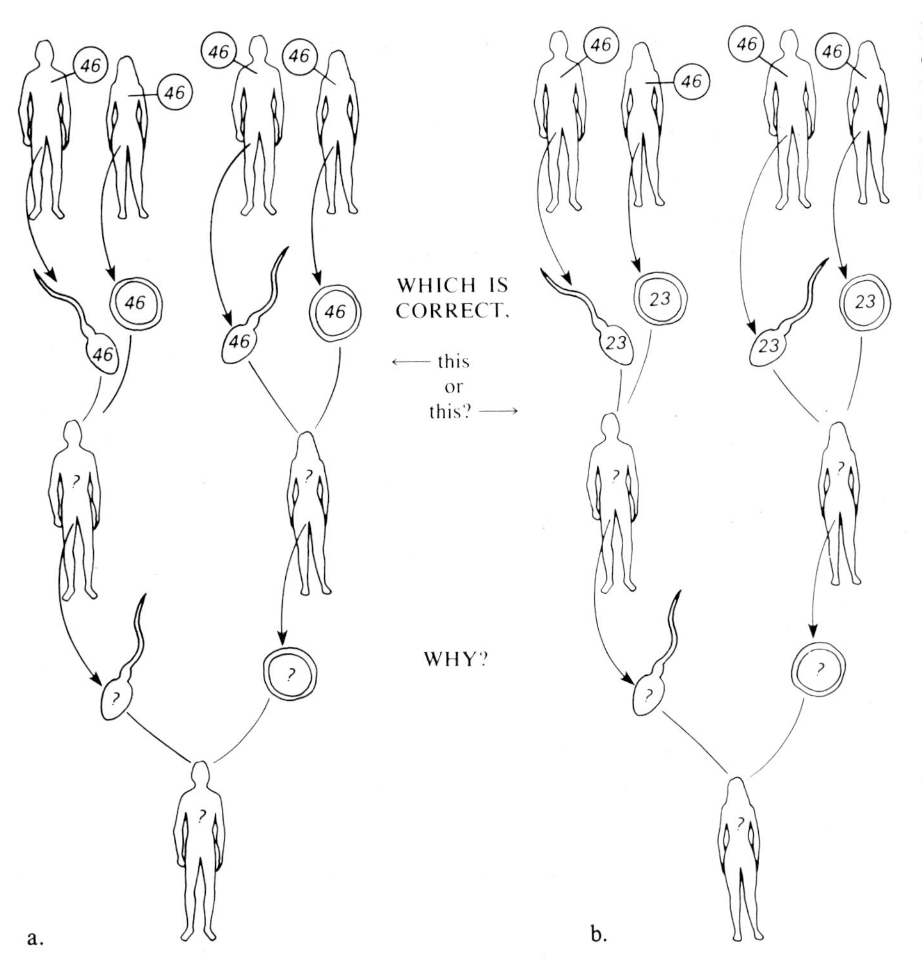

Figure 8.15 Two patterns for chromosome distribution during the sexual part of the life cycle of mammals. Is pattern a or b correct? Courtesy of W. B. Saunders Company and Biological Sciences Curriculum Study, Inc.

8.21 Plants have a slightly different life cycle pattern than animals, yet the distribution pattern of chromosomes during mitosis and meiosis is essentially the same. Only slight variations occur.

Each plant species has two plant forms that are part of the life cycle (figure 8.16).

The diploid ($2n$) plant form is called a sporophyte because it produces haploid ($1n$) *spores* by meiosis. These spores produce by mitosis give rise to the alternate plant form called a gametophyte, which makes gametes. Gametes join together to form a zygote, which gives rise to a sporophyte plant by mitosis to complete the life cycle. A plant body may be obtained from either a ____________ or a ____________. How many sets of chromosomes are in a spore? ______

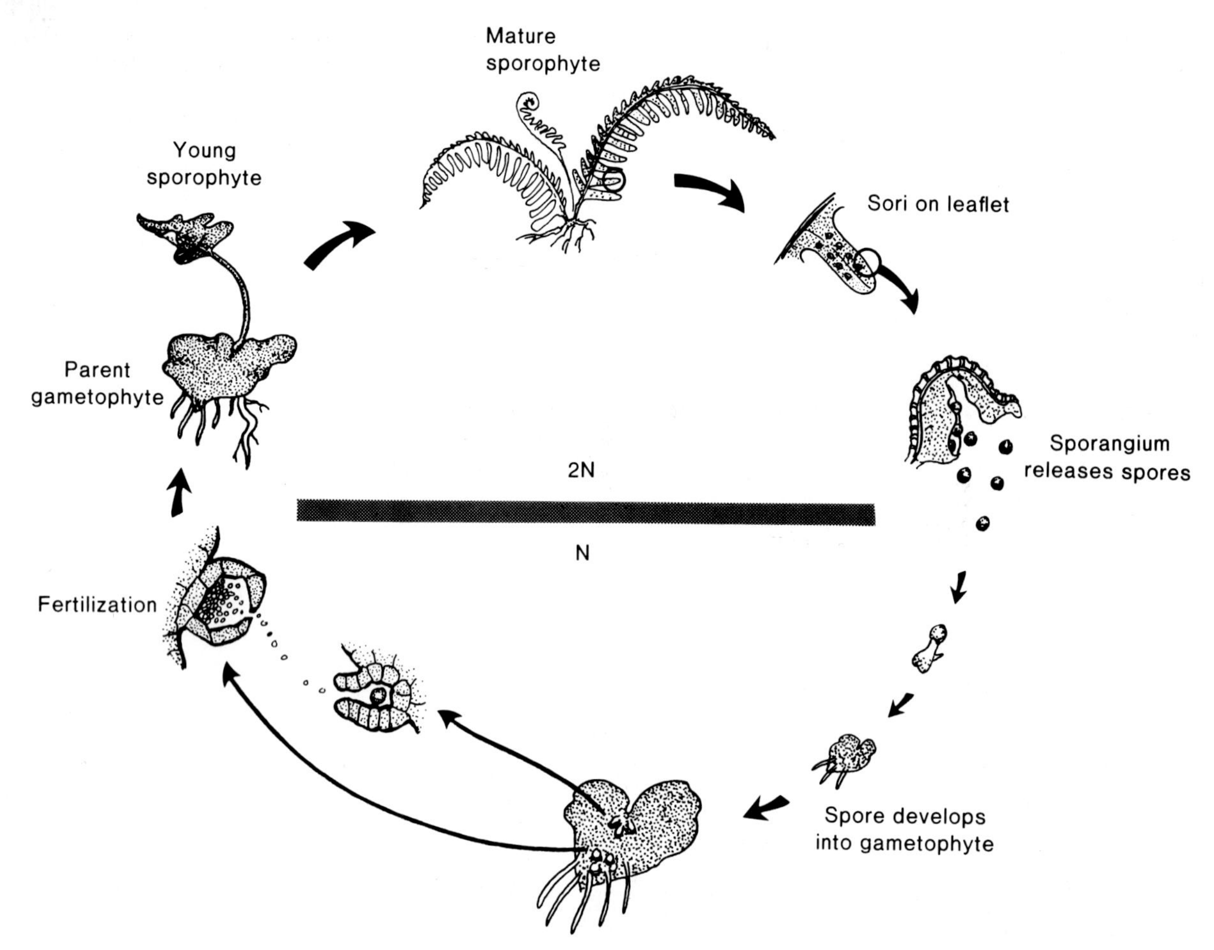

Figure 8.16 Life cycle of a fern plant showing two adult plant bodies, gametophyte and sporophyte.

8.22 Where does meiosis occur, and what are the products? Study figure 8.17 and complete data table 8.3 in the logbook.

Data Table 8.3 Meiosis in Animals and Plants

Organism	Location	Products
1. Human	Testes/Ovaries	Sperm/Eggs
2.		
3.		
4.		

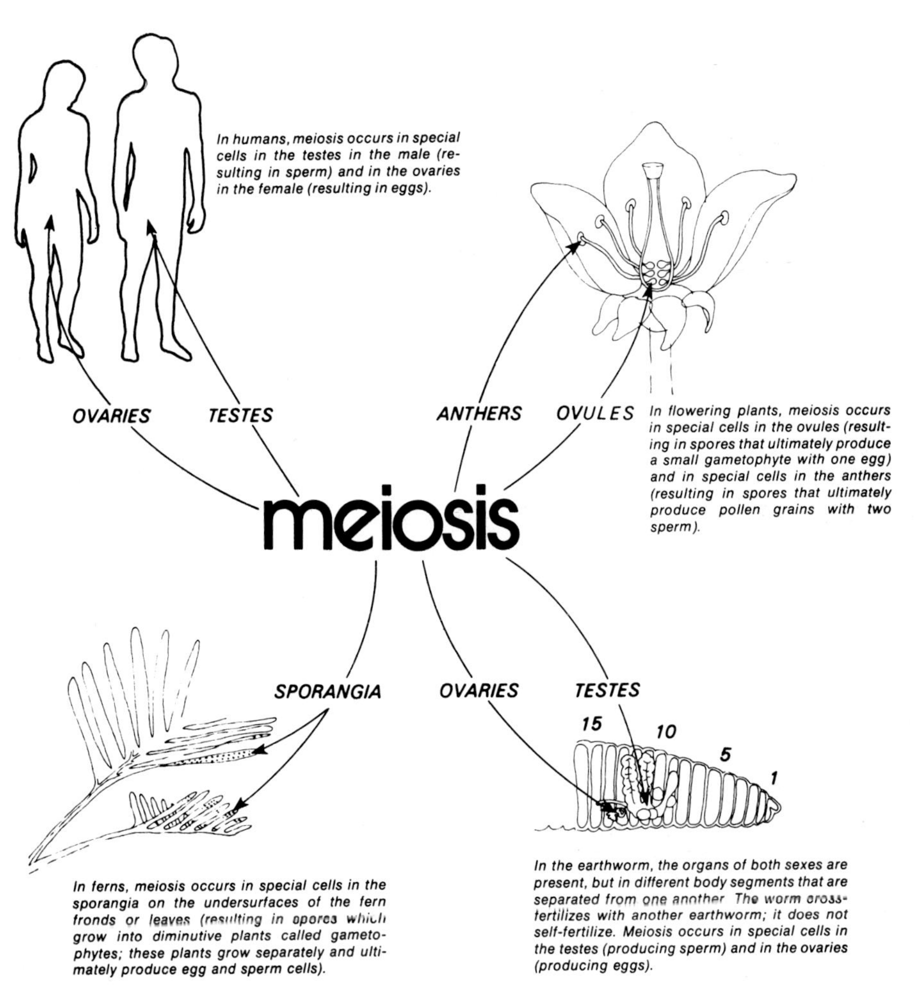

Figure 8.17 Meiosis in animals and plants showing sites of occurrance and meiotic products. Courtesy of W. B. Saunders Company and Biological Sciences Curriculum Study, Inc.

8.23 How is the genetic material of the chromosomes distributed from generation to generation by meiosis? To answer this question requires that you understand how 2 sets of chromosomes in a diploid cell are halved (reduced) to 1 set of chromosomes in a haploid gamete cell (2 sets to 1 set). Therefore, when two gametes are joined (each with one set of chromosomes), the zygote will have 2 sets of chromosomes. The cyclic pattern is $2n$ to $1n$ to $2n$. Meiosis makes it possible to *maintain a constant chromosome number* per cell generation after generation.

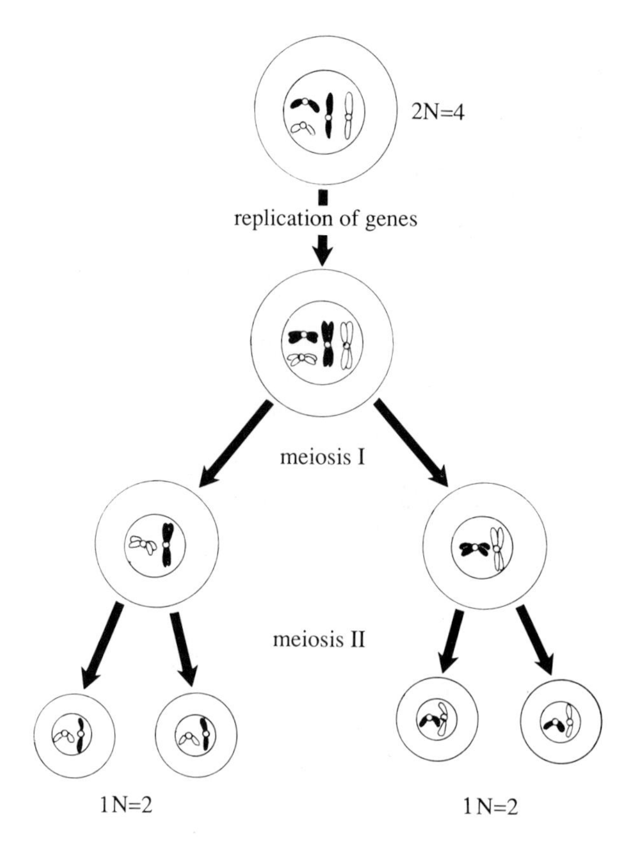

Figure 8.18 Nuclear divisions of meiosis.

8.24 Meiosis consists of two nuclear division phases called meiosis I and meiosis II (figure 8.18).

How many daughter cells are produced from two meiotic nuclear divisions? ______ What are these daughter cells called? ______ Observe that the synthesis (replication) of genes and chromosomes occurs before meiosis I. How many chromatids are present per each original chromosome? ______ In figure 8.18 the diploid cell ($2n = 4$) contains homologous pairs of chromosomes in the nucleus. How many homologous pairs do you observe? ______ During meiosis I do the chromatids or homologous pairs separate from each other? ______ Are chromatids or homologous pairs separated from each other in meiosis II? ______

8.25 The stages of meiosis I and meiosis II are shown in figure 8.19. Write a description in your logbook of each stage of meiosis I and meiosis II. In the description, describe what happens during nuclear division and cytokinesis.

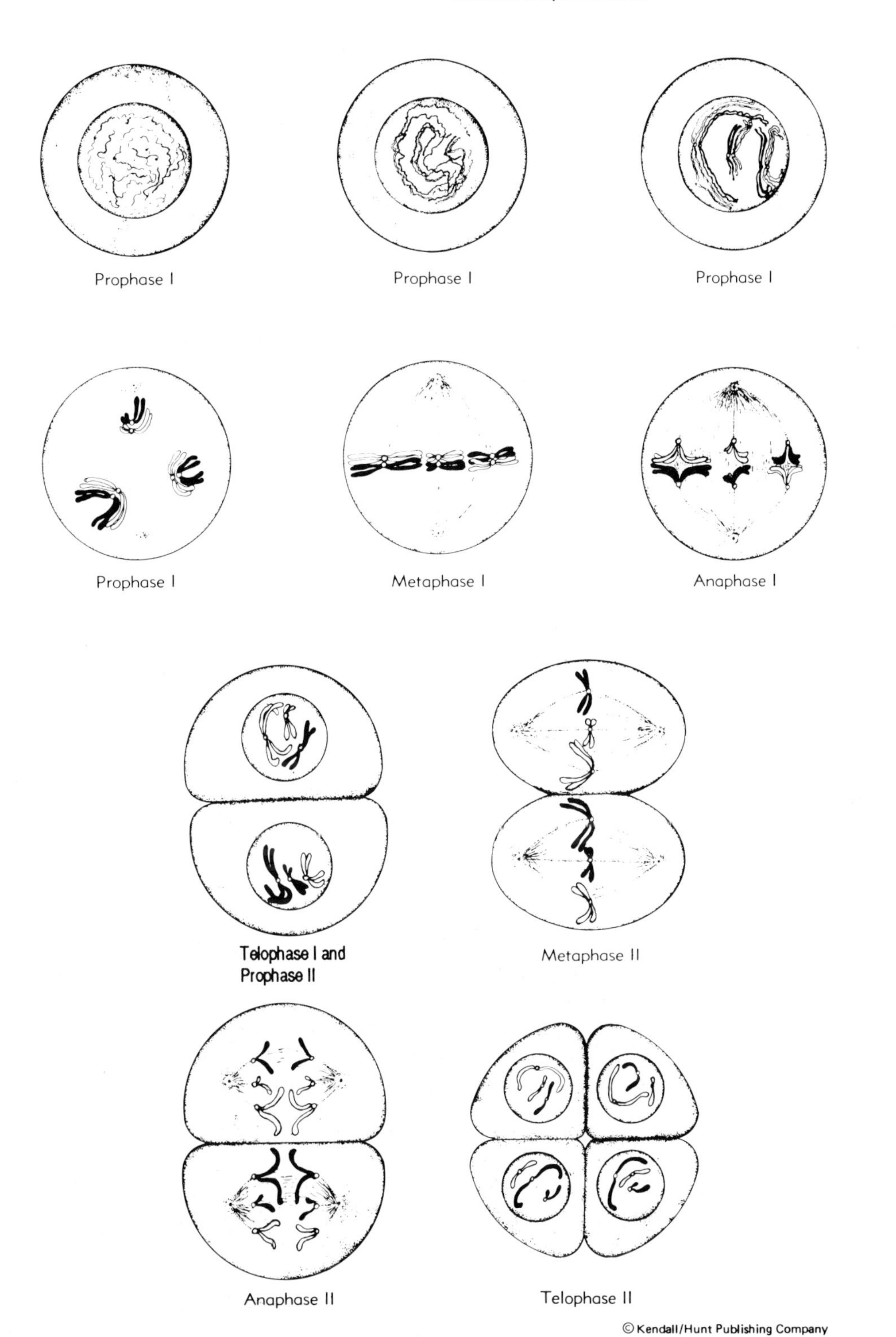

Figure 8.19 Stages of meiosis I and meiosis II.

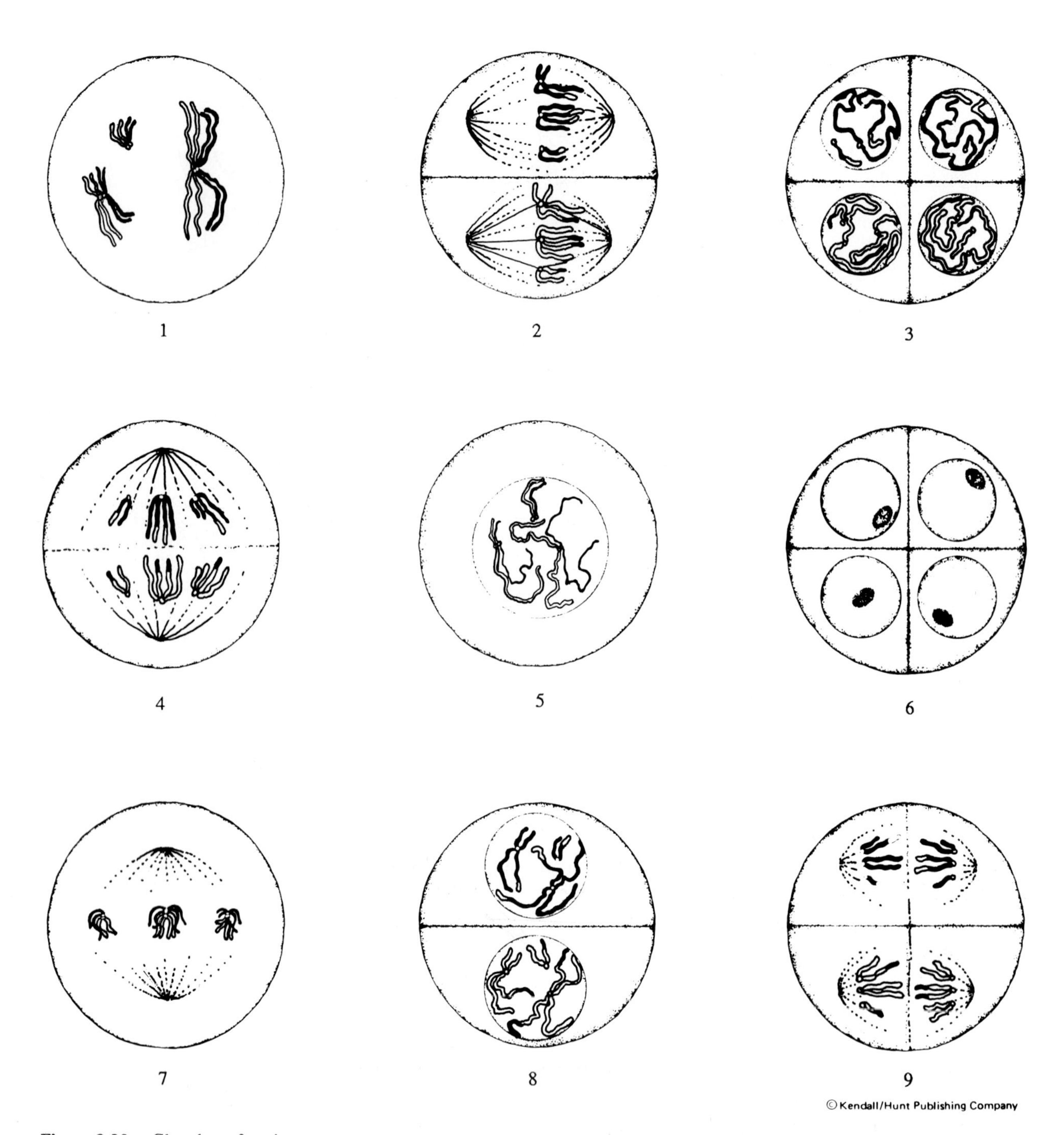

© Kendall/Hunt Publishing Company

Figure 8.20 Sketches of various stages of meiosis.

8.26 Rearrange the number of the cells number 1, 5, 4, and 7 in figure 8.20 in the order you think the steps of meiosis I and meiosis II might occur. Record the sequence in your logbook. ________

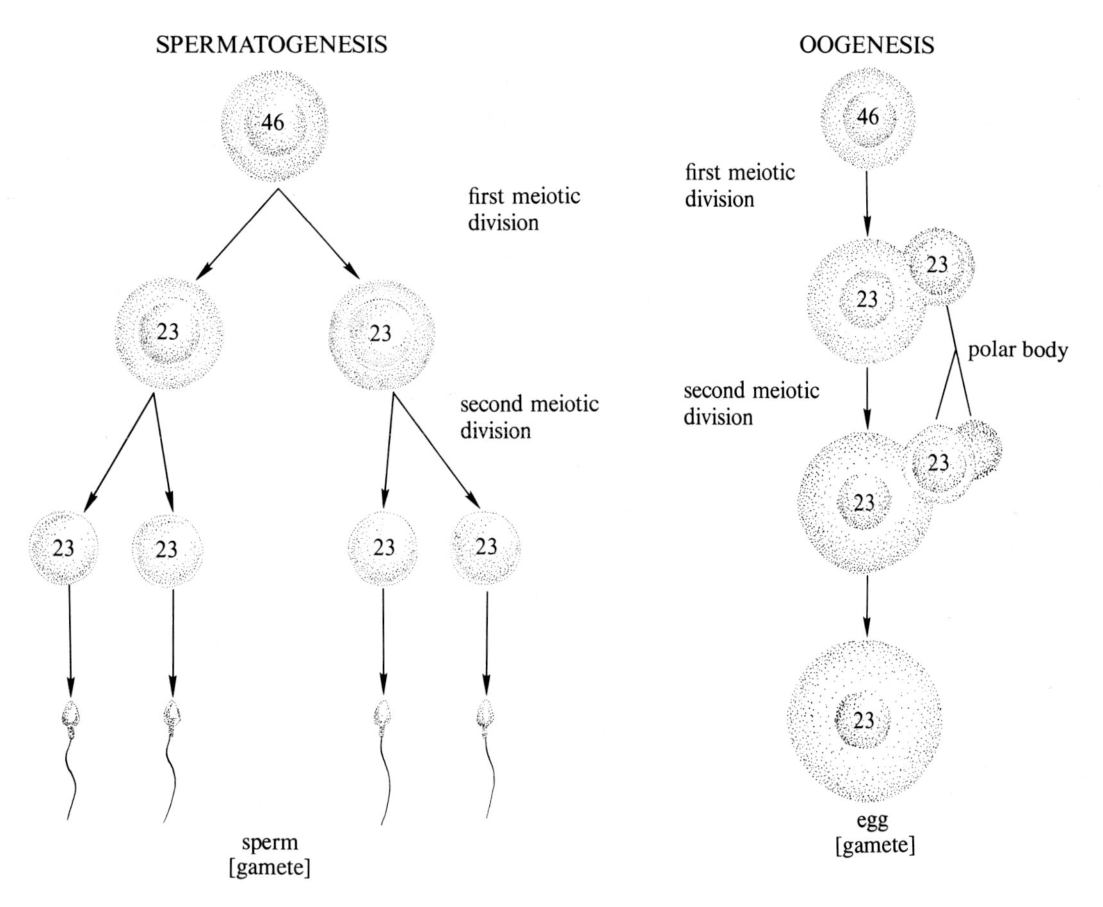

Figure 8.21 Meiosis in humans.

8.27 Meiosis in humans is shown in figure 8.21. Write the name of the process that produces sperm in the logbook. ________
In oogenesis the female produces during meiosis I two cells of unequal size. The smaller one is called a **polar body.** After meiosis II, there are at least two polar bodies produced. All polar bodies are nonfunctional.

During meiosis I, the homologous look-alike chromosomes (with paired chromatids) come together (**synapsis**) in an alignment (*tetrad*) that permits sister homologous chromatids to join together and exchange (*cross over*) genetic information between homologous chromatids (figure 8.22). This exchange of gene information provides a mechanism for the random rearrangement of genes.

The significance of meiosis, then, is both the reduction of chromosome sets ($2n$ to $1n$) and the opportunity for introducing genetic variation in a species of plant or animal by rearranging the alignment (order) of genes.

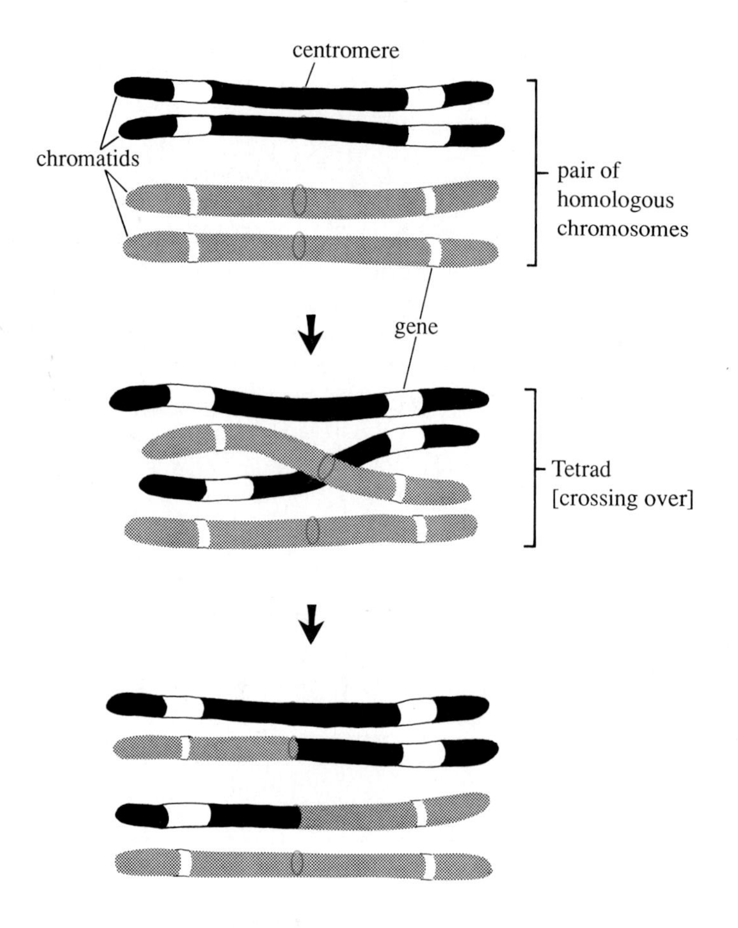

Figure 8.22 Exchange of genetic information between homologous chromatids.

8.28 The steps in the life cycle of a plant are symbolized in the pattern of figure 8.23. Plant 1 represents the sporophyte plant ($n = 2$) and plant 2 represents the gametophyte plant ($n = 1$). Use the following terms and label the remaining parts of the diagram in your log-book: zygote, diploid, haploid, gamete (egg, sperm), meiosis, *A* arrowed line. You will find out later what MC represents.

8.29 Symbolize chromosomes with strips of colored construction paper. Cut three blue strips of different size and shape and three red strips to match exactly the size and shape of the blue strips. Letter your strips as shown below. Your strips might look like the following in size and shape, but they do not have to. (It will help if they are not bigger than the illustrations.)

Blue set	A	b	G	R
Red set	a	B	g	r

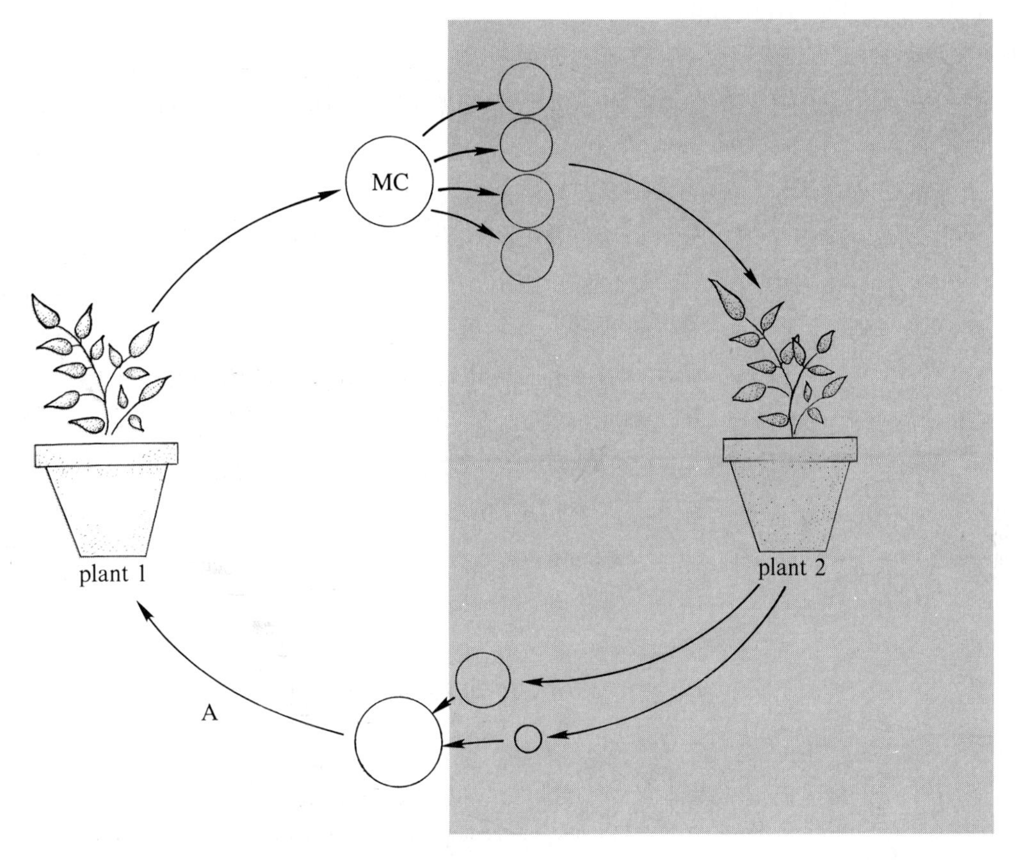

Figure 8.23 Steps in the life cycle of a flowering plant.

Now redraw the diagram of figure 8.24 on a clean sheet of white paper. Make the size of the parts large enough to contain your paper chromosomes. Place all the chromosomes together in the zygote part of the diagram to symbolize **fertilization.**

1. How many *sets* of chromosomes are present in the zygote?

   ________ n = ________

2. How many chromosomes are present? ________
3. How many homologous pairs are present in the zygote?

   ____

4. Name by letters one homologous pair. ____________

Move your chromosomes through one complete mitotic division. Cut additional paper strips to obtain duplicate chromatids.

5. How many chromosomes are present in each sporophyte plant cell? ________ Suppose a sporophyte plant contained ten thousand cells. How many chromosomes would be found in the nucleus of each cell if no meiosis occurred? ________

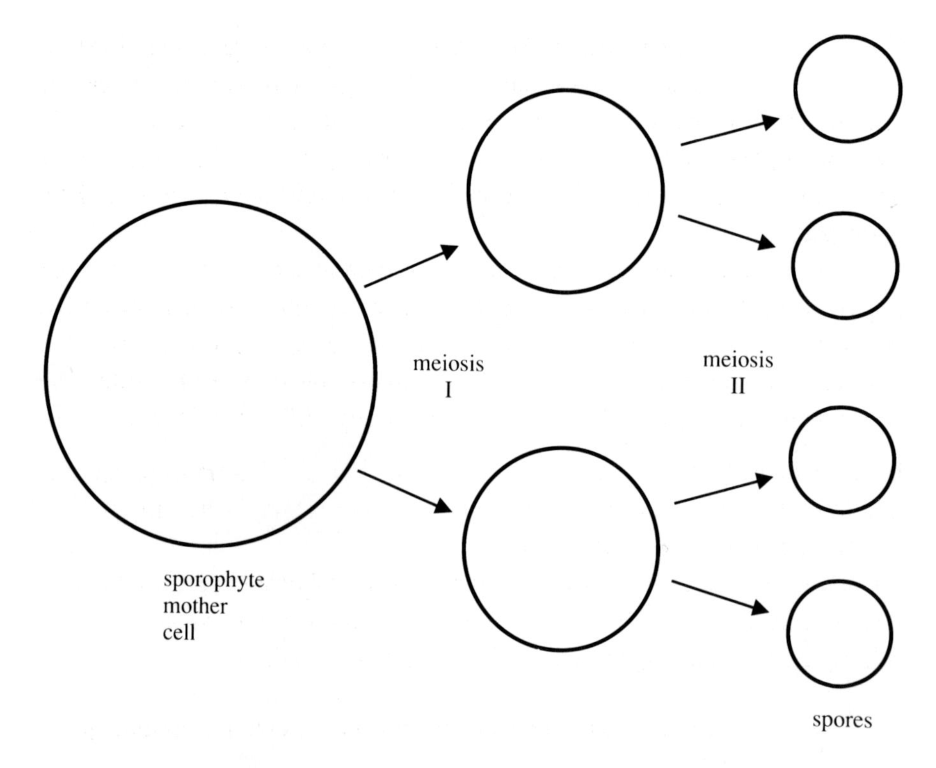

Figure 8.24 Production of spores in plants by meiosis I and meiosis II.

A sporophyte plant makes spores using meiosis to distribute the chromosomes to the spores. This meiotic process occurs in cells called sporophyte *mother cells* (MC). Use figure 8.24, your redrawing of it, and figure 8.18 to move the strips of chromosomes through one meiotic division.

Hint: Assume the interphase mother cell contains duplicated DNA and chromosomes. Therefore, start by placing the red and blue duplicated chromosomes in the MC of your redrawn diagram figure 8.23.

6. How many chromatids are in the MC? ________
7. How many chromatids are in each cell at the end of meiosis I? ________ At the end of meiosis II? ________
8. How many spores contain only three at the end of meiosis II? ________
9. How many spores contain only one color of chromosome at the end of meiosis II? ________
10. Can you see that meiosis I separates homologous look-alike pairs of chromosomes from each other? ________ (yes/no)
11. Can you see that meiosis II separates the chromatids? ________ (yes/no)

If you answered no to either or both of these questions, check with your teacher.

8.30 During meiosis I, the chromosomes may be rearranged so that each set of maternal (from the egg) and paternal (from the sperm) chromosomes in the diploid parent cell do not necessarily remain together in the gametes (animals) and spores (plants). This provides for *variation* in the genetic information found in each gamete produced by each parent.

Let us study the rearrangement of chromosomes in meiosis. Place your paper chromosomes with each duplicated set on the MC in figure 8.24. Align these 2 sets of chromosomes along the dotted line (symbolizes the equatorial plane at meiosis I). All the chromosomes may not fit within the MC when you do this. That is okay. How many different alignment combinations can be obtained? ________ *Hint:* Use the letters to figure this one out, and note the chromosomes can be *rearranged* so that not all blue are on one side of the dotted line.

Do the meiotic chromosome manipulation of meiosis, noting the distribution of the letters (symbolizing genes) *A* and *a*. This will allow you to follow the possible alignments of this homologous pair.

1. What different combination of these genes can be obtained in the spores?
   Answer: AbGR, Abgr and aBgr, aBGR.
   Now follow the gene symbols *G* and *g*.
2. What different combinations of these genes can be obtained?
   ______, ______ and ______, ______
3. How many different gametophyte plants can be obtained?
   ____

8.31 Further genetic variation can occur at meiosis I when chromosomal look-alike partners (homologues) may physically exchange (cross over) parts of their chromosome structure. This also creates gametes or spores with different combinations of genetic information. Both meiotic I events lead to considerable variation in the individual organism.

If at synapsis in meiosis I the red and blue homologues containing *A* and *b* (blue) and *a* and *B* (red) exchange parts by breaking in the middle of each homologue, then the following spore (1n) combinations would occur after meiosis I and II: *ABRG, ABgr* and *abgr, abGR*. How is this combination variation different from the combination variation obtained by homologous pair alignment?

8.32 Sex determination: Will it be a boy or a girl? Do some reading about so-called sex chromosomes (*X* and *Y*), and then answer the questions in your logbook.

1. How many *X* chromosomes are present in the male cell? ______ the female cell? ________
2. How many chromsomes are in the male cell? ________ the female cell? ________
3. The female gamete produced by meiosis will contain (*X* only, *Y* only, or either an *X* or *Y*). ____________
4. The male gamete produced by meiosis will contain (*X* only, *Y* only, or either an *X* or *Y*). ____________
5. Which gamete will determine the sex of the baby (egg sperm)? ____________

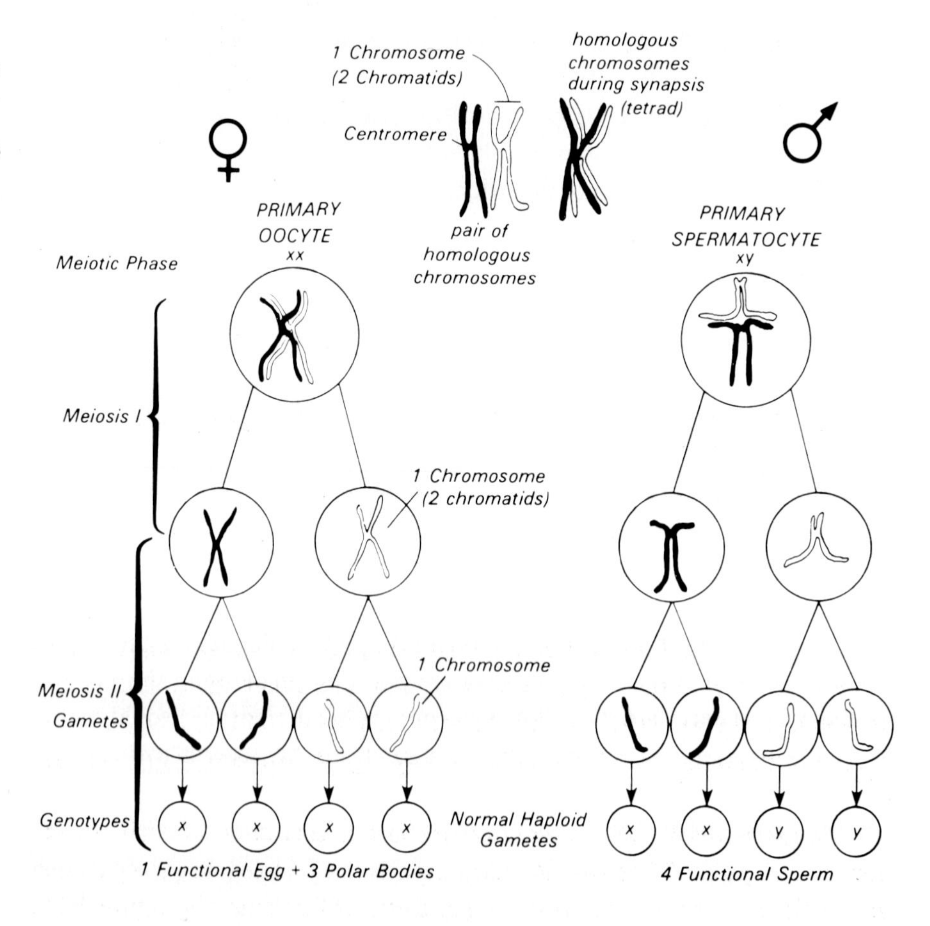

Figure 8.25 Pattern of chromosome separation in female and male humans during meiosis. Courtesy of W. B. Saunders Company and Biological Sciences Curriculum Study, Inc.

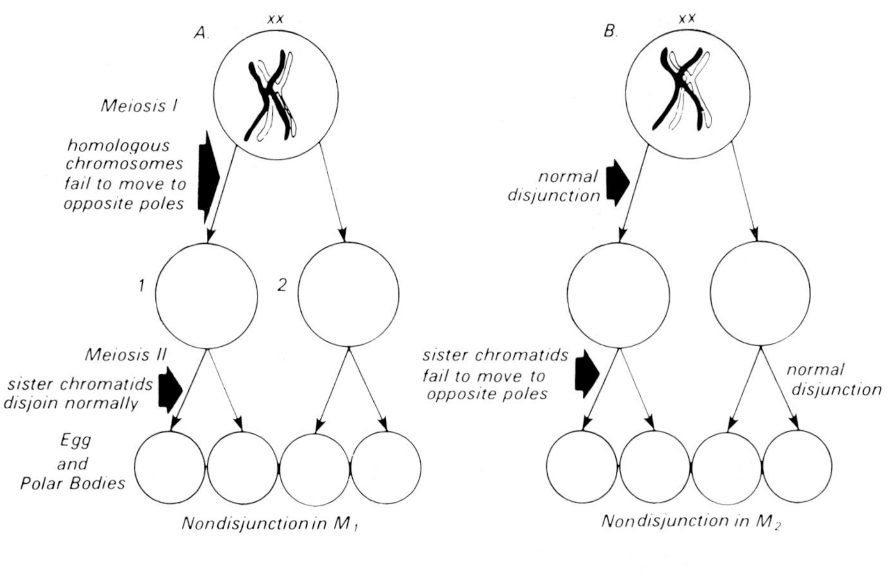

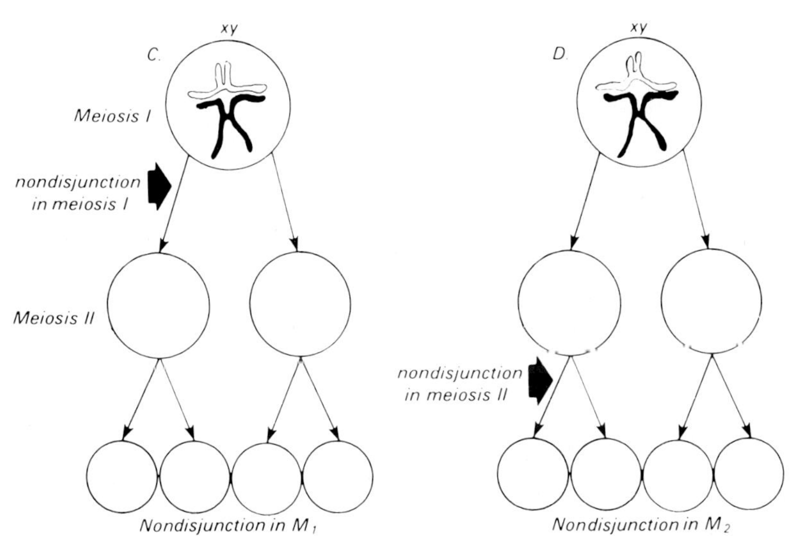

Figure 8.26 Abnormal patterns of chromosome separation during meiosis. Courtesy of W. B. Saunders Company and Biological Sciences Curriculum Study, Inc.

8.33 By now you should understand that chromosomes carry genetic information in genes, and these genes are transmitted from the parent to the offspring in gametes. You also learned that genes affect the individual traits of an organism through protein synthesis. Sometimes in the production of gametes by meiosis, chromosomes are subject to accidents in their movements. These accidents may result in the production of an abnormal individual characterized by physical defects and/or mental deficiency.

What happens when chromosomes fail to separate normally during meiosis? Figure 8.25 shows the normal pattern of chromosome separation in female and male humans during meiosis. After studying figure 8.25, complete in your logbook the diagrams in figure 8.26 by *drawing* the chromosomes.

If the diploid cells of an organism carries more than 2 sets of chromosomes, the condition is called **polyploidy. Aneuploidy** is a diploid cell that contains an excess or deficiency of individual chromosomes. This condition arises from *nondisjunction.* The best known case of aneuploidy is Down's syndrome (until recently called *mongolism*). There are a number of other abnormal aneuploidic conditions resulting from nondisjunction of the sex chromosomes.

8.34

Cellular reproduction is a continuous process. It appears to be a complex process, but once you have the steps mastered, the process does not seem quite so foreign. To review this chapter, use the study points below. When you feel you have mastered the material related to the study point, place a check (✓) in the appropriate box.

On completion of this chapter, you should be able to do the following:

☐ 1. Distinguish between the two main cell cycle phases.

☐ 2. Identify the four phases of mitosis and discuss what is happening in the cell at each phase.

☐ 3. Distinguish between asexual and sexual reproduction and give examples of each.

☐ 4. Distinguish between diploid and haploid cellular situations.

☐ 5. Discuss the stages of meiosis.

☐ 6. Compare and contrast polyploidy and aneuploidy.

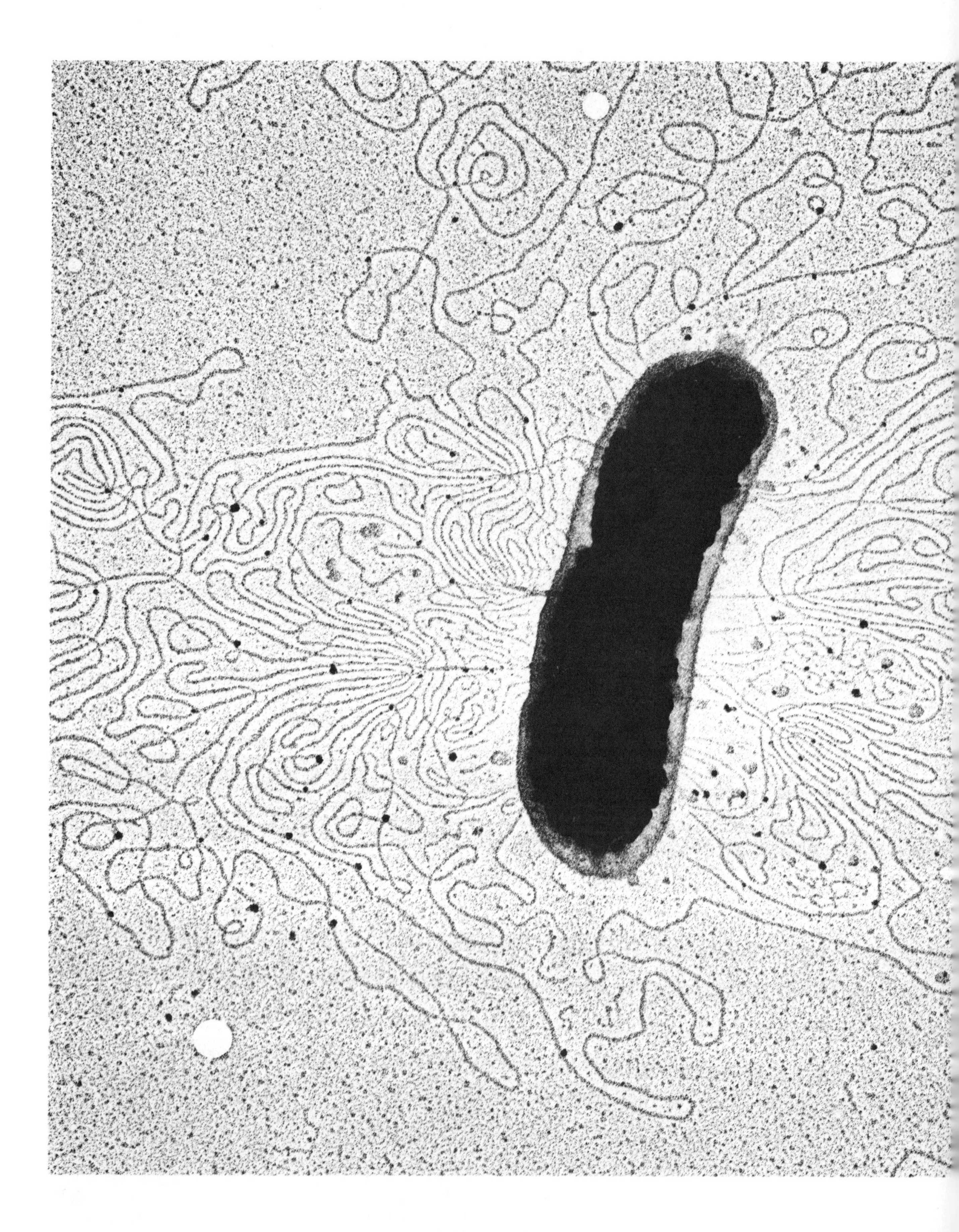

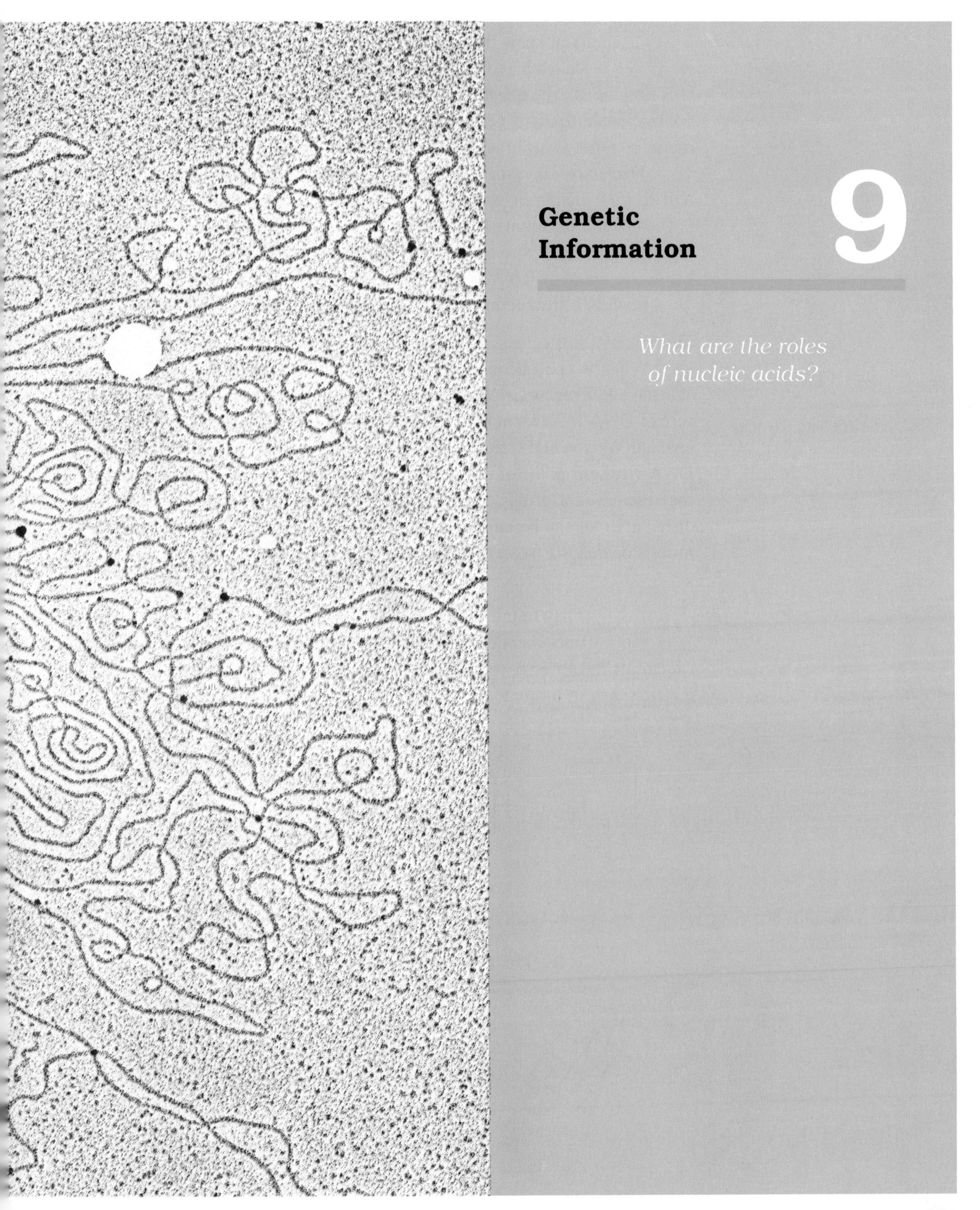

Genetic Information

9

What are the roles of nucleic acids?

9.1 You know that chromosomes contain genetic information, also called hereditary material, in chemical units called **genes.** Now it is time to ask the question, "What is the chemical nature of the gene?" Many people thought genes were protein, but in 1928, Fred Griffith and again in 1940, Oswald Avery demonstrated that genes are **nucleic acids.**

There are two types of nucleic acids: **DNA** (deoxyribonucleic acid) and **RNA** (ribonucleic acids). DNA, located in eukaryotic chromosomes, is the genetic information passed from one cell to another during mitosis and meiosis. Genes also are found in bacteria and viruses. In a *few* viruses, genes are made of RNA.

In what cell structure(s) of eukaryotic plants and animals do you find DNA? ________________

What is the chemical structure of nucleic acids? In 1953, James Watson and Francis Crick told the scientific world that the chemical structure of DNA was two *helical strands* twisted together. This discovery then enabled Watson and Crick to predict how DNA was replicated during the DNA synthesis phase of the life cycle of a cell. It is in this interphase period that the amount of nuclear DNA per cell doubles in preparation for the distribution of the hereditary material by chromosomal movement during mitosis or meiosis (see chapter 8).

9.2 They found that each strand of DNA is made up of four distinct chemical compounds called **nucleotides** (figure 9.1). Each nucleotide has three parts: (*a*) *nitrogen base* (named by a letter and a geometric shape in figure 9.1), (*b*) *sugar,* and (*c*) *phosphate* group.

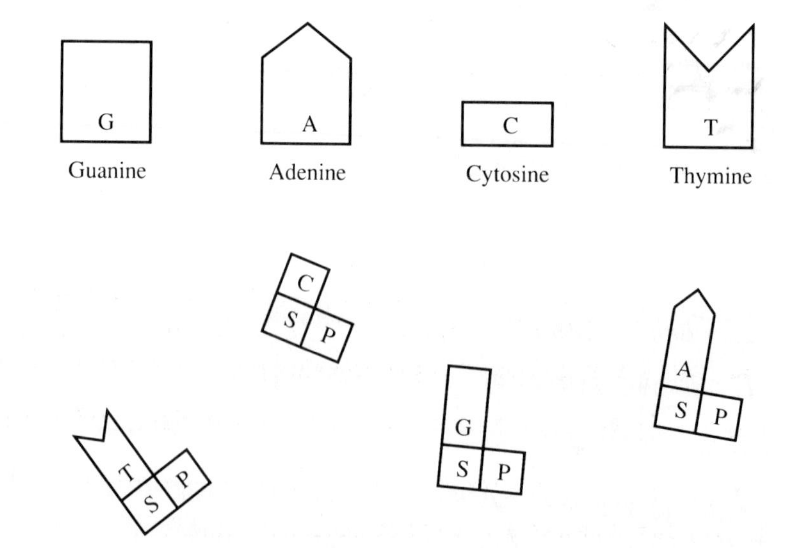

Figure 9.1 Parts of nucleotides called nitrogen bases.

Figure 9.2 Three parts of a nucleotide. S = sugar P = phosphate group.

Note there are four different nitrogen bases that are always directly connected to a sugar (S). The sugar-nitrogen base is connected to a phosphate group (P) (figure 9.2). The nucleotide is the building block monomer molecule of DNA. Figure 9.3 illustrates four nucleotides joined together to form a single strand (chain). The chain is twisted to form a helical structure. However, for our initial studies we will look at it as being a straight linear chain. The process of joining the nucleotide monomers together is called a condensation reaction (see chapter 5).

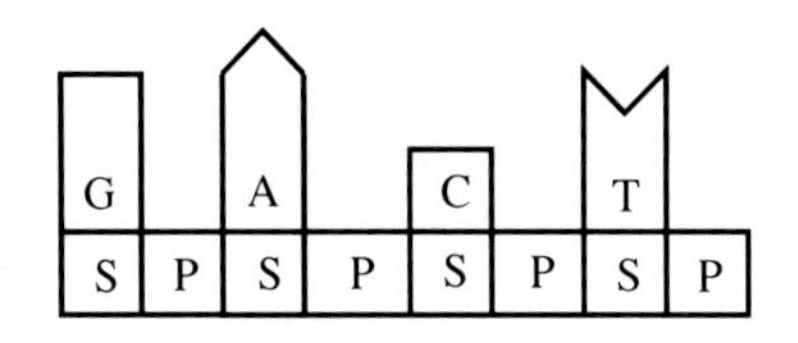

Figure 9.3 Four nucleotides joined together.

9.3 The (S) stands for the sugar *deoxyribose,* and the (P) stands for the *phosphate* group. From the basis of what you now know, a nucleotide consists of three parts. List them in your logbook.

1. ______________________
2. ______________________
3. ______________________

DNA receives its name, **deoxyribonucleic acid,** from the presence of the sugar called ______________.

9.4 If you notice the geometry of each nitrogen base (figure 9.1), you should realize that two of the bases will match up like pieces of a puzzle. Guanine unites only with cytosine. You make the remaining matches between the nitrogen bases.

Guanine: ______cytosine______

Adenine: ______________________

Cytosine: ______________________

Thymine: ______________________

9.5 Figure 9.4 shows what is called the DNA ladder. For your purpose, it is easier to use the DNA ladder as a visual model rather than its actual third dimensional shape. Remember, as you noted above, only two nitrogen bases can join together.

Based on the diagram in figure 9.4 and 9.7, the uprights (sides of the ladder) are made of ______________ and ______________. The rungs (steps) of the ladder are made of ______________.

Figure 9.4 DNA ladder model of two strands of nucleotides joined together at the nitrogen bases.

	S	P	S	P	S	P	S	P	S	P	S	P	S	P	S	P
	G		A		C		T		G		A		C		G	
	C		T		G		A		C		T		G		C	
P	S	P	S	P	S	P	S	P	S	P	S	P	S	P	S	

9.6 Names have been given for similar nitrogen bases: Guanine and adenine are called *purines,* and the nitrogen bases cytosine and thymine are called *pyrimidines.* As you have already noted, nitrogen bases bond to a single counterpart. The bond that forms between two bases is a weak *hydrogen bond.* This is represented in figure 9.4 at the line where the two matching nitrogen bases join. Label the weak hydrogen bond in figure 9.4 of your logbook.

9.7 The structure of DNA (figure 9.5) is summarized as follows:

- The DNA molecule is made up of two strands wound about one another in a double helix.
- Each strand is made up of a chain of nucleotides.
- A nucleotide is composed of a phosphate group, a deoxyribose sugar, and a nitrogen-containing base (purine or pyrimidine).
- The nucleotides of each strand are linked together by joining the phosphate group of one nucleotide to the sugar group of a second nucleotide.
- The two strands are held together by weak hydrogen bonds between each purine and its "partner" pyrimidine.

Label the following on figure 9.5 in your logbook: sugar-phosphate backbone, a paired nitrogen base (purine joined to pyrimidine), and a hydrogen bond.

9.8 While Watson and Crick were doing their work, other biologists had discovered that cellular DNA synthesis occurs just prior to mitosis and meiosis. Figure 9.6 helps show where DNA synthesis occurs in relationship to mitosis. Label figure 9.6 in your logbook at the place where chromatids separate during mitosis. Which phase in figure 9.6 would you expect DNA replication to occur? ______________________

In your logbook make a diagram similar to the one in figure 9.6 that represents meiosis.

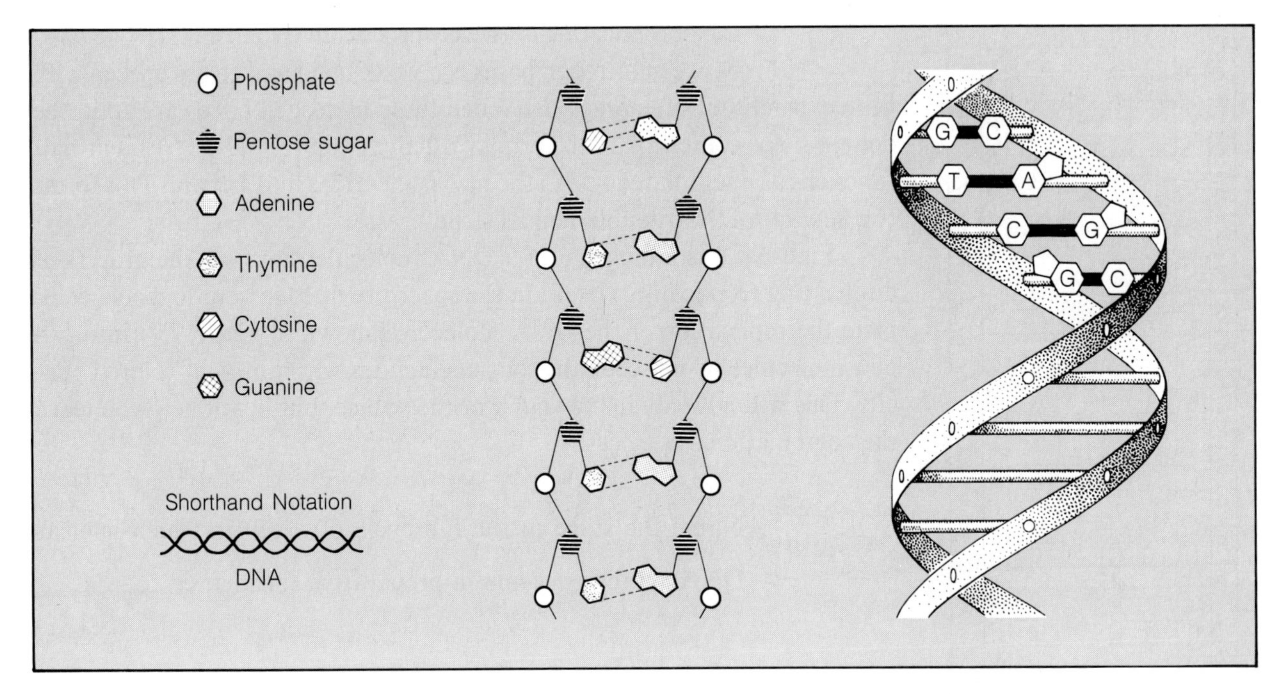

Figure 9.5 Structural diagrams of DNA.

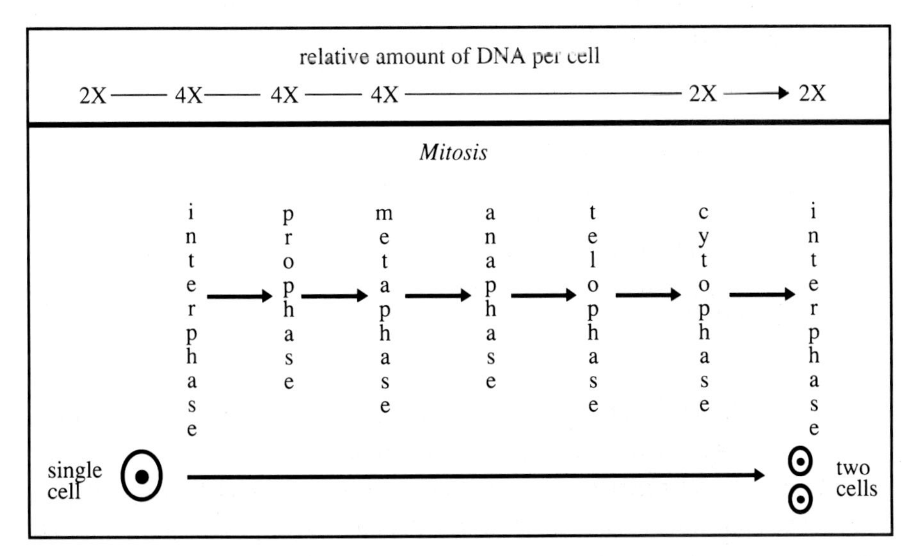

Figure 9.6 Relative amount of DNA at interphase and mitosis.

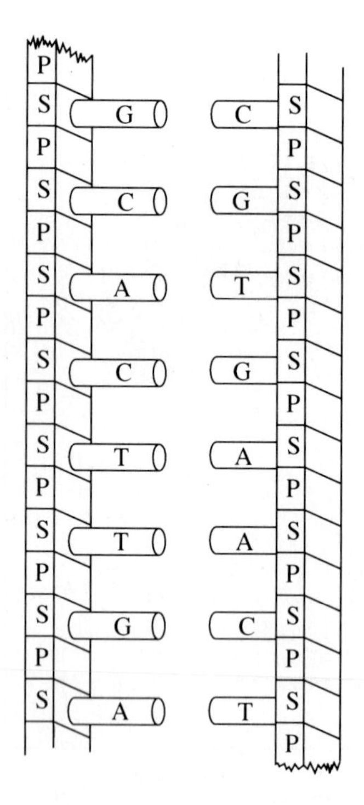

Figure 9.7 Ladder model of DNA showing the two separated nucleotide strands.

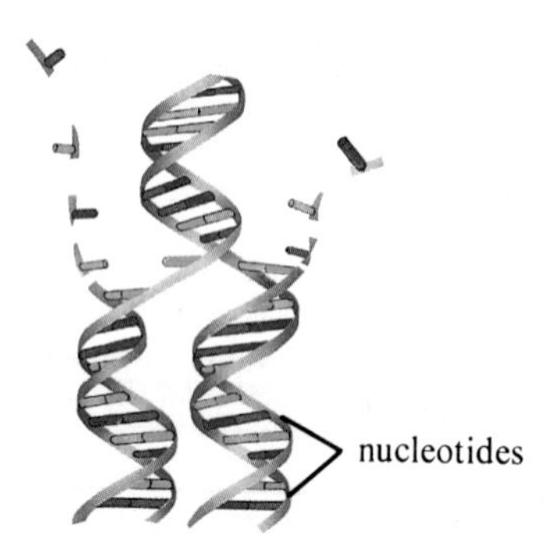

Figure 9.8 Duplication of a DNA molecule.

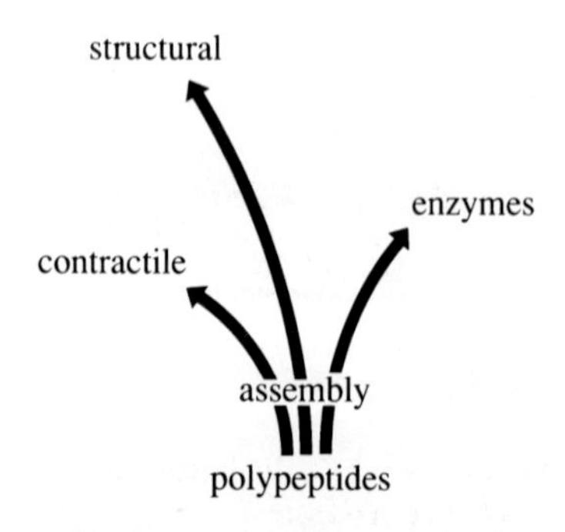

Figure 9.9 Assemble of polypeptides into different proteins.

9.9 DNA replication is an easy process to visualize. (*A*) The bond between nitrogen bases is a weak bond and easily broken with little expenditure of energy. (*B*) When these hydrogen bonds are split, the double helix comes apart to form two halves (figure 9.7). (*C*) Each half functions as a template to align the new nucleotides into a chain. This forms two separate DNA double-helical structures.

Figure 9.8 is a diagram of a DNA molecule that is in the process of duplicating (replicating) itself. In the space provided in your logbook, complete the replication of the DNA molecule (shown in figure 9.7) into two new molecules. Color code the four nucleotides with different colored pencils. This will not only make your work look nice, but it will help you learn the combination sequences.

9.10 Name the stage in the life cycle of a cell in which nuclear DNA synthesis occurs in preparation for meiosis. ________

9.11 Are the two new DNA molecules alike? ________ Describe briefly in your logbook why it is important for the replication of DNA to form identical DNA molecules.

In summary, you have studied how the *flow of information* occurs from cell-to-cell during mitosis and meiosis. You have learned in this chapter how DNA replication occurs in preparation for making diploid daughter cells (mitosis) and for making haploid daughter cells (meiosis).

9.12 Living organisms, as noted in chapter 7, are made of matter organized and maintained through its interaction with energy. You also know from chapter 7 that living systems interact with energy and transform it into a form that can be used to perform work. These energy transformations that occur in living cells are called cellular metabolism. How do cells manage these cellular activities? The answer is they manufacture and use protein molecules called **enzymes.** Where do cells obtain the molecules that are found in all cellular membranes? The answer is they manufacture these proteins and insert them into existing membranes. Where do some cells obtain the contractile proteins they use to move about (ameoba, leukocytes), pump fluids (smooth muscle cells), or move the skeletal framework of mammals (skeletal muscle cells). The answer is they manufacture these proteins. A part of the manufacturing process often requires the assembly of structured proteins, enzymatic proteins, and contractile proteins from polypeptides (figure 9.9).

9.13 You already know that a protein molecule may be a single polypeptide or several polypeptides joined together (see chapter 5). You also learned that protein molecules are macromolecules manufactured at the ribosomes and are products of the hereditary information stored in the DNA. You are aware that proteins are manufactured by joining amino acids together with a covalent bond (in this specific case, a peptide bond). A chain of amino acids is called a **polypeptide.**

DNA is important to the manufacture of proteins because DNA contains the information for joining the amino acids together at the ribosome. However, the manufacture of protein occurs in the cytoplasm, and all the DNA (except for a small amount in chloroplasts and mitochondria) is located in the nucleus.

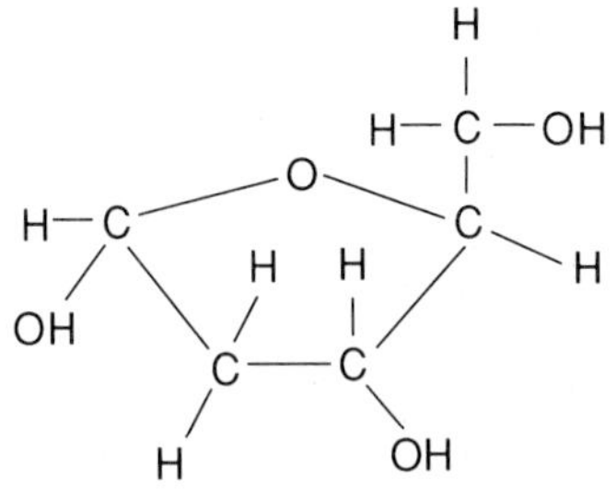

DNA sugar

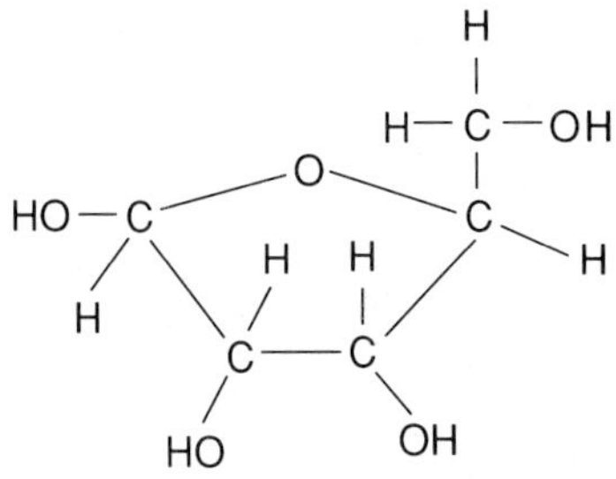

RNA sugar

Figure 9.10 Nucleic acid sugars.

9.14 Biochemical studies have shown that genetic information "flows" from the genes in the nucleus to the cytoplasm by a *ribonucleic acid* (RNA) messenger. A knowledge of the chemical structure of RNA is important to understand how DNA controls the manufacture of protein molecules. One difference between DNA and RNA is visible in the organic structure of the two *sugars* that make up DNA and RNA. Observe figure 9.10 and circle on your logbook figure, the structural difference between the two sugars.

Another difference between DNA and RNA is that thymine is not found in RNA. Instead, a similar base, *uracil,* is present. One other significant factor should be mentioned. That is, most of the DNA is found in the nucleus of the cell, and RNA is found both in the nucleus and in the cytoplasm of the cell.

RNA is a single strand (chain) of nucleotides. Study figure 9.11 and circle in your logbook the chemical parts of the RNA molecule that are different from DNA.

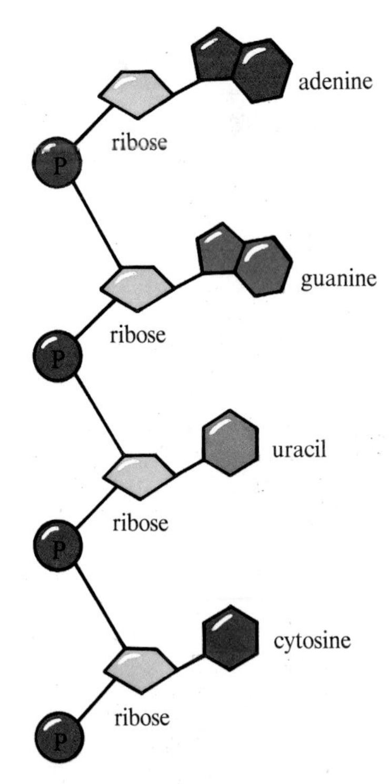

Figure 9.11 A single strand of RNA nucleotides.

9.15 Now let us learn how the genetic information, stored in sets of nucleotides in DNA, is used by living cells to make proteins.

First, you must understand that there are three major kinds of RNA: messenger RNA (mRNA), transfer RNA (tRNA), and ribosomal RNA (rRNA). You may have already guessed the role of rRNA. It helps form the ribosome, and it helps put together the amino acids. Second, all three RNAs are made from different DNA genes. For your study, you will learn

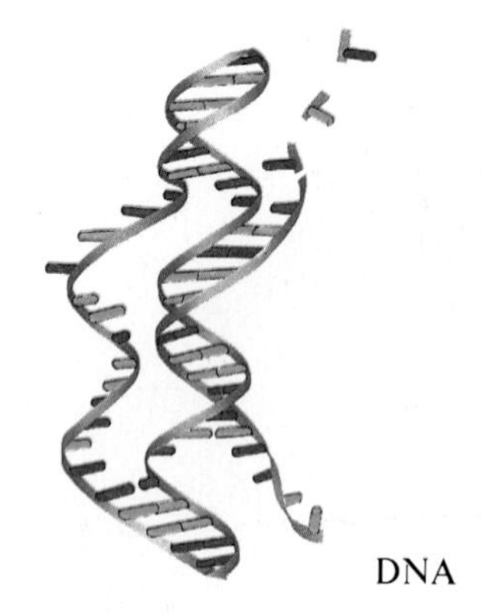

Figure 9.12 Transcription of information from DNA to RNA.

about the production of mRNA from DNA. Third, messenger RNA is formed from only one strand of a DNA double helix by a process called **transcription** (figure 9.12). Transcription is the process during which the information of DNA strand is transferred to a RNA strand.

Certain chemical rules must be obeyed for the transcription process to occur. First, the mRNA strand must be a *complementary copy* of one DNA template strand. Second, in DNA, each strand contains a complementary copy of nucleotides that are the complement of the other. That is, thymine (T) is always hydrogen bonded (base paired) with adenine (A), and guanine (G) is always paired with cytosine (C). You can observe this in figures 9.4 and 9.5. In data table 9.1 in your logbook, list the DNA complement (counterpart).

Data Table 9.1 DNA Complement

DNA Alphabet	DNA Counterpart
A (adenine)	T (thymine)
C (cytosine)	
G (quanine)	
T (thymine)	

9.16 Observe in figure 9.8 that when the replication of both DNA strands occurs, a complementary copy of the template strands is produced. Messenger RNA is a complementary copy of a single DNA template strand.

Complete the DNA sense strand and the messenger RNA nucleotide sequence in data table 9.2 in your logbook by writing the complementary mRNA nucleotide letter opposite the template (sense strand) DNA nucleotide letter.

Data Table 9.2 Building Messenger RNA

DNA	A	A	T	G	C	G	T	G	G	(Sense Strand)
	|	|	|	|	|	|	|	|	|	
	T	T	A	C	G	C	A	C	C	(Sense Strand)

DNA A A
mRNA ______________________________

Figure 9.12 shows the transcription process in detail. Observe that the mRNA strand is a complementary copy of one DNA sense strand. Also observe that only a part of the DNA double helix comes apart during transcription, unlike that of DNA replication. Write in your logbook the complete nucleotide sequence for the *DNA template strand* and its complementary mRNA copy shown in figure 9.12. Use data table 9.1 as a guide for writing your nucleotide sequence.

9.17 Transcription in the nucleus is the first step in the manufacture of proteins. This step results in a *linear* sequence of DNA nucleotides being transcribed into a *complementary linear* sequence of RNA nucleotides. The second step is the *translation* of the message (sequence of mRNA nucleotides) in the cytoplasm (figure 9.13). Prior to the second step all RNA messages are chemically altered in the cytoplasm by adding nucleotides to the ends of the RNA strand. Many RNA messages are also shortened by removing non-coding segments called **introns;** the remaining coding nucleotide segments called **exons** are joined.

Translation of messenger RNA is a *decoding* step that requires tRNA and rRNA. Protein is similar to RNA because a protein molecule is made of monomers linearly aligned in a specific sequence. The decoding process requires that a linear sequence of mRNA monomer nucleotides be decoded into a linear sequence of monomer amino acids.

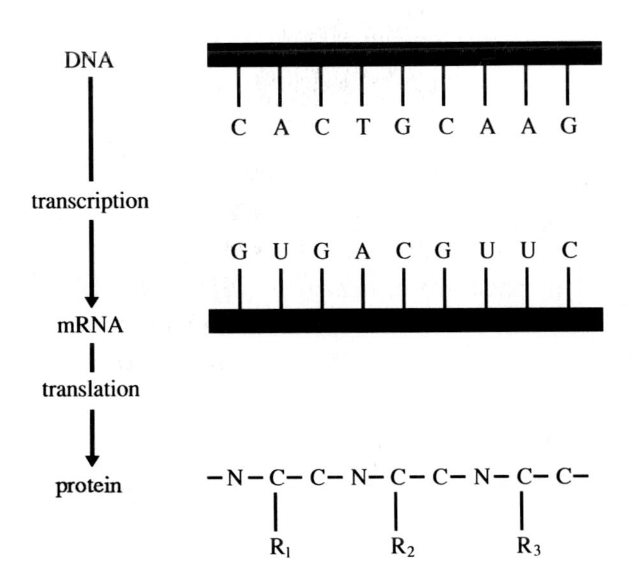

Figure 9.13 Flow of information from DNA to protein. Transcription and translation.

Data Table 9.3 Alphabet Code

DNA Code Word (Nucleus)	RNA Code Word-Codon (Cytoplasm)	Amino Acid (Cytoplasm)
ACC	UGG	tryptophan
ACG		glycine
AGC		arginine
TTA		asparagine
ATG		histidine
CGC		valine

9.18 Let us take a look at the code and how is it decoded. Each amino acid is represented in the DNA sense strand by a sequence of three nucleotides. Each nucleotide sequence can be written as an alphabet set of three letters. The letters represent the four different nucleotides.

Knowing which set of letters of the nucleotide alphabet represent each amino acid allows you to write the DNA alphabet code word for the amino acid. For example, in data table 9.3 several amino acids that are found in the cytoplasm are listed along with their DNA nucleotide alphabet. *ACC* is the alphabet code for *a*denine *c*ytosine *c*ytosine.

During transcription, the genetic information of DNA is coded into a RNA form. Complete data table 9.3 in your logbook by writing the RNA alphabet code for each DNA alphabet code. *Please note* that the RNA alphabet code is the complement to the DNA alphabet code.

9.19 The coded RNA message (mRNA) is obtained from one strand of the DNA double helix. A typical portion of a single helical strand of mRNA might look like the one in figure 9.14. Is the messenger RNA strand in figure 9.14 coded from DNA strand number 1 or number 2? ________. Number ________ is the sense strand.

When the mRNA shown in figure 9.12 is decoded in the cytoplasm, how many amino acids would be joined together in a polypeptide chain?

____

9.20 Scientists have decifered the genetic alphabet code for all the amino acids. Data table 9.4 shows all the possible mRNA alphabets (*codons*) for each amino acid. Note that data table 9.4 contains the complementary mRNA codons for the amino acids. Use data table 9.4 to name in your logbook the amino acids that would be assembled into a

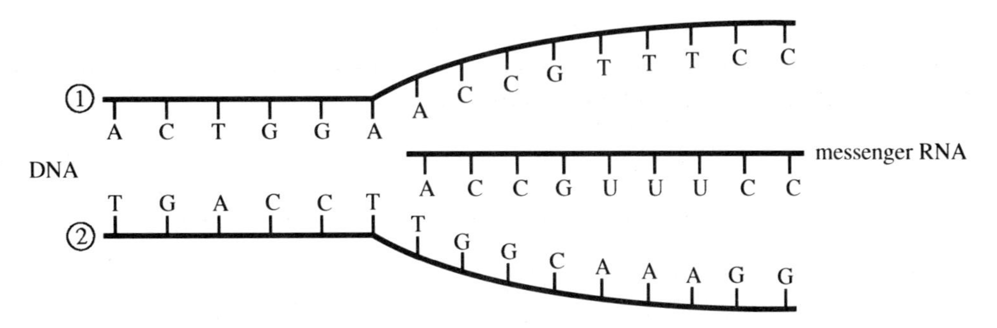

Figure 9.14 Transcription. Flow of information from a single strand of DNA to a single strand of RNA.

polypeptide according to the mRNA illustrated in figure 9.14. ________, ________________, ________________. *Hint:* You must decode the mRNA by reading the alphabet right-to-left.

Data Table 9.4 RNA Alphabet Codes

First Position	Second Position: *U*	*C*	*A*	*G*	Third Position
U	UUU Phe	UCU	UAU Tyr	UGU Cys	U
	UUC	UCC Ser	UAC	UGC	C
	UUA Leu	UCA	UAA STOP	UGA STOP	A
	UUG	UCG	UAG	UGG Trp	G
C	CUU	CCU	CAU His	CGU	U
	CUC Leu	CCC Pro	CAC	CGC Arg	C
	CUA	CCA	CAA Gin	CGA	A
	CUG	CCG	CAG	CGG	G
A	AUU	ACU	AAU Asn	AGU Ser	U
	AUC lie	ACC Thr	AAC	AGC	C
	AUA	ACA	AAA Lys	AGA Arg	A
	AUG Met	ACG	AAG	AGG	G
G	GUU	GCU	GAU Asp	GGU	U
	GUC Val	GCC Ala	GAC	GGC Gly	C
	GUA	GCA	GAA Glu	GGA	A
	GUG	GCG	GAG	GGG	G

9.21 Now decode the mRNA molecule in figure 9.12. Start your decoding at the upper part (the free end) of the mRNA strand. List the amino acids in sequence (the order of placement in the polypeptide) coded by the mRNA of figure 9.12. Also list any *stop* messages that you may decode. In addition to the *stop* signal, there are nucleotide alphabet code signals that mark the end of the message and, therefore, mark the end of assembling amino acids. There are three nucleotide alphabets (UUA, UAG and UGA) that function similarly to a period at the end of a sentence. These signal the end of the decoding process and the completed polypeptide is released from the ribosome.

9.22 Figure 9.15 shows the process of translation at the ribosome. Note the role for transfer RNA (tRNA) is (*a*) to transport the amino acids to the ribosome and mRNA strand for assembly, and (*b*) to decode the mRNA nucleotide alphabet. How is the messenger RNA decoded by the transfer RNA molecule? Let us review. You know that chromosomal DNA transfers its genetic information (stored as a sequence of alphabet nucleotides) to messenger RNA. mRNA flows to the ribosome in the cytoplasm, where amino acids are joined together in the sequence dictated by mRNA. mRNA is a complementary DNA blueprint.

The decoder of mRNA is a transfer RNA-amino-acid complex molecule. Transfer RNAs are small uniformly-shaped molecules. Located at one end of the tRNA molecule is an attachment site for an amino acid. Each amino acid has its specific tRNA molecule to which it becomes at-

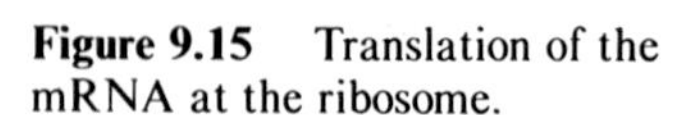

Figure 9.15 Translation of the mRNA at the ribosome.

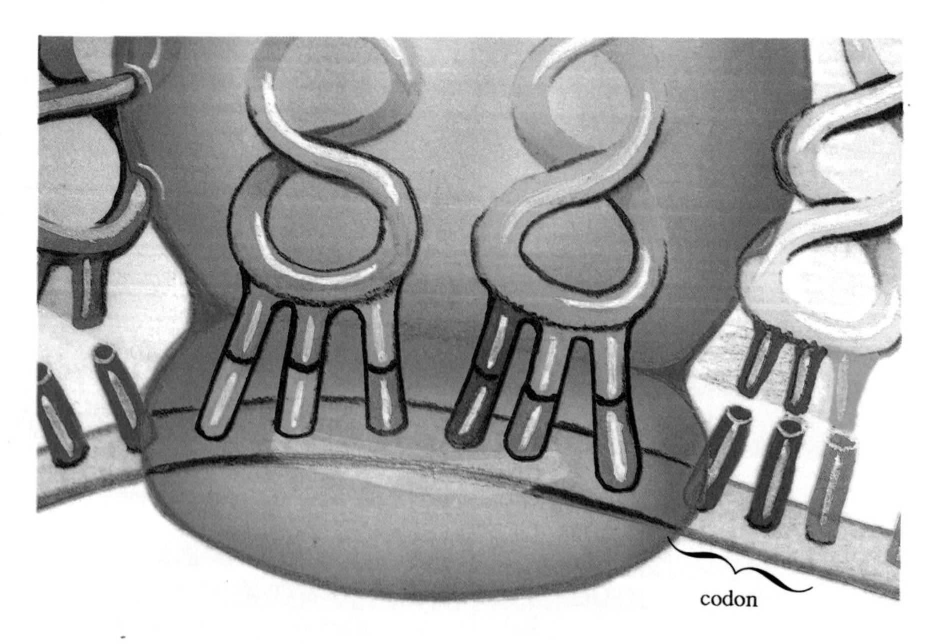

tached. At the opposite end of the tRNA molecule, a set of three nucleotides is located. These three nucleotides are called an *anticodon.* Each anticodon has a complementary *codon* in a messenger RNA molecule.

9.23 In figure 9.14 the messenger RNA codons are (reading right-to-left)—CCU UAU CCA. What are the tRNA anticodons for each codon? *Please note:* the anticodon for CCU is GGA. What are the other two anticodons? ____________ and ____________

When the decoding process takes place, the anticodon of a tRNA-amino-acid complex must find its complementary codon in the messenger RNA. The binding of a tRNA-amino-acid complex occurs only if the anticodon is complementary to the codon. If the tRNA-amino-acid complex has a GGA anticodon, what amino acid is bound to the tRNA? ________ Hundreds of tRNAs, each carrying one specific amino acid, decode the genetic message of an RNA molecule. This results in hundreds of amino acids joined together to form a polypeptide (figure 9.16).

A single gene contains enough information to assemble all the amino acids for a single polypeptide. This is known as the *one-gene-one-polypeptide* hypothesis. Figure 9.16 summarizes the overall process of protein manufacture.

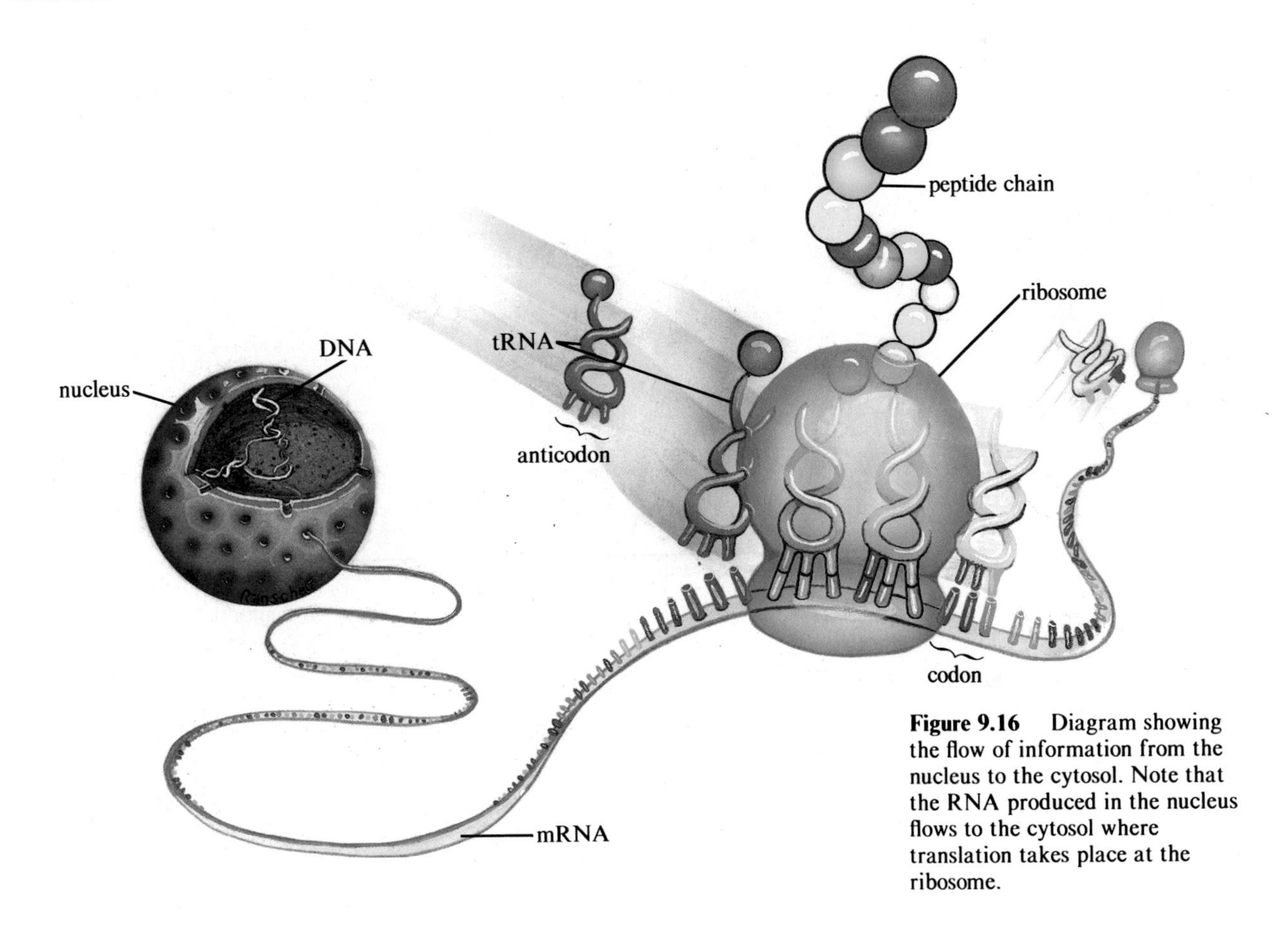

Figure 9.16 Diagram showing the flow of information from the nucleus to the cytosol. Note that the RNA produced in the nucleus flows to the cytosol where translation takes place at the ribosome.

9.24 Write a summary of the flow of genetic information from DNA to RNA to polypeptide on a separate sheet of paper. Attach your report to your logbook for this chapter.

9.25 It is the exact duplication of DNA that ensures genetic traits to be transmitted from parent to offspring. This is important if the offspring is to resemble its parents. If for some reason there is a mixup in the DNA code in the gamete producing cells, this change becomes permanent in the offspring and could be passed on to future generations. A change in the DNA code is referred to as a **mutation.** The effect of a mutation is an abnormal or missing protein.

An example of the effect of a mutation is seen in the disease phenylketonuria. PKU is the result of a missing enzyme. A genetic mutation in the DNA causes an abnormal enzyme to be made. This results in a mental abnormality.

Over 1,500 human diseases are genetically inherited. These defects are contained in the sex cells of some males and females. When the gametes are produced by meiosis, *some* of the gametes produced may contain defective genes. Many of these inherited diseases you may already know, such as diabetes, PKU, cystic fibrosis, Tay-Sachs, and sickle-cell anemia.

Figure 9.17 Effects of sickle-cell anemia. a. Normal red blood cells. b. Abnormal (sickled) red blood cells. c. Amino acid sequence of a small portion of the hemoglobin.

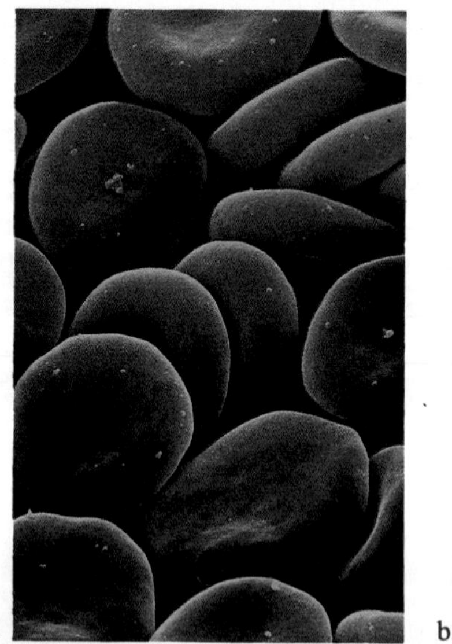
a.

b.

c. **Normal** Valine — Hisidine — Leucine — Threonine — Proline — Glutamic acid — Glutamic acid -----

[1] [2] [3] [4] [5] [6] [7]

Sickle-cell anemia Valine — Hisidine — Leucine — Threonine — Proline — Valine — Glutamic acid -----

Sickle-cell anemia is a very interesting and extremely serious genetically inherited disease that is caused by a changed hemoglobin molecule. Hemoglobin is found in red blood cells and carries oxygen to the cells throughout the body.

The presence of the changed hemoglobin may cause the red blood cells to change shape, or to "sickle" (figure 9.17*b*). This may cause blockage of the small blood vessels and prevent the normal flow of blood. The result is pain and damage to internal organs.

Sickle-cell anemia is a serious disease and can shorten the life span by 35 to 40 years. The sickle-cell trait is not a disease and normally causes none of the problems produced by sickle-cell anemia. But the sickle-cell trait does affect health when oxygen supply is low, such as periods of high levels of air pollution.

More than 2 million people worldwide are affected by sickle-cell anemia. In the United States, about one of every six hundred African-Americans is born with sickle-cell anemia. Sickle-cell anemia and the sickle-cell trait are also found among Spanish-speaking people and in other families whose origins are countries such as Greece, Italy, and Turkey.

A hemoglobin molecule is made of four polypeptides, each approximately 150 amino acids long. The amino acid sequence has been determined and a part of this sequence is shown in figure 9.17*c*. Analysis of normal hemoglobin and sickle-cell anemia hemoglobin showed a single difference existed between these two types of blood. How does sickle-cell anemia hemoglobin differs from normal hemoglobin (see figure 9.17*c*)?

____

This change in a single amino acid represents a mutation in the DNA alphabet code. This mutation (and other mutations) are inherited through sexual reproduction. Sickle-cell anemia is an alteration of a single nucleotide pair within the gene coding for one kind of hemoglobin polypeptides.

9.26 Normal hemoglobin (H) is a dominate trait over sickle hemoglobin (h), which is recessive. A person with the sickle-cell trait has a (Hh) genotype, and a person with sickle-cell anemia has a (hh) genotype. You may want to review chapter 8 before answering the following questions in your logbook.

1. If a person with normal hemoglobin and a person with the sickle-cell trait have children, what percentage of their children will have normal hemoglobin? ____________
2. What percentage of their children will have the sickle-cell trait? ____________

3. What percentage of their children will have sickle-cell anemia?

   ____

4. If a person with normal hemoglobin and a person with sickle-cell anemia have children, what percentage of their children will have normal hemoglobin? ____________
5. What percentage of their children will have the sickle-cell trait?

   ____

6. What percentage of their children will have sickle-cell anemia?

   ____

9.27 The results of chromosomal study have led to a better understanding of the causes of mutations and genetic maladays. The role of the genetic counselor is to help prospective parents discover potential problems. Two such problems have been mentioned, namely, PKU (phenylketonuria) and sickle-cell anemia. Other hereditary problems occurring in human beings are also listed in data table 9.5.

Data Table 9.5 Hereditary Problems

Trait	Description
Cataract	Clouding of the lens
Color Blindness	Inability to distinguish between red and green colors
Muscular dystrophy	Degeneration of muscles
Sickle-cell anemia	Inability of red blood cells to carry oxygen
Polydactyly	Having extra fingers and/or extra toes

Write a report on one of the above hereditary problems and attach it to this chapter in your logbook.

9.28 Genetic engineering is a set of techniques that permit scientists to isolate a single gene (remove the gene from all the rest), make new gene combinations (bypass meiosis), transfer the gene into another cell or organism (bypass fertilization), and to produce large quantities of their specific gene and its product (proteins). These techniques were first used to genetically modify bacteria DNA. Today genetic-engineering scientists can move genes into the eukaryotic cells of plants and animals. The process often starts by placing a gene into the plasmid DNA of a bacterium (figure 9.18).

Genes *cloned* (making many exact copies) in this way can be introduced into plant and animal cells.

Because this technology permits the rearrangement of genetic information, useful applications have arisen or are on the horizon of availability. Already the pharmaceutical, chemical, and food industries have used this technology to produce products such as human insulin, human growth hormone, and a vaccine against cattle foot-and-mouth disease. With list of cloned plant genes lengthening yearly, increase in yields of important crop plants is being speculated. Genetically engineering plants resistant against pathogens is already accomplished. Cloned plant genes also permits molecular biologists to study gene response to environmental and biological stress.

a.

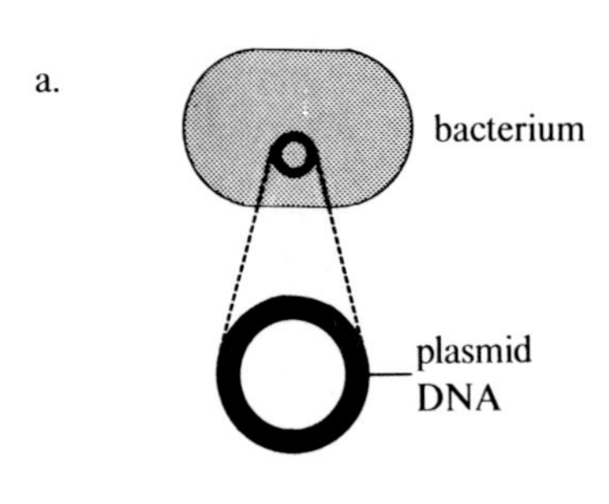

b.

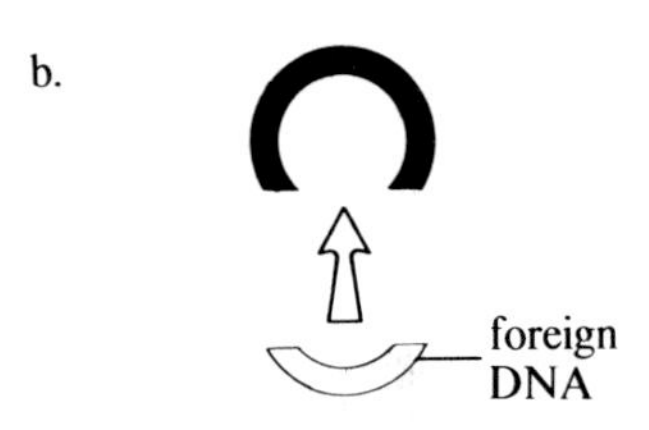

c.

d.

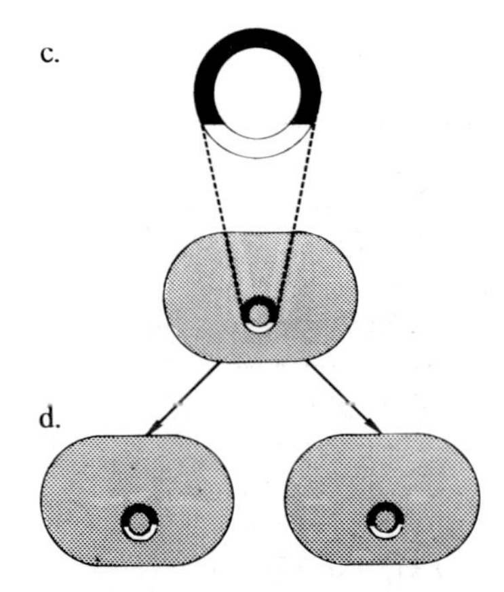

Figure 9.18 Genetic engineering. a. Plasmid DNA is removed from bacteria. b. Foreign DNA is removed from another cell and joined to the plasmid DNA. c. The hybrid DNA is reinserted into bacteria. d. Normal replication of the bacteria and its DNA hybrid.

9.29 What are the dangers? Even with benefits of applied genetic engineering, some persons have questioned the risks of this technology. Most experts agree that the hazards of basic research are minimal. However, no one is completely certain about the long-range consequences of creating new genetic combinations.

There are three major concerns:

- Genetically engineered organisms might have potentially harmful effects on human health and the environment.
- The impact on evolution is unknown.
- This technological capacity might affect the well-being of human beings.

For now, genetic engineering research proceeds with caution under federal guidelines.

9.30 Scientific knowledge about DNA and RNA is relatively new. No doubt more discoveries will be forthcoming. To review this chapter, look at the study points below. When you feel you have mastered the information related to the study point, check (✓) the appropriate box in your logbook.

On completion of this chapter, you should be able to do the following:

- ☐ 1. Identify the sequence of events leading to the discovery of DNA.
- ☐ 2. Identify the components of DNA and RNA.
- ☐ 3. Create a "DNA ladder," including each of the four nitrogen bases and their base counterparts.
- ☐ 4. Identify the steps for DNA replication.
- ☐ 5. Discuss the process of transcription.
- ☐ 6. Discuss the process of translation.
- ☐ 7. Recognize some of the hereditary problems that occur from gene alterations.
- ☐ 8. Recognize the given three major concerns of genetic engineering.

Mendelian Genetics

10

What did Gregor Mendel do for us?

Figure 10.1 Gregor Mendel, the "father of genetics."

10.1

Genetics is the science that deals with genes and variation. **Heredity** is defined as the transmission, through reproduction, of factors that cause offspring to resemble their parent or parents. Therefore, a definition of genetics may be more complete if it includes the study of both heredity and variation. Whatever definition you use, the science of genetics has had its ups and downs.

The nineteenth-century figure who has ultimately been named the "father of genetics" was Gregor Mendel (1822–1884).

Mendel received his early horticulture training on his father's farm. At the age of twenty-one, he entered the monastery at Brunn to be ordained a priest three years later.

While at the monastery, Mendel studied garden peas to discover what was later to be called the "laws of heredity." In all, Mendel selected seven traits of garden peas to study. He called the traits *unit characteristics*. The traits that Mendel studied are listed along with the alternative appearance. They are the following:

- The stem was either tall or short.
- The dried seed pod was either not constricted or constricted (inflated).
- The ripe seeds were either smooth and round, or rough and wrinkled.
- The seed pod was either green or yellow.
- The flowers were either axial or terminal to the stem.
- The flowers were either purple or white.
- The seed coat of the peas were either yellow or green.

Studying his experiment, one trait at a time, Mendel asked, "What sort of offspring would result if the pea plants were cross-fertilized artificially?" The results were graphic. All plants bore smooth seeds with yellow seed coat color. The pods were all inflated and green. The flowers were all axial and purple, and the stems were all tall.

Mendel recognized that one trait seemed to remain hidden, and one appeared each time. He used the word **dominant** to refer to the characteristic that was the most potent. The hidden characteristic was called **recessive.** Since he used pure strains of plants, he called the mixed plants **hybrids.** (Mendel's second question was naturally referring to the fate of the recessive traits.) What happened to the recessive trait? Data table 2 illustrates the results of the F_2 (second filial) generation when the F_1 hybrids were permitted to self-fertilize. The symbolism used in data table 10.2 may need some explanation. A ♂ means male, and a ♀ means female. P stands for

Data Table 10.1 The Seven Traits Studied in Mendel's Famous Pea Plant Research.

TRAIT	CHARACTERISTICS	
STEM LENGTH	 Tall	 Short
SEED SHAPE	Round	 Wrinkled
POD SHAPE	 Inflated	 Constricted
FLOWER POSITION	 Axial	 Terminal
SEED COLOR	Yellow	 Green
FLOWER COLOR	 White	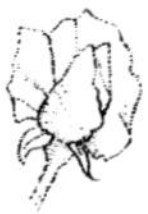 Purple
POD COLOR	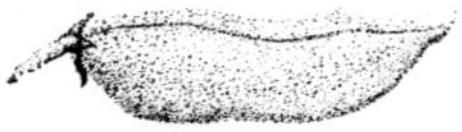 Green	 Yellow

Data Table 10.2 Crossing of Traits and the Results in the F_1 and F_2 Generations

No.	P_1 Cross ♂ + ♀	F_1 Plants	F_2 Plants	Ratio
1	Long + short stems	all long	787 long 277 short	2.84:1
2	Inflated + constricted pods	all inflated	882 inflated 229 constricted	
3	Round + wrinkled seeds	all round	5774 round 1850 wrinkled	
4	Yellow + green seeds	all yellow	6022 yellow 2001 green	
5	Axial + terminal flowers	all axial	651 axial 207 terminal	
6	Purple flowers + white flowers	all purple	705 purple 224 white	
7	Green + yellow pods	all green	428 green 152 yellow	

parent, F_1 means the first filial generation, F_2 means the second filial generation. What, then, would the symbol for the third filial generation be? ________________ What filial generation are you from your grandparents? ____________ Study data table 10.2.

10.2 Data table 10.2 illustrates the results from Mendel's experiments. Look at these results carefully. Determine the ratio of dominant-to-recessive traits for each trait and record this ratio in the data table. Do not round off the ratio numbers to whole numbers; instead, round off decimal notation to hundredths, or two decimal places. The first one is done for you, so you have an example to follow.

10.3 If you round off the ratios to whole numbers, is there a similar ratio for each of the seven traits? ________ What is it? ____________________________ Is the dominant trait numerically larger than or numerically smaller than the recessive trait? ________________

10.4 From his work Mendel was able to formulate several laws that needed to be addressed. His first law is called the *law of segregation.* From his data, Mendel proposed that a hereditary character (trait), such as tallness, is determined by a pair of factors. When a gamete is formed, one member of the pair is segregated only to become matched with another member again when sperm and egg unite.

Figure 10.2 illustrates Mendel's law of segregation. Note that each pea plant has two **alleles,** one for tallness and one for shortness. The dominant allele is represented with a capital letter. The recessive allele is indicated with a small letter. One allele is found on each **homologous chromosome** at a particular site or **gene locus.** As the two segregated alleles join during the fertilization of the egg by the sperm, they match up. If a dominant allele is present, the dominant trait will show. If no dominant trait is present, the recessive character will be allowed to show.

Two dominant (TT) or two recessive (tt) alleles are said to be purebred, or **homozygous** dominant or recessive. One dominant allele and one recessive allele is said to be **heterozygous** (Tt) dominant. By writing traits in this fashion, you can read genetic manuscripts and easily recognize the **genotype** and/or **phenotypes** being reviewed. The genetic makeup of an organism is known as the genotype. This is different from phenotype which is an observable characteristic. For example, TT is the homozygous genotype for tall.

1. Write the homozygous recessive trait for shortness. __________
2. Write the heterozygous dominant trait for tall. ____________
3. List whether the following genotypes are homozygous dominant, homozygous recessive, or heterozygous dominant.

TT = ________________________

Ww = ________________________

aa = ________________________

AA = ________________________

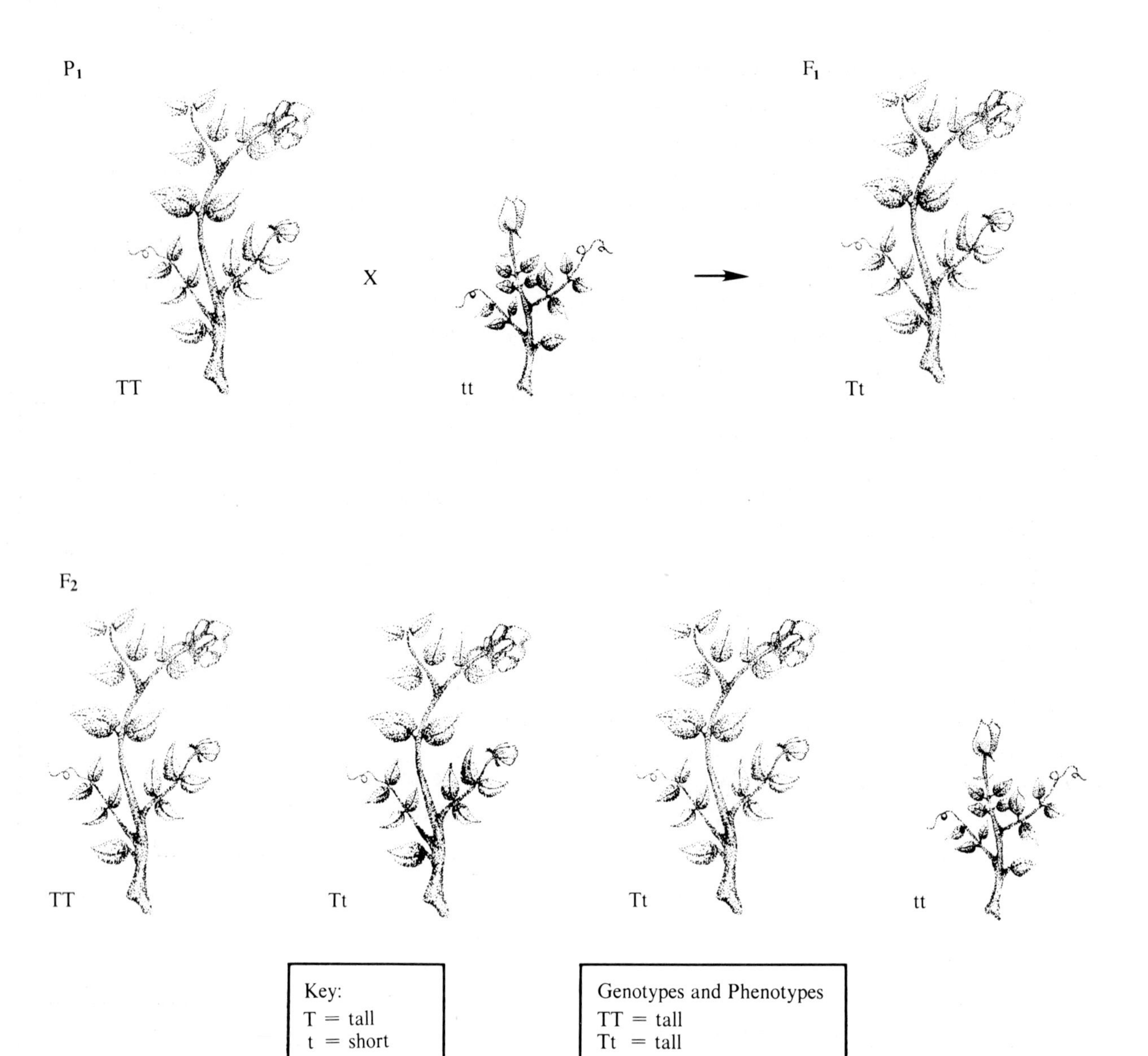

Figure 10.2 Mendel's law of segregation is illustrated here. Note that although the short plant in F_1 doesn't seem to be seen, it reappears when like F_1 plants are crossed yielding the F_2 generation.

10.5 A parent shows the genotype Ee. The alleles for the **gametes** are E and e, which the parent obtained from his/her parents. Once gametes are known for each parent, you can calculate what the genotypes for each offspring will be. A good method for doing this is using a **Punnett square,** a means for solving genetic problems introduced by R. R. Punnett, a poultry geneticist.

The Punnett square was designed to show the genetic possibilities of offspring. It is best that you become acquainted with the Punnett square step-by-step. You will begin with the Punnett square using a *single trait.*

The Punnett square is a series of boxes, as illustrated in figure 10.3.

The top shall be identified as the position for male traits, carried by the sperm gamete, and the left side shall be identified as the position for female traits, carried by the egg gamete, as illustrated in figure 10.3.

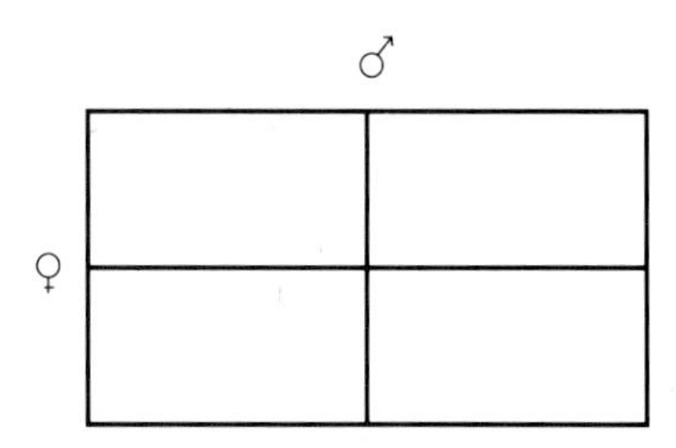

Figure 10.3 The format for a typical Punnett square.

If capital T stands for tall plants and small t represents short plants, the crossing of them could result in an F_1 generation. The parents are *homozygous.*

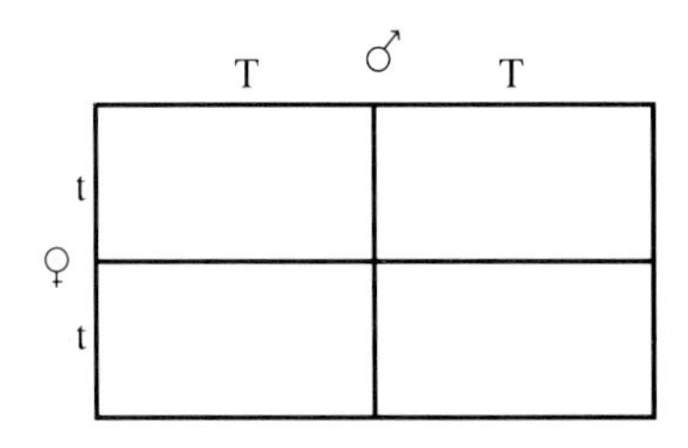

Figure 10.4 Placement of the alleles on a Punnett square.

To determine the F_1 generation, merely cross multiply the alleles.

Note: The dominant trait is always written first. T is dominant over t. The plant would be tall, but each offspring would carry genetic information for shortness.

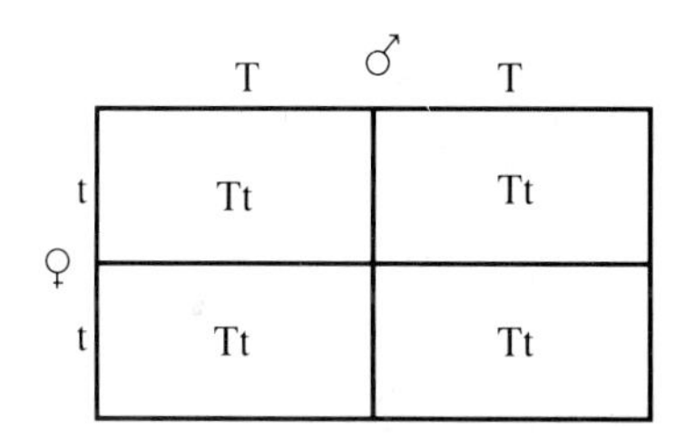

Figure 10.5 Solving a genetic problem with a Punnett square.

If you were to cross a male and female heterozygous pair to produce a F_2 generation, the results would begin to resemble Mendel's findings. The Punnett square is set up for you to complete below.

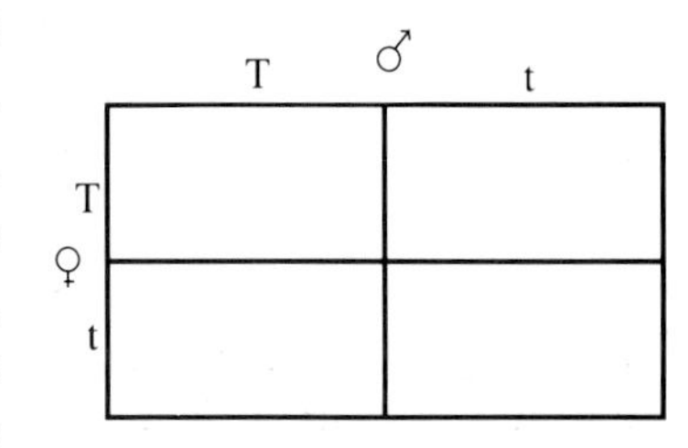

Figure 10.6 Work this problem in your logbook.

10.6 While each of the cells in figure 10.6 can be considered a separate, single *genotype,* the way in which an individual looks may be similar to others, even though the genotypes are different. For example, in figure 10.6 you have one-fourth of the individuals as homozygous tall. At the same time, one-half of the individuals will be heterozygous tall. But, tall is tall and, therefore, three-fourths of the individuals will be tall. The fact that these organisms all look tall is called the *phenotype.* The phrase *look-alike* is often used with phenotypes. In the example just used, the phenotype ratio is 3:1 with three individuals showing TT or Tt and one individual showing a homozygous recessive genotype. The genotype ratio for this example is 1:2:1. Do you see how these ratios were obtained?

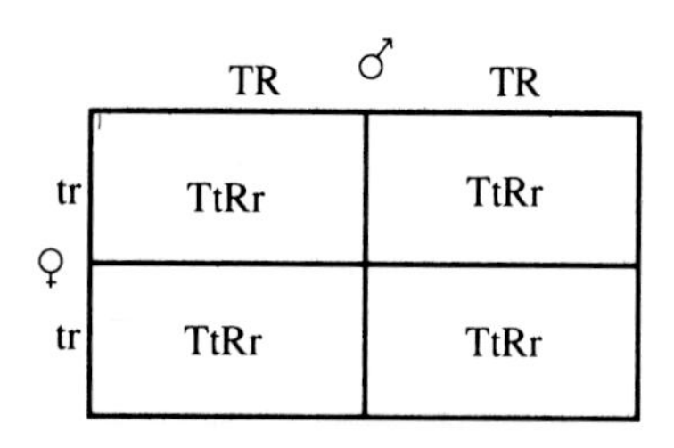

Figure 10.7 A solved genetic problem using two alleles. Note this problem illustrates parents with homozygous traits.

10.7 Mendel's second law, the *law of independent assortment,* states that members of each pair of factors segregate independently of all other pairs. This is readily seen when two genetic traits are studied at the same time. To apply two traits at one time, the Punnett square becomes a bit more involved. Let us check this out. The traits that will be used are plant height and seed shape. These are keyed below.

T = Dominant tall, and t = recessive short

R = Dominant round, and r = recessive wrinkled

The F_1 generation is worked out for you in figure 10.7. Note the same basic arrangement, as with a single trait, is used. Note further the male parent was homozygous dominant for both traits, and the female was homozygous recessive. The F_1 generation will be (heterozygous, homozygous). ________________

If two F_1 individuals are crossed, TtRr × TtRr, the results will be quite stunning. The Punnett square for this problem is set up for you. You will need to fill in the possible gamete combinations for each parent before completing the Punnett square.

10.8 Now that you have completed the Punnett square, list the individual genotypes: ________________________________

____

What is the phenotype ratio for the above Punnett square? ________

Figure 10.8 A genetic problem for you to solve in your logbook. This Punnett square is set up for parents with two heterozygous traits. Do you understand why this problem is set up this way?

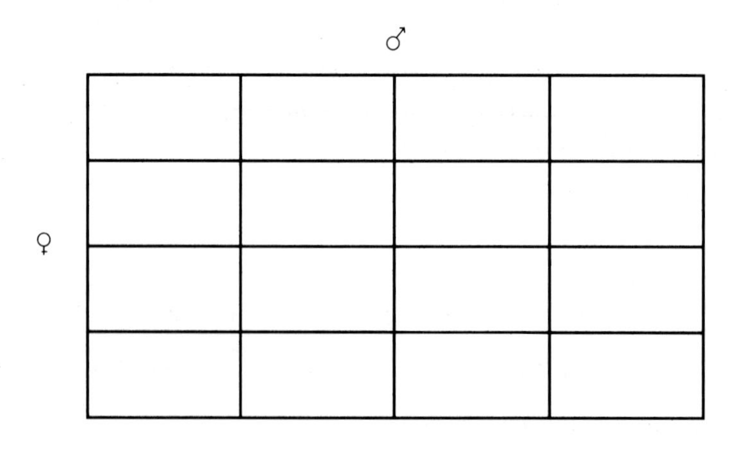

10.9 Many times the Punnett square method of determining the genetic traits of the offspring is not as easy to use as the algebraic form. The following is the algebraic approach to solving Mendel's genetic problems. If a hybrid is crossed with a pure recessive, the ratio would be 1 : 1. If we use T for tall and t for short, this can be illustrated as in figure 10.9.

Punnett square

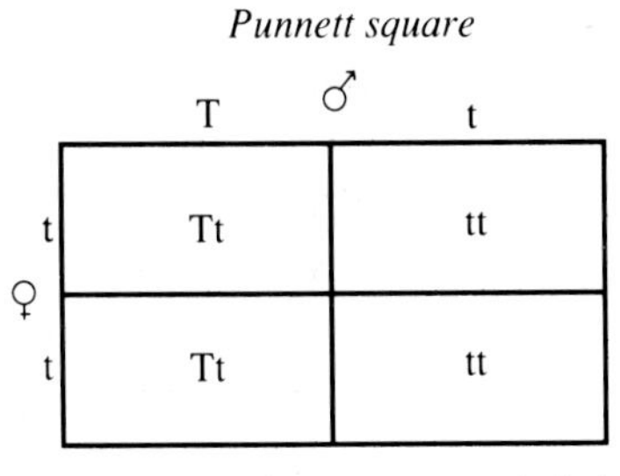

You can see the genotype ratio is 1:1

Algebraic method

[T + t]
x [t + t]

Tt + tt
Tt + tt

2 Tt + 2 tt

Genotype ratio = 1:1

Figure 10.9 Comparing two methods for solving the same genetic problem. Note one parent has a homozygous trait.

You can see the genotype ratio is 1 : 1.

By applying the same methods but using a two hybrid cross, the results would be as those seen in figure 10.10.

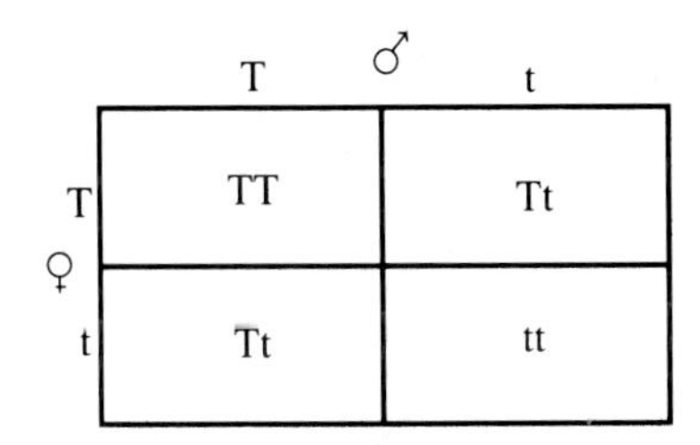

Genotype ratio = 1:2:1

[T + t]
x [T + t]

TT + Tt
Tt + tt

TT + 2 Tt + 2 tt

Genotype ratio = 1:2:1

Figure 10.10 Comparing two methods for solving the same genetic problem. Note both parents are heterozygous for the trait.

You solved a two trait problem above with both parents being heterozygous. Using the algebraic method, solve the same problem. Use the same letter symbols (TtRr). (*Hint:* Work each trait first, and then multiply the products.) Show your work on a separate sheet of paper and attach it to your chapter for evaluation.

10.10 What is the phenotype ratio of the problem you just completed algebraically? ________ Does this ratio agree with the phenotype ratio you listed in 10.8? ________

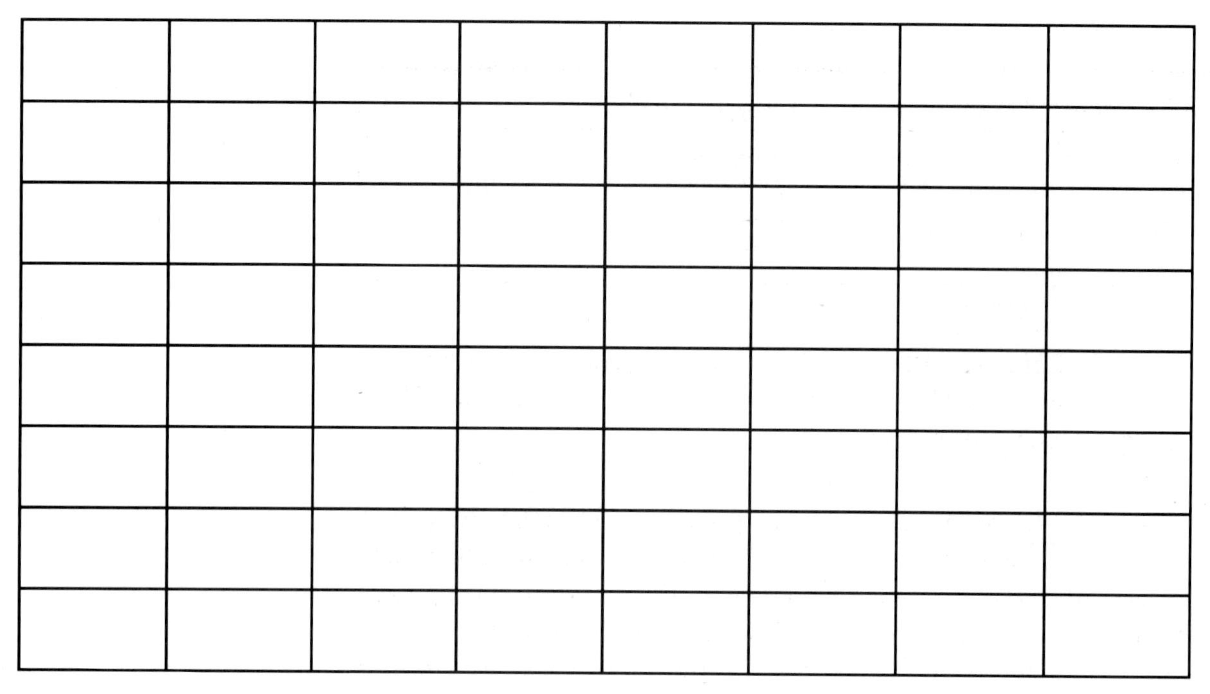

Figure 10.11 Punnett square for a trihybrid cross. Solve this problem with the grid given to you in your logbook.

10.11 You have completed a monohybrid and a dihybrid cross. A trihybrid cross is also possible. Using three of Mendel's pea traits, complete the trihybrid cross of heterozygous parents for seed coat color, flower color, and flower position (YyPpAa) in figure 10.11.

10.12 What is the phenotype ratio of the above trihybrid cross?

____

10.13 Genetics, as we know it today, goes beyond heredity according to Mendel. Up to this point, all of the **autosomal** situations, involving any chromosome other than a sex chromosome have been based on traits that had just two alleles. There are *multiple allele* situations. Blood types may have three alleles per gene.

The normal phenotypes of blood are A, B, AB, and O. However, the possible genotypes for each phenotype suggests that A and B are dominant to each other or codominant, and both are dominant to O. The only way a person with type O blood may come by that genotype is $i^{O}i^{O}$.

Using parents with $I^{A}i^{O}$ and $I^{B}i^{O}$ blood types, show the possible genotypes for offspring. (You may wish to use a Punnett square to show your results.)

Data Table 10.3 Blood Groups

Phenotypes Blood Types	Possible Genotypes
A	I^AI^A, I^Ai^O
B	I^BI^B, I^Bi^O
AB	I^AI^B
O	i^Oi^O

10.14 Data table 10.3 reflects the possible genotype-phenotype possibilities. Compare your answer in 10.13 to this table. What seems to be unique to that mating?

10.15 In some instances, there seems to be a lack of dominance with some traits. This suggests *incomplete dominance.* Some animals and plants show incomplete dominance when they are mated. Using a Punnett square, cross a red bull (H^R) with a white heifer (H^w). It is okay to use more than one letter. Assume the cattle to be homozygous to their respective traits.

10.16 You should have all of the offspring as being H^RH^w. In this situation, the red and white colors mix to make a roan color, which is neither red nor white, a visible sign of incomplete dominance. Complete a F_2 generation cross. What is the phenotype ratio of the offspring?

____

10.17 *Epistasis* is the ability of a gene to mask the effect of a different gene on a different locus point. Deafness is a human trait that is sometimes controlled by epistatic conditions. Individuals with H or E alleles have normal hearing while the presence of the DD alleles cause deafness regardless of any other gene possibilities. Assume parents to be HHEeDd and HhEeDd. What are the possible genotypes for this cross? You will need to use a Punnett square to solve this problem. What is the percent of individuals that would have a deafness disability from the results of this Punnett square? ________ What is the phenotype ratio for this problem? ________

10.18 Autosomes are all chromosomes other than sex chromosomes. Sex chromosomes identify the sex of the individual. The sex chromosomes that indicate a female are (*XX*). The male has one of the *X* chromosomes replaced with a *Y* chromosome (*XY*). The possibility of having a child born male or female is ________ . Show this percentage with the use of a Punnett square.

Were 50 percent of the offspring male? ________ female? ________

10.19 The *Y* chromosome carries no sex-linked genes. Sex-linked genes have alleles on the *X* chromosomes only. Eye color in the fruit fly Drosophila melanogaster is a sex-linked trait. Let X^R stand for red eye color, and X^r stand for white eye color. The fact that the trait is shown with an *X* means it is sex-linked. Solve the problem where a red-eyed male mates with a heterozygous red-eyed female.

What are the results of this crossing?

10.20 As mentioned above, a dominant trait in fruit flies is the red eye color. White eyes are recessive. A red-eyed male is mated to three females. The results are as follows:

1. Fly A, white eyes, has mostly red-eyed offspring.
2. Fly B, white eyes, has some white-eyed offspring.
3. Fly C, red eyes, has some white-eyed offspring.

Determine the genotypes of each parent including the male fly. Indicate the genotypes by symbols (X^R meaning red eye and X^r meaning white eye) in data table 10.4.

Data Table 10.4 Determining the Eye Color of Drosophila melanogaster

Individual	Genotype
Fly A	
Fly B	
Fly C	
Male fly	

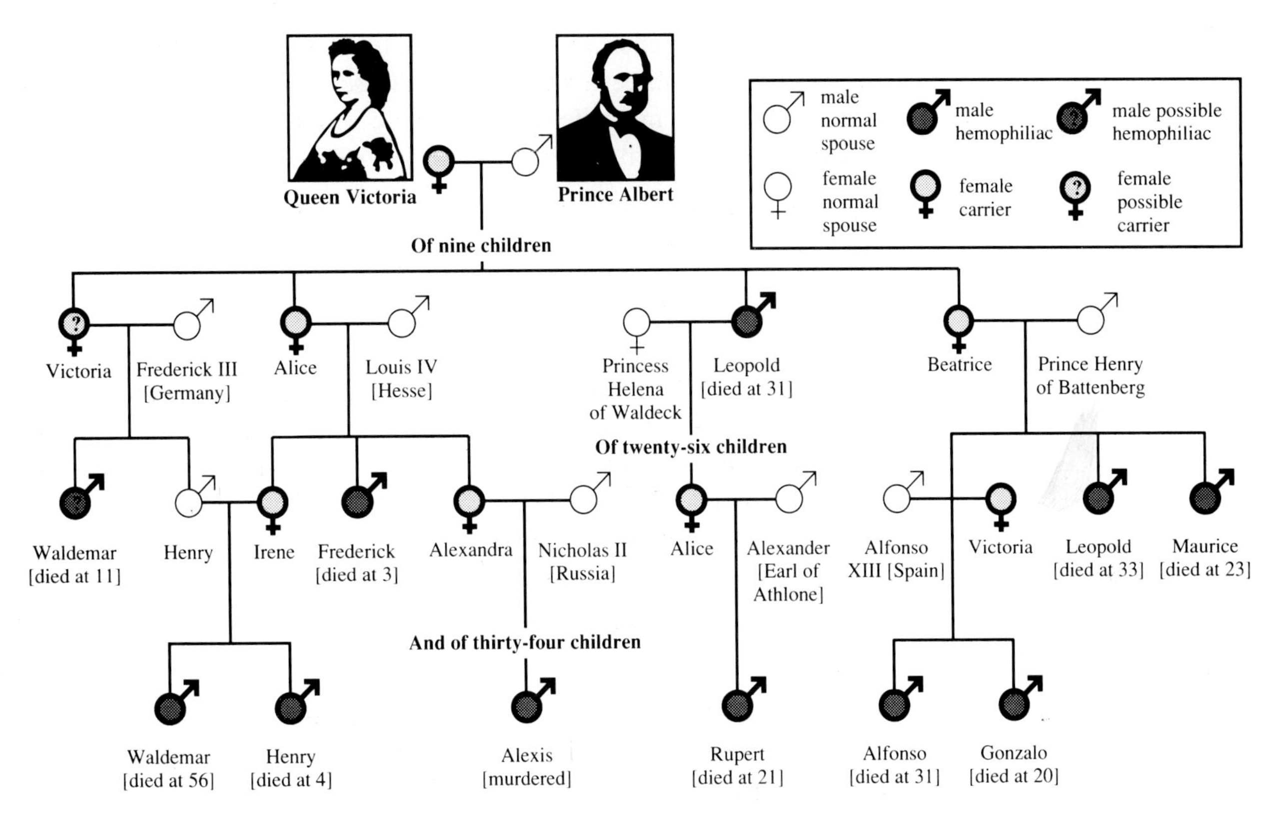

Figure 10.12 Pedigree of the Royal Family of England, exclusive of the present family, which was not affected by hemophilia.

10.21 There are many sex-linked traits associated with humans. Colorblindness is a recessive sex-linked trait. The Punnett square below (figure 10.13) illustrates a cross between a normal female and colorblind male.

1. Are any of the female offspring colorblind? ______
2. Are any of the male offspring colorblind? ______
3. Are the female offspring heterozygous or homozygous? ______
4. Are the male offspring heterozygous or homozygous? ______
5. What would the offspring genotypes be if a heterozygous female were to mate with a normal vision male? Colorblind male? Show your work below.

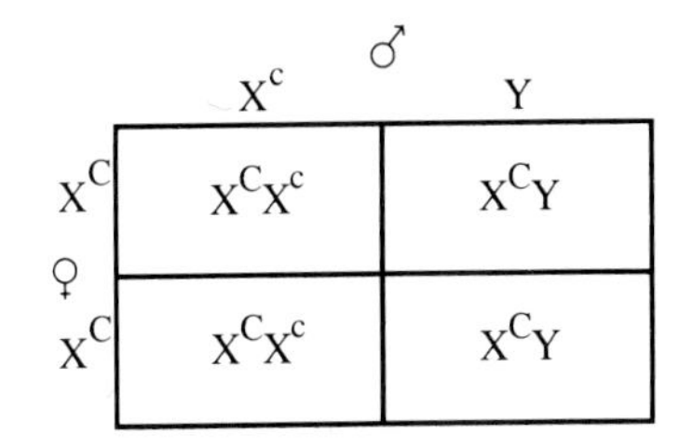

♀ \ ♂	X^c	Y
X^C	X^CX^c	X^CY
X^C	X^CX^c	X^CY

Figure 10.13 Punnett square showing the results of a cross between a color-blind male and a homozygous, normal vision female.

10.22 Hemophilia is probably one of the best examples of a sex-linked genetic disease. The history of the royal family of England is the best example of this genetic disease and how it is propagated. Figure 10.12 depicts a simplified version of the royal family. The present family has had no trace of this disease.

Notice that women are only carriers of hemophilia. Rarely is a female afflicted with this disease. Using a Punnett square, show the only way a female could be hemophilic (approximately 1 per billion). Let X^H be the normal allele and X^h be the afflicted allele for hemophilia.

10.23 Supposing you are a genetic counselor and a married couple has come to you to determine the possibilities for their offspring to be hemophilic. Show the genotypes and phenotypes that are possible with each situation. Again let X^H be the normal allele and X^h be the afflicted allele for hemophilia. Show your work and give your final answers on a separate sheet of paper.

Situation 1: Normal male + normal female. No history of hemophilia.

Situation 2: Normal male + normal female. The female's family has a history of hemophilia.

Situation 3: Normal female + male with hemophilia. The female's family has a history of hemophilia.

10.24 There is a trait that is easy for you to check to determine whether you have it or whether you are free of it. This trait is the ability to taste PTC (phenylthiocarbimide). Tasters are able to distinguish a definite bitter flavor while nontasters sense nothing. Tasting is dominant. PTC paper can be tasted by three-fourths of the American population while the remaining one-fourth cannot taste the paper. Get enough PTC paper so each member of your family can be tested. Simply place a piece of PTC paper on your tongue. The rest will come naturally. Record your reaction to PTC paper and that of each member of your family in data table 10.5. Are your parents homozygous tasters, heterozygous tasters or homozygous nontasters? ________________ Explain.

Data Table 10.5 Results of Families Ability to Taste PTC.

Relation	Taster (yes/no)

10.25 Heterozygous parents for *albinism,* a trait for lack of skin pigment, have a one-in-four chance of producing an albino child. Determine the genotype of an albino male who is a taster but has a nontaster mother. Discuss the rationale for your answer.

10.26 What is the genotype of a woman who is a nontaster, normal pigmented skin, but whose father is an albino taster and whose mother has homozygous traits for normal skin color and tasting. Discuss your answer.

10.27 If the male in section 10.25 and the female in section 10.26 were to marry, what would be the possible genotypes of their offspring? Show your calculations.

10.28 Data table 10.6 lists a dozen genetic traits. Take a look at your classmates and others in your school. Literally poll them to determine which traits are dominant and which are recessive.

Data Table 10.6 Genetic Traits

Genetic Trait	Tally Sheet Dominant Trait	Recessive Trait
Freckles	Present	Not present
Hair color	Not red	Red
Widows peak	Present	Not present
Tongue rolling	Able to do	Not able to do
Dimples	Present	Not present
Hair of middle joint	Present	Not present
Hair texture	Curly	Straight
Hand preference	Right-handed	Left-handed
Ear lobe	Free	Attached

10.29 Carefully read each statement below. One or more of the statements are genetic principles, and one or more are false statements or superstitions. Check the appropriate box for each statement indicating whether or not you think it is a genetic truth.

Statement	True	False
1. Acquired characteristics, such as mechanical or painting or mathematical skills, may be inherited.		
2. The father always determines the sex of a child.		
3. Genes of the offspring are received as 50 percent ratio from the parents.		

10.30 There have been many concepts presented in this chapter that are based on Mendel's experiences. You may need to review this chapter carefully so you feel confident of your understanding of the concepts. Review the study points below. If you feel you can relate specifically to the study point, place a check (✓) in the corresponding box.

On completion of this chapter, you should be able to do the following:

☐ 1. Distinguish between dominant and recessive traits.

☐ 2. Determine genotype and phenotype ratios using Punnett squares.

☐ 3. Determine entire filial generations with Punnett squares.

☐ 4. Explain Mendel's law of segregation.

☐ 5. Explain Mendel's law of independent assortment.

☐ 6. Explain the meaning of codominant traits.

☐ 7. Explain how sex-linked traits are transferred and accepted.

☐ 8. Explain how multiple alleles react similarly to codominant situations.

☐ 9. Explain what happens with incomplete dominance.

Genetic Ratios/ Genetic Poker

11

How are genetics and poker similar?

11.1 If you flip a coin in the air and allow it to land on the floor, the probability it will turn up heads is 50 percent. Likewise, the probability of the coin turning up tails is ________ percent.

Using a single coin, flip it 50 times. Record a tally mark in the appropriate place in data table 11.1 after each flip of the coin. How many times should fifty flips turn up heads? ________ tails? ________ What ratio do you predict will be the number of times each will turn up? ____

Data Table 11.1 Heads or Tails

Heads	**Tails**
Total	Total

What is the actual ratio from your data? ________ Does the actual ratio match your prediction? ________

11.2 The percentage of error can be calculated. It tells you how far off you are from expected answers. To calculate the percentage of error, subtract the number you got from the accepted number (25) and divide the accepted number (25) into the difference. Multiply your division answer by one hundred. This will give you an answer based on percent. The mathematical formula for this calculation would look like the following:

$$\frac{\text{Difference}}{\text{Accepted}} \times 100 = \text{percent of error}$$

Show your calculations, in the space provided in your logbook, for calculating the percentage of error for the data obtained in data table 11.1.

11.3 Read all of section 11.3 before you begin. If you were to put fifty pennies into a box heads up and shake it up and down three times, you would expect ________ (number) pennies to be remaining heads up. If you were to remove the pennies showing tails and repeat the procedure, you would expect ________ pennies to show heads up.

Data Table 11.2 Shaking the Pennies

Trial No.	Prediction	Actual
1		
2		
3		
4		
5		
6		
7		
8		
9		
10		
11		
12		
13		
14		
15		
16		

Now, obtain a box containing fifty pennies. Place all of the pennies heads up and shake the box up and down three times. Count the pennies showing heads and record this number in data table 11.2. Remove the pennies showing tails, and continue the process with those pennies that remain. Continue this process keeping the conditions as similar as possible until you have removed all of the pennies, recording the results after each shaking sequence. Remember to remove those pennies that have turned over. *Make your complete predictions in data table 11.2 before you do any shaking.*

11.4 How did your predictions in data table 11.2 hold up against the actual data you collected?

Why were there variations? Could you have done better in controlling the variations?

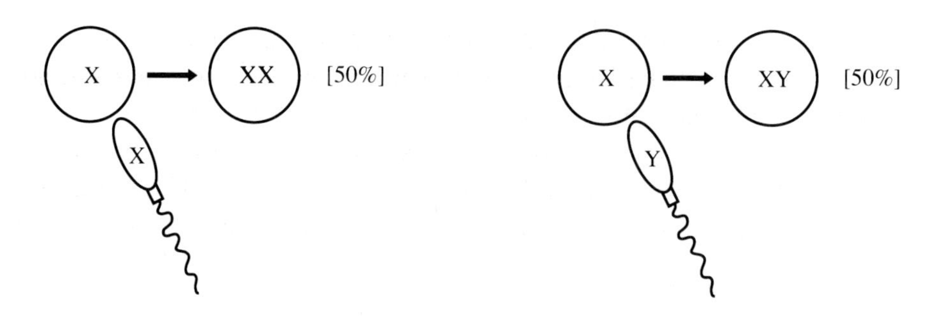

Figure 11.1 The probability of sex determination by fertilization.

11.5 While flipping coins and shaking boxes of coins is fun and easy, it hardly has anything to do with genetics, or does it? Percentages are a part of the study of genetics, as are probabilities.

You can determine the sex of offspring by knowing which sperm cell will fertilize the egg cell. You studied this point in chapter 10. Sex chromosomes are different for female and male. The female chromosomes are identical. The female egg contains two ***X* chromosomes.** The male, on the other hand, has different chromosomes. The male carries an ***X* chromosome** and a ***Y* chromosome.** The determination of the sex of the offspring is directly related to which sperm cell fertilizes the egg.

If you follow figure 11.1, you will be able to follow sex determination by fertilization.

Since you now know what the outcome can be, what is the percentage of an offspring being a male? ________ female? ________ This seems somewhat similar to flipping a coin.

11.6 In chapter 10 you became aware of blood genotypes. If type O is a recessive trait and A and B are dominant to O but not dominant over each other (codominant), then what blood type would a person have if the father is type O and the mother is type A? ________ Why?

Would the same be true if the type A parent were type B? ________

11.7 A father is typed and found to have type AB blood. A mother is typed and found to be AO. What is the percentage of the offspring having type A blood? ________

What are the chances (percent) of an offspring having type B blood?

____

11.8 Probabilities enable you to estimate the chances for a particular prediction to come true. However, predictions will not always come true. Certain outside influences may affect the results.

If you were to use a deck of cards and take only the four 9s from the deck, the probability of your drawing the 9 of spades would be one chance in four. What would be your chance of drawing a 9 from the entire deck?

________ Explain.

11.9 Dice are interesting devices for looking at probability. Two dice are rolled simultaneously. The probability of getting one die to have a 1 would be ________. The probability of having the second die roll to a 1 would be ________. The probability of both die rolling so both show a 1 and, therefore, getting "snake eyes" is ________.

11.10 In a family with four children, three of which are boys and one a girl, the possibilities for order of children born are: boy-boy-boy-girl, boy-girl-boy-boy, boy-boy-girl-boy, or girl-boy-boy-boy. If you use symbols B for boy and G for girl, the order would look like this: BBBG, BGBB, BBGB, GBBB.

The probability for any one of the combinations would be:

BBBG $(\frac{1}{2} \times \frac{1}{2} \times \frac{1}{2} \times \frac{1}{2}) = 1/16$

GBBB $(\frac{1}{2} \times \frac{1}{2} \times \frac{1}{2} \times \frac{1}{2}) = 1/16$

BGBB $(\frac{1}{2} \times \frac{1}{2} \times \frac{1}{2} \times \frac{1}{2}) = 1/16$

BBGB $(\frac{1}{2} \times \frac{1}{2} \times \frac{1}{2} \times \frac{1}{2}) = 1/16$

If you were to figure the probability, but not specify any order, the probability would be:

$$\frac{1}{16} + \frac{1}{16} + \frac{1}{16} + \frac{1}{16} = \frac{4}{16} \text{ or } \frac{1}{4}$$

Using a similar example, as above, calculate the probability for each order if there were five children in a family having three girls and two boys. Include the symbolism for the orders and the calculations for each.

What would be the probability in the above example if there were no order specified for the calculation? Show your work.

Data Table 11.3 Spades and Clubs

Plays	TT—Two Spades	tt—Two Clubs	Tt—Club & Spade
1–16			
17–32			
33–48			
49–64			

11.11 The Chi-square test is a statistical test that is useful in genetics. This test enables observed variations to be analyzed mathematically. The Chi-square test allows for comparisons to be made between expectations and actual results. The function of this test is to compare obtained results with expected results on the basis of chance.

Genetic ratios normally are not found to be a precise 3:1, 2:1, or 9:3:3:1. The reason for this is the number of variations due to chance. To show this more clearly obtain two decks of cards from your teacher. Remove all the jokers. Separate each deck into two respective piles of black and red cards. You should have four piles, two black and two red. For this part you will need only the two piles of black cards. Shuffle each pile and place them face down.

Place a single card from each pile face up simultaneously. Record whether the two cards are both spades, clubs, or a spade and a club in data table 11.3. After sixteen plays, reshuffle the piles and repeat the process three more times for a total of sixty-four plays, recording each play by a tally mark in the appropriate space.

11.12 Assume the spades are dominant for tallness and the clubs recessive for shortness. How many tall plants did you have after sixty-four plays? ________ Short? ________

If your data had shown exactly a 3:1 ratio, how many tall plants would you have expected? ________ short? ________

Is your data identical to the expected? ________ Explain the reason for any difference.

11.13 The purpose of the Chi-square test is to determine if experimental data provides an acceptable approximation of an expected ratio. Using your data from data table 11.3, complete data table 11.4.

Data Table 11.4 Chi-square for One Trait

	Tall (TT or Tt)	**Short (tt)**
Actual total of sixty-four plays (a)		
Expected total of sixty-four plays (e)		
Deviation from expected ($a - e = d$)		
Deviation squared ($d \times d = d^2$)		
$\frac{\text{Deviation squared}}{\text{expected}}$ $\left(\frac{d^2}{e}\right)$		
Chi-square $= \frac{d^2}{e}$ for tall $+ \frac{d^2}{e}$ for short $= X^2$ (in decimal)		

Use the Chi-square test (data table 11.8) at the end of this chapter to determine the probability of your results. As you look at this table of numbers, note the column for degrees of freedom. The row you pick from this column is determined by subtracting one (1) from the number of columns of data you are collecting. In the case of data table 11.4, there are two columns. Therefore your degree of freedom row would be one, or $2 - 1 = 1$.

Read across this row until you find your Chi-square value, or approximate value. Note the number heading your column. If necessary, your teacher may need to show you how to interpolate for a more precise answer, but for our work, a close approximation is adequate. Assume your Chi-square value was .148. With one degree of freedom, this figure lies in the 0.70 column.

Values between 0.50 and 0.99 would indicate any variation of the results would be due to chance. As this figure continues towards 0.99, the greater the probability the variation between expected and actual results is due to chance. Likewise, as the X_2 value continues towards .001, the least likely the results were due to chance. What results do you get with your data?

Does this value indicate the variation was due to chance? ________
Mendel had determined an approximate 3:1 ratio of tall and short plants in the F_1 generation. Is your ratio similar? ________

11.14 In the previous investigation you were only referring to one trait, tall or short. To expand to two traits, you will need to use both the black and red cards. You will continue to use sixteen plays at one time up to a total of sixty-four plays, just as you have been doing. You will record tally marks of what is happening as you did in the previous

Data Table 11.5 Spades, Clubs, Hearts and Diamonds

Plays	**Two Spades TT**			**Two Clubs tt**			**Club and Spade**		
	Two Hearts AA	**Two Diamonds aa**	**Heart and Diamond Aa**	**Two Hearts AA**	**Two Diamonds aa**	**Heart and Diamond Aa**	**Two Hearts AA**	**Two Diamonds aa**	**Heart and Diamond Aa**
1–16									
17–32									
33–48									
49–64									

investigation. For example, if the black cards turned over are two clubs, go to the two-clubs column of data table 11.5. Then turn up two red cards simultaneously. If you have turned up a heart and a diamond, under that heading in the Two Clubs column, you will record your tally mark. Repeat this process for the remainder of the plays. This investigation will illustrate a **dihybrid** cross as opposed to a **monohybrid** cross from the previous investigation. Record all of your results in data table 11.5.

11.15 If spades are dominant for tallness and clubs recessive for shortness, and if hearts are dominant for axial flowers and diamonds are recessive for terminal flowers, how many of each possible combination do you have? tall, axial? ______ ; short, axial? ______ ; tall, terminal? ________ ; short, terminal? ________ . What ratio would you expect to obtain with two traits? ________ Did you? ________ Mendel did get a 9:3:3:1 ratio. Are your results similar to his? ______

11.16 Complete data table 11.6 using your results in data table 11.5. Using the Chi-square table, determine the probability of your results with three degrees of freedom. What is the probability of your results for a dihybrid cross? ________________

Data Table 11.6 Chi-Square for Two Traits

	TA	tA	Ta	ta
Actual total of sixty-four plays (a)				
Expected total of sixty-four plays (e)				
Deviation from expected ($a - e = d$)				
Deviation squared ($d \times d = d^2$)				
$\frac{\text{Deviation square}}{\text{expected}} \left(\frac{d^2}{e}\right)$				
Chi-square $= \frac{d^2}{e}$ for each $+ \frac{d^2}{e}$ for each $= X^2$				

Data Table 11.7 Chi-square Values

Degrees of freedom (N)	P=.99	.95	.90	.80	.70	.50	.30	.20	.10	.05	.02	.01	.001
1	0.00015	0.0039	0.016	0.064	0.15	0.46	1.07	1.64	2.71	3.84	5.41	6.64	10.8
2	0.020	0.10	0.21	0.45	0.71	1.39	2.41	3.22	4.61	6.00	7.82	9.21	13.8
3	0.16	0.35	0.58	1.01	1.42	2.37	3.67	4.64	6.25	7.82	9.84	11.34	16.3
4	0.30	0.71	1.06	1.65	2.20	3.36	4.88	5.99	7.78	9.49	11.67	13.28	18.5
5	0.55	1.15	1.61	2.34	3.00	4.35	6.06	7.29	9.24	11.07	13.39	15.09	20.5
6	0.87	1.64	2.20	3.07	3.83	5.35	7.23	8.56	10.65	12.59	15.03	16.81	22.5
7	1.24	2.17	2.83	3.82	4.67	6.35	8.38	9.80	12.02	14.07	16.62	18.48	24.3
8	1.65	2.73	3.50	4.59	5.53	7.34	9.52	11.03	13.36	15.51	18.17	20.10	
9	2.09	3.33	4.17	5.38	6.39	8.34	10.66	12.24	14.68	16.92	19.68	21.67	
10	2.56	3.94	4.87	6.18	7.27	9.34	11.78	13.44	15.99	18.31	21.16	23.21	
15	5.23	7.26		10.31		14.34		19.31		24.99		30.58	
20	8.26	10.85		14.58		19.34		25.04		31.41		37.57	
25	11.52	14.61		18.94		24.34		30.68		37.65		44.31	
30	15.0	18.50		23.36		29.34		36.25		43.77		50.90	

Data Table 11.8 Normal Curve Tally Sheet

	1	2	3	4	5	6	7	8	9	10	11	12	13
1–25													
26–50													
51–75													
76–100													
101–125													
126–150													
151–175													
176–200													
201–225													
226–250													
251–275													
276–300													

11.17 Some statistical problems require data to be projected similar to that of a normal curve. Large population tallies would be such a use for this type of statistical measure.

If a large population of students were to take an examination, the majority would receive the letter grade of C. This is true if a *normal curve* is used to determine a grade. Fewer, but equal numbers, would get a B and a D, and the smallest number, but also an equal number of students, would get an A or an F grade. An example of a normal curve is illustrated below.

Can a normal curve be constructed from legitimate data? Get the normal curve game, which includes a game board and twenty-five marbles. Run enough data so you can complete a graph from the data. Release the marbles in the same fashion to assure little variation. You will need at least three hundred pieces of data to make your data significant. This is one time when "the more the better" philosophy really works. Tally your data in data table 11.8.

Figure 11.2 A normal curve is perfectly symmetrical. Such a curve, however, is difficult to attain due to variations in attainable data.

11.18 Graph the results of the data that you have collected above on a piece of graph paper. Graph the slot number on the (*X*) or horizontal axis and the number of times per slot on the (*Y*) or vertical axis. Graph paper is available in your logbook.

1. Does the graph from your data resemble figure 11.2? ______
2. Are there any points that do not fit on the curve? __________
3. Propose some reasons for this.

11.19 Having a statistical background enables you to see more clearly the framework for genetic studies. This chapter is somewhat difficult but fun to complete. Review the study points for this chapter. Place a check (✓) in the box when you have mastered the study point.

On completion of this chapter, you should be able to do the following:

☐ 1. Determine the probabilities for various combinations of offspring.

☐ 2. Use the Chi-square table effectively.

☐ 3. Explain the probabilities of a monohybrid and dihybrid cross using Chi-square statistical procedures.

☐ 4. Explain the use of a normal curve.

An Introduction to Evolution

12

What ever happened to life?

12.1

Sexual reproduction gives claim to an essential feature, namely, the formation of a **zygote.** The single cell zygote forms from the uniting of two sex cells, or gametes, the egg (or ova) and the sperm (or spermatozoa).

Our ancestors were not knowledgeable about sexual reproduction and heredity. Aristotle, who is called the "Father of Biology," was unable to observe reproduction as precisely as many of his other biological observations. He believed that blood was the key to reproduction, and his influence continued for centuries.

During the Renaissance period more exact information became available. William Harvey (1650) studied reproduction by using pregnant animals. By dissecting pregnant animals at various stages of gestation, he determined the development of the young animals yet to be born. Through this work, Harvey realized that the egg was a very important part of the reproduction of offspring. He made no direct conclusions about the existence of the sperm or about conception, however.

Figure 12.5 Homunculus

Anton van Leeuwenhoek, along with his first microscope, was able to study the human egg. Since the sperm cell is still thousands of times smaller than the egg cell, it was not detected with earlier microscopes. In the 1700s, when scientists were able to study sperm cells, they noticed a tail and a head. The head was described as having a small man in it, so the sperm cell was called *homunculus,* Latin for "little man." The illustration below compares a drawing of a homunculus and a sperm cell. By using your imagination, you should be able to see what looks like a homunculus in the actual sperm cell.

Even with the discovery of the sperm cell, the idea of conception was still not fully understood. Some believed the sperm cell was wholly responsible; others believed the "all egg" theory, identified by Harvey.

In the middle 1700's, Pierre Louis de Maupertius stated that the independent egg and sperm theories were incorrect. He claimed that individuals developed with characteristics from each parent. It was not, of course, until the middle 1800s that science was able to combine the information about the role of sperm cells and egg cells into usable theories. For it was during this period of time that Gregor Mendel, an Augustinian monk, Charles Darwin, a naturalist, and others began to formulate their theories of heredity, natural selection, and organic evolution and to apply them to the living world as we know it. The information that these scientists gathered and reported greatly altered the understanding of life science during the next one hundred years.

12.2 In 1809, Jean Baptiste de Lamarck, a French biologist, proposed an explanation for the continued development and change of species. Lamarck felt that an organ of an animal increased in size if the organ was used. He also felt an organ would atrophy if it were not used. As a result, certain organs developed and were passed on to offspring, while others disappeared completely. Lamarck's ideas were widely accepted.

A most often used example of Lamarck's theory involves the early development of the giraffe. According to Lamarck's theory of use and disuse of organs, giraffes probably had short necks originally. The short neck allowed the animal to reach grass for food. However, as grass became scarce, the need to attain tree foliage for food demanded that giraffes acquire longer necks. Those giraffes able to acquire this trait survived and passed the trait on to their offspring. Giraffes unable to acquire the trait for a longer neck did not survive.

12.3 Some thirty years after Lamarck's proposal, Charles Darwin made a memorable trip as a naturalist aboard the H.M.S. Beagle. The Beagle was bound for the Galapagos Islands, off the west coast of South America, so Darwin was able to study the wildlife there. Because of these observations, Darwin was able to write his theory of evolution in a book entitled *The Origin of Species by Means of Natural Selection.*

The theory of natural selection proposes the following factors to account for evolutionary change: overproduction, the struggle for existence, variation, the survival of the fittest, and inheritance.

If we compare the Lamarck's giraffe theory and Darwin's, there are distinct differences. First, Lamarck felt that animals grew the organs to compete for survival. Darwin said instead that the shorter giraffes died out but those that were naturally taller were able to survive. Thus, the genetic inheritance for tallness was passed on to their offspring. The genetic inheritance for shortness was eradicated because the parent stock died out as food became inaccessible to them.

The major criticism of Darwin's work was that it did not explain satisfactorily how variations occurred. Nor did his theory distinguish between variations caused by heredity and variations caused as a result of environmental and geographical changes. You know that hereditary variations were later explained by Gregor Mendel. Environmental and/or geographical changes needed further clarification.

12.4

In 1901, Hugo De Vries made certain modifications of Darwin's theory. De Vries called his theory the **mutation theory.** Essentially, he proposed that mutations were the basic cause of variations.

De Vries would explain that mutations brought about by normal genetic flow resulted in giraffes that had abnormally long necks and longer legs. These mutations became the "normal" giraffe because the shorter animals died out. In other words, the mutation was the source of the variation.

Fellow scientists' main criticisms of De Vries' theory were twofold. First, De Vries did not discuss the cause of the mutation. And second, there had never been a mutation of such magnitude (or population size) to enable observers to recognize the appearance of a new species distinct from the "normal" species. The scientific world continued to argue over the work of Darwin, Mendel, and De Vries.

Today, however, there is a modern theory of evolution that takes into account the best of Darwin, Mendel, and De Vries. Still using the giraffe as our model, the modern theory explains the change as follows:

1. The ancestors of the giraffe were smaller, with shorter necks and shorter legs. This enabled them to graze on low-level vegetation, possibly competing with other grazing animals. In essence, there may have been too many animals competing for the available food, a concept known as *overproduction.*
2. When food is scarce, the members of an entire population become involved in a *struggle for existence.* In the case of the giraffe, the competition was for low tree branches and their leaves.
3. The *variations* amongst the animals that were competing became increasingly important. Those giraffes with slightly longer legs and necks could reach the leaves on upper branches more easily, so they had more food available than shorter giraffes. As a result, the taller animals survived. The favorable genes for longer legs and necks were passed on to the offspring. The longer necks and legs could be considered a mutation from the "normal," as De Vries would say. The modern theory calls this the *survival of the best-adapted.* Darwin called it "survival of the fittest" or "natural selection."
4. Because the mutations were, in fact, variations in *heredity,* they were passed on from generation to generation. The end result is the modern giraffe. Is the giraffe still evolving? Probably not, but who is to say? Evolution usually takes a long time, or it may be continually ongoing.

Ecologists speak of animals having a niche. It is conceivable that giraffes are not evolving anymore because they are not competing with any other animals in their niche.

12.5

When members of the same species live in the same area, they are referred to as a **population.** When there is a change in the genetic makeup of a population, scientists say evolution has occurred, or is occurring, in that species. By studying the **gene pool** of a population, population geneticists are able to determine whether or not evolution is occurring within a given species.

A population can be studied mathematically to show how a gene pool and the gene frequency occur. The results produce substantial statistical data to support the existence or non-existence of an evolutionary trend. The principle that is used is called the **Hardy-Weinberg Principle.**

In 1908, Godfrey Hardy and Wilfred Weinberg showed mathematically how a gene pool of a given population does not change. When placed under specific conditions, genotypes maintain a constant ratio in a population. This occurs generation after generation. Of course, specific criteria must be met. These criteria include the following:

1. Mating must be random.
2. The genetics of a given population stays within the gene pool. Therefore, individuals cannot migrate out of the gene pool of the study population, nor can any outside individuals migrate into the study population. The term used to describe the outward flow of species is emigration. The converse of emigration is immigration.
3. No mutations can occur.
4. A large sample must be studied.

The fruit fly *Drosophila melanogaster* will be used to help you see how the Hardy-Weinberg Principle is applied.

If W and w are the allelic genes for red and white eyes of the fruit fly, we can assume that both types exist in fruit fly populations. One would expect normal Mendelian genetic principles to apply as the fruit flies mate to produce offspring. If 100 males and 100 females, each having the following genotypes, were allowed to reproduce, the results of the gene frequency should be easy to calculate.

Male and female genotypes:

36 males and females, each having a genotype homozygous dominant for red eyes

____

48 males and females, each having a genotype heterozygous dominant for red eyes

____

16 males and females, each having a genotype homozygous recessive for white eyes

____

Using the given symbols for red eyes and white eyes, fill in the above blanks with the appropriate genotype.

To allow easy calculations, assume each parent produces 10 gametes. Data Table 12.1 then establishes the gene frequencies of eye color in the female *Drosophila melanogaster.*

Since there are 100 individuals, the data can conveniently be expressed in percentages. Therefore, 60 percent of the individuals have the dominant trait for eye color and 40 percent of the individuals have the recessive trait.

Data Table 12.1 Gene Frequencies of Percent of Total Eye Color for Female *Drosophila melanogaster*

# of individuals	Genotype	W Gametes	w Gametes	Gamete Total	% of Total
36	homozygous dominant	$36 \times 10 = 360$	- - - - -	360	36
48	heterozygous dominant	$\frac{48 \times 10}{2} = 240$	$\frac{48 \times 10}{2} = 240$	480	48
16	homozygous recessive	- - - - -	160	160	16
= 100		600	400	1000	100

Since it is possible that male *Drosophila m.* show the same results as those given for females in Data Table 12.1, we can then say both females and males have equal percentages of dominant and recessive traits. Therefore, from these parents, offspring would have the following gene frequency:

		♂ W (.60)	♂ w (.40)
♀	W (.60)	WW (.36)	Ww (.24)
♀	w (.40)	Ww (.24)	ww (.16)

Therefore:
WW = .36

Ww = .24 + .24 = .48

ww = .16

In the above Punnett square, a predictable 1:2:1 Mendelian ratio is indicated. The Punnett square also shows that 36 percent of the offspring are homozygous dominant, 48 percent are heterozygous dominant and 16 percent are homozygous recessive. How do these figures compare with those established in Data Table 12.1?

In short, the Hardy-Weinberg Principle supports the idea that matings, when occurring randomly, when the population is sufficiently large, and when no mutations interfere, cause the gene pool frequencies to remain essentially constant from one generation to another.

This is called **genetic equilibrium.** If we apply genetic equilibrium to the classic giraffe example, we might assume today's giraffes are in a state of genetic equilibrium since no known changes seem to be occurring.

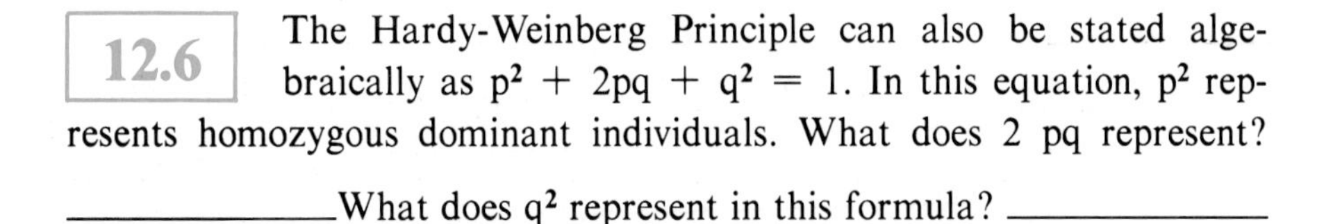

12.6 The Hardy-Weinberg Principle can also be stated algebraically as $p^2 + 2pq + q^2 = 1$. In this equation, p^2 represents homozygous dominant individuals. What does 2 pq represent? ____________ What does q^2 represent in this formula? ____________

12.7 If you use the above algebraical formula with the data given in the Punnett square above, you should find the formula to be true. Transfer the data from the Punnett square into the formula. Determine if the algebraic formula works. Show your work.

12.8

Knowing something about the Hardy-Weinberg Principle is necessary for population geneticists. But the principle has its flaws. For example, one of the criteria for using this principle is random mating. In the wild, however, mating is seldom random. Male animals compete for the right to mate. The most colorful male birds attract the females. Deer species combat each other for the right to mate with a specific doe. Many animals in herds have a single male that dominates the entire herd. In such cases, only the dominant male mates with the females.

Another part of the principle states that individuals of a species must stay in the population so the gene pool remains intact. Normally, individuals move freely, and as a result, move into and leave study populations regularly. The coming and going or migration of individuals changes the gene pool. Those changes affect the validity of the data when using the Hardy-Weinberg Principle.

Yet, migration is a natural behavior of animals, as well as a major evolutionary force. In fact, migration is just one of four major forces of evolution. The other three forces are genetic drift, mutation, and natural selection. All four forces have one thing in common: each is capable of causing evolution in a species over an extended period of time. Using a reference book, discuss each of these forces of evolution.

migration:

genetic drift:

mutation:

natural selection:

12.9

The search for evolutionary evidence continues. In their search, scientists have identified five types of evidence that support the theory of evolution.

1. Living things have unique, as well as similar, characteristics. These characteristics allowed an orderly taxonomic scheme to be developed more than one hundred years ago, one that continues to be effective. You will remember the categories used for the taxonomic scheme: kingdom, phylum, class, order, family, genus, and species. The relationships between living things is evident in this taxonomic scheme, and such relationships suggest evolutionary development.

2. There appears to be evidence of a common ancestry based on morphological structures. Organs of the various body systems are quite similar in terms of structure and location, as well as embryonic development. There is also evidence that vestigial organs, such as the appendix, also point to some common ancestor.
3. Comparative biochemistry between species suggests an evolutionary relationship.
4. The similarity of vertebrate development at the embryonic stage suggests that many vertebrates evolved from a common ancestry.
5. The formation of genetic hybrids suggests that species can change. This tends to support the idea that evolution can give rise to variations in a species.

12.10 The science world has accepted the concepts that support evolution. There are still many questions, however, that need to be answered. Only time and continued research will allow the theories of evolution to be more fully understood. More research is needed to get beyond homologous structures (structures that have a similar evolutionary origin) and analogous structures (structures that have similar function but not similar evolutionary origins).

Review the materials in this chapter. When you can complete the task in each of the study points below, check (✓) the box. When you complete Chapter 12, you should be able to do the following:

☐ 1. Discuss the differences of the evolution of the giraffe as proposed by Lamarck and Darwin.

☐ 2. Identify how De Vries' mutation theory was similar to Darwin's theory of natural selection.

☐ 3. Apply the modern theory of evolution to the evolution of the giraffe.

☐ 4. List the criteria for using the Hardy-Weinberg Principle.

☐ 5. Apply the Hardy-Weinberg Principle to a hypothetical situation.

☐ 6. Define the four forces of evolution.

☐ 7. List the five sets of evidence that support the theory of evolution.

Plants and the Environment 13

How do plants meet problems of existence on Earth?

13.1

You are an expression of life on Earth, the third planet from the sun in our solar system, within our galaxy. Everything about your body reflects the specific physical and chemical conditions of our planet, its gravitational force, its solar radiation, its atmosphere, its chemical elements, and, perhaps more than anything else, its abundance of water.

What does a **terrestrial organism** (one that lives on the land, rather than in the water) need to survive? You are a terrestrial organism, and you have a lot in common with rosebushes and radishes, which also are terrestrial organisms. You are an animal and they are plants, but your body and their bodies have similar needs to live.

In this chapter we will look briefly at the problems of survival on Earth, especially as they relate to plants. We will look at the structure of the plant body and see that organs are adaptive responses to environmental challenges. We also will formulate and test hypotheses, take measurements, gather data, and do simple calculations. In this chapter and the next few, we will emphasize **vascular plants,** those plants that have specialized **vascular tissues** to conduct or transport fluids, minerals, and food throughout the plant body.

There are millions of plant species on Earth, and other millions that have lived here but are now extinct. Living plant species are grouped into several major categories, based on structural and reproductive characteristics, as shown in data table 13.1.

Data Table 13.1 Classification of Plants into Major Groups

Nonvascular plants. Plants that do not have specialized tissues for conducting water and dissolved nutrients.

- Mosses (figure 13.1)
- Liverworts
- Quillworts

Vascular plants. Plants that have specialized conducting tissues called xylem and phloem.

- **Nonseed-Producing Vascular Plants**
  - Club mosses
  - Horsetails and scouring rushes
  - Ferns and their relatives (figure 13.2)
- **Seed-Producing Plants**
  - Gymnosperms: Conifers and their relatives
  - Angiosperms: Flowering plants (figure 13.3)

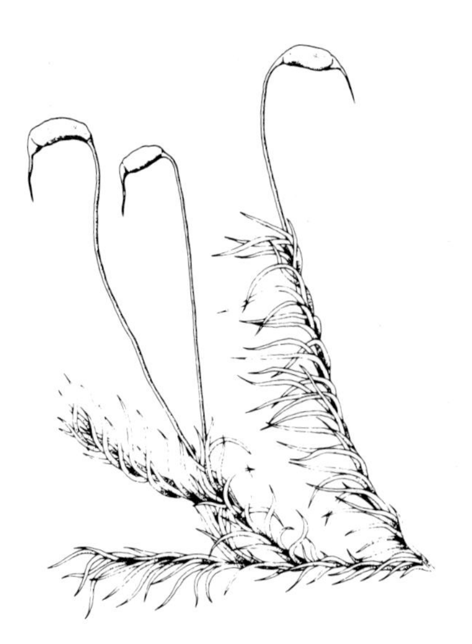

Figure 13.1 Moss. A nonvascular plant.

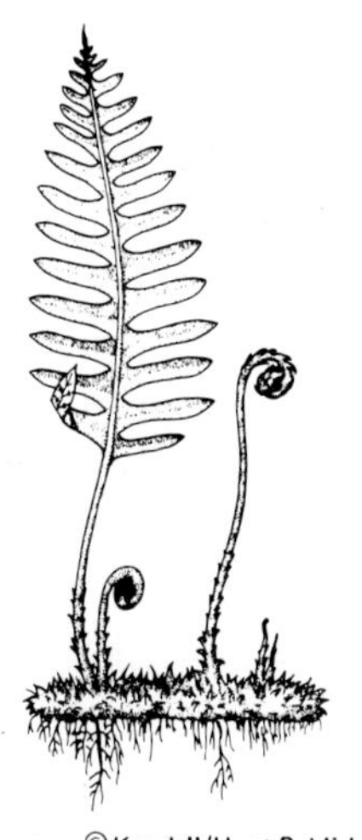

© Kendall/Hunt Publishing Company

Figure 13.2 Fern. A nonseed-producing vascular plant.

Figure 13.3 The dandelion, a seed-producing plant. Seed-producing plants are the most complex of the vascular plants, and this complexity is related to their high degree of adaptation for living in many kinds of environments, especially on land. In the plant kingdom, they have the most complex body forms, the most complex tissues, and the most complex life cycles, including the ability to produce seeds. Seed-producing plants are categorized either as gymnosperms or angiosperms. The dandelion is a flowering plant, or angiosperm. Like other angiosperms, its seeds develop within fruits. By contrast, the seeds of gymnosperms develop in cones. Most gymnosperms are coniferous trees, but the category also includes several unusual species.

13.2

Three-fourths of Earth's surface is covered by water. In these waters, photosynthetic protists (diatoms, phytoflagellates, and algae) are responsible for almost all of the photosynthetic production. On land, however, most photosynthesis is performed by plants, which are *complex, multicellular, eukaryotic, chlorophyll-containing organisms.* (Review earlier chapters if you do not remember these concepts.) Plant cells are surrounded by **cell walls.** A plant body is composed of thousands of cells, glued together into tissues by substances called **pectins.** As you will see, specialized tissue types perform specialized functions within the plant body.

The study of plants is called **botany,** and a scientist who studies plants is called a *botanist,* or *plant scientist.* There are many kinds of careers for plant scientists (data table 13.2).

Data Table 13.2 Career Opportunities in the Plant Sciences

All disciplines
Training as a biological scientist, agricultural scientist, or medical scientist can find expression in a broad range of careers in teaching, research, industry, business, and direct care of people, their animals, and their plants. In addition to careers regarded as "typical" for people with biology training, there also are opportunities for science writers, illustrators, science librarians, photographers, technicians, and specialists in computers and scientific equipment. The possibilities are many and varied, and a preference for biology, botany, or zoology is not a limitation.

Biological scientist
Such a scientist works mainly in a specialized discipline, or area, and may be one of the following: botanist, microbiologist, ecologist, naturalist, geneticist, plant pathologist, systematist, taxonomist, plant physiologist, morphologist, anatomist, paleobotanist, photobiologist, developmental botanist, tropical biologist, marine biologist, freshwater biologist, mycologist, phycologist, virologist, pharmacologist, toxicologist, cytogeneticist, or bioengineer.

Agricultural scientist
A scientist who works in an applied, agriculture-related area may be one of the following: agronomist, field crop manager, soil scientist, soil conservationist, plant ecologist, plant physiologist, agrigeneticist, plant breeder, range manager, food technologist, horticulturist, forester, forestry technician, plant quarantine and pest control inspector, turf manager, county extension agent, farmer, or farm manager.

Medical scientists and health care professionals
These specialists work mainly in human and domestic animal biology, but some plant science-related fields also are available, such as public and environmental health, pharmacology, nutrition, bacteriology, and medical mycology.

Ecologists are scientists (some of them botanists) who study the interactions of organisms with their environments. This science is called **ecology.** Ecologists understand that *each place on Earth has a specific combination of interacting physical conditions and organisms* (data table 13.3). These interacting components make up functioning ecological systems called **ecosystems.**

Data Table 13.3 Physical and Biological Components of the Environment

Physical Factors of the Environment

Radiation. Visible light, infrared and ultraviolet radiation, and other kinds of radiation (X ray, gamma ray, and microwave).

Atmosphere. Gases, humidity, particulate matter, and pollutants.

Substrate. Soil or other surface on which an organism lives (texture, composition).

Water. Abundance, ionic composition (gases, minerals, and pH), and other chemicals in solution.

Biological Factors of the Environment

Organisms of the same species that are present and have potential effects on an organism. These effects may be cooperative (as in reproductive functions), but often are competitive (competition for food, water, space, light, and other essential environmental factors).

Organisms of different species that are present and have potential effects on an organism. These effects may be beneficial, as in providing habitat, symbiotic relationships, or recycling of nutrients. Other kinds of interactions are not mutually beneficial. Often the interactions are food related, with one organism providing food for the other through predation (plant eaters and flesh eaters), parasitism, or pathological attack. There also may be competition among members of different species for food, water, space, light, and so forth.

By now you may be asking, "What does ecology have to do with the structure of plants?" The answer is, "Everything!" *Photosynthetic organisms produce food that other organisms must have to live and grow,* and understanding their structure is a key to understanding their tremendous role in feeding both themselves and other organisms. A few ecosystems

Figure 13.4 Cactus. Members of the cactus family live in arid New World deserts. They typically have protective spines and ridged stems filled with water storage tissues, but lack evaporation-prone leaves.

Figure 13.5 Euphorb. Plants in the euphorb family live in arid Old World regions. They have adaptive resemblances to cacti, even though they are not closely related. Leaves, as shown here, may develop when moisture conditions are more favorable.

derive nutrition from bacterial synthesis in warm water surrounding volcanic vents in the ocean floor. Most of Earth's ecosystems, however, run on solar energy harnessed by photosynthesis.

1. What are the specific physical (nonliving, or abiotic) components of the ecosystem that makes up your yard at home?
2. What are the specific biological (living, or biotic) components of the ecosystem that makes up your yard at home?

13.3 In looking at all of Earth's ecosystems, it becomes clear that *similar kinds of organisms live wherever physical conditions are similar* (figures 13.4 and 13.5). We call these general kinds of ecosystems **biomes.** Data table 13.4 lists the major biomes of our planet.

1. What physical factor seems to be the one that most distinguishes one biome from another?

Without water, life as we know it is not possible. *Availability of water, sunlight, and a favorable temperature range are the most important factors in determining the kinds of organisms that can live in a particular place.* Where sunlight and water are the most favorable to support lush plant growth, there also is the greatest abundance of animal and microbial life.

2. Which biomes support the greatest density and diversity of living things?
3. Which biomes are the most sparsely populated by organisms?
4. Why are plants a critical component of every ecosystem?
5. Why is it true to say that the plants present at a site determine the other life forms that are present?
6. What have you eaten today that was derived, directly or indirectly, from plants?
7. What features of your classroom are derived from plants, either directly or by manufacturing processes? (Remember to include the wood of your pencil and the paper on which these words are printed.)

Data Table 13.4 Earth's Major Biomes: Kinds of Ecosystems

a. Arctic and alpine environments
b. Boreal coniferous forests
c. Temperate deciduous forests
d. Warm-climate evergreen forests (mainly conifers)
e. Tropical rain forests (mainly broadleafed plants; jungle)
f. Grasslands (prairies, plains)
g. Deserts
h. Montane coniferous forests
i. Temperate montane forests and coniferous rain forests
j. Marine environments
Earth's marine environments cover about 71% of its surface and are, in reality, *highly* varied, depending on latitude, depth, proximity to shorelines, and so on. Marine environments usually are distinctly stratified (in layers) in regard to light and temperature gradients. (Surface waters are usually brighter and warmer.) As a result, organism distribution also is stratified. Some of the major kinds of marine environments are: coastal (from shoreline to continental shelf), coral reef, open sea (lighted zone), abyss (dark zone, including ocean bottom), and underwater volcanic vents.
k. Special environments
While not usually identified as biomes, as such, the following kinds of environments are varied and significant: estuaries, wetlands, freshwater ponds and lakes, freshwater streams and rivers (river systems), and salt water lakes.

13.4

In preceding chapters you learned that **photosynthesis** is the process by which green plants convert water, carbon dioxide, and solar energy into carbohydrates. Also, you learned how plants and other organisms use food carbohydrates (sugars and starches) to release energy (respiration) and to build other kinds of molecules (organic synthesis). The primary food storage carbohydrate produced by green plants is **starch.** The primary structural carbohydrate is **cellulose,** the main component of plant cell walls.

As you learned previously, plant cells tend to absorb water, causing a buildup of water pressure inside the cells. The pressure of the intracellular solution against a cell membrane (and cell wall) is called **turgor.** Cells and tissues that are plump and swollen due to turgor are said to be *turgid.*

In **herbaceous plants** (those that do not produce woody stems and roots), turgor is the primary means of support. Such plants have a water-based, or *hydrostatic,* skeleton. Visualize in your mind the strained bulging of thousands of individual plant cells within the cellulose boxes of their cell walls, and you will have an image of what causes herbaceous plant tissues to be stiff and crisp. By contrast, tissues of **woody plants** are supported by a skeleton of reinforced, hardened cell walls, as we will examine in more detail in the next chapter.

Now look at the living plants your teacher has provided. Plant A has not been watered for several days. Plant B has been watered within the past 24 hours.

1. Construct a hypothesis: What do you think has caused plant A to become so wilted?
2. Make another hypothesis: Why do you think the stems of plant B are firm and its leaves fully expanded?

Scientists use their own knowledge and experience, along with the knowledge of others they read about in research publications, to decide if a hypothesis is reasonable. They then test it. They measure and record results that help them to interpret the probable acceptability of the hypothesis.

Assume you wanted to *test* the two hypotheses you stated above.

3. What could you do to cause plant B to become like plant A?
4. What could you do to cause plant A to become like plant B?
5. What outcomes (results) would you predict from these tests?
6. Make another hypothesis: What do you think would happen if both plants were deprived of water permanently?
7. How would you test this hypothesis?
8. Do you think it is necessary to perform such a test yourself? Why or why not?

Obviously, the relationship between plant tissues and water is a critical factor in a plant's structure, as well as its survival. Dependence on water is a major factor that has shaped plants structurally, physiologically, and reproductively to survival on Earth.

13.5 If you were to reduce the lifetime of an individual organism to one summary statement, you could say that *the overall problem of existence is to grow, develop, and survive to reproductive maturity.* Grain crop plants, for instance, rush from seed to flowering to seed within a growing season. Their vegetative bodies die and their living seeds are harvested. The seeds are planted to obtain a new generation, and so on, year after year. Along the way, each generation both takes from and contributes to its environment.

1. What specific environmental factors are critical factors or barriers to an individual plant's survival to reproductive maturity? (It may be easier to answer this question if you think of one particular garden plant or forest tree.)

You will recall that those factors that are related to physical aspects of the environment are called *physical* or *abiotic environmental factors.* (Refer back to data table 13.3.)

2. Put a (P) in front of each physical factor you have listed above.

Those factors that are related to organisms in the environment are called *biological* or *biotic environmental factors.* Biological factors may be more difficult to list. They include predation (feeding on) and competition, as well as such beneficial interactions as reproduction, dispersal of pollen and seeds, and nutrient recycling. (Refer back to data table 13.3.)

3. Put a (B) in front of any biological factors you have listed above.

Keep in mind these specific environmental factors as we continue our discussion of plants. You will see how plant bodies are adapted in many ways to cope with their environments while obtaining what they need to survive.

13.6 Plant cells are nearly identical in cytoplasmic structure to animal cells. The combination of cells into a *plant body,* however, results in an overall body form that is quite different from that of multicellular animals. The body of a vascular plant is arranged into two major divisions: a **shoot system** and a **root system** (figure 13.6).

The *shoot system* includes stems, leaves, and organs for sexual reproduction. Some specialized asexual reproductive structures also may be produced by the shoot system, as we will investigate in a later chapter. In addition, roots may form from parts of the shoot system.

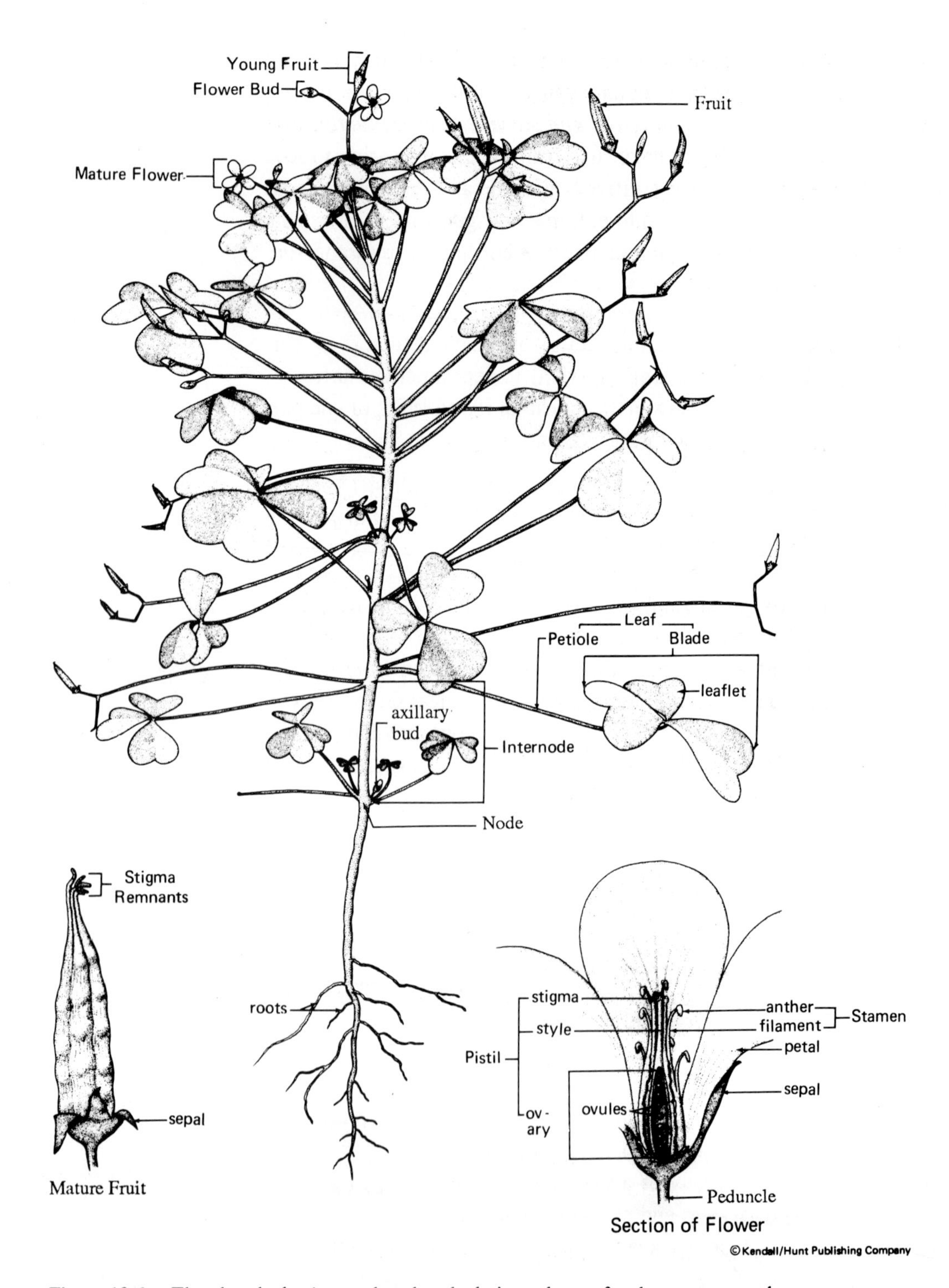

Figure 13.6 The plant body. A vascular plant body is made up of a shoot system and a root system. The shoot system of a flowering plant, such as this *Oxalis,* consists of a stem and its products: leaves, branch stems (from buds), and flowers. Fruits and seeds develop from flowers. Compare this drawing to the diagrammatical plant body in figure 13.8 to gain an overall functional perspective.

The *root system* is made up of older, stronger roots that join the stem, as well as younger, finer roots that continue to grow and penetrate the soil. The mature roots provide strong anchorage and the younger roots absorb water and nutrients. Young branch roots can grow from mature roots, enlarging the scope of the root system.

How do these systems grow and develop? *A plant body increases in length by growth at its tips.* This growth at the tips (apices) is called **apical growth.** Specifically, young shoots originate from buds on existing shoots, and young roots originate (mainly) from the farthest ends of existing roots. Because new growth occurs at the tips, the most mature plant tissues are located nearest the center of the plant body.

Plant stems and roots also increase in girth as they mature. This second type of growth is accomplished by internal layers of **cambium** tissues.

This pattern of development is quite different from the way an animal body develops. Can you imagine yourself "becoming" a tree? Think of the main axis of your body (your trunk) remaining the same length, but getting thicker. At the same time, imagine your arms and legs (your limbs) continuing to elongate and branch upward and downward. Furthermore, as a tree you continue to grow your entire life, *and* are able to regenerate body parts to replace those that break off along the way.

1. Based on what you already know about plants, in what ways does it seem to you that plant bodies are different in form from animal bodies?

13.7

Plants face problems of life similar to those of animals, but must meet them in different ways, because a plant is bound its entire life to the spot of earth where the seed from which it grew fell or was planted. Plants are not free to wander in search of food, water, shade, sunlight, or mates. They cannot hunt prey. They cannot hide or run to escape storms, fires, floods, predators, bulldozers, or chainsaws.

The fact that a plant is anchored, however, does not mean that it cannot seek what it needs from the environment. Its shoot system of stems and leaves grows upwards in response to light, and the leaves become arranged into a mosaic pattern that fills all the lighted spaces of the shoot system. If shaded by an adjacent plant or a human-made structure, the shoot system will lean toward the brightest spot. This growth response toward light is called **positive phototropism** (figure 13.7).

Figure 13.7 Phototropism. Plant growth is responsive to environmental influences, such as the direction and intensity of light. Plants orient their leaf surfaces toward a light source, and the youngest tissues are the most sensitive. Judging from the growth response you see in this photograph, from which direction is the brightest light coming?

Underground, the root system branches and spreads, filling the soil with a structural system for anchorage and for absorbing water and dissolved nutrients. It may astonish you to know that the bulk of a plant's root system is about the same as its shoot system. Look at a large tree the next time you are outside and try to imagine the extent of its root system. Young roots tend to grow toward gravity (**positive geotropism**) and toward water (**positive hydrotropism**).

1. Sketch a "balanced" cartoon plant, labelling its shoot system, the soil line, and its root system.

13.8 Let us now do an exercise designed to give you both a feel for the extent of the shoot system and some practice in measurement and calculation. Examine the assortment of plants your teacher has provided. Each kind of plant has a somewhat different growth form, with longer or shorter stems and different arrangements of leaf attachment to the stems. Work individually or in teams according to your teacher's instructions to complete this activity. Choose one plant, and use a metric ruler for your measurements. Do not remove leaves from the plant unless you are instructed to do so.

The first part of the investigation will be to define a "typical" mature leaf. The second part will be to use that leaf to estimate total mature leaf area. Examine a leaf. You can see it has two major regions, a thin leafstalk, or **petiole,** and a broad, flat **blade.**

1. What is the name of your plant?

2. Measure the length and maximum width of the blades of the five largest leaves. Do not remove the leaves and do not include the leafstalk (petiole) in your measurement. In your logbook write your measurements in data table 13.5.

Data Table 13.5 Sample Leaf Dimensions

Length in cm		**Width in cm**	
Average =	cm	Average =	cm

3. Now calculate the average length and average maximum width of your five leaves and record them in the data table.
4. Can you determine the leaf area by simply multiplying the average length and average maximum width figures?
5. Why would such a calculation be inaccurate?

13.9 Now find a leaf that closely matches the average length and average maximum width dimensions you have calculated. This leaf will represent a "typical" mature leaf. Remove the leaf carefully.

1. Follow the instructions to complete parts a–d in data table 13.6 in your logbook.

Data Table 13.6 Approximating Total Mature Leaf Area.

a. Sketch the outline of the sample leaf on the grid portion of data table 13.6, indicating as accurately as possible its true size and shape.

b. Total leaf area in number of grid units = ________

c. Calculate leaf area in metric units:

________ grid units $\times$ 25 mm^2 = ________ mm^2 leaf area

Conversion of leaf area to cm^2:

________ mm^2 / 100 = ________ cm^2

d. (__ leaves) $\times$ (__ cm^2) = ____ cm^2 total mature leaf area

b. After you have sketched the leaf outline, count the number of whole- and half-grid units enclosed within the leaf blade outline and record the total number of units under your sketch. (In computer imaging, such grid units are called *pixels,* and they are calculated automatically. Today, however, you must be the computer!)

c. Each grid unit here is 5 mm $\times$ 5 mm = 25 mm^2. Therefore, multiply the number of units covered by your leaf blade times 25 to calculate the leaf area in mm^2. Convert your mm^2 figure to cm^2 and record it below the leaf outline. (Remember that 100 mm^2 = 1 cm^2.)

d. Next, count the number of mature leaves present on your selected plant. (Include the one you removed in your count.) Multiply the number of mature leaves times the per leaf area in cm^2 that you calculated above. Your total is an estimate of the total mature leaf area of the plant.

Measure out an equivalent cm^2 area on your work table or a portion of the floor to provide a graphic display of the estimated leaf area of your plant.

2. What are its dimensions?

3. Are you surprised by its extent? Explain.

It should be apparent to you that this is an imprecise way of determining the total leaf area, which would take into account other leaf categories by age and size. You have had an opportunity, however, to use measurement and calculation in a scientific way to arrive at your approximation of total mature leaf area.

13.10 A final important aspect of any scientific test is to determine *sources of error,* factors that can cause a test to be inaccurate or biased. In an actual, controlled experiment, you would subject the results to statistical analysis, which would take into account probability for error and predict whether or not the results could be considered reliably accurate. Your teacher may wish you to perform such an analysis as an extension of this activity.

1. What are some sources of error in the investigation you conducted in 13.8 and 13.9?

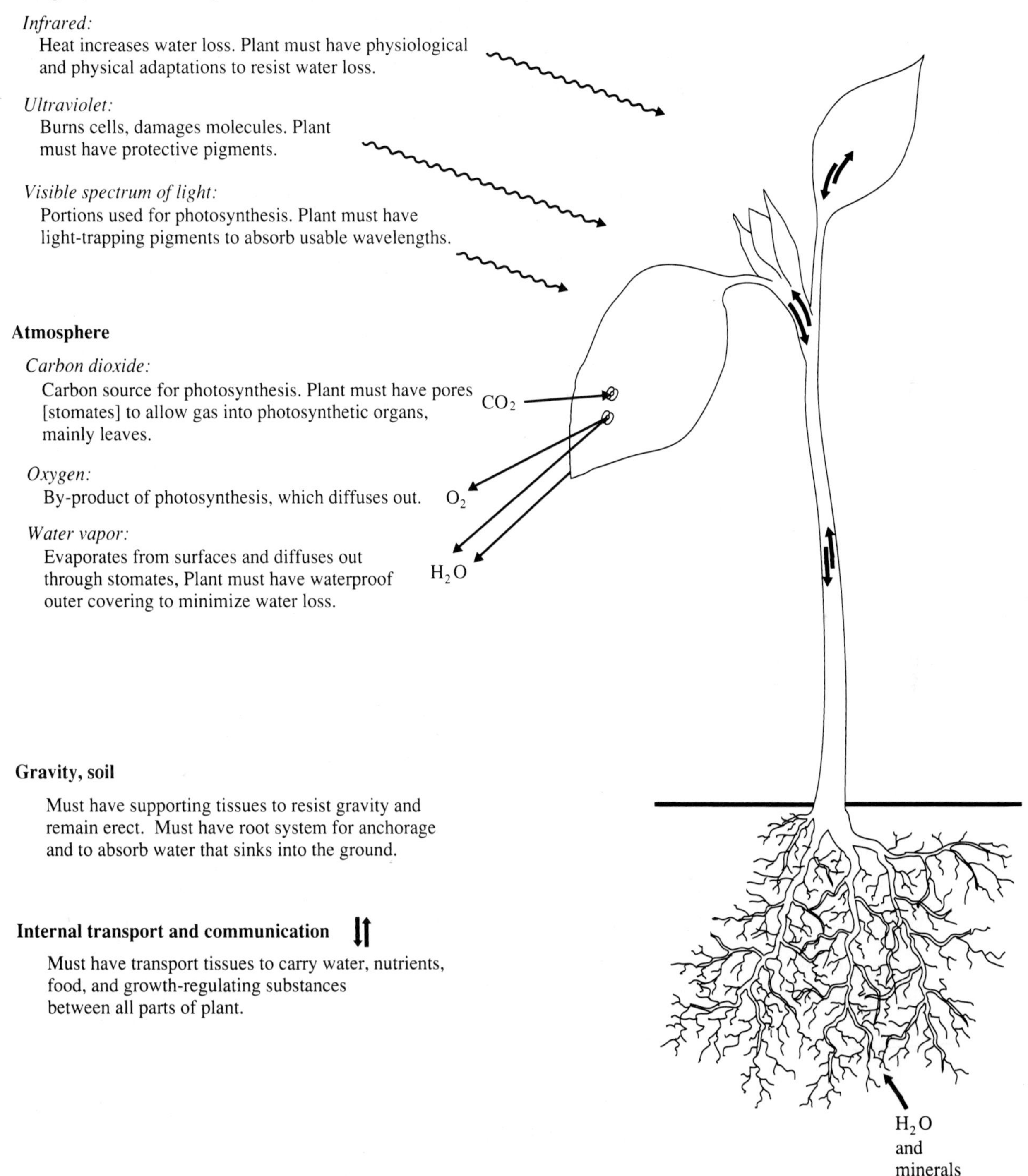

Figure 13.8 Environmental challenges and the plant body. The body of a terrestrial plant is adapted to take and process what it needs from the environment and, at the same time, to protect itself from environmental hazards. Some of the major environmental interactions are depicted here. This schematic understanding of the plant body will be useful to you during the rest of your study of plants.

13.11 Now that you have some appreciation of the effective leaf area of your plant, we will consider its balance with the rest of the plant body.

The last investigation was designed *for* you. This time, however, you will design your own investigation. Use what you have learned about the scientific method in this course to design an experiment that will give you answers to the following questions: (*a*) What is the relative mass of the shoot system compared to the root system? (*b*) What is the proportion of leaves, stems, and roots to the entire plant body?

First, *establish the objectives of your experiments.* In this case, the questions above have established your objectives. Given these objectives, you need to restate them as a set of **hypotheses,** tentative explanations or expectations that you will test. With the general observations and knowledge you already have, complete the hypotheses below.

1. It is my hypothesis that the live weight of the root system of this plant is (equal to, more than, less than) the live weight of its shoot system.
   What is your basis for this hypothesis?
2. It is my hypothesis that the live weight of the leaves is (equal to, more than, less than) the live weight of the stems.
   What is your basis for this hypothesis?

Second, *determine the form you want your data to take.* In this case, you are collecting measurements of live (wet) weight. In the metric system you will be measuring in grams.

Third, *set up your methods.*

3. How (by what methods) will you obtain your data (measurements)?
4. What equipment will you need, and how will you get it?
5. In what form will you record your data? Set up a chart to show the headings you would use.
6. How will you analyze (evaluate, summarize, and compare) your data?

Finally, *identify the potential sources of error in your experiment, and suggest ways to minimize (control) them.* For instance, live weight consists mainly of water. To get a more accurate measurement of actual plant substance (biomass), dry weight often is used.

7. What sources of error would you anticipate, and how would you try to control them, reducing their influence on your experiment?

13.12 This brief introduction establishes a context for your understanding of how plants have adapted for life on Earth (figure 13.8). Before you continue, examine the study points. If you understand a study point, place a check mark in the appropriate box in your logbook. If you do not understand the study point, you need to return to the text of the chapter and review the concepts.

On completion of this chapter, you should be able to do the following:

☐ 1. Identify some careers in the plant sciences.

☐ 2. Understand physical and biological environmental factors as challenges and conditions to which plants have become adapted.

☐ 3. Name the major biomes of Earth and understand the critical role of plants in any ecosystem.

☐ 4. Understand the role of turgor in plant support.

☐ 5. Understand the concepts of plant body elongation from tips, and of increase in girth.

☐ 6. Recognize the roles played by the major plant organs.

☐ 7. Have an appreciation for the effective leaf area of a plant.

☐ 8. Identify sources of error in an experiment.

☐ 9. Outline four major components of experimental design.

Seed Structure and Germination 14

What is a seed and how does the young plant emerge?

14.1

Each seed-producing plant you have ever observed, handled, or eaten originated as a microscopic **embryo,** tucked within the nourishing, protective tissues of a **seed** (figure 14.1). What is a seed? By definition, a seed develops from an ovule (maternal tissue) in a cone or flower, and contains an embryonic plant. The embryo part of the seed develops from a zygote, the cell that is formed when a sperm nucleus unites with an egg nucleus. (The details of plant reproduction are discussed in chapter 20.) Most kinds of seeds are characterized by a period of **dormancy,** in which growth is prevented and metabolism occurs very slowly.

What are the parts of a *dormant* seed? A hard **seed coat** protects internal tissues by reducing water loss. It also reduces susceptibility of the seed to physical damage and to bacterial and fungal diseases. The food supply for **germination** (the process by which seedlings emerge from the seed) and early seedling growth is packed either in storage tissue of maternal origin or in embryonic storage tissue. The **embryo** (embryonic plant) itself consists of immature tissues that are capable of generating a mature plant body by mitosis. Finally, the cells of the seed contain important substances called **growth regulators,** mainly *hormones.* Some of these growth regulators are *inhibitors* that maintain seed dormancy; others are *growth promoters* that encourage cell division and enlargement.

Imagine what it is like to be an embryonic plant. You are nestled snugly inside a protective capsule, surrounded by food-filled cells that yield nourishment in response to the demands of your body. You are in a suspended, restful state. What is the urgency to emerge? What could happen to you there?

For one thing, you are one of hundreds to thousands of seeds produced on a plant to ensure that even a small percentage of you may live to reproduce the next generation. You are at the base of many food chains, sought after by seed-eating birds, mammals (including humans), and insects.

For another thing, the timing of your species may be such that you cannot survive winter in the seed, but must establish yourself as an independent plant soon after release from your parent. Silver maple fruits, for instance, are shed abundantly in the spring, and contain seeds that must germinate immediately, before they dehydrate. Most kinds of temperate climate seeds, however, are equipped to *overwinter* in a dormant condition for at least one season. Garden vegetable and flower seeds manage this nicely.

Let us assume you are one such seed, a tiny seed that dropped from the dried fruit husk and rolled into a crack in the soil at the end of the previous summer. Winter is over and the soil has been soaked repeatedly by melting snow and/or spring rains. The soil is warmed daily by the sun. It is time for you to germinate. How do you do it?

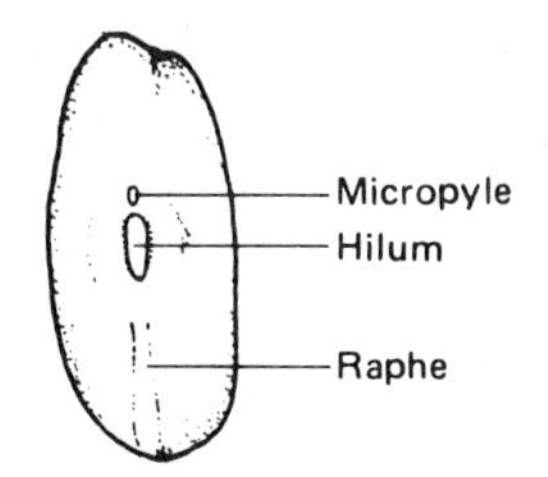

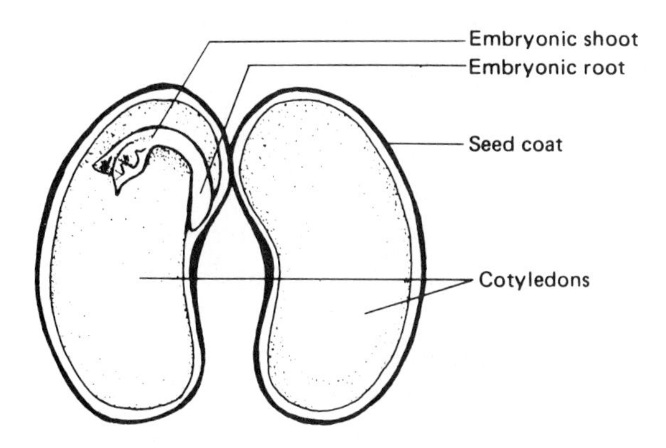

Figure 14.1 External and internal structure of a bean seed. The external features include the micropyle, through which the egg was fertilized, the hilum where the seed was attached to the inside of the bean pod, a seamlike raphe, and the protective seed coat. Inside the seed, the bean embryo is made up of nutrient-filled cotyledons (seed leaves) and the embryonic organs and tissues that will give rise to the shoot and root systems.

Water is responsible for getting germination started. Snowmelt and rainwater have washed over you day after day, gradually rinsing away enough inhibitors to release your tissues from dormancy. Like Sleeping Beauty, you have received the "kiss" that awakens you from a long sleep. With the reduced influence of inhibitors, your tissues become responsive to the presence of growth promoters. Imbibition occurs; water penetrates the seed and internal tissues begin to swell. Mitosis begins in the immature tissues of the embryo. More enzymes and hormones are produced, food is mobilized faster, and still more water is absorbed into the seed. Cellular enlargement occurs, mainly because water enters developing cells. As a result, physical pressure builds up, pressing outward against the surrounding seed coat. Eventually the seed coat ruptures, freeing the seedling to emerge, root first.

Once out of the seed, the germinating seedling faces an extremely vulnerable period. It must quickly establish a root system to draw water from the soil. Failure to do this during the first several hours tends to be fatal. In fact, one important role of inhibitors in the seed is to keep the seed dormant until sufficient water is present in the soil, as determined by the sufficient soaking to rinse away the inhibitors!

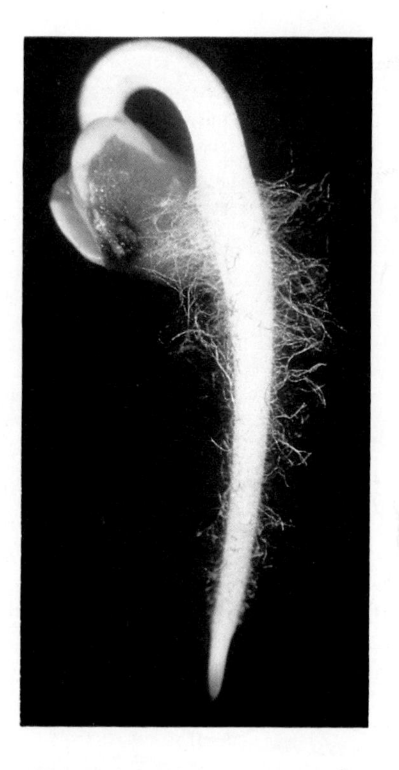

Figure 14.2 Germinating seedling. The stem and root structures grow rapidly as the root thrusts its way into the soil, establishing contact with soil moisture through its many, delicate root hairs. The emerging stem has formed a hook that breaks upward through the soil surface, while the young leaves and apical meristem are still protected by the cotyledons.

Obviously, a seedling also must quickly establish a shoot system capable of conducting photosynthesis before seed food reserves are depleted. How can you get your tender shoot apex up through heavy, rough soil without it becoming crushed? Seedlings cannot fold arms over their heads as they are pushed upwards, but they may be protected in one of several ways. The burst of primary root growth and initial stem elongation may push the shoot straight upward with the seed coat or even the fruit wall (as with a sunflower) still cupped over it. Or, the embryonic stem may form a **hook,** a hairpin bend that breaks through the soil, dragging the shoot tip along with it (figure 14.2). The hook straightens after emergence above the soil line. Finally, the cotyledons remain clamped around the shoot tip until after emergence. They then bend back, exposing the tiny true leaves, which expand and begin to photosynthesize. The cotyledons gradually wither as their storage tissue is used up; eventually, they fall off. With this introduction, let us move into an examination of seed structure, function, and germination.

14.2 Today (*Day 1*) you will set up an investigation of seed germination. Each of the next five to eight class days you will spend the first 5 to 10 minutes of the class period observing the progress of this investigation, according to the instructions in parts 14.5 and 14.6. Then you are to go on to other segments of this chapter.

Data Table 14.1 Experimental Setups for Germination Experiments

Setup Number	Kind of Seed	Nature of Environmental Variables
1	mung bean	no water, bright indirect light, room temperature
2	mung bean	water, bright indirect light, room temperature
3	mung bean	water, darkness, room temperature
4	mung bean	water, darkness, cold temperature
5	radish	no water, bright indirect light, room temperature
6	radish	water, bright indirect light, room temperature
7	radish	water, darkness, room temperature
8	radish	water, darkness, cold temperature
9	alfalfa	no water, bright indirect light, room temperature
10	alfalfa	water, bright indirect light, room temperature
11	alfalfa	water, darkness, room temperature
12	alfalfa	water, darkness, cold temperature
13	lettuce	no water, bright and room temperature: well-lit site that is not in direct sunlight or near a heat register
14	lettuce	water, bright and warm: well-lit site near a heat register or radiator; or in an incubator
15	lettuce	water, bright and room temperature: well-lit site that is not in direct sunlight or near a heat register
16	lettuce	water, dark and room temperature: enclose Petri dish completely in aluminum foil; place next to #15 setup
17	lettuce	water, dark and cold: Petri dish enclosed in aluminum foil, in a refrigerator

You will observe the germination of four kinds of seeds: mung bean, radish, alfalfa, and lettuce, recording how long germination takes and how seedlings emerge and develop. You also will test the effects on germination of three environmental variables—water, light, and temperature.

The mung bean, radish, and alfalfa seeds will be grown in germinating chambers made of glass jars. The lettuce seeds will be grown in Petri dishes. Data table 14.1 identifies all of the experimental setups. Your teacher will divide the class into four or more teams for these activities. The detailed instructions for Day 1 are presented in sections 14.3 and 14.4.

14.3 *Day 1 instructions for mung bean, radish, and alfalfa teams:* Set aside two dozen of your seeds in a small container or plastic bag. This will be the "no water" sample. Use a marker to label the container according to the seed variety you have been assigned.

Next, obtain three very clean quart-size glass jars. Using a wax marker, label each jar according to the numbers presented in data table 14.1. For each jar you will need a clean piece of nylon hosiery or cheesecloth about twice as large as the mouth of the jar. You also will need a sturdy rubber band or canning jar ring. This equipment will serve as a *germination chamber* in which to grow your seedlings (sprouts) to test the effects of light and temperature.

If you are using mung bean seeds, put about ¼ cup of seeds in each jar. If you are using radish seeds, put about 2 tablespoons of seeds in each jar. If you are using alfalfa seeds, put about ¼ cup of seeds in each jar. Then secure the piece of hosiery or cheesecloth over the top of each jar, using a rubber band or canning ring.

Next, *gently* run cold tap water into each jar, through the fabric cover. Fill the jar and rinse the seeds for 10 to 20 seconds. Then gently pour off half of the water so the seeds can soak overnight. (The jars will be half full of water.) Try not to leave seeds stuck to the cover. Place the jars in their designated experimental sites. Have your teacher check your setups.

Use section 14.5 instructions for subsequent days of the experiment.

14.4 *Day 1 instructions for the lettuce team(s):* You will need five very clean Petri dishes, each lined with four layers of sterile filter paper, about one hundred leaf lettuce seeds, water that has been sterilized by boiling, transparent tape, a wax marker, and aluminum foil. Use a wax marker to label each dish on the *side* of the lid (not the top).

Leave one dish dry and flood the other four with sterilized water. Put the covers on all of them. Allow the filter paper to soak for 10 minutes, and then pour off the standing water. Do not squeeze the paper or otherwise try to get out all of the water that clings between the paper and the container.

Next, arrange about twenty lettuce seeds, evenly spaced, on top of the filter paper in each Petri dish. Place the lid on the Petri dish and tape it closed in four places, taking care not to tilt it.

The labels will be 13, 14, 15, 16, and 17, as listed in data table 14.1. Wrap Petri dishes 16 and 17 completely with aluminum foil so no light can get in. Mark the outside of the foil wrappers also. Have your teacher check your Petri dishes, then put them in their specified locations.

Use section 14.6 instructions for subsequent days of the experiment.

14.5 Instructions are given here for the daily observation of germination setups of mung beans, radishes, and alfalfa. Use data tables 14.2, 14.3, 14.4, and 14.5 to record your own answers *and those of the other teams.* Also note each day the progress of the seeds that were not given water. Examine the seeds closely each day to report changes in appearance, such as: increased plumpness, first sign of root growth and shoot growth, length of emerging root, length of emerging shoot, development of leaf pigmentation (color), and any other changes you notice. You may find it helpful to make small sketches, to help you remember the actual appearance of seeds and seedlings at each stage.

Day 2: Without removing the cover, carefully pour off the water in which the seeds soaked. Gently rinse the seeds by running cold tap water through the cover, then pour off the excess water. From now on, the seeds should be kept moist but should not be covered with water. If they are submerged in water, the living tissues will not get enough oxygen to grow rapidly; also, the anaerobic conditions in the water will encourage growth of decomposition bacteria and the seeds will rot.

Remove the cover and take out three or four seeds to measure them and observe any changes. Record your information in the appropriate data table. Then share your data with the rest of the class. This procedure will allow each member of the class to have a complete record of information. Replace the cover and return the jars to their designated sites, except now place the jars on their sides.

Days 3 and 4: Rinse, observe, record and share data about the sprouting seeds. Place the jars on their sides in their designated sites.

Day 5: Rinse the sprouts a final time, examine them, record and share data. After you have completed your data tables, have your teacher look at them.

Finally, use one or more of each kind of sprout to complete section 14.10 of this chapter. Be sure to complete your observations today, even if you have to wait until another class period to write your conclusions. Notice that you will be sharing your sprouts with all of the other members of your class. Once everyone has finished with the observations, the sprouts may be eaten.

14.6 Instructions are given here for the lettuce experiment observations, *Days 2 through 5 or longer.* Lettuce seeds germinate more slowly than the three other kinds of seeds we are using. Also, the rate of seedling development may be affected by your classroom temperatures and lighting. Therefore, your instructor may wish for you to extend this experiment to 8 or 10 days to maximize your results. Be sure to check with your instructor before terminating the experiment!

After you have made and recorded your own observations each day, record the cumulative class data for the other kinds of seeds on data tables 14.2, 14.3, 14.4, and 14.5 of your logbook. On *Day 5,* do section 14.10 *with your classmates,* after first reading the last paragraph of section 14.5 above.

Data Table 14.2 Alfalfa, Radish, and Mung Bean Germination (Record Data in Log Book)

Environment 1, 5, or 9: No Water, Room Temperature, Light					
Alfalfa seeds	**Day 1**	**Day 2**	**Day 3**	**Day 4**	**Day 5**
Seed width (in mm)			——	——	——
Seed length (in mm)			——	——	——
Have the seeds enlarged					
Seed coats intact or split					
Length of emerging root					
Length of emerging shoot					
Chlorophyll present and where					
Radish seeds					
Seed width (in mm)			——	——	——
Seed length (in mm)			——	——	——
Have the seeds enlarged					
Seed coats intact or split					
Length of emerging root					
Length of emerging shoot					
Chlorophyll present and where					
Mung bean seeds					
Seed width (in mm)			——	——	——
Seed length (in mm)			——	——	——
Have the seeds enlarged					
Seed coats intact or split					
Length of emerging root					
Length of emerging shoot					
Chlorophyll present and where					

Data Table 14.3 Alfalfa, Radish, and Mung Bean Germination (Record Data in Log Book)

Environment 2, 6, or 10: Water, Room Temperature, Light					
Alfalfa seeds	**Day 1**	**Day 2**	**Day 3**	**Day 4**	**Day 5**
Seed width (in mm)			———	———	———
Seed length (in mm)			———	———	———
Have the seeds enlarged					
Seed coats intact or split					
Length of emerging root					
Length of emerging shoot					
Chlorophyll present and where					
Radish seeds					
Seed width (in mm)			———	———	———
Seed length (in mm)			———	———	———
Have the seeds enlarged					
Seed coats intact or split					
Length of emerging root					
Length of emerging shoot					
Chlorophyll present and where					
Mung bean seeds					
Seed width (in mm)			———	———	———
Seed length (in mm)			———	———	———
Have the seeds enlarged					
Seed coats intact or split					
Length of emerging root					
Length of emerging shoot					
Chlorophyll present and where					

Data Table 14.4 Alfalfa, Radish, and Mung Bean Germination (Record Data in Log Book)

Environment 3, 7, or 11: Water, Room Temperature, Dark					
Alfalfa seeds	**Day 1**	**Day 2**	**Day 3**	**Day 4**	**Day 5**
Seed width (in mm)			———	———	———
Seed length (in mm)			———	———	———
Have the seeds enlarged					
Seed coats intact or split					
Length of emerging root					
Length of emerging shoot					
Chlorophyll present and where					
Radish seeds					
Seed width (in mm)			———	———	———
Seed length (in mm)			———	———	———
Have the seeds enlarged					
Seed coats intact or split					
Length of emerging root					
Length of emerging shoot					
Chlorophyll present and where					
Mung bean seeds					
Seed width (in mm)			———	———	———
Seed length (in mm)			———	———	———
Have the seeds enlarged					
Seed coats intact or split					
Length of emerging root					
Length of emerging shoot					
Chlorophyll present and where					

Data Table 14.5 Alfalfa, Radish, and Mung Bean Germination (Record Data in Log Book)

Environment 4, 8, or 12: Water, Cold Temperature, Dark					
Alfalfa seeds	**Day 1**	**Day 2**	**Day 3**	**Day 4**	**Day 5**
Seed width (in mm)			———	———	———
Seed length (in mm)			———	———	———
Have the seeds enlarged					
Seed coats intact or split					
Length of emerging root					
Length of emerging shoot					
Chlorophyll present and where					
Radish seeds					
Seed width (in mm)			———	———	———
Seed length (in mm)			———	———	———
Have the seeds enlarged					
Seed coats intact or split					
Length of emerging root					
Length of emerging shoot					
Chlorophyll present and where					
Mung bean seeds					
Seed width (in mm)			———	———	———
Seed length (in mm)			———	———	———
Have the seeds enlarged					
Seed coats intact or split					
Length of emerging root					
Length of emerging shoot					
Chlorophyll present and where					

When you look at the Petri dishes, do not remove their tops. The overlapping Petri dish halves keep mold spores from getting in and also minimize evaporation of water. You will be able to make your observations through the glass. Be sure to rewrap the aluminum foil snugly after you observe the seeds that are in dark conditions so no light can penetrate. If the foil tears, replace it with new foil.

Examine the seeds closely each day to report changes in appearance, such as: increased plumpness, first sign of root growth and shoot growth, length of emerging root, length of emerging shoot, development of leaf pigmentation (color), and any other changes you notice. You may find it helpful to make small sketches, to help you remember the actual appearance of seeds and seedlings at each stage.

1. Record your daily observations in your logbook.

It will be important for you to determine the *germination percentages* for each set of variables. On the *last day* of your observations, count the number of seeds that germinated in each Petri dish and divide that number by the total number of seeds in each Petri dish. Multiply the answer by 100 to convert the number to a percentage. A summary of this calculation is included in data table 14.6.

2. Summarize your final observations and percentages on data table 14.6 in your logbook.

Your observations of the physical appearance of seedlings *should include:* presence or absence of roots and shoots; length in mm of roots and shoots; brightness of chlorophyll (yellow to dark green), if it is present; other differences you may notice among the several setups.

Then complete the activity in section 14.11.

14.7 In flowering plants, seeds mature within a **fruit,** which develops by growth and ripening of the **ovary** of a flower (figure 14.3). In practical terms, you could think of a seed as an embryonic plant packed in a protective case along with its own lunch. Several kinds of flowering plant seeds are available for your observation, including those of radish, bean, sunflower, peanut, and corn plants. Obtain one of each kind of seed to observe at your desk.

Examine the peanut first. The peanut shell is the dried wall of the fruit. Remove a seed from a peanut.

1. How many seeds are in the fruit?
2. Do other members of your class have a different number of seeds, and if so, how many?

Data Table 14.6 Lettuce Seed Germination Results (Record Data in Log Book)

Date ________ Day # ________

Physical appearances and germination percentages at end of experimental treatments.

$$\frac{\text{Number of germinated seeds}}{\text{Number of seeds in Petri dish}} \times 100 = \text{percentage}$$

13. *No water:*
    Germination %: ________
    Appearance:

14. *Bright and warm:*
    Germination %: ________
    Appearance:

15. *Bright and room temperature:*
    Germination %: ________
    Appearance:

16. *Dark and room temperature:*
    Germination %: ________
    Appearance:

17. *Dark and cold temperature:*
    Germination %: ________
    Appearance:

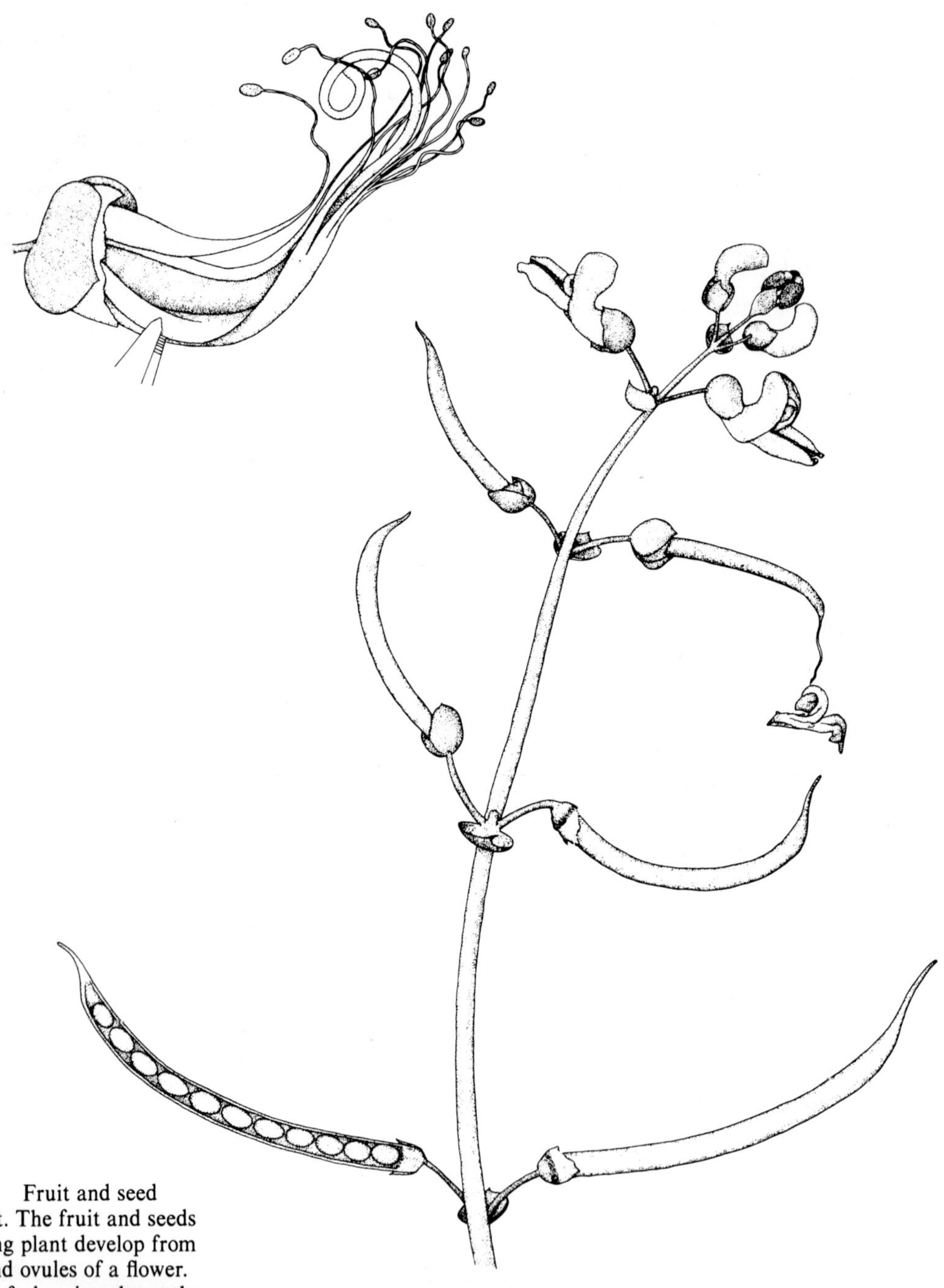

Figure 14.3 Fruit and seed development. The fruit and seeds of a flowering plant develop from the ovary and ovules of a flower. The upper-left drawing shows the ovary of a legume flower, and the larger drawing shows developing fruits (pods) of different ages—they enlarge as they ripen. The most mature pod is opened to show the fully developed seeds.

©Kendall/Hunt Publishing Company

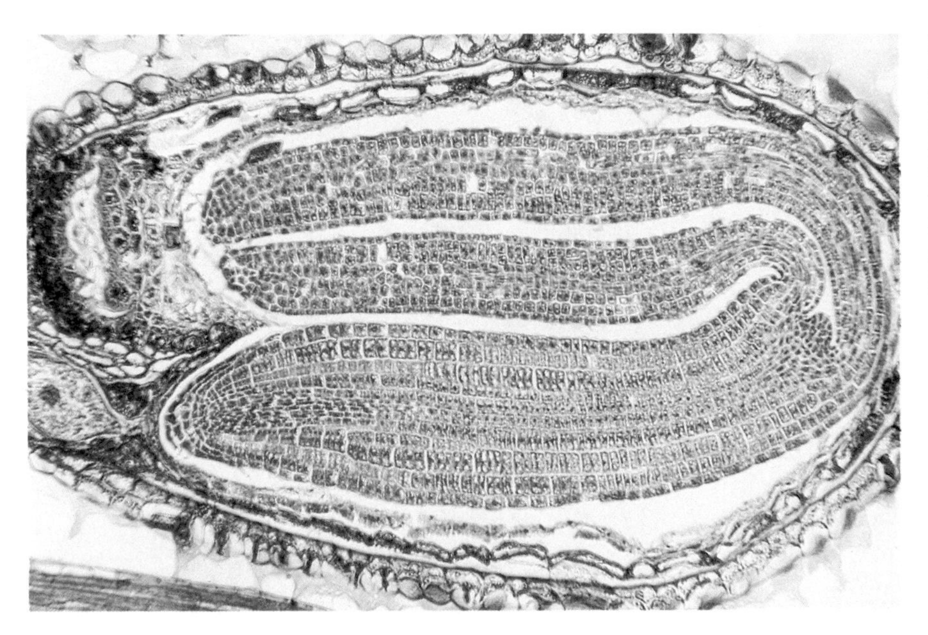

Figure 14.4 Microscopic section of a shepherd's purse seed. In this tiny seed, the two cotyledons are folded to one side of the rest of the embryo, rather than enclosing it as they do in the bean seed (fig. 14.1). Even in a seed as tiny as this, there is a fully formed plant embryo that can grow and mature into an adult plant.

Wrapped loosely around each peanut is a papery hull, which represents the **seed coat.** For contrast, examine the radish seed, which has a hard, thickened seed coat. Carefully open the two halves of a peanut seed. The main axis of the small **embryo** is clearly visible, including the *embryonic root, stem, and leaves.* The two large halves of the seed also are part of the peanut embryo. They are called the seed leaves, or **cotyledons.** The cotyledons have two major functions: (*a*) they store food to nourish the germinating seedling until it is mature enough to draw nutrients from the soil and to carry out photosynthesis; (*b*) they physically protect the delicate shoot tip as it is shoved upward through the soil during germination.

How do cotyledons develop? During seed development a nutrient storage tissue called **endosperm** (inner seed) develops, surrounding the embryo. The endosperm forms a large part of many kinds of seeds, such as corn and other grasses. However, in many flowering plants, including the peanut and shepherd's purse, the endosperm is digested by the developing embryo, which then stores the digested food in its cotyledons (figure 14.4).

Flowering plants are divided into two main groups, the **monocots** (monocotyledons) and **dicots** (dicotyledons) (figure 14.5). Look closely at the long version of each of these two names, and you will see that the *-cot-* part of each word refers to the cotyledons.

3. If *mono-* means "single" and *di-* means "two," how many cotyledons do you suppose are found on monocot embryos?
   on dicot embryos?

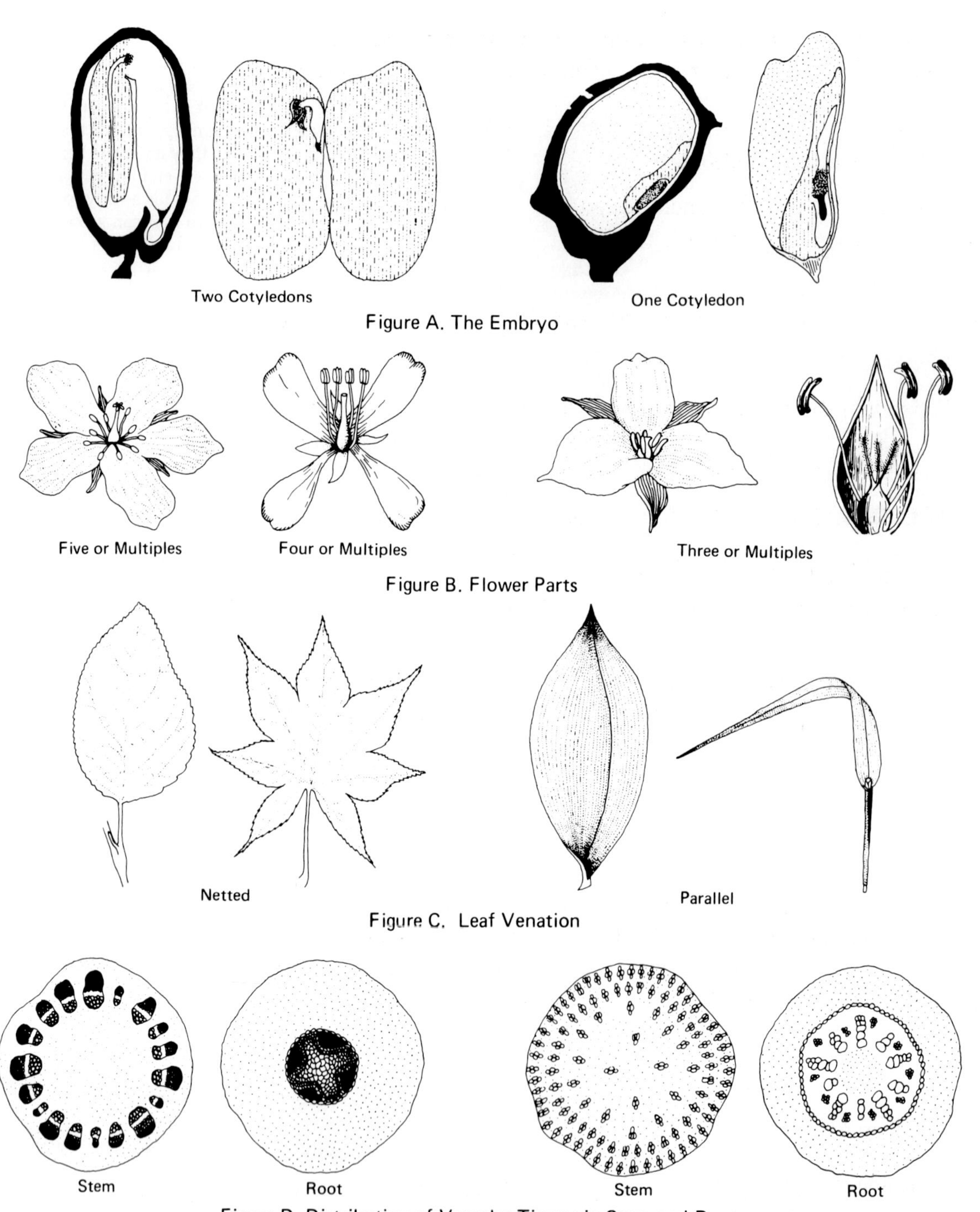

© Kendall/Hunt Publishing Company

Figure 14.5 Comparison of monocot and dicot characteristics. Flowering plants are classified into two major subgroups. A monocotyledonous plant has a single seed leaf that enfolds the embryo, whereas a dicotyledonous plant has two seed leaves. There are other significant differences in the structure of flowers, leaves, stems, and roots. You will study these differences in greater detail in later chapters.

Grasses, onions, and lilies are examples of monocots. Beans, sunflowers, and most of the commonly eaten fruits and vegetables are dicots.

4. Based on your observation of the seed, would you say the peanut is a monocot or a dicot?
5. Sketch the opened peanut seed and label the four main parts of the embryo. Then you may eat it.

14.8 Examine the sunflower "seed" and then carefully crack it open.

1. Based on what you observe about the outer shell and its contents, do you think that what we call a sunflower "seed" is actually a *seed* or a *fruit?* State the basis for your opinion.
2. Is the seed covered by a seed coat?
3. What other seed parts can you identify?
4. What part(s) of the sunflower "seed" do we eat?

14.9 Examine both dry and soaked corn kernels. The dry kernel still is in a dormant condition. Soaking has activated enzymes in the other kernel, beginning the germination process.

1. Which of the two kernels is plumper and why?
2. What is the physical process of soaking up water called?

The outer covering, or *hull,* that clings tightly to the outside of the corn kernel is actually the fruit wall. The seed coat is present, but not obvious. By definition, a corn kernel is a fruit, not a seed. Specifically, it is a kind of fruit called a *grain.* Look at figure 14.6, showing a corn kernel bisected lengthwise. You can see that the tiny embryo is located near the pointed end of the kernel. Most of the seed interior, however, is filled with endosperm. Using a razor blade, *carefully* cut open a fresh or soaked corn kernel and look for the embryo.

3. Is it easy to distinguish the embryo from the endosperm? Explain.
4. What do you hypothesize is the primary storage substance in the endosperm?

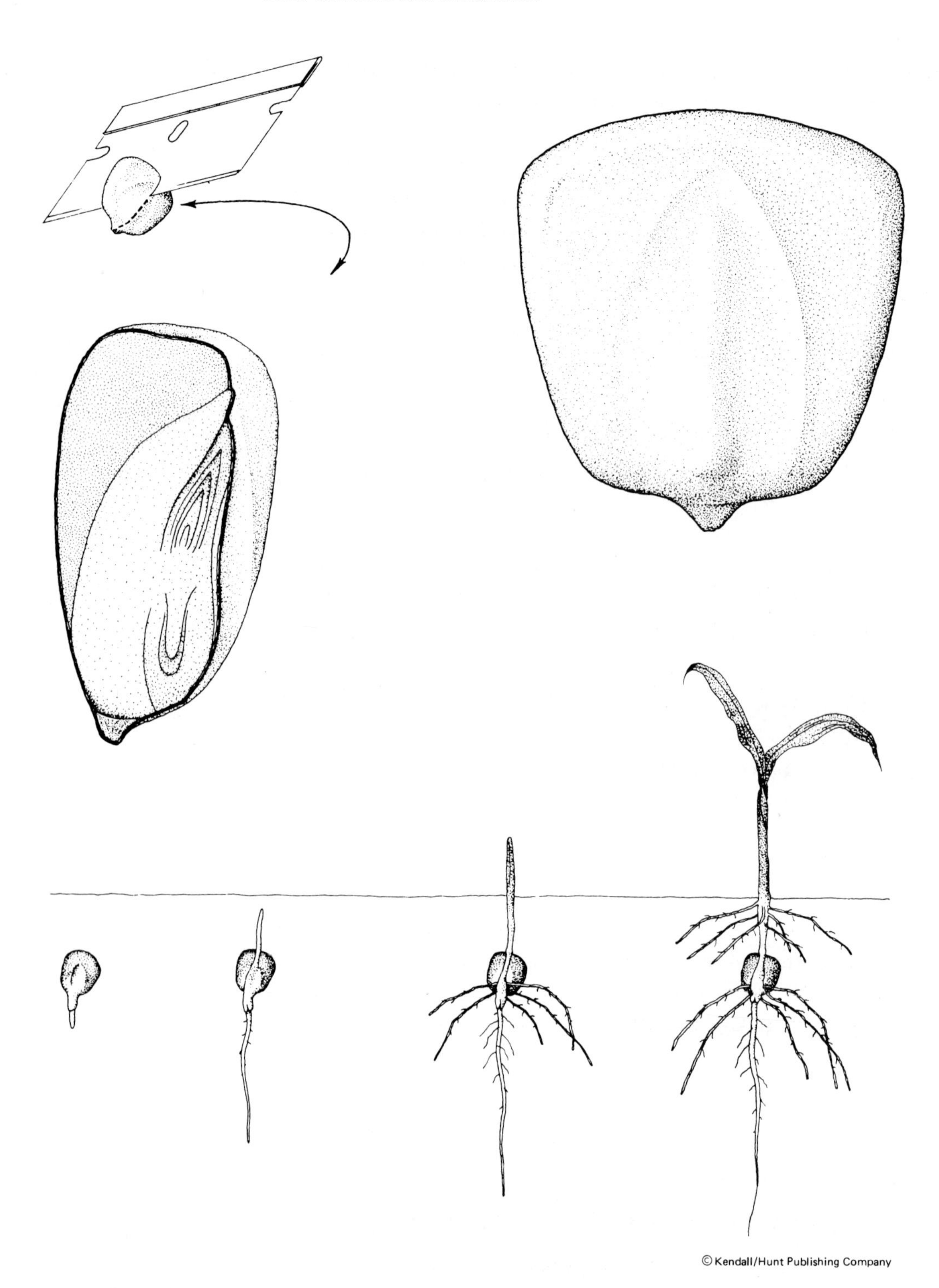

© Kendall/Hunt Publishing Company

Figure 14.6 Corn kernel structure and germination. In the bisected grain, the sheathed embryonic leaves and the shoot apex they surround resemble the flame of a candle. The embryonic root is surrounded by a space. The root and shoot are embedded in a layer of tissue that will produce hormones during germination. The cotyledon is a pad of tissue cupped against the rest of the embryo. Compare the corn germination sequence to bean germination, described earlier.

Many kinds of chemical stains and indicators can be used to help us see greater detail in biological specimens. The corn kernel, for instance, may appear to be a homogeneous, creamy white. To distinguish the embryo from the endosperm, we can stain either the embryo or the endosperm.

Put a drop of iodine solution on a clean glass microscope slide, then place the cut surfaces of the corn kernel down in the iodine. Leave them in the iodine for at least 3 minutes, turn them over, gently blot the iodine from the cut surfaces, and observe them.

5. How can you now distinguish the endosperm tissue from the embryo?
6. What specific reaction has occurred?
7. Does this reaction confirm your hypothesis about endosperm storage?

When grains are processed into flour, the embryo part of the kernel, called the *germ*, is removed. The endosperm then is ground up (milled) and sifted to remove fragments of the fruit wall/seed coat.

8. Based on this information and your own observations, what can you conclude is the major component of flour?
9. Why do you think grains are such an important source of food for humans and domestic animals?

14.10 On *Day 5* of your germination project, complete your own data table and then examine the most vigorous alfalfa, radish, and mung bean seedlings. *Reminder:* Do this activity even if you are on the lettuce team. You must place the sprouts on a wet paper towel or in water to keep them from drying out while you examine them. A magnifying glass or binocular microscope will aid your examination greatly.

1. How can you distinguish the transition region between the young stem and the young root?
2. What part(s) of the seedlings have developed chlorophyll?
3. Sketch one each of an alfalfa, radish, and mung bean seedling and label the following structures: primary root, main stem axis, cotyledons, true leaves, shoot apex, and root apex.

As your final observation in these experiments, eat some of your sprouts, plain or with salad dressing.

4. How would you describe the texture of succulent young seedlings?
5. How would you describe the distinctive flavors of alfalfa, radish, and mung bean seedlings?

Based on the introductory description of germination and your own observations, answer the following questions about the seeds you have germinated. If there were differences among the alfalfa, radish, and mung bean seeds, please note them.

6. Which species germinated most quickly? most slowly?
7. Did your seedlings emerge root-first or shoot-first?
8. Did the seed coats remain attached to the seedling root or to the shoot?
9. How long did the seed coats remain attached?
10. At any time did you notice the presence of a hook? Please explain.

14.11 Let us now analyze some of the *environmental variables* that can affect germination. *Review your complete data* for the alfalfa, radish, mung bean, and lettuce seeds exposed to four sets of environmental conditions as the basis for answering the following questions.

Plants need certain things from their environment, including air, water, an array of specific minerals, light, and favorable temperatures. In an experimental situation, each factor is considered an *experimental variable.* In a given experiment, however it is wise to *vary* only one or two factors and to *control* (keep constant) the other factors.

1. Why is it important to limit the number of variables in an experiment?
2. In your germination activity, which environmental factors were the same (constant) for all of the seeds?
3. Which environmental factors were variables for these experiments?

4. What influence, if any, did the presence or absence of light have on germination? State your conclusions separately for each species.
5. What influence, if any, did temperature have on germination? State your conclusions separately for each species.
6. What significant roles does water play in germination?
7. What was the result of withholding water from the seeds?

14.12 In this unit you were introduced to seed structure and germination, and you had an opportunity to conduct an extended experiment. Before you continue to the next chapter, review the study points below. Check off the points you understand. Return to the text to review any activities and concepts you do not understand.

On completion of this chapter, you should be able to do the following:

- ☐ 1. Describe seed and embryo structure.
- ☐ 2. Understand the several phases of germination.
- ☐ 3. Understand the role of growth inhibitors in dormancy.
- ☐ 4. Germinate edible seedlings.
- ☐ 5. Understand the general effects of light and temperature on germination of selected seeds.
- ☐ 6. Keep data records, including both objective and subjective observations.

Plant Tissues 15

How do specialized tissues meet the needs of a plant?

15.1 Organisms whose bodies are composed of many cells are called **multicellular organisms.** Some simple multicellular organisms have bodies that are composed of a loose aggregation of specialized cell types. Most multicellular organisms, however, have true **tissues,** groups of cells that are specialized to perform specific functions in the body of the organism.

You are familiar with animal bodies. Feel the lining of your mouth with your tongue. The smooth epithelial tissue there protects underlying cells from air and other substances that enter your mouth. Muscle tissue enables your eyes to move, following these words across the page. Nervous tissue enables your brain to sense and interpret the words you see.

What about plant bodies? They also are made up of specialized tissues. Plant tissues carry out specific roles in a plant body, enabling it to manufacture and store food, grow, obtain and distribute what it needs from the environment, and to resist such environmental hazards as gravity, wind, temperature extremes, and organisms that would feed on its tissues. Plant bodies also produce growth-regulating substances and a variety of other secretions.

15.2 You will recall that plant cells are nearly identical in cytoplasmic structure to animal cells. Plant tissues, however, differ greatly from animal tissues. The most basic distinction of plant tissues is that each kind of tissue has its own supporting skeleton, formed by many cell walls. By contrast, animal cells have only a delicate outer cell membrane; therefore, most animal tissues are soft. Since both plant and animal bodies meet similar environmental challenges to survival, however, their tissues can be classified in functional categories that are analogous to some of the animal tissue categories (data table 15.1). Figures 15.1 and 15.2

Data Table 15.1 Functional Categories of Plant Tissues

Category	Tissue Name	Functions
Covering		On young stems, leaves, fruits:
	Epidermis	Protects against water loss and invasion by microbes. Permits gas exchange between atmosphere and inner tissues.
		On roots:
	Epidermis	Protects against abrasion from soil particles and microbe invasion. Absorbs water and minerals.
		On woody plants:
	Cork	Protects, cushions, and insulates inner tissues.

Data Table 15.1 Continued

Category	Tissue Name	Functions
Support	*Parenchyma and Collenchyma*	Hold herbaceous shoot system upright, maximizing exposure to sun, air, and pollen.
	Sclerenchyma, including fibers and sclereids	Provides flexible toughness, keeping conducting tissues from becoming crushed or kinked, yet allowing plant to bend; also forms hard parts, such as shells, seed coats.
	*Xylem**	Extensive woody development provides physical and physiological support for large, long-lived plant body.
Vascular transport	*Xylem* and Phloem**	Movement of fluid solutions throughout plant body, carrying water, raw materials, and food (mainly sugars) between sites of absorption, organic synthesis (including photosynthesis), storage sites, and sites of energy demand.
Growth	*Meristems, including apical meristems, vascular cambium, and cork cambium*	Specialized tissues in which cell division occurs, giving rise to new body cells that mature and differentiate into specific kinds of cells/tissues.
Growth regulation	*Parenchyma, possibly other*	Actively growing sites are primary production sites for hormones that regulate growth, development, and responses to environmental stimuli.
Food manufacture	*Parenchyma*	Photosynthesis in parenchyma cells that contain chloroplasts (called chlorenchyma cells).
Food storage	*Parenchyma*	Storage of excess food molecules, particularly as starch.
Reproduction	*(Discussed in chapter 20)*	Specialized tissues in which meiosis occurs, giving rise to haploid spores, which germinate and grow into gamete-producing (sexual) forms of the life cycle.
		Asexual or vegetative reproduction by formation of roots on separate shoots or leaves and by formation of specialized vegetative structures.

**Xylem and phloem are complex tissues* (composed of several cell types) including conducting cells, sclerenchyma cells, and parenchyma cells.

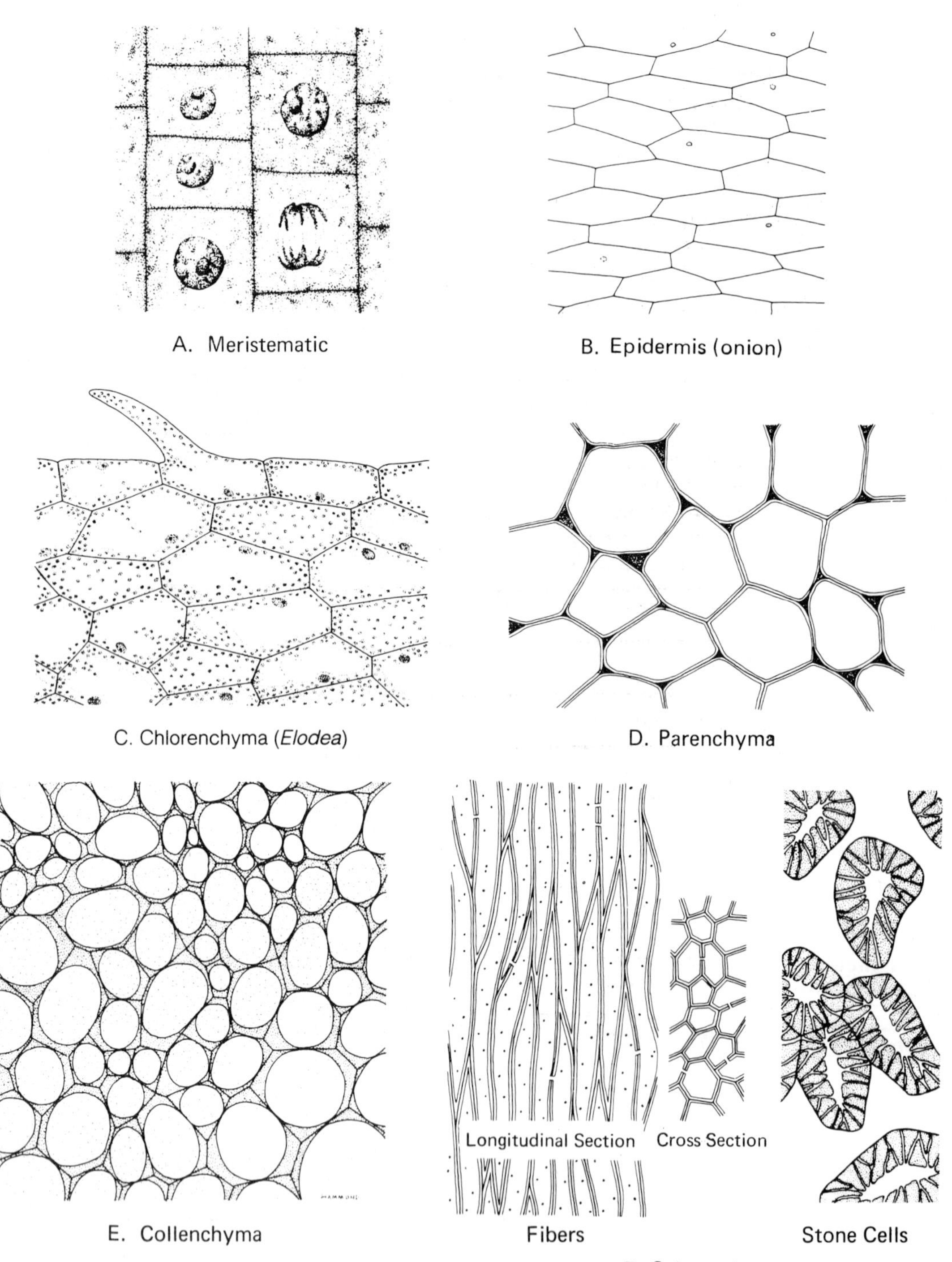

© Kendall/Hunt Publishing Company

Figure 15.1 Some simple plant tissues. A simple plant tissue is composed of one main cell type. In these tissues, note the way cells are fitted together, their shapes, the relative cell wall thicknesses, and the presence or absence of such special features as dividing nuclei, chloroplasts, and intercellular spaces.

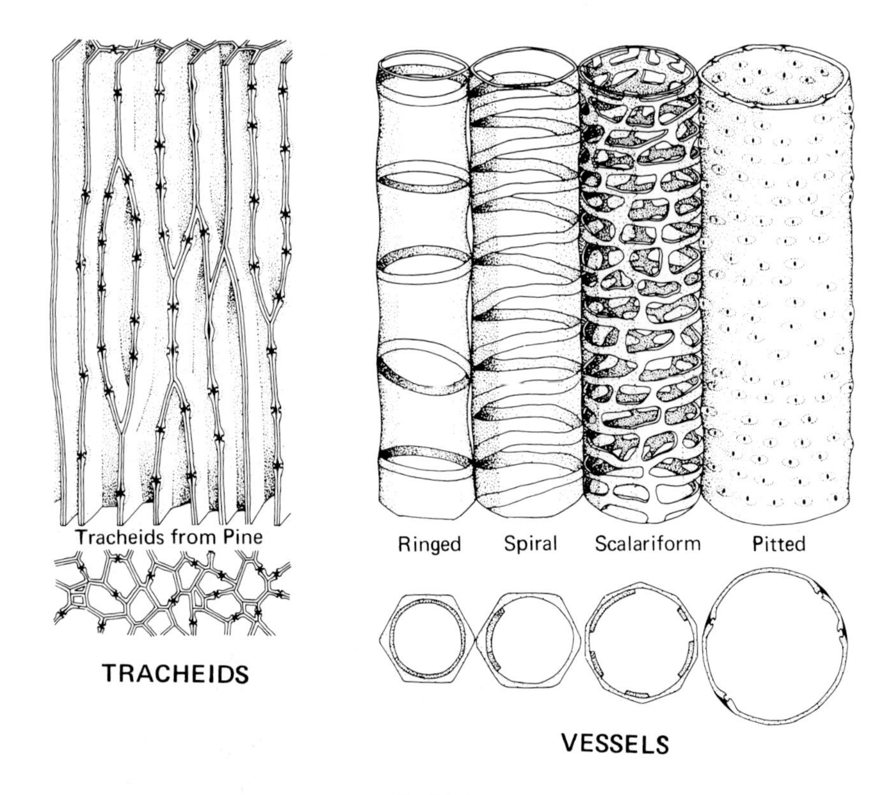

A. Xylem

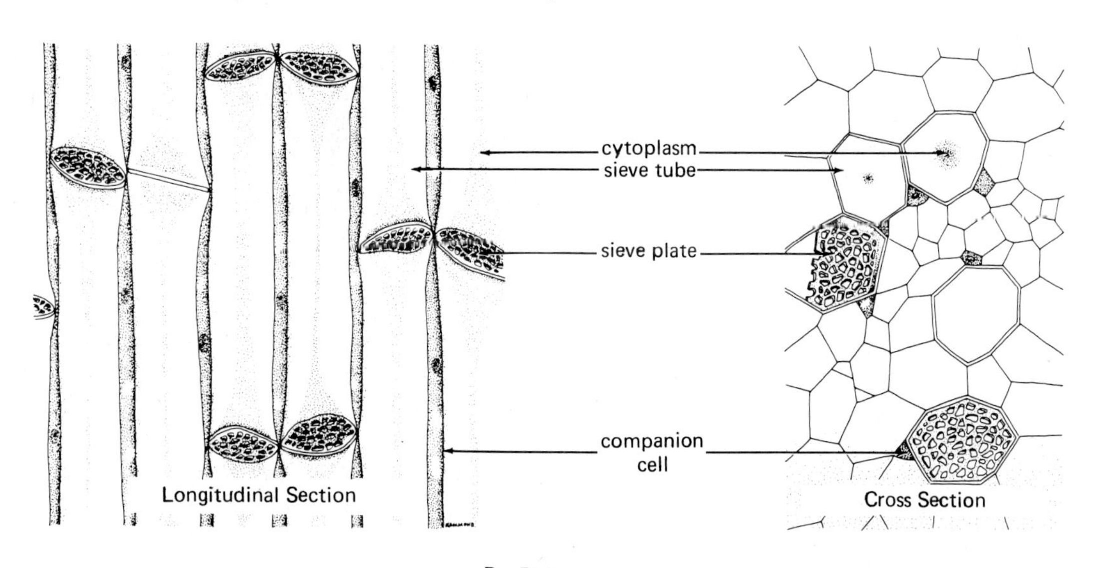

B. Phloem

© Kendall/Hunt Publishing Company

Figure 15.2 Xylem and phloem conducting cells. Tracheids and vessels carry water and dissolved minerals upward from the roots of a plant. Conifers have only tracheids, whereas flowering plants have both tracheids and vessels. Sieve tubes conduct sugar solutions both upward and downward. Xylem and phloem tissues also have parenchyma and sclerenchyma cells; they are complex tissues.

illustrate the major plant cell and tissue types. Although the illustrations are not all drawn to the same scale, you can tell that cell types vary in structure. These differences are related to differences in the functions they perform in the plant body.

The plant body is a mosaic of these tissue types, because each plant **organ** is made up of several different types of tissues, another way in which complex plant and animal bodies are similar. Also, the organs of the plant body are interconnected by vascular tissues, which transport vital fluids.

You will notice that many plant tissue names end in *-enchyma,* which is derived from a word meaning "poured in." The ending refers to the way cells are packed together in tissues. The first part of each tissue name refers to a distinguishing feature of the tissue. Data table 15.2 shows some of the word roots used in botanical naming, and understanding them will help you to understand the terms used in later units and in other reference materials.

1. What is a fundamental difference between plant and animal cells and, therefore, their tissues?
2. What plant tissues are responsible for protecting the plant body from its external environment?
3. Which functional category is represented in plants but *not* in animals?
4. What tissue type has the greatest variety of functions, and what are these functions?
5. What support tissue is found only in herbaceous stems?
6. What tissues transport (conduct) solutions throughout the plant body?
7. What is a *complex* tissue?
8. What is the function of meristems (meristematic tissues)?
9. Why do most plant tissue names end in *-enchyma?*

Data Table 15.2 Word Roots for Some Terms Commonly Used in Botany

(*L* means the word was derived from Latin; *Gr* means the word was derived from Greek; *Sp* means the word was derived from Spanish.)

Word Root	Origin	Relevant Translation
Terms relating mainly to the overall plant body:		
api, apex	L, *apex*	tip or summit
axil	L, *axilla*	armpit
cambium	L, *cambiare*	to exchange
chloro	Gr, *chloros*	green
colla	Gr, *colla*	glue
cork	Sp, *alcorque*	cork
derm	Gr, *derma*	skin
epi	Gr, *epi*	upon, on the outside
meristem	Gr, *meristos*	divided
parenchyma	Gr, *parenchyma*	
para		beside
enchyma		in-poured
phloem	Gr, *phloios*	bark
phyll	Gr, *phyllum*	leaf
phyto, phyte	Gr, *phyton*	plant
scler	Gr, *scleros*	hard
stoma	Gr, *stoma*	mouth
vessel	L, *vascellum*	hollow or concave container
xylem	Gr, *xylon*	wood
Terms relating to sexual reproduction:		
anther	Gr, *anthos*	flower
carpel	Gr, *karpos*	fruit
ovule	L, *ovum*	egg
petal	Gr, *petalon*	leaf
pistil	L, *pistillum*	pestle
pollen	L, *pollen*	fine dust
sepal	Gr, *skepas*	covering
stamen	L, *stamen*	thread of life

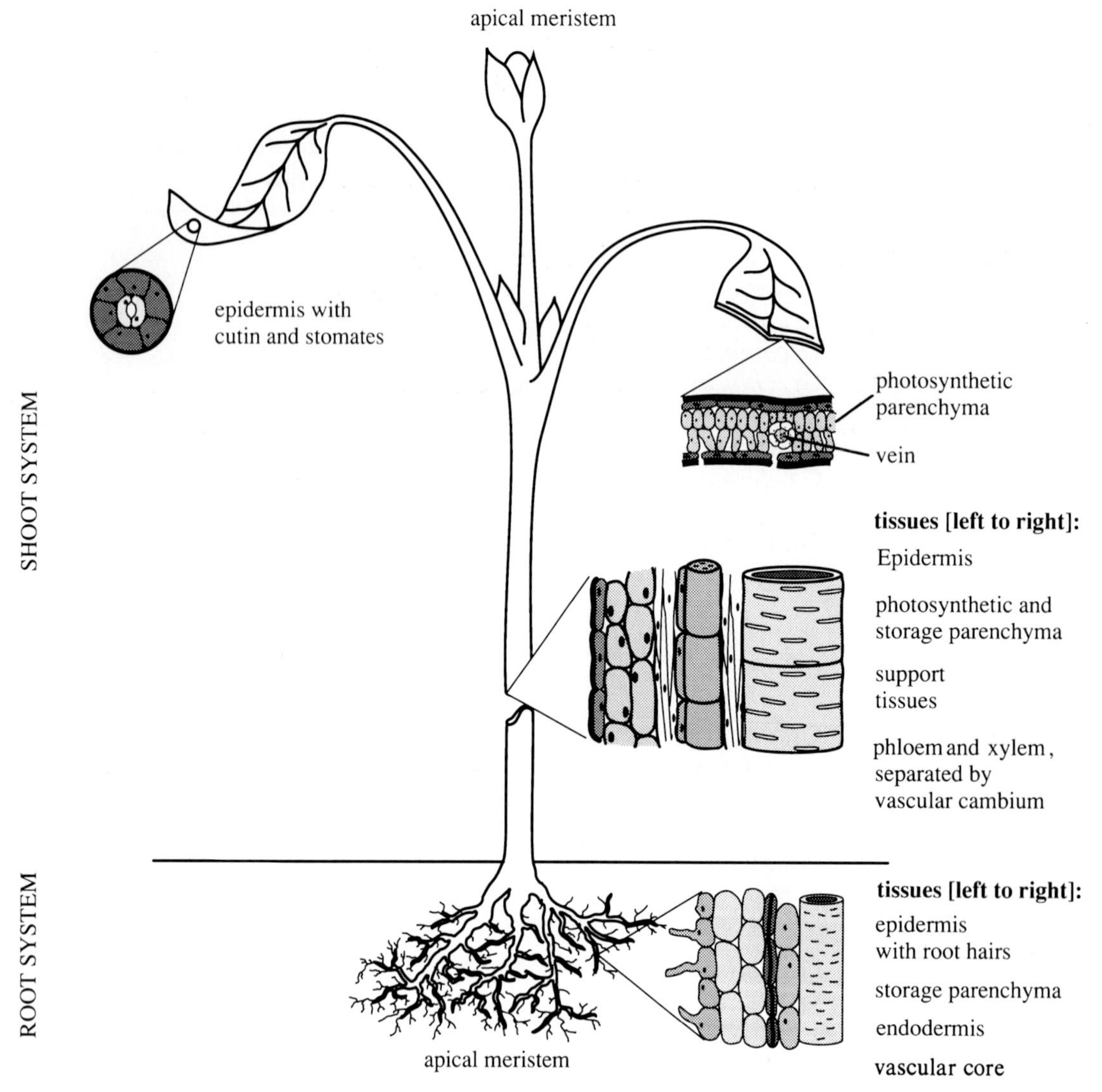

Figure 15.3 Tissue distribution in a plant body. This *highly* schematic drawing shows some of the tissues that make up leaves, stems, and roots. Compare it to data table 15.1 to understand the roles of the various tissues in the life of a plant.

Figure 15.3 shows how some of these tissues are distributed in the body of an **herbaceous** plant, one that does not develop a long-lasting, permanent body with woody tissues. In a woody plant body, the youngest parts of the stems and roots are at their tips. They resemble the shoot and root tips of herbaceous plants in tissue types and in their arrangements.

15.3 This chapter has presented a brief introduction to plant tissues to prepare you for investigations of plant organs. Review the study points below, and be sure you understand them before checking them off.

On completion of this chapter, you should be able to do the following:

☐ 1. Understand the major functions of the several plant tissue types, so you will think about their functions each time you observe them in a particular plant organ.

☐ 2. Say and spell the tissue names, so they will be part of your biological vocabulary.

Herbaceous Stems

16

What is an herbaceous stem and how does it grow?

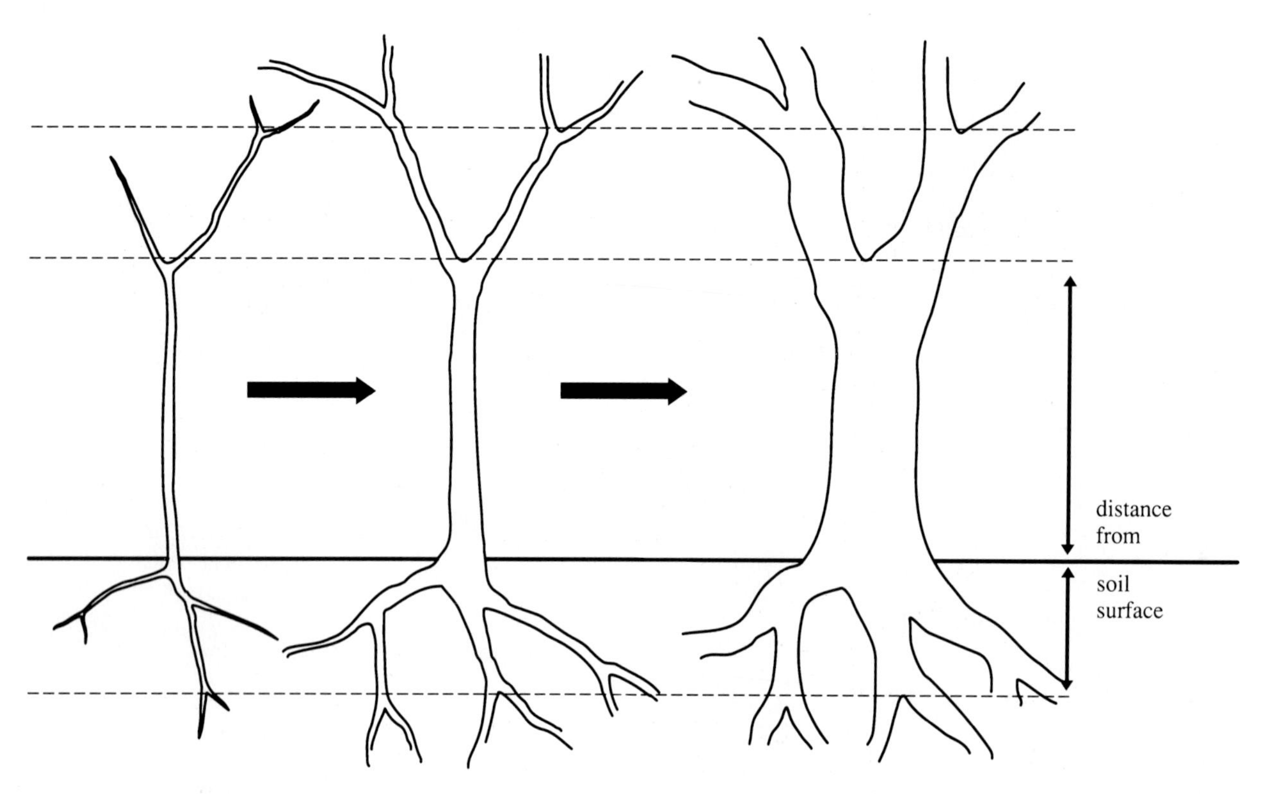

Figure 16.1 Dimensions of stem growth. During its lifetime, a plant stem grows and matures. It begins as the delicate herbaceous stem of the seedling. In many species, the stem maintains its herbaceous characteristics, but in others it develops extensive wood and bark. Whether herbaceous or woody, a stem grows in two dimensions. It grows *in length* (height) by adding new tissues at the stem tips. It grows *in girth* (thickness) by the development of xylem, phloem, and cork tissues. This illustration is intended to illustrate that, no matter how tall or how old a plant becomes, the mature tissues of a stem are not responsible for increase in height. Thus, in this case, the branches of a tree remain at the same height above the ground as the tree ages. The same growth principles are seen in roots.

16.1

What is a stem? The **stem** is the most complex plant organ and has the greatest variety of functions. It is responsible not only for support, transport, and storage, but also for the generation of other organs. Thus, you can think of the stem as *the parent organ of the entire shoot system.* How does a stem grow? It increases in length by adding new cells at its tip, and it increases in diameter (girth) by adding new cells to its conducting tissues (figure 16.1).

Examine the herbaceous plants your teacher has provided, and look at figure 16.2. Each shoot consists of the stem and any leaves, buds, cones, flowers, or fruits that the stem has produced. Notice there is a main stem plus many branches, or sideshoots. Branches that bear only leaves are called *vegetative shoots,* and those that bear flowers are called *flowering shoots.*

As directed by your teacher, appoint one member of your group to cut off a sideshoot (carefully) and take it to your desk to study it. Do not break or remove any structures from your shoot until you have examined them as directed.

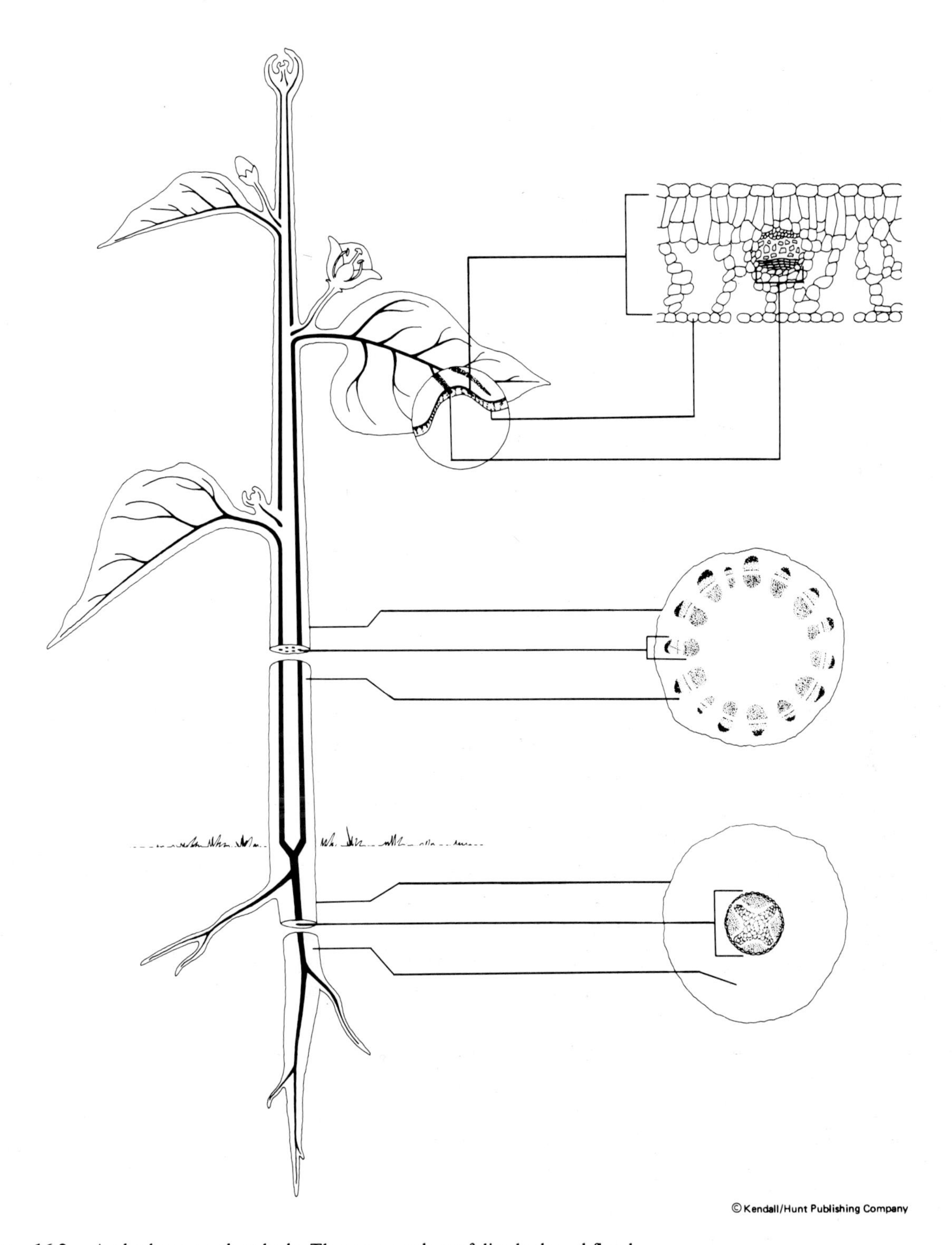

© Kendall/Hunt Publishing Company

Figure 16.2 An herbaceous plant body. The stem produces foliar buds and floral buds, from which leaves and flowers develop, respectively. Nodes are the points along the stem at which buds arise. Vascular tissues are continuous from the roots to the tips of the stem-produced structures. In the root they are arranged as a vascular core; in the stem they form vascular bundles that are continuous with veins in the leaves and flowers.

First, examine how the leaves are attached to the stem. The places on the stems where leaves are attached are called **nodes,** and the spaces between nodes are called **internodes.** The way leaves are attached to the stem is a genetic characteristic that can be used to help identify and classify plant species. In many plants there is a single leaf at each node, so the leaves are attached to the stem in a spiral, or alternating way. This is called an *alternate leaf arrangement.* In many others, two leaves come out opposite each other, an *opposite leaf arrangement.* A third arrangement is for three or more leaves to originate at a single node, a *whorled leaf arrangement.*

1. What is the name of the plant you are examining?
2. What kind of leaf arrangement does it have?
3. Make a simple sketch to show how the leaves are arranged on the shoot.

Examine the shoot more closely. The point of leaf attachment to the stem is called the **leaf axil.** Within each leaf axil there is a **bud,** which will give rise to a sideshoot. Because it is in a leaf axil, such a bud is called an **axillary bud.**

Look at the tip of any vegetative shoot and you will see a cluster of tiny, immature leaves. Dissect them away carefully and you will expose the delicate bud at the shoot apex, the **terminal** or **apical bud.** (Apex is a word that means the "tip" or "peak" of something.) The *apical meristem* is a tiny pad of delicate growth tissue located at the apex of each shoot (bud). In the following activities, you will examine its microscopic structure to learn about the nature of plant growth. Figures 16.3 and 16.4 show different views of shoot apical meristem.

4. Return to figure 16.2 and label the following structures in your logbook: node, internode, leaf, flower, axillary bud, and apical bud. Be sure to use neat, horizontal lines in your labeling.

16.2

A plant shoot becomes longer because new cells arise by cell division of the apical meristem. Buds also are developed from the apical meristem. As noted above, each bud has its own apical meristem. In a vegetative shoot, nodes begin to form around the sides of the apical meristem, differentiating into *primordial leaves* and axillary buds. In a flowering shoot, floral buds are formed instead of vegetative buds. Floral buds contain *primordial flowers.*

Figure 16.3 Celery stem apex. Peel the leafstalks of celery away until you get to the short, conical stem to which they are attached, and you will expose the stem apex. To get the magnified view afforded here, however, requires a scanning electron microscope. The dome-shaped apical meristem is surrounded by primordial leaves. The leaf blades are already starting to differentiate into their divided form and have tiny epidermal hairs at their bases.

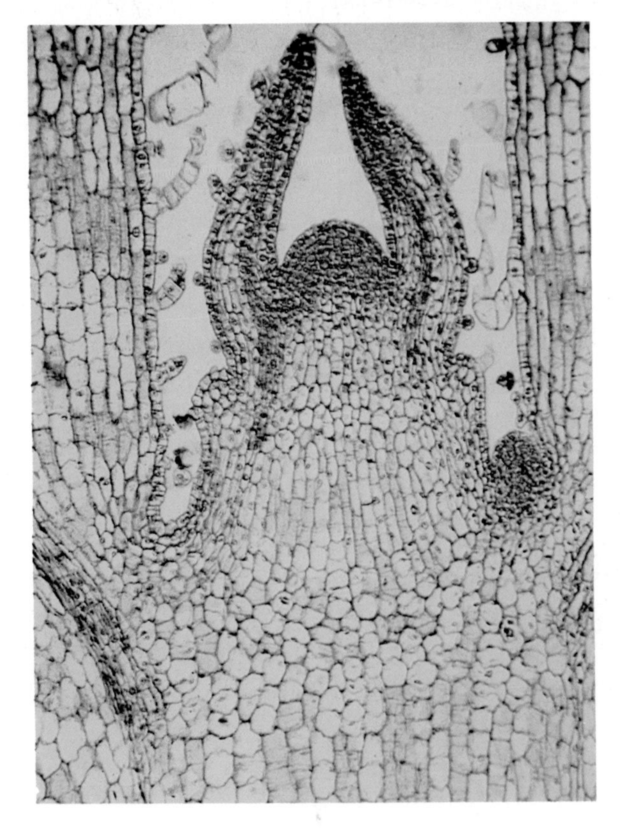

Figure 16.4 *Coleus* stem apex. *Coleus* is a popular houseplant with variegated leaves. Compare this view of a stem apex to that in figure 16.3. Sectioning the stem tip into thin slices allows us to see how the cells change as they grow and mature. Here you can see the apical meristem as a dome of small, dense cells—looking a bit like a head between two upraised arms, which are primordial leaves. Older, more mature cells are larger and show some differentiation into tissue types. Vascular tissues will develop from the strands of darker cells in the young leaves; a more mature vascular strand is visible in the lower left corner. Epidermal hairs and an axillary bud also are visible.

From your teacher, obtain a microscope slide showing a longitudinal section of a *Coleus* or *Elodea* stem tip, and identify apical meristem, newly formed stem tissues, primordial leaves, and developing axillary buds. At this young stage of development, most of the cells are in the process of enlarging, but do not show much differentiation.

1. How would you describe the shape and dimensions of the young cells?

2. List the products of the apical meristem.

16.3 Now we will examine the differentiated tissues of the stem, for which figure 16.5 should be helpful. Refer to it as you make your microscopic observations. *Remember to use good microscopy techniques, always beginning your observation with low power and always having the low-power objective in place when moving a slide to or from the microscope stage.* This technique lessens the chances of breaking a microscope slide or damaging the lens objective.

Before you take a microscopic trip to discover stem anatomy, you need to take a "mind trip" to establish a context. This exercise in imagination will help you to understand how each stem tissue and structure relates to the whole stem in three dimensions.

The concentrically arranged tissue regions that are apparent in the thin slice you will view are actually vertical *columns* of tissue. To help you imagine how they are arranged, think about what the outer skin of the stem would look like if all of the tissues inside it were removed. It would be apparent to you that the epidermis is a *cylindrical sheath* that clothes the full length of the stem. Next, mentally suspend vascular "strings" vertically within the tube of epidermis. Finally, imagine injecting foam to fill in all of the space between the suspended strings and the outer sheath. Allow the foam to become semisolid. The individual foam bubbles resemble parenchyma cells, and the collective mass of foam resembles parenchyma tissue. Congratulations, you have just created an herbaceous stem model!

Now place a microscope slide of a young dicot stem on an unprinted area of this page and look at it closely. The discs that you see are extremely thin slices of a young herbaceous stem, cut so thinly they are transparent and only one-cell layer thick.

1. Are the discs on your slide circular or squarish?

2. How many millimeters wide/thick was the stem from which the slices were taken?

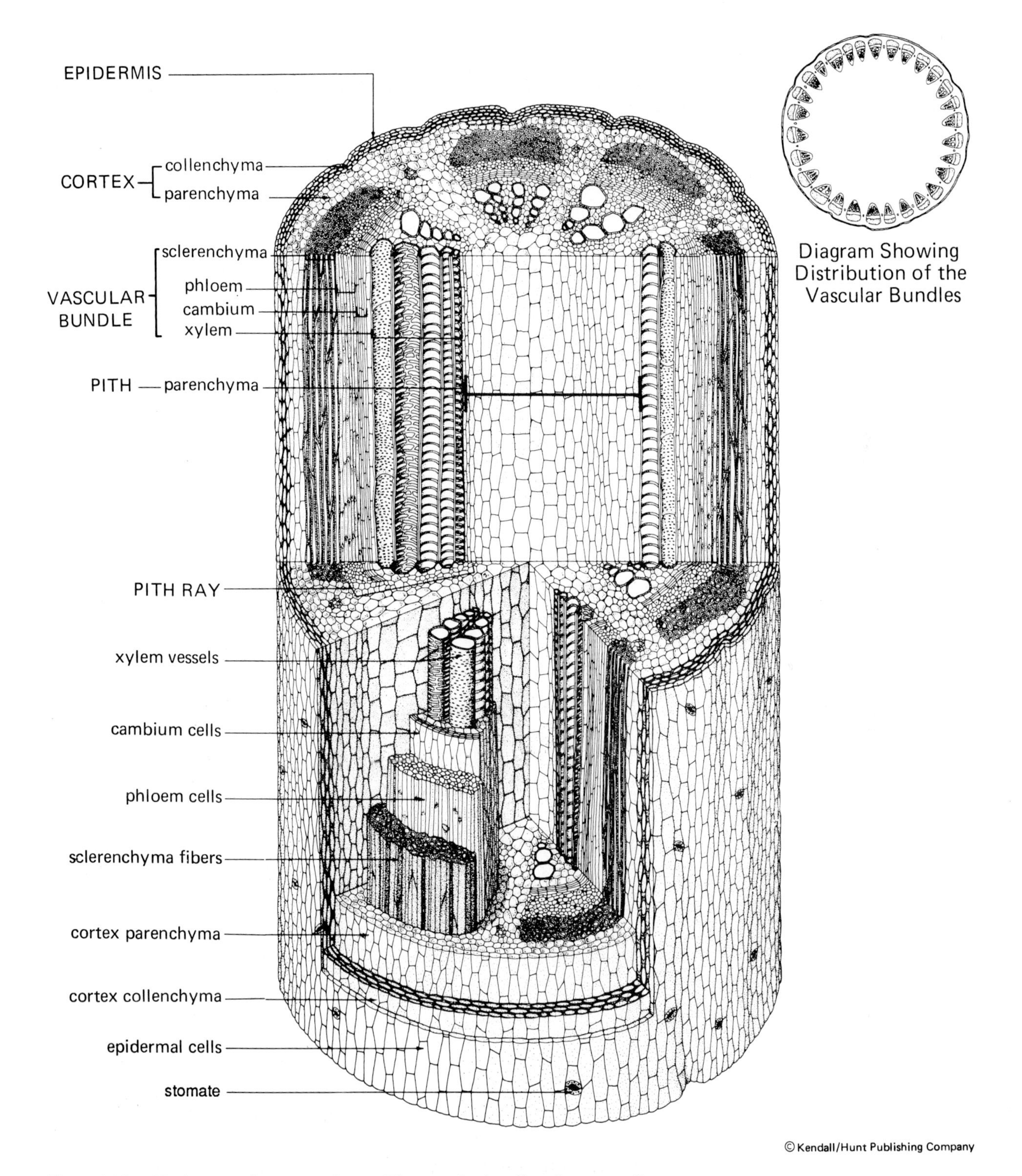

© Kendall/Hunt Publishing Company

Figure 16.5 Herbaceous dicot stem tissues. The vascular bundles of a young dicot stem are arranged in a ring. They are surrounded by the parenchyma tissue of the cortex, pith rays, and pith. The stem depicted here has a solid pith, but some stems are hollow in the center. The stem is reinforced by thick-walled, living collenchyma tissue. The vascular bundles are reinforced by thick-walled sclerenchyma tissue.

Most plants have round stems. If your specimen is a radish (*Raphanus*) or sunflower (*Helianthus*) seedling, the stem sections will look like circles. If it is from a young alfalfa (*Medicago*) stem, it will be squarish. One group of plants, the sedges, even have stems that are triangular in shape (sedges have edges). Like leaf attachment, stem shape can be used to help identify and classify plants.

Keeping your imagined stem model in mind, examine your slide with the microscope, beginning with the low-power objective. As you examine the slide, refer back to figure 16.5.

3. Do the vascular bundles seem to be randomly arranged or are they arranged in a pattern? If in a pattern, how would you describe it?

4. What kind of symmetry (asymmetry, radial, or bilateral) is evident?

5. How is the parenchyma tissue arranged in relation to the vascular bundles and the epidermis?

6. Continuing to observe with the low-power objective, sketch an outline of your stem cross section. Show the correct number, shape, and arrangement of the vascular bundles. You do not need to show individual cells.

16.4 Carefully change to the high-power objective to observe finer detail. As you know, if you are properly focused using low power, you will be nearly in focus when you switch to high power. Use only the fine adjustment knob when the high-power objective is in place.

1. What colors are present in your slide?

2. Where do you think the colors came from?

Most tissues are *unpigmented,* lacking pigments to give them color. Plant tissues that do have pigments (chlorophyll, carotenoids, and anthocyanin) become bleached out by the preservatives and solvents used to prepare microscope slides. Therefore, many kinds of chemical *stains* are used to make cellular structures more visible. Stains are applied after tissue sections have been mounted to the glass slide; then the coverslip is glued on.

Different stains penetrate and are absorbed by different chemical substances within cells. A bluish-purple stain (methylene blue) commonly is used to set off the cell nucleus. For most botanical specimens, a green stain (fast green) is used to highlight cell walls that are composed mainly of cellulose and pectins. A red stain (safranin) commonly is used to highlight cell walls that are coated or impregnated with **secondary substances** to make them hard or waterproof.

Cell wall secondary substances include the following:

Lignins	harden cell walls, as in fibers, vessels, and tracheids.
Cutin	protects external epidermal cell surface from water loss.
Suberin	seals cork cell walls against air and water movement.

3. On your slide, which cells have the thickest walls, those that are stained green or those that are stained red?
4. What can you infer about the chemical composition of the green-stained cell walls?
5. What can you infer about the chemical composition of the red-stained cell walls?

16.5 Look around the outside of the stem section, where you will see a layer of **epidermis,** which is composed of flattened, tile-like cells.

1. Why do the epidermal cells look like flat rectangles on your microscope slide? (Remember the orientation of your specimen.)

The light-pink outer covering that surrounds the stem (it may have been torn in places during preparation of the slide) has no walls or other cellular structures, so you can tell it is noncellular, a secretion. This outer covering is the **cuticle,** composed of cutin.

2. What cells secreted the cuticle?
3. What does the cuticle do for the stem?

Next look at the center of the stem section, where the cells seem to be of a uniform type. (If you have an alfalfa stem, it may be hollow in the very center, so examine the cells surrounding the cavity.) The central tissue area of the stem is called the **pith.** Switch between low- and high-power magnification as needed to clarify your observations.

4. What tissue type makes up the pith and surrounds the vascular bundles?
5. Are the parenchyma cells larger or smaller than most of the other cells present?
6. Do the parenchyma cells seem to have a specialized shape, or do they appear to derive their shape from the way they are packed in among their neighbors? Describe their shape as appropriately as you can.
7. Sketch a cluster of five adjacent parenchyma cells. Resist the temptation to draw only circles; pay attention to the actual angles and thicknesses of the cell walls.
8. If numerous oval or egg-shaped objects are present inside the parenchyma cells, they are **starch grains.** What does the presence of starch grains tell you about one function of parenchyma tissue?

Look for a blue-stained **nucleus** in each parenchyma cell. Some cells seem not to have a nucleus, but you know that each living parenchyma cell should have one.

9. Explain this apparent absence of a nucleus in some cells in your section.

16.6 You have eaten celery leafstalks, so you know about celery "strings," which are **vascular bundles.** Now imagine the inside of an herbaceous stem, with similar "strings" running its length.

1. If it looks like a string *lengthwise,* what does a vascular bundle look like in *cross section?*

Return to the low-power objective, if you have not already done so, and position your slide so you can see the ring of vascular bundles in your stem cross section (figure 16.6). Each bundle is composed of an outer ridge

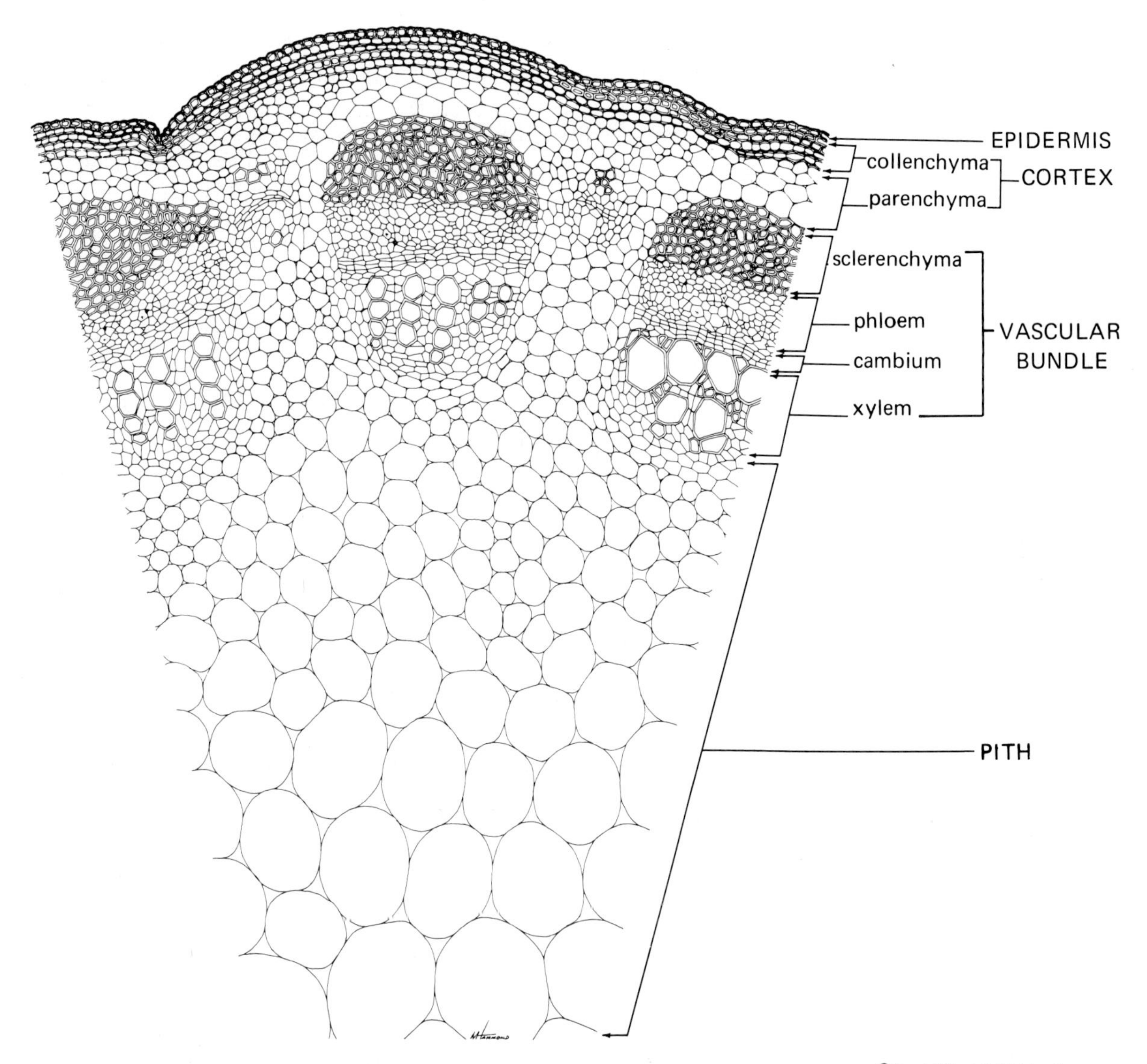

© Kendall/Hunt Publishing Company

Figure 16.6 Dicot vascular bundles. This drawing shows how tissues are arranged in the vascular bundles of an herbaceous dicot stem. Keep in mind that the bundles are long, vertical strings. A layer of sclerenchyma fibers makes up the outer region. A vascular cambium zone adds new phloem and xylem tissue to each vascular bundle. The addition of new phloem and xylem is called secondary growth.

of tough, thick-walled *fibers* (sclerenchyma tissue) that help to support and protect the vascular tissues. In your slide, the sectioned ridge of fibers looks like a "hat" atop the bundle; therefore, it is called a **bundle cap.**

2. Do you think the phrase *bundle cap* gives an accurate suggestion of how the fibers relate to the whole vascular bundle longitudinally (remember the "string")? Explain your answer.
3. What color are the fibers stained, and what can you infer about their cell wall composition?

4. Sketch a group of five adjacent sclerenchyma fibers as they appear in your cross section.

5. Refer back to figure 16.5 to understand how the bundle fibers look when seen lengthwise. Describe the shape and cell wall characteristics of a single fiber from a vascular bundle.

Was your description scientific? Did it portray an accurate image of the fiber's relative length, width, shape, wall thickness, and the diameter of its internal chamber (lumen)? Did you see a nucleus in any of the fibers? It would be most unusual if you did, because fibers grow and differentiate until their walls become thick and hard. At that time the cytoplasm and nucleus degenerate, so that the functionally mature fiber is a dead skeletal element. The fibers will be mentioned again in connection with woody tissues.

Inside the bundle cap is **phloem** tissue, which conducts dissolved food up and down the plant body. Phloem is composed of several cell types, including sclerenchyma cells, parenchyma cells, conducting cells, and their companion cells.

In flowering plants the conducting cells of the phloem are joined end-to-end, and are separated by a perforated cell wall that looks like a strainer, or sieve; consequently it is called a **sieve plate.** Sieve plates allow solutions to pass from cell to cell. The connected cells form a conducting tube called a **sieve tube,** and each conducting cell that forms it is called a **sieve tube member** (or sieve tube element). Perforations on the sidewalls of sieve tube members also allow fluids to move sideways between adjacent cells.

Each sieve tube member has living, functional cytoplasm, but is *enucleate,* lacking a nucleus. You know what a nucleus is supposed to do for a cell, so you must ask yourself, "How can a sieve tube member regulate its activities? What controls its metabolism?" The answer to both questions is that it has a helper cell. Each sieve tube member has a small, nucleated **companion cell** to assist it and direct its metabolism. The sieve tube member and its companion cell are in intimate communication via membrane-lined passages through the plasmodesmata of the cell walls. Like sieve tube members, companion cells are elongate.

Look again at your slide, using high power to examine the phloem tissue, which is stained green. Next to each sieve tube member you should be able to see a smaller companion cell that looks boxlike in cross section. The combination of larger sieve tube members and small, squarish companion cells is a good way to identify phloem tissue. Not all of the sectioned companion cells will appear to have a nucleus.

6. Can you hypothesize why this is so?

Identify sieve tube members on your slide, using low-power magnification to locate them and high power to observe them more closely. As you observe, compare your view to figure 16.5 to understand the longitudinal arrangement of the cells.

7. From their thickness and the way they are stained, what can you infer about the chemical composition of their cell walls?
8. What is the general shape of the sieve tube in cross section?
9. Can you find evidence of a nucleus in any of the sieve tube members?
10. Sketch a sieve tube member and its companion cell, as they appear in cross section on your slide. Show their relative sizes, shapes, and wall thicknesses.

Look for a sieve plate, or part of one. If you do find one, tell your teacher so others may see it. Not all slides may have one visible.

16.7 Efficient water transport is critical to the survival of a plant, because plant shoots lose enormous amounts of water vapor each day by **transpiration,** water evaporation from plant surfaces. Water also is used as a hydrogen source during photosynthesis. The function of **xylem** tissue is to transport water and dissolved substances (minerals, hormones, and so forth) upward through the plant body.

Carefully return to low-power magnification and look at a vascular bundle. The largest cells in the vascular bundle are **xylem vessel elements,** barrel-shaped, open-ended cells that are arranged end-to-end to form vessels. **Vessels** are large, hard-walled tubes that carry a continuous stream of water molecules from the roots to the rest of the plant. Mature vessels have no cytoplasm—the cells *die* to become functional. Vessel sidewalls are perforated, permitting water movement between adjacent cells.

1. How many large vessel elements are present in the bundle you are examining?
2. From their thickness and their staining, what can you infer about the chemical and physical characteristics of vessel cell walls?

Compare your view to figure 16.5.

Smaller xylem conducting cells called **tracheids** also may be present in the vascular bundles. They are thick-walled, and are more angular in cross section and narrower in diameter than the vessel members. Tracheids have perforated end walls and sidewalls, but the perforations are not as complex as those of the vessels. Tracheids do not form continuous tubes.

Tracheids are a more primitive kind of plumbing than are vessels, but they are, nonetheless, very effective. Conifers, such as pines and firs, have tracheids in their wood, but not vessels. Flowering plant xylem has both tracheids and vessels, and the vessels are highly developed.

3. What xylem cell types conduct water vertically in angiosperms?
4. What xylem cell type conducts water vertically in conifers?
5. Do you see fibers in the xylem tissue on your slide? How can you tell they are fibers?
6. Do you see parenchyma cells in the xylem tissue on your slide? (They may be more boxlike in appearance than those you saw in the pith.)
7. Based on your observations, and comparing them to what you learned about phloem tissue, would you say that xylem is a complex tissue (composed of more than one cell type) or a simple tissue (composed of a single cell type)? Explain your response.

Now examine the region that is a boundary *between* the phloem and xylem. You will see a line of thin-walled, flattened cells. The line is called **vascular cambium** tissue. The vascular cambium is made of growth (meristematic) cells that produce new phloem, xylem, and vascular cambium cells. Some new cells are produced on the phloem side of the vascular cambium, and some are produced on the xylem side (figure 16.7). Therefore, the youngest, most immature xylem and phloem cells are always right next to the vascular cambium on both sides. Most vascular cambium cells are elongated, like the cells they produce, as shown in figure 16.5.

8. What is the function of the vascular cambium?
9. Cells produced on the outer side (farther from the stem center) of the vascular cambium mature into what tissue?
10. Cells produced on the inner side of the vascular cambium mature into what tissue?

If your stem is slightly older, you may see a layer of vascular cambium extending *between* adjacent vascular bundles. This sidewise development of vascular cambium occurs when the cambial cells divide longitudinally along a plane that is *radial* in respect to the stem center (see figure 16.7). Eventually the vascular cambium may form a continuous *cylinder* of xylem, vascular cambium, and phloem (figure 16.8). Your teacher may want you to examine a slide that resembles figure 16.8.

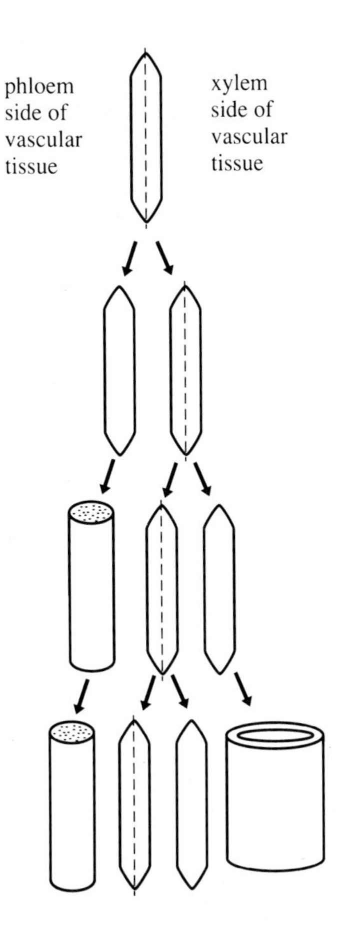

1. vascular cambium cell divides lengthwise
2. one daughter cell differentiates into a phloem cell, the other becomes a vascular cambium cell
3. vascular cambium cell divides lengthwise
4. one daughter cell differentiates into a xylem cell, the other becomes a vascular cambium cell
5. vascular cambium cell divides lengthwise
6. one daughter cell differentiates into a xylem cell, the other becomes a vascular cambium cell
7. and so on!

Figure 16.7 Xylem and phloem formation. Each time a vascular cambium cell divides, one of the resulting cells differentiates into either a phloem or xylem cell, and the other one becomes a new vascular cambium cell. This illustration traces the results of three divisions, with the potential for a fourth division indicated. The illustration had to be arranged vertically, according to time frame, but you should try to visualize how it might occur, over time, *in a single horizontal plane,* adding thickness to the vascular tissues.

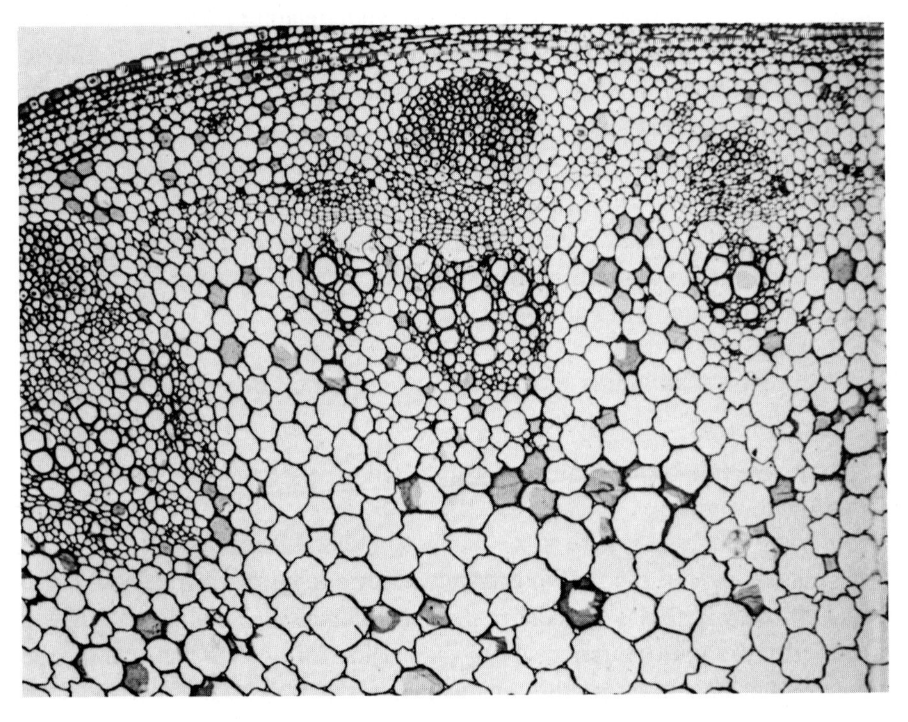

Figure 16.8 Fusion of vascular bundles. In this micrograph you can see a band of tissues between the vascular bundles that differs in appearance from the surrounding parenchyma cells. The band is formed by parenchyma cells that are dividing and differentiating into cambial cells. Once the cambium layer is formed, typical phloem and xylem will be produced, as you can see here by the presence of a tiny, new vascular bundle that has formed beside the original bundle. This process eventually forms a complete cylinder made of concentric rings of phloem, vascular cambium, and xylem. This transition occurs in some herbaceous plant stems as well as in stems that will eventually become totally woody in arrangement (next chapter).

The production of a continuous vascular cambium ring, *and* the subsequent production of additional xylem and phloem is called **secondary growth,** which you will examine in the next chapter. By contrast, **primary growth** is the result of maturation of tissues produced by apical meristem tissue. The herbaceous plant body is formed mainly by primary growth, having limited secondary growth. Secondary growth is a major characteristic of woody plants.

16.8 Examine the tissue surrounding the vascular bundles. It should look familiar to you. The regions between the bundles are called **pith rays,** because they radiate from the pith. The region between the vascular bundles and the epidermis is called the **cortex.**

1. What tissue type forms the pith rays and cortex?
2. Based on the tissue type that predominates, what would you hypothesize to be the functions of the pith rays and cortex?
3. What physical evidence supports your hypothesis?

Now put the low-power objective in place, remove the microscope slide, and return it to your teacher as directed.

16.9 Now let us return to a stem you can hold in your hands and transfer some of the microscopic information to a whole structure. Obtain a white (Irish) potato from your teacher. Yes, a potato is a kind of stem. The potato **tuber** is the enlarged tip of a **rhizome,** a kind of stem that grows horizontally beneath the soil surface. To prepare the potato tubers for this activity, your teacher has placed them in a lighted environment for about one week.

Look closely at the exterior of the potato tuber. One end shows dried wisps of vascular tissue where the tuber was removed from the rest of the stem. Nodes and internodes also are present, identified by the "eyes" of the potato tuber, which are clusters of buds at the nodes.

1. How many buds are present at each node? Be sure to look all the way around the tuber.
2. Sketch your potato tuber and add labels to show nodes and internodes.

The outside of the potato is protected by a layer of **cork** tissue, which was formed by a special meristem called **cork cambium.** Think for a moment about why the corky outer covering had to be formed. When young, the tuber was protected by its thin epidermis, but as it grew and expanded, it became too large for the epidermis to stretch around it. However, a layer of cork cambium developed in the outer part of the cortex and began producing cork cells. When the epidermis began to split from the stress, the interior of the stem was protected by the tougher, suberized (tissue protected by suberin) cork tissue. Continual cork production enabled the tuber to continue to expand because its "skin" expanded with it.

3. What color is the outside covering of the potato?

4. What kind of tissue are you eating when you eat potato "skins"?

5. What color are bottle corks and tree bark? (Bottle corks are made from the bark of the cork oak.)

6. What can you infer about the major tissue component of tree bark?

Gently scrape away a patch of the corky skin of the potato, remembering that it has been exposed to light for several days.

7. What color is the tissue just beneath the cork?

8. What pigment is responsible for this color?

9. What kind of tissue is this?

10. What special process is occurring in this pigmented tissue?

Place a knife straight across the tuber and slice away one end. Examine the interior tissue arrangement. The two outermost tissue regions are the corky skin and then the cortex. The outermost cortical cells are pigmented, while the rest are white. In fact, the whole interior of the potato is white.

11. Why do you think the potato's interior is white?

Now test your hypothesis. Scrape some of the milky fluid from the cut surface of the tuber and smear it onto a clean, blank microscope slide. Even with no magnification you probably can see that the fluid is grainy. Add a tiny amount of water and a coverslip and then observe the slide with your microscope.

12. What do you think the ovoid bodies you see are?

Leaving the slide and coverslip in place, carefully put a small drop of iodine solution next to the outside edge of the coverslip. Observe the color change that takes place when the iodine solution contacts the ovoid bodies.

13. Does this chemical test confirm your original identification of the ovoid bodies?

14. Based on your observations, what do you infer the primary function of the potato tuber to be?

16.10 You may know that some plants produce chemical substances that are toxic to animals. The production of toxic or distasteful substances is a plant's way of protecting itself from being eaten by animals, a characteristic arrived at by adaptation via natural selection.

Many plants produce substances that are directly poisonous or that provoke an **allergic reaction.** A substance that provokes an allergic response is called an **allergen.** Allergic responses are of several types. Sensitive persons can experience skin irritation from touching substances to which they are allergic. For instance, poison oak, poison ivy, and poison sumac are North American plants that produce an irritating resinous substance on their exterior surfaces. That substance causes an *allergic dermatitis* (skin inflammation) in humans and some other animals. Some people develop red, itchy bumps called *hives* (urticaria) after ingestion of specific allergens. *Hay fever* is a common allergic response to plant pollen or fungal spore allergens. *Asthma* and *severe shock* (anaphylaxis) are extreme, life-threatening allergic responses.

Many people are allergic to foods that are perfectly safe for most of us, including wheat, eggs, and certain fruits and vegetables. Ironically, many of humankind's most valuable food plants produce toxins. All domesticated and wild members of the nightshade family (Solanaceae) produce toxic alkaloids, including the potato, tomato, and eggplant. The alkaloids poison the nervous system, primarily, and are most highly concentrated in the green tissues of these plants. Therefore, their foliage must *never* be eaten by humans or livestock. The fruits of wild nightshades and of potato plants are very toxic. The ripe fruits of tomatoes and eggplants and the tubers of white potatoes contain relatively low alkaloid concentrations, so are safe to eat. However, you probably have heard warnings about eating "green" potatoes, and this toxicity is why!

Obtain a paper towel or facial tissue before doing this activity, and *do not do the activity if you have an allergy to potatoes, tomatoes, or eggplants.* Cut out a small piece of white, inner potato tuber tissue and taste it. It will have a definite potato taste, with a very slightly bitter aftertaste. Next, cut out a small sliver of the green potato tuber tissue. Taste it, but do not swallow it. This sample will taste more bitter because of the higher concentration of alkaloids, mainly one called *solanine.* Now spit out your sample into the towel or tissue and discard it. If you accidentally swallow some of it, it will not hurt you in such a small amount unless you are unusually sensitive to it.

16.11 So far we have examined the tissue arrangement in herbaceous stems of flowering plants that are classified as **dicots.** Now let us look at **monocot** stems (figure 16.9). Monocot and dicot stems have the same functions in the life of the plant, including support, conduction, storage, and production of leafy and flowering shoots. The arrangement and growth characteristics of the stems differ, however, and we will look briefly at these characteristics.

Grasses are good examples of monocots, and corn is a grass. Examine the fresh or dried cornstalks your teacher has provided. Notice that the outer covering is hard, rough, and fibrous. Identify *nodes* and *internodes.* Notice that the corn leaves (as those of other grasses) form a **sheath** around the node. **Adventitious roots** often form around nodes, especially near the ground. Adventitious roots are those that arise from other than root tissues; for example, from leaf or stem tissues. In corn, particularly, whorls of strong adventitious roots from the lowest nodes form **prop roots** that help to support the plant.

1. Sketch a segment of the cornstalk (stem) showing a node with adventitious roots, plus part of the internode on either side.

2. Examine the interior tissues where the stalk has been cut or broken. What tissue type fills the stem interior?

If the cornstalk is fresh, this tissue will be soft and juicy. Scrape some of the pulp onto your *clean* finger and taste its sweetness. Even if the cornstalk is dried, some of the sweetness may remain. High fructose syrup (corn syrup) is obtained from processing corn stalks and grains.

3. What are the dense "strings" that are arranged longitudinally within the stem?

Corn stems have a solid interior, but many grasses have hollow stems, as you may know if you ever have pulled out a grass stalk to chew on or have looked at straw. (Where do you think humans got the idea to make paper "straws?")

4. What food carbohydrates are stored in the cornstalk?

5. Can you think of another large grass with sweet, pulpy stems that have economic value? (Aloha . . .)

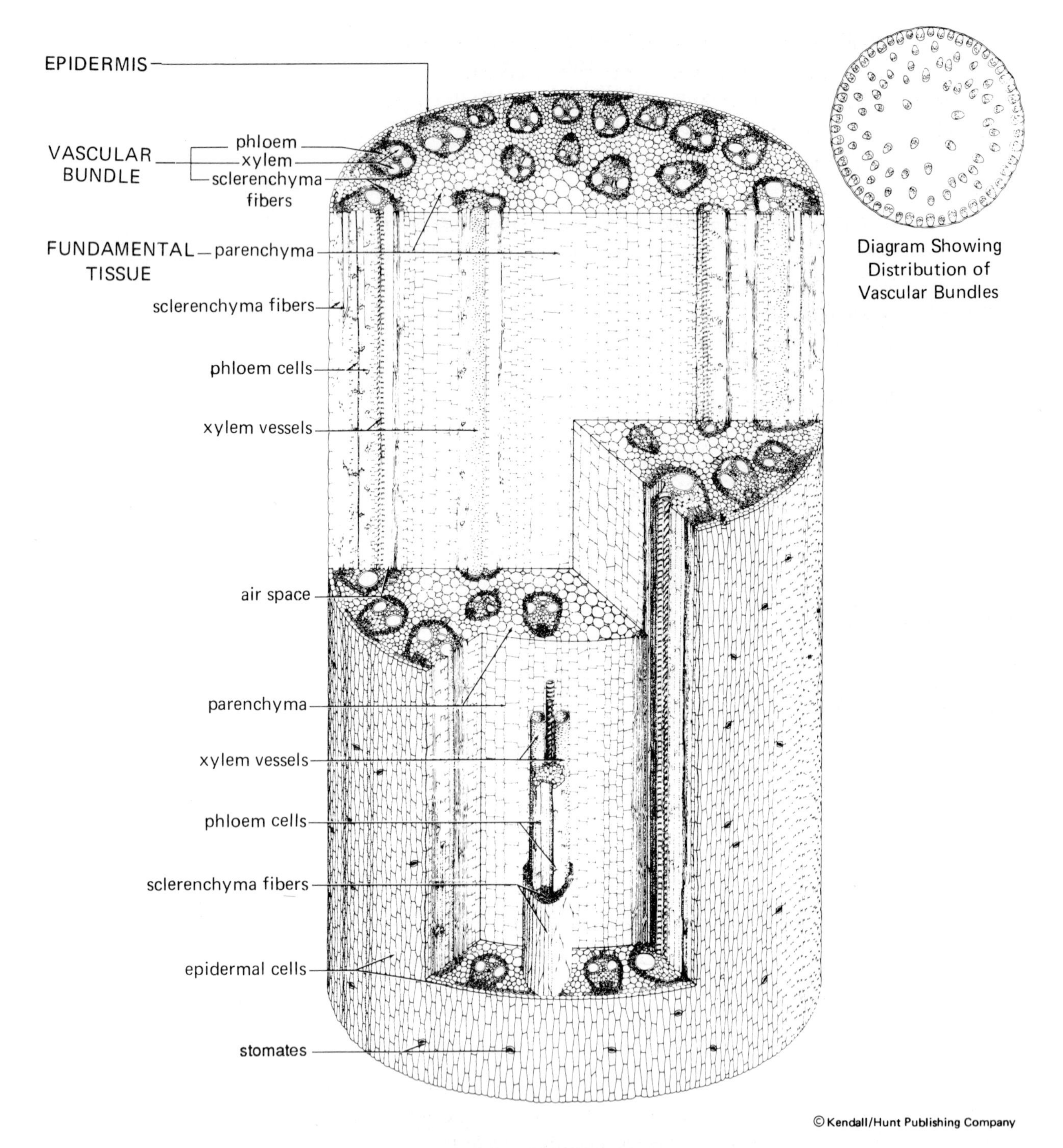

© Kendall/Hunt Publishing Company

Figure 16.9 A typical monocot stem. The vascular bundles of a monocot stem have a scattered arrangement, rather than being arranged in a ring. This is related to the fact that, with few exceptions, monocot stems do not become woody, but remain herbaceous. Their scattered vascular bundles remain independent of one another, in contrast to the pattern seen in dicots.

16.12 Obtain a microscope slide that has cross sections of corn (*Zea*) stems. Position your corn stem slide so the outer layers of the stem are in the center, focus in low power, and then change to the high-power objective. Identify the multicellular **epidermis** (which has *stomates* you may be able to see), underlain by sclerenchyma fibers, which have thick, lignified cell walls.

View the slide with low power and move it around to observe the scattered arrangement of the vascular bundles. Unlike dicots, there are no distinct areas that can be called pith and cortex, so the parenchyma is called simply **ground parenchyma.**

1. Does monocot parenchyma tissue appear to differ significantly from dicot parenchyma tissue?

2. Sketch a schematic diagram of the corn stem section, showing vascular bundle arrangement. (You need not worry about showing individual cells in this sketch.) Label the epidermis, ground parenchyma, and a vascular bundle.

Center a single vascular bundle in your field of vision and use both low- and high-power magnification to clarify detail. The xylem and phloem are arranged differently within a monocot bundle than in a dicot bundle. The bundle may seem to have a "face." In the monocot vascular bundle the largest cells form the "eyes" and "mouth" of the section. Usually, there is a torn-looking space near the "mouth," a result of physical stresses during growth. Monocot vascular bundles have a **bundle sheath** of parenchyma cells that assist in transferring substances into and out of the conducting cells. Refer to figure 16.10 and compare it to your vascular bundle.

3. How many large xylem vessels are apparent in your bundle?

4. How would you describe the shape of the phloem conducting cells and companion cells in cross section?

5. Do you see a vascular cambium layer? Explain.

Unlike stems of dicots, most monocot stems do *not* undergo secondary growth. The xylem and phloem in the vascular bundle are only primary tissues, those that originated from the apical meristem. Therefore, vascular cambium usually is missing from monocot vascular bundles.

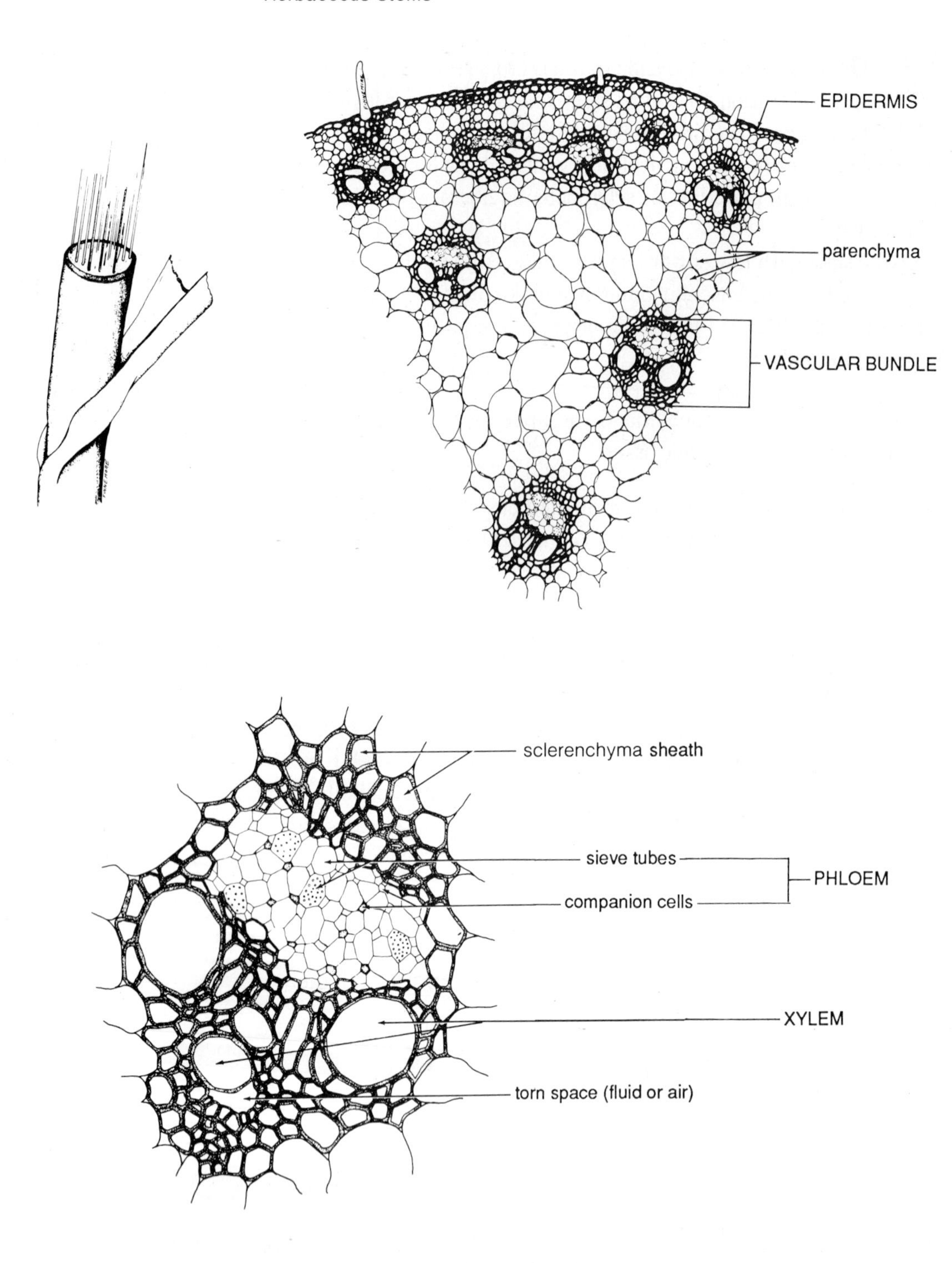

© Kendall/Hunt Publishing Company

Figure 16.10 Monocot vascular bundles. Three degrees of magnification show: an overview of how the vascular bundles are arranged in the corn stem (upper left), a closer view that shows the bundles surrounded by ground parenchyma (upper right), and cellular detail in a single vascular bundle (lower).

Figure 16.11 Bamboo: a woody monocot. Bamboos are the largest, toughest monocots. Their hard, woody stems are an important component of various subtropical and tropical ecosystems, and have many human uses. Bamboo stems have hollow centers, with outer walls that are strongly reinforced with sclerenchyma fibers as well as xylem and phloem. They show a different secondary growth pattern than do woody dicot stems. Bamboos, like other grasses, have meristematic tissues at their nodes, solid discs of tissue that partition the hollow stem into chambers.

Figure 16.12 Corn stem nodes and internodes. Corn is a large grass that also shows clear nodes and internodes. Unlike the bamboo, the internodes are not completely hollow, but are filled instead with soft, sugar-rich storage parenchyma tissue. Monocot nodes not only are sites of growth in length of the stem and of leaf production, but also are sites of adventitious root formation, as shown by the prop roots that typically form on corn plants.

If secondary growth occurs in monocots, it is at the nodes. Bamboo, unlike most monocots, has hard, woody stems (figure 16.11), but is a good example of nodal growth. The joints in the wood are its nodes, and the disc that crosses each node retains some meristematic activity. Thus, the bamboo shoot can elongate at its nodes, as well as at its tip. Edible bamboo shoots are tender shoots that are collected just as they begin to emerge from the ground, and they often are used in Asian cooking. You can see the nodes and internodes clearly in monocot stems that are sliced lengthwise (figure 16.12).

16.13 Now check your stem diagnostic skills. Figures 16.13 and 16.14 are pieces of a broccoli stem and an asparagus stem. Observe the tissue arrangements.

1. Is broccoli a monocot or a dicot? How can you tell?
2. Is asparagus a monocot or a dicot? How can you tell?

16.14 Many plants have stems that are specialized for reproduction or food storage. Think about the circumstances under which such specializations might be useful to the plant. Plants that are in temperate climates usually go through an extended period of dormancy. The dormant period may occur in winter, when conditions are too cold and water is frozen, or it may occur during the hottest, driest part of the year. During the dormant period, photosynthesis may not occur at all—think about all of the trees that lose their leaves during winter. Some metabolism still occurs in stems and roots, but at a lower rate.

As a plant comes out of dormancy, there is a sudden burst of growth and energy use. The plant must rely on stored food resources to meet its metabolic needs until the new leaves are fully productive.

Have you noticed how grasses tend to spread up over the edges of sidewalks, or how strawberry plants produce young plants on "runners"? Both of these examples show how plants reproduce by forming specialized

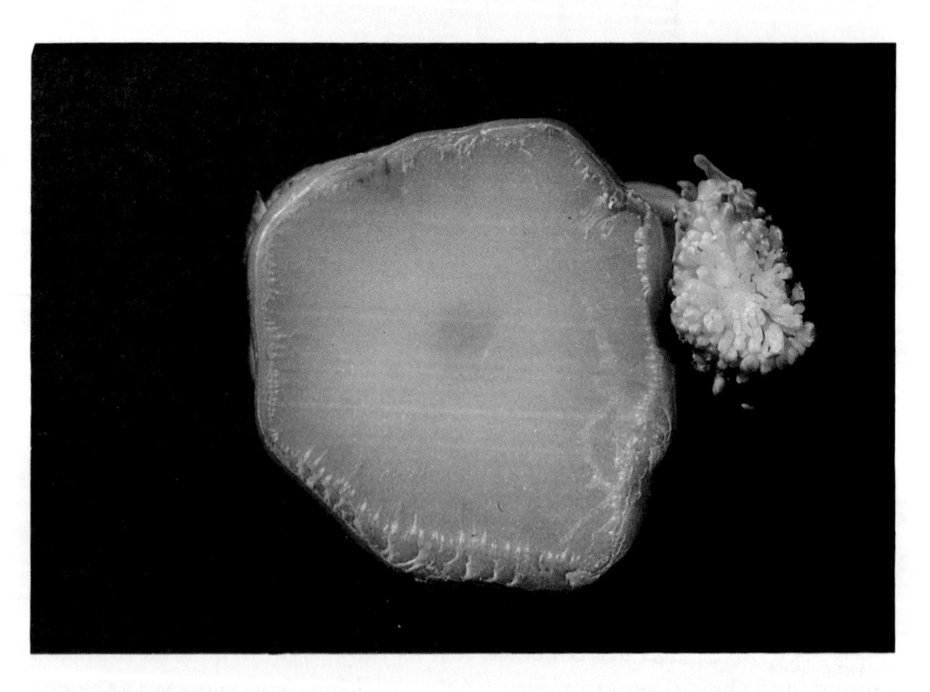

Figure 16.13 Broccoli stem. Judging from the arrangement of stem tissues, would you conclude that broccoli is a monocot or a dicot? This stem segment happens to have a small offshoot that has arisen at the node below where this stem was cut off.

Figure 16.14 Asparagus stem. Judging from the arrangement of stem tissues, would you conclude that asparagus is a monocot or a dicot? This is an older asparagus stem, so the vascular bundles are much tougher than those in the young sprouts we eat.

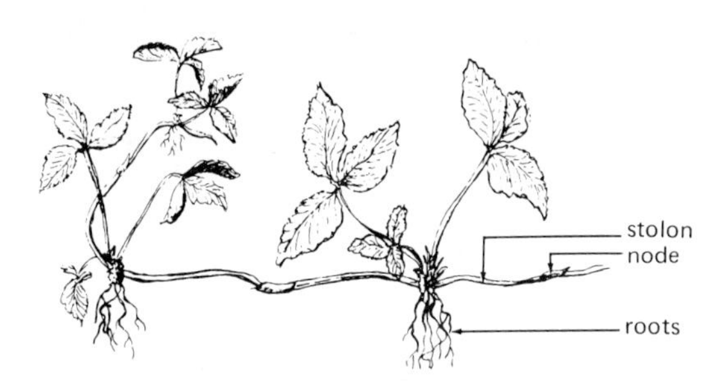

Figure 16.15 Strawberry runners. Many plants produce runners (stolons), horizontal stems that grow across the soil surface and develop new shoot and root systems at their nodes. This is an important means of vegetative or asexual propagation in such plants. Once the new root system has become established, the stolon can be severed and each plant is a new, independent organism.

horizontal stems, from which grow adventitious roots (figure 16.15). Figure 16.16 shows several examples of specialized food storage and reproductive stems. The potato tuber is an example of a stem that is specialized both for food storage and reproduction. Potatoes are grown from pieces of tubers, rather than from seeds. The new shoot and root systems develop from nodes on the tuber pieces, and are nourished by starch that is stored in the tuber until the new plant body is self-sustaining.

16.15 In this chapter you have been introduced to the structure of herbaceous stems, in which the mature tissues are the result of primary growth. Before going on to the study of woody tissues, use the study points listed below to be sure you have grasped the information and concepts you will need. If you do not understand a study point, review the chapter text once more.

Figure 16.16 Specialized stems. Stems in this plate are specialized for food storage and asexual propagation. They include rhizomes (grass, iris, ginger), a tuber (potato), a corm (gladiolus), and bulbs (onion, hyacinth, lily). Note that most of the storage tissue in a bulb is actually *leaf* tissue. Chapter 20 deals in more detail with the reproductive aspects of specialized stems.

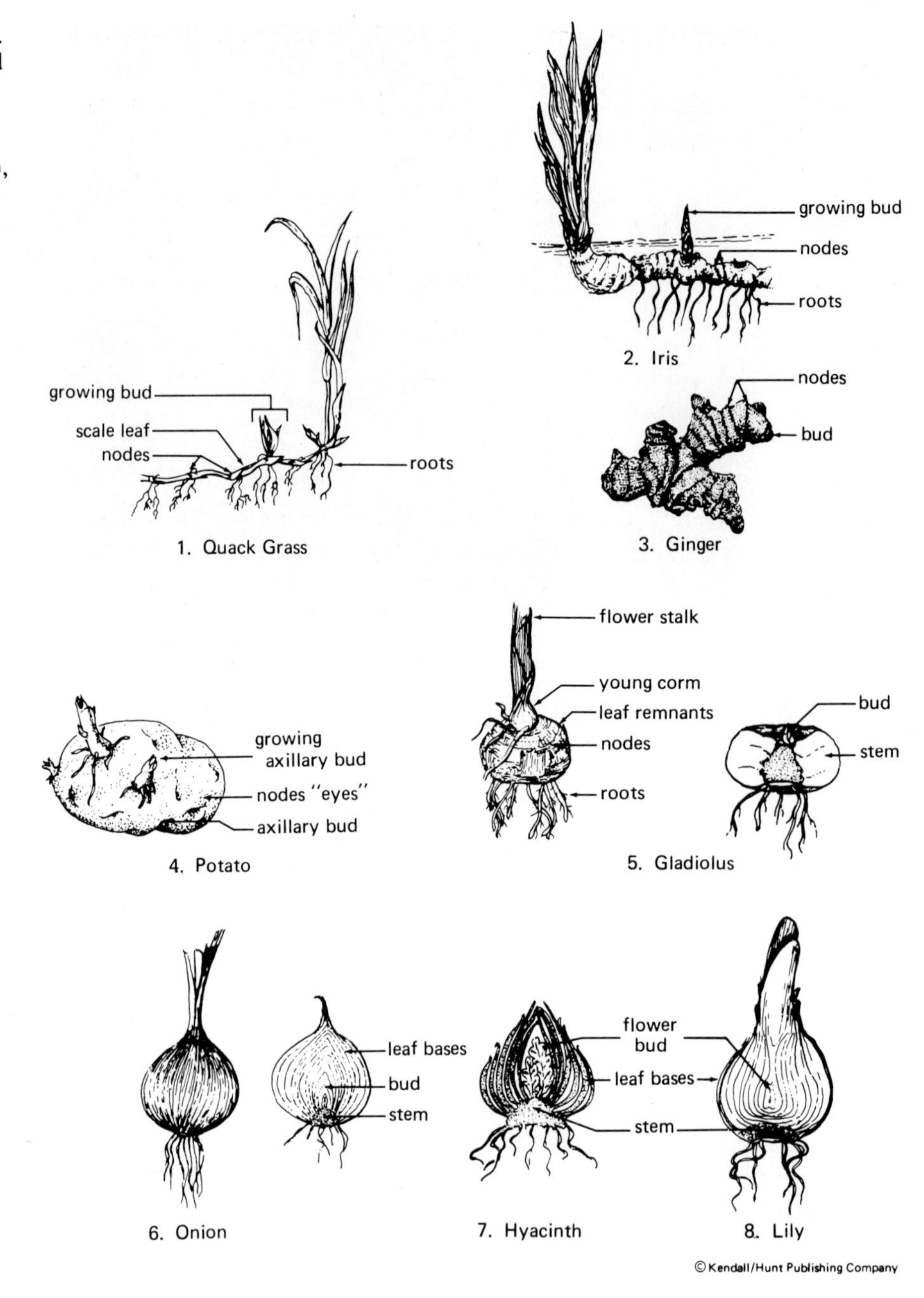

On completion of this chapter, you should be able to do the following:

☐ 1. Understand the several functions of stems and why the stem can be called the "parent" of the shoot system.

☐ 2. Identify node structures, including leaves, flowers, buds, and (in some plants) roots.

☐ 3. Recognize tissues of dicot stems, their arrangements, and their functions.

☐ 4. Recognize tissues of monocot stems, their arrangements, and their functions.

☐ 5. Understand vascular bundle structure.

☐ 6. Consistently relate the physical structure of specific cell and tissue types to the roles they perform in the life of a plant.

Woody Stems 17

What is a woody stem and how does it grow?

17.1 Imagine what it would have been like to have lived one thousand years ago. What would you have eaten, worn as clothes, and used as a shelter? What would you have used as furnishings and essential household utensils? How would you have foraged, hunted, and defended yourself? How would you have warmed yourself and cooked?

You might have imagined yourself living at the time when North America was occupied by the several Native American civilizations that lived intimately with the widely varied environments of the continent. You may have recalled pictures of wood-beamed cliff dwellings, pole-and-hide tipis, wood-and-earth hogans, sodhouses, or log longhouses. Did you also envision mats, baskets, and water jugs woven of plant materials, moccasins stuffed with cattail fluff for insulation, packframes, and camp fires?

To what extent would you have relied on *woody plants* as a source of food, fiber, construction materials, fuel, and medicines? For example, salicylic acid (aspirin), long used to reduce fever and relieve pain, has been extracted from the inner bark of willows for this purpose for thousands of years. The point is, of course, that humans did and still do depend on woody plants for many useful purposes. It is an ancient relationship, dating back at least to the time when the first humans foraged, used pointed sticks to dig for food, and used twigs and branches for shelter and to fuel cooking fires.

Woody shrubs and trees grow as *forests,* which are important to our planet in many ways. We cut trees for lumber, plywood, and paneling. We tear logs into chips and then cook them in a chemical stew to make paper and to extract lignins, resins, gums, turpentine, and other useful products. The forest products industry is a traditional mainstay of the American economy.

Furthermore, forests are "oxygen factories" for the biosphere. Through photosynthesis they renew the atmosphere's oxygen supply. At the same time, they remove large quantities of carbon dioxide produced by cellular respiration and combustion of wood, coal, gasoline, and other fuels. **Photosynthesis** is the only natural process that has this tremendous ability to exchange atmospheric carbon dioxide and oxygen. Woody plants convert so much atmospheric carbon dioxide to biomass (their bodies) that forests also are major *carbon reservoirs* in the biosphere.

Forests also retain moisture in upper soil levels and in the atmosphere, maintaining conditions that are suitable for many other lifeforms. In fact, forests have a major impact on world climate and soil patterns. Wherever in the world uncontrolled deforestation has occurred, drought and loss of

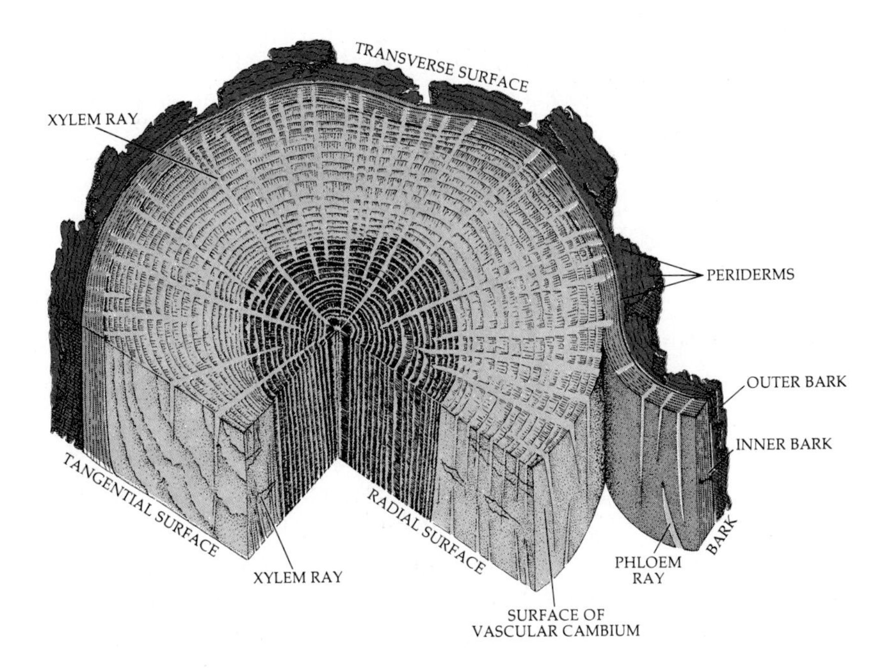

Figure 17.1 A mature woody stem. This drawing of a woody dicot stem shows the relationship of the **wood** (xylem) to the vascular cambium and the **bark** (phloem and cork). It also shows three ways the stem can be cut or sectioned, giving three different views of the wood grain and, at the microscopic level, the cells of the stem tissues. (See also figures 17.7, 17.8, and 17.9.) From Raven-Evert-Eichhorn, *Biology of Plants* 4th Edition, Worth Publishers, New York, 1986, page 455. Reprinted with permission.

topsoil have followed. The disastrous desertification of agricultural land in northern Africa and the resulting famine conditions were predicted at least a decade before they finally reached critical proportions in 1984 and 1985.

In many ways, we take woody plants for granted, not really appreciating their complex roles in our lives and in the biosphere as a whole. In chapter 16 we explored the development of herbaceous stems. In this chapter we will look at the development and activities of long-lived woody stems—wood with a bark wrapper (figure 17.1).

17.2 You should keep in mind several questions as you look at the materials provided in this chapter and attempt to apply your understandings to your experiences outside the classroom. Begin by suggesting answers to the questions below.

1. What are some advantages of having a long-lasting, woody body over having a short-lived herbaceous body?
2. What are some disadvantages?
3. What functions do you think the wood performs in the life of a plant?
4. What do you think are the functions of the bark?

Figure 17.2 Buds. This maple twig has a cluster of buds at its apex. It also has buds that are located at the nodes, called axillary buds because each arises in the axil or angle that is formed between the stem and a point of leaf attachment. Buds on woody plants are formed during the summer growing season, rest through the winter in a state of dormancy, then begin growing in the spring. In maples, the large apical bud is a floral bud—it will produce a flowering shoot. By contrast, the smaller buds beside it and the axillary buds are foliar buds—they will produce leafy (foliage-bearing) shoots.

17.3 Look at the samples of woody twigs your teacher has provided, and select one to study at your work station. Depending on the time of year, leaves may or may not be present. Identify the **nodes** and **internodes** on the twig.

1. Is the leaf arrangement on your twig *alternate, opposite,* or *whorled?*

2. Can you think of some useful reasons for being able to recognize different kinds of leaf arrangement? If so, what?

Next identify the **apical buds** and **axillary buds.** You may be able to distinguish between *foliage buds* (vegetative buds) and *flower buds* (figure 17.2). Flower buds usually are larger and rounder. It is easier to recognize them in the spring, when the buds are swelling. *In woody plants, foliage and flower buds for the next year are formed at the end of the summer growing season.* They then must survive winter conditions in a resting, or dormant, state. Woody plant buds are covered by hardened **bud scales** that protect them from water loss during dormancy.

3. Sketch a bud from your twig and show the detail of the overlapping bud scales.

Now gently peel away the bud scales, taking care not to crush or tear the embryonic tissues they protect.

4. Describe a single bud scale: transparent or opaque, smooth and shiny or fuzzy, colored or clear, and other characteristics you may notice.
5. Is the bud a foliage bud or flower bud, and how can you tell?
6. Are the interior bud tissues green, and, if so, how did light enter the bud?
7. Remembering your earlier study of herbaceous buds, name the kinds of tissues and embryonic structures you would expect to be present.

17.4 As a winter/spring class project, place some freshly cut winter twigs from pussy willow plants in a vase. Set the vase in a lighted but moderately cool place in the classroom and observe changes that take place over the next few weeks. Change the water at least once a week to keep it from becoming putrid and slimy from bacterial growth. As the buds begin to open, the fuzzy bud scales will open and shoots will emerge.

The floral shoots will contain either male organs (yellow-tipped stamens) or female organs (vaselike pistils). Foliage shoots will produce, obviously, leaves.

This project works best in the spring, when the willows already have begun the physiological changes associated with the return of the growing season. Otherwise, the influence of dormancy-causing hormones in the twigs will be much stronger than the influence of growth-promoting hormones that are present, affecting your observations.

17.5 When your skin is cut or torn, it heals itself by forming scar tissue. Woody plants also produce scar tissue, a layer of cork. Your woody twig has scars that show where leaves and bud scales formerly were attached. Furthermore, each scar is like an individual "fingerprint," showing the exact outline and vascular bundle arrangement of the structure that was attached (figure 17.3).

How are these several kinds of scars formed? The falling away and replacement of stem structures is a natural process, and the plant body makes *advance* preparations to lose leaves, bud scales, flowers, and fruits. The stem prepares for the loss of attached structures by forming a special cork cell layer, the **abscission layer,** as part of the entire process leading up to the loss of these structures. The process is called **abscission,** meaning "cutting away from." (Think of *ab* as in abduction, and *scise* as in scissors.) Abscission is a nondamaging, self-cutting way for stems to release old structures, including leaves, bud scales, flowers, and ripened fruits. The abscission layer permits structures to fall from their own weight, breaking at the weakened abscission layer and leaving behind a dry corky scar.

Look at the large **leaf scars,** where leaves have been attached to the twig. You can see the shape of the leaf petiole where it was attached. Inside the scar you can see individual **bundle scars,** where tiny vascular bundles extended from the xylem and phloem of the stem into the leaf.

1. Sketch a leaf scar, accurately portraying its shape and the number and position of the vascular bundle scars.
2. Do all of the leaf scars on your twig have a consistent number and arrangement of bundle scars?
3. Look at twigs from two other species. How do the leaf-scar shapes and bundle arrangements differ from those on your original twig?
4. Why do you think physical differences, such as node and leaf scar characteristics, can be used consistently to identify species? Consider your answer carefully.

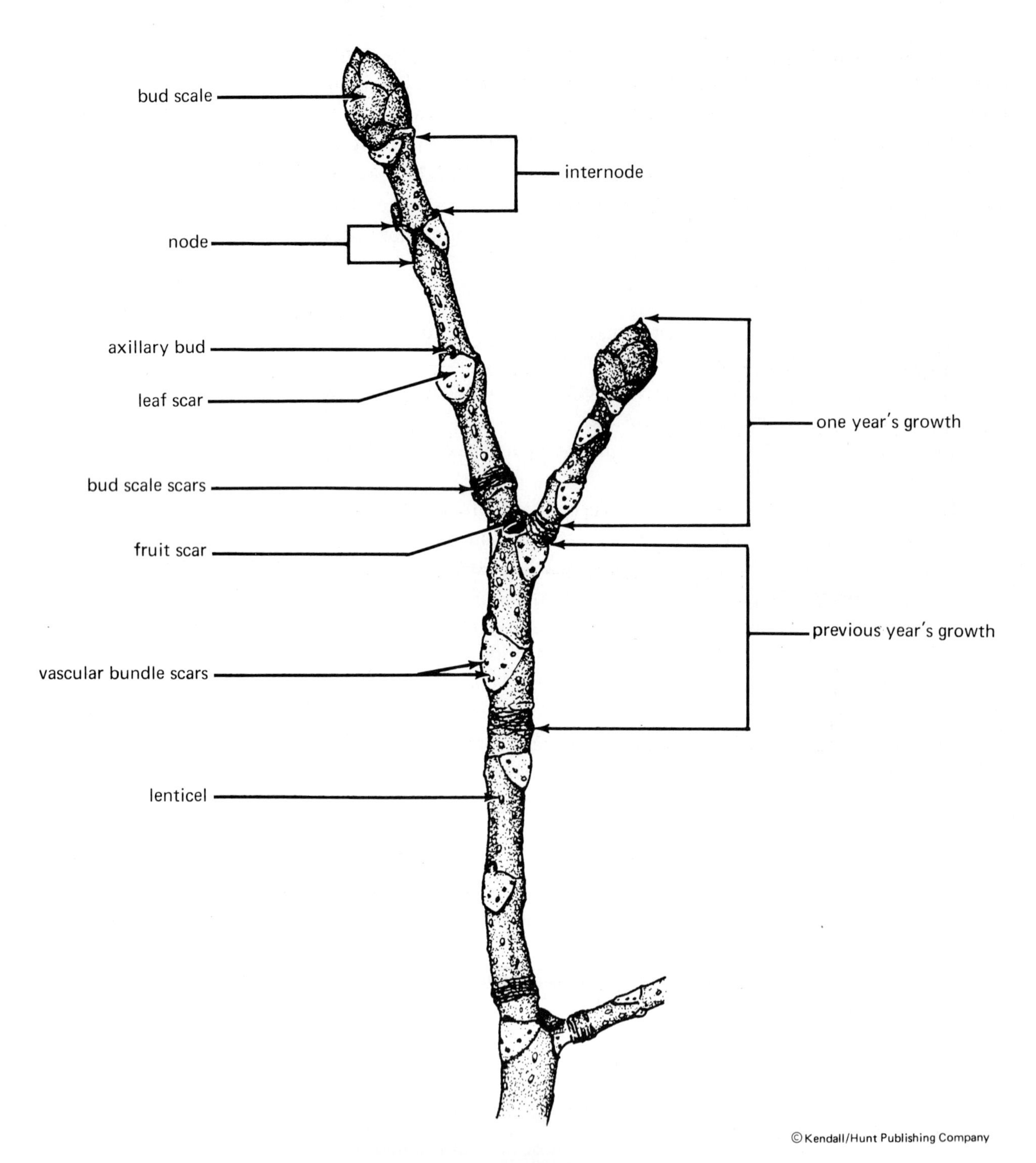

© Kendall/Hunt Publishing Company

Figure 17.3 Woody twig characteristics. The external structures of a woody twig tell a story. Bud scale scars mark past annual growth. Leaf scars tell leaf arrangement, shape of the petiole, and distribution of the vascular bundles in the petiole. Buds mark the promise of the coming season's growth. Lenticels allow some gas exchange between stem tissues and the air.

Examine your twig again. Each new season's growth emerges from a bud that has been protected by bud scales. Therefore, rings of **bud scale scars** mark the beginning of a year's growth (figure 17.3). Beginning from the tip of your twig, find the bud scale scars from the most recent growing season. Then work your way up the twig (away from the tip), looking for bud scale scars from previous growing seasons.

5. How many growing seasons are represented on the twig section you have?

Environmental factors affect how vigorously a shoot grows in a season. They affect both the number of cells that are produced and how much the cells expand (how large they become). You can reason, for instance, that a shoot will grow more vigorously when there is plenty of spring and summer moisture, and less vigorously during drought conditions.

6. Measure your twig and record the length in cm (or mm) of shoot growth that occurred during the three most recent growing seasons. If you can distinguish fourth and fifth growing seasons, record that data also.
7. What hypotheses can you make about environmental conditions of the past three-to-five growing seasons, based on your observations?
8. Do you have sufficient evidence to have confidence in your hypotheses?
9. How could you test them?
10. What kind of meristematic tissue forms corky scar tissue in the abscission layer that forms between stems and the structures they shed?

17.6 Plants also can respond to *wounding,* which is physical damage to tissues. They have both "first aid" and permanent repair mechanisms. When a plant first is wounded, cytoplasm and/or sap may ooze from the wound. As it dries, it glazes over the exposed tissues. This is a simple passive phenomenon. On the other hand, many plant species produce *special substances* that help to protect wounded tissues. These

Figure 17.4 Latex. Pencil tree or milkbush produces a milky latex that becomes gummy as it hardens. The plant in the photograph has been deliberately wounded to cause the latex to flow. The latex of this plant and other members of the euphorb family contains toxic substances that can produce mild to severe skin irritation. By contrast, latex produced by the common dandelion is nontoxic.

substances are **gum, pitch,** and **latex.** They help to keep disease-causing bacteria and fungi from starting to grow in the wound, and also help to keep the underlying tissues from dehydrating while they repair themselves. Humans have found many uses for gum, pitch, and latex.

Some plants, such as cherry trees, exude *gums,* sticky, water-soluble substances that harden to a clear, shiny, often tinted form. Plant gums often are used as thickeners and binders in processed foods (for example, xanthan, guar, and locust bean gums). You probably eat some every day in baked goods, cereals, puddings, dairy products, and beverages.

1. This evening, look at food container labels in your home and list the products that contain plant gums.

Conifers exude *pitch,* which is thick, sticky, and aromatic—the Christmas tree smell. Pitch contains several *resinous substances,* which not only plug the wound but also are antiseptic (discourage growth of bacteria and fungi). Pitch is not water-soluble. Because it is high in hydrocarbons, pitch burns readily and rapidly, whether it fuels a forest fire or is used to start a campfire. Rosin, used by dancers, gymnasts, and players of stringed instruments, comes from pitch (smell it sometime). Other resins make masking tape sticky. Organic solvents, especially turpentine, and a variety of other useful substances are extracted from pitch.

Latex is a milky substance exuded from some plants (for example, dandelions, milkweeds, and rubber plants). When it hardens, it becomes rubbery. Natural (nonsynthetic) *rubber* is made from rubber trees, whose trunks are slashed to collect the latex. The chewy ingredient of chewing gum is *chicle,* a plant latex. Latex from many tropical and subtropical plants contains irritating, toxic substances that help to discourage herbivores (figure 17.4). Milkweed and poinsettia latexes are mildly irritating to the skin. Latex from the manchineel tree that grows in Florida is dangerously potent: it can blister the skin and blind the eyes.

Permanent wound repair of woody stems is slower and much more complicated than simple cork formation or the release of special substances. It involves filling in the exposed space with new vascular cambium, wood, and bark over one or more growing seasons. Figure 17.5 shows a healing wound in a tree trunk. You will understand the repair process better after the discussion of woody stem growth that follows.

17.7 The formation of new vascular cambium, xylem, phloem, cork cambium, and cork is called **secondary growth,** and it is most prominent in woody plants. Secondary growth contributes to the thickness of a stem, usually determined by *girth* measurement (like taking your waist or hip measurement). **Wood** and **bark** are produced by secondary growth.

Figure 17.5 Wound repair. This tree was damaged some time ago. A large patch of bark was lost and the underlying wood was exposed. The exposed wood surface now is dry and dead. Vascular cambium from the area surrounding the wound has been producing new tissues that are gradually overgrowing the wound. In time, the exposed patch of wood will be completely covered. Until that happens, however, it continues to be exposed to weathering and attack by insects and fungi.

Are you having difficulty visualizing how plants grow? Plant growth is special; the concepts are not that complicated, just different from animal growth (figure 17.6). For one thing, you already know that plants can grow throughout their lifetimes by the mitotic production of cells at their shoot tips, root tips, and cambial layers. This means that in each growing season, brand-new embryonic tissues form that mature into new, young shoots, roots, and vascular tissues. Think about these questions: Can animals keep growing throughout their lives? Do we generate new body parts, or are we stuck with what we have?

Plants also are capable of significant **regeneration.** If a twig is pruned or broken away from a branch, it will be replaced by a shoot from another bud. Likewise, a shrub or tree that has been cut down may send up new shoots from the stump that remains. Some prairie oaks exist as brushy thickets of shoots that have regenerated from hundreds-of-years-old stumps that were burned to the ground by successive prairie fires. Can we survive such drastic disruptions and regenerate lost body parts?

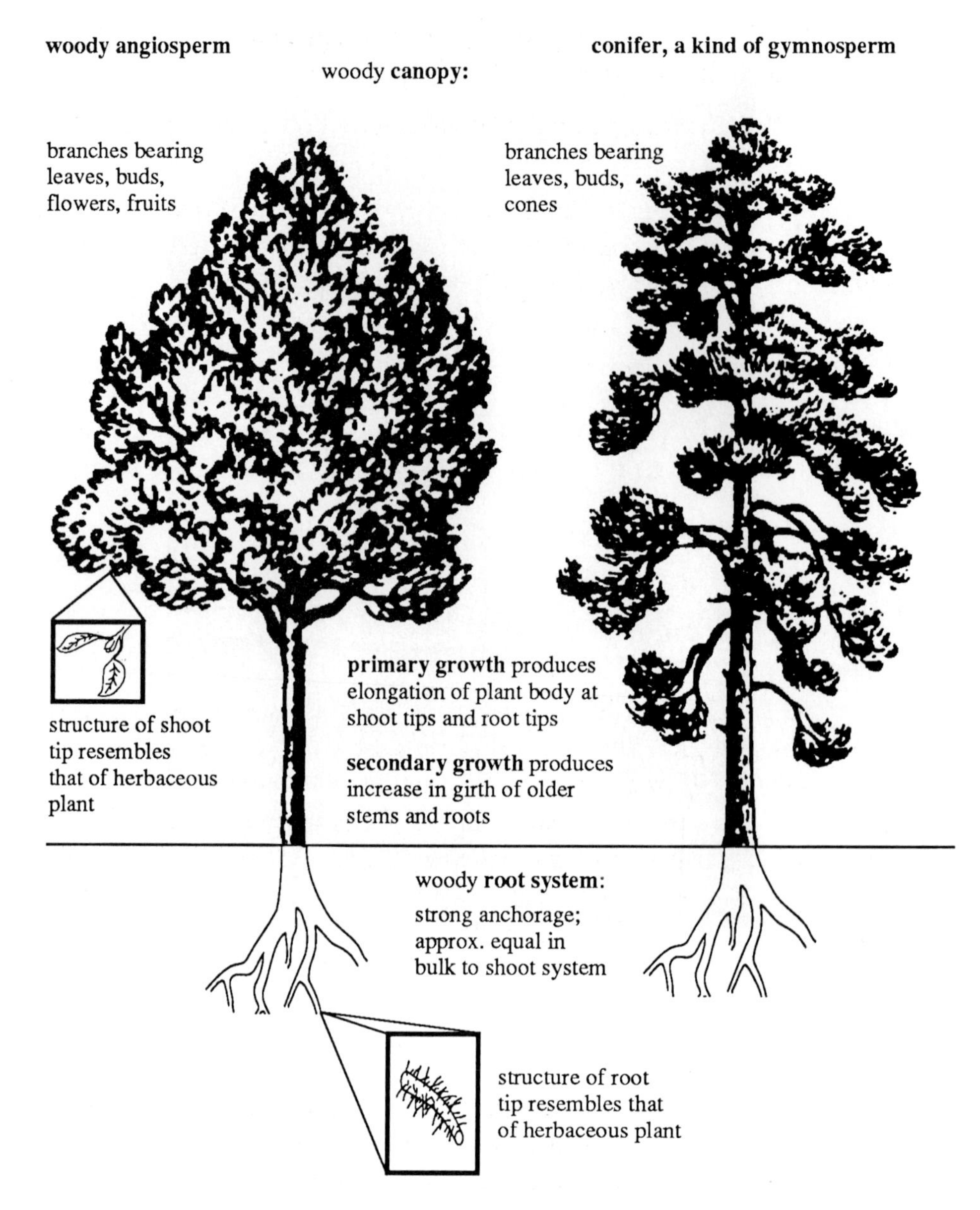

Figure 17.6 Basic tree structure. As trees grow, they become taller and fuller as a result of apical growth, growth produced by apical meristem in buds. An existing twig elongates by growth from the terminal bud. New shoots, leaves/needles, flowers/cones are produced from buds that arise on existing twigs. The meristematic and maturation processes of stems thus are parent processes to the entire canopy.

Furthermore, plant growth is provided only by the *meristematic* tissues. By contrast, mitotic processes in mature animals usually are devoted only to repair and renewal processes once physical maturation has occurred.

Before we begin the next activity, answer the following questions to be sure you understand how plant growth occurs.

1. Of primary and secondary growth, which kind increases the length of a stem (or root?)
2. What kind of meristematic tissue is involved in primary growth?

3. Of primary and secondary growth, which kind increases the girth of a stem (or root)?

4. What meristematic tissues are involved in secondary growth?

17.8 In previous sections you examined herbaceous stems, and it was mentioned that very young shoots of woody plants resemble herbaceous stems. Think about it. Of course they start out that way, because they develop from apical meristem! First we will look at some fully developed woody stems, and then we will discuss how they arrived at that state.

Examine the woody stem cross sections your teacher has provided. Identify the wood and the bark.

1. The wood is made up of what complex tissue type?

2. The bark is made up of what two tissue types?

3. What meristem separates the wood from the bark?

The wood shows obvious *annular rings* that usually correspond to the growing seasons, so they often are called **annual rings.** Each ring consists of two regions. The widest part is softer, lighter tissue containing large cells produced during the most active part of the growing season, spring and early summer, when water is most abundant. Toward the end of the growing season, the number of cells produced and their size become reduced, resulting in production of a thinner, denser, and darker line.

4. Why do you suppose water availability influences the size of the cells that are produced at different times of the year?

The width of the annular ring in a given year also is influenced directly by environmental conditions. Adverse conditions, such as inadequate light, inadequate moisture, and crowding by adjacent plants, can result in tightly spaced rings. The suppression may be evident on only one side of the stem; perhaps this is where it was growing next to another tree or building. A common pattern of variation in ring thickness in forest trees is rapid early growth characterized by thick rings, followed by a period of suppression, probably as a result of crowding, followed by another period of thicker rings after the forest was thinned out by fire or cutting.

As you can see, the structure of the xylem is a historical record of environmental conditions that were present during the lifetime of a tree. Comparative studies of living and fossilized wood aid our understanding of climatic changes that have occurred in North America through millions of years. By studying the beams supporting ancient Native American pueblos, researchers also were able to infer which years had good harvest seasons, and which were marked by drought, helping us to learn more about both the original inhabitants of our continent and the environmental conditions they endured.

5. Using the number of annular rings as a guide, determine the approximate age of some of the stem sections at the time of death.
6. What was the age at death of the youngest stem sample present?
7. Was the oldest stem more than a century old?

Next, choose one stem slice to take back to your work station and use it as a basis for answering the following questions.

8. Are all of the xylem rings the same thickness?
9. Can you see any evidence of suppression on one side of the stem, and what can you infer from this evidence?
10. Is the wood softer in the spring wood or in the summer wood of each ring? How does this relate to cell size?

If you look very closely or with a magnifying glass, you also may be able to see a few rings in the bark. However, the bark is continually stretched and split by the pressure of xylem expansion, so its structure is very irregular. The thin, dense lines in the bark correspond to deceased cork cambium regions. Most of the older bark consists of crushed layers of phloem and cork. You will find a more detailed presentation of bark at the end of this unit.

11. Use your imagination and the information you have to write a brief "life story" for the stem you are observing. (Tree, "This Is Your Life!")

Woody stems also are characterized by the presence of **wood rays,** vertical stacks of living parenchyma cells that conduct substances between the stem center and its outer regions. In cross section, they look like radiating spokes of a wheel. Identify wood rays in your samples.

Finally, you may notice that the outermost few years of xylem tissue are a lighter color than the interior (older) xylem. The interior of the stem is the **heartwood,** and its darker color comes from substances stored there, such as minerals and assorted organic substances. The outer cylinder of xylem is the **sapwood,** which is the active conducting part of the stem.

12. Is most of the wood in the tree sections heartwood or sapwood?
13. Based on this observation, do you infer that most of the wood in the tree sections is conducting or nonconducting in function?

17.9

Most of the trees logged in the United States and Canada are converted to lumber or some kind of paper product. Perhaps you have worked with wood yourself, or have observed some building project. How is a tree cut to convert it to lumber? What gives different kinds of wood their distinctive *wood grain patterns?* Why does the wood grain look different on the wide part of a board than on the ends and sides? The answer lies in the way the xylem tissues, particularly the annular rings, are oriented.

The nature of the xylem tissue also dictates the *structural characteristics* of each kind of wood: its softness or hardness, strength, flexibility, porosity, insulating characteristics, and so on. Substances that are stored in the parenchyma cells of the xylem determine other wood characteristics, such as color, flammability, and natural resistance to decay.

Examine the wood blocks your teacher has provided and try to imagine how each block was positioned when it was part of a tree. First look at the ends, which really are cross sections of the xylem tissues.

1. Can you see evidence of annular rings in the xylem?
2. How many seasons did it take the tree to grow the chunk of wood you hold in your hand?

Next, look at the edges of the block, to see the lengthwise relationship of the annular rings to the wood grain. Finally, examine the broad surface of the block, which exhibits the grain pattern you see most often. It is the view seen on exposed boards and on veneers, such as paneling and plywood.

17.10 Now that you are familiar with the macroscopic view of wood and how it is oriented to the overall structure of the stem, look at the microscopic view of a cross section of a young woody stem in figure 17.7.

1. Use what you have learned about plant tissues and stem structure to label the following structures:
   (*a*) bark, (*b*) cork, (*c*) phloem, (*d*) vascular cambium zone, (*e*) wood, (*f*) annular rings, (*g*) wood rays, (*h*) spring wood, (*i*) summer wood, (*j*) vessel, (*k*) tracheid, (*l*) fiber, and (*m*) pith.

Think again about the relationship of the microscopic view you have just labelled to how woody stems are in all three dimensions. Figure 17.8 shows a section of conifer wood, cut in three different ways, as were your wood blocks. A **cross section** cuts across (perpendicular to) the long axis of the stem. Figure 17.7 was a cross section. A **radial section** is a longitudinal section that cuts from the center to the outside of the stem, *along* the rays. If you ever have split wood with an axe, you may have found that this is the easiest plane along which to split because the rays are relatively soft. A **tangential section** is a longitudinal section that cuts *across* the rays instead of with them. Most wooden boards are tangential slices of stem. Refer also to figure 17.1 for perspective.

Compare the cross section view in figure 17.8 to the figure 17.7 view. Study the appearance of the vascular rays in both figures. Now look at figure 17.9, a three-dimensional drawing that shows the details of how cells are arranged in the wood of a dicot. Notice that the cross sectional view allows you to see how wide a cell is (its diameter), how large its interior space is, and how thick its cell walls are. By comparison, both radial and tangential views show you how the cells look lengthwise. Using all three microscopic views enables you to construct a mental concept of what the cells actually look like. This is an example of how scientists are able to use different but complementary sources of information to arrive at explanations.

Also, notice that seeing three views gives you a clearer idea of how tissues and structures are arranged. The vascular rays are the most striking example. They appear to be thin, radiating lines when viewed in cross section or tangential section. A radial section, however, provides an opportunity to see how they look from the side, revealing them as vertical stacks or plates of parenchyma cells.

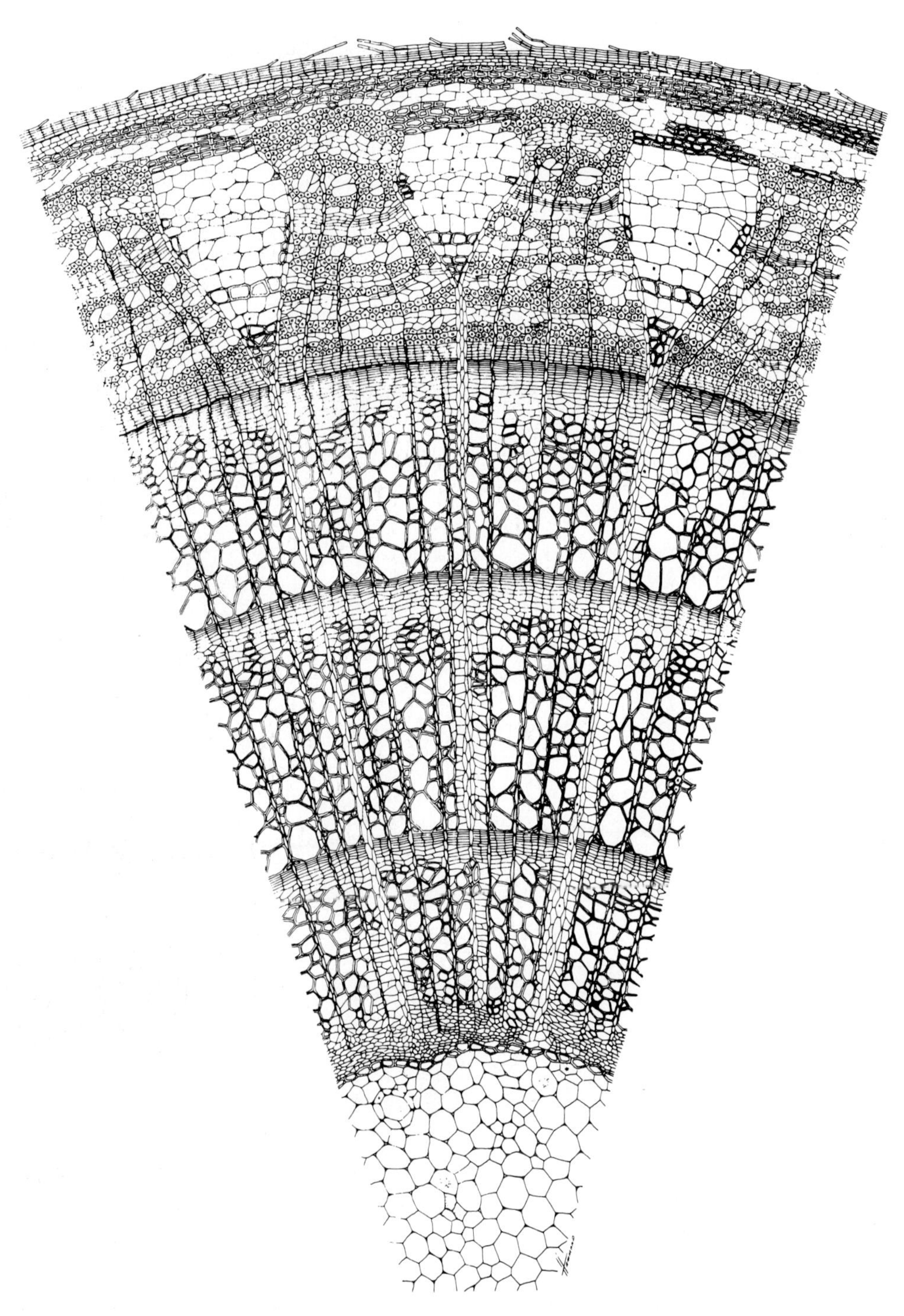

© Kendall/Hunt Publishing Company

Figure 17.7 Young woody dicot stem. In the transition from an herbaceous to woody state, the stem has retained its central pith, but has added three seasons of secondary growth. The outward expansion has stretched the earlier phloem layers, causing them to split. Parenchyma tissue has filled in the resulting wedge-shaped spaces. The epidermis has been replaced by cork, which also is being stretched. A deeper cork layer has formed across the top of the former bundle caps, and yet another layer of cork cambium is apparent about ⅓ of the way into the older phloem. It will extend across the parenchyma wedges, producing a new cork layer.

Figure 17.8 Wood in 3-D. This block of conifer wood shows how tissues are oriented in three dimensions within a woody stem, and gives an anatomical perspective to familiar ways lumber is cut. Unlike dicots, conifers have no vessels in their xylem, so their wood lacks the porous look of many dicots (such as the one in the previous figure). The arrangement of the vascular rays can be seen clearly, however, and is similar among woods. The rays are stacks or plates of parenchyma cells that radiate from the center to the outside of the stem, as can be seen in either the cross section (top) or radial section (right side). The broad surface of most boards is a tangential section (left side), in which the depth and width of the rays can be seen; *i.e.,* how many cell layers tall and wide they are.

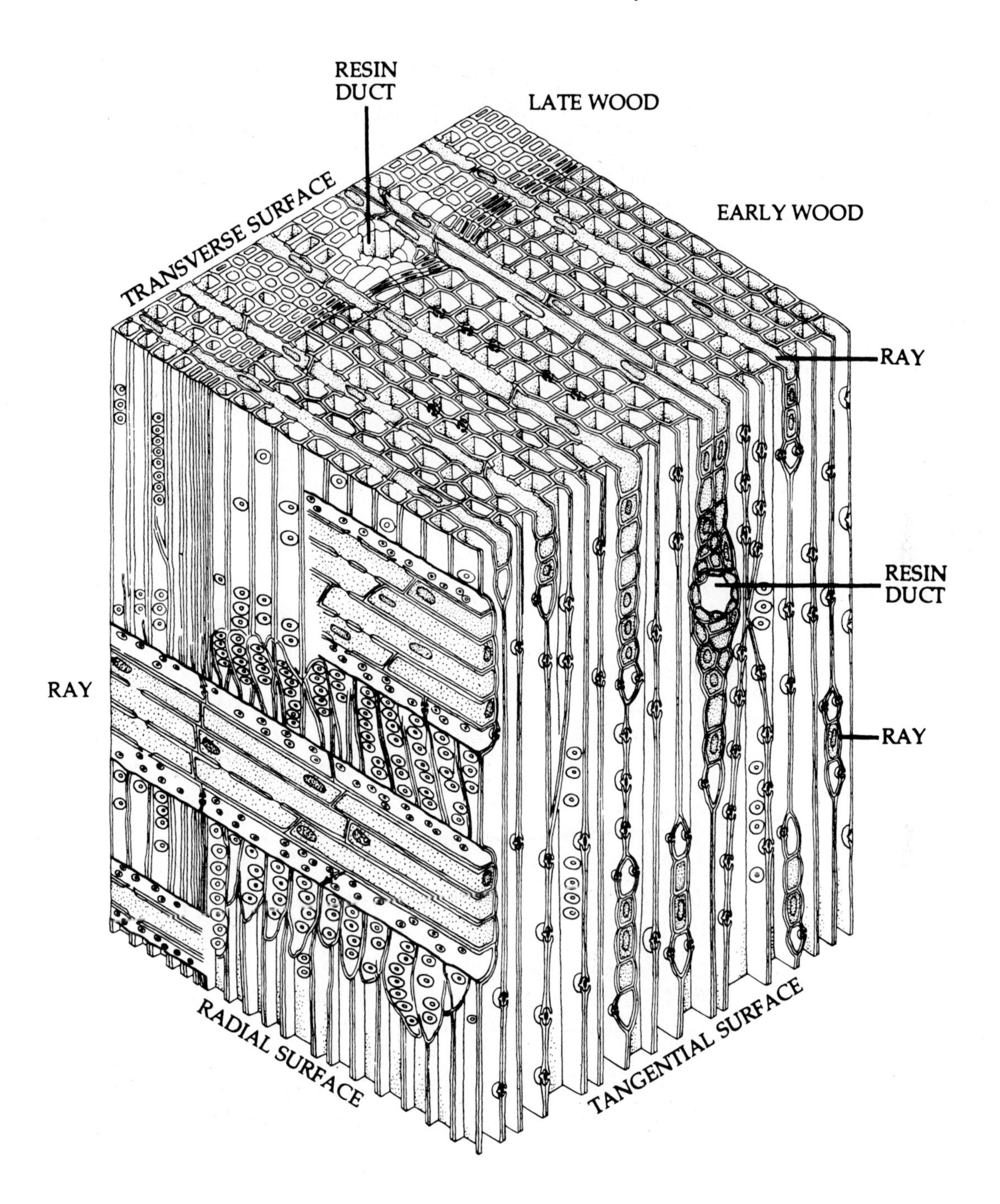

Figure 17.9 Cellular view of conifer wood. Compare this figure to figure 17.8 and see if you can match up these labelled structures with corresponding features of figure 17.8. From Raven-Evert-Eichhorn, *Biology of Plants* 4th Edition, Worth Publishers, New York, 1986, page 459. Reprinted with permission.

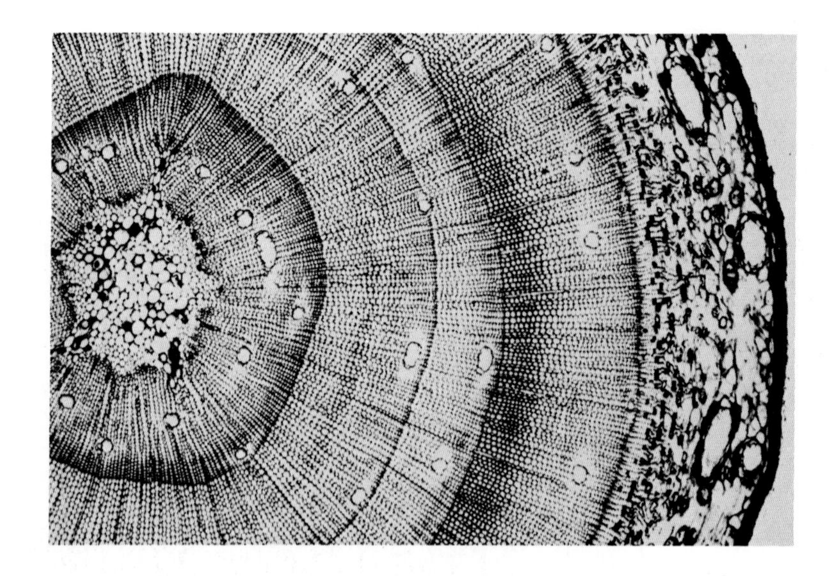

Figure 17.10 Young conifer stem. Similarities and differences between angiosperms and gymnosperms are apparent when you compare this pine stem cross section to figure 17.7. Pith is present, but the wood is composed only of tracheids and fibers; no vessels are present. Cell-lined resin ducts are present in the wood, and larger, resin-filled cavities are present in the bark. You may have experienced these aromatic resins yourself, getting "pitch" on your hands or clothing while handling a Christmas tree or conifer wood. The cambial zone is apparent as a dense layer of cells between the youngest wood and the youngest phloem. In both this photograph and the figure 17.7 drawing you can see, strikingly, that the layer of living, functioning phloem cells is quite thin.

17.11 The two main branches of seed-producing plants, the angiosperms (flowering plants) and gymnosperms (mainly coniferous trees) have some fundamental differences. Xylem structure is one of these differences. Examine figure 17.10, which shows a cross section of a young pine stem. The most obvious difference is that the xylem is composed mainly of tracheids and fibers. Vessels are notably absent. Transport of water up the xylem occurs by cell-to-cell transfer through **pits** in the tracheid walls.

Coniferous trees produce aromatic resins—you know them as pitch. In this figure you can see large **resin cavities** in the bark and smaller **resin ducts** or **canals** in the wood. The resin cavities are pockets or "blisters" in the bark. The resin ducts are elongated structures.

Figure 17.11 shows cross, radial, and tangential views of the microscopic structure of pine wood. Use these three views to build a mental picture of tracheids in gymnosperm wood, comparing them to previous figures.

1. How would you describe the shape of tracheids, in terms of width, length, cell wall thickness, and size of the internal cavity?
2. What is the function of the pits that are visible in the radial and tangential views?

Figure 17.13 Bark structure. The bark of a mature tree contains layers of cork and crushed phloem tissues. Old cork cambium layers give it a ringed appearance. For all practical purposes, the phloem tissue is only an important structural feature of the innermost bark, next to the cambium. It's a bit amazing to realize that such a thin layer of living phloem is responsible for transporting dissolved food and hormones between the root and shoot systems of even the largest tree. Most of the bark is cork. The corky bark protects the inner tissues from dehydration, extreme temperatures, predators and pathogens, mechanical damage, and fire.

Try to imagine the stresses and logistics associated with this growth pattern. Think of a circle of people holding hands, facing outward, to represent the vascular cambium. When the vascular cambium divides, it makes a new layer of xylem cells that must stretch all the way around the previous layer of cells. To simulate this action, imagine each person in the circle taking a step forward without anyone letting go. The cambium divides again, forming another ring of xylem. Imagine each person taking another step forward. Can they do it without pulling apart? How about the vascular cambium and the outermost ring of xylem cells? How do they compensate for this outward expansion? Sooner or later, your ring of hand-holding people can expand the circle only if new persons are inserted into the circle. The same is true in the cambium! By longitudinal division on a radial plane, new cambial cells are inserted into the ring.

4. What three kinds of stem tissue are formed by mitotic division of the vascular cambium?

17.13

What are the growth stresses on the tissues of the bark, which also is stretched by the expanding xylem? The layers of phloem tissues are squeezed and become crushed between the hard outer bark layers and the xylem. The bark stretches and cracks in response to the expanding girth of the stem. Successively deeper **cork cambium** layers form to maintain a protective seal between the air and the live tissues at the base of the cracks (figure 17.13).

Figure 17.14 Woody dicot bark. Compare this photomicrograph of a young woody stem to the drawing in figure 17.7, and relate the description provided there to the actual tissues you can see in this illustration. You can see the line of cambium that separates the wood from the bark. In the bark, you can see the thick outer cork tissue, the cellular detail of the conducting part of the phloem tissue, the wedges of parenchyma tissue that make up the phloem rays, and areas of cells that are becoming cork cambium.

Figure 17.14 shows how parenchyma and cork tissues are produced for a time, filling in the gaps caused by the outward-pushing growth forces. Older bark is composed only of dead cells, and is sloughed off in flakes, strips, or chunks. Only those phloem layers *nearest* the vascular cambium contain living, functional cells.

Remembering the function of the phloem, can you guess what happens to a woody stem when a complete ring of bark is removed from the stem? This type of injury is called **girdling,** and it can prove fatal for a plant if the main stem is girdled. If phloem transport is completely cut off, the food supply to the roots is virtually eliminated. Girdling that occurs in the winter can prevent movement of stored food upward to the emerging shoot tissues. Either way, tissue starvation is a result of girdling. A girdled stem may be able to regenerate new phloem if the vascular cambium is not completely removed or killed by subsequent dehydration.

1. What tissue separates the bark from the wood?
2. Where are the youngest bark cells found, nearer to or farther from the wood?
3. What other meristematic tissue is found in the bark, and what kind of tissue does it produce?
4. What are some functions of the bark?
5. Can you think of economic uses for bark?

17.14 This chapter has offered views of woody tissue from microscopic, macroscopic, and economic perspectives. Review the study points listed below and check them off if you feel you have mastered them. If you have not, return to the text of the chapter to study them again before continuing to the next activity.

On completion of this chapter, you should be able to do the following:

☐ 1. Understand several ways in which woody plants are important in the biosphere.

☐ 2. Understand the concept of secondary growth.

☐ 3. Recognize external characteristics of woody twigs and be able to infer certain environmental information from them.

☐ 4. Understand the functions and processes of abscission and wound healing.

☐ 5. Understand the functional and structural characteristics of bark and wood and the roles of primary and secondary growth in the formation of the plant body.

☐ 6. Know how to infer past environmental conditions by "reading" the record of annular rings in a woody stem, particularly the main stem, or trunk.

Leaves and Roots 18

How do leaves and roots perform their functions?

18.1 Green is the color of photosynthesis. Poetically speaking, you even could say that green is the color of life on Earth. What fundamental concept does that statement represent? Green is the color of chlorophyll, the pigment responsible for capturing solar energy so it can be converted to food energy to support life. The color green resides in plants and in photosynthetic microorganisms. Following this logic, it is obvious that photosynthetic organisms, such as plants, are the critical energy intermediaries on which all life, including human life, depends.

All green parts of a plant are able to conduct photosynthesis because they contain chlorophyll; however, **leaves** are specifically adapted to be efficient sites of photosynthesis. That is their "job" in the life of a plant. In the first part of this chapter we will look at how leaves are adapted to perform that role.

18.2 We will begin with some questions you already should be able to answer.

1. What is the primary function of a leaf?
2. What is the predominant color of leaves, and what substance gives them that color?
3. What major plant organ gives rise to the leaves?
4. Are leaf cells the direct result of mitosis in the apical meristem or in the vascular cambium?

18.3 Let us compare the structure of a typical, broad, flat, dicot leaf to the ingredients needed for photosynthesis. *First,* light must be present. *Second,* chlorophyll must be present or the leaf cannot convert solar energy to carbohydrate (food) energy. *Third,* the leaf must have carbon dioxide as a source of carbon and oxygen atoms. *Fourth,* the leaf must have water as a source of hydrogen atoms and electrons and as an indispensible component of the cytoplasm. *Fifth,* enzyme systems must be present to catalyze and regulate photosynthesis *and* the subsequent conversion and transport of photosynthetic products, mainly sugars and starches.

1. What five major ingredients must be present for photosynthesis to occur? (Two of them are part of the cell.)
2. Where does the plant get each of these ingredients?

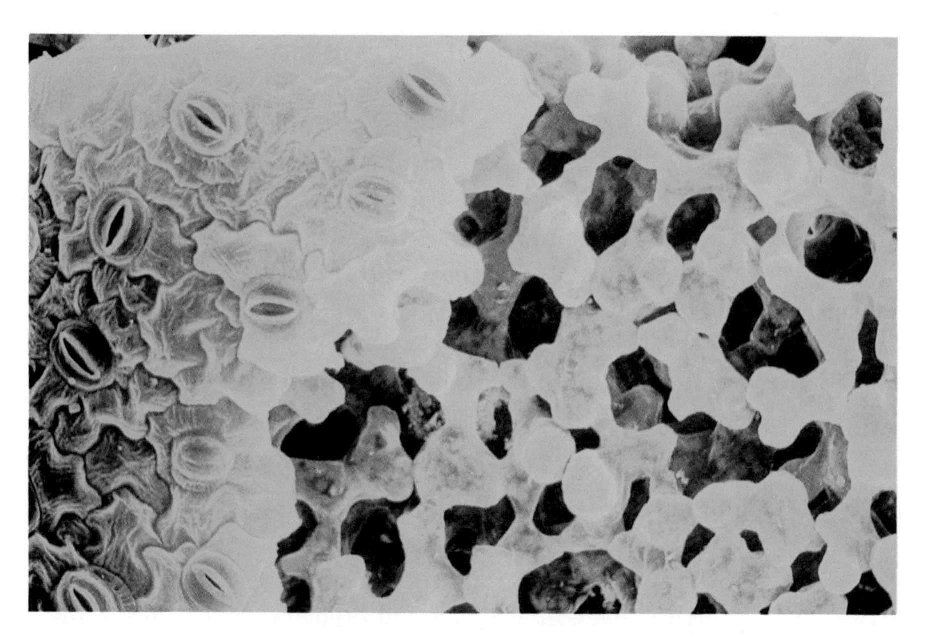

Figure 18.1 Inside a leaf. This scanning electron microscope view was prepared at an angle, enabling you to see epidermal tissue on the left edge and chlorenchyma tissue on the right. Numerous *stomates,* through which gasses and water vapor diffuse, are visible in the epidermis. Chlorenchyma cells in this leaf region are lobed, creating a large amount of air space. The result is a "spongy" appearance; therefore, this leaf region is called *spongy mesophyll.* (Compare this view to figures 18.2 and 18.3.) Carbon dioxide enters the leaf through the stomates, becomes dissolved on the wet cell surfaces, then is absorbed into the cells, where it is used for photosynthesis. Can you visualize how the structure of the spongy mesophyll maximizes such absorption?

3. What parts of plants have you seen that are green?
4. Make a hypothesis. Why do you think the broad, thin structure of a leaf is an adaptation for photosynthesis? Discuss it with your lab partner(s) to come up with what seem to be the best answers.

18.4

Chlorophyll is contained within specialized, complex cytoplasmic organelles called **chloroplasts,** as you learned earlier. Chlorophyll develops in the chloroplasts of large, thin-walled **parenchyma cells** that are exposed to light. Most of the leaf is chlorophyll-containing parenchyma tissue, called **chlorenchyma** tissue, for short (figure 18.1). Chlorenchyma cells make up most of the thickness of the leaf, its **mesophyll** (mid-leaf) part.

We will look first at a fairly simple leaf. Put a drop of water on a clean microscope slide and take it to where your teacher has a sample of the aquatic plant called *Elodea.* Use forceps to *carefully* remove a single leaf from the tip of a shoot. Do not crush the leaf. (A smaller, younger leaf will be easier to observe than a thicker, older leaf.) *Immediately* place the leaf in the water droplet on your slide. (Do not let the leaf dry out.)

Place a cover slip over the leaf and observe your slide preparation using the low power of the microscope. Focus carefully and move the slide around as needed to find a thin area where you can see individual cells clearly. The **cell walls** will be easy to distinguish, as will the clear interior space, a fluid-filled **central vacuole.** The **cytoplasm,** bounded by the cell membrane, is pressed into a thin, clear layer between the cell wall and central vacuole. The contents of the central vacuole are contained within the **vacuolar membrane** (also called the tonoplast).

1. What is the name of the membrane that separates the cytoplasm from the cell wall?
2. What is the name of the membrane that separates the interior cytoplasm from the central vacuole?

The cytoplasm is a complex concoction of organic substances, minerals, and organelles. Rich in enzymes and other proteins, it has a semiviscous consistency, similar to thin, raw egg white. If you are careful and lucky, you may be able to see the slightly amber **nucleus,** which often is lodged in a corner formed by the cell wall. If you do find a nucleus, share your good fortune with your teacher and classmates.

3. What are the ovoid green organelles you can see?
4. Do they appear to be moving? If so, they are evidence of the constant circulation within the cytoplasm.
5. Sketch three adjacent cells, showing their shapes, the thickness of the cell walls that separate and attach them together, the central vacuole, and some chloroplasts. Your drawing does not need to be fancy, but you should try to keep the different parts in proportion to one another.

18.5 Next, we will look at a more complex leaf, one from a terrestrial plant. Refer to figures 18.2 and 18.3 as you make your observations and read the subsequent discussions. Examine the microscope slide that has dicot leaf cross sections. Use both low and high power to help you see greater detail. Answer the questions below based on your observations, but then leave the slide in place while you read the next segment.

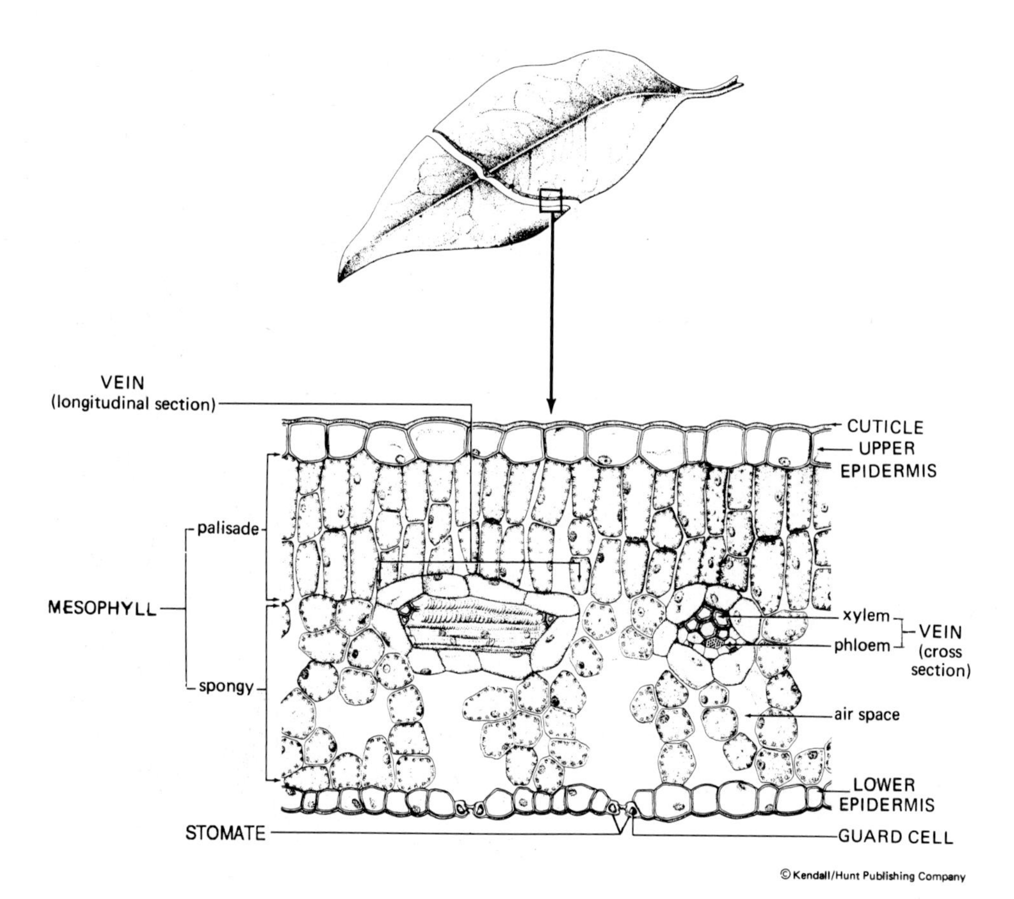

© Kendall/Hunt Publishing Company

Figure 18.2 Cross section of a dicot leaf. This drawing will help you to identify regions and structures on your microscope slide. The leaf is composed of a photosynthetic middle—the *mesophyll*—wrapped in and protected by epidermis. Water and minerals are brought to the leaf in the veins. Carbon dioxide diffuses into the leaf via stomates. Oxygen that is produced during photosynthesis and water vapor lost from cell surfaces diffuse out of the leaf through the stomates.

1. How many cell layers thick is the leaf? (If the thickness is variable, you can express your response as a range of numbers.)
2. Explain why the cell walls are red and green on your microscope slide.
3. Are all the cells you see the same size and shape?
4. Are there differences in cell wall thickness among the different kinds of cells?
5. Try to name some of the cell or tissue types that are present, identifying them by cell wall thickness and presence or absence of cytoplasmic structures.
6. In what tissue region do you detect large spaces between the cells?

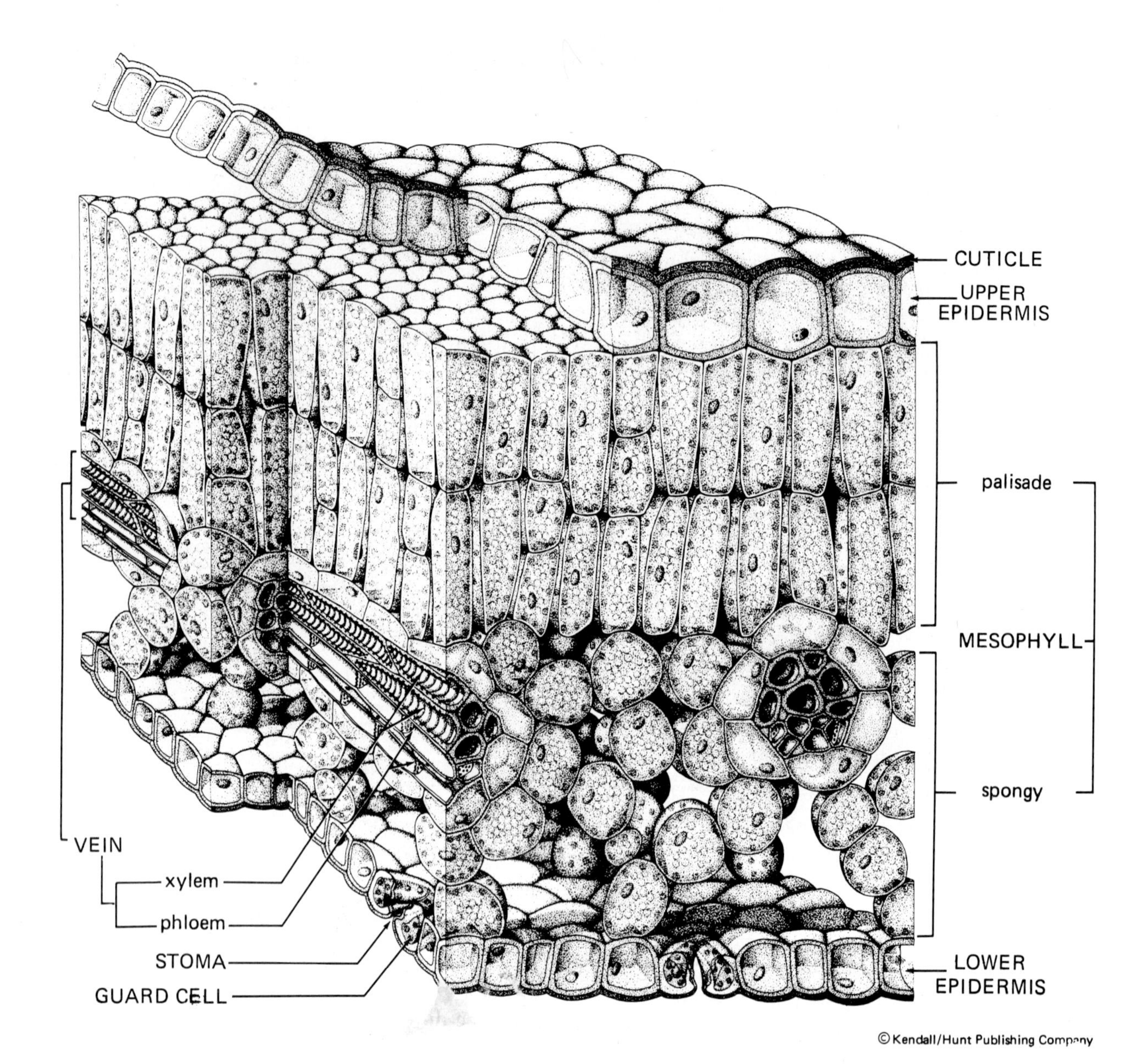

Figure 18.3 3-D view of a dicot leaf. Combining this view with those presented in the preceding figures should give you a good conception of what the inside of a leaf is like, although this drawing does not capture the lobed nature of the spongy mesophyll cells. Notice that specialized, nonphotosynthetic cells surround the veins. They aid in loading carbohydrates into the veins from surrounding photosynthetic cells.

18.6

The shoot system of a terrestrial plant is filled with water, yet it is surrounded by drier air. The natural tendency is for water to evaporate from the plant tissues, a process called **transpiration.** Dry cells are dead cells! To protect the leaf from desiccation (drying out), the outside leaf surface is wrapped in cutin-covered (cutinized) **epidermis.** The thickness of the cuticle varies from species to species. *Waxes* may be present outside the cuticle, as you may know if you ever have polished an apple.

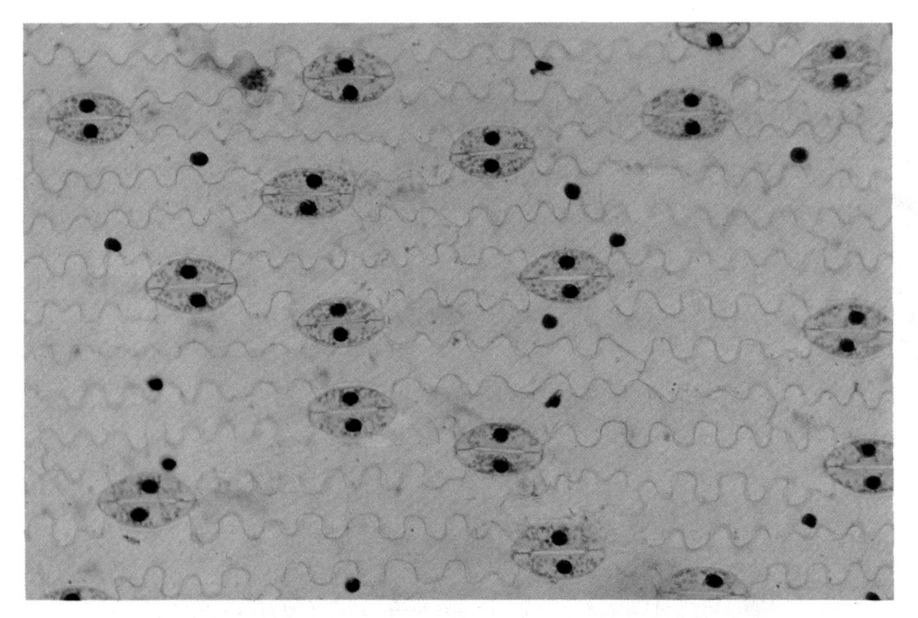

Figure 18.4 Leaf epidermis. Cells often have interesting and beautiful shapes, as this slide shows. Epidermal tissue is particularly variable, as it includes surface cells, guard cells, and epidermal hairs of several types. Basic epidermal cells are flat and fit together like tiles, forming a sealed surface. The cells of this leaf are particularly interesting, with interlocked, undulating edges. Some nuclei are visible. This view also shows that guard cells are the only epidermal cells to contain chloroplasts.

Imagine yourself wrapped in a continuous sheath of waxed paper. You will not dry out, but how will you breathe? Would you poke holes in the covering? Some oxygen and carbon dioxide can pass through the leaf cuticle, but not enough to support photosynthesis. Plants solve the problem of getting sufficient air by having **stomates** built into the epidermis (figure 18.4). Each stomate is composed of two photosynthetic **guard cells** that open and close a tiny pore.

1. What gas must diffuse into the leaf from the atmosphere in order for photosynthesis to occur?
2. What major gases diffuse out of the leaf into the atmosphere? (*Hint:* one of them is a by-product of photosynthesis and the other is the vapor form of a liquid substance.)
3. What is the name of the structures through which most of the gas molecules pass?
4. What term describes evaporative water loss from plant surfaces?
5. How might evaporative water loss be beneficial to the leaf on a warm, sunny day?

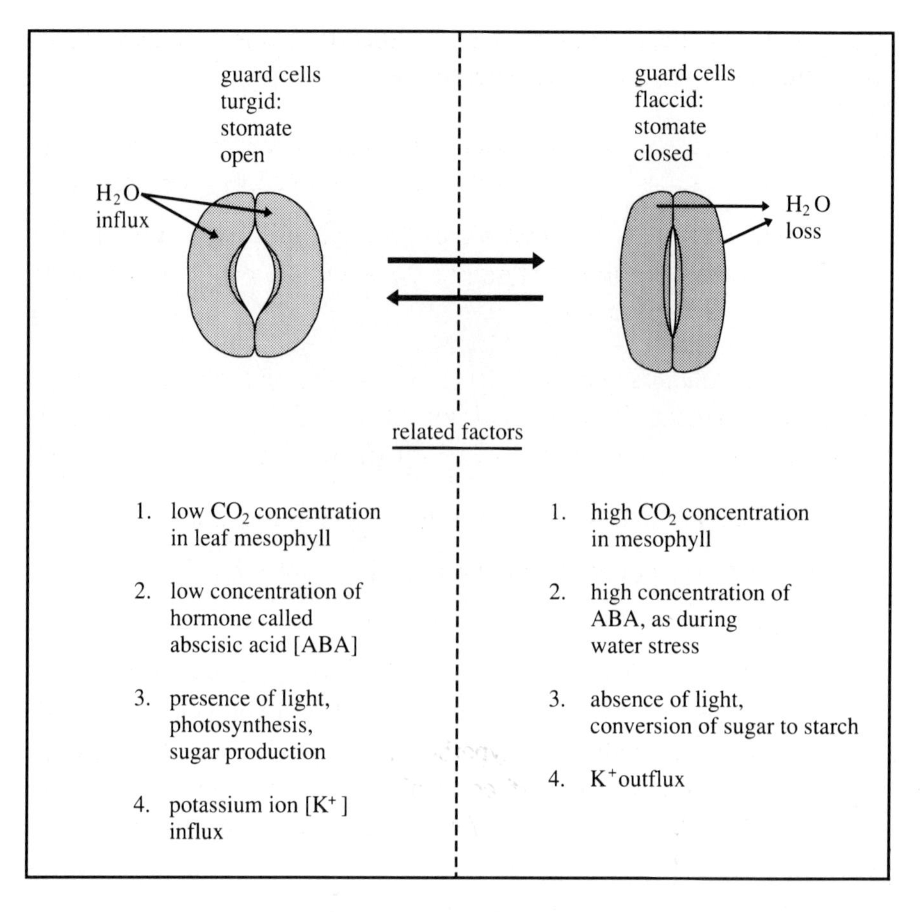

Figure 18.5 Stomate opening and closing. The pore (stoma) of a stomate is opened when the surrounding guard cells swell and bow outward. It is closed when the guard cells become flaccid. Although turgor change is the operative factor of stomatal opening and closing, several other factors of the cellular environment also are involved.

Guard cells usually are the only cells of the epidermis that have chloroplasts, a fact that seems to be related to their ability to regulate the stomate opening (figure 18.5). The cell wall that lines the pore is thicker than the cell walls that are attached to adjacent epidermal cells. When the guard cells are turgid, the thickened region does not expand, but the rest of the cell does. As a result, the cells bow outward and the pore is opened. When the guard cells are flaccid, the pore is closed. At least one hormone helps to regulate the stomatal opening.

6. Under what conditions would it be beneficial for the plant to have the stomates closed?

7. Are such conditions also those that would tend to cause the guard cells to become more flaccid?

8. What happens to the stomatal opening when the guard cells become flaccid?

Look at the epidermis on your leaf section. Remember that you will see stomates in *section,* rather than whole.

9. Sketch a sectioned stomate from your slide, showing both guard cells. Label the guard cells, stomatal pore, and chloroplasts.
10. Move the slide around while observing with low power. Do there seem to be more stomates on the upper surface of the leaf or the lower surface?

If you cannot find enough stomates on your slide to make this generalization, combine information from other students' observations to increase the amount of data and record the data.

11. Number of stomates in upper leaf surface sections.
12. Number of stomates in lower leaf surface sections.
13. If there seems to be a difference in stomatal distribution between the upper and lower leaf surfaces, suggest one or more hypotheses to explain the difference.
14. Make an additional hypothesis. In aquatic plants whose leaves float on the water surface, would you expect stomates to be found on the top or bottom leaf surface? Explain your reasoning.

18.7 **Epidermal hairs** are extensions of epidermal cells. They are found on leaves, stems, fruits, and flowers, and they vary greatly in structure, color, and chemical content. *Hairlike* epidermal hairs make plant surfaces fuzzy, as with the leaves of mullein, African violets, purple passion plant, and mint geranium. It is thought that a forest of epidermal hairs reduces air flow over the leaf surface, decreasing water vapor loss. It also may discourage animals, especially insects, from feeding on the leaves.

Mints and many other plants have *glandular hairs,* epidermal hairs that contain aromatic, oily, or toxic substances that discourage potential herbivores. Crushing the glandular hairs releases these substances. Some glandular hairs produce an **exudate,** something that is secreted to the hair surface. The sundew and Venus's flytrap, both carnivorous plants, have sticky hairs that lure and trap insects.

Perhaps you have had an unpleasant "brush" with a stinging nettle, which has sharp, brittle, *stinging hairs*. When you touch the hairs, they stab your skin and shatter, releasing irritating chemicals into the tiny puncture wounds. A painful, burning sensation results!

The various kinds of epidermal hairs are interesting, even beautiful. Using a magnifying glass or a binocular microscope, look at the several kinds of leaves your teacher has provided.

1. Which plant(s) have colorless epidermal hairs?
2. Which plant(s) have pigmented epidermal hairs?
3. Which plant(s) have glandular epidermal hairs?

18.8

The leaf *mesophyll* is composed mainly of chlorenchyma tissue. The mesophyll has many *air spaces*. Try to imagine what it would be like to become miniaturized enough to crawl into a leaf through a stomate. You would be surrounded by glistening mesophyll cells enclosed in water-saturated, fibrous cell walls. You could grasp the cellulose fibers for support, but would need to be careful about putting a foot through the more delicate wall areas and into the cytoplasm. Solar radiation absorbed into the leaf would cause the atmosphere in the leaf air spaces to be hot and saturated with humidity. After sloshing around over and between warm, wet parenchyma cells that are vibrant with biochemical activity (might they "hum" with energy?), you would probably find yourself eagerly exiting through another stomate to the evaporative cool of the external atmosphere!

Keeping in mind this miniature view, examine the mesophyll shown on your microscope slide. What you see there is a thin section, but try to visualize in your mind what the mesophyll might have looked like in three dimensions. Notice the relatively thin cell walls, the many chloroplasts, and the intercellular (between-cell) air spaces; look for nuclei.

The mesophyll in a "typical" dicot leaf consists of two layers of differently shaped parenchyma cells. The upper part is made up of tall, columnar cells. These cells are densely packed, maximizing the number of chloroplasts available to trap light. The lower region is spongy, made up of irregularly shaped cells having larger air spaces. The lobed cells have large surface areas, enhancing carbon dioxide absorption, oxygen release, and water evaporation. (Remember your observation about stomate distribution?) The mesophyll is penetrated by many **veins,** the *vascular bundles* of the leaf. You will examine veins in section 18.9.

1. Sketch part of your leaf section, showing its tissue regions and the location of at least one vascular bundle. You do not need to show great cellular detail, unless otherwise instructed. Label the following structures in your drawing: upper epidermis, mesophyll, vascular bundle (vein), lower epidermis, stomate, and epidermal hair (if any are present).
2. Draw one or two examples of columnar mesophyll cells and spongy mesophyll cells.
3. In what specific ways is the structure of the mesophyll adapted for efficient photosynthesis?

18.9

So far you have learned that leaf structure is related to photosynthesis in the following ways: (*a*) the epidermis reduces water loss while permitting carbon dioxide to diffuse into the leaf interior; (*b*) oxygen and water vapor diffuse out of the leaf; (*c*) parenchyma cells in the mesophyll contain chloroplasts that trap light energy to be used in photosynthesis; (*d*) air spaces and the shape of mesophyll cells enhance environmental gas exchanges.

The remaining raw material for photosynthesis is water. Where does it come from? *Veins* carry solutions between the leaf and the rest of the plant body. Leaf veins branch off the stem vascular tissues. Veins are seen easily in most leaves. Figure 18.6 is a photograph of a leaf skeleton. The soft parenchyma tissues and epidermis have been removed and only the tough vascular tissues remain.

Figure 18.6 A skeletonized leaf. Chemical removal of all of the soft tissues of the leaf reveals the full extent of the vascular system.

Vein *arrangement* is another characteristic that distinguishes the dicots and monocots. In monocot leaves, such as those of grasses, lilies, irises, and daffodils, the leaf veins are parallel to one another. In dicots, the veins branch, forming a network.

1. Is the leaf in figure 18.6 from a monocot or a dicot?
2. Examine the leaves your teacher has provided and identify which plants are monocots and which are dicots.

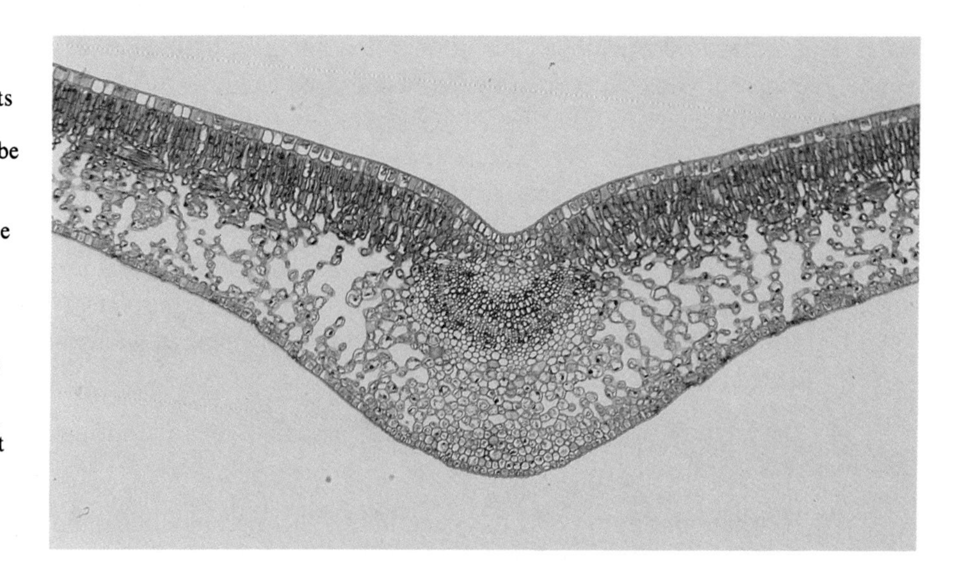

Figure 18.7 Dicot leaf cross section. This section passes through the *midrib* of the leaf, its thickened, central ridge. The tissue regions of the leaf should be familiar to you by now. For instance, you can identify upper and lower epidermis and compare stomate distribution on the two surfaces. Furthermore, the differentiation of the mesophyll into palisade and spongy layers enables you to conclude that this illustration is oriented with the top side of the leaf in an upward position. Relate this arrangement to leaf function: densely packed palisade cells maximize chloroplast exposure to light; loosely packed spongy cells maximize carbon dioxide absorption. The thick midrib includes a prominent vein, with xylem, phloem, and supportive tissues. You also can identify several small veins—compare their appearance here to the figure 18.3 drawing to help you understand their structure and various orientations.

Again place the dicot leaf slide on the microscope stage and move it around until you locate one of the larger veins. The vein contains xylem and phloem tissues, all surrounded by a parenchyma cell sheath (figures 18.3 and 18.7).

3. Is the xylem on the upper or lower side of the vein?

Imagine a thin sliver or string of the vascular tissue of a stem entering a leaf petiole. (Remember bundle scars on stems?) The string of conducting tissue is a vascular bundle, containing xylem, vascular cambium, and phloem. Since the xylem was *inside* the cambium and phloem of the *stem,* it is on the *topside* of the vascular bundle that enters the leaf. Be sure you are able to visualize this relationship.

Identify the following cell types in the vein: (*a*) xylem conducting cells, (*b*) phloem conducting cells, and (*c*) fibers.

4. What substances are carried to the leaf in the vein xylem?
5. What substances are carried from the leaf in the vein phloem?
6. What is the function of the fibers in a vein?

18.10 Prepare a wet mount of a *very* thin slice of dicot leaf. To do this place a piece of fresh leaf on a wet glass microscope slide, and slice off several extremely thin cross sections. *Be careful not to cut yourself or leave the razor blade lying loose on the counter.* Add more water, if needed, and place a cover slip over the sections. Examine your preparation under the microscope. Compare what you can see in this fresh section of a living leaf to what you could see in the permanent slide.

1. What tissue regions can you identify and what are their colors?
2. What cellular structures can you identify?
3. What are the advantages of observing living cells?
4. What are the advantages of observing cells that have been chemically preserved and stained?

18.11 Perhaps your teacher will want you to section a thicker leaf. Plants that are adapted to life in Earth's widespread arid climates often have water storage tissues. Such plants are called *succulents.* Many dry climate plants also have a multicellular epidermis, one that is more than a single layer thick. The succulent jade plant (*Crassula*) and its relatives store moisture in their thick leaves. Examine the surface of the leaves that are provided with a magnifying glass or binocular microscope to see the smooth, waxy surface.

Prepare and examine a wet mount of a very thin slice of a jade plant leaf. Then answer these questions.

1. What tissue region takes up most of the interior of the leaf?
2. Are the parenchyma cells present larger than those of the prepared leaf section? Hypothesize a reason for your observation.
3. Does the inner mesophyll seem to have as much chlorophyll as the outer regions?
4. What do you hypothesize to be the main function of the clearer mesophyll tissue?

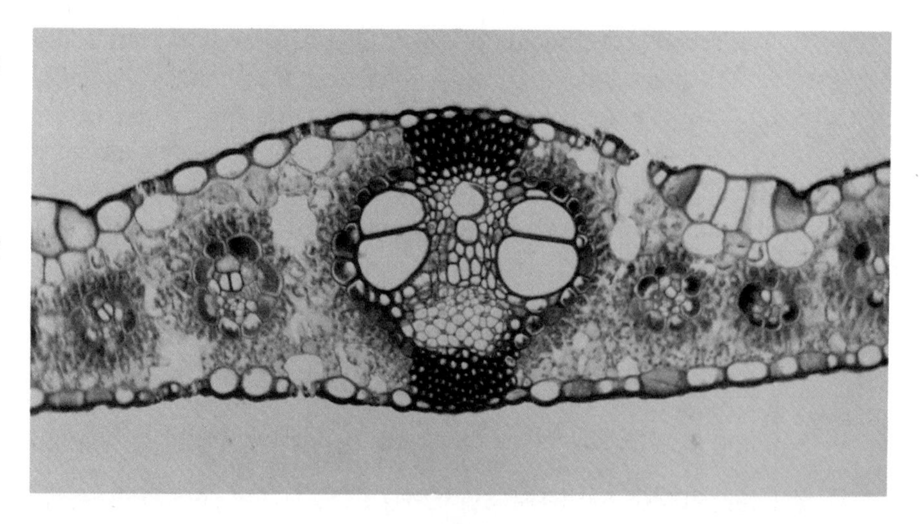

Figure 18.8 Monocot leaf cross section. Corn leaf mesophyll is not differentiated into layers. The large vein in the center of the photograph includes several vessels. The phloem is centered between and below the two pairs of large vessels. Sclerenchyma fibers strengthen the outer sides of the bundle. You can see stomates on both epidermal surfaces. Notice the row of five large, bulliform ("bubble-shaped") cells in the upper right part of the epidermis. Grass leaves roll when they dry out because these special cells lose water and shrink. The shrinkage of many bulliform cells along the entire upper epidermal surface shortens this surface, causing the leaf to roll inward. When water is restored, the bulliform cells fill and the leaves unroll.

18.12 Examine a microscope slide containing sections of a corn (*Zea*) leaf, such as the one shown in figure 18.8. Study the epidermis, mesophyll, and veins.

1. How would you describe the shape of corn mesophyll cells?
2. Is the mesophyll separated into palisade and spongy layers?
3. Do the veins seem to have fewer or more xylem cells than the dicot veins you examined?
4. Are the parenchyma cells surrounding the veins larger or smaller than they appeared in the dicot slide?

In grasses, parenchyma cells around the veins often are enlarged. They actively assist in the transfer of substances between the veins and the mesophyll.

18.13 Plant tissues require large amounts of water to function normally. In addition, they need water as a raw ingredient of photosynthesis. However, as you have seen in the previous sections, the plant body loses water continuously from the shoot system, and it must constantly be replenished. Most vascular plants get all of their water from the environment via the root system. Minerals and other substances are absorbed along with the water. In addition to absorbing water and nutrients, roots anchor and support a plant.

The root system originates with the single embryonic root that is present in the seed. However, a single root is insufficient to serve an entire plant, so a system of branching roots develops. Where do the rest of the roots come from? They have two sources: (*a*) new roots arise as branches of existing roots; or, (*b*) they may arise from the stem.

The tissues of the root system are continuous with those of the shoot system, although tissue arrangement in roots and stems differs. A region of transition is present where the shoot and root systems blend together. Remember how growth occurs? The shoot apex and root apex begin growing in their opposite directions even before germination. The products of the shoot apical meristem differentiate into a shoot kind of organization, and cells produced by the root apical meristem differentiate into a root kind of organization. In this section of chapter 18 we will examine root growth, organization, and function.

18.14 What do you already know about roots? Have you seen roots and root systems firsthand? If you have transplanted flower or vegetable seedlings, you have seen root systems and know the importance of not tearing tender young roots. If you have pulled weeds, you also have seen root systems. If you have shopped for vegetables, you have seen roots. What are the external characteristics of roots? Your teacher has provided an assortment of root-bearing plants. Look at them after you have read this introduction.

When you pull up a small clump of grass and then a dandelion or sunflower, you can see that they have distinctly different kinds of root systems (figure 18.9). The grass root system consists of numerous fine roots; it is called a **fibrous root system.** The dandelion, on the other hand, has one major root from which smaller *branch roots* grow, a **tap root system.**

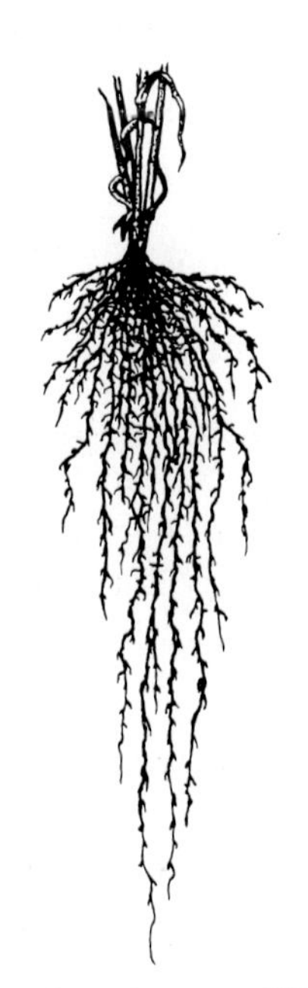

Fibrous Root System of Wheat

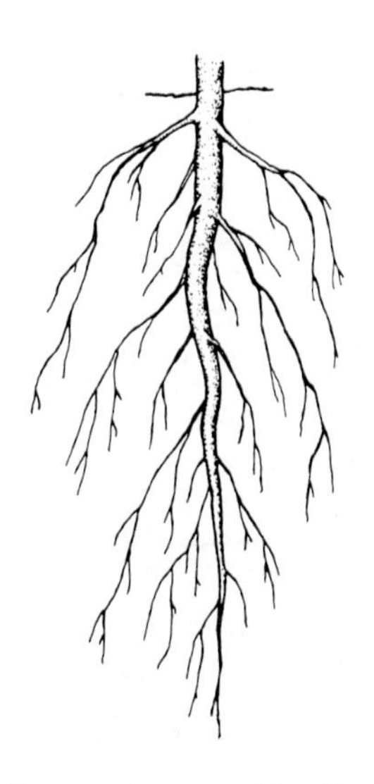

Taproot System of Sunflower

Figure 18.9 Root systems. Grasses typically have a fibrous root system composed of many small roots. Many dicots have a tap root system, consisting of one main root and a number of smaller branch roots.

Roots often are major sites of food storage for plants that live more than a year (figure 18.10). Humans rely on **storage roots** as important food sources. Storage roots have an abundance of parenchyma tissue, making them much plumper than nonstorage roots.

Roots also can develop from stem or leaf tissues. They are called **adventitious roots** (figure 18.11). Most houseplants are propagated from pieces of a parent plant (cuttings, see chapter 20), by formation of adventitious roots.

1. Name at least four kinds of plants that produce edible storage roots. (Do not include any tubers—they are stems!)
2. Can you think of another vegetable that usually comes with the roots attached? (*Hint:* you trim them off before eating the shoot system.)
3. What are adventitious roots, and can you see any advantage to a plant's being able to produce adventitious roots?

18.15 In our study of seeds and germination, you saw how plant embryos are arranged in seeds and how they emerge from the seeds during germination. You also had an opportunity to appreciate how rapidly mitosis in the various meristematic tissues gives rise to a plant body consisting of millions of individual cells. Examine the onion bulb that has been placed over water for several days. This is the kind of setup used to grow the onion roots that were made into the onion root-tip slides you looked at during your investigation of mitosis in plant tissues. These young roots are the result of primary growth.

1. What is the meaning of the last statement?
2. Do the roots seem to be of uniform diameter?
3. Would you hypothesize the roots to be approximately equal in age and maturity?
4. Does the onion have a fibrous or tap root system?
5. Are branch roots evident?
6. Identify the root cap on each onion root and describe it here.

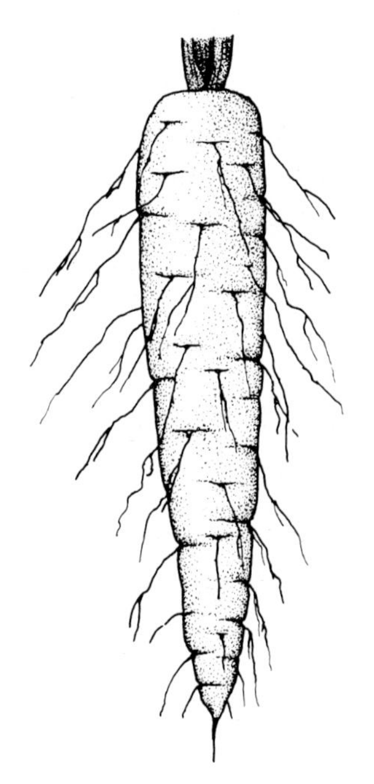

Food Storage Root of Carrot

Figure 18.10 Storage roots. Plants that grow for more than one season often store food reserves in fleshy root tissues. Stored food is mobilized in the spring to support new growth.

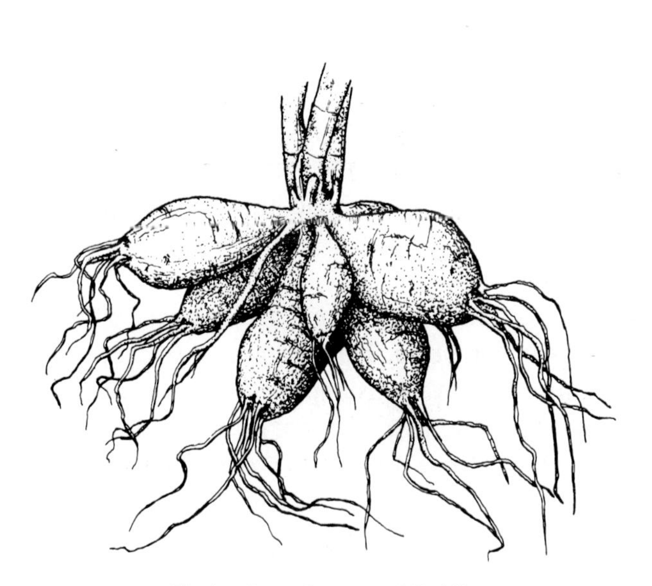

Fleshy Root System of Dahlia

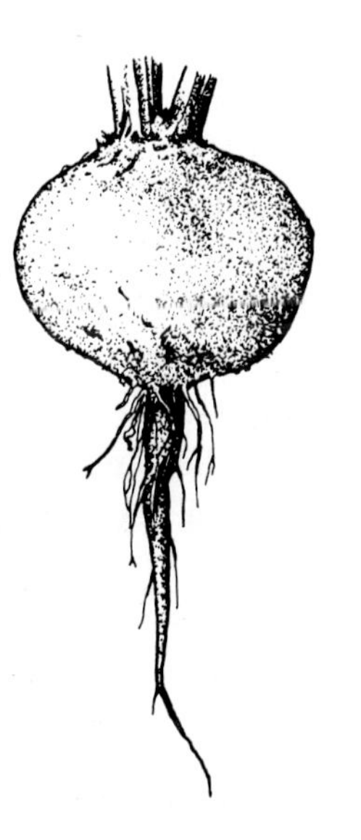

Fleshy Taproot System of Beet

© Kendall/Hunt Publishing Company

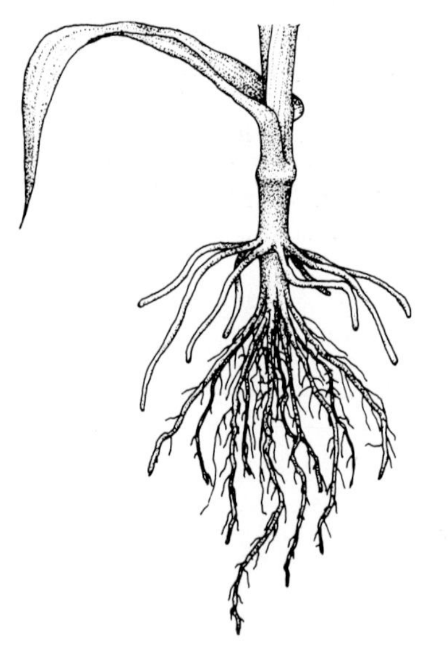

Prop Roots of Corn

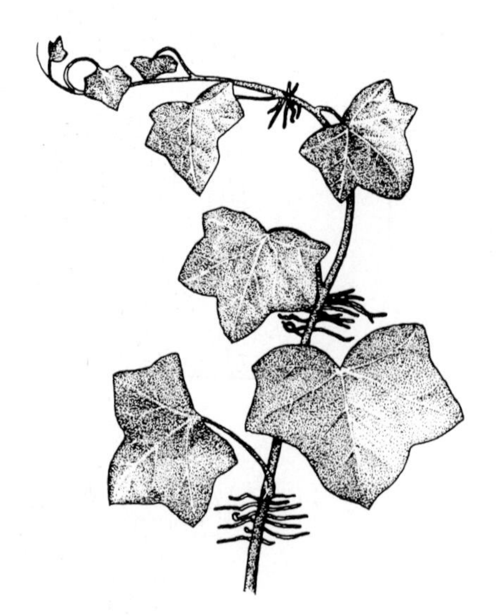

Aerial Roots of English Ivy

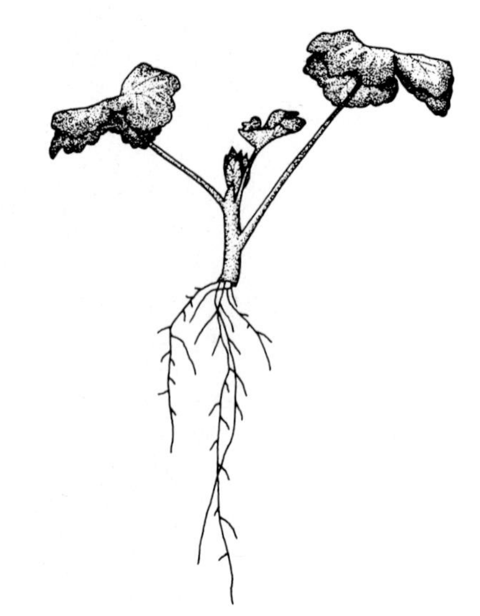

Adventitious Roots on Cutting of Cultivated Geranium

© Kendall/Hunt Publishing Company

Figure 18.11 Adventitious roots. Adventitious roots develop on stems or leaves. Corn and some other plants develop *prop roots* that begin above the soil level and help to support the shoot system. *Aerial roots* help to anchor climbing plants, such as ivy and Virginia creeper. Adventitious roots also are an important means of vegetative (asexual) propagation, enabling broken-off parts of a shoot system to develop into new individual plants.

18.16 Obtain a microscope slide having longitudinal sections of onion (*Allium*) root tip. Use a millimeter ruler to measure the length of a section before you put the slide under the microscope. Examine the root tip sections under the microscope, beginning with low-power magnification. One end of each section is flat, where the root tip was sliced off to make the slide. Focus your attention on the other end, the rounded, growing tip. Compare what you see on your slide to figure 18.12. Notice, especially, the major *developmental regions* of the root tip.

The **root cap** resembles a thimble that covers and protects the tiny region of tender mitotic cells at the root tip. This role is a significant challenge, and you can see that its outermost cells appear to be degenerating. The outer cells of the root cap break down as the root is propelled through the soil by growth and expansion of newly formed tissues. Visualize the growth forces involved. A growing shoot pushes through air—no problem there. A growing root pushes through the soil, a rough encounter for fragile cells. (Can you imagine yourself being driven headfirst through the soil?) Degenerating root cap cells form a slippery layer that helps to lubricate this passage. New root cap cells are added continuously by the apical meristem.

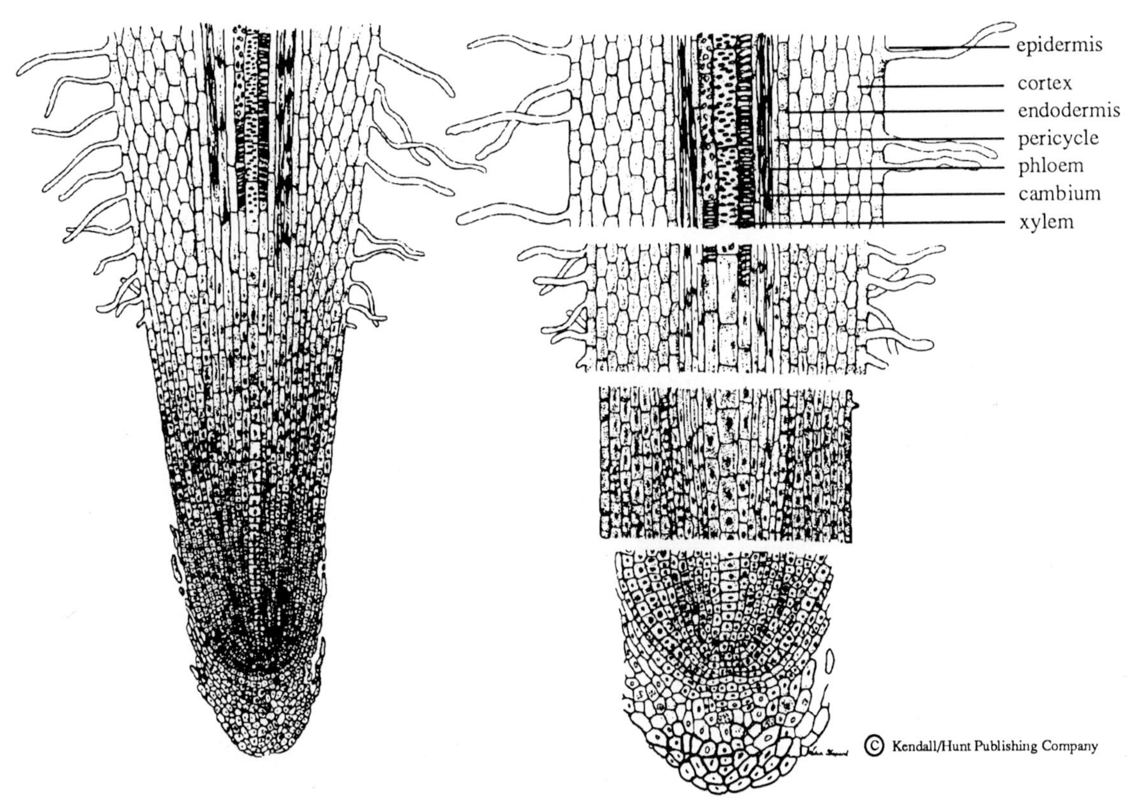

Figure 18.12 The root apex. In this drawing and on your slide, you can see progressive origin, enlargement, and maturation of cells in the root tip. Chromosomes can be seen in various stages of mitosis in the meristematic cells, which look relatively small, thin-walled, and boxlike. The apical meristem of a root tip produces cells in two directions. (1) Those cells that are produced toward the main body of the root enlarge and differentiate into the functional tissues of the root. Tubular extensions of the epidermal cells, called *root hairs,* mark the region of the young root in which the tissues are beginning to differentiate. (2) Those cells that are produced toward the soil form a *root cap* that protects and lubricates the delicate apical meristem as the root grows through the soil.

Recently, another important role has been attributed to root cap cells, that of sensing the direction of gravitational force. Tiny grains inside root cap cells settle downward, influencing the orientation of root growth toward the source of gravity. As a result, roots usually grow downward!

The **apical meristem** cells make up a meristematic region between the root cap and the newly formed cells of the root body. They are small cells with prominent nuclei, very thin cell walls, and no major vacuoles. In many of the cells you can see stained chromosomes, captured as they were engaged in the intricate ballet of mitosis. *The apical meristem produces new cells in both directions.* Cells produced toward the tip of the root add to the root cap. Cells produced away from the tip of the root mature into the tissues of the root body.

The next stage of growth after mitosis is *cellular enlargement.* The young cells expand in response to the osmotic uptake of water. As the cells swell, additional cytoplasm and cell walls are formed. The root tip region immediately behind the apical meristem consists of cells that are beginning to enlarge. Upward from these youngest cells are ranks of progressively older cells in various stages of expansion. The root tip always shows this region of tender cells that are enlarging but have not yet differentiated into different cell types. This region of the root tip, therefore, is called the *region* or *zone of cell enlargement.*

Think of yourself as a very young root cell for a moment. You were formed by mitosis. Then you absorbed water and got bigger, as did the other cells beside you that are about the same age. What about the ages of cells that are above and below you?

1. Is the cell "above" you (closer to the main plant body) older or younger than you?
2. Is the cell "below" you (closer to the root apex) older or younger than you?
3. As the root tip continues to grow, do you remain the same distance from the apical meristem, or does it move farther and farther away from you?

Even as the mitotic action becomes increasingly distant from you, you keep changing, maturing. If you are an outermost cell, you become a flat, tilelike epidermal cell, with a root hair. If you are an internal cell, you differentiate into some other kind of specialized cell—perhaps a parenchyma cell or a xylem vessel element. You are part of an entire "neighborhood" of maturing and differentiating cells: *a region* or *zone of maturation and differentiation.* If you have successfully imagined this fantasy, you have a feeling for the dynamics of growth in a root tip.

Move your microscope slide to an area where **root hairs** are present, marking the beginning of the maturation zone. Why do you think we use root hairs to mark the zone? The presence of a root hair identifies a functionally differentiated epidermal cell. Root hairs are an easily observed marker that gives, by inference, a clue that other tissues also are maturing.

Just inside the functional epidermal cells, you should be able to see a layer of maturing parenchyma cells. Finally, there is a core of elongated cells that will give rise to the vascular cambium, xylem, and phloem.

4. Label the following regions and structures on figure 18.12 in your logbook: (*a*) root cap, (*b*) apical meristem, (*c*) enlargement region, (*d*) differentiation region, (*e*) root hair. (Use good labeling technique.)

Once again, examine the whole onion with its roots suspended in water, and measure a typical root. How does its length compare to the length of the growing tip you have just examined? Remember this: no matter *how* long a root grows, new primary growth occurs only at its very tip!

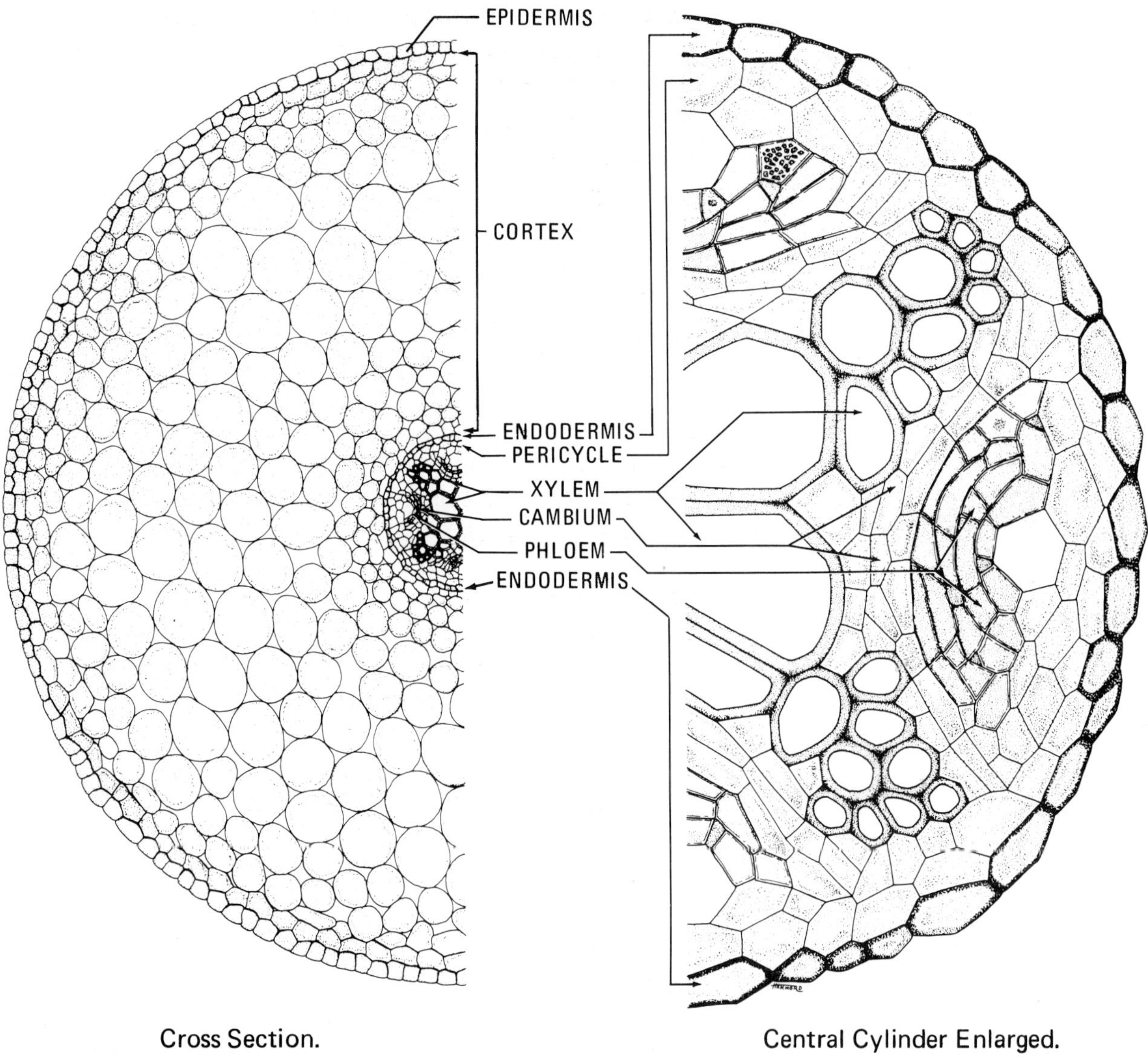

© Kendall/Hunt Publishing Company

Figure 18.13 Young dicot root. The arrangement of functional tissues in a young dicot root is quite different from tissue arrangement in a stem. Xylem and phloem form a central *vascular core,* surrounded by the parenchyma cells of the *cortex.* The innermost layer of cortical cells, the endodermis, helps to regulate molecular traffic into the vascular core. The outermost layer of the vascular core, the pericycle, is meristematic.

18.17 Now that you understand the *origin* of root tissues, we will go down to the microscopic level to observe the details of the differentiated tissues resulting from primary growth. Obtain the microscope slide that has cross sections of a very young dicot root. Figure 18.13 will help you to identify structures that are visible on your slide. Use the following discussion and descriptions to identify and understand the functions of the epidermis, cortex, endodermis, pericycle, vascular cylinder, xylem, phloem, and vascular cambium.

Figure 18.14 Roots and the soil. In this scanning electron microscope view, tiny soil particles look like monstrous boulders in comparison to the delicate root hairs. Consider the functional relationship between the soil particles and the root hairs. Water molecules stick together (cohesion) and they also tend to stick to the soil particle surfaces (adhesion). As a result, a thin film of water is held between soil particles, and this water is that which is available to plant root hairs. Thus, the "thirst" of both a tiny lettuce seedling and a massive oak tree is satisfied by the molecule-by-molecule osmosis of water through the multitudinous, fragile root hairs that are found only near the growing tips of young roots.

Look at the slide with low-power magnification, and you will see that the tissue arrangement is quite different from that of an herbaceous stem. Both organs have an **epidermis,** but the root epidermis is not heavily cutinized and does not have stomates. It is adapted for *absorbing* water, not preventing water loss! Therefore, many epidermal cells have tubular extensions, the *root hairs,* that grow outward through the small, water-filled spaces between soil particles (figure 18.14). The delicate root hairs, present only on the youngest roots of a plant, are responsible for most of the water uptake for the entire plant, whether it is a radish or an oak tree.

The root **cortex** is several cell layers thick, forming a cylinder of storage tissue inside the epidermis. Its innermost cell layer is the **endodermis.** The endodermis separates the cortex from the vascular tissues that make up the **core** of a root. Endodermis means, literally, "inner skin," because endodermis cells have a thickened, waterproofed strip (the Casparian strip) on four sides, where they are joined to other endodermis cells. The strip keeps water from passing *between* the endodermis cells and into the vascular core.

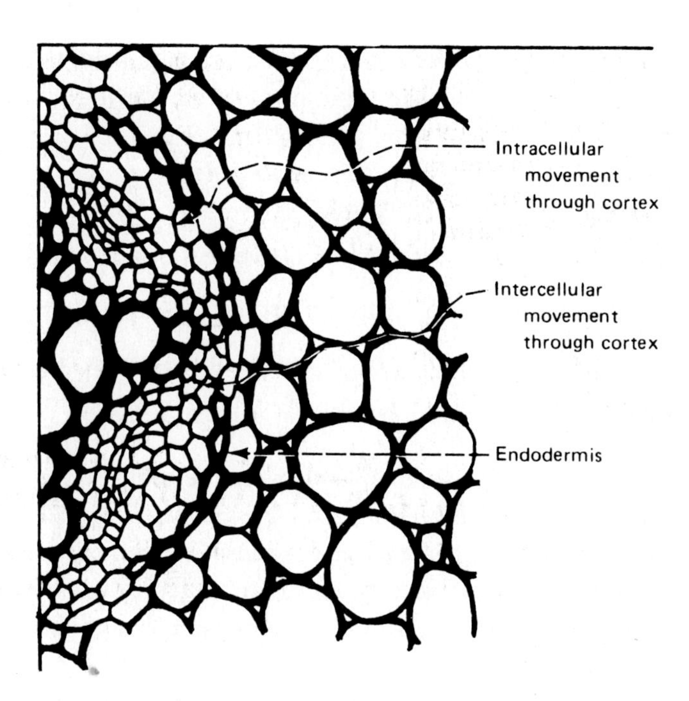

Figure 18.15 Water movement in roots. Water that is absorbed through the epidermis moves laterally (radially) toward the vascular core. Once it enters the xylem, it then can be transported upward. An analogy might be of a person walking along a corridor, then taking an elevator. Water and dissolved substances can move either between or through the cells of the cortex until they get to the endodermis. Molecules can then enter the vascular core only by passing through the living membranes of endodermal cells, so are selectively taken in or excluded.

The strip is visible on your slide because it absorbs more stain. The cell walls that *face* the cortex and the root core, however, are not sealed.

Why would a root have an inner layer of "skin" to seal off the vascular tissues? After all, a water solution must get to the vascular core so it can be taken up to the rest of the plant! To hypothesize an answer to this question, we will look at how water does move from the outside to the center of the plant (figure 18.15). The soil contains water, in which are dissolved all sorts of substances. Some of these substances are useful plant nutrients, such as nitrates, iron, magnesium, and calcium. However, the soil solution also contains substances that may be disruptive to normal metabolic processes. Water that is absorbed through the epidermal cells and then travels *intracellularly,* through the cells of the cortex before reaching the endodermis, is subject to a sorting process. What has done the sorting? Living cell membranes are able to transport or exclude many specific molecules.

However, it also is possible for water to leak into the root between epidermal cells, or through injured cells, and move through the cortex *intercellularly,* without having to pass through living cell membranes. The endodermis acts as a kind of "final filter," assuring that any solution that enters the vascular cylinder has first been osmotically screened by passage

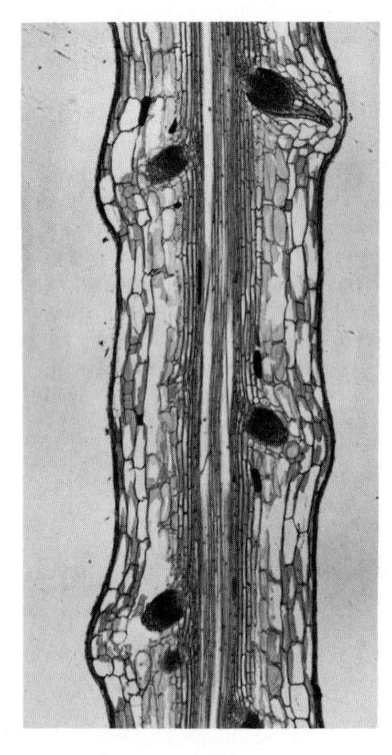

Figure 18.16 Branch root origin. Root systems consist of main roots that produce many branch roots. The branch roots originate from a special meristematic root tissue, the *pericycle,* which is the outermost cell layer of the vascular core. In this micrograph you can see a longitudinal view of the various tissue layers that were shown in cross section in figure 18.13. You also can see several branch roots forming. The structure of a branch root tip is the same as the root tip previously described, in which the apical meristem divides to produce not only root body tissues but also a root cap. Branch roots grow through the cortex before emerging into the soil.

through the cell membranes of the endodermis. It also helps to exclude air bubbles that might interrupt the columns of water moving upward in the xylem.

1. What is the outer tissue layer of the root called?
2. Are root hairs present on any of the sections of your slide?
3. What kind of tissue forms the cortex?
4. What do you think the ovoid bodies in the cortex cells could be?
5. What does the innermost ring of cortical cells do?

Move the microscope slide so you can see the central core of the root, which is the **vascular cylinder.** (Botanists also call it a *stele.*) How does this arrangement differ from the herbaceous stem slide you examined? First, there is no pith in the root. Second, there are no individual vascular bundles. Instead of pith, the centermost structure in a young dicot root is a lobed column of **xylem tissue.** In cross section, the lobes may look like a cross or star. **Phloem tissue** is nestled in the grooves between the lobes of xylem tissue. Roots add xylem and phloem as they grow; therefore, **vascular cambium** first develops between the xylem and phloem tissue. As xylem and phloem are added, the xylem loses its lobed character, becoming rounded.

Like shoots, roots form branching systems. Where do **branch roots** come from? The outermost layer of cells of the vascular cylinder is called the **pericycle;** it lies just *inside* the endodermis. The pericycle is a special meristematic root tissue that gives rise to vascular cambium at the tips of the xylem lobes *and* gives rise to the branch roots. You can see it on your slide as a layer of thin-walled cells just inside the endodermis.

Figure 18.16 shows tiny branch roots growing outward from the vascular cylinder and through the cortex, heading for the external environment. You can see that they grow just the same way other young roots grow, from the apical meristem, with the same primary growth characteristics.

Your teacher may have a slide or an illustration of a more mature root, in which you can see the results of *secondary growth.* In woody plants, the roots grow to resemble woody stems. Expansion causes loss of the epidermis and cortex, and cork cambium layers develop. The result is a bark-covered, woody root, complete with annual rings! Woody roots are often important storage sites and, consequently, have proportionately more parenchyma tissue than do woody stems. (From where do you suppose all those sugars in maple sap come in the spring?)

18.18 Keep in mind the microscopic anatomy you have just examined as you do this activity. Your teacher has provided some fresh carrot roots for you to examine. First, look at the external morphology of a carrot root, which is adapted for food storage. At the top of the root you can see where the shoot system was (or is) attached. The leaves arise from a very short, compact stem, forming a clustered shoot system.

1. Remembering the kinds of pigments that are present in plants, what pigment is mainly responsible for the orange color of the carrot root?

The root is covered with *epidermis*. Along its length you can see dimpled areas or wrinkles where small *branch roots* have emerged from the main root. On store-bought carrots, the branch roots were torn off when the root was harvested and washed for market. However, by looking closely you may be able to see their stubs.

2. Is it likely that branch roots have root hairs? Explain.
3. What do you think is the function of the branch roots?
4. What are three functions of the main (large) carrot root?
5. Does the carrot have a fibrous root system or a tap root system?

Using a very sharp vegetable peeler or other cutting instrument, make some delicately thin cross sections of a carrot root, placing them immediately onto a wet microscope slide. Avoid sectioning at an oblique angle or the tissue orientation will not be a perfect cross section. After you have made several sections, select the best two sections for examination. Examine the carrot root cross sections with a magnifying glass or a binocular microscope. Note the central *core*.

6. What is the anatomical name of the root core?
7. What is the name of the region between the epidermis and the vascular cylinder, and what is its main function?
8. What complex tissue types do you think might be found in the vascular cylinder?
9. What do you think is the function of the vascular cylinder?
10. Is the vascular cylinder darker or lighter in color than the outer root tissues?

11. Based on your own experience in eating raw carrots and in what you know about the structure of vascular tissues, how can you explain the fact that the vascular cylinder is tougher than the outer tissues?

12. Based on your own experience in eating raw carrots and in what you know about storage parenchyma tissue, how can you explain the fact that the cortex tissue tastes sweet?

In your sections, or those of a neighbor, you should be able to see occasional spiky looking extensions of the vascular cylinder into the cortex. These extensions are the vascular cylinders of branch roots. Try to visualize the physical relationship of the branch roots to the main root. Remember, the branch roots do not arise as surface structures, but have their origin in the outer tissues of the vascular cylinder.

13. How do the origins of branch roots and branch shoots differ?

You can also see **vascular rays** that extend from the center of the vascular cylinder to its perimeter. As in wood, root vascular rays are vertical plates of parenchyma tissue. Whereas xylem and phloem conducting cells move water solutions along the longitudinal axis of the plant body, vascular rays move substances radially, between the outer and inner tissues of the root.

14. Sketch one of your carrot root cross-sections (actual size or larger) in the space below, and label the following regions and structures: (*a*) epidermis, (*b*) cortex, (*c*) vascular cylinder, (*d*) branch root origin, and (*e*) vascular ray.

18.19 Your teacher has cut several carrot root segments in half lengthwise. Use the vegetable peeler to obtain a thin, *longitudinal* section of the root, placing it immediately on a wet microscope slide. Examine the strip of root with backlighting; then look at it with a magnifying glass or binocular microscope. You can see all of the structures visible in the cross sections, but from a different perspective. The main differences will be two: (*a*) you can see more branch roots coming off the vascular cylinder, and (*b*) you can see that the vascular rays are deep bands of parenchyma, not just thin lines as they appear in cross section.

1. Sketch a portion of the lengthwise section (actual size or larger) in the space below and label the same structures you labeled in activity 14 of section 18.18.

18.20 In this chapter you have combined macroscopic and microscopic observation techniques to learn how leaves are adapted as organs for photosynthesis, and how roots are adapted for their several roles. Review the study points listed below and check them off if you feel you have mastered them. If you have not, return to the text of the chapter and your notes to study them again before continuing to the next activity.

On completion of this chapter, you should be able to do the following:

☐ 1. Explain how all of the ingredients for photosynthesis get into a leaf.

☐ 2. Understand the basic structure of dicot and monocot leaves.

☐ 3. Understand the environmental conditions in a photosynthesizing leaf.

☐ 4. Prepare fresh leaf sections for microscopic study.

☐ 5. State the several functions of root systems.

☐ 6. Recognize fibrous and tap root systems and understand their separate advantages.

☐ 7. Explain how root cells are formed and mature in a root tip.

☐ 8. Identify the arrangement of tissues in a dicot root that has undergonc primary growth.

☐ 9. Understand the continuity of the vascular tissues between the root, stem, and leaves.

☐ 10. Prepare fresh root sections for microscopic examination.

Fluid Transport Throughout the Plant Body 19

How do solutions move within the plant body?

19.1

While learning about the external and internal structure of stems, leaves, and roots, you also learned about the distribution of the vascular tissues. They are continuous from the tip of the smallest root to the edge of the farthest leaf. The vascular tissues carry a solution that is mostly water, plus sugars, amino acids, mineral ions, hormones, and a variety of other substances. In this chapter we will look at the physical mechanisms of plant transport.

We still do not know *all* of the scientific details about fluid transport in plants, but we do know enough to present logical, tested hypotheses. With what you already know, you should be able to understand these hypotheses readily. As with many biological processes, the easiest way to understand xylem and phloem transport is to think in terms of cause and effect. Keep cause and effect in mind as you complete the following activities, referring to figure 19.1.

First, consider the **xylem.** The main function of xylem tissue is to move a water solution from the root system to the shoot system. What does your brain tell you about the orientation of this movement? Conduction in the xylem is analogous to an elevator that only goes *up.*

Next, consider the **phloem.** The main function of phloem tissue is to move nutrients, especially sugars, from sites of production to storage sites, or from storage tissues to sites of need. Most sugars are produced, you will recall, in the leaves. They are transported out of the leaves to the rest of the plant body, where they supply *energy* for respiration and *molecular fragments* for growth. They also are moved to *storage* parenchyma tissues, such as pith or ground parenchyma (in stems), cortex (in stems and roots), vascular rays (in woody stems), and developing fruits and seeds. When the plant has need of nutrients stored in roots or stems, the nutrients can be carried upward via the phloem. Conduction in the phloem, therefore, is analogous to an elevator that can go either *up* or *down.*

Finally, consider lateral and radial transport. The elongated conducting cells of the xylem and phloem conduct *vertically.* Only a few body cells actually touch these conducting tissues, yet all cells are nourished by them. How does this occur? Water, nutrients, hormones, and other substances are transported *radially* and *laterally* by *cell-to-cell transfers.* Parenchyma tissue, such as that in the **vascular rays,** mobilizes substances between the vascular tissues and outer tissues. Cell-to-cell exchanges occur by diffusion and other membrane transport mechanisms, as you studied in earlier chapters.

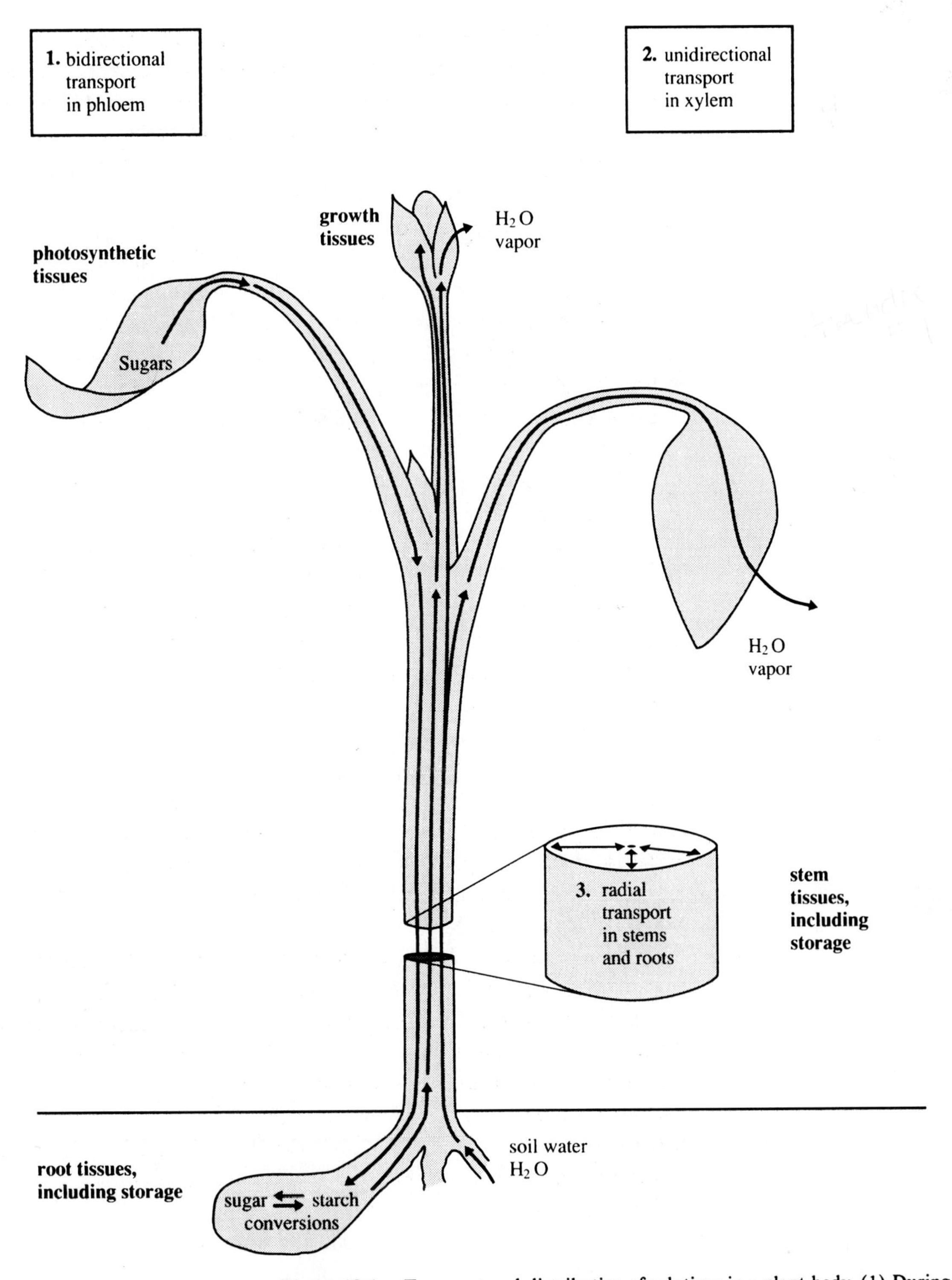

Figure 19.1 Transport and distribution of solutions in a plant body. (1) During photosynthesis, phloem transport is mainly *downward* to tissues that need food for metabolism, and to storage sites. At other times, food needs of the plant body are met by movement of sugars *upward,* from storage sites. (2) Water and substances absorbed through the roots are transported *upward,* through the xylem. (3) Solutions also are moved *radially,* nourishing and permitting storage in body cells that are not in direct contact with the conducting cells.

The major quest of research on internal transport processes of plants has been to understand the physical and biochemical mechanisms of xylem and phloem transport. These are some of the questions that must be answered:

What are the driving forces that can pull and/or push water from roots that may be buried 20 meters or more below the soil surface to leaves that may be 20 meters or more above the soil surface?

How can forces generated in the body of a plant counteract the powerful force of gravity? If you have ever carried a bucketful of water, you know how heavy it is.

What are the driving forces of phloem transport?

What physical and biological forces enable phloem transport to occur in two directions?

Demonstrating the upward movement of a solution is easy. Observe the white carnations and celery stalks that have been standing in a dye solution and compare them to those that have been standing in plain water.

1. How do the dye-treated materials differ in appearance from the nontreated materials?
2. What, specifically, has caused this difference?
3. What structures can be observed more easily in the dye-treated materials?
4. Make an hypothesis about how the dye traveled from the container solution to the top of the plant. (Imagine the pathway taken by a single molecule of dye.)

Using a razor blade, carefully prepare a few incredibly thin cross sections of dyed celery stalk, placing them directly onto a wet microscope slide. Add a coverslip and then observe the thinnest section with a microscope.

5. Does your microscopic observation support your expectation that the solution was transported in the xylem? If so, how? If not, propose an explanation for the discrepancy between what you expected and what you actually observed.
6. Based on your observations and your hypothesis, what can you infer about the movement of water in the nontreated materials?

19.2 What physical forces are involved in xylem transport? O **motic pressure** causes water to enter root cells. As water fuses into the vascular cylinder it pushes upward along the path of le resistance, through the xylem vessels and/or tracheids. The upward pressure of water caused by osmotic absorption is called **root pressure.** Root pressue is demonstrated by the upward movement of tree sap in the spring. Maple sap to make maple syrup and candy is obtained by pounding special taps into the wood in the spring, right before the trees begin to flower. Sap flows out through these taps, driven by root pressure.

Root pressure alone, however, cannot account for the total picture of water transport in plants. The primary driving force of xylem transport during the active growing season is **transpiration,** water loss from plant surfaces, especially leaf surfaces. Remember the inside of a leaf? Water vapor in the humid air spaces escapes through the stomates and is replaced by water from cell surfaces. Each molecule of water that evaporates from the *surface* of a parenchyma cell is replaced by a water molecule from *inside* the cell. That water molecule is replaced, in turn, by a water molecule from an *adjacent* cell. The transfer continues, from cell to cell, until eventually the supply chain of water molecules can be traced back to the vein itself.

Try to imagine a whole string of water molecules "holding hands," from a parenchyma cell surrounded by air spaces, through the other cells it touches, then all along a xylem tube to the root (figure 19.2). When a molecule at the top pulls, the whole chain is moved up! This explanation is, in essence, the **transpiration-pull** theory of xylem transport. The greater the water loss by transpiration, the more rapid the xylem transport.

1. What kind of weather most favors water loss by transpiration?
2. During what kind of weather do outdoor plants need to be watered most frequently? Why?
3. What happens to plants that have inadequate soil moisture under such conditions?

The transpiration-pull theory is one in which biologists have great confidence. However, there are other physical factors besides evaporation that are integral parts of the process. You can begin to explore these other factors by answering the questions below. Your teacher may want you to try the experiments the questions suggest.

4. Do you think you could sit up in a tree (or on a ladder or inside an upper-story window) and suck water through a length of plastic tubing from a container located at ground level?

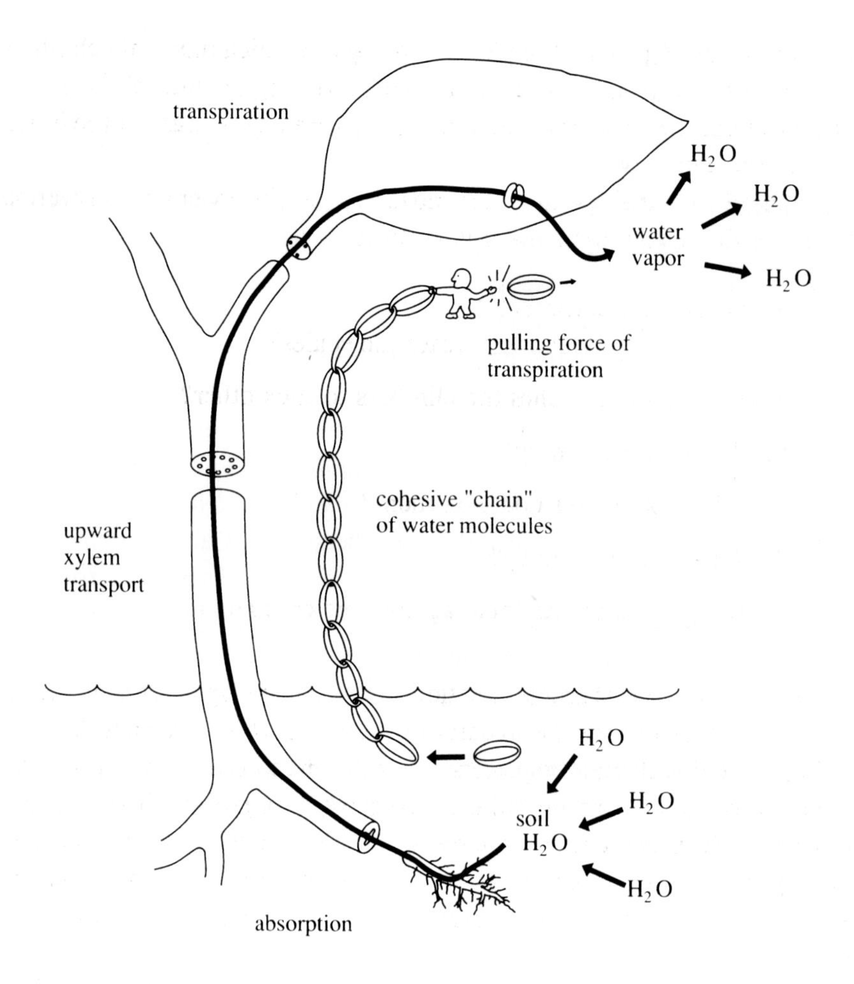

Figure 19.2 Transpiration-pull model of xylem transport. Water molecules are drawn through the xylem tracheids and/or vessels in a continuous stream, held together by cohesion. Most of the time, the driving force of the process is transpiration. As water molecules are vaporized from leaf surfaces and escape into the atmosphere, they are replaced by water molecules that are "pulled into place" behind them. The supply of water molecules is maintained by osmotic absorption of water into the roots. Even in a very tall tree, the column of water is able to resist the downward pull of gravity because of (a) cohesion of the water molecules and (b) adhesion of the water molecules to the conducting cell walls.

5. What would be your main source of resistance?
6. Do you think it would be easier to suck water through tubing having a wider diameter or thinner diameter?

We mentioned above that water molecules "hold hands"; they tend to stick together. This is obvious when a droplet of water beads on a tabletop, or on your skin, instead of spreading out uniformly across the surface. The tendency for like molecules to stick together is called **cohesion.** The cohesive attraction among water molecules is what allows them to be pulled up through the vessels in a steady stream.

The narrowness of the vessels also aids in water transport, by placing a large internal surface area in contact with the stream of water molecules. In addition to their attraction for each other, water molecules tend to stick to the walls of the vessels. The tendency for unlike molecules to stick together is called **adhesion.** The adhesive attraction between the water molecules and the vessel walls helps the chain (stream) of water molecules to resist the force of gravity. Imagine an analogy in which mountain climbers who are roped together are being pulled up through a rock chimney. As they move up, they also brace their arms, legs, and backs against the inside walls of the chimney.

Relate this analogy to water movement in the xylem by answering these questions. The first answer is given for you.

7. What are the climbers?
   (Answer: water molecules)
8. What force attaches the climbers to each other?
9. What pulls them up?
10. What force pulls down on them?
11. What is the chimney?
12. What force braces them against the chimney walls?

19.3

Earlier chapters mentioned phloem transport in context, relating it to the structure of stems, leaves, and roots. The introduction to this chapter emphasized that the direction of phloem transport varies. During active photosynthesis and growth, sugars and hormones are transported from the leaves to the rest of the plant body, a *downward* movement. In the spring, when perennial plants come out of dormancy, sugars and other substances are transported from storage tissues to the developing buds, an *upward* movement.

Demonstrating phloem transport is not as easy as demonstrating xylem transport because it involves substances that are produced by living cells, rather than substances that are absorbed from the environment and subsequently transported. However, more sophisticated techniques based on the same general methods can be used.

In the dye-uptake experiment, we used a visible dye to mark the water. To follow phloem transport, scientists have used the same principle, using a *radioactive isotope* of the carbon atom to mark (label) sugars produced in the leaves. With what you know about photosynthesis, you should be able to guess how the radioactive isotope of carbon is supplied to the leaf for incorporation into sugar molecules: as carbon dioxide (CO_2) made from the carbon isotope.

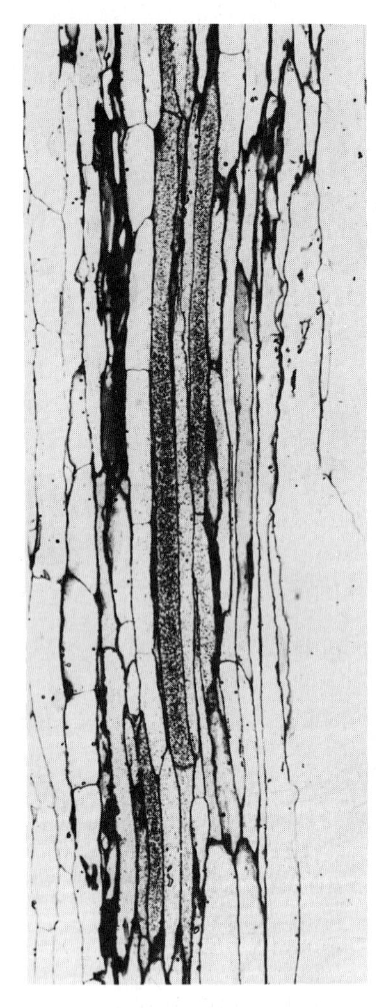

Figure 19.3 Radioactively labelled sugars in sieve tubes. Carbon dioxide containing a radioactive carbon isotope was provided to this plant, which incorporated the carbon atoms into sugars during photosynthesis. Later, the plant tissue was harvested, thinly sliced, and exposed to photographic film. The release of radioactive particles caused the labelled sugar molecules to show up as black dots in the photograph, demonstrating that sugars are, indeed, transported in phloem tissue.

Unlike dye, it is not possible to see radioactive sugar molecules directly. However, it is possible to detect them photographically, because the radiation they emit causes a change in photographic film. Figure 19.3 is a photograph made from film that has been exposed to shoots containing labeled sugar molecules, which show up as tiny dots of radioactivity.

1. What are several reasons this method of demonstrating phloem transport has been explained to you in writing and photographs rather than being given to you to do as an experimental activity?

Another ingenious method of analyzing the composition of phloem cell contents is by use of small insects called aphids. Aphids feed by piercing plant tissues with their sharp, sucking mouthparts, called stylets. Some species probe around intercellularly, sucking out fluids from parenchyma tissues. Other species penetrate individual cells and withdraw the contents of the cell vacuoles. Still other species are specific phloem feeders, working their stylets into the sieve tubes and withdrawing the fluid that is being transported: they feed directly from the sugar pipeline!

By instantly freezing phloem-feeding aphids before they could withdraw their stylets, then carefully cutting the aphid bodies away from their stylets, botanists were able to withdraw sieve tube contents through the hypodermic channels established by the aphids (figure 19.4). Analysis of phloem contents at various sites permits comparisons and tracking of substances. This is a good example of the creative thinking that is a part of scientific research.

Currently, the **pressure flow hypothesis** of phloem transport seems to be our most acceptable explanation. The major forces involved are *diffusion, osmosis,* and *active transport.*

Refer to figure 19.5 as a model for understanding the pressure flow hypothesis, matching parts of the figure to the stages described below. You may assume that phloem transport is downward from a leaf in this explanation.

Sugars are produced in chlorenchyma cells. Some sugars are transformed immediately into starch grains within the chloroplasts as a temporary means of storage while awaiting transport from the cells. Conversion of excess sugars to starch helps to maintain a stable osmotic concentration in the cytoplasm. Some sugars diffuse into the intercellular fluid and from cell to cell following a diffusion gradient.

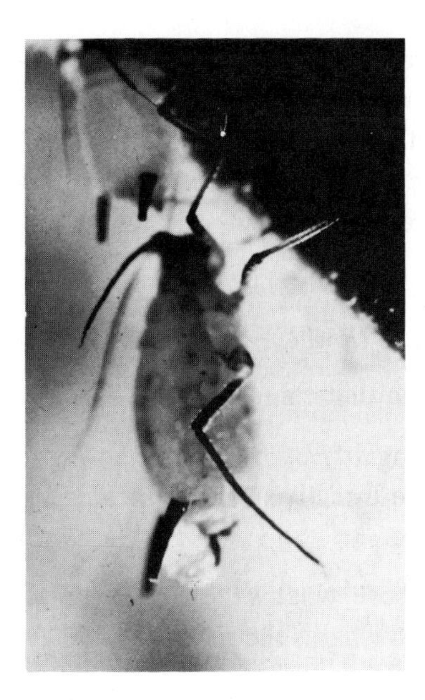

Figure 19.4 Aphids as research collaborators. Aphids feed by piercing plant tissues with their hypodermic-like mouthparts. Excess sugar solution is released as "honeydew" droplets, shown here. Some aphid species feed directly within the sieve tubes, and their stylets thus form a delicate and precise instrument for collecting sieve tube contents for scientific analysis. (See text.)

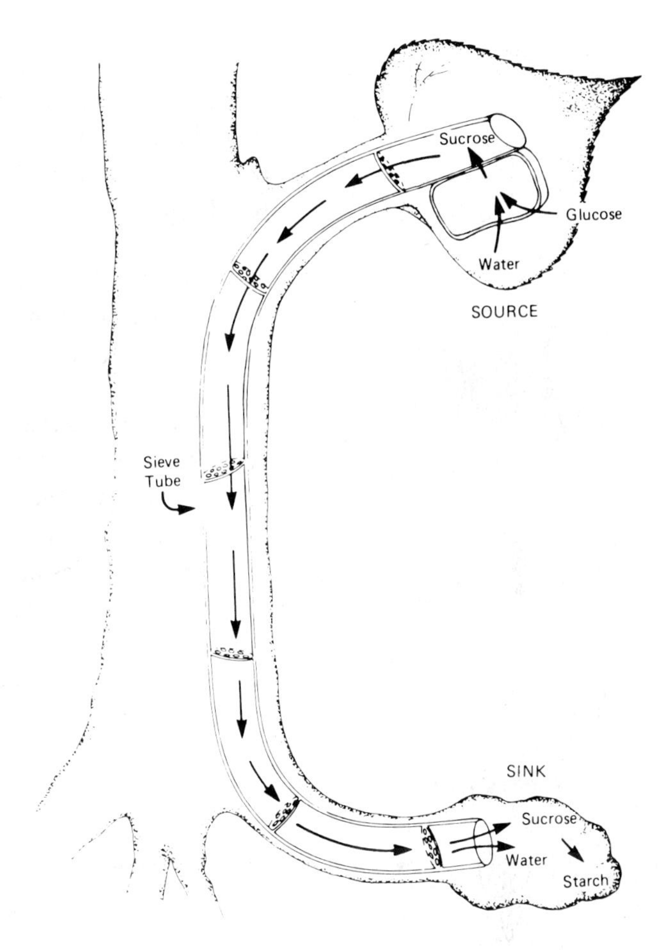

Figure 19.5 The pressure flow model of phloem transport. Diffusable sugars, such as glucose, are *loaded* into sieve tubes by active transport, then are converted to sucrose, which does not diffuse back into the surrounding cells. The high sugar concentration causes water to diffuse into the sieve cells from surrounding cells all along the way, building up pressure in the pipeline, and so pushing its contents along. At the same time, sugars are being *unloaded* at sites of need or storage, creating an osmotic tendency for water also to leave the sieve tubes. Thus, it is thought, a mass flow of sieve tube contents toward the unloading site (sink) occurs. This drawing shows leaf-to-root transport, but transport also can occur in the other direction.

Sugars from the chlorenchyma cells and intercellular fluid are loaded into the sieve tubes of the veins. Since the sugar concentration in the sieve tube is already higher than in the leaf cells, the process of *phloem loading* is carried out by *active transport,* against a diffusion gradient. The parenchyma cells that form a sheath around the veins are especially adapted for active transport of substances.

The higher concentration of sugars in the sieve tubes creates an osmotic gradient. As a result, many water molecules enter the sieve tubes from surrounding cells. The resulting influx of water creates a surge of movement, called mass flow, down the pipeline.

Farther down the sieve tube, sugar molecules are *unloaded* into storage tissues, where they are converted to starch or sugar molecules that are too large to diffuse through the cell membranes (such as sucrose). A combination of diffusion and active transport is involved.

The constant removal of sugars from the phloem stream helps to maintain diffusion gradients that favor continued downward flow of the sieve tube contents. In tissue regions where the sieve tube contents become more dilute than the surrounding tissues, water molecules respond to the osmotic gradient. They reenter the xylem and are recycled—they catch the "up" elevator!

2. What is meant by the term *pressure flow?*
3. Name 2 physical forces/phenomena that contribute positively to phloem transport.
4. Name 3 cellular processes that are directly involved in phloem transport, and explain their roles.
5. In phloem transport, as described above, what is the carbohydrate *source* (where they are formed), and what is the *sink* (where they are deposited)?

19.4

In chapter 19 you have learned the physical and biological bases of fluid transport throughout the plant body. Review the following study points and check off those you have mastered. If you cannot check off a point, return to the chapter contents and study it again before continuing to the next chapter.

On completion of this chapter, you should be able to do the following:

☐ 1. State the specific functions of xylem and phloem transport.

☐ 2. Understand the role of cell-to-cell transfers in the distribution of substances in a plant body.

☐ 3. Explain the transpiration-pull theory of xylem transport, including identification of the physical forces involved.

☐ 4. Explain the pressure flow hypothesis of phloem transport, including identification of the physical forces and biological processes involved.

Plant Reproduction 20

How do plants reproduce?

20.1 Each individual organism is the "latest model" of a species line that goes back thousands, even millions, of years. The inherited species legacy of the many years is present as DNA molecules found in every cell. Therefore, reproduction is a climactic event in each generation, ensuring survival of the species.

One characteristic of organisms is that they are able to **reproduce.** More specifically, *members of each species produce new members of the same species by passing on parental DNA to individuals of the next generation.* As you know from earlier chapters, specific DNA sequences are the genes that control all aspects of an organism's life, including its biochemical composition, physical structure, environmental tolerances, and behavior. Each species is distinctive because of the way its DNA operates. You are a member of the species *Homo sapiens.* As the most recent representative of a long line of ancestors, what is the role of an individual organism in the survival of the species?

First, consider what is involved in *one* lifetime. A typical individual lifetime progresses through several stages: *inception, embryonic* state, *juvenile* growth period, period of vigorous *reproductive maturity, senescence* (period of degenerative metabolic changes and slow growth rate), and, finally, *death.* The early growth phase is the most rapid (figure 20.1). In species terms, the most significant phase of the lifetime is the reproductive phase.

You are familiar with the vertebrate reproductive pattern, which is sexual. In **sexual reproduction** of animals, a female parent produces a haploid egg cell, which combines with a haploid sperm cell produced by a male parent to produce a zygote. The zygote is the first diploid cell of a new individual's body. Like vertebrates, plants carry out sexual reproduction. *Unlike* vertebrates, however, the life cycle of a plant species has two **alternating generations.** Furthermore, plants also carry out **asexual reproduction,** giving rise to new individuals without syngamy (fusion of gametes).

In this chapter, we will look briefly at how plants carry out the transfer of inherited information through reproduction.

20.2 Plant life cycle generations are represented by two kinds of organisms: a **sporophyte** and a **gametophyte.** In vascular plants, such as ferns, gymnosperms, and angiosperms, the plant form you recognize is the sporophyte; therefore, the sporophyte is referred to as the *dominant phase* of the life cycle of vascular plants. The gametophytes of vascular plants are small and inconspicuous. In fact, the gametophytes of conifers and flowering plants are not even free-living individuals. They grow inside

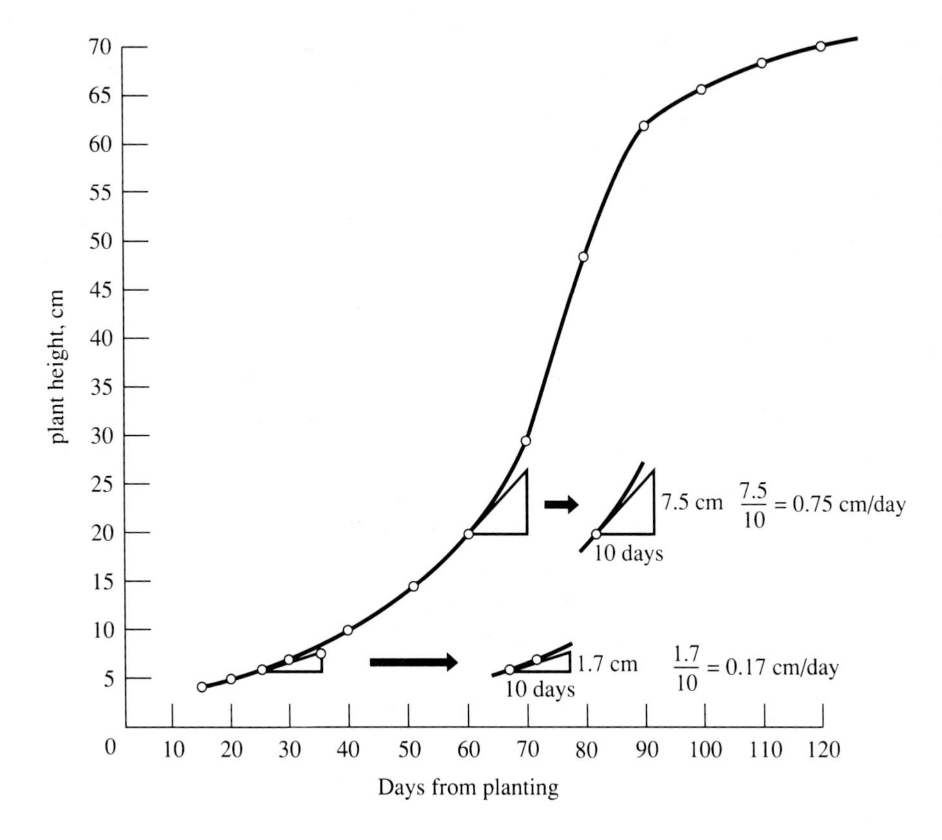

Figure 20.1 A plant growth curve, from seedling to maturity. Early growth of a seed plant is the most rapid, as the plant develops a body that is capable of drawing needed resources from the environment. As mature vegetative growth is reached, resources are diverted to food storage, to the production of reproductive structures, such as cones or flowers, and, eventually, to seed development.

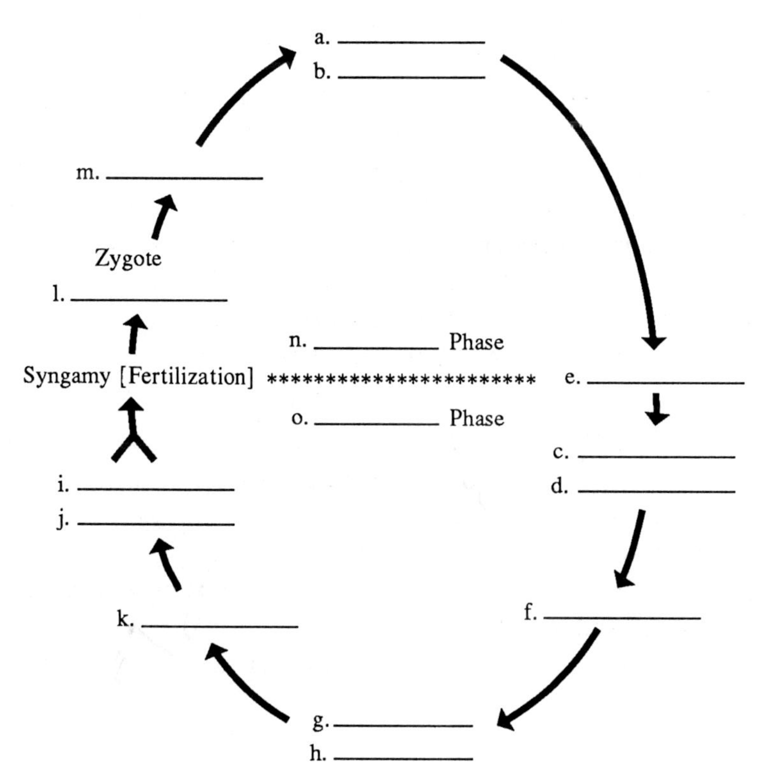

Figure 20.2 Generalized plant life cycle. A larger version of this diagram is found in the student logbook. Follow the instructions in the text to complete the diagram in your logbook.

special structures that are part of the sporophyte. In mosses and liverworts, the gametophyte is the *dominant phase* of the life cycle, and the small sporophyte grows attached to the body of the gametophyte.

What, exactly, is a sporophyte and what is a gametophyte? First, *sporophyte* means "spore plant," because sporophytes produce nonsexual reproductive cells called **spores.** *Gametophyte* means "gamete plant," because gametophytes produce sexual reproductive cells, the **gametes.** Second, the definitions of these phases also is based on how many sets of chromosomes are present. The cells of a "typical" sporophyte are **diploid,** having two sets of chromosomes. Gametophyte cells have *half* the chromosomal complement of the sporophyte; therefore, a "typical" gametophyte is **haploid.** Both spores and gametes in such a "typical" life cycle are haploid. (If you do not remember clearly what is meant by *chromosome sets, diploid,* and *haploid,* refer back to chapter 8 to refresh your understanding.)

How do the separate sporophyte and gametophyte generations fit into the life cycle of a plant, and what is the significance of these alternating generations? To answer these questions, you need to understand the life cycle story. Examine figure 20.2, a diagram of a generalized plant life cycle. Use the above information, and your understanding of meiosis, mitosis, and syngamy to construct this story by answering questions and completing the labels for figure 20.2 in your logbook.

We will use the sporophyte as the starting point in our plant life cycle story.

1. Write *sporophyte* in blank *a* in the diagram.
2. Write $2n$ in blank *b*.
3. What kind of reproductive cell (spore or gamete) is produced by the sporophyte? Write your answer in blank *c*.
4. Is the blank *c* cell n or $2n$? Write your answer in blank *d*.
5. What kind of nuclear division (meiosis or mitosis) occurs in the sporophyte body to produce the blank *c* cell? Write your answer in blank *e*.

A spore is a single cell, yet it gives rise to a multicellular plant body, the gametophyte.

6. What kind of nuclear division is responsible for this growth? Write your answer in blank *f*, and write *gametophyte* in blank *g*.
7. Is the gametophyte haploid or diploid? Write n or $2n$ in blank *h*.

8. What kind of reproductive cell (spore or gamete) is produced by the gametophyte? Write your answer in blank *i*.
9. Is it haploid or diploid? Write your answer in blank *j*.
10. What kind of nuclear division occurs in the gametophyte body to produce the blank *i* cell? Write your answer in blank *k*.
11. How many of these special reproductive cells are necessary to produce a zygote?
12. Is this kind of reproduction sexual or asexual?
13. Is the zygote n or $2n$? Write your answer in blank *l*.

The **zygote** is the first cell of the sporophyte generation. The zygote contains the combined DNA of its parent cells, which it duplicates and then passes on to two new cells when it divides. They, in turn, faithfully duplicate the DNA and pass it on when they divide. Eventually, a multicellular body is formed with each cell containing all of the species information. In plants, the earliest sporophyte body is the **embryo,** an immature stage of sporophyte development. In many plants the embryo enters a period of dormancy, awaiting the stimulus of appropriate environmental conditions in which to continue its growth and maturation.

14. What kind of nuclear division is responsible for the multicellularity of a plant body? Write your answer in blank *m*.
15. Finally, write in blanks *n* and *o* which half of the life cycle is the *diploid phase* and which half is the *haploid phase*.

In our life cycle story, we are back where we started, at the sporophyte. This *pattern* is the same for all plants, although the *details* may differ. Reflect for a moment on how this pattern compares to an animal life cycle, as represented by vertebrates such as humans.

16. Does the human life cycle include alternating, different persons, one that produces spores followed by one that produces gametes?
17. Which generation is absent from the human life cycle?
18. What kind of reproductive cell is absent?
19. Does your life cycle include both diploid and haploid phases? Explain.
20. What *process* is responsible for *halving* the number of chromosome sets in a life cycle?
21. What *process* is responsible for *restoring* the paired chromosomal condition in a life cycle?

You now know how the sporophyte and gametophyte generations fit into a plant life cycle, but you still need to grasp the *significance* of these alternating generations. The plant life cycle is not just different in an arbitrary way from the vertebrate life cycle. It is *fundamentally* different, a different evolutionary response to the problem of reproduction. Assuming, therefore, that natural selection resulted in the alternating-generation type of life cycle in plants, can you think of its advantages? Put yourself in the place of a plant. Consider some of the special features and problems in the life of an immobile organism.

Except for the seed plants (gymnosperms and angiosperms), plants release spores directly into the environment, where they germinate and grow into free-living gametophytes. Spores are an *excellent* means of reproduction. They are tiny, lightweight, and easily dispersed by wind or water. Therefore, spores are an important means of species distribution. Also, such plants typically produce thousands, even millions, of spores in a season. Many spores do not end up in hospitable environments for gametophyte growth; however, their sheer numbers vastly increase the potential for successful continuation of the life cycle. A further advantage of spores is that they do not require combination with any other cell in order to give rise to a new individual, which makes reproduction much simpler just in logistical terms alone! Also, a spore is protected by a desiccation-resistant case, enabling the spore to survive for years, if necessary, until it is stimulated to germinate by exposure to favorable environmental conditions.

Why does *any* life cycle have both a diploid phase and a haploid phase? How do these phases relate to natural selection? Meiosis that leads to spore production distributes chromosomes of only one parent into the spores. The generation produced thus is a continuation only of genes present in the parent plant. Syngamy, however, combines chromosomes (and therefore genes) from two separate cells, possibly from two different parents.

The unique genetic composition of a zygote thus produced then must survive the tests of the environment through maturation and survival of the sporophyte before its genetic combinations can be passed on in the next generation of spores. Like many spores and young gametophytes, many zygotes and young sporophytes die, some because of what we might think of as plain bad luck, rather than some fundamental genetic inferiority. For instance, they may end up on a rock, be killed by an animal or fungus, or shrivel up during a dry spell. Even these misfortunes, however, are part of natural selection, which is random in nature.

20.3 Mosses (figure 20.3) and their relatives, the liverworts, are classified as *bryophytes* (moss plants). They are the major plant group in which the gametophyte is the dominant generation. Examine the moss plants your instructor has provided as you read through this activity. Even large moss plants are small in comparison with plants having a dominant sporophyte stage.

Bryophytes tend to grow as mats on wet rocks, soil, or tree trunks. They require moist environments for at least part of the year, though most species can survive dry periods by entering dormancy. Why do bryophytes require moist environments? In the first place, they do not have vascular tissues; therefore, they do not have *true* roots, stems, and leaves with efficient water absorption and transport. In the second place, water is needed as a medium for sexual reproduction. The male sex organs of bryophytes release sperm cells directly into the environment. This release is effective only when water is present because the sperm cells must swim into vase-shaped female organs, each of which usually contains a single egg cell. Rainy weather is ideal for this activity, creating an aquatic film within the plant mat. Raindrops even help, splattering sperm-bearing droplets over plant surfaces.

Figure 20.3 Moss plants. This photograph shows both the gametophyte and sporophyte stages of the life cycle. The low-growing green plant is the gametophyte, which produces gametes. The erect structures, tipped by spore cases (sporangia), are sporophytes. Each sporophyte is a mature, spore-producing plant that began life as a zygote that was formed when an egg cell and a sperm cell united. In turn, each spore that is produced by the sporophyte is capable of growing into a new gametophyte.

1. Based on what you know about plant life cycles, what process occurs after a sperm cell contacts an egg cell?
2. What is the name of the cell that results from this process?

Figure 20.4 shows some stages in a moss life cycle. As you can see, the sporophyte begins its growth while it is still in the female structure. As it continues to grow, it sends a small finger of tissue down into the gametophytic tissue, thereby becoming a "friendly" parasite on its own parent, from which it derives water and nourishment throughout its entire existence. This is an efficient arrangement, requiring a minimum of growth and survival effort on the part of the sporophyte, and it is quickly able to mature and produce spores. The spores develop in an expanded **sporangium,** or spore case, at the tip of the sporophyte.

3. Examine your moss specimen closely. Identify the gametophyte and any sporophytes that may be present. Carefully remove an individual gametophyte "branch" with the sporophyte attached and sketch it in the space below. Label: (*a*) *gametophyte,* (*b*) *sporophyte,* and (*c*) *spore case.*

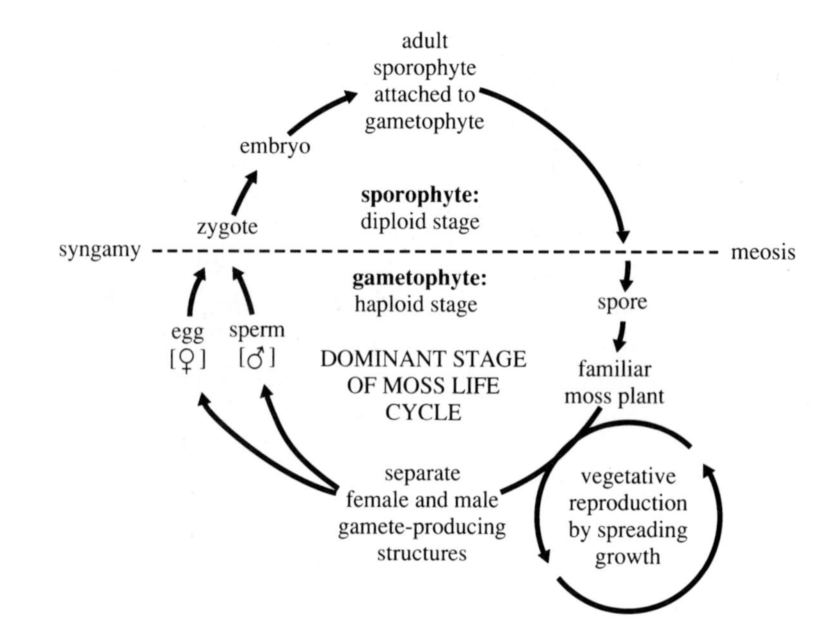

Figure 20.4 Moss life cycle. Mosses, liverworts, and their relatives are classified as bryophytes. They have a gametophyte-dominant life cycle, meaning that the gametophyte is the conspicuous form, the form that is most noticeable, most independent, and that persists from year to year. This life cycle demonstrates the same basic features as the generalized plant cycle you worked on in part 20.2. The diagram also shows that mosses can produce new plants by vegetative reproduction, forming clones (numbers of asexually reproduced, genetically identical individuals).

20.4

Life cycles of seed plants are much more familiar to you than the life cycles of bryophytes and of nonseed-forming vascular plants (ferns, horsetails, and club mosses). Flowering plants, in particular, are of interest to humans because we appreciate the beauty of flowers and rely on the fruits and seeds they produce for much of our own food. Many plant species die the same season they first produce seeds, an abbreviated lifetime indeed compared to trees and shrubs that continue to flower and fruit even after they have become senescent.

What are flowers, and why do so many of them look and smell so lovely? Obtain a single gladiolus blossom. Cut the flower away from the flowering stalk very carefully and do not leave any part of the flower attached to the stalk. Compare your flower to figure 20.5 to identify its several components, which are arranged in whorls around a short stem. The showiest floral structures are the **sepals** and **petals,** both of which are modified leaves. Cut away one sepal and one petal (at their bases) so you can see inside the flower.

1. Are the gladiolus sepals and petals similar or different in appearance?
2. How can you tell them apart?

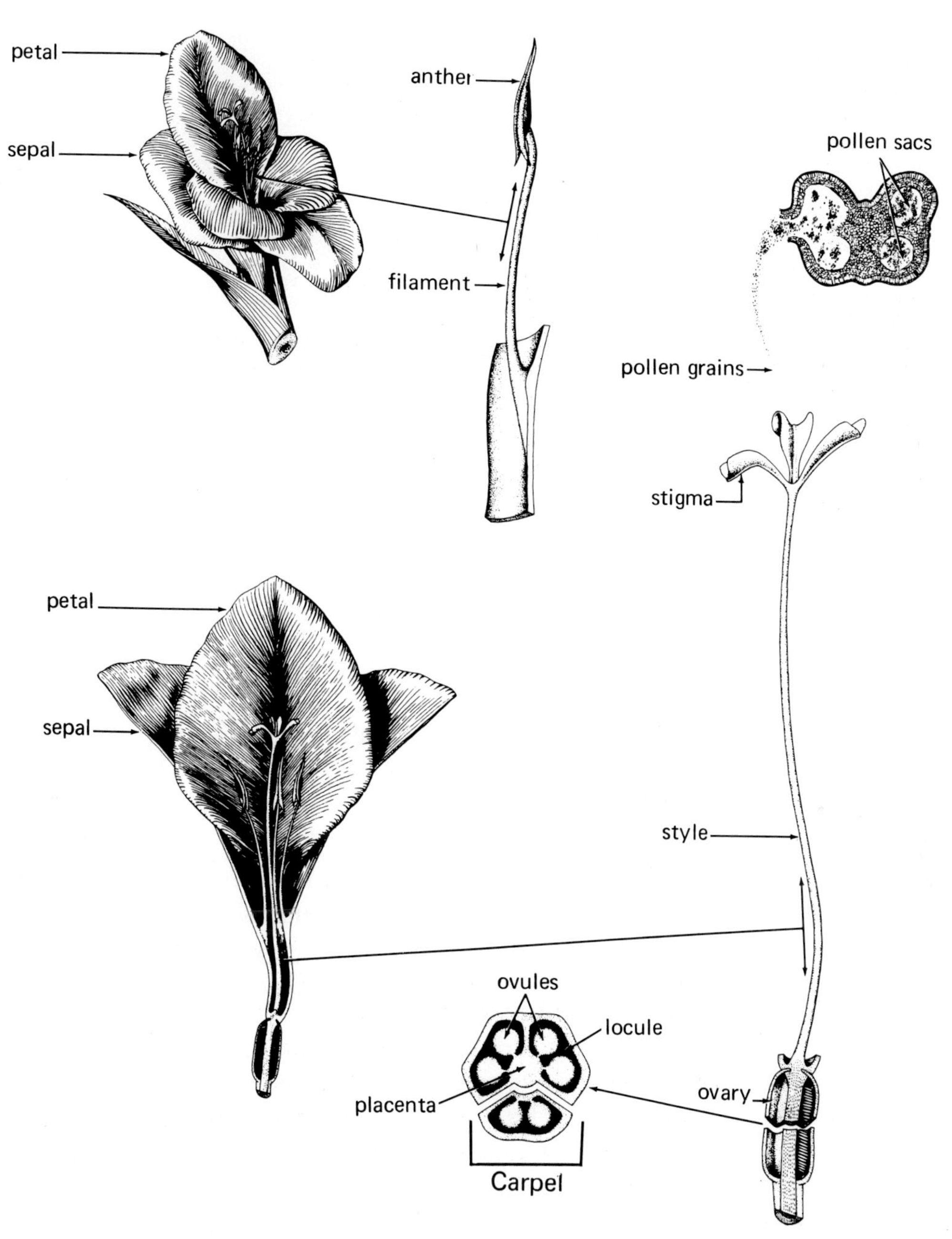

© Kendall/Hunt Publishing Company

Figure 20.5 The gladiolus flower. Floral parts are arranged in concentric rings, or whorls. From outermost to innermost, they are the sepals, petals, stamens, and pistil(s). The stamens and pistils are reproductive structures, whereas the sepals and petals are accessory structures that are often specialized to attract pollinators.

Figure 20.6 Ragweed pollen. Pollen grains are the male gametophytes of flowering plants. Each pollen grain contains a small amount of cytoplasm, a vegetative (growth-directing) nucleus, and a sperm-producing nucleus. The pollen grain is surrounded by a protective, desiccation-resistant coat, the outer surface of which is often sculptured or sticky. Sticky pollen, such as that of goldenrod, is transferred on the leg hairs of insects. Ragweed pollen is dry and spiky, and is spread by the wind. How might its external structure be related to this means of dispersal?

3. Judging from the vein arrangement in the sepals and petals, would you deduce that gladioli are monocots or dicots?
4. Does your flower have a fragrance?
5. Does it have any **nectaries?** These can be identified by the presence of a sugary exudate called **nectar.**

Several stalklike structures are attached near the bases of the petals. These are the male reproductive structures, the **stamens.** Each stamen has an expanded tip, the **anther,** containing sacs within which **pollen** (figure 20.6) is produced. A pollen grain is the *male gametophyte* of a seed plant, and it contains two **sperm nuclei.**

6. How many stamens are present?

If the anthers in your flower are mature, pollen grains may already have been released. Examine the anthers carefully, using magnification, if available.

7. Is pollen present?
8. What color is it?
9. Is it dry and powdery, or does it tend to stick to your fingers?
10. How does/did the pollen get out of the anthers?

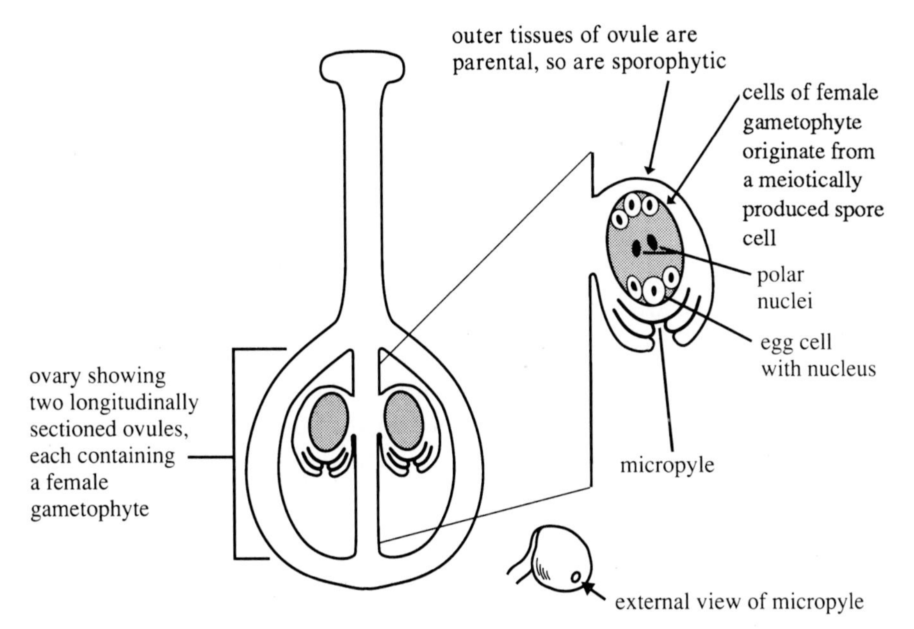

Figure 20.7 Ovary and ovules. An ovary contains one or more chambers (locules) in which the ovules develop. Each ovule is made of maternal, sporophytic tissues and an enclosed, dependent gametophyte. The outer ovule layers are penetrated by a small shaft, the micropyle ("little hole"), through which the pollen tube of the male gametophyte enters the ovule. The gametophyte of many flowering plants consists of only seven cells, the largest of which has two nuclei.

Now examine the structure in the center of the flower. It is the **pistil,** the female floral structure. The pistil of this flower is formed by the fusion of separate **carpels** during its development. You can see evidence of this fusion by examining the pistil for external seams, the number of lobes at its tip, and the number of chambers inside the expanded **ovary** at its base.

Carefully remove the entire pistil from the flower, examine it, and then slice the ovary crosswise to reveal its chambers.

11. How many chambers are present?
12. From this information, what can you infer about the number of carpels represented?
13. What do you hypothesize the small white objects attached to the walls of the ovary to be?

If you said "egg," you were close; they are **ovules.** An ovule is a multicellular structure in which the *female gametophyte* is developed. Figure 20.7 shows the multicellular female gametophyte encased in sporophytic tissue.

14. Why do you think this is referred to as a "captive" gametophyte?

Now return your attention to the external characteristics of the pistil. The tip of the pistil is called the **stigma,** and it is modified to receive pollen grains during **pollination,** the transfer of pollen from anthers to stigma. Stigmas often are sticky, bumpy, or covered with epidermal hairs to help trap pollen grains. The stigma is attached to the ovary by a stalk, called the **style.**

20.5 Your instructor may wish for you to repeat your observations on other flowers that are provided. If so, use separate sheets of notebook paper to answer the same questions about other flowers.

20.6 Many flowering plants are pollinated by wind. Humans who suffer pollen allergies are affected mainly by these wind-borne pollens, which tend to be light and nonsticky. The stigmas of wind-pollinated flowers often are covered in epidermal hairs.

On the other hand, many flowering plants expend a great deal of growth energy to encourage pollination by animals, especially insects. If you were a bee, what would cause you to be attracted to a flower? First, the flower would have to be a source of something you need for survival, and second, the flower would have to attract you to it. Based on what you already know about flowers, answer the questions below.

1. What (specifically) do flowers have to offer animal pollinators that is beneficial to the animals?
2. In what ways do flowers advertise themselves to animal pollinators?

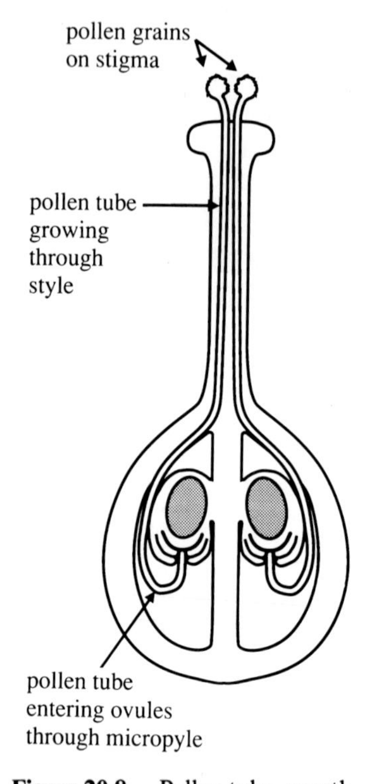

Figure 20.8 Pollen tube growth. After pollen grains land on the stigma of a receptive flower, they begin to germinate. The growth of a hollow pollen tube is directed by the tube nucleus. Eventually, the end of the pollen tube enters the micropyle. It is then ready to grow through the last layer of sporophytic tissue to enter the female gametophyte.

What happens *after* pollination? More to the point, how does syngamy occur? Consider the physical conditions involved. The sperm nuclei cannot swim to the egg nuclei, because they are inside a desiccation-resistant pollen coat and the egg nucleus is embedded within tissue. Also, the style is between the pollen grain and ovule.

The answer is that the pollen grain *germinates,* sending a slender **pollen tube** growing through the style tissue toward the ovary. Growth of the pollen tube is directed by a special nucleus called the **tube nucleus.** When the pollen tube reaches the ovary, it curves toward an ovule, entering the ovule by a small opening called the **micropyle** (little pore) (figure 20.8). The *sperm nuclei* are released after the end of the pollen tube disintegrates (figure 20.9). One sperm nucleus fuses with the egg nucleus, forming a zygote. The other sperm nucleus fuses with a gametophytic nucleus, forming a cell that will give rise to the **endosperm** tissue to nourish the embryo as it develops and subsequently germinates.

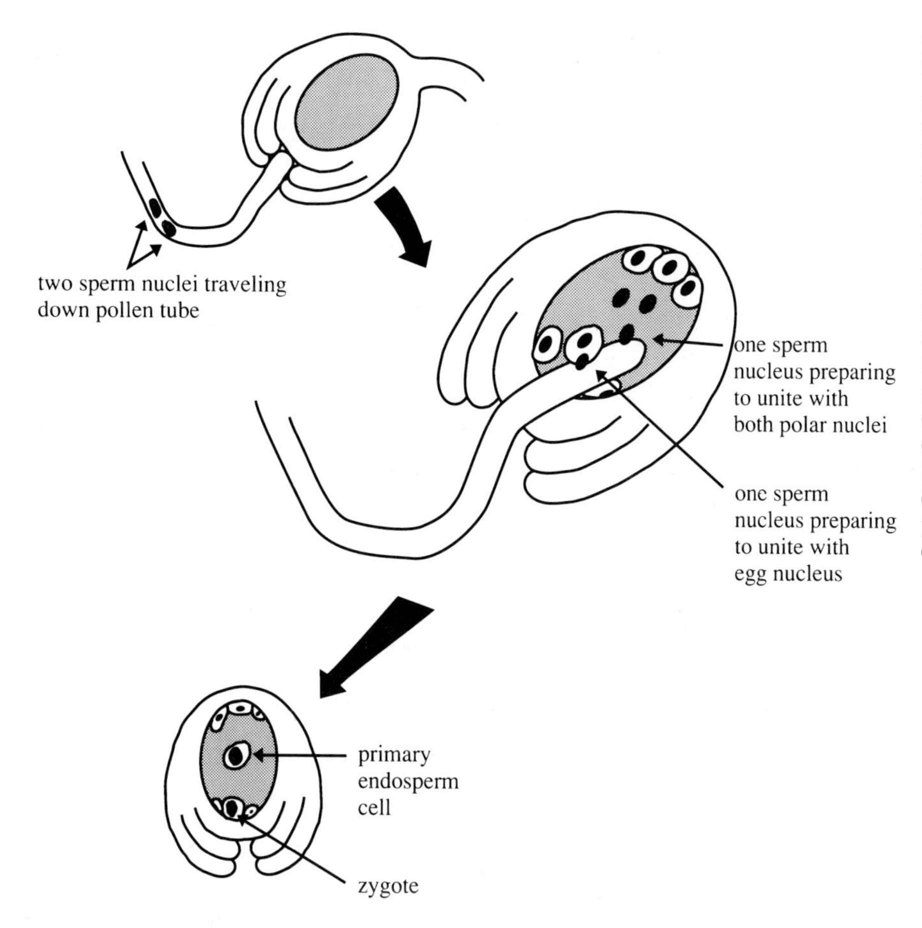

Figure 20.9 Double fertilization in flowering plants. As the pollen tube digests its way through the style, two sperm nuclei make their way down the tube. After entering the female gametophyte, the end of the tube weakens and permits the sperm nuclei to get out. One sperm nucleus fuses with the egg nucleus, forming a zygote, which is the beginning of the new sporophyte. The other sperm nucleus unites with both polar nuclei, forming a primary endosperm cell. The primary endosperm cell is thus triploid ($3n$). It gives rise to the endosperm, a nutritive tissue that provides food for the developing embryonic sporophyte.

This utilization of both sperm nuclei is a special feature of flowering plant reproduction, referred to as **double fertilization.** In conifers, by contrast, a sperm nucleus unites with an egg nucleus, but the endosperm is formed by female gametophyte tissue only.

3. How is seed plant reproduction superior to the method seen in bryophytes as an adaptation to terrestrial environments in which free water is not available for gamete transfer?
4. What functional parallels can you see between seed plant reproduction and terrestrial vertebrate reproduction in regard to internal syngamy and embryo development? What do you think is responsible for such parallelism?

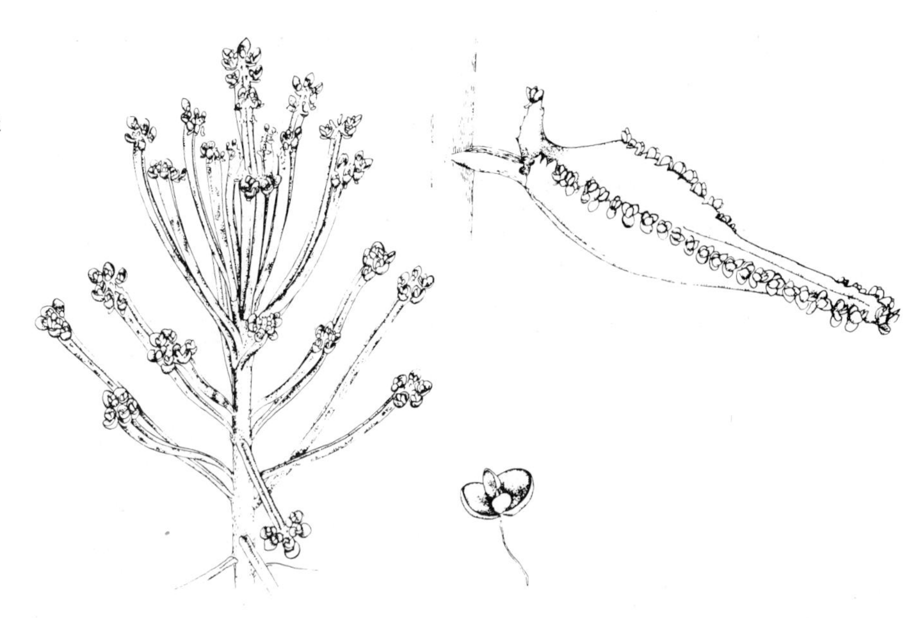

Figure 20.10 Vegetative reproduction of plantlets. *Kalanchöe* and *Bryophyllum* reproduce by seed production, but also by the production of hundreds of tiny plantlets along their leaf margins. Each plantlet is potentially able to grow into a new plant. All of the plantlets produced by a single parent plant make up a clone.

20.7

Earlier chapters mentioned that **adventitious roots** can form on stems or leaves, a basis for asexual (vegetative) reproduction. In the next few activities we will focus on the question, "How do plants reproduce asexually?" For the first activity you will need a small plant container filled to within 4 cm of the rim with potting soil.

Vegetative reproduction is the production of a new plant body from a part of the old one. Many plant species have *natural* means of vegetative reproduction, producing *miniature plantlets, bulbs, offshoots,* or other such structures that give rise to new shoots (figures 20.10 and 20.11). Examine a *Kalanchöe* or *Bryophyllum* plant, sometimes called airplane plant or mother-of-millions.

1. What asexual reproductive structures are present?
2. From what plant organ have they arisen?
3. Carefully remove four small plantlets and return to your station. Examine them carefully, and sketch one in the space provided below. Label the organs you can identify.
4. Is each structure a *complete* (all vegetative parts present) plant body in miniature? Explain.

Now moisten the soil in your potting container and place the plantlets on the soil, pressing them in only *slightly*.

5. What do you think will happen to each of these plantlets over the next 2 weeks (provided that you keep the soil moist)?
6. Does each plantlet have the potential to become a full-sized reproductive adult like its parent?
7. What is/are the advantage(s) to the plant to be able to reproduce in this way?

Figure 20.11 Vegetative reproduction of bulblets. Some monocots produce small bulbs from their bases or on their leaves. Note the presence of some young bulblets on the left and one that has established its own root system on the right. Is this group of plants a clone?

If a jade plant is present, carefully remove an *older* leaf, one that is located closer to the base of a stem rather than its tip. The older leaf will break away from the stem more easily without damage to the stem. Lay the leaf on the soil surface along with the plantlets from the previous activity, then loosely tent the top of the container with a plastic sandwich bag. After several weeks, a tiny new plant will form at the end of the leaf, where it formerly was attached to the stem.

8. How does this sort of leaf propagation differ from the production of a miniature plantlet?

Kalanchöe, Bryophyllum, and jade plant all are examples of *succulent plants,* plants that have special water-storage tissues. Their leaves are plump and juicy, enabling them to draw moisture from their own tissues during dry periods.

9. In what kind of climate do you think succulent plants naturally grow?

20.8 For this activity you will need a sharp knife or pair of scissors, a Zip-Loc® plastic bag, two paper towels, and assorted plants provided by your instructor. First, fold the paper towels to lie flat along one side of the plastic bag; then moisten the towels, pouring off any excess water.

Cut off an approximately 10-cm-long stem tip of Swedish ivy, and return to your work station to examine it and apply some of what you know about shoot structure.

1. How would you describe the shape of the stem in cross section?
2. What kind of leaf arrangement is present?
3. Does handling the stem impart a fragrance to your fingers? How would you describe it?

All of the above characteristics are typical of members of the mint family, not the ivy family.

4. Based on your own observations, therefore, what might you conclude about the taxonomic classification of Swedish ivy?

Examine the leaf undersurface, which is richly populated with orange, jewel-like glandular hairs, and then gently rub a finger across it.

5. What happens to your finger where it rubbed the leaf?
6. What is the source of the pigmented fluid?

Next, look along the stem surface for small protrusions. Beneath each tiny bump is an immature **adventitious root.** Most plants are capable of asexual reproduction by the formation of adventitious roots from stem tissues. Even woody twigs and branches can grow adventitious roots under favorable conditions. The ability to grow new plants from *cuttings* removed from the body of a parent plant is the basis for much of the horticultural industry. The majority of popular house plants are propagated in this way.

What are the advantages of vegetative propagation? In the first place, it is a means of obtaining new individuals that are genetically identical to the parent. In nature, a plant that is especially well adapted to where it is growing can give rise to a **clone,** a whole population of identically well-adapted offspring produced by natural vegetative reproduction. From a horticultural standpoint, it means that when a particularly beautiful, vigorous, or disease-resistant plant is developed, it can be perpetuated indefinitely. A second advantage is that cuttings usually are easier and faster means of propagation than growing seeds. You are going to demonstrate that now.

Carefully trim off all leaves on your Swedish ivy cutting except those at the last 2 to 4 cm of the shoot. Then place your cutting on the moistened paper towel in the plastic bag.

The kind of cutting you have just made is called a *stem tip cutting,* because it includes the stem apex. Another kind of stem cutting is the *stem segment cutting,* which is a segment of stem containing one or more nodes. Return to the Swedish ivy plant and remove a 10-cm segment of stem from which the tip *already has been removed.* No leaves need be present on the stem, just nodes, which will give rise to new shoots. Place your stem segment cutting in the plastic bag.

Coleus is a popular houseplant noted both for its vigorous growth and its brightly colored leaves. Prepare a stem tip cutting of one or more kinds of *Coleus.* Be sure there are no large leaves left on the cuttings before you place them in the bag, as they will take up space, use energy, and be extremely susceptible to fungal growth.

Finally, seal the bag and tape or pin it to a cabinet, wall, bulletin board, or other place designated by your teacher. Do *not* place it in direct sunlight or near a heater.

7. Why not?

Over the next several days, observe the growth of the adventitious roots, roots hairs and all.

8. Date and record your observations in your logbook and turn them in to your teacher in 7 to 8 days. Then you may take your cuttings home to plant them.

Of course you also can start the cuttings directly in soil; however, starting them in plastic enables you to observe the roots as they grow.

9. Can you think of two reasons to provide a sealed, moisture-retaining rooting environment?
10. What food sources can the tissues of the cutting rely on before new shoot and root systems are established?

20.9 In this chapter you have been introduced to the general pattern of plant life cycles, in which there are two distinct generations, the gametophyte generation and the sporophyte generation. Review the study points itemized below and check off those you feel you have mastered. If necessary, return to the text to review points you do not understand.

On completing this chapter, you should be able to do the following:

☐ 1. Sketch a characteristic growth curve, identifying the several stages of a lifetime.

☐ 2. State clearly the differences between a sporophyte and a gametophyte in terms of function, chromosomal condition, and the processes of meiosis and syngamy.

☐ 3. Understand the importance of spores in a plant life cycle.

☐ 4. Name a major plant group in which the gametophyte is the dominant generation.

☐ 5. Know that the plants most familiar to you are seed plants, which have a dominant sporophyte generation and a captive gametophyte generation.

☐ 6. Understand the relationships between pollen grains and male gametes, and between ovules and female gametes.

☐ 7. Explain double fertilization as it occurs in the flowering plants.

☐ 8. Recognize that the pattern of pollen transfer, followed by internal syngamy and embryo development, is an effective adaptation to reproduction on dry land.

☐ 9. Explain some environmental and genetic implications of vegetative reproduction.

☐ 10. Grow plants from cuttings and from miniature plantlets with an understanding of environmental needs that must be met to assure success.

The Nonseed-Forming Plants and the Fungi

21

What are the non-seed plants and fungi?

Figure 21.1 Fossilized horsetails. Horsetails and other primitive kinds of vascular plants were abundant in ancient swamps and forests, and many were preserved as fossils. We know that these groups appeared earlier than the seed plants because their fossils are found in older rock formations than are the seed plants. The approximate age of fossils can be verified by several kinds of scientific dating techniques.

21.1

By now you have been well-introduced to the seed plants, with passing references to other plant groups, most recently the bryophytes. Bryophytes lack vascular tissues, but the other plant groups have them. The vascular plants are the club mosses, horsetails and scouring rushes, whisk ferns, ferns, and seed plants. Although they are presently overshadowed in number of species and complexity by the seed plants, the other vascular plant groups have a rich fossil history (figure 21.1).

Like seed plants, the other vascular plants have a *sporophyte-dominant* life cycle. Unlike seed plants, they have *free-living gametophytes* that release sperm cells into a film of water, through which they must swim toward egg cells held within vase-shaped female organs. Their sexual reproduction process, therefore, has a very vulnerable stage. The gametophytes of these plants usually are small, inconspicuous, photosynthetic, and grow in moist conditions.

In this chapter we will look briefly at nonseed-forming plant groups, plus fungi. Fungi, you will recall, are classified in their own kingdom because they differ strikingly in both structure and nutritional mode from plants. Space permits only a short chapter here, but we encourage you to explore additional references in your classroom and library.

21.2

You examined the *moss* plant body in chapter 20. If your teacher has additional specimens of mosses and liverworts present, examine them. The body of a nonvascular plant, undifferentiated in terms of vascular tissues, is called a **thallus.** As you learned earlier, this term also is used to describe the bodies of multicullular algae. Where the bryophyte plant body contacts the surface on which it grows (its **substrate**), you may see slender extensions called **rhizoids** that help to attach the plant to the substrate and that also may be able to absorb water.

1. Lacking vascular tissues, the rhizoid cannot transfer water through vessels or tracheids; therefore, how do you hypothesize that any fluid might be able to travel from the rhizoids to other parts of the plant body?
2. Are the leaflike parts of the moss thallus truly leaves? Justify your answer.
3. Are the bryophyte plant bodies you are examining gametophytes or sporophytes?

To early herbalists, plants that looked like an organ of the human body were assumed to have medicinal properties associated with that organ. The thallus of most *liverworts* is flat, lobed, and has a surface pattern that resembles that of liver. This appearance may be the reason they were named "liver plants" (wort means "plant"). Examine the specimen or photographs that have been provided. Both sexual and asexual reproductive structures may be present. Special multicellular discs called **gemmae** are produced (by mitosis) in small craterlike cups on the liverwort surface (figure 21.2). When released from their cups, gemmae grow into new gametophytes.

Figure 21.2 Liverwort gemmae cup with gemmae. Multicellular discs called *gemmae* are produced by the liverwort gametophyte. When dispersed, each gemma can grow into a new gametophyte. This is a type of asexual (vegetative) reproduction.

4. What environmental factors might serve to disperse gemmae from the parent plant?
5. Are the gemmae a means of sexual or asexual reproduction?

Figure 21.3 Sexual reproductive structures of liverworts. Like mosses, liverworts have a gametophyte-dominant life cycle and are classified as bryophytes. The flat gametophyte body produces tall, umbrella-like structures that bear egg-producing archegonia, as well as shorter rosette-shaped structures that bear sperm-producing antheridia. In this photograph you can see numerous female structures. A good view of a male structure is provided in the lower left corner of the photograph.

In liverworts, male and female reproductive organs develop in clusters atop stalks formed by the gametophyte body (figure 21.3). The top of the male reproductive stalk is a flat *splash platform* that functions as follows: flagellated sperm cells are released when water is present, and the spattering of raindrops disperses sperm-laden droplets. Some of these droplets land on female reproductive stalks, which have down-curved spokes. The openings of the female organs are on the underside of the spokes (figure 21.4). Water landing on the spokes runs down and becomes suspended around them, creating hanging pools in which sperm cells can survive until they have swum into the female organs.

Each zygote that results from syngamy undergoes mitotic division and differentiation to grow into a microscopic sporophyte, which remains attached to its parent (figure 21.5). From then on the growth pattern is similar to that of mosses, except upside-down!

6. How do you hypothesize that the liverwort sporophyte obtains water and nourishment?

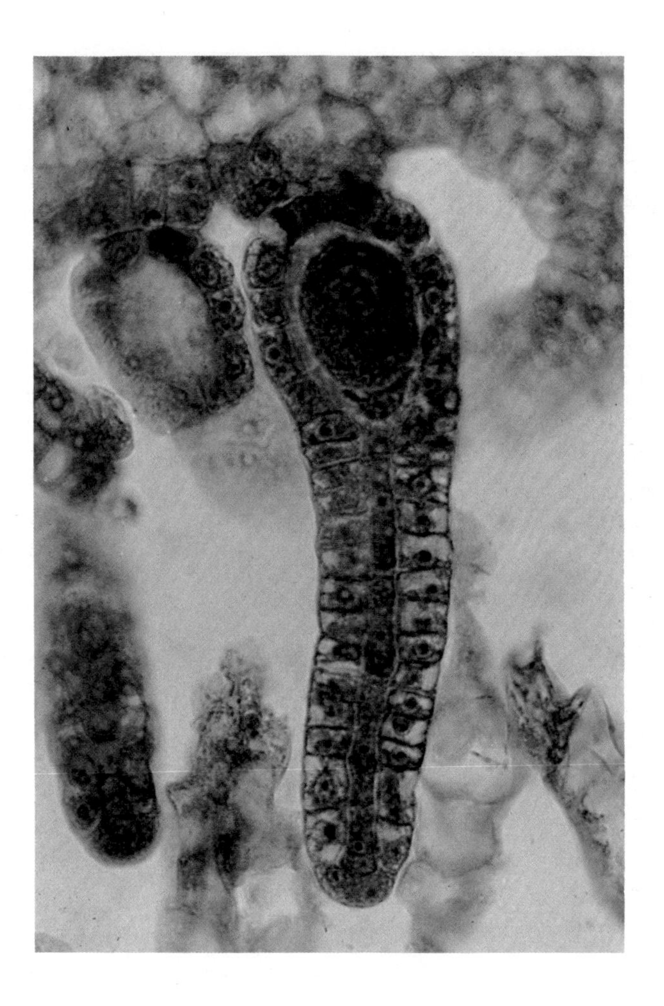

Figure 21.4 Bryophyte archegonium. This excellent view of a liverwort archegonium shows its single egg cell. Multiflagellated sperm cells swim to the archegonium and then through its slim passageway to the egg. When a sperm cell nucleus unites with the egg cell nucleus a zygote is formed—the beginning of the next sporophyte generation.

7. What kind of reproductive cells does the liverwort sporophyte produce?
8. Are they *n* or 2*n*?
9. When they germinate, what do they become?
10. In what kinds of habitats would you expect to find liverworts?

Figure 21.5 Liverwort sporophytes. As in mosses, liverwort sporophytes grow attached to the gametophyte body. Here you can see small, bulbous sporophytes embedded in the underside of the archegonia-producing structure.

21.3

Ferns are vascular plants, most of which grow as a horizontal stem, or **rhizome,** from which roots extend downward and leaves extend upward (figure 21.6). Ferns, along with the other nonseed-forming vascular plants, have a rich fossil record (figure 21.7). Coal, for instance, is made of the compressed and chemically altered bodies of plants, many of them ferns, that lived hundreds of million years ago. Ferns with

© Kendall/Hunt Publishing Company

Figure 21.6 Fern sporophyte. Ferns have a sporophyte-dominant life cycle and have vascular tissues. The fern body usually consists of a horizontal stem, a rhizome, that grows at or just beneath the soil surface. Leaves arise from nodes along the rhizome. Adventitious roots also grow from the rhizome. Note that the leaf of this species is a compound leaf, one that is divided into numerous leaflets. Sporangia are clustered on the undersides of the leaflets, as shown more clearly in figures 21.8 and 21.9.

Figure 21.7 A reconstructed ancient forest. Forests of the Coal Age were dominated by primitive vascular plants, including both low-growing and tree-sized ferns. By comparison, our modern forests are dominated by seed-producing trees and shrubs. The exhibit shown here has been reconstructed from fossil evidence. It features ferns, plus some kinds of vascular plants that have been extinct for millions of years.

which you may be familiar in the wild or as houseplants grow in ankle-high to waist-high clumps. Some ferns are much smaller. The largest are the tree ferns of tropical climates, which grow up to 20 m tall. Most ferns occupy moist habitats, and some even are aquatic.

1. Examine fern specimens your instructor has provided. Identify the rhizome and leaves. Where do the roots arise?
2. How can you identify the nodes on the rhizome?

The rhizome of many ferns is covered with small dry scales, giving it a fuzzy appearance. Most ferns have leaves that are divided into **leaflets.** Examine the back of the leaflets for brown structures, which are clusters of **sporangia** (spore cases). In some ferns, the spore case clusters are round; in other species they are strips on the leaf backs, or they are enclosed in folds of the leaf edge (figures 21.8 and 21.9).

3. In the space below, sketch a smaller-than-life-sized fern leaf, identifying its point of attachment to the rhizome, leaflets, and clusters of spore cases.
4. Record the following measurements to accompany your sketch: leaf length; leaflet length; spore case diameter (if round) or other dimensions.

According to your teacher's instructions, prepare a wet mount of a single sporangium or obtain a prepared slide. Examine your slide with low-power magnification.

5. Are spores present within the sporangium?
6. Is the sporangium closed or open?

Figure 21.8 Fern sporangia. In this example, each sporangial cluster (sorus; *pl.*, sori) is covered by a thin tissue flap. When the air is relatively dry, sporangial cases snap open to release the spores. The tiny spores are easily carried on air currents. Those that land in favorable, moist sites germinate into delicate, photosynthetic gametophytes. Spores are an important means of species dispersal. Because of their resistant outer cases, they can remain viable for long periods of time.

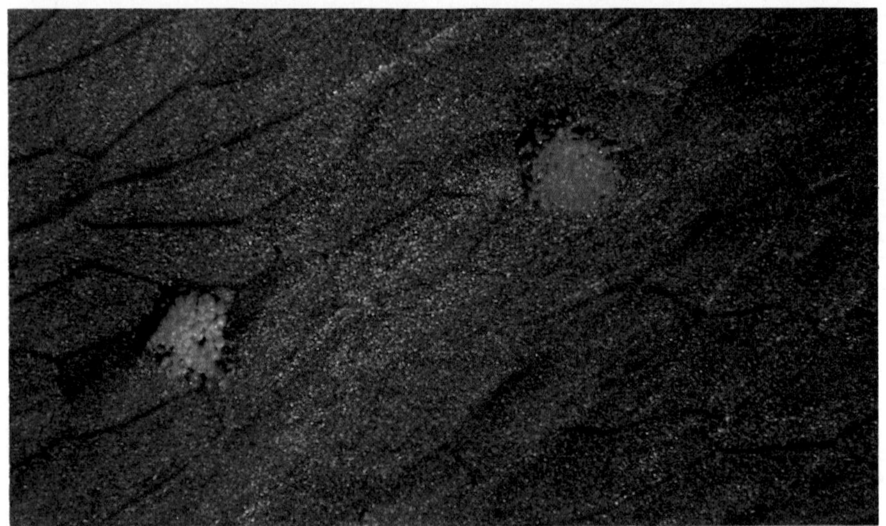

Figure 21.9 Fern sporangia. These sporangial clusters (sori) are not covered by a tissue flap.

Figure 21.10 Spores within a fern sporangium. The row of prominent cells along the left and upper edge of this sporangium is part of a mechanism that tears open the sporangium, beginning with the weak spot on the right side.

These fern sporangia are able to throw their spores into the environment by a catapulting action. Notice the "backbone" of cells up the back and over the top of the sporangium shown in figure 21.10.

7. Are their cell walls thinner toward the inside or the outside of the spore case?

During dry conditions that are suitable for spore release, water evaporates from these cells through the thin cell walls, causing the entire band of cells to shrink and tighten. The cells cannot contract, however, on their

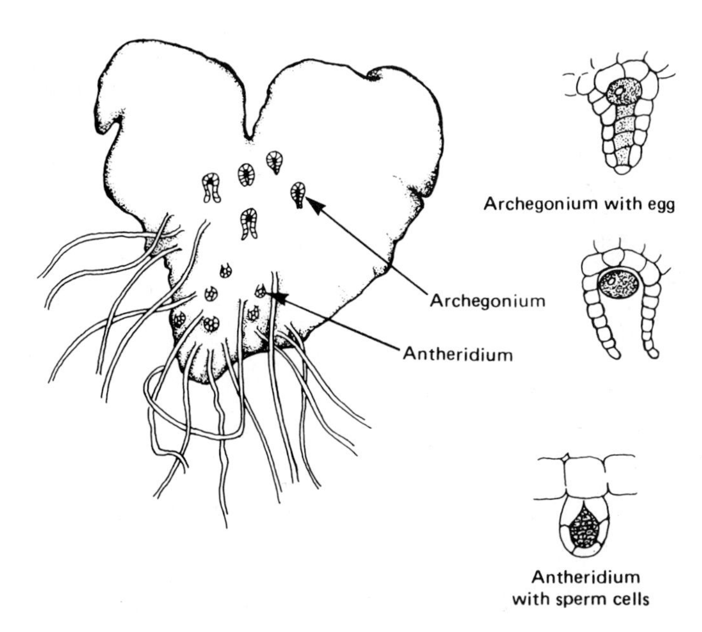

Figure 21.11 Fern gametophyte. The fern gametophyte is a delicate, green thallus that is often less than a centimeter in diameter. Some species have separate sexes, but the example shown here produces archegonia and antheridia on the same plant. Sperm cells have many flagella, and swim toward the archegonia along a gradient of sugary solution produced by the archegonia. Egg and sperm unite to form a zygote, which divides by mitosis to produce an embryonic sporophyte, which grows and matures. Initially the young sporophyte remains attached to the gametophyte, but eventually the small gametophyte dies and the sporophyte continues to grow into a familiar fern plant.

thickened side. As a result, the tightness is concentrated on the outside of the band of cells until the sporangium eventually rips open at an opposing weak spot. The explosive force of this rupture causes spores to be flung out of the case. If you can find some sporangia that have not yet opened on the living fern, you can observe this activity by letting them dry out on your slide as you watch. If you would like to try this, do not use a coverslip and use little or no water. Watch closely because the final snap occurs rapidly and unpredictably.

8. What does each spore have the potential to become?
9. By what kind of nuclear division are the fern spores formed?
10. What generation of the fern plant have you been observing?

Figure 21.11 shows a fern *gametophyte,* bearing male and female organs. As mentioned earlier, water must be present for the flagellated sperm that are produced to swim to the female organs.

21.4 The remaining groups of vascular plants probably are less familiar to you than the seed plants and the ferns. For our purposes, we will only briefly survey them here.

Horsetails and *scouring rushes* (*Equisetum*) (figure 21.12) often grow on sandy banks, such as along roadcuts, railroad tracks, and the edges of coastal dune vegetation. They have hollow, jointed stems, and they lack functional leaves. Their spore cases are clustered in a clublike structure at

Figure 21.12 Horsetail. Horsetails are vascular plants with whorled, jointed leaves. They often grow in poor, sandy soils.

Figure 21.13 Horsetail strobili. The sporangia of horsetails and scouring rushes are clustered in cones, or strobili (*sing.,* strobilus). When the spores are mature, you can tap such strobili and release clouds of yellow spores. As in ferns, the spores of horsetails and scouring rushes germinate into small, independent, photosynthetic gametophytes. Water is required to allow sperm cells to swim to the eggs, and the delicate gametophytes quickly perish when moisture is no longer available. The sporophytes, on the other hand, are tough and able to withstand dry conditions. Thus, we can say that the water-dependent sexual phase is the weak link in the life cycle, in terms of full adaptation to terrestrial environments. The same is true of bryophytes, ferns, whisk ferns, and club mosses.

the tip of the plant, called a **cone,** or **strobilus** (figure 21.13), a structure that is akin to the cones and flowers of seed plants. Their bodies are impregnated with silica, giving them a gritty texture.

1. Knowing this, can you guess how scouring rushes got their name?

The *whisk fern* (*Psilotum*) is a "living fossil," a remnant of Earth's earlier history. Like *Equisetum* species, *Psilotum* has its own distinctive body form. It has slender, leafless, green stems that are dichotomously branched (figure 21.14). (Dichotomous means two-branching; you may remember using a dichotomous key to identify plants or insects.) Small, three-parted sporangia are nestled in nodes along the stems. As the spores mature, they become bright yellow, showing through the clear sporangial walls. The attractiveness of the whisklike vegetation, accented by brilliant yellow spore cases, makes *Psilotum* an attractive houseplant.

Figure 21.14 Whisk fern sporophyte, with sporangia. The whisk fern, *Psilotum,* is a primitive vascular plant with a life cycle pattern like that of ferns and horsetails. Its delicate, green gametophyte is dependent upon the presence of free water, both for survival and for the completion of sexual reproduction.

Club mosses (*Lycopodium, Selaginella*) are so named because they have a mosslike growth form and also produce cones (strobili) (figure 21.15). Some club mosses are valued as decorations, such as princess pine, reindeer moss, and the so-called resurrection plant. The bright yellow spores of *Lycopodium* have had industrial uses as a dry lubricant, filler, and even as a body powder.

Selaginella demonstrates a significant stage in the modification of vascular plant life cycles as carried forward into the seed plants: the production of *two kinds of spores,* instead of a single kind. What does that mean? It means that one kind of spore is produced that grows only into male gametophytes, and a second kind of spore is produced that grows only into female gametophytes. How does that relate to the seed plants? Recall our exploration of flowering plants. In the anther, only male gametophytes (the pollen grains) are produced. In the ovule, only female gametophytes are produced. This is a refinement of the same *pattern* seen in this club moss.

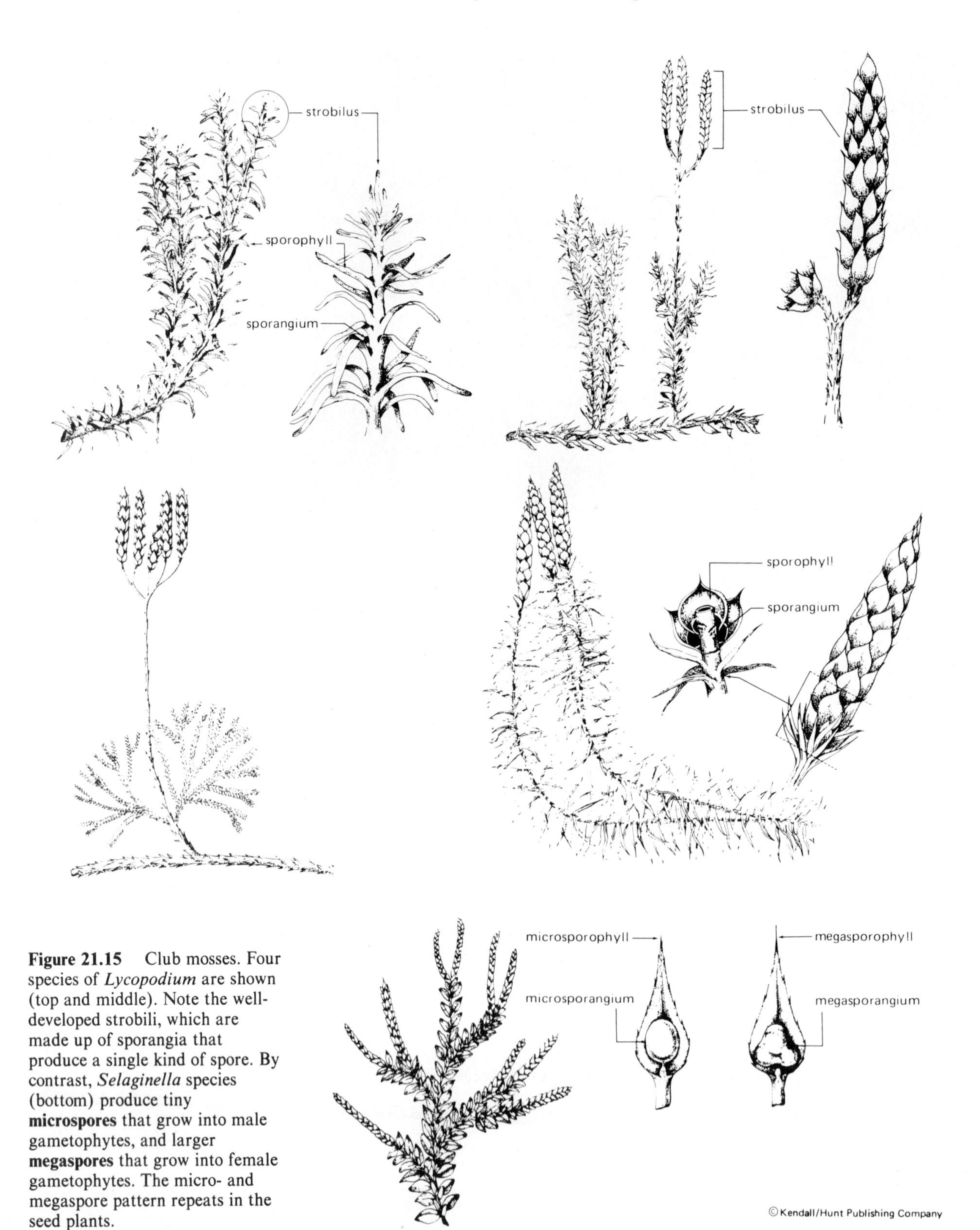

Figure 21.15 Club mosses. Four species of *Lycopodium* are shown (top and middle). Note the well-developed strobili, which are made up of sporangia that produce a single kind of spore. By contrast, *Selaginella* species (bottom) produce tiny **microspores** that grow into male gametophytes, and larger **megaspores** that grow into female gametophytes. The micro- and megaspore pattern repeats in the seed plants.

© Kendall/Hunt Publishing Company

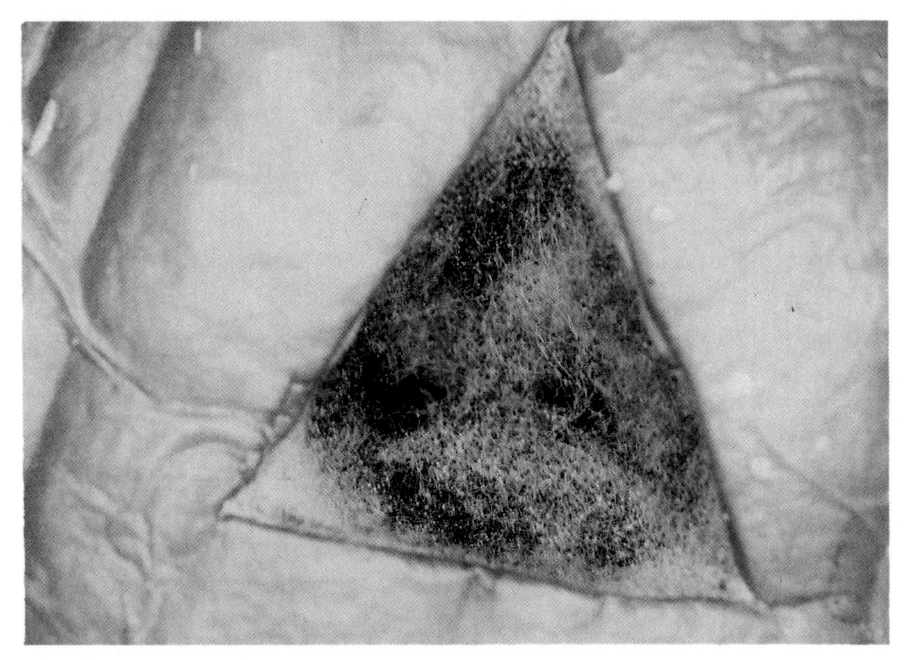

Figure 21.16 Fungal mycelium. The body of a fungus is constructed entirely differently than that of plants. It is composed of numerous filaments called hyphae, which form a collective mat or web, called a mycelium. Also unlike plants, fungi do not carry out photosynthesis. They obtain nourishment by digesting the living or dead tissues of other organisms, such as this pumpkin. Normally, the outer coverings of living plants and animals protect them from fungal attack. The inner tissues of this pumpkin were exposed, however, when it was carved into a jack-o-lantern.

21.5 The *fungus* body is totally unlike that of plants. The body of most fungi is called a **mycelium,** a mass of filaments that are called **hyphae** (sing., *hypha*) (figure 21.16). Except for reproductive structures, there is little differentiation of cells into tissues. The cell walls are composed of various polysaccharides, including cellulose (common to plants) and chitin (a substance also produced by many animals). Fungi also differ from plants in being heterotrophic, feeding by absorption.

Obtain a culture or prepared slide of black bread mold (*Rhizopus*) and examine it with a hand magnifier or binocular microscope. If using a slide, you may look at it with low-power magnification under a monocular microscope.

1. Describe the appearance of the mycelium.
2. What is the name of each body filament?
3. Do the filaments seem to be solid or hollow?
4. Are crosswalls evident?
5. How do these filaments compare to plant tissues and the filamentous algae you studied?

Fungi reproduce by vegetative spread of the mycelium and by spores. They are extremely important members of ecosystems because they penetrate and digest dead plant and animal remains, beginning the breakdown process for recycling of the nutrients locked up in these remains. Some fungi have economic uses, whereas many others are parasites and pathogens. Athlete's foot and ringworm are two common fungal diseases of humans, in which the mycelium grows in the outer layers of the skin.

Return to your bread mold and look for small balls at the tips of some hyphae. These are sporangia.

6. What cells do they produce?

Obtain a cultured mushroom to examine (figure 21.17). Identify its *stalk, cap,* and *gills.* Pull the cap away from the stalk, then pull apart the cap so you can see the arrangement of gills more clearly. Microscopic spore-producing structures line the surfaces of the gills. Use your fingers to tear the stalk lengthwise.

7. Does the stalk seem to be composed of filaments?
8. What are the filaments called?

When you see a **mushroom, bracket fungus** or **puffball** growing wild (figure 21.18), you are seeing the fungus equivalent to the "tip of the iceberg." Most of the fungus body is a webby mycelium that branches throughout its substrate, whether that substrate is a dead tree or forest floor. The visible part is a special reproductive structure that develops only to produce and disseminate spores. Therefore, picking a mushroom does not kill the fungus. Cultivated mushrooms can be harvested again and again from the same mycelium.

9. In terms of crop production and harvesting, how does mushroom cultivation differ from cultivation of fruit-bearing plants?

The cultivated mushrooms sold in grocery stores are edible members of the genus *Agaricus. However, you should never eat unidentified mushrooms!* Although there are many edible (and delicious) species, expert identification is required. Some mushrooms are toxic enough to cause cramps, vomiting, diarrhea, fever, chills, hallucinations, convulsions, coma, permanent cell damage, or even death.

Yeasts are another important kind of fungus. Some species are used in human food production and industrial processes that involve fermentation. Major by-products of yeast fermentation are alcohol and carbon

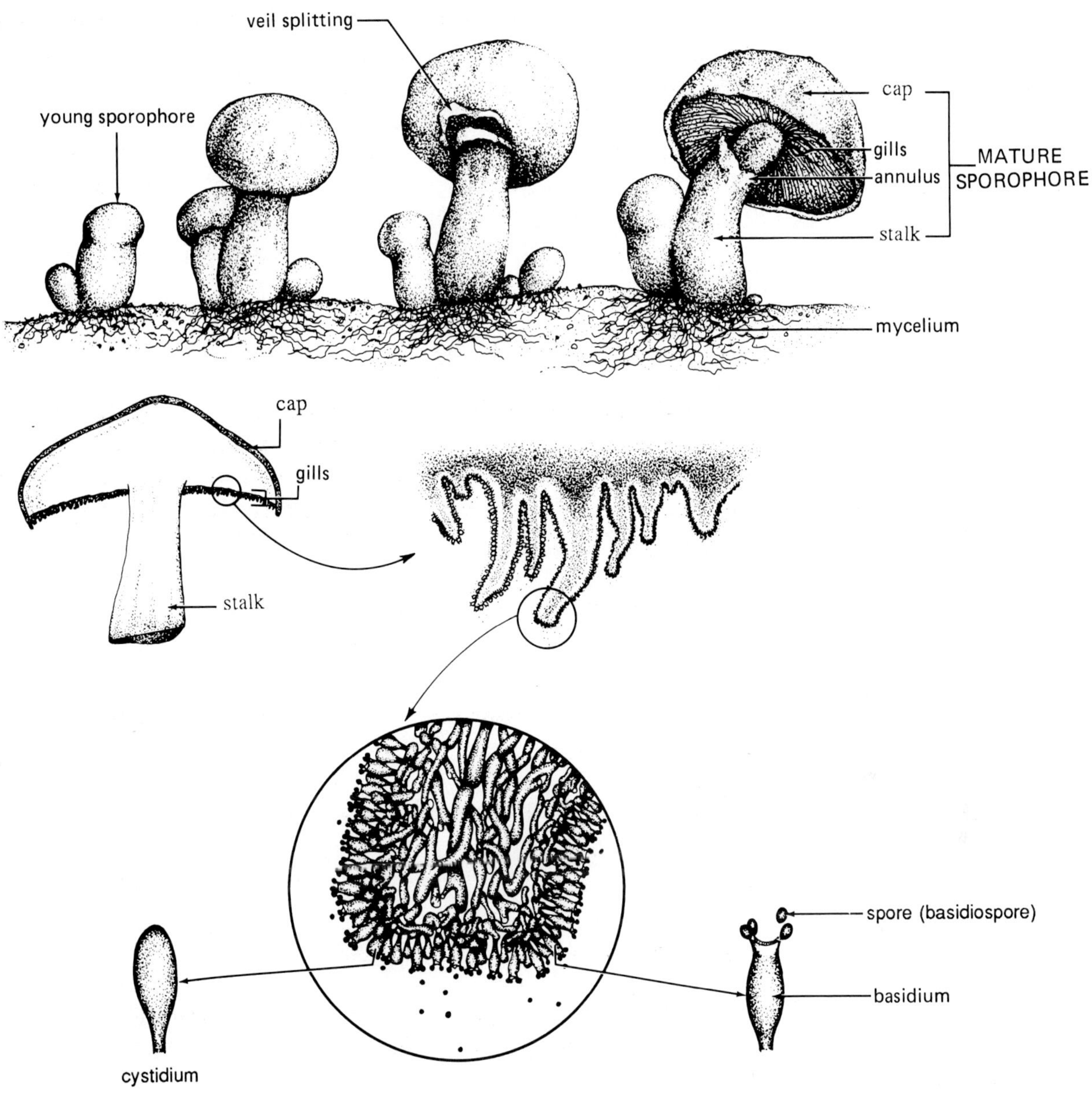

© Kendall/Hunt Publishing Company

Figure 21.17 Mushroom growth and reproduction. The familiar mushroom is only the spore-producing part of the total fungus body, which consists mainly of the food-obtaining mycelium.

dioxide. Alcoholic beverages produced by fermentation are, therefore, naturally carbonated until they are refined and distilled. Yeast breads rely on the carbon dioxide bubbles produced by living yeast cells to cause dough "to rise"; the alcohol evaporates during baking. Yeast undergo a type of asexual reproduction called **budding,** whereby miniature yeast cells are pinched off from the body of a mature cell (figure 21.19).

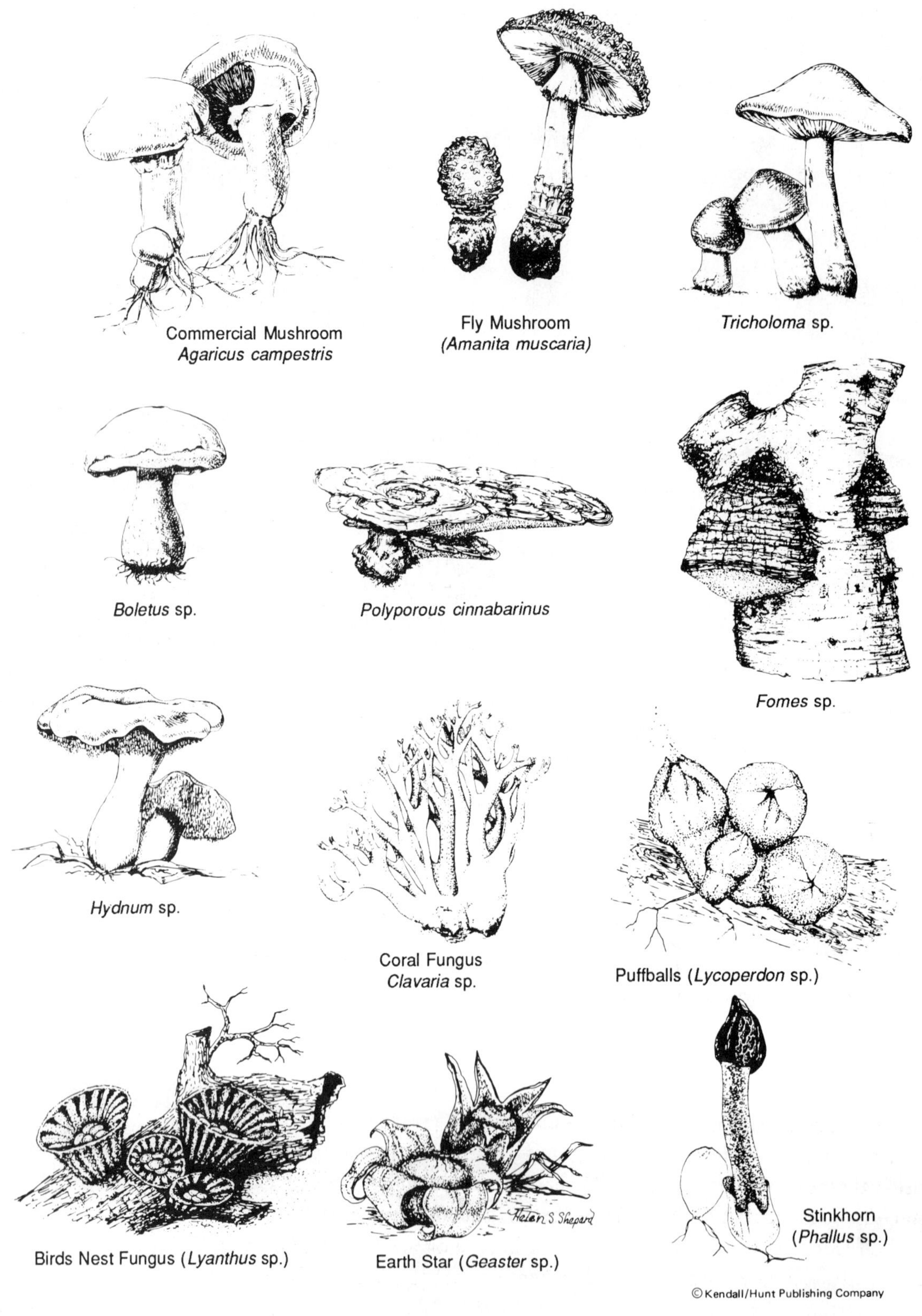

© Kendall/Hunt Publishing Company

Figure 21.18 Some kinds of fungi. Mushrooms are the most familiar fungal body form, but this plate also shows puffballs and some other interesting forms. Some of these species are edible, but most are not. At least one, the fly *Amanita,* can cause a very painful death.

The life cycles of fungi are varied, so it is not useful to generalize about them. It is interesting, however, to know that life cycles of some of the parasitic and pathogenic species (figure 21.20) involve more than one **host** organism. Wheat rust is a serious disease of wheat plants that spends one phase of its life cycle growing on barberry plants. Apple-cedar rust attacks apple trees, spending a different phase of its life cycle on cedar trees. The dual life cycle enables a fungus to reproduce on both hosts, *greatly multiplying the number of spores produced.* Are you ready for even more complication? Many fungi produce more than one kind of spore, even on the same host!

Table 21.1 identifies some of the major categories of fungi, and some examples.

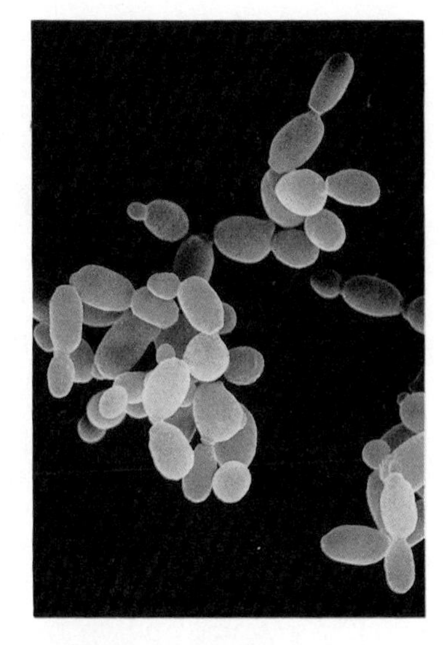

Figure 21.19 Yeast. Yeast cells are fungi for which we have important human uses. They carry out a type of respiration called fermentation, which releases carbon dioxide gas and ethyl alcohol. Baker's yeast is used to make raised dough, and brewer's yeast is used to produce fermented beverages. Yeasts reproduce asexually by budding, as can be seen in this micrograph.

Data Table 21.1 Major Groups of Fungi*

Group Name	Examples
Zygomycota (zygote fungi)	Black bread mold
Ascomycota (sac fungi)	Yeasts, water molds, bread molds, truffles, and morels
Basidiomycota (club fungi)	Mushrooms, cup fungi, bracket fungi, puffballs, jelly fungi, rusts, and smuts
Deuteromycota (imperfect fungi)	
This is an artificial classification category for about twenty-five thousand species of fungi whose life cycles are imperfectly known. It includes *Penicillium,* source of the antibiotic called penicillin, and many other species of fungi of great medical and economic importance.	
Mycophycophyta	Lichens

*Slime molds once were classified with the fungi, but they now are classified with the protists. Therefore, they are not listed in this table.

21.6

Lichens are classified as fungi, but their bodies are jointly made up of a fungus and photosynthetic algae or cyanobacteria. The lichen body is, therefore, a *symbiotic partnership.* The fungus component of a lichen usually is an ascomycote.

Lichens grow mainly in polar and temperate climates, requiring alternating damp and dry conditions. They live on soil, tree bark, bare rocks, and in the sea-spray zone of rocky coasts. Because they are the first kind of organism to populate bare rock sites, lichens are called *pioneer plants.*

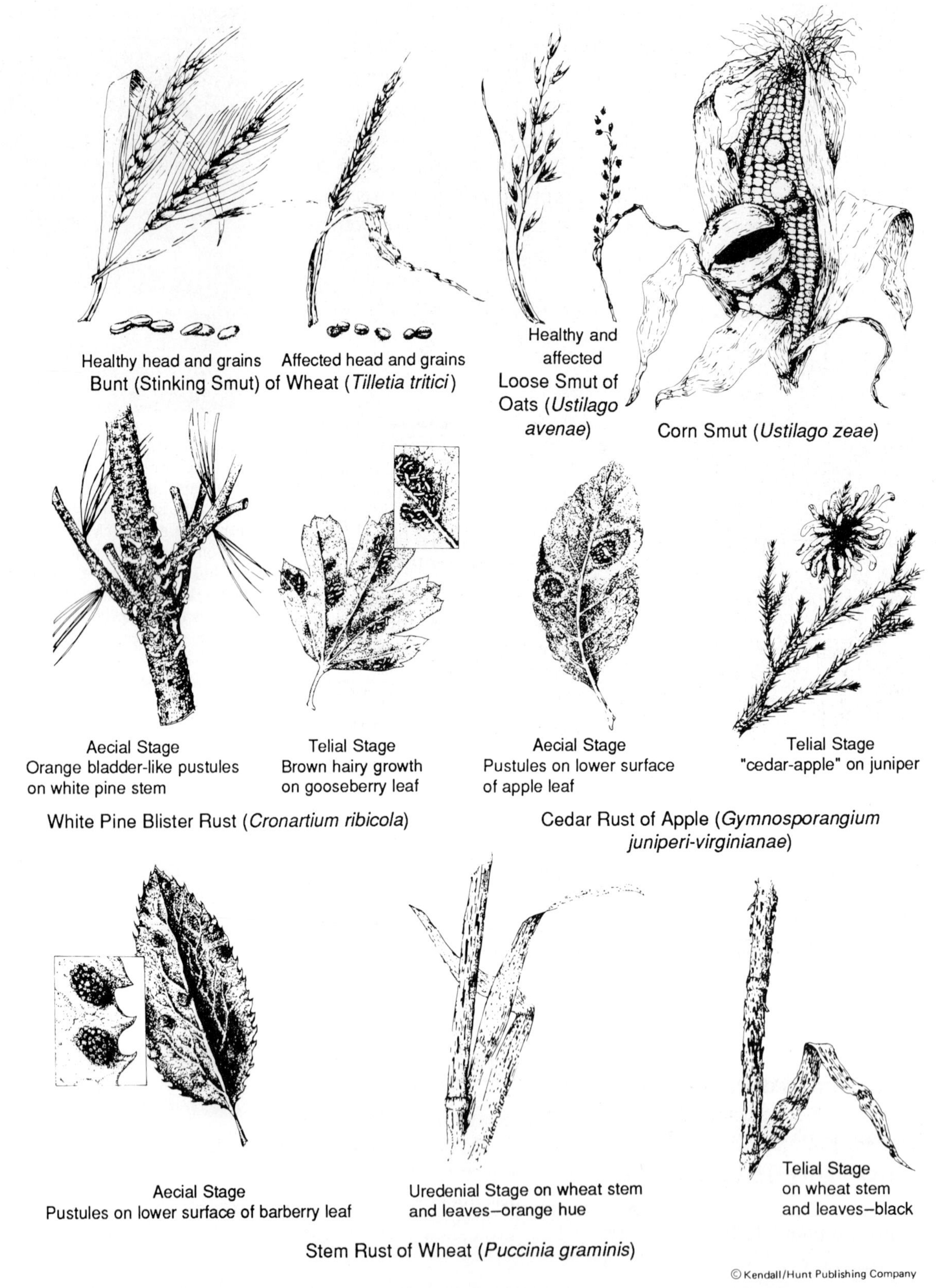

© Kendall/Hunt Publishing Company

Figure 21.20 Pathogenic fungi. Rusts and smuts are kinds of fungi that obtain nourishment from living plant tissues. The species shown here are serious agricultural pests. Rusts and smuts often have multi-parted life cycles involving two or more host species.

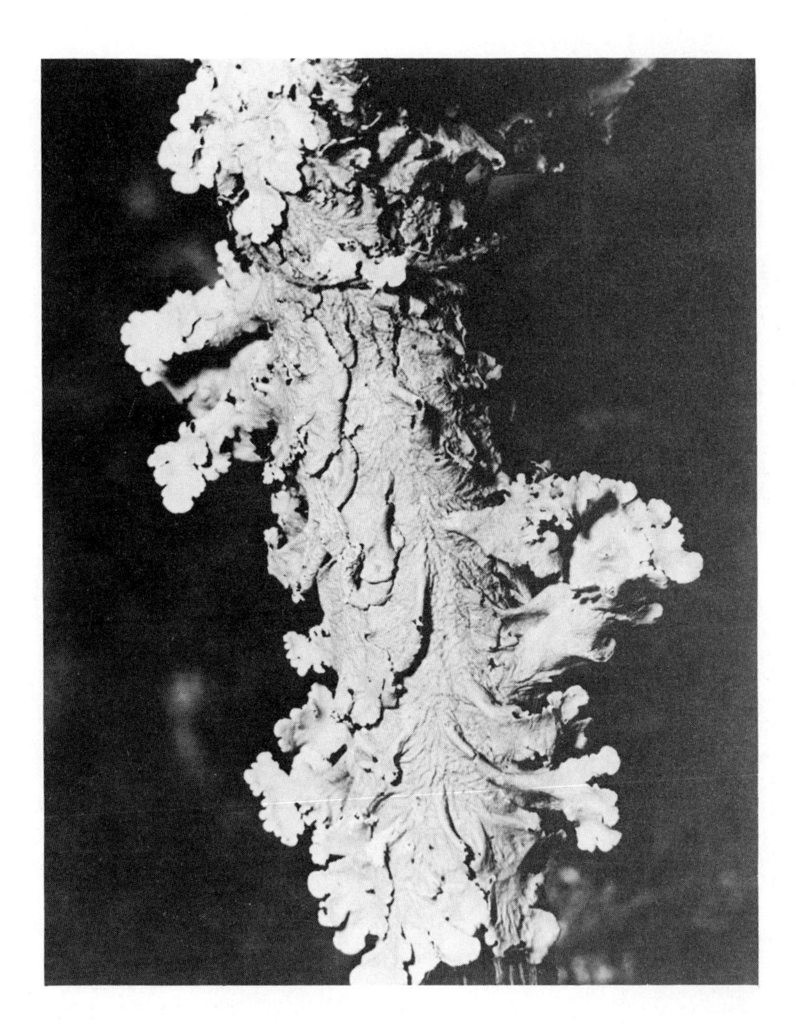

Figure 21.21 A leafy (foliose) lichen. The lichen body is a mutualistic association of a fungal mycelium with photosynthetic bluegreen bacteria or green algae, depending on the species. The broad thallus shown here is often found in moist forests.

Lichens are extremely slow growing; some that have been measured on gravestones may grow as little as a few millimeters in 100 years. Lichens are very susceptible to air pollution, as demonstrated by the lack of living lichens in and near some industrial areas.

Figures 21.21, 21.22, and 21.23 show the three different body forms among lichens: leafy (foliose), spiky (fruticose), and encrusting (crustose).

21.7

In this chapter you have learned a few of the major characteristics of members of the plant kingdom that are not seed producers. The observations and answers you have provided were designed to reinforce, through practice and application, the concepts of plant reproduction that were introduced in chapter 20.

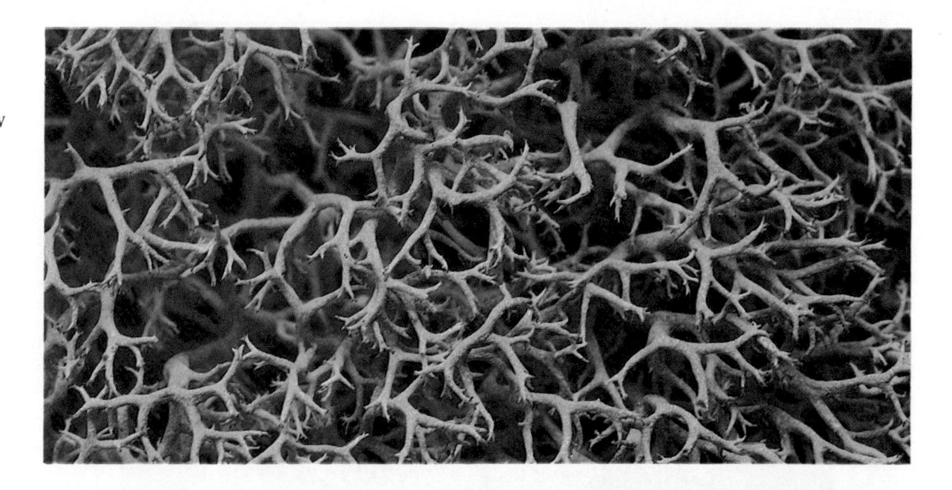

Figure 21.22 A spiky (fruticose) lichen. The spiky growth form is often characteristic of lichens that grow in drier habitats. Some spiky lichens grow profusely on the dead branches of trees in coniferous forests. Other species grow on sandy soil—you may be familiar with the red-topped "British soldier" lichens.

Figure 21.23 An encrusting (crustose) lichen. Crustose lichens are often found growing where nothing else can, as on the surfaces of rocks, masonry, and gravestones. Some crustose lichen patches may be hundreds of years old.

Review the study points listed below and check off those you understand. Review the text to clarify any points you do not understand before going on to the next chapter.

On completion of this chapter, you should be able to do the following:

- ☐ 1. Name the seven major plant groups, six of which are characterized by the presence of vascular tissues and a sporophyte-dominant life cycle.
- ☐ 2. Explain why the pattern of seed plant sexual reproduction is a more refined adaptation to terrestrial life than is the release of gametes directly into the environment as seen in bryophytes and the non-seed forming vascular plants.
- ☐ 3. Identify the two main kinds of bryophytes.
- ☐ 4. Identify the major structural features of ferns.
- ☐ 5. Identify horsetails, club mosses, and the whisk fern.
- ☐ 6. Explain, fluently, the life cycle pattern that is common among plants.
- ☐ 7. Understand several ways in which fungi differ from plants.
- ☐ 8. Name several important roles fungi play in the biosphere, including interactions with humans and their activities.
- ☐ 9. Understand the relationship of hyphae to a mycelium, and of reproductive structures to a mycelium.
- ☐ 10. Name and recognize several kinds of fungi.
- ☐ 11. Explain what a lichen is.

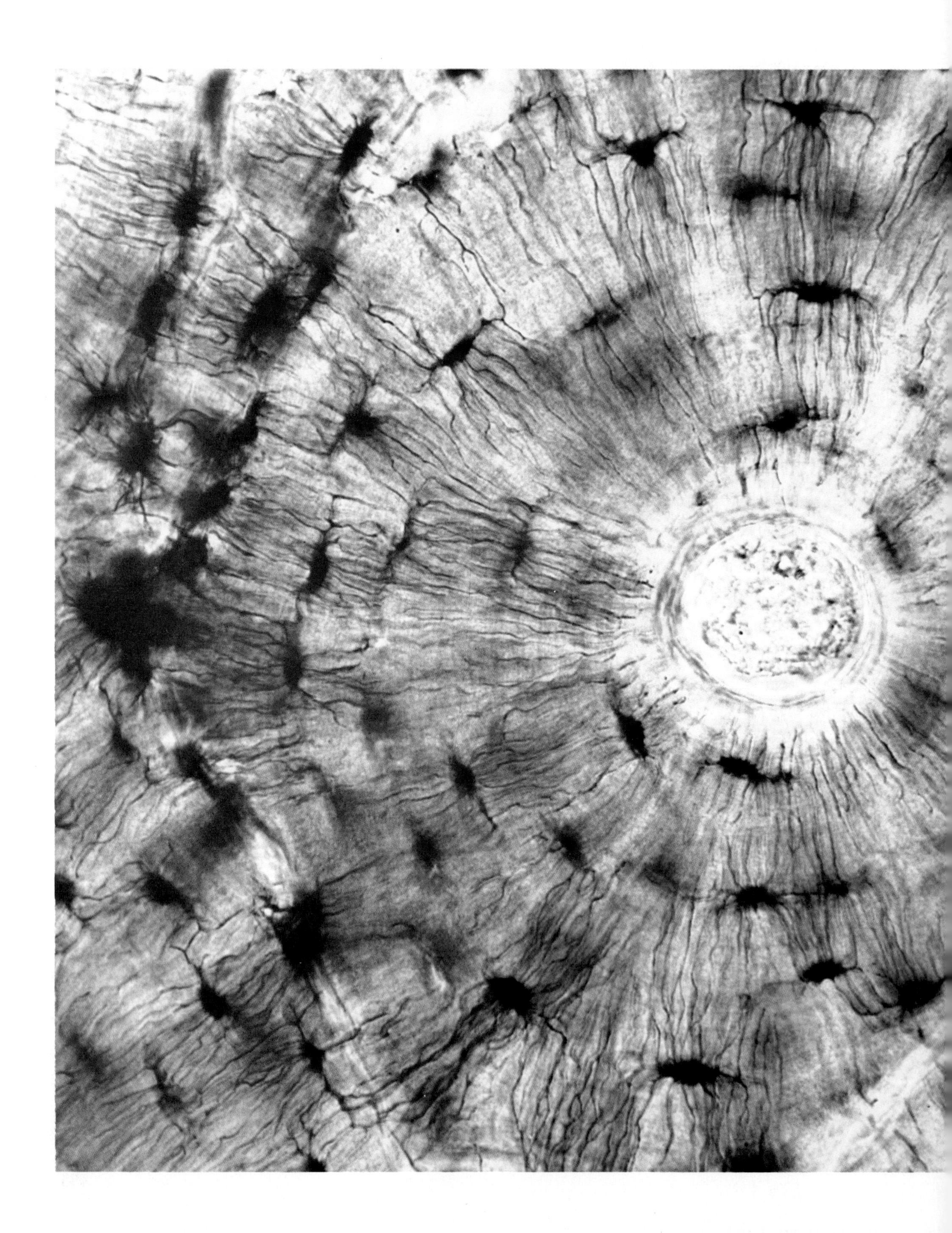

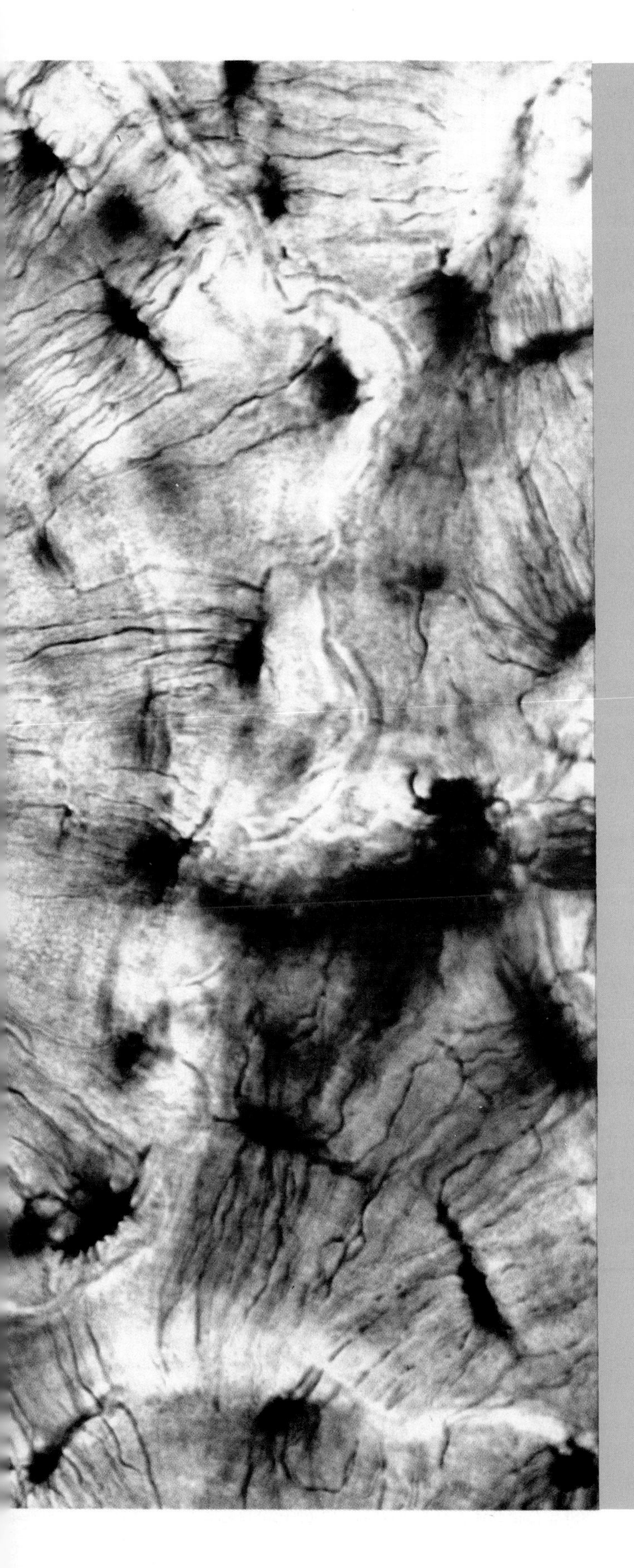

Human Anatomy: The Body Plan 22

What is your body plan?

22.1 This chapter will introduce you to the study of human anatomy and its corresponding physiology. Essentially, **anatomy** is the science of the structure or parts of the human body. **Physiology** is the science dealing with the function of the structures of the human body.

Eleven body systems are typically used to describe the human form. The eleven systems are listed below. To check your present background about these systems, match the description in the right column with the system in the left column by writing the correct letter in the blank provided. The same matching program is located in your log book.

	System		Description
____	1. integumentary	*a.*	Production of sperm and eggs
____	2. skeletal	*b.*	Removal of nitrogenous wastes from the body
____	3. muscular	*c.*	Interchange of oxygen and carbon dioxide within the lungs
____	4. respiratory	*d.*	Secretion of hormones
____	5. circulatory	*e.*	Regulates all of the systems of the body
____	6. digestive	*f.*	Support system for the entire body
____	7. endocrine	*g.*	The protective surface covering of the body
____	8. nervous	*h.*	Provides the ability to move, grasp, lift, and so forth
____	9. reproductive	*i.*	Transport system throughout the body
____	10. urinary	*j.*	Breakdown of food into usable products for the body
____	11. lymphatic	*k.*	Cleans the body of foreign material

22.2 All body systems are made of many organs. For example, the heart is an organ of the circulatory system. The stomach is an organ of the digestive system. The skin is the primary organ of the integumentary system. Organs are made of tissues, and tissues are made of specific cells.

There are four distinct types of tissues found in the body. They are the **epithelial, connective, muscle,** and **nerve** tissues. All tissues are equally important, because each has a function to perform that is *necessary* to the well-being of the whole body.

The study of tissues here is rather extensive. It will involve numerous, careful observations of tissues at the cellular level. As each tissue is described below, you are to observe a prepared slide of the tissue with a compound microscope. All observations should first be made with the low-power objective and then, when possible, with the high-dry objective.

You are to illustrate, color, and label the cell parts of each tissue. Viewing field circles are provided for your illustrations. Attempt to illustrate several cells of each tissue. For reference use, it is okay to increase the size of the cells you illustrate. The first tissues you will study will be epithelial tissues.

The epithelial tissues make up all coverings of the body, line all body cavities, and form the glands of the body. Epithelial tissues always have an exposed surface that is free of cellular contact. They also have a basement membrane that anchors the epithelial tissue to the underlying connective tissue.

There are five basic types of epithelial tissues. They are: *simple squamous epithelium, stratified squamous epithelium, cuboidal epithelium, columnar epithelium,* and *ciliated columnar epithelium.* Epithelial tissue is classified, first of all, on the basis of whether it is arranged in a single layer (simple) or more than a single layer (stratified). Secondly, epithelial tissue is classified according to the shape of the individual cells: squamous (flat), cuboidal (like a cube), or columnar (tall, like a column).

Get prepared slides now and complete the first part of this activity. Fulfill the requirements as stated above.

22.3 Connective tissues function as a support system for many organs and they connect many organs to one another. Connective tissue has an abundance of intercellular material called **matrix,** which helps distinguish the various types of connective tissues. Complete this section in a similar manner to section 22.2.

22.4 There are three types of muscle tissue that you need to know. These should be studied in the same manner as the two previous sections.

Data Table 22.1 Epithelial Tissue

Tissue	Information	Illustration	Slide illustration
Simple squamous epithelium	A single layer making up the linings of the peritoneum, pleural membranes and the circulatory vessels.	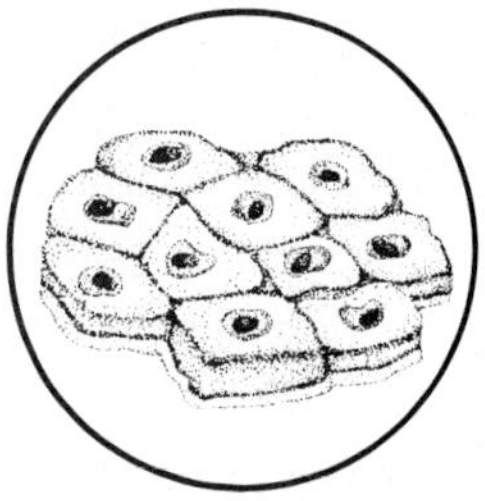	
Stratified squamous epithelium	Layered squamous epithelial cells are found in the lining of the mouth, that is, the cheek cells you looked at in section 4.2. The outer layer of skin (epidermic) is an example of these cells.	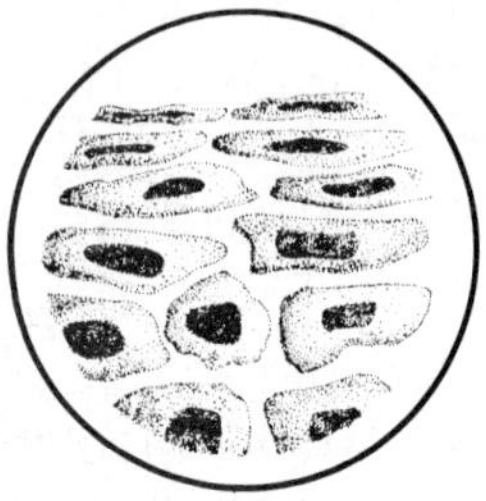	
Simple columnar epithelium	These are elongated cells that line the digestive tract. Goblet cells are typical of plain columnar tissue. The goblet cells secrete mucus that protects the surrounding cells of the tissue.	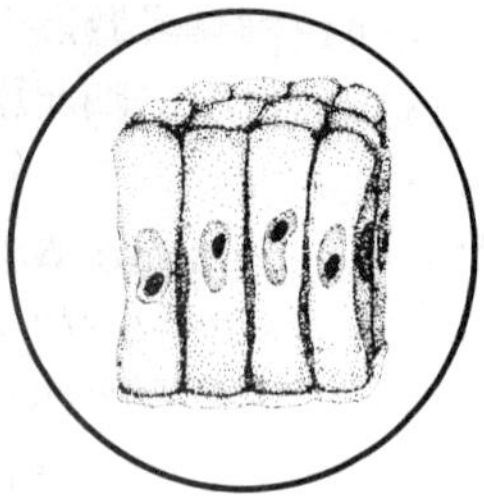	
Ciliated columnar epithelium	These columnar cells have cilia or hairlike projections protruding from their exposed surfaces. This type of tissue lines the trachea, bronchial tubes, and pharynx.	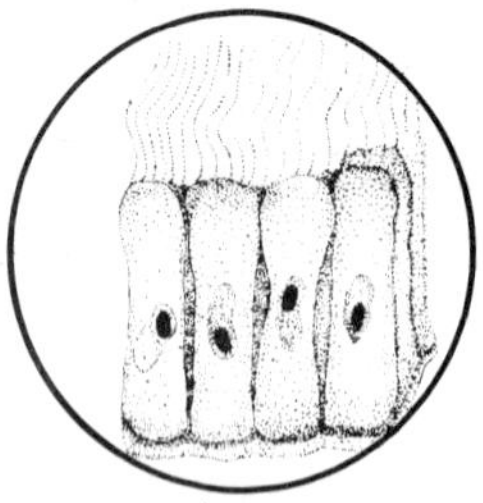	
Cuboidal epithelium	The name implies that these cells are cube-shaped. They function as secretory cells. Cuboidal cells are found in such glands as the thyroid gland and the salivary glands.	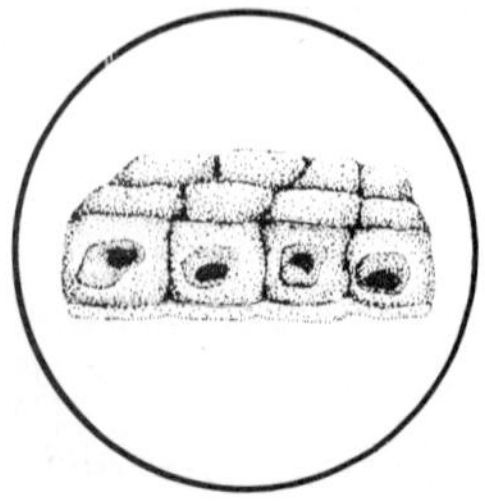	

Data Table 22.2 Connective Tissue

Tissue	Information	Illustration	Slide Illustration
Areolar connective tissue	Areolar tissue contains yellow and white fibers between large cells called fibroblasts. This connective tissue attaches the skin to the body.		
Fibrous connective tissue	This tissue is extremely strong, deriving its strength from the masses of white fibers. This connective tissue makes up the tendons and ligaments of the body.		
Reticular tissue	Reticular tissue gives the lymph nodes support.	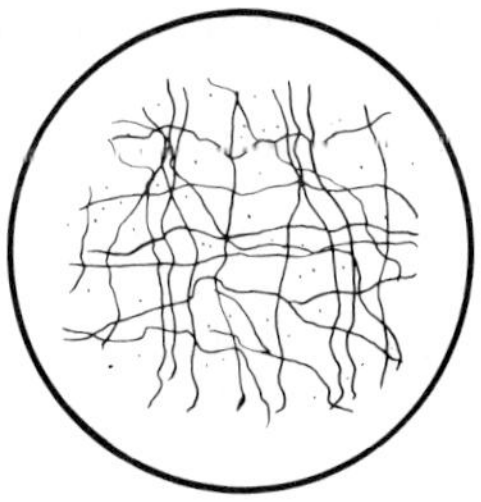	
Adipose tissue	These cells are especially designed for the storage of fat in large vacuoles. This type of cell can be found nearly anywhere in the body, either singly or in large groups.	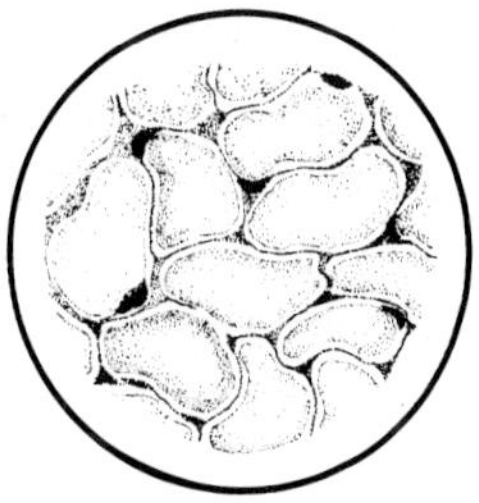	

Data Table 22.2 *Continued*

Tissue	Information	Illustration	Slide Illustration
Hyaline cartilage	Hyaline cartilage covers the articulating surfaces of the bone materials and forms the basic inner structure of the nose and rib cartilages. This material is, in essence, densely packed white fibers.		
Elastic cartilage	Similar to hyaline cartilage but more flexible, this material contains more yellow fibers. Elastic cartilage is found in the external ear, eustachian tube, and in parts of the larynx.		
Bone	Bone gives support to the body and is made of a dense, rigid matrix. Throughout bone material there are passageways called **Haversian canals** through which blood vessels run.		
	Surrounding each Haversian canal are many **osteocytes.** These are mature cells connected by **canaliculi.**	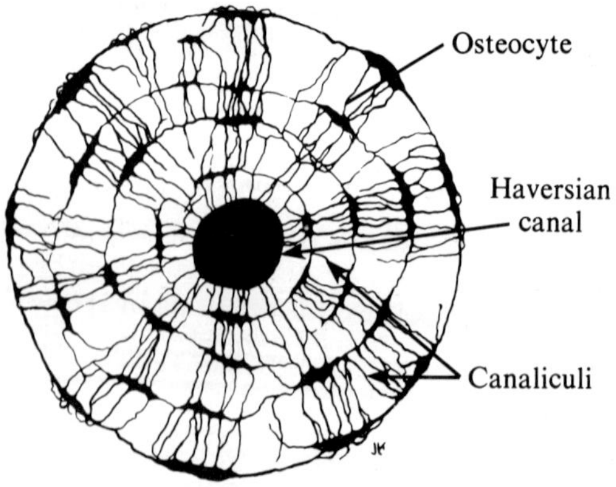 Osteocyte Haversian canal Canaliculi	

Data Table 22.3 Muscle Tissue

Tissue	Information	Illustration	Slide Illustration
Striated muscle tissue	This type of muscle tissue forms the body muscles. These are the muscles that allow voluntary movement of the arm, leg, trunk, neck, and so forth. Note the striations and the nuclei.		
Smooth muscle tissue	This tissue makes up the muscle of the visceral organs, that is, the stomach, intestines, and so forth. Smooth muscle lacks striae. Smooth muscle tissue also has fewer nuclei.		
Cardiac muscle tissue	This is the muscle of the heart. This tissue has less distinct striations, fewer nuclei, and dark, broad bands called intercalated disks.		

22.5 Nervous tissue carries nerve impulses to all parts of the body, thus regulating body functions. There are two types of neurons that make up the nervous system—motor neurons and sensory neurons.

Neurons that carry messages from organs to the spinal cord and the brain are called **sensory neurons.** Neurons that then carry an action message back to muscles are called **motor neurons.** The end result is an action regarding the received stimulus. View the slide representing one of the many neurons and illustrate it as you have done in the previous exercises.

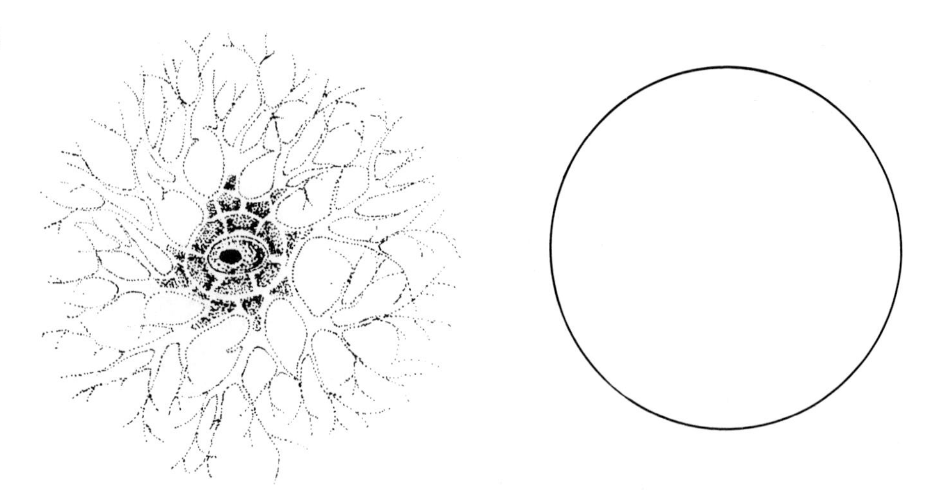

Figure 22.1 A typical nerve cell called a neuron.

22.6 Being able to recognize the cells, tissues, and systems is only a beginning. This background information is necessary, however, for future studies. In order to communicate further and effectively about anatomy, you also need to know the relative position of one anatomical structure to another. Study the terms below for relative body positions and the discussion given for each. When you have the terms mastered, check (✓) the box and continue to the next.

Relative position	*Discussion*
☐ Superior and inferior	If a structure is above another structure, it is *superior* to that structure. A structure below another structure is said to be *inferior.*
☐ Anterior and posterior	If any structure is in front of another, it is considered to be *anterior.* If it is behind the structure, it is located *posteriorly.*
☐ Dorsal and ventral	In order to use these terms accurately, you must visualize a four-legged animal or one that is normally in a horizontal position. The back of a dog would be the **dorsal** side, while the stomach would be considered **ventral.**

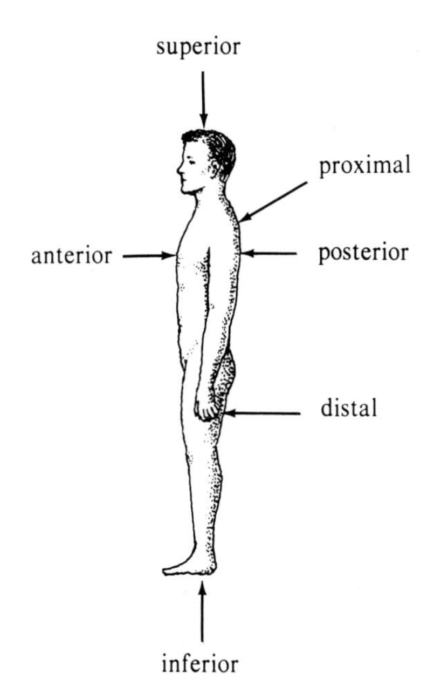

Figure 22.2 Terms used to explain the relative positions. Compare this figure to Figure 22.3.

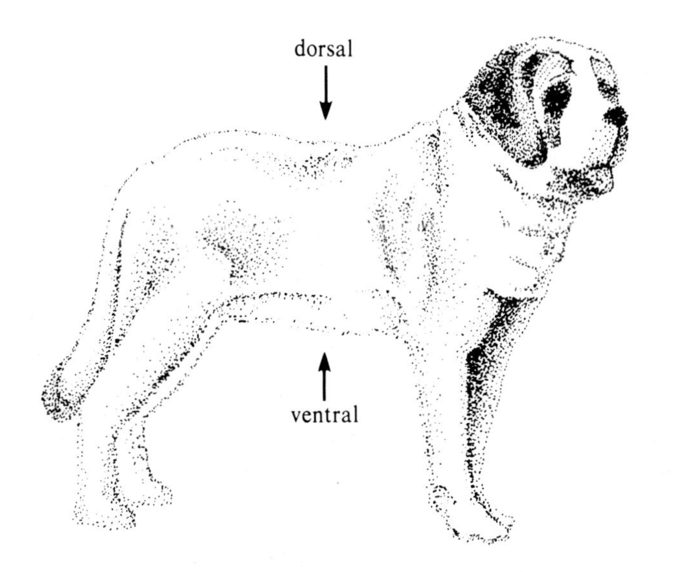

Figure 22.3 With positions taken on four legs, dorsal replaces posterior and ventral replaces anterior.

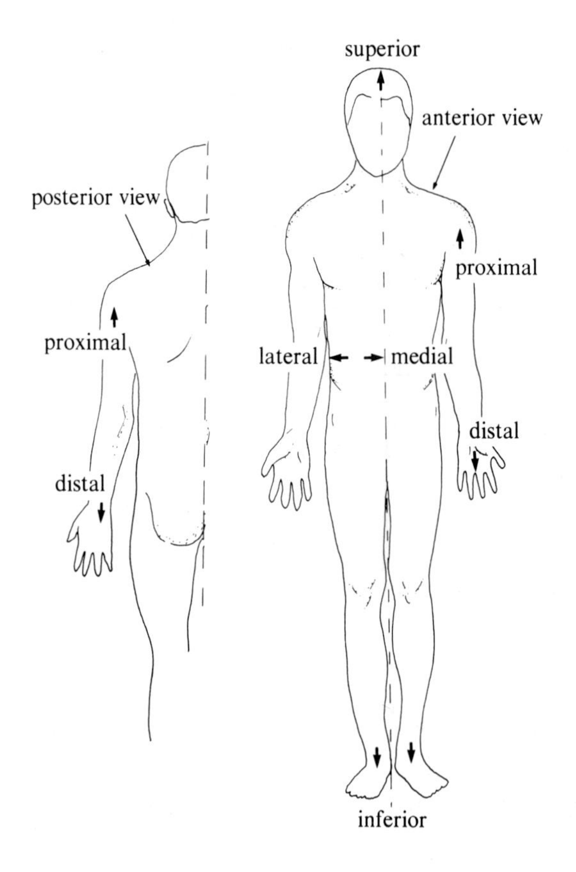

Figure 22.4 Relative body positions. Compare this figure to Figure 22.2.

☐ Proximal and distal

Both terms are used in discussions of legs, feet, arms, and hands. The part that is closest to the point where the limb attaches to the main trunk of the body is the **proximal** part. The part that is furthest away is considered **distal.** (*Hint:* Remember distance.) See figure 22.4.

☐ Medial and lateral

If you were to place your hand on your breast bone, you would be identifying the middle of your body, or the median line. If an anatomical structure is close to the median line, that structure is called **medial.** Likewise, the surface farthest away from the median line is considered **lateral.**

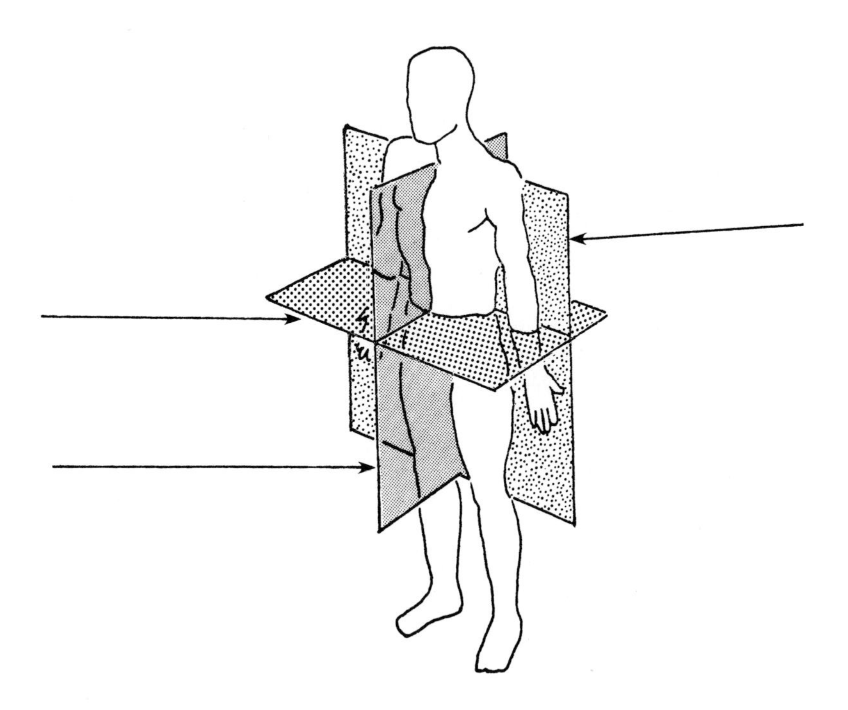

Figure 22.5 The planes of the body.

Do you have these terms in mind? Good! Now let us apply them. Identify the following phrases as to whether they indicate superior or inferior, anterior or posterior, dorsal or ventral, proximal or distal, medial or lateral.

1. Last digit of finger ______
2. Palm side of hand in relation to backside when the arm is hanging at the side ______
3. Nose in relation to shoulder ______
4. Elbow in relation to the wrist ______
5. Foot in relation to waist ______
6. Chest (standing upright) ______
7. Back (standing upright) ______
8. Outer surface of arm ______

22.7 In studying the human body, we frequently make use of sections through various planes of the body as a means of locating structures. Like the terms that place structures in relative positions, the following terms are used to describe these planes. Label figure 22.5 properly from the following descriptions.

Figure 22.6 Specific cavities of the body.

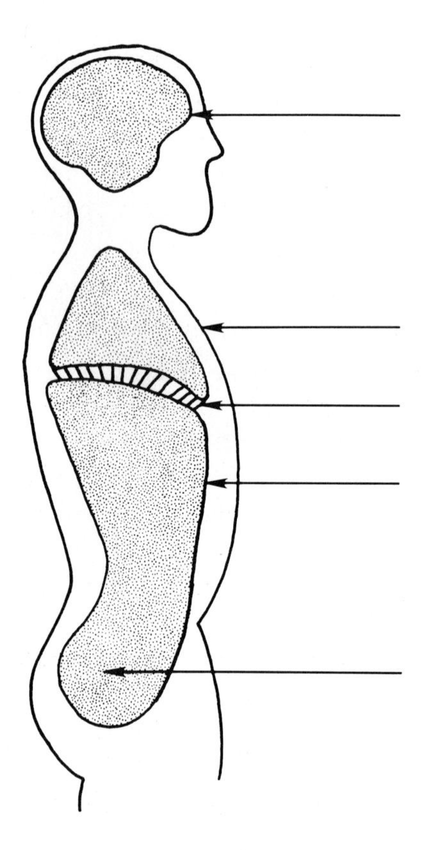

A **transverse plane,** cross section divides the body into upper and lower segments (portions). A **frontal plane,** or *coronal* plane, divides the body into front and back portions. A **longitudinal plane,** or *sagittal* plane, divides the body into right and left halves. A midsaggital plane divides the body into equal halves.

22.8 Figure 22.6 illustrates specific cavities in the body. In order to be effective with any study of anatomy, an understanding of the body cavities is needed. The body cavities are described below. Label each cavity from the descriptions.

The **thoracic** cavity contains the lungs and heart. It is separated from the abdominal cavity by a strong muscle called the **diaphragm.** The **abdominal** cavity, which is inferior to the thoracic cavity, contains the stomach, small intestines, liver, and other organs. Inferior to but continuous with the abdominal cavity is the **pelvic** cavity, which contains the urinary bladder, colon, and large intestine. The two cavities together are called the **abdomino pelvic** cavity. The **cranial** cavity, which is the most superior cavity of the body, contains the brain.

22.9 Each of the cavities is lined with a protective membrane. For example, the thoracic cavity is protected by the **plural membrane,** while the heart, also a part of the thoracic cavity, is protected by the **pericardium.** The organs of the abdomino-pelvic cavity are surrounded by the **peritoneum.** Three **meninges** surround the brain.

22.10 The background you now have should allow you to be more successful with the remaining chapters. You will note, however, that this chapter had a considerable amount of vocabulary. Before you go on, review the vocabulary by writing, in your own words, a working definition of each boldfaced or italicized word in this chapter. Do this on a separate sheet of paper and attach it to this chapter. Then review the study points below. Place a check (✓) in the box when you are certain you understand the objective.

On completion of this chapter, you should be able to do the following:

☐ 1. List the eleven body systems and give the main function of each.

☐ 2. Identify the five basic types of epithelial tissues.

☐ 3. Identify the various types of connective tissues.

☐ 4. Identify neurons.

☐ 5. Distinguish between the various relative positions of the body.

☐ 6. Identify and explain the planes that divide the body into sections.

☐ 7. Identify specific cavities of the body.

☐ 8. Identify the protective membranes that surround each cavity.

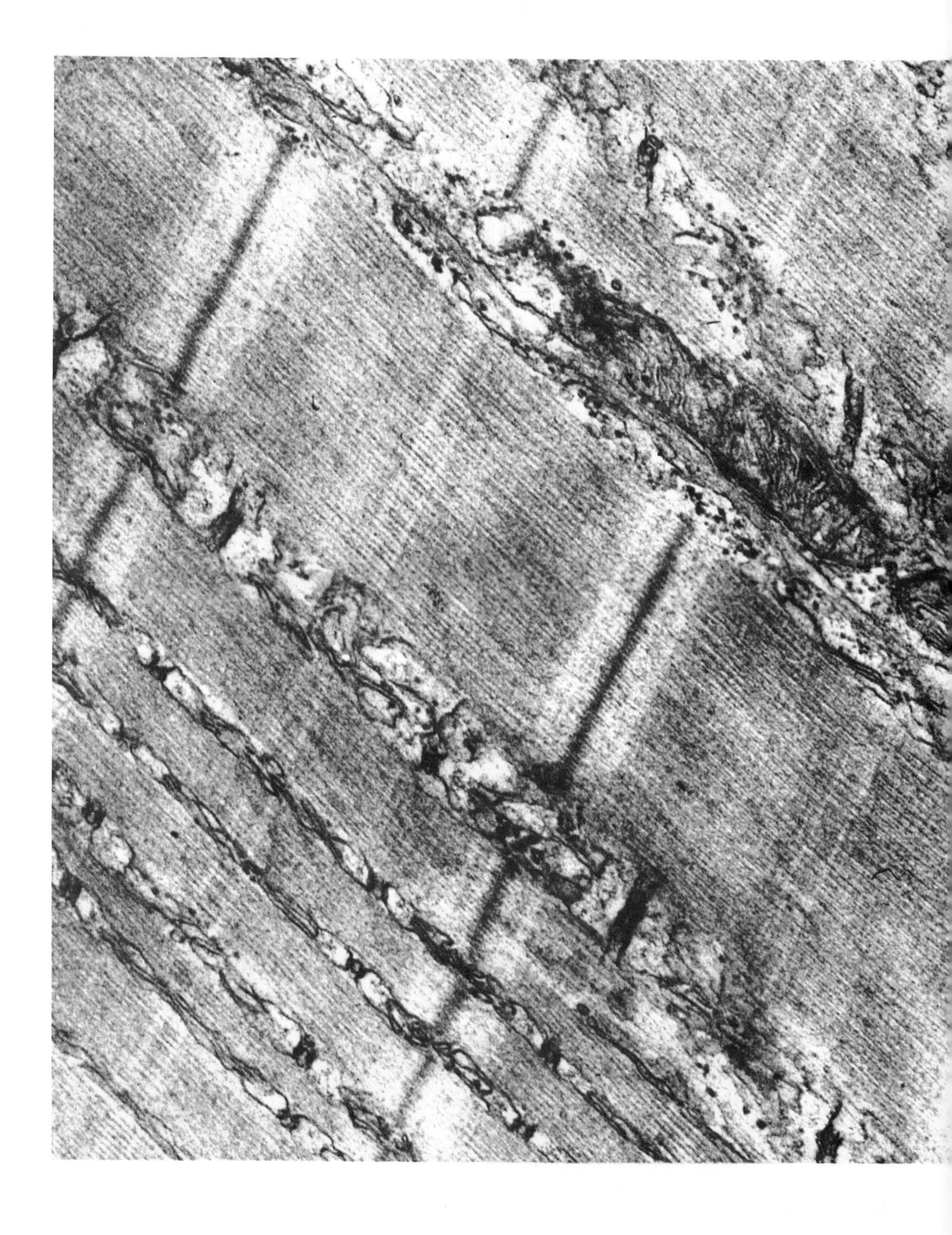

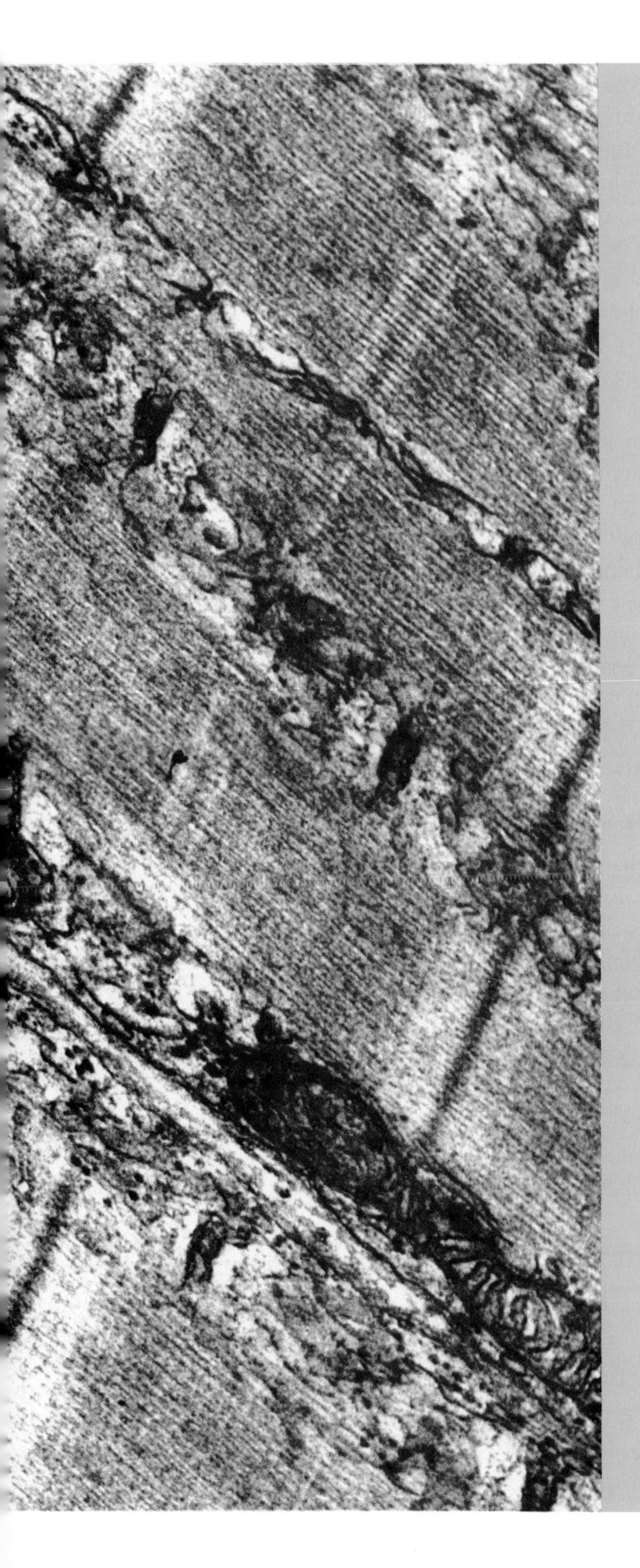

The Skeletal and Muscle Systems and Articulations 23

What are the skeletal and muscle systems and how do they effect articulation?

23.1 The skeletal system has four basic functions. One such function is *support*. Compared to the framework of a house or the steel girders of a building, the skeletal system is made of a strong, nonflexible substance called *bone matrix*.

A second function of the skeletal system is *protection*. The skeletal system encloses many of the vital organs of the body, such as the heart, lungs, spinal cord, and brain. The encasement of these organs gives these organs protection from injury.

A third function of the skeletal system is *movement*. When muscles, attached to two different adjacent bones and separated by a joint, contract, movement occurs.

A fourth function of the skeletal system involves a process called **hemopoiesis,** the process of forming new blood cells. Red blood cells are produced in the marrow of spongy bone in nearly all long bones, ribs, and vertebrae. *Red marrow,* called **myeloid** tissue, actively produces millions of red blood cells per minute because it is necessary to replace the old cells at the same rate as their removal. The life of a red blood cell is relatively short, probably less than 120 days.

The shaft of long bones, called the **diaphysis,** has a central cavity called the **medullary cavity.** This cavity is filled with yellow marrow. Yellow marrow consists mainly of fat cells. If the body requires more red blood cells than is being produced by the red marrow, yellow marrow can be converted to red marrow for that purpose. Refer to figure 23.1 to see the initial structure of a bone. Be familiar with each part that is labeled.

The ends of the long bones are called **epiphyses** and consist of cancellous or spongy bone. Each bone is covered with a tough connective tissue layer called the **periosteum.** The periosteum is primarily involved in the growth and repair of bone.

Bone is the hardest of all connective tissues due to the large amounts of calcium salts that make up its matrix. Bone cells are called **osteocytes.** Concentric layers of bone matrix and osteocytes around cylinder-shaped tubes called **Haversian canals** make up a Haversian system.

Haversian canals contain blood vessels that carry blood to the osteocytes. Osteocytes communicate with each other by many cytoplasmic canals called **canaliculi.** (See figure 23.2.)

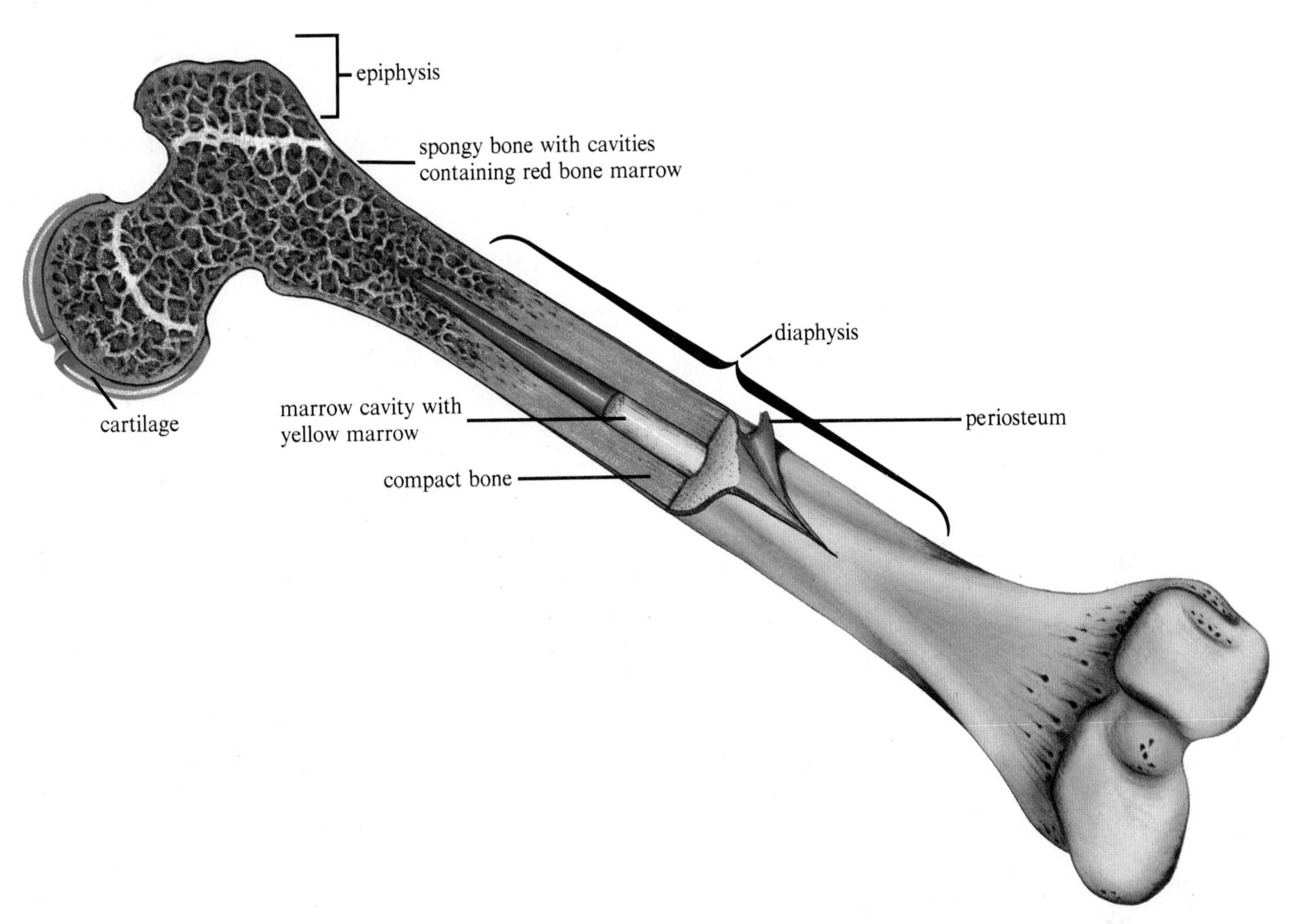

Figure 23.1 Basic anatomy of bones. Like the understructure of a building, your bones are your support structure.

23.2 In total, there are 206 bones in the adult skeletal system. Originally, at birth, there were 270 bones, but 64 bones have fused together. The skeletal system is divided, for convenience, into two major parts. One part, the **axial skeleton,** includes the skull, vertebral column, and ribs. The second part, the **appendicular** skeleton, consists of the bones of the upper and lower extremities, pectoral girdle, and pelvic girdle.

The bones of the skeletal system exist in four general categories, which are listed below. Check the box (✓) when you have mastered each. Refer to figure 23.3 for the location of the bones.

Data Table 23.1 The Bones of the Skeletal System Categorized According to Shape

☐	Long bones	The long bones consist of the femur, tibia, fibula, humerus, radius ulna, and the phalanges.
☐	Short bones	Included in this group are the wrist and ankle bones, which are called the carpals and tarsals, respectively.
☐	Flat bones	The flat bones include the ribs, scapula, and several of the cranial bones.
☐	Irregular bones	This group of bones makes up the remainder of the bones that are not listed in the above categories. The vertebrae, the sacrum, the coccyx, and the mandible are examples.

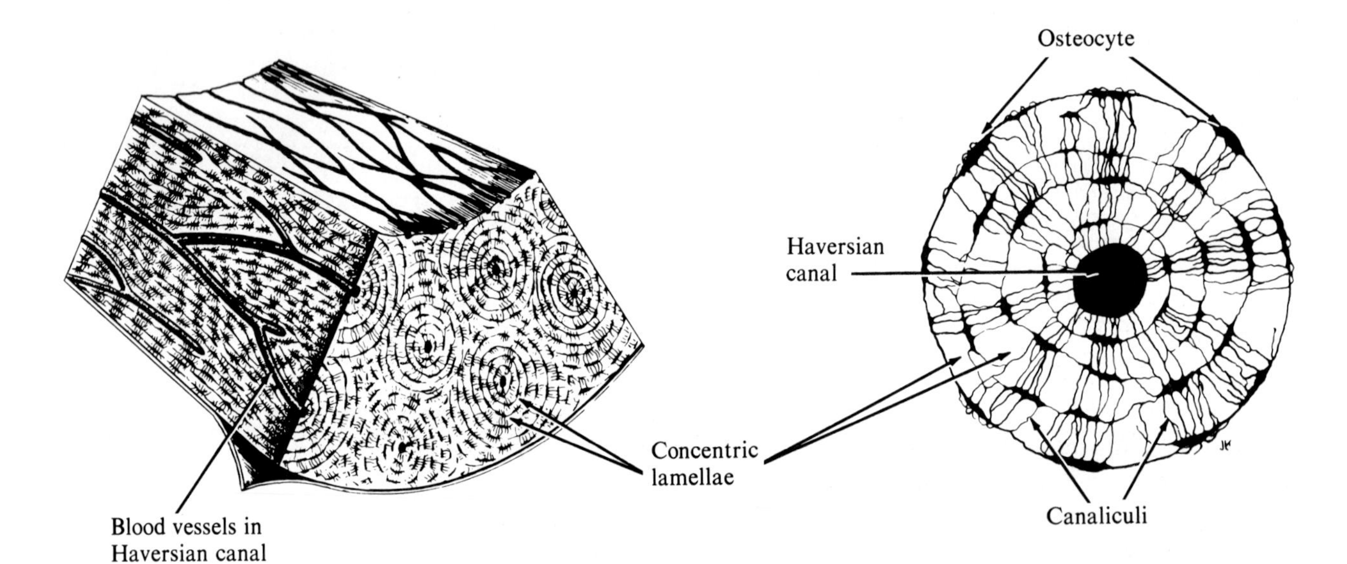

Figure 23.2 Histology of bone tissue. Note the many Haversian systems made up of osteocytes, canaliculi, lamellae, and Haversian canals. A Haversian system is often referred to as an osteon. Osteocytes are mature cells that no longer produce new bone tissue. They do, however, support ongoing cellular activities. The many canaliculi form an intricate transport network throughout bone to accommodate nutrient needs and removal of waste products. (Courtesy of James Koevenig.)

Data Table 23.2 Structural Terms Associated with the Skeletal System

	Term	Definition
☐	1. Foramen	A hole in a bone allowing for the passage of nerves and blood vessels.
☐	2. Fossa	A slight depression on the surface of a bone.
☐	3. Fissure	A narrow slit in a bone.
☐	4. Meatus	A canal-like opening through a bone; considered to be more tubelike than a foramen.
☐	5. Sinus	A cavity in a bone.
	Terms that relate mainly to **processes,** which are considered to be any prominence on a bone, are listed below.	
☐	1. Condyle	A rounded head of a bone for articulation with another bone.
☐	2. Crest	A ridge along a bone for muscle attachment.
☐	3. Head	A rounded projection that is supported on the neck of a bone for articulation.
☐	4. Spine	A sharp process for muscle attachment.
☐	5. Trochanter	A large roughened process on a bone for muscle attachment.
☐	6. Tubercle	A small rounded process for muscle attachment.
☐	7. Tuberosity	A large rounded process for muscle attachment.

23.3 Bones have *holes* and *processes* and contain *cavities* and *depressions*. Specific terms that relate to these holes, cavities, processes, and depressions are listed in data table 23.2. Study them carefully. Each is important when identifying a part of a bone. Place a check (✓) in the box when you have mastered the term and its definition.

While many of these terms seem unfamiliar to you at this time, they will become more familiar as they are used. To learn the names of the bones and their parts will take some time. You will be expected to identify bones verbally, as well as in a written exam. This is a big order, and you will find it will take a lot of work. If you have real bones available, use them while you study the various parts of the skeletal system.

23.4 The muscle system is comprised of three types of muscle tissue. Smooth and cardiac muscle tissues make up many of the internal organs, which will be discussed in later chapters. We will only be concerned with striated muscle tissue at this time. Striated muscles are also called skeletal muscles or voluntary muscles because of their association with the skeleton and types of movement.

Figure 23.3 Basic skeletal anatomy.

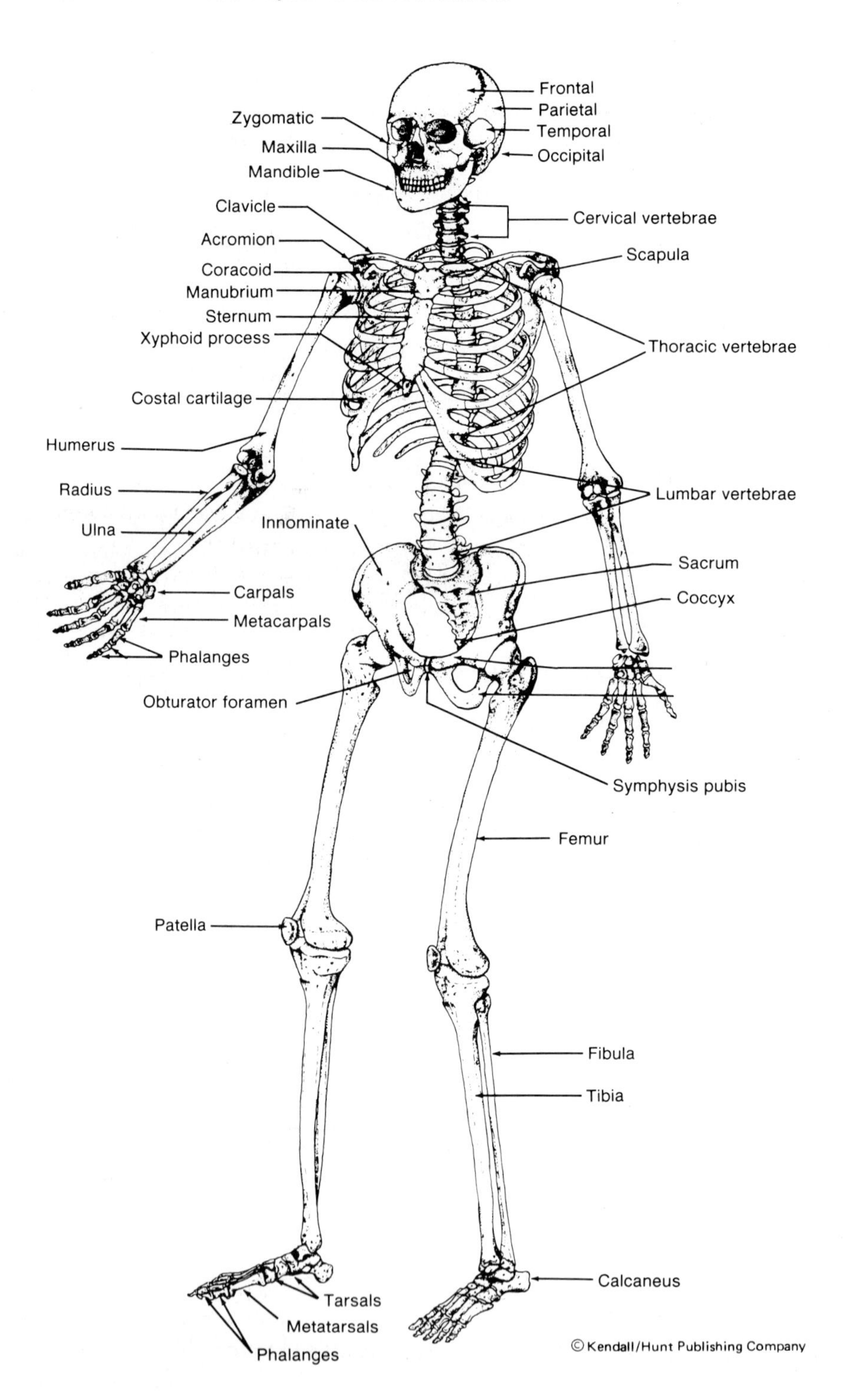

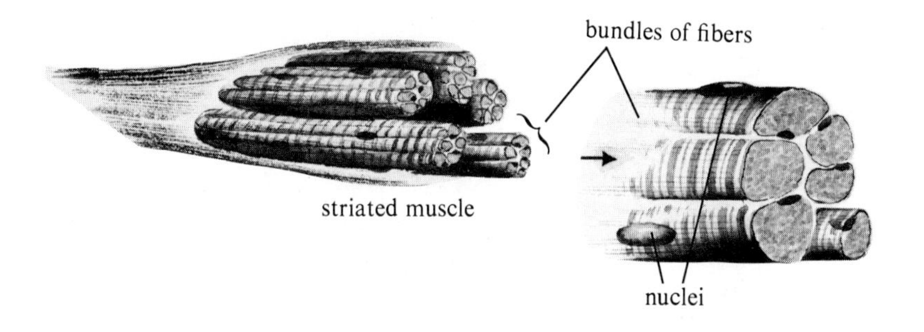

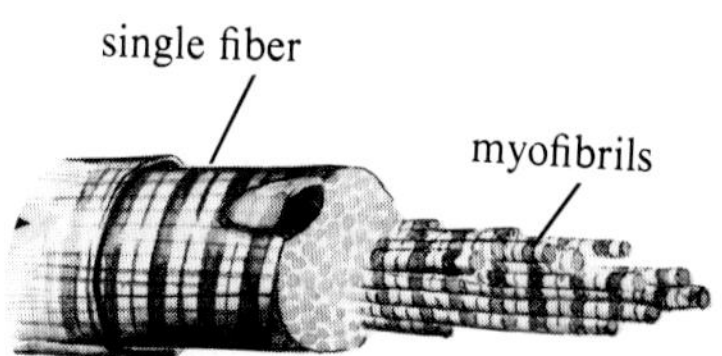

Figure 23.4 A generalized composite of muscle tissue. Courtesy of James Koevenig.

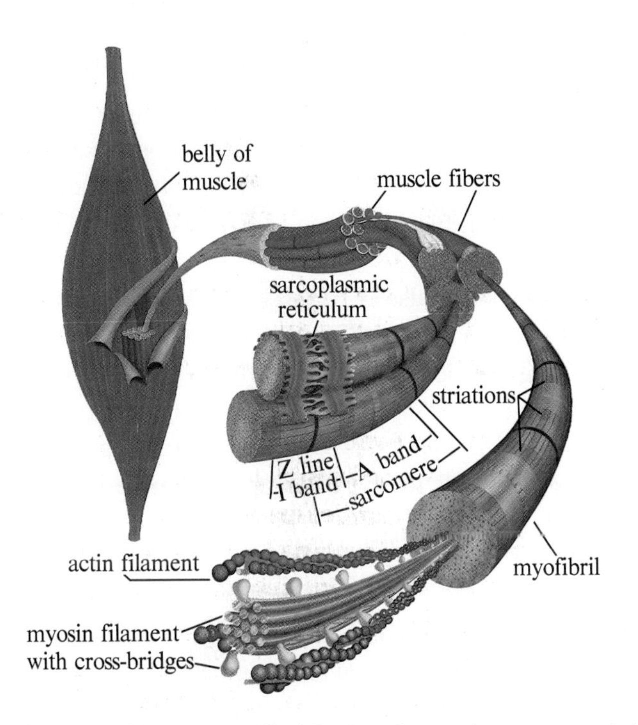

Figure 23.5 A second composite of muscle tissue. Compare this rendering with Figure 23.4. Note the positioning of the components of a **sarcomere,** the contractile unit from Z line to Z line.

Striated muscle fibers are cylindrical and contain many nuclei. Striated muscle fibers equal individual cells. The *cytoplasm,* or **sarcoplasm,** found in each fiber contains numerous mitochondria and **myofibrils** that are instrumental in muscle contraction.

Figure 23.4 illustrates the parts of striated muscle. Refer to it as required while you study the mechanics of skeletal muscle contractions.

Muscle contractions occur only after muscle tissue is stimulated by nerve cells called **motor neurons. Myosin filaments** and **actin filaments,** which are muscle proteins, play a major role in muscle contractions. The larger myosin filaments are found only in the A bands of a myofibril, while the smaller actin filaments are located in both the A band and the I band (refer to figure 23.5). *Note:* The region of the A band lacking actin filaments is known as the H zone. Note also that actin filaments do not cross

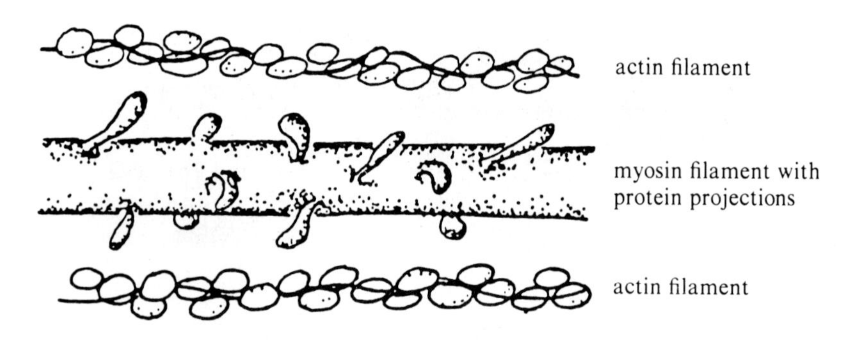

Figure 23.6 The contraction of muscles occurs as these filaments work together.

the Z line, but rather anchor to this region and extend into the A band. The distance from one Z line to the next is termed a **sarcomere.**

During a contraction, the I band is shortened and the H zone disappears when the actin and myosin filaments slide over one another. The overall effect is the shortening of the muscle tissue resulting in a contraction.

Refer to figure 23.6 regarding the tiny protein projections called **cross-bridges** on the myosin filament. These extensions from the myosin filament extend toward the actin filaments and play a valuable role in the contraction of muscle fibers.

When a nervous impulse is received by a muscle fiber, calcium ions (Ca^{++}) are released from the sarcoplasm of the muscle tissue. The result is the formation of an electrostatic bond between actin and myosin producing **actomyosin,** a neutral compound. The cross-bridges are temporarily attached to the actin filament. The cross-bridges collapse and the actin filaments are pulled towards the center of the sacomere, thus shortening the muscle.

Since the actin filaments are anchored to the Z lines of the muscle fibers, they slide over the myosin filaments in the contraction or shortening of the muscle, creating tension. This is why muscles feel tight when contracted. Draw your fist towards you to pronounce your bicep. Feel the tightness? The process described above goes on all the time. Every time your body moves, muscles are carrying out this process.

When the muscle is relaxed as a result of removing the neural stimulation, the calcium ions are removed, and the muscle returns to a resting state. However, even though the muscle tissue has relaxed, it is always contracted to some extent, a condition called **muscle tone.** Muscle tone is important to maintain, but it may be lost if the nerves are impaired in any way.

Figure 23.7 illustrates the general stage of muscle tissue contraction. Discuss what is occurring.

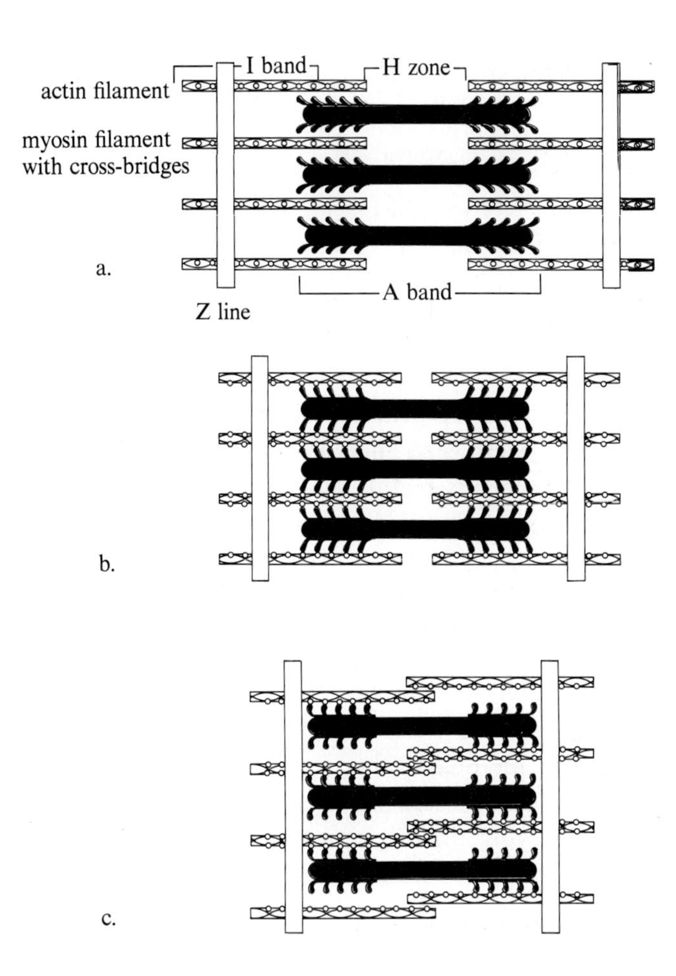

Figure 23.7 The mechanism for muscle contraction: (a) actin and myosin filaments at a relaxed state; (b) cross-bridges attach to the actin filaments and essentially pull the actin filaments toward each other causing (c) a shortening of the muscle. The estimated time for the entire process is 1/40th of a second.

23.5 Muscle contraction requires considerable energy to be expended in order to do the work involved. This energy produced by certain cell components is in the form of specific high energy ATP compounds.

Each cell of muscle tissue contains numerous **mitochondria** that produce the ATP molecules. Myosin filaments contain an enzyme called ATPase, which causes ATP to break down into ADP and P. The energy that held the ADP and P together is released and used for muscle activity by allowing the reaction to occur between actin and myosin filaments.

$$\underset{\text{triphosphate)}}{\underset{\text{(Adenosine}}{\text{ATP}}} \xrightarrow[]{\substack{\textit{ATPase} \\ \textit{(enzyme)}}} \underset{\text{diphosphate)}}{\underset{\text{(Adenosine}}{\text{ADP}}} + \underset{\text{(Phosphate)}}{\text{P}} + \underset{\text{released}}{\text{Energy}}$$

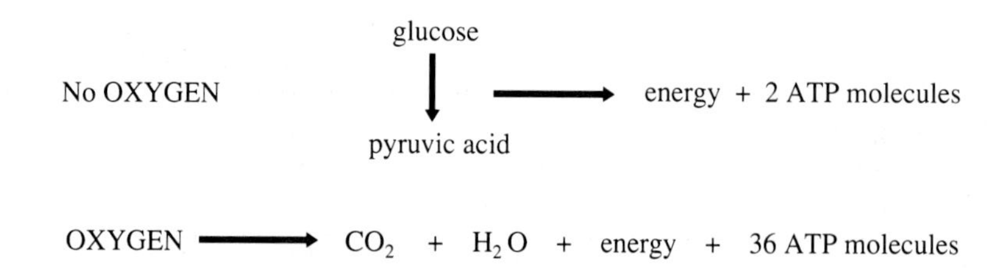

Figure 23.8 Generic formula for ATP use.

Oxygen is required for this reaction to occur. Oxygen is obtained by respiration and is carried to the cells of the body by hemoglobin, found in red blood cells.

Figure 23.8 illustrates a generic formula for ATP use. In this diagram the top part of the system is an **anaerobic** reaction, where minimal ATP is produced. Where oxygen is readily available, an **aerobic** reaction occurs, providing 36 ATP units. Normal muscle activity would be represented by the second reaction.

When would the first reaction be prevalent? Explain.

23.6 Muscles have an **origin** and an **insertion.** The origin is the end of the muscle that is fixed or relatively immovable. The insertion is the end of the muscle that is attached and capable of moving a joint. Occasionally, some muscles have more than one origin. Complete data table 23.3 by identifying the origin, insertion, and action of each muscle listed. You will need to use reference materials.

The movement of a part of the body is accomplished by the use of many muscles, not just a single one. One muscle, however, is determined to be the **prime mover,** or the *main muscle.* The muscles that assist the main muscle are called **synergists.** Muscles that have an opposing action to the prime mover are called **antagonists.** It should be noted that muscles come in pairs, and for every action by one muscle, there is an opposing muscle doing an opposite action. For example, the prime mover in flexing your arm is the *biceps brachii.* The antagonist to this muscle is the *triceps brachii,* which allows you to extend your arm.

23.7 Figures 23.9a and b illustrate anterior and posterior views of the surface muscles. Study them so you know the muscle on sight. Remember, these are just the surface muscles and do not illustrate the many underlying muscles of the body.

Data Table 23.3 Selected Muscles of the Human Body

Name of Muscle	Origin	Insertion	Activity
1. Trapezius			
2. Deltoid			
3. Biceps brachii			
4. Triceps brachii			
5. External oblique			
6. Rectus femoris			
7. Vastus lateralis			
8. Gastrocnemius			
9. Soleus			
10. Latissimus dorsi			
11. Biceps femoris			
12. Gluteus maximus			
13. External oblique			
14. Sternocleidomastoid			
15. Masseter			

Figure 23.9a Human muscle system, anterior view.

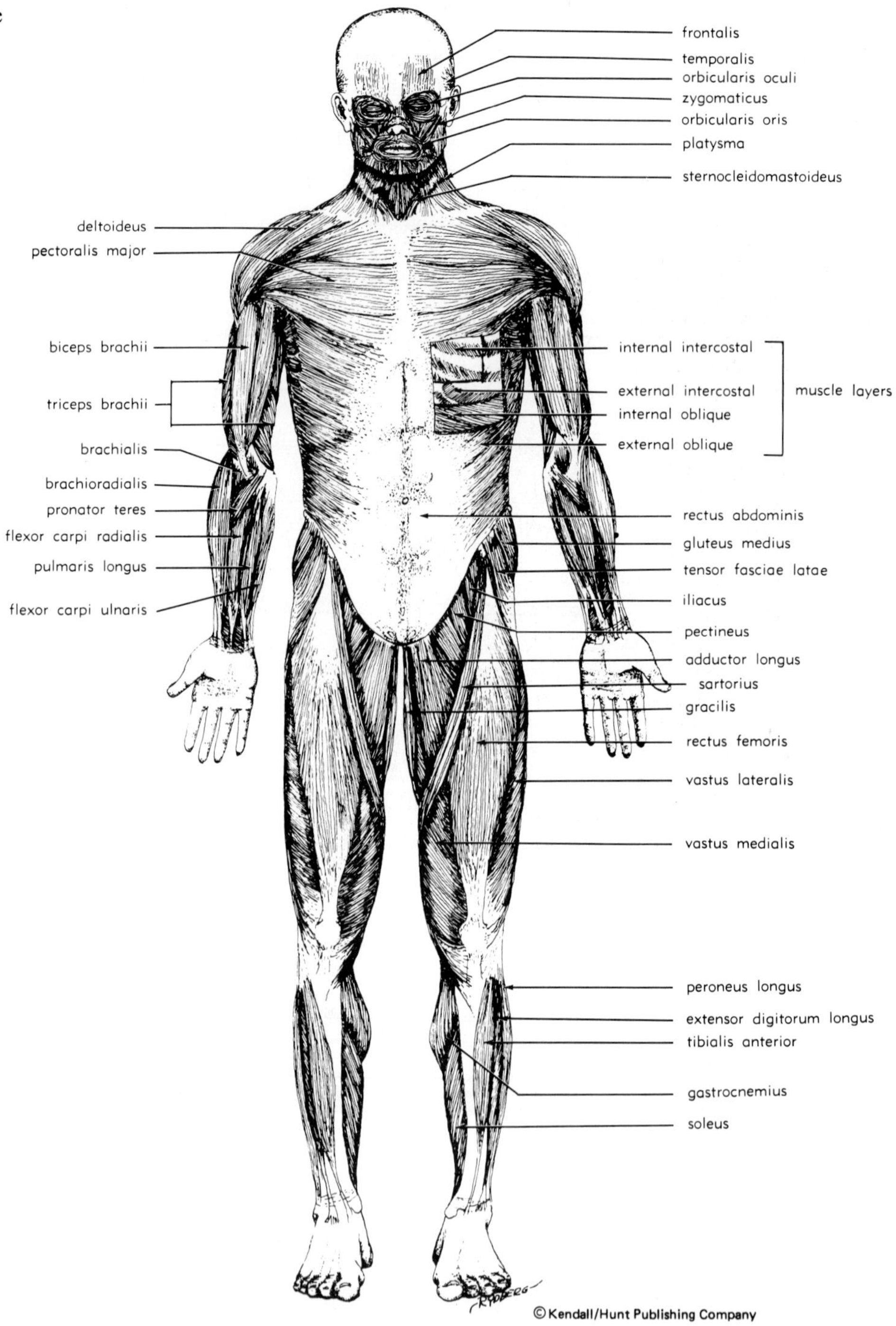

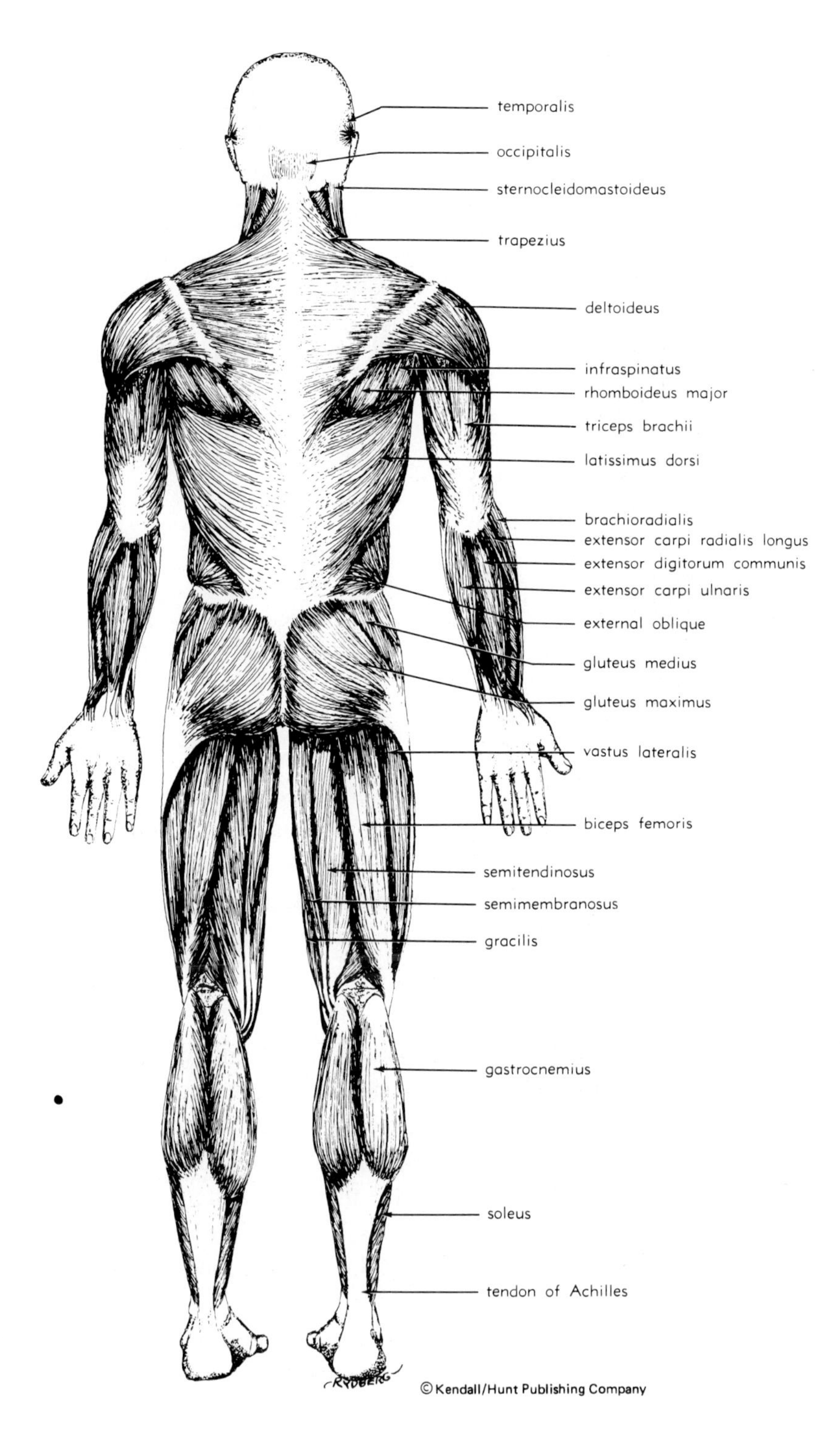

Figure 23.9b Human muscle system, posterior view.

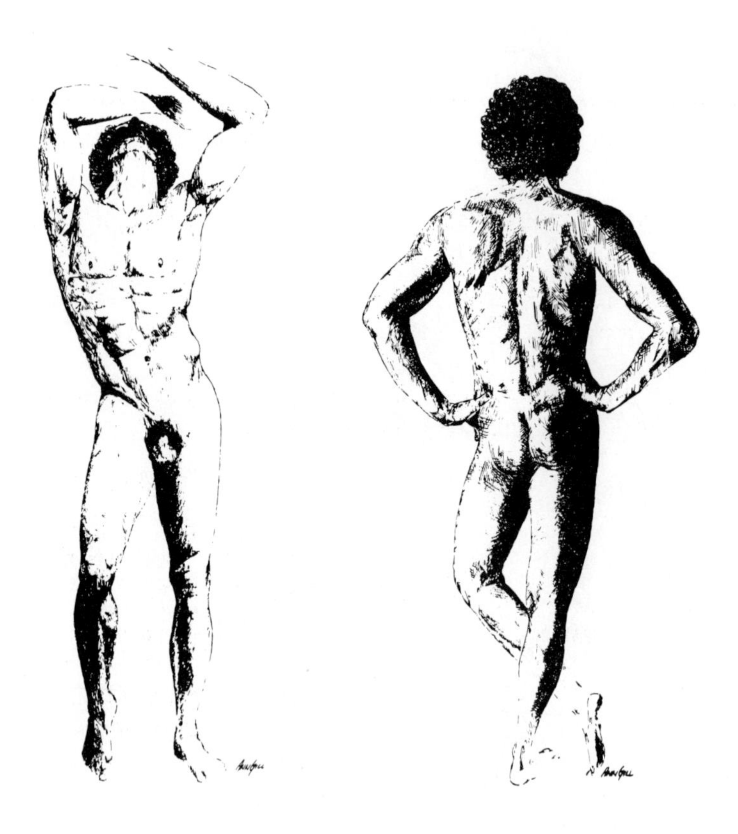

Figure 23.10 Dorsal and ventral view of surface muscles.

23.8 Having studied the surface muscles as they appear anatomically, look now at figure 23.10. How many muscles can you distinguish? Label the noticeable muscles with a blue pen. When you have done this, label those muscles that you cannot distinguish, but note their approximate location with a red pen.

23.9 The term **articulation** literally means a joining position of two bones. In some cases, the articulation of two bones allows considerable movement; in other instances, there is virtually no movement. The importance of identifying with articulations is significant because of the association of the skeletal system with the muscular system.

Three general classes of articulations based on the degree of movement in each are presented in data table 23.4. Place a check (✓) in the box as you master each articulation.

Each of the joint categories are necessary because of different needs that exist in various bone connections.

Data Table 23.4 Articulations

☐	1. Synarthroses (immovable)	Joints of this type have no movement.
☐	2. Amphiarthroses	Joints of this type have limited movement.
☐	3. Diarthroses (freely movable)	Joints of this type have a high degree of movement.

If we take a closer look at the different joint classifications, we will find subclasses. There are two subclasses to *synarthroses.*

☐	1. Suture	Sutures are found mainly in the skull and are characterized by two bones joined together by fibrous tissue.
☐	2. Syndesmoses	In this form of articulation, bones are joined together by ligaments. An example woud be the articulation between the distal ends of the radius and the ulna.

Joints that tend to be only slightly movable or have minimal articulation are called **amphiarthroses.** This category has two subclasses.

☐	1. Symphyses	These are joints where a bone is joined together by fibrocartilage, such as the pubic symphysis.
☐	2. Synchondrosis	The joining of the epiphyses and the diaphysis of developing long bones are examples of synchondrosis. Normally, the two boney surfaces are connected by cartilagenous tissue.

Data Table 23.4 *Continued*

The third joint category is called **diarthroses.** These joints are characterized by the free motion they are capable of accommodating.
The characteristics of diarthroses are:

1. A cavity encapsulated with fibrous cartilage and reinforced with ligaments.
2. Opposing bones are covered with cartilage that provides a smooth gliding surface. This surface operates, in part, by the action of **synovial fluid.** Synovial fluid is produced by the synovial membrane that covers the articulating surfaces of the bones of a joint. It lubricates the articulating surfaces, thus minimizing friction. While under pressure in a joint, synovial fluid keeps the surfaces of the two bones from touching. The **viscosity** of synovial fluid is often affected by temperature. This could account for the tendency of certain joints to feel stiff during cold weather.

Examples of joints that are categorized as diarthroses are

1. the ball and socket joint of the hip and the femur;
2. the hinge joint of the elbow, where the radius and ulna unite with the humerus;
3. the pivot point of the skull on the axis;
4. the saddle joint of the thumb;
5. the gliding joint of vertebrae and the intervertebral discs; and
6. the condyloid joints between metacarpals and phalanges.

23.10 The ability to move an arm or a leg is a combination of action between the joint, muscles, tendons, and ligaments functioning in harmony. Since there is little or no movement in synarthrodial and amphiarthrodial joints, the majority of movements are associated with diarthrodial or synovial joints.

Fourteen basic types of motion that may be accomplished by various synovial joints are presented in data table 23.5. Figure 23.11 supports data table 23.5. As they are described, you should do the movement, using your arms and/or legs. Place a check (✓) in each box as you master the concept.

23.11 The examples given above can be tried by you as you review the descriptions. You should note that other synovial joints can render the same action as those described above and that this listing is not all-inclusive.

Data Table 23.5 Body Movements

	Movement	Description
☐	1. Abduction	Lift your arm until it is extended straight out from the shoulder and away from the midline of the body.
☐	2. Adduction	This is the opposite of abduction. The lowering of an extended arm to your side is adduction.
☐	3. Pronation	With your arm hanging down to your side, turn your hand so your thumb is near your body.
☐	4. Supination	The opposite of pronation is accomplished by turning your palm out so the thumb is pointing away from the body.
☐	5. Flexion	Make a fist and draw your forearm towards you. This motion is flexion.
☐	6. Extension	The opposite of flexion is extension, or movement of your forearm away from you.
☐	7. Rotation	This movement is accomplished by turning your head at its axis.
☐	8. Circumduction	Many times a fast-ball pitcher makes a complete rotation of his pitching arm before releasing the ball. This is circumduction. Very few joints are capable of giving this much motion.
☐	9. Retraction	This motion is accomplished by opening the mouth.
☐	10. Protraction	The opposite action of retraction would be closing the mouth.
☐	11. Inversion	This type of movement is accomplished by moving the foot from its flat position so that an inward motion is completed. The large toe is turned upward by this motion executed at the ankle joint.
☐	12. Eversion	Eversion is the opposite of inversion and is accomplished by turning the foot outward.
☐	13. Dorsiflexion	This motion also affects the foot. It is accomplished by pulling the toes back when the foot is extended in front of you as you sit with your leg elevated.
☐	14. Plantar flexion	This motion is the opposite to dorsiflexion, accomplished by pointing your toes.

Joints are the junctions between two bones. Together with muscles they result in movement. Having completed a study of the skeletal system, muscular system, and articulation, you are now ready to take a brief look at the system that enables motion to happen in the first place, namely, the nervous system. But first, check over the study points for this chapter. When you know the study point listed, place a check (✓) in the box.

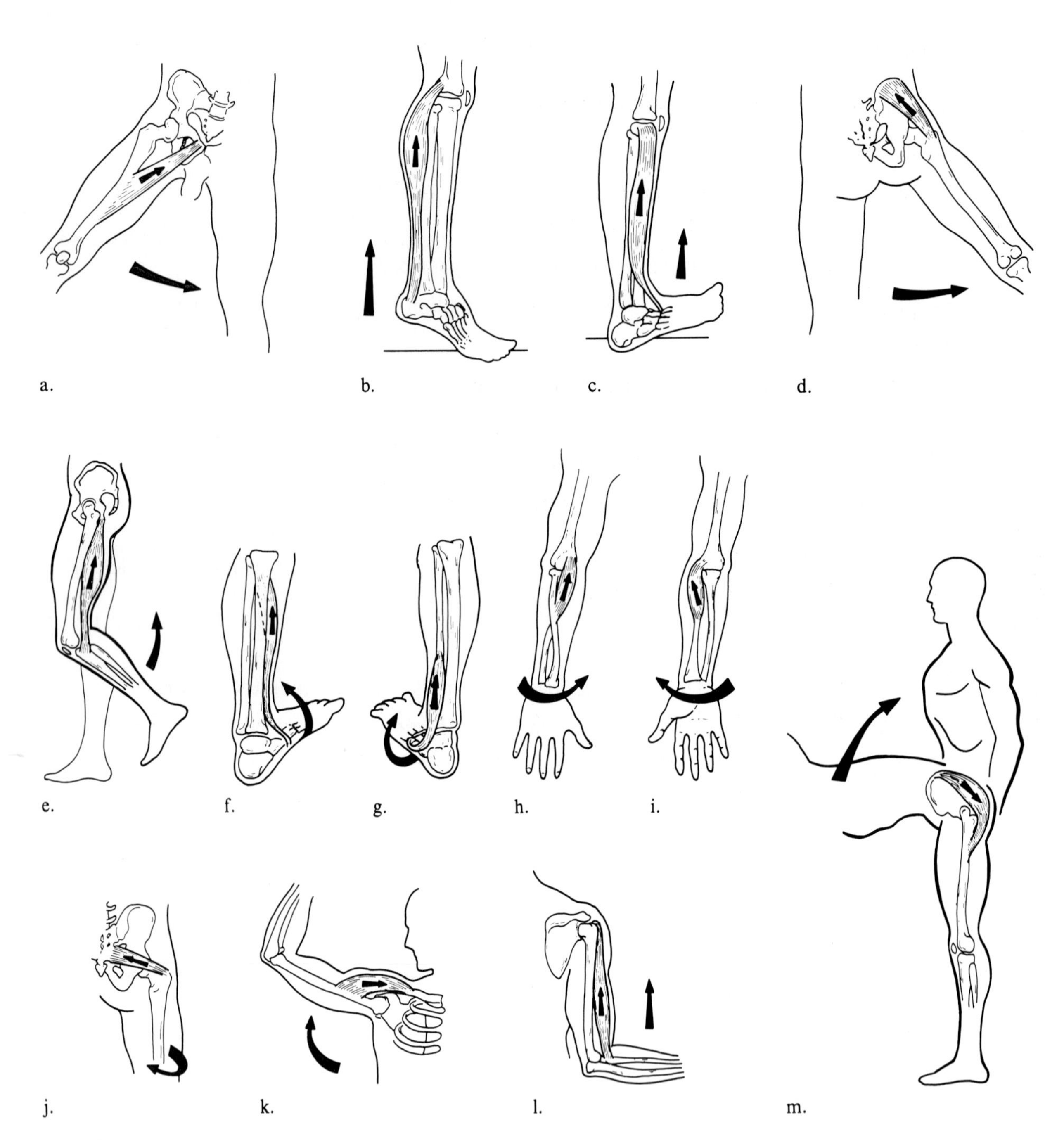

Figure 23.11 Body movements due to joining articulations.

On completion of this chapter, you should be able to do the following:

- ☐ 1. Identify the four functions of the skeletal system.
- ☐ 2. Identify the basic components of a typical long bone.
- ☐ 3. Distinguish between and recognize assigned long, short, flat, and irregular bones of the skeletal system.
- ☐ 4. Identify the various components of bone tissue.
- ☐ 5. Identify and recognize assigned processes, cavities, and depressions of bones.
- ☐ 6. Identify and explain the sequence and structure of muscle tissue.
- ☐ 7. Explain the process of muscle contractions.
- ☐ 8. Explain the difference between an origin and an insertion.
- ☐ 9. Identify assigned muscles.
- ☐ 10. List the origin, insertion, and action of specific muscles.
- ☐ 11. Identify and explain the various articulations and joint classifications.
- ☐ 12. Identify six diarthroses joints.
- ☐ 13. Explain the fourteen basic motions that are accomplished with various diarthroses joints.

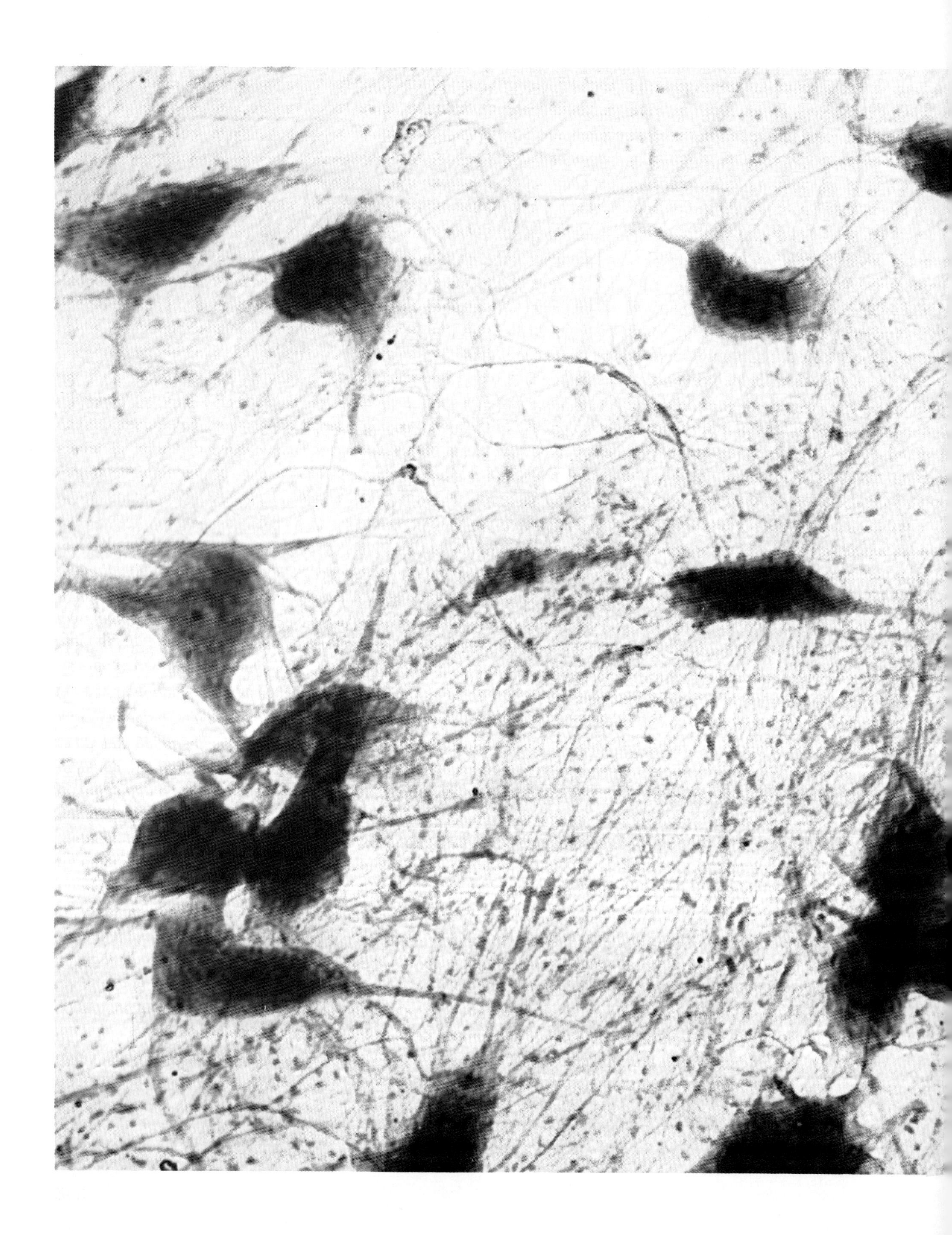

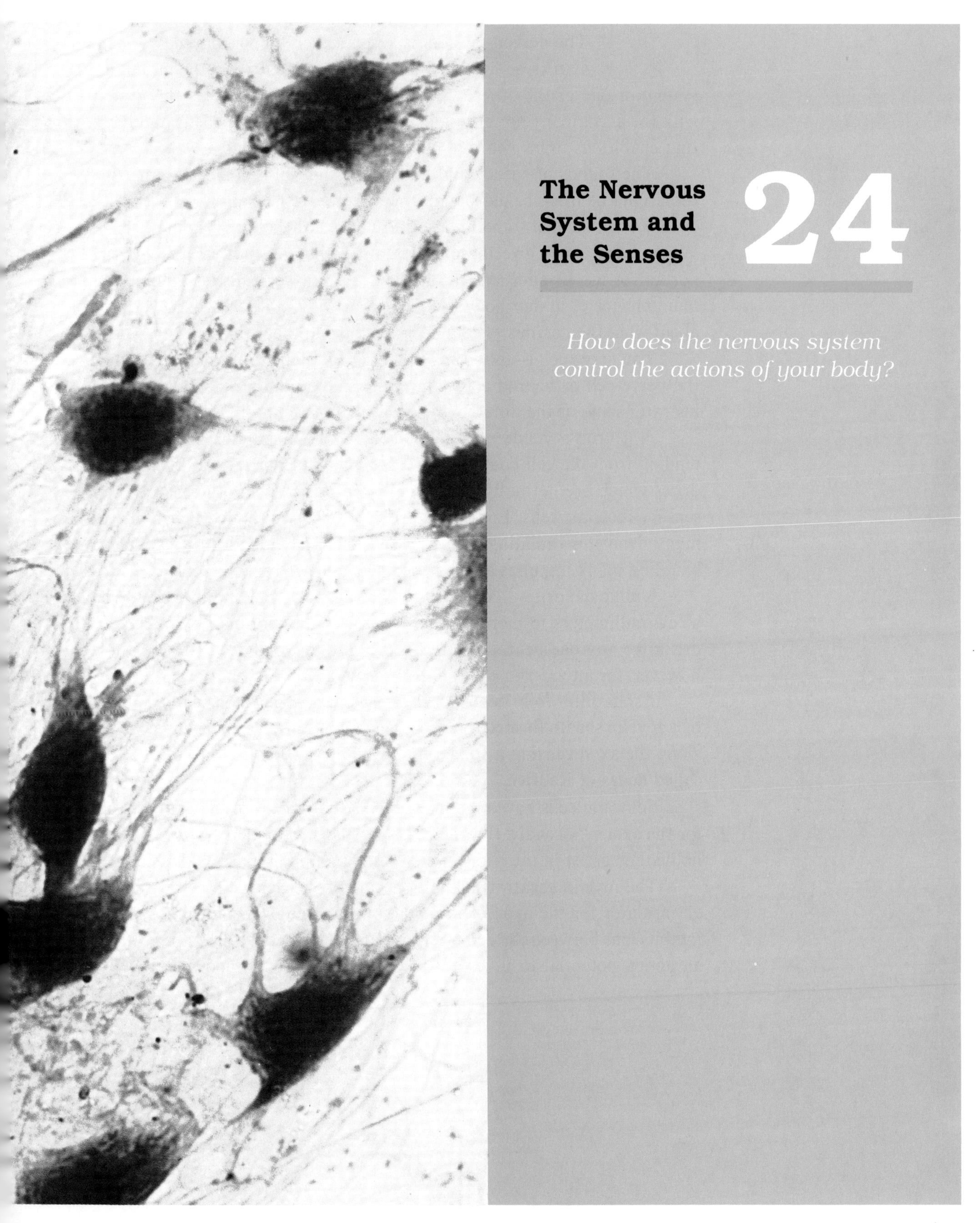

The Nervous System and the Senses 24

How does the nervous system control the actions of your body?

24.1 The nervous system is an extremely important body system. Can it be qualified as more important than the circulatory system or one of the other systems? Probably not. But, it is this system, with the accompanying senses of sight, hearing, smelling, taste, and touch, that enables us to be aware of our external surroundings.

The sounds of speech and music, the sight of a golden summer day, the aroma of fresh baked bread, the sweetness of cookies fresh from the oven, and the feel of something soft are ours to behold through our senses. We probably don't give these senses much thought, but each is very important to us. In addition, the nervous system carefully coordinates the maintenance of all our body functions. For the most part, you never think about what the nervous system does for you and that is okay, its automatic.

The nerve cell is unlike any other cell. Specific to the nervous system, the nerve cell or neuron comes in many shapes and sizes. Figure 24.1 illustrates some of the different types of neurons found in the human body.

A neuron is made up of a cell body, which contains a nucleus. Extending from the cell body are processes called **dendrites** and **axons.** Dendrites which are normally short, receive impulses from other neurons, or, sensory receptor cells. Look at figure 24.2. You will notice that there are many dendrites radiating from the cell body. The dendrite is capable of handling many impulses because of this extensive branching.

Unlike dendrites, axons are typically longer, ranging in length from a few millimeters to more than a meter. While there may be many dendrites for any one neuron, there is normally just one axon. It is possible, however, for an axon to have several secondary branches.

Axons differ from dendrites in another way. They are usually enclosed by a myelin sheath formed by specialized **Schwann cells.** Schwann cells gap along the axon causing it to look lumpy. The gaps have a name. They are called **nodes of Ranvier.**

Schwann cells have several functions. For one, they provide nutrition for the axons. Secondly, they help damaged axons to mend themselves. Finally, they assist in the transmission of nerve impulses.

The myelin sheath is actually a part of the Schwann cell. It acts as an insulator for the axon. As impulses flow through the axon, the sheath inhibits ions between the cell interior and exterior from interacting causing an action potential to be established. However, at the nodes of Ranvier,

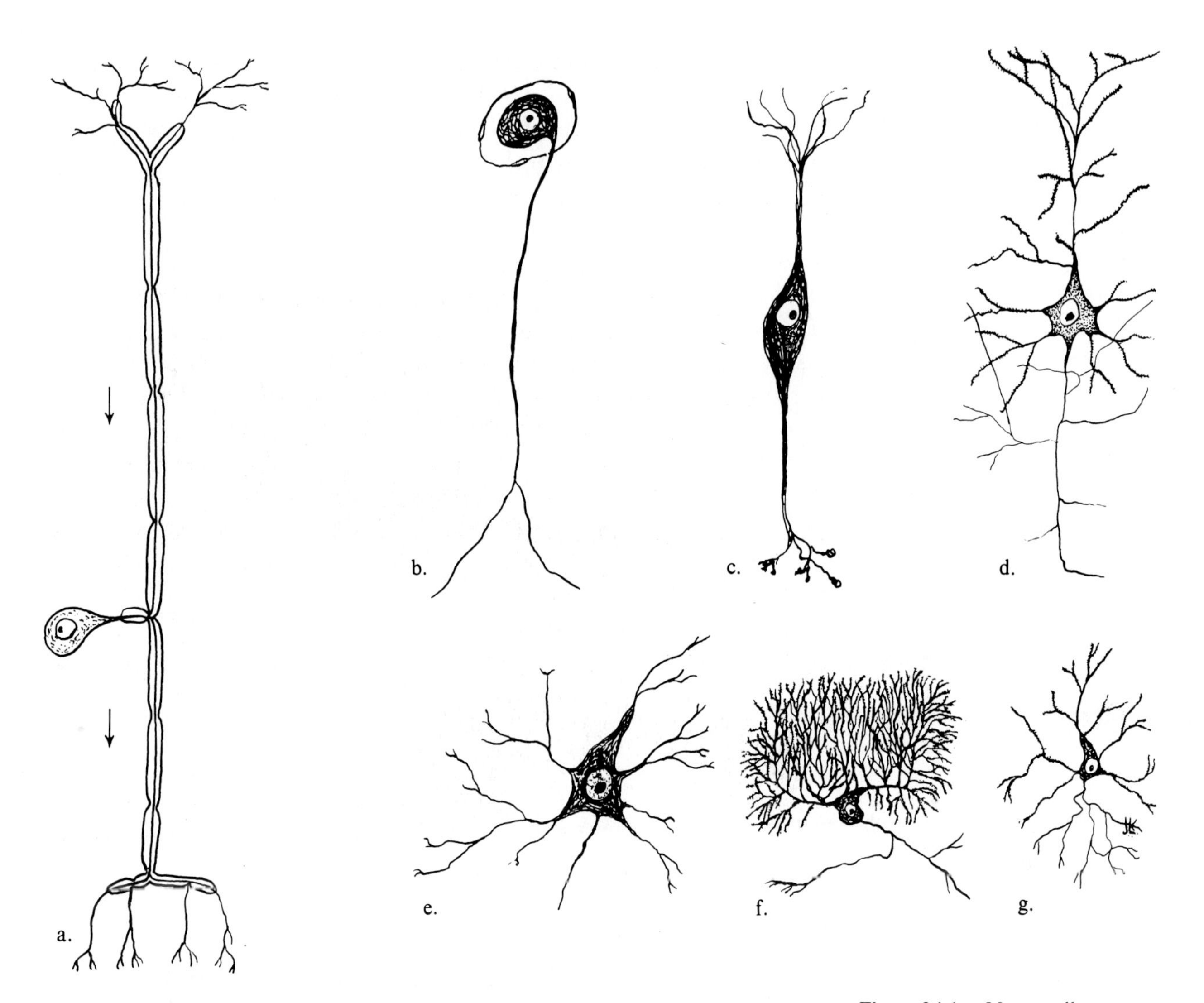

there is no myelinated covering. So, at these points, an ion exchange is possible. The result is the membrane, at the nodes of Ranvier, is depolarized and an action potential is set off. The nerve impulse, then, must leap the gap causing rapid impulse conduction. Scientists have found that myelinated nerves are able to transmit a nerve impulse twelve times faster than an unmyelinated nerve. Or another way of putting it, the unmyelinated dendrites do not transmit nerve impulses as fast as the myelinated axons.

Figure 24.1 Nerve cells: (a) Sensory neuron, (b) Unipolar cell, (c) Bipolar cell, (d) Pyramidal multipolar cell, (e) Star shaped multipolar cell, (f) Purkinje's cell, and (g) Granule cell. (Courtesy of James Koevenig.)

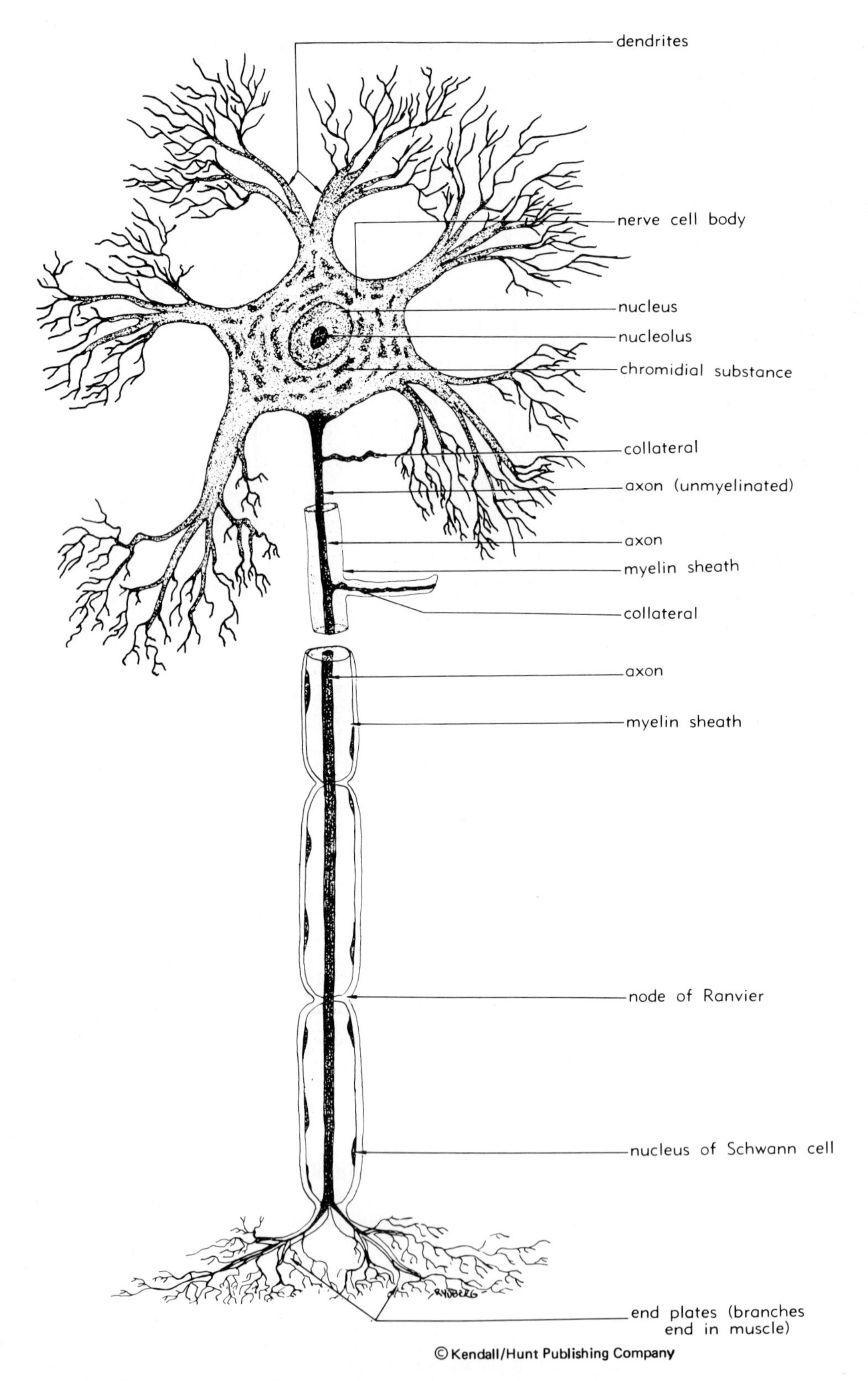

© Kendall/Hunt Publishing Company

Figure 24.2 Anatomy of a motor neuron. Compare this neuron to the sensory neuron illustrated in figure 24.1. How are they similar? different?

As the impulse travels through the axon, it ultimately reaches the end of the axon. At that point the impulse must jump from the axon of one neuron to the dendrites of another. The gap between the axon and the dendrites is called a **synapse.** The signal traverses the *synaptic junction,* not as an electrical spark, but by means of chemicals called **neurotransmitters.** The neurotransmitters travel the synaptic gap triggering the nerve impulse in another neuron. The time involved for neurotransmitters to become established and cross a synaptic junction is about 1/40th of a second. Some of the more common neurotransmitters are epinephrine (adrenalin), norepinephrine, and acetylcholine.

Nerve impulses are received and carried to the spinal cord or brain for a reaction. That impulse is carried by the nerve cells as described above. These nerve cells are called *afferent sensory neurons.* The reaction of the body is translated back to a gland or muscle by *efferent motor neurons.* The entire process is known as a **reflex arc.** The doctor tapping your knee, to determine the presence of reflexes is a good example of the reflex arc and the movement of a nerve impulse. Figure 24.3 illustrates the movement of an impulse from reception to the spinal cord and back to a muscle for reaction. The knee jerk type of action involves only two neurons.

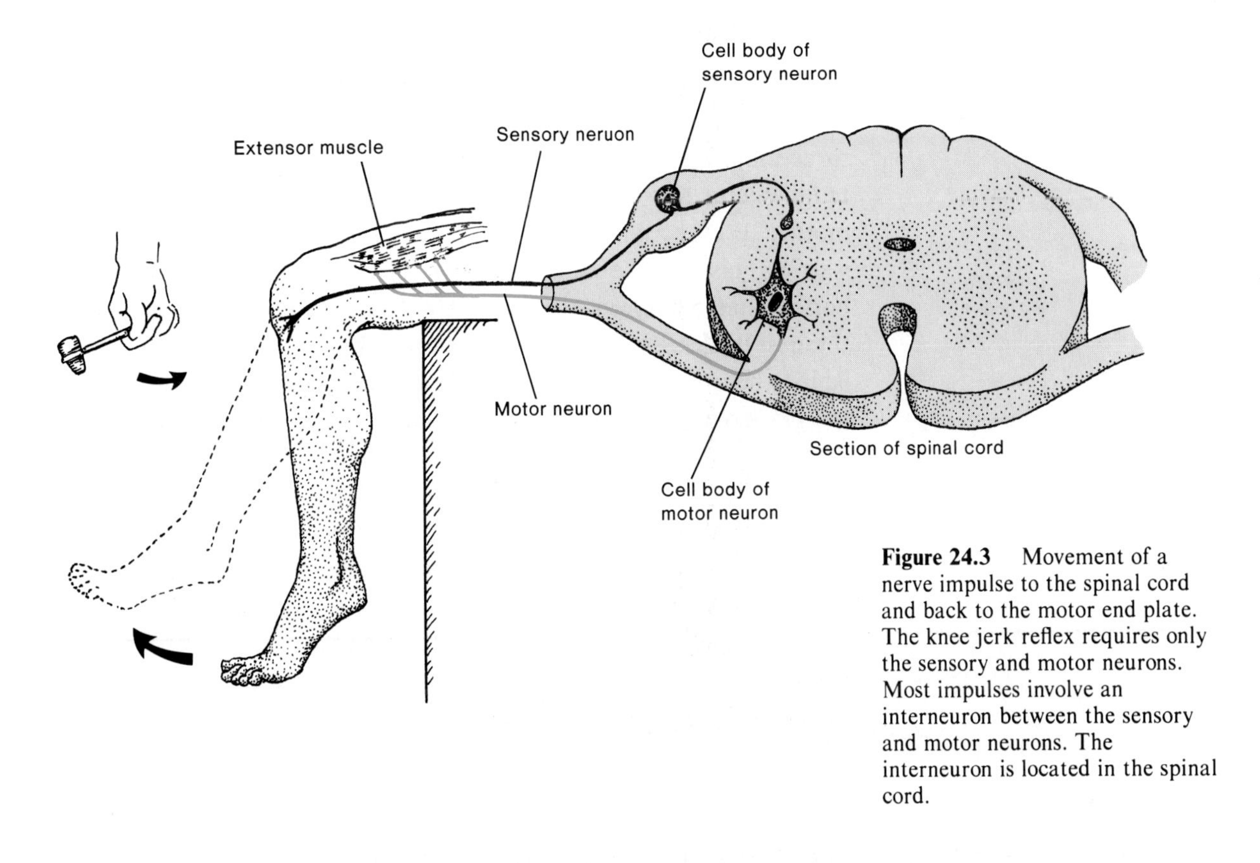

Figure 24.3 Movement of a nerve impulse to the spinal cord and back to the motor end plate. The knee jerk reflex requires only the sensory and motor neurons. Most impulses involve an interneuron between the sensory and motor neurons. The interneuron is located in the spinal cord.

Some impulses, however, involve three neurons, the sensory neuron, the motor neuron and an interneuron which is found in the spinal cord. The interneuron lies between the sensory neuron and the motor neuron. It receives an impulse from a sensory neuron, and, in turn, passes the impulse on to the motor neuron.

Your teacher has placed some prepared slides of neurons on the supply table. Get one of the slides and view it with the aid of a microscope. Illustrate the nerve cell that you see.

Label all parts that are visible.

24.2 There is more to the nervous system than just neurons. We need to now look at what constitutes the *central nervous system* and the *peripheral nervous system.* Dividing the nervous system into these two categories provides us a means of looking at this complex system in a more organized manner.

The central nervous system (CNS), is made up of the brain and spinal cord. The peripheral nervous system is made up of the spinal and cranial nerves. Let's look at the central nervous system first.

Central Nervous System

The main organ of the central nervous system is the brain. Figure 24.4a illustrates the brain. Encased in the skull, the brain is surrounded by three tough membranes called **meninges.** The outermost meninx, made of tough white fibrous connective tissue, is called the **dura mater** ("hard mother"). Directly below the dura mater is the **arachnoid layer** ("spider's web"). This weblike mater offers additional protection to the brain. The bottom meninx layer is the **pia mater** ("soft mother"). The pia mater is the thinnest of the layers, and contains many blood vessels which aids in nourishing the underlying cells of the brain. Between the arachnoid layer and the pia mater exists a watery fluid called **cerebrospinal fluid.** Along with hair, skin, and the skull, the main function of these layers is to protect the brain from injury.

The brain is made up of four basic parts: the cerebrum, the cerebellum, the diencephalon, and the brain stem. (Refer to figure 24.4b). Each part of the brain has an anatomy of its own with respective functions.

The cerebrum is the largest part of the brain, bearing four main lobes. Data table 24.1 identifies each lobe and its function. Using figure 24.4a and b as an aid check (✓) the appropriate box when you know where each lobe is located and its respective function.

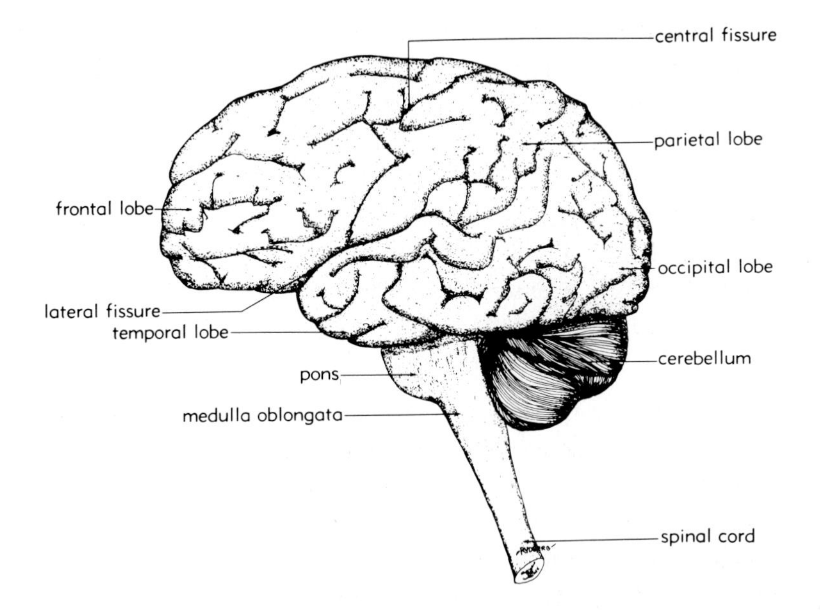

Figure 24.4a Basic anatomy of the human brain.

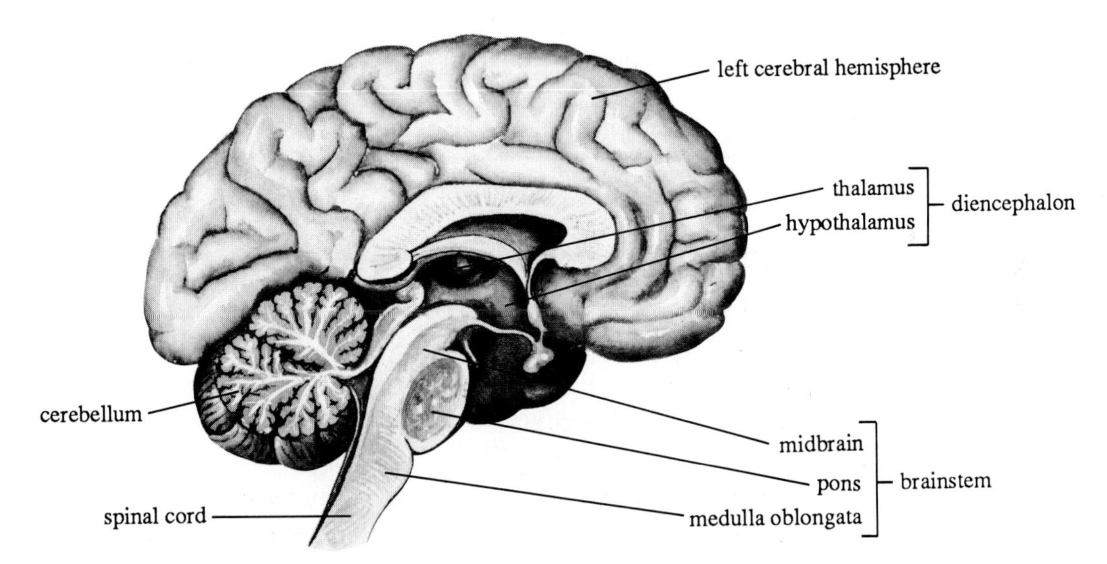

Figure 24.4b Sagittal section of the human brain showing the four basic regions. (Courtesy of James Koevenig.)

Data Table 24.1 The Four Lobes of the Cerebrum

Lobe	Function
☐ 1. Frontal	Maintains control of voluntary muscles; also the region where complex planning and problem-solving take place.
☐ 2. Occipital	Center for all visual senses.
☐ 3. Parietal	The senses of touch, pressure, and pain from the skin; also the area responsible for speech.
☐ 4. Temporal	Responsible for the sense of hearing; also the memorization of music and visual senses.

24.3 The brain stem and diencephalon are also shown in figure 24.4b. The main parts of the brain stem and the diencephalon, and their respective function are found in data table 24.2. Check (✓) each box as you become familiar with the location and function of each part.

Data Table 24.2 Specific Regions of the Brain Stem

Brain Part	Function
☐ 1. Diencephalon made up of	
☐ *a.* Thalamus	Receives sensory impulses and connects them to the cerebral cortex.
☐ *b.* Hypothalamus	Regulates heart rate and blood pressure; also regulates body temperature, sleep, and body weight.
☐ 2. Brain stem made up of	
☐ *a.* Midbrain	Serves as a reflex center, forming a cerebral pathway to higher levels of the brain; maintenance of auditory senses and posture are also carried out here.
☐ *b.* Pons	Aids in regulating the rate of breathing.
☐ *c.* Medulla oblongata	Controls vital parts of the body including the digestive tract, the circulatory system, respiration, and the vasomotor system.

24.4 The cerebellum, located below the occipital lobe of the cerebrum, serves to coordinate all skeletal muscle movements, as well as posture.

Peripheral Nervous System

24.5 From the central nervous system we can now move to the peripheral system. The peripheral nervous system is comprised of the spinal and cranial nerves. As the name implies, the cranial nerves are located near or within the brain. These nerves specialize in determining sensations or in carrying out a specific activity, or both. The spinal nerves are directly involved with stimulating glands to discharge their respective secretions. The cranial nerves and the spinal nerves service the two subsystems of the peripheral nervous system, namely, the *somatic nervous system* and the *autonomic nervous system.*

The somatic nervous system relates to our conscious actions and reflexes. In other words, it is the somatic system you voluntarily control. Scratch your nose. Clap your hands. Wink your right eye. The fact that you can do these things is a result of the somatic nervous system, a system you have control over.

The autonomic system, on the other hand, controls your unconscious activities. You simply do not have any conscious control over the autonomic system. Think about it. Do you control your heartbeat, digestion of food, or visual perceptions? These things are automatic. You merely ride along, with the autonomic system doing for you, all of the time. While you control winking, you usually don't control blinking. Blinking is a reflex that happens thousands of times each day with you rarely acknowledging the fact that you have blinked at all.

The autonomic system may be further divided into two other subsystems: the *sympathetic* and *parasympathetic nervous systems.* Each acts in opposition to one another. The sympathetic system is capable of increasing heart rate, dilating the pupils of the eyes, diverting more blood from the stomach and intestinal tract to the brain and skeletal muscles, and a host of other activities.

The parasympathetic nervous system, on the other hand, works just the opposite of the sympathetic system. The parasympathetic system slows the heart rate, constricts the pupils of the eyes and relaxes the flow of blood to the brain and muscles so that the intestinal tract can behave normally.

Figure 24.5 Five spinal nerve regions showing the 31 spinal nerves that service the body.

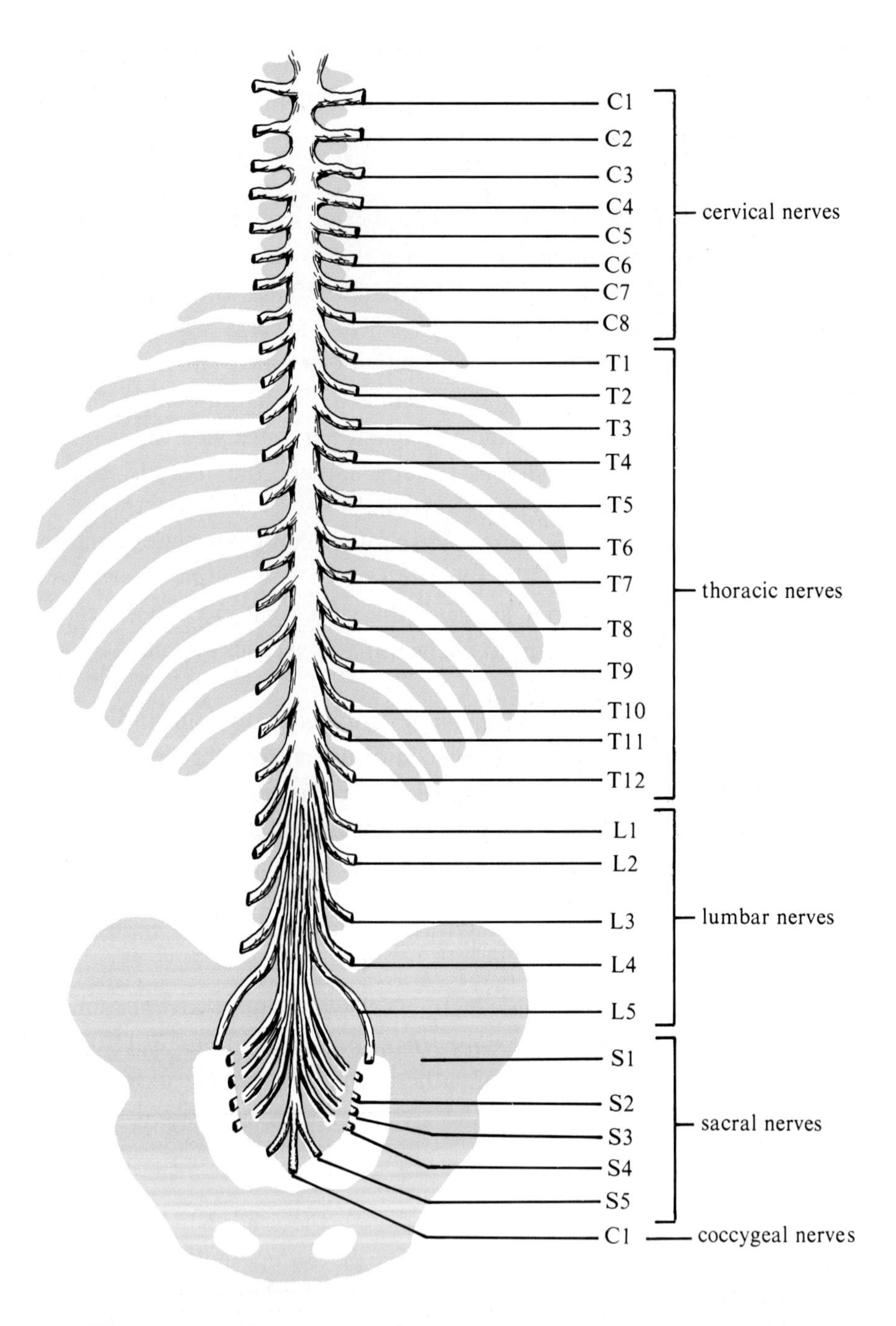

The sympathetic and parasympathetic systems seem to keep each other in balance. If a stressful situation should arise, however, the sympathetic system dominates. Sometimes called the "fight or flight" system, the sympathetic system gives you the edge in athletics, stage performances, or wherever and whenever you need to muster your best.

The sympathetic and parasympathetic systems are serviced by thirty-one pairs of spinal nerves that have been identified by their origin in relation to their position with the vertebral column (see Figure 24.5), and twelve cranial nerves (see Figure 24.6).

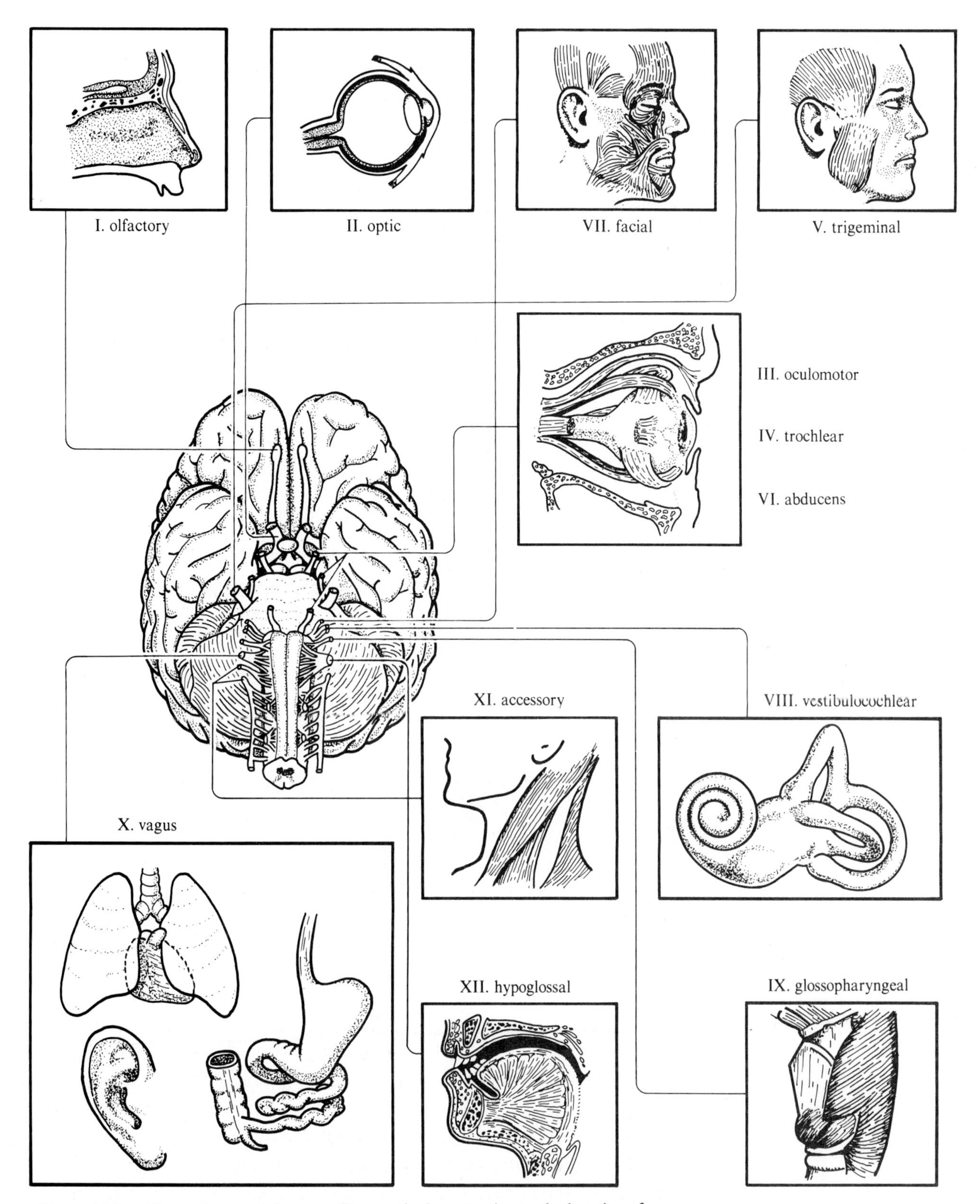

Figure 24.6 The twelve cranial nerves. Pay particular attention to the location of numbers III, IV, and VI.

While each nerve is important, many biologists believe the cranial nerves are the most important. Such a position comes from the fact that these nerves keep your brain in touch with your surrounding environment. Data table 24.3 identifies the twelve cranial nerves and their respective functions. Check (✓) each box when you are satisfied you know each nerve and its function. In order to help you remember the specific and static order of the cranial nerves, use this jingle to retain the first letter of the cranial nerves: "On old Olympus towering top, a family, very good, viewed armadillo hops".

Data Table 24.3 Twelve Cranial Nerves

Cranial Nerve	Function
☐ I. Olfactory	Smell.
☐ II. Optic	Vision.
☐ III. Oculomotor	Raises eyelids, moves eyes, and adjusts amount of light entering the eye.
☐ IV. Trochlear	Moves the eyes (eye movement).
☐ V. Trigeminal	Controls muscles of mastication: sends impulses to lower teeth, lower jaw, upper teeth, upper mouth parts, and tear glands in the eyes.
☐ VI. Abducens	Moves the eyes (eye movement).
☐ VII. Facial	Controls taste, facial expressions, and salivary glands.
☐ VIII. Vestibulocochlear	Hearing and equilibrium.
☐ IX. Glossopharyngeal	Impulses to pharynx, tonsils, tongue, and salivary glands.
☐ X. Vagus	Impulses to heart and smooth muscles of visceral organs.
☐ XI. Accessory	Impulses to muscles of neck and back.
☐ XII. Hypoglossal	Impulses to muscles to move the tongue.

24.6 Sensory System

The sensory systems have long been associated with the nervous system. There are essentially five sensory systems, each reacting to specific stimuli. The sensory systems are made up of highly specialized and developed *sensory receptors*. These receptors have the ability to trigger nerve impulses that contribute to the way in which we understand the world around us. Everything we know about our world comes to us through our senses. Everything we know depends, to some extent, on the manner by which our senses work.

Smell

The remainder of this chapter will focus on the five senses. The first sense we will take a look at is the sense of smell. Linked to the sensation of smell, the nose takes in, registers and deciphers the many smells that confront us daily. While many animals have a keener sense of smell than humans, we are able to detect countless thousands of different aromas.

The term *olfactory* refers to smell. The olfactory receptor cells, which project through the cribriform plate of the skull, are partly responsible for the detection of the sensation called smell. These highly specialized cells connect to the *olfactory bulb* via sensory nerve fibers.

Odors could be thought of as particles mixed with air or other gases. Some consider odors a gas. When an odor/gas enters the nasal cavity, it dissolves in the mucus lining. The odor/gas/mucus solution stimulates the olfactory cells in the olfactory bulb which connects to the olfactory nerve. Impulses are then carried by the olfactory nerve to the olfactory portion of the cerebral cortex. There, the impulses are to "read".

Only a small area in the roof of the human nasal cavity, the so called *regio olfactoria,* is provided with the olfactory mucous membrane, which contains sensory cells specialized to detect odors. (Refer to Figure 24.7.) In normal respiration, only a small proportion of the odor-bearing air drawn into the lungs comes in contact with this membrane. Therefore, to enhance the amount of air to come in contact with the membrane, sniffing is used. Sniffing has the effect of increasing the proportion of air that makes contact with sensory cells.

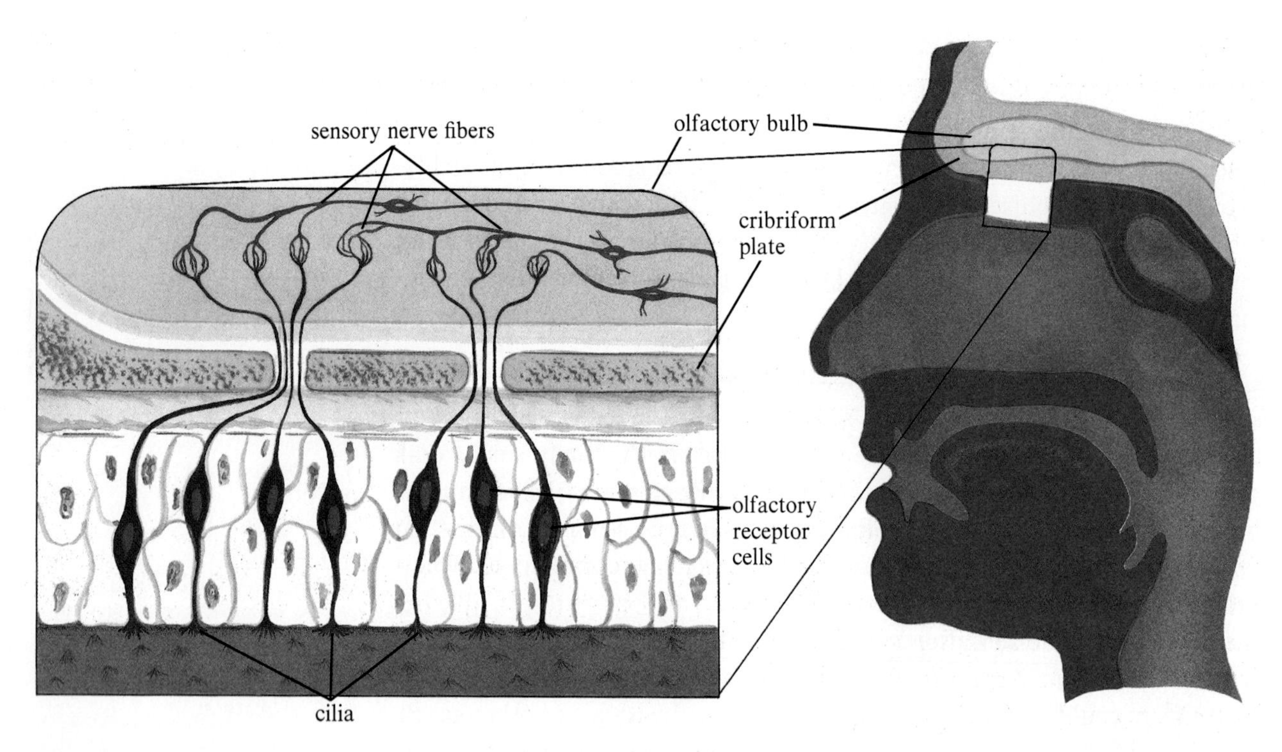

Figure 24.7 Location of, and the support anatomy for the sense of smell.

The acuteness of the sense of smell varies greatly from one animal species to another. Humans have a rather limited range of detection of odors compared to other animals. Some dogs can smell millions of times more acutely than humans. These animals live in a smell-oriented world of which humans have only the dimmest conception. Pigs, for example, have been used for a long time to sniff out the prized truffle which lies buried in the soil. Dogs are used to detect the presence of explosives and narcotics.

Data table 24.4 lists the five basic odors that can be smelled. In addition to the name, there is a short description of each odor listed. There are five bottles on the supply table. They are listed by number and not name. You are to determine which bottle goes to which smell. (Caution: Whenever you smell any substance, you should fan your hand over the substance and carefully sniff the air. Never place your nose directly over a substance to smell without first determining its strength and noxious characteristics.)

Data Table 24.4 Basic Odors

Odor	Description	Number of bottle
a. Camphorous	Like camphor	______
b. Floral	Smells like flowers	______
c. Ester	Candy smell	______
d. Pungent	Sharp odor	______
e. Putrid	Scent of decaying meat	______

24.7 Taste

The sense of taste is the responsibility of the tongue, and to some extent, the palate, throat and tonsils. Specialized taste cells, along with support cells, form organs of taste called **taste buds.** Taste cells open to the surface of the tongue by means of *taste pores.* Tiny hairs that are attached to the taste cells project from the pores. It is these hairs that are believed to be responsible in determining the taste sensation. (See figure 24.8.)

There are four major taste sensation areas on the tongue. Each seems to be capable of determining a specific taste. Figure 24.9 illustrates the receptor sites for four of the basic tastes. However, you should know that scientists actually list five major taste sensations. Data table 24.5 lists these sensations. The metallic taste is not readily detected from the taste buds located on the tongue, but taste sensitive areas for the metallic taste do exist on the palate and the superior regions in the back of the mouth.

Data Table 24.5 Basic Tastes

Taste	Description
a. Sweet	As produced by sugar
b. Sour	As produced by acidic foods such as a lemon
c. Bitter	As produced by coffee
d. Salty	As produced by salt
e. Metallic	As produced by metals

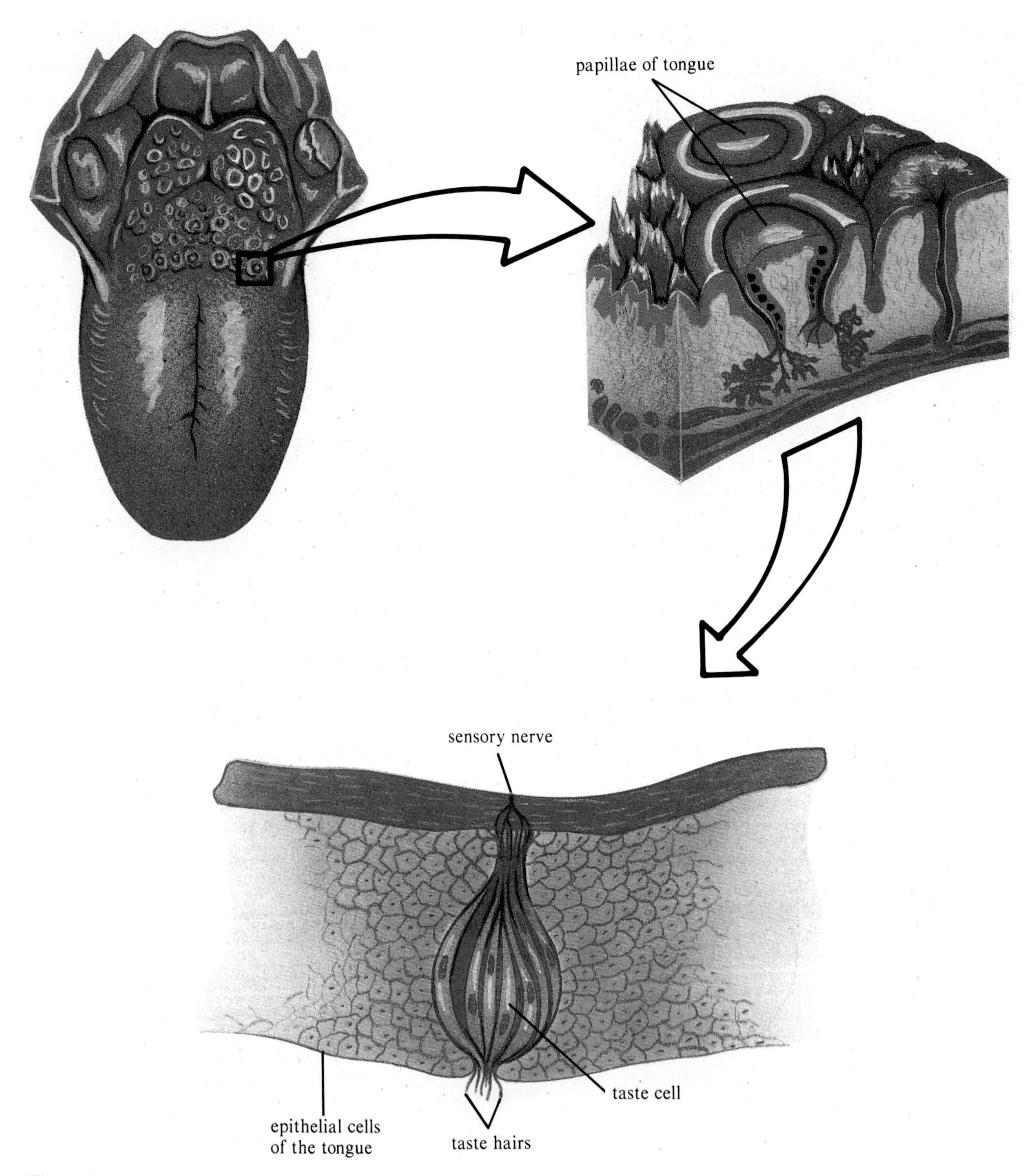

Figure 24.8 Anatomy associated with the sense of taste.

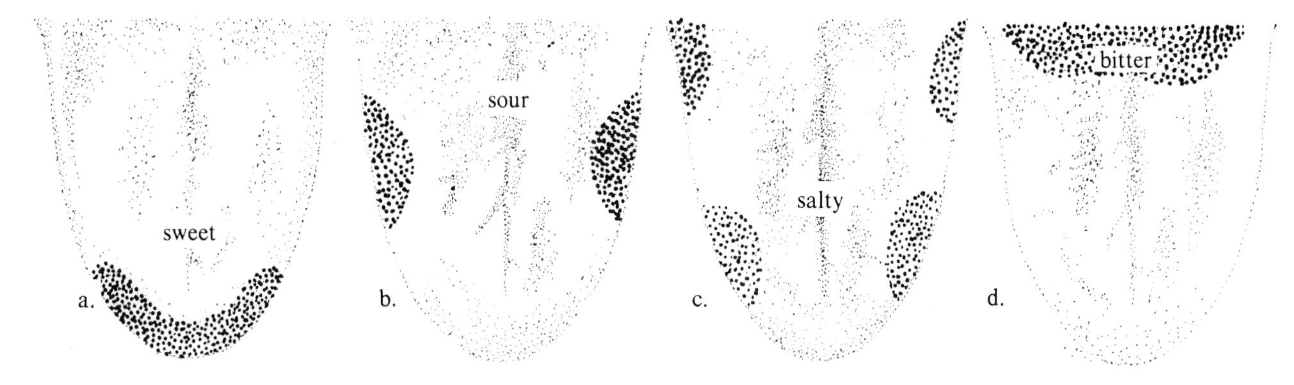

Figure 24.9 Four taste sensations can be identified on the tongue in specific locations.

Hearing

24.8

The school bell rings breaking the sounds of lecture from a sing song voice. School's out. Time to go. The first thing that gets turned on in the car is the radio. The best song in the entire world is playing. Now the ignition is connected and the car starts. Ah, the hum of a well tuned engine. A friend honks at you. He waves, then yells some instructions. Sounds, all sounds. As you read the next few paragraphs, follow the progression of sound by locating its position in figures 24.10 and 24.11.

The sense of hearing begins with the sounds, but unless your hearing apparatus is working, you hear nothing. The sense of hearing begins by sound waves being channeled into the *auditory canal* by the *pinna.* The sound waves strike the *tympanic membrane,* causing it to vibrate. The vibrations, could be seen with an instrument called an osciliscope. The same vibrations are mimiced by the typanic membrane.

As the tympanic membrane begins to vibrate, the *middle ear* bones or ossicles, the malleus, incus and stapes move accordingly. The vibration of the stapes at the *oval window* causes wave motion in the fluid contained in the cochlea, which is located in the *inner ear.* As the wave motion progresses through the fluid found in the cochlea, it starts the *basilar membrane* vibrating. These vibrations are in turn, registered by the *organ of Corti.* Keeping in mind that the *organ of Corti* is located within the *cochlea,* it is here that thousands of tiny hairs are located. These tiny hairs are attached to literally thousands of nerves that are sensitive to the motion of the hairs. The nerves translate the hair motion into pitch and volume via the *vestibulocochlear nerve.*

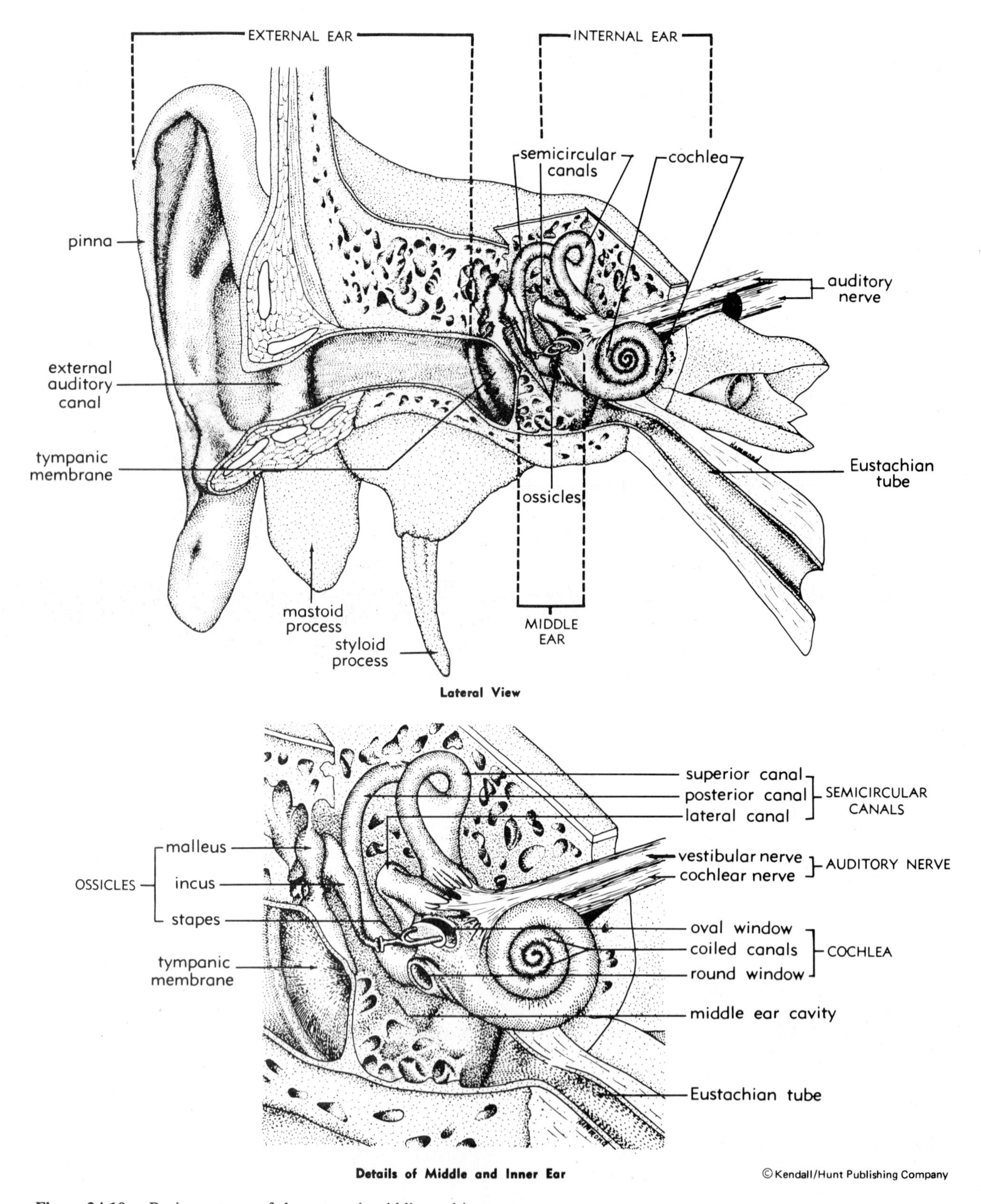

Figure 24.10 Basic anatomy of the external, middle, and inner ear.

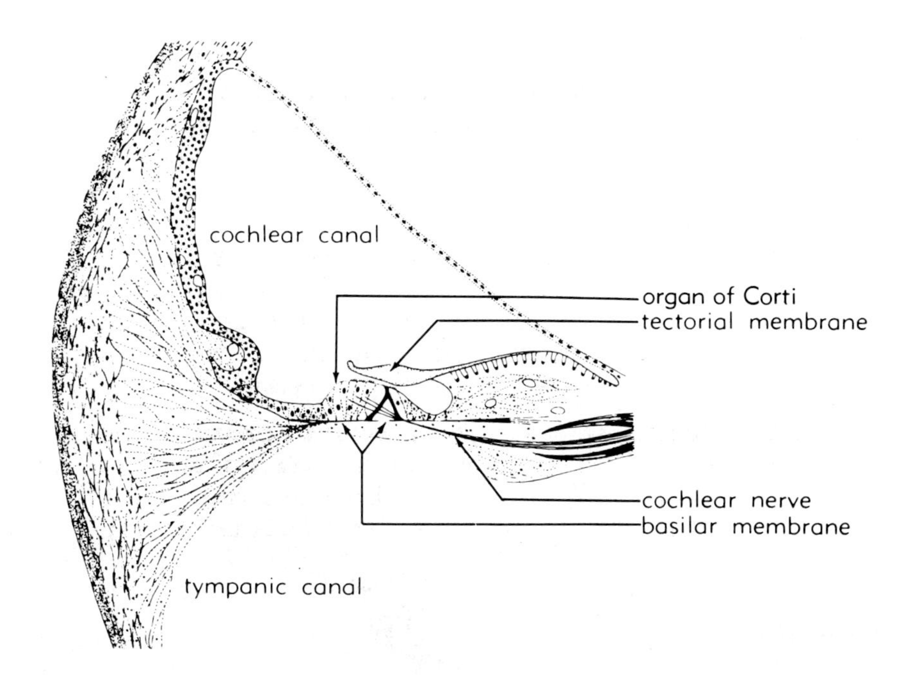

Figure 24.11 Section through basal turn of cochlea showing location of the organ of Corti.

Like the sense of smell, many animals use sound to their advantage. Humans have a nominal range of sound, but we are usually not capable of hearing sounds below 20 cycles per second nor higher than 20,000 cycles per second. Dogs, bats and dolphins are capable of hearing sounds considerably higher than humans. Bats and dolphins create high pitched sounds so they can use these sounds to position themselves in relation to surrounding objects.

24.9 Let's look a little closer at some sounds. As you listen to your own voice, you have learned the sounds you make and how you sound. Yet, when you listen to yourself on a magnetic tape, your first reaction is that this is not me. Wrong. It is you. You simply sound different to yourself because, much of the sound you hear is conducted through the bone of your skull. This sound is different than what you hear when your voice is played back to you from a tape recorder. The following short activity will give you a better understanding of this conduction of sound.

You will need a tuning fork. Start the tuning fork vibrating by striking it on the heel of your hand. Listen to the sound. Put the tuning fork closer to and further away from your ear. This is the sound of the tuning fork. Now repeat the process, but place the nonvibrating end of the fork on your mastoid process, just behind your ear and listen to the sound. Are you hearing a similar or different sound? ____________ Explain.

24.10 Place the vibrating fork on your forehead. Again, listen carefully to the sound being transmitted. In the latter two parts of this experiment, you are actually feeling as well as hearing the sound waves being developed. What are the differences of the sound in the three locations?

24.11 The *semicircular canals* are another part of the inner ear, but have a different function. The semicircular canals are responsible for our sense of balance. The canals can be seen in figure 24.10. They lie at right angles to one another and occupy a different plane in space. Containing many receptors, the canals are filled with a fluid similar to the fluid within the cochlea. When your head changes position, this fluid moves and stimulates the receptors. With this stimulation, the receptors start impulses that go through a branch of the auditory nerve to the cerebellum. As a result of this stimulus, the brain becomes aware of changes in the head's position. It is critical that such information be dealt with so that you don't fall as you move around each day with your definite acrobatic motions. You should note that the superior and posterior canals are positioned vertically. The lateral canal has a horizontal placement.

Sight

24.12 Another sense is your sense of sight. As you look around, you are aware of what is around you because of this sense. The sense of sight, accomplished with the eyes, is considered by many as the single most important sense.

As light enters the eyes, it follows a specific path. Figure 24.12 illustrates the path of light into the eye. As light passes through the eye it travels through several *refractive substances.* The cornea is the main light-refracting part of the eye. The vitreous humor, lens, and aqueous humor are other light-refracting parts. Besides light-refracting, the lens also is the major part of the eye for focusing so that the object you look at can be plainly seen.

Objects that are close or far away come into focus because of the adjusting lens of the eye. The ciliary muscles, attached to the lens, pull on the lens causing it to flatten. When this happens, objects that are far away can be seen in focus. As the ciliary muscles relax, the lens thickens so that objects that are close are seen in focus.

Problems in focusing are shown in figure 24.13. Such problems with refraction of light by the lens includes *nearsightedness* or **myopia,** *farsightedness* or **hyperopia,** and *astigmatism.* Your teacher has placed a Snellen chart in the room. Check your vision with the Snellen chart. Obtain any further instructions from your teacher. What is your vision using the Snellen chart? ____________

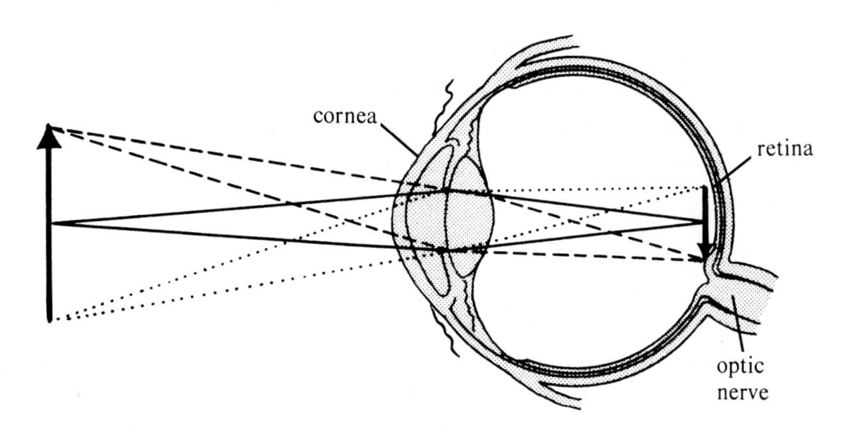

Figure 24.12 The path of light through the eye to the retina. Courtesy of James Koevenig.

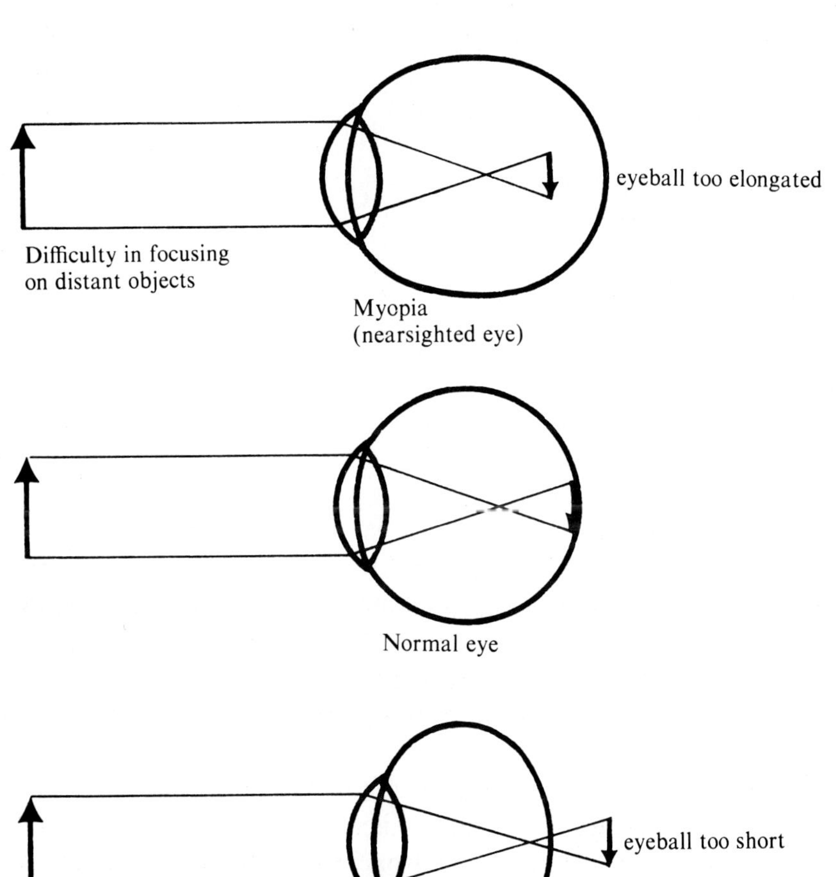

Figure 24.13 Common eye/lens problems. Courtesy of James Koevenig.

24.13 Your teacher has also put up an astigmatism chart. If you have an irregularity in your lens, certain lines on this chart will appear to be more defined. Check your vision with this chart. Do you have the symptoms of an astigmatism? ____________ Remember, only a certified optometrist or opthamologist can officially determine your exact eye situation.

24.14 As light passes through the lens and on to the retina, it strikes special cells called **rods** and **cones.** See figure 24.14. Cones detect fine detail, register bright light, and detect color. The primary colors: red, yellow, green and blue are picked up by the cones. All of the other colors are products of mixing these colors but are also detectable by the cones. You will recall from chapter 10 that an hereditary disorder is colorblindness. Many suffer from red and green colorblindness or combinations of these colors.

To determine your color receptivity, obtain a colorblind chart from your teacher. You will need to have a partner assist you. Determine whether or not you have a colorblind situation. Are you colorblind or have some sort of problem determining color situations as contained in the test? ______ Don't be afraid to admit to a problem. It hasn't interferred with your success so far. I have a green/brown problem.

24.15 The other part of the retina are the rods. Rods respond to dim light. Rods contain a special pigment called *visual purple* or more acurately, **rhodopsin.** Rhodopsin increases within the rods as less and less light becomes available. As the concentration of rhodopsin increases, vision in the lack of light becomes enhanced. You experience this when you enter a darkened theatre.

Rhodopsin is made up of a protein called *opsonin* and another chemical similar to carotene. This chemical is capable of being converted to vitamin A. If vitamin A is lacking, rhodopsin is unable to effectively assist true vision in the dark. This condition is known as *nightblindness.*

Whether during the light time or dark time, light is passed to the retina and there it is focused on the retina and transferred to the *optic nerve.* The *optic nerve* sends the image to the brain for righting, definition of color and determining the sense of distance. All of this essentially happening at the same time, and with no apparent voluntary direction on your part.

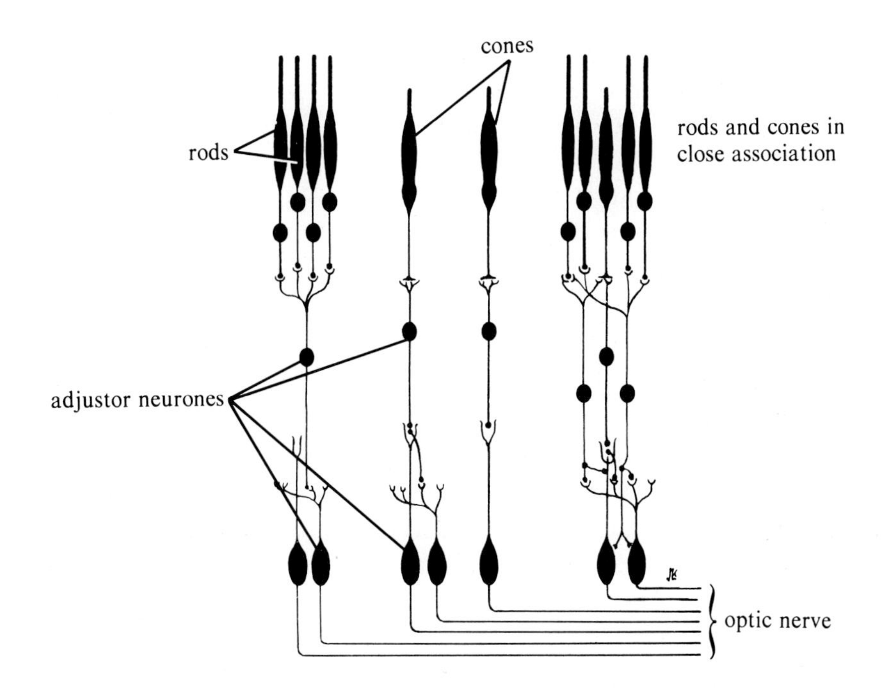

Figure 24.14 Rods and cones of the retina. (Courtesy of James Koevenig.)

Where the optic nerve joins with the retina, no rods or cones are present. As a result, no light sensitivity exists. This point is called the **blind spot.**

You have probably heard someone say, "she/he has a blind spot". This means that she/he does not see all the facts. Of course, most of us have mental blind spots at one time or another in our lives. But did you know that everyone has a true *physical blind spot*—one for each eye?

In the following activity, you will investigate your own blind spots. You and your partner will need a centimeter-scale ruler and figure 24.15.

a. Cover your left eye with your hand and hold figure 24.15 at arm's length. *Stare at the cross with your right eye only.* Slowly bring the paper closer toward your face.

b. At the instant the spot disappears, stop moving the paper toward you. Have your partner measure the distance in centimeters, from your right eye to the paper. Record this measurement in data table 24.6. Continue moving the paper closer to your face (staring only at the cross) until the spot reappears. Have your partner measure this distance also. Record this distance in data table 24.6.

c. Repeat the experiment with your left eye (covering the right). This time, however, stare at the *spot,* not the cross. Record your measurements in data table 24.6.

Figure 24.15 Symbols used for determining the blind spot.

d. Complete data table 24.6 to determine your total blind distance. Was your total blind distance more or less than your partner's? ________ Find out how the entire class did on this experiment. You should find differences in the total blind distance.

As mentioned at the beginning of this section a person's blind spot is anatomically on the *retina*. This is a place where no *photoreceptor cells* exist.

Data Table 24.6 Blind Spot Distance for Each Eye

	Right	**Left**
Disappearing distance		
Reappearing distance		
Total blind distance (TBD) (disappearing minus reappearing)		

24.16 Just to the right of the blind spot is a pit called the **fovea centralis.** The *fovea centralis* is the center of the *macula lutea* or *yellow spot.* This is the region of the eye where critical vision occurs. You need to be aware of the fact that the macula lutea is made up entirely of cones. Therefore, this region will be the most sensitive to bright light, fine detail and color. Study the anatomy of the eye in figure 24.16 carefully. Locate each of the anatomical parts of the eye mentioned so far. Better yet, let's dissect an eye so that you can "see" the parts first hand. You will need a pig or cow eye and proper dissecting materials. Your teacher may wish to give you some additional instructions before you begin.

a. Study the external features of the eye carefully. Find and identify everything possible.
b. Make an incision in the eye by making an incision halfway back and parallel to the lens. You will be cutting through the outer scleroid coat and middle choroid coat. Both are tough protective tissues of the eye. The inner *retina* is quite fragile.

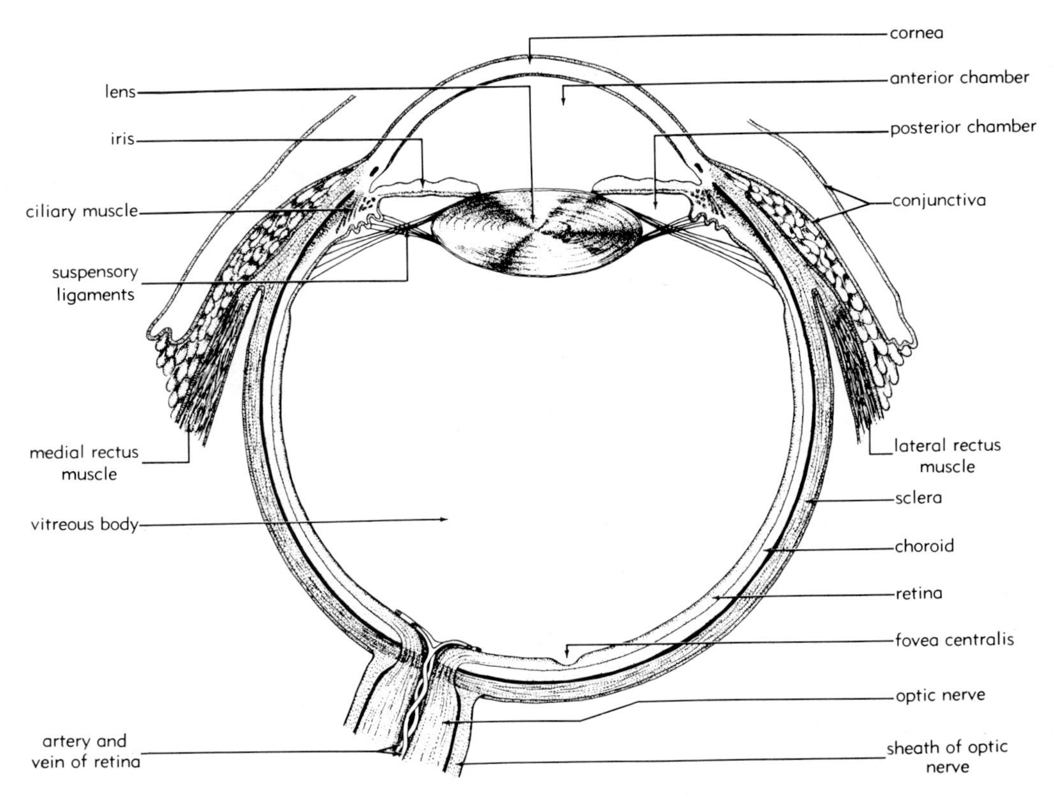

Figure 24.16 Basic anatomy of the eye.

c. Having made your incision, push from the bottom of the eye forcing the *vitreous humor* out of the eye. It will resemble egg white. The lens and the *ciliary muscles* will become noticeable as you invert the inside of the eye outside. Study them both carefully. Put the lens on a piece of paper where there is some printing. What do you see? __________

d. Look at each part of the eye carefully. Study the eye with the assistance of figure 24.17. When you feel you have done a thorough job with this dissection, clean up your lab space and dispose of the eye by your teachers instructions.

24.17 Touch

The final sensory system that we will study is sometimes identified as a body system unto its own. We will treat this system here because the purpose of the system is detection of the surroundings through the nerve endings in the skin. The *integumentary system* includes the skin, hair, nails, and the oil and sweat glands. Figure 24.17 illustrates a cross section of skin.

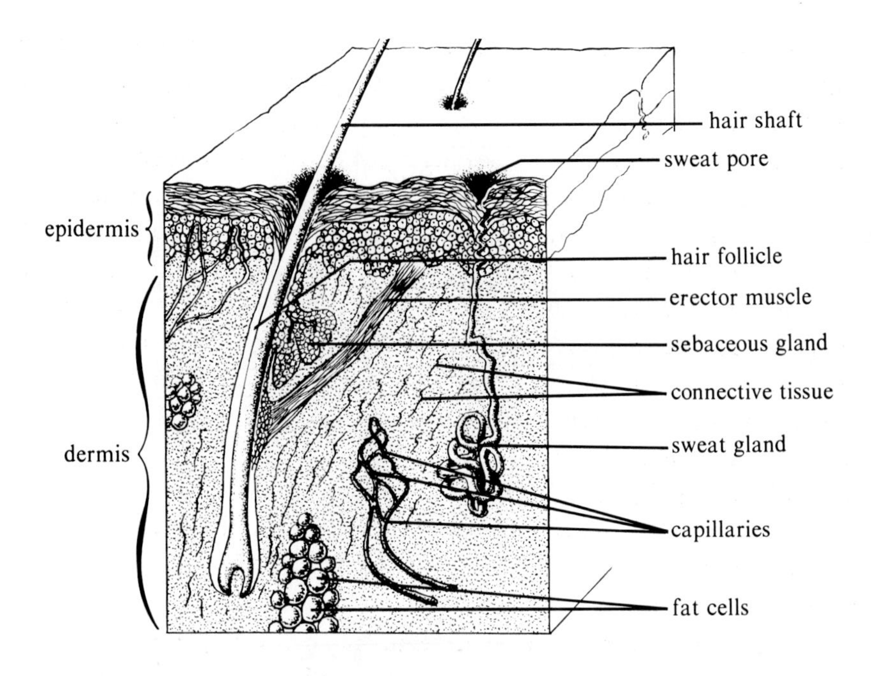

Figure 24.17 Cross section of skin illustrating selected anatomy. Courtesy of James Koevenig.

Skin is made of five layers of cells in the *epidermis*. These stratified layers include:

1. Stratum corneum
2. Stratum lucidum
3. Stratum granulosum
4. Stratum spinosum
5. Stratum germinativum

As the cells of the outermost layer of skin, the stratum corneum as lost each day by the hundreds of thousands, new cells replace them. New epidermal cells are constantly being made at the germinativum level. As cells progress upward to the outermost level, they become hardened by a process called **keratinization.** Nails and hair also undergo this hardening process.

Just below the epidermis is the *dermis*. Throughout the dermis and reaching into the epidermis exist many nerve endings. These nerves give us a tactile sense of our environment. Touch also is our register of pain, pressure, and changes in temperature.

The sensory nerves of the skin are distributed unevenly over the skin area and lie at different depths in the skin. For instance, if you pull a string gently over the palm of your hand or across your lips, you will feel this sensation because of the *Meissner corpuscles*. These receptors are close to the surface of the skin. They respond to light touch.

Pacinian corpuscles respond to pressure. These receptors lie deeper in the skin. If you press the eraser end of your pencil against your skin, you feel the eraser as pressure. Since the Pacinian corpuscles lie deeper in the skin, the pressure stimulus is a stronger sensation than a touch stimulus. You may think there is no difference between touch and pressure, but without this difference, you would be unable to pick up objects.

Heat and cold stimuli are detected by different receptors. This is an interesting, protective adaptation of the body. Actually, cold is not an active condition. Cold results from a lowering of heat energy. If both great heat and intense cold were to stimulate a single receptor, you would be unable to react to either. However, since some receptors are stimulated by heat and others by the absence of it, you can react to both conditions. *Krause corpuscles* respond to cold. *Ruffini corpuscles* respond to heat.

The sensation of pain is also a protective device of the body. *Free* nerve endings throughout the skin react to mechanical, thermal, electrical, and chemical stimuli that can register as pain.

A common sensory perception is one in which stimulation in one part of the body gives rise to a sensation (frequently pain) that seems to be localized in some different or remote part of the body. You may have experienced an example of this *referred pain phenomenon* when swallowing something very cold, such as ice cream. Often a feeling of sharp discomfort is felt somewhere else, possibly in the forehead region or the back of the neck.

The occurrence of referred pain seems to be related to common nerve pathways that are used by sensory impulses coming from both skin areas and visceral organs (organs within the body cavity). Consider pain impulses from the heart (associated with heart attacks) that seem to be conducted over the same nerve paths as those coming from the skin of the shoulder and the inside of the left arm. As a result, the cerebral cortex may incorrectly interpret the source of the impulses as being the shoulder or arm rather than the heart.

The following activity will demonstrate the referred pain phenomenon. Place your elbow in a dish of ice cold water (10° C) and note the progression of sensation that you experience. At first you will feel some discomfort in the region of the elbow. In a couple of minutes the sensation will be felt elsewhere. Keep your elbow immersed in the water 2 to 3 minutes. Where do you feel the referred sensation? ____________________

The *ulnar nerve* supplies the ring and little fingers, and the lateral side of the hand. The ulnar nerve is responsible for transferring the sensation in this experiment.

24.18 The skin is a receptor site for sensations. Many of you like to get a suntan in the summer. Some of you are unable to tan and find the sun more of a foe. Darkening of the skin or tanning is accomplished by the presence of *melanocytes* in the skin. These cells are capable of producing a pigment called **melanin.** Melanin is capable of capturing the harmful infrared rays of the sun. In so doing, the *melanocytes* darken causing the skin to take on the tanned appearance. Everyone does this to some degree depending on the number of *melanocytes* present. Albino people, however, have few or no *melanocytes*. Therefore, to these people, the sun can be quite harmful.

24.19 Skin has a few extensions to identify with, but we should make note of those few. The most prominent feature on skin is the hair. Hair can be found on most parts of the body except the palms of the hand and the soles of the feet. Each *hair shaft* has a root embedded in the skin forming a *hair follicle. Sebaceous glands* secrete oil to keep the hair soft. Around each hair follicle is a group of muscles called the *arrector pili muscles*. These muscles are capable of causing the hair shafts to contract and the skin area around each hair shaft to rise, causing a condition known as *"goose pimples."*

The purpose of active arrector pili muscles is to aid in heat retention to the skin. Sweat glands, on the other hand, aid in heat loss. As needed, the body will give off fluids to the skin via the *sweat ducts*. As the fluid rests on the skin it evaporates. Evaporation is a cooling process so the process assists the cooling of the skin.

24.20

Before continuing, review this chapter carefully. The study points should help. Read each study point. If you can readily identify with the study point, place a check (✓) in the box.

On completion of this chapter, you should be able to do the following:

☐ 1. Distinguish between the central nervous system and peripheral nervous system.

☐ 2. Distinguish between the autonomic and somatic systems, and the sympathetic and parasympathetic systems subdivisions.

☐ 3. Discuss the reflex arc.

☐ 4. Identify and relate the function of the parts of the brain.

☐ 5. Identify and give the function of each of the twelve cranial nerves.

☐ 6. Discuss the anatomy related to and the physiology of the sense of smell.

☐ 7. Discuss the anatomy related to and the physiology of the sense of taste.

☐ 8. Discuss the anatomy related to and the physiology of the sense of sight.

☐ 9. Discuss the anatomy related to and the physiology of the sense of touch.

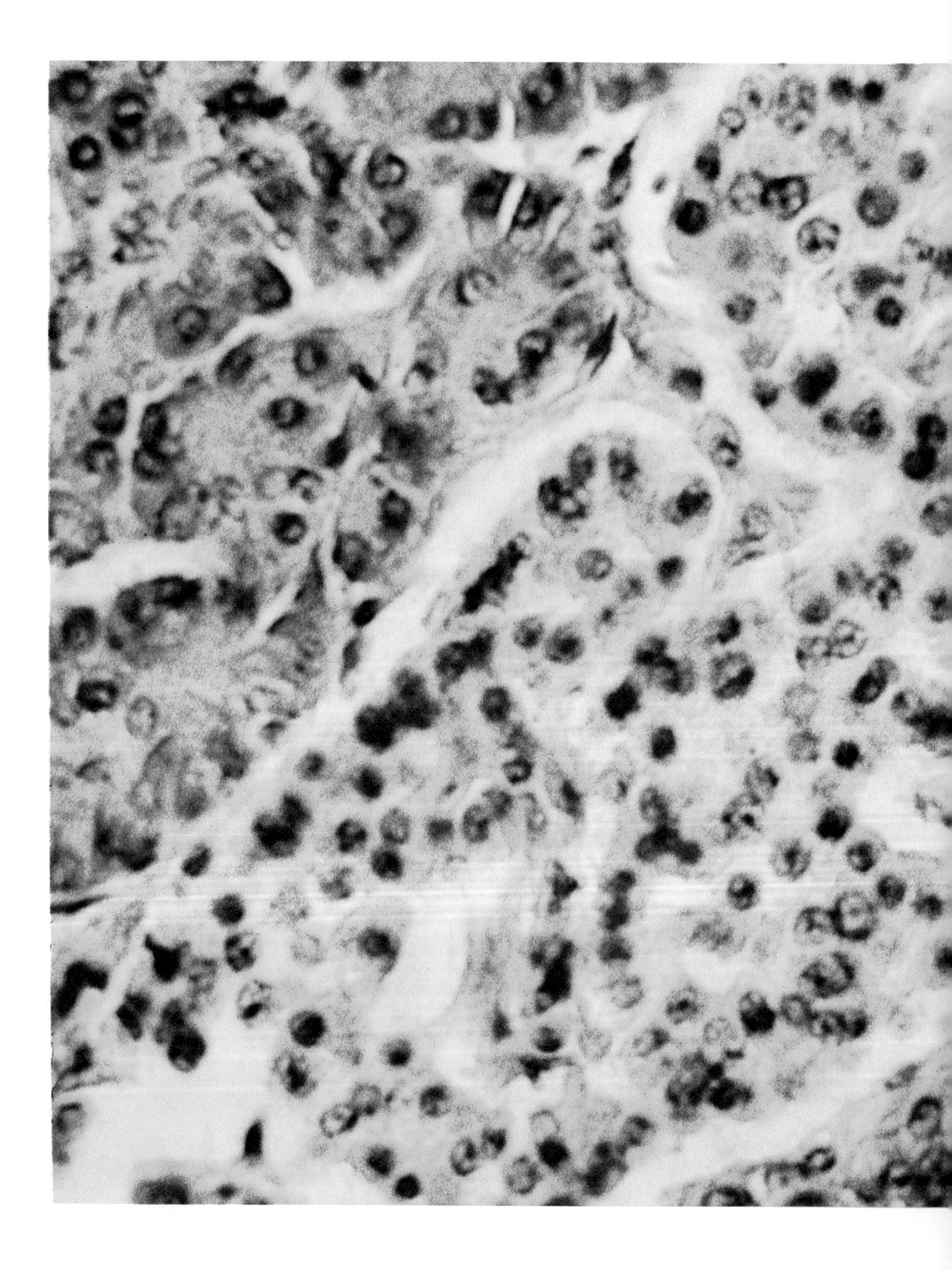

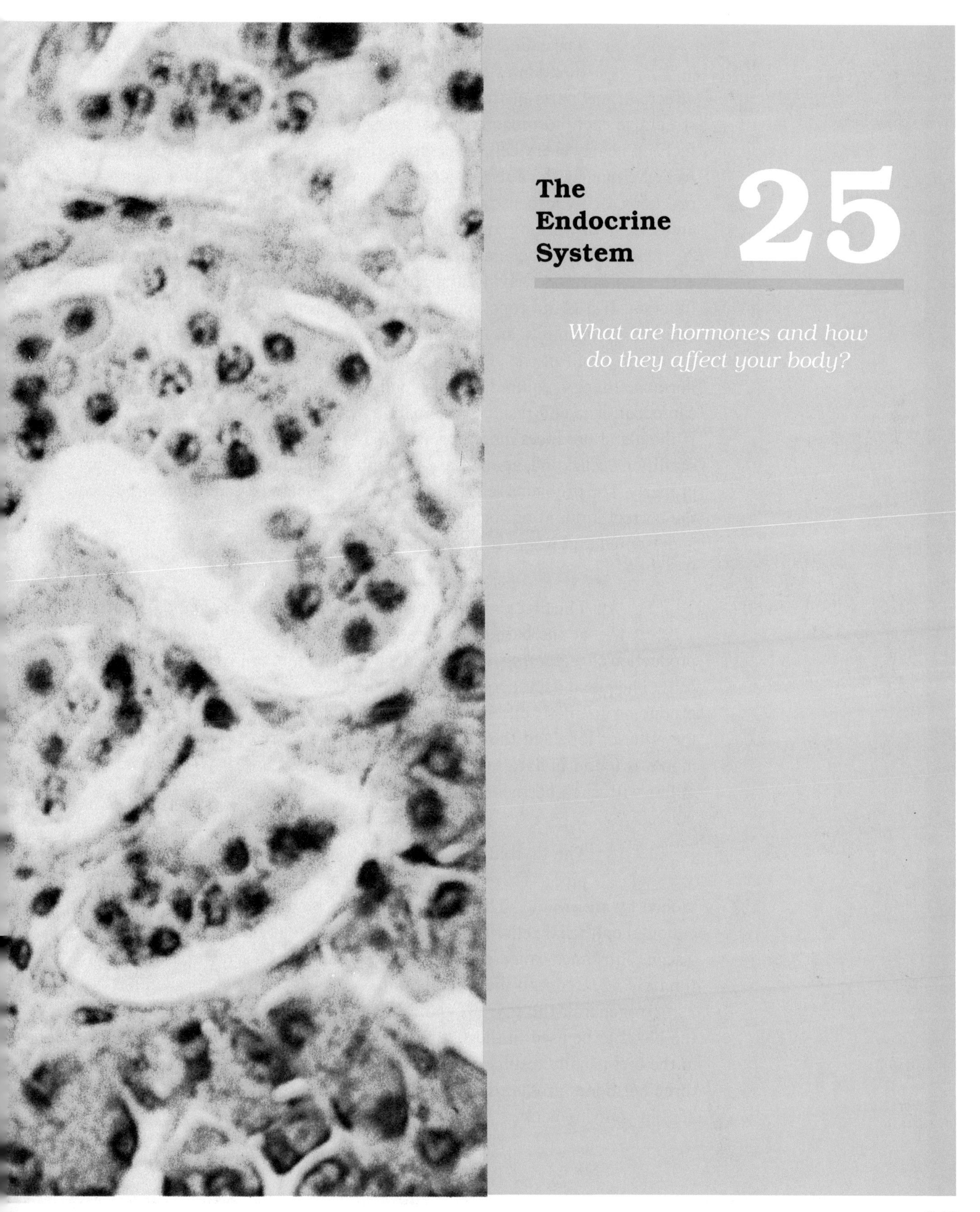

The Endocrine System 25

What are hormones and how do they affect your body?

25.1

The endocrine system consists of ductless glands. The general function of each endocrine gland is to secrete **hormones** that affect various parts of the body. The hormones are carried by the bloodstream so each hormone not only has specific effects, but also general effects on nearly every cell of the body. The release of the hormones is controlled either by the autonomic nervous system through complex chemical reactions, or in response to tropic hormones released from the **hypothalamus,** or both working together.

Hormones are chemically identified as **amines, proteins,** or **steroids,** with each hormone having a specific molecular structure, thus a specific function. If a gland functions properly, the proper amount of hormone is released according to an individual's **metabolic rate.** A malfunction of the gland can result in one of two opposite conditions. If a gland produces less hormone than is required, a **hypo** condition (for instance, hypothyroidism) can occur. The end result is the inability of the body to function properly. If the gland produces more hormone than is required by the body, a **hyper** condition occurs and, again, the end result is the body's inability to function properly. The problems related to the endocrine glands' inability to produce the correct amount of hormone will be discussed as each endocrine gland is presented and where it is appropriate.

25.2

The first gland for you to study is the pituitary gland. Located at the base of the brain, the pituitary gland is controlled by production of *regulating factors* by the hypothalamus or by specialized nerve cells. The gland itself actually consists of two separate types of tissue, each producing distinctly different types of hormones. These two parts are called the anterior lobe and the posterior lobe. A summary of the pituitary hormones is found in data table 25.1. Check (✓) each box as you become familiar with each hormone and its function.

25.3

The thyroid gland is located adjacent to the trachea (windpipe). There are two lobes, one on either side, that are attached by an isthmus. The gland is made of *secretory follicles* lined with cuboidal epithelial cells. The follicles of the thyroid gland are filled with a viscous substance called **colloid,** which is the stored hormones that are ultimately released into the blood stream.

In general, the thyroid gland has the ability to remove iodine from the blood to be used in the synthesis of its hormones. Thus, lack of iodine in the diet usually results in a deficiency of certain thyroid hormones. The three hormones produced by the thyroid gland are discussed in data table 25.2 in your log book.

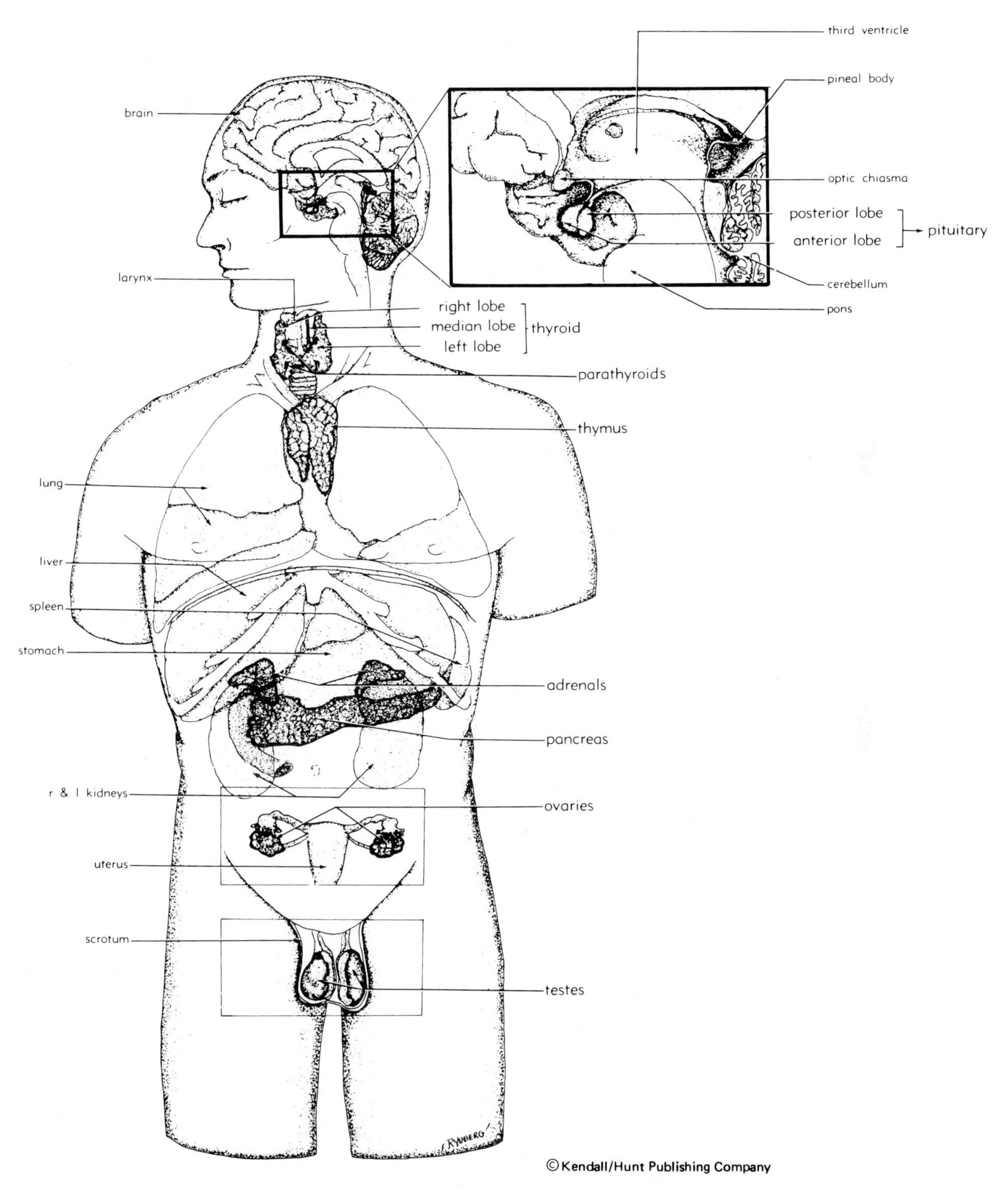

© Kendall/Hunt Publishing Company

Figure 25.1 Relative positions of the glands of the endocrine system.

Data Table 25.1 Function of the Pituitary Gland Hormones

Anterior Lobe: The anterior lobe, made of epithelial cells, secretes GH, TSH, ACTH, FSH, LH, and prolactin.

Hormone	Function
☐ 1. GH (growth hormone) or STH (somatotropin)	This hormone regulates the growth of the body by controlling the growth of the tissues. If hyposecreted, **dwarfism** will result. The hypersecretion of this hormone causes **gigantism.**
☐ 2. TSH (thyroid-stimulating hormone or thyrotropin)	The primary function of this hormone regulates the secretion of thyroid hormones. *Hypothyroidism* and *hyperthyroidism* occur when small or large amounts, respectively, are secreted. The results of these conditions will be discussed with the thyroid gland.
☐ 3. ACTH (adrenocorticotropic hormone)	The major function of this hormone is to regulate the cortex of the adrenal glands.
Gonadotropic hormones	
☐ 1. FSH (follicle-stimulating hormone)	FSH in the female regulates the growth of the Graafian follicle in the ovary and the estrogen levels of the body. In males, FSH stimulates the production of sperm cells.
☐ 2. LH (lutenizing hormone)	LH stimulates the development of the corpus luteum in the female. The corpus luteum is a secretive organ that produces progesterone. In the male, this hormone is called ICSH (interstitial-cell-stimulating hormone) and is responsible for production of certain male hormones (androgens), which are responsible for secondary sex characteristics.
☐ 3. Prolactin or LTH (luteotropin hormone)	Prolactin stimulates the mammary glands to produce milk and helps maintain the corpus luteum.

Posterior Lobe: The posterior lobe is responsible for two hormones. The posterior lobe is made up of special nerve cells called **pituicytes** rather than the epithelial cells that make up the anterior lobe. The two hormones are discussed below.

Hormone	Function
☐ 1. ADH (antidiuretic hormone)	ADH regulates the amount of urine that is produced and ultimately excreted. A decrease in ADH causes **diabetes insipidus.**
☐ 2. Oxytocin	Oxytocin is secreted at the later stages of pregnancy and aids in childbirth by causing uterine contractions (labor). Oxytocin also causes milk to be ejected from the mammary glands as a baby nurses.

Data Table 25.2 Function of the Thyroid Gland Hormones

☐	1. Thyroxine and triiodothyronine	Both of these hormones govern the rate at which carbohydrates give up energy, regulate growth, and increase protein synthesis. In growing, a deficiency of these hormones results in **cretinism.** The levels of hormones of the thyroid gland are closely regulated by the TSH of the pituitary. In adults, hypothyroidism results in **myedema,** characterized by low basal metabolism, a gain in weight due to excess fluid, and lowered mentality. **Hyperthyroidism** may result in the formation of a **goiter.** If excessive TSH is secreted by the pituitary gland or there is an iodine deficiency, the thyroid gland becomes overactive and enlarges, forming the goiter. Iodine intake can be maintained by using iodized salt in your daily diet.
☐	2. Calcitonin	Calcitonin regulates the amount of calcium in the blood by inhibiting the amount of calcium that is released from bones.

25.4

The parathyroid glands are small glands located on the thyroid glands. The four small glands secrete a single hormone called PTH (parathyroid hormone, or parathormone). The function of this hormone is to increase the blood calcium concentration and decrease the blood phosphate concentration. The regulation of calcium is important because calcium is involved in muscle contractions, membrane permeability, and blood clotting.

Hyperparathyroidism causes an increase of PTH to be secreted, with the end result being a softening of the bone tissue, which could lead to deformity. Excess calcium can collect in the kidneys causing kidney stones to form.

Hypoparathyroidism can cause a lowering in blood calcium concentration resulting in a condition known as **hypocalcemia.** Low-calcium levels affect nerve fibers that control muscle reactions resulting in muscle spasms. In severe cases, death may result.

25.5

The adrenal glands are located on the superior end of the kidney and consist of two regions, the *medulla* and the *cortex*. The two regions and their respective hormones are discussed in data table 25.3 in your log book.

Data Table 25.3 Function of the Adrenal Gland Hormones

Medulla

Hormone	Function
☐ 1. Epinephrine	Epinephrine increases blood pressure due to dilation of the coronary vessels. It also causes an increase in blood sugar, blood fatty acids, ACTH, and TSH.
☐ 2. Norepinephrine	This hormone complements the actions of epinephrine. The reverse of epinephrine occurs.

Cortex

Hormone	Function
☐ 1. Mineralocorticoids	These hormones regulate salt absorption by various tissues.
☐ 2. Glucocorticoids	These hormones assist in the breakdown of fats, proteins, and carbohydrates.
☐ 3. Adrenal sex hormones	Very little is known at this time regarding these hormones. Addison's disease can occur as a result of hyposecretion of cortical hormones. A hypersecretion may result in a condition called Cushing's disease.

25.6 The pancreas is located adjacent to the stomach and is part of the small intestine. The endocrine portion of the gland consists of small concentrations of glandular tissue called the *Islets of Langerhans*. The islets have two basic cell types: alpha cells secrete the hormone **glucagon;** beta cells secrete the hormone **insulin.** Failure to produce insulin in an adequate amount results in a disorder called **diabetes mellitus.**

Glucagon regulates the liver's ability to convert glycogen into glucose. This action can cause a rise in the blood glucose level. Insulin opposes the action of glucagon by decreasing the amount of blood sugar. Hyperinsulinism can cause a coma, or a condition called insulin shock. Hypoinsulinism results in a need for additional blood glucose.

25.7 The **ovaries** and the **testes** produce gonadal hormones that regulate the secondary sex characteristics in the female and male, respectively. The testes produce testosterone, which is necessary for the formation of viable sperm cells in the male. The ovaries produce estrogens and progesterones, which regulate secondary sex characteristics, the menstrual cycle, and pregnancy.

25.8 The thymus gland and the pineal gland are both glands of uncertain functions because little is known about them. The thymus gland is thought to secrete **thymosin,** which acts to affect white blood cell production. The pineal gland, however, is known to secrete a hormone called **melatonin,** which is thought to stimulate the hypothalamus to affect the pituitary gland.

25.9 The endocrine system has an important role to play for normal human development and function. It might be of interest to you to study one of the disorders in greater detail. Discuss the possibility of writing a paper about Addison's disease, Cushing's disease, diabetes mellitus, or a different hormone-related disorder with your teacher. If you write such a paper or if one is assigned, attach it to this chapter and hand both in at the same time for evaluation.

25.10 The endocrine system is complex, to say the least. The failure of any gland can cause a problem in hormonal balance. Most of the time, however, this does not happen.

Review the study points for this short, but important chapter. Check (✓) each study point when you are satisfied you know the concepts paralleling the objective.

On completion of this chapter, you should be able to do the following:

☐ 1. List the nine glands of the endocrine system.

☐ 2. Discuss the hormones and respective functions of each endocrine gland.

☐ *a.* pituitary

☐ *b.* thyroid

☐ *c.* parathyroid

☐ *d.* adrenal

☐ *e.* pancreas

☐ *f.* ovaries

☐ *g.* testes

☐ *h.* thymus

☐ *i.* pineal

The Digestive System 26

What is the digestive system and how do enzymes work?

26.1 The digestive system consists of a continuous tube beginning with the mouth and ending at the anal orifice. This tube is modified throughout its length to form specialized organs that perform certain specific functions involved with the mechanical and chemical breakdown of food. This system involves taking in food and converting it to a useful form so the cells of the body can use the foodstuffs for various body activities and functions.

The digestive system begins with the mouth or **buccal** (oral) **cavity.** Food enters the oral cavity and is physically broken down to allow it to be digested more quickly. Physical breakdown occurs with chewing.

Chemical breakdown begins in the mouth by the action of enzymes. Enzymes are organic catalysts that break large particles of food into smaller and smaller particles. The enzymes themselves are not changed, but merely help the action to occur. In the mouth, the enzyme that is present is in saliva. Saliva is produced by three sets of salivary glands, which secrete an enzyme called **ptyalin,** or **salivary amylase.** The function of this enzyme is to begin breaking down starchy foods. In addition to the enzymic effect that saliva has on food, it is also a lubricant for the mouth and throat.

Enzymes enable chemical reactions to occur. There are many enzymes that act to break down foodstuffs, produce energy and waste products. When energy is produced, the reaction is called **exothermic,** or exergonic.

Opposite to this process are enzymic actions that synthesize or build. Because energy is required to carry out building processes, the reactions are called **endothermic,** or endergonic. The investigation below will illustrate both types of reactions.

Pour 5 mL of 3 M hydrochloric acid (HCl) in a small test tube.

Health or hazard warning!

Student Note: The capital *M* in 3 M refers to molar concentration. This is a basic unit of measurement just as feet, pounds, and gallons. The higher the number, the greater the concentration.

Record the temperature of the acid in data table 26.1.

26.2 Carefully pour 5 mL of 3 M NaOH into the same test tube. Gently stir for 1 minute. Record the final temperature of the solution in data table 26.1.

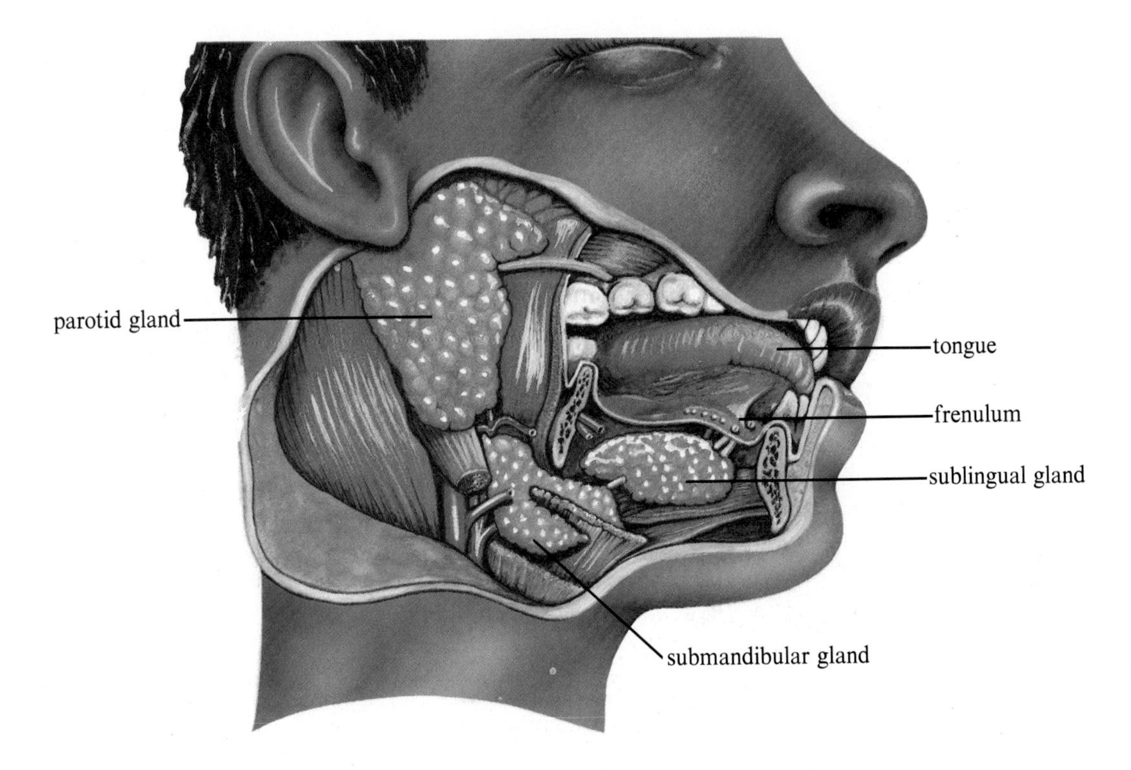

Figure 26.1 Anatomy of the mouth region showing the location of the salivary glands.

Data Table 26.1 Exothermic Reaction

Initial temperature = __________
Final temperature = __________
Difference in temperature = __________

1. Did the temperature of the solution change? ________
2. Does a change in temperature indicate that a chemical reaction has taken place? ________
3. Was heat energy present during the reaction? ________
4. Is this data quantitative or qualitative? ________

26.3 It was mentioned above that there are two types of chemical reactions related to heat energy. To illustrate the second type of chemical reaction, an endothermic reaction, put 10 mL of distilled water into a 20 × 150 mm test tube. The test tube should be clean. Record the temperature of the distilled water in data table 26.2. To the distilled water, add 1 gram of NH_4Cl (ammonium chloride). Stir continuously with a stirring rod until the ammonium chloride has dissolved. Record the final temperature, after a minute has passed, in data table 26.2.

Data Table 26.2 Endothermic Reaction

Initial temperature =	____________
Final temperature =	____________
Difference in temperature =	____________

1. What happened to the temperature of the solution?

   ____

2. How are the results of data table 26.1 similar to or different from the results of data table 26.2?

26.4 The first investigation, in which you used sodium hydroxide, (NaOH) is an example of an (exothermic reaction/endothermic reaction) ______________. The second investigation, in which you used ammonium chloride (NH_4Cl), is an example of an (exothermic reaction/endothermic reaction) ______________. Keep in mind that cells are "factory units" of the human body system, and it is in the cell where many of the chemical reactions take place. Some cellular reactions require energy input while other cellular reactions release energy.

26.5 Enzymes are protein catalysts. They speed up chemical reactions in organisms. Enzymes may be grouped by the location of their activity. *Endoenzymes* function inside of cells where they are actually made. *Exoenzymes,* however, are also made inside the cell, but are secreted outside the cell where they perform their functional (catalytic) role.

Exoenzymes specifically react with the processes of digestion. Exoenzyme digestion begins in the mouth and continues in the stomach from the chemical action of gastric juices. After a period of time, the churned, partially digested food is passed on to the small intestine. It is here that final exoenzyme digestion takes place.

26.6 The presence of *exo*enzymes permits cells of organisms to use the food that is ingested. At the same time, the individual cells are able to perform only those reactions for which *endo*enzymes are available. Let us continue.

Hydrogen peroxide is an extremely active chemical. It is used as a bleach for hair styling and often is used to wash out cuts and scrapes.

The formula for hydrogen peroxide is H_2O_2: similar to water but different. Hydrogen peroxide is found in some cells as a byproduct of cellular reactions. It is very toxic. A cellular endoenzyme breaks H_2O_2 into H_2 and O_2, and thus prevents cellular death. The chemical equation for this reaction is as follows:

$$2H_2O_2 \xrightarrow[\text{Enzyme}]{\text{Catalyst}} 2H_2O + O_2$$

Let us investigate this in liver cells. Get a small piece of liver. Place the liver on a watch glass and add 15 drops of 3 percent H_2O_2 onto its surface. Describe what happens.

26.7 Place a second piece of liver in boiling water. Boil the liver for several minutes. Remove from the water and repeat the procedure you used in section 26.6. Describe what happens.

26.8 What are the differences (if any) in the reactions of the two test tubes? Explain these differences.

26.9 The endoenzyme that is found in liver cells is called **catalase.** It was catalase that caused the bubbling action with the hydrogen peroxide. According to the chemical equation in section 26.6, hydrogen peroxide is broken down into water and oxygen in the presence of a catalyst. Are your results supportive of this statement? ________

26.10 Materials such as food that enter the cell are used by the cell as its source of energy, as building blocks for maintenance within the cell, and/or as building blocks for cell secretions such as enzymes. Temperature influences enzyme activity. The pH level also influences enzyme activity. Usually enzymes are most effective at an optimum temperature and pH.

The molecules that enzymes affect are called **substrates.** The substrate is a specific chemical substance that the enzyme is going to change. Enzyme molecules are usually large, and substrate molecules are usually smaller. It is believed that the geometric configuration of the enzyme molecule complements the shape (geometric configuration) of the substrate molecule. Some enzymes may have their shape changed by the substrate

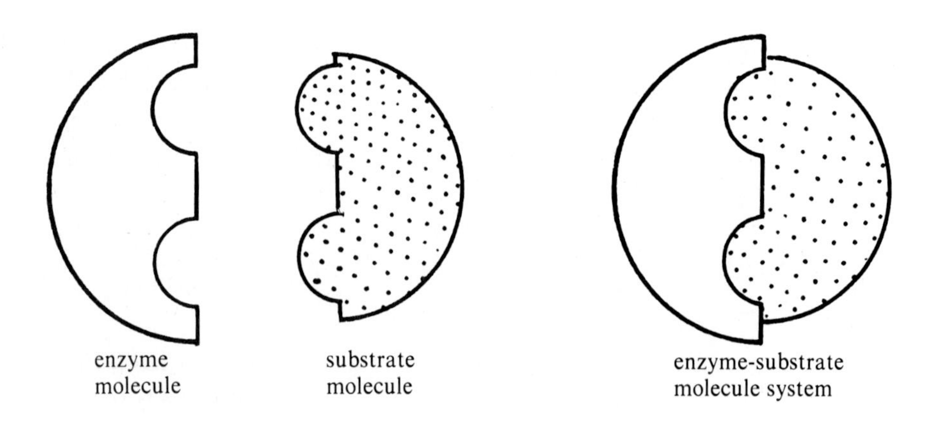

Figure 26.2 An enzyme-substrate molecule system.

molecule in order to be complementary to the substrate. This means that each enzyme fits to a specific substrate just as a key fits only one lock. Figure 26.2 illustrates an enzyme-substrate system.

As you can see, the enzyme molecule here is larger than the substrate molecule. The substrate molecule and the enzyme molecule match up perfectly, similar to the pieces of a jigsaw puzzle, or as stated earlier, to a lock-and-key model.

Not every enzyme has the proper shape to react structurally with every substrate chemical. In fact, enzymes are quite specific in terms of which substrate they react with. This characteristic is known as enzyme *specificity.* The two main types of enzyme specificity are (*a*) *absolute specificity,* where a particular enzyme reacts with only a particular chemical compound (substrate), and (*b*) *absolute group specificity,* which simply means that a particular enzyme will react with a particular group of chemicals, such as the alcohols.

Many things can **inhibit** enzymes from reacting; temperature is one of these. Of course, this goes beyond normal inhibiting. Excessive temperature destroys the enzymes. *Inhibitors* that are not destructive exist in living tissue. When an inhibitor comes in contact with the substrate of the enzyme, the enzyme substrate complex formation is stopped. Figure 26.3 illustrates this phenomenon.

To illustrate the inhibiting effect, obtain a small piece of liver and place it in a petri dish. Record everything you can observe. Note the texture, color, toughness, and other characteristics of the liver.

Cover the liver with mercuric chloride. Put the lid on the dish, label it, and set it aside overnight. Because you are working with a mercury compound, wash your hands with soap and water.

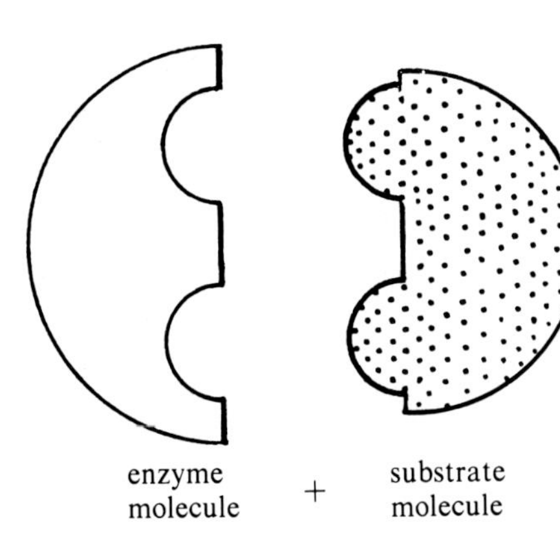

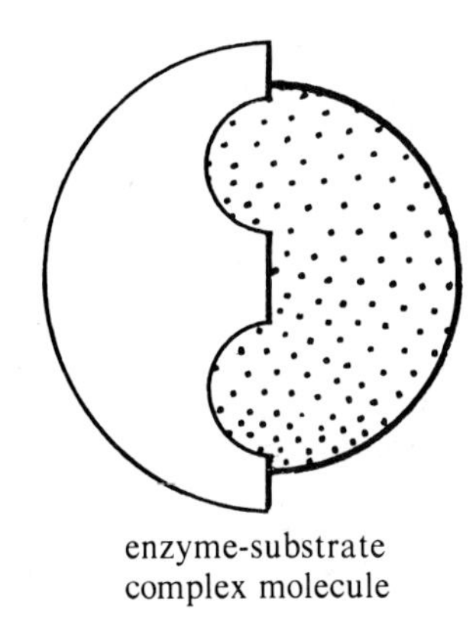

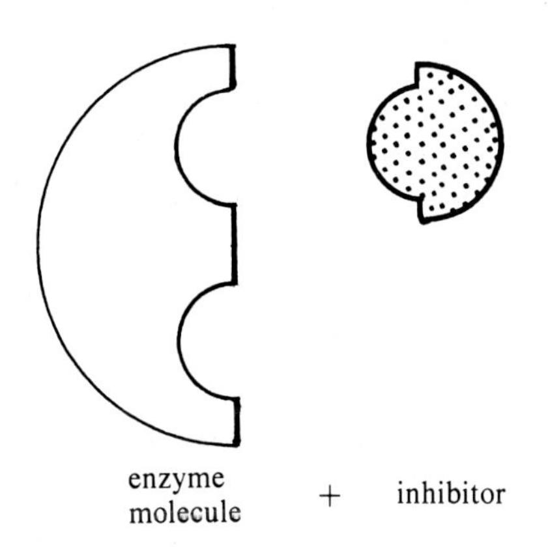

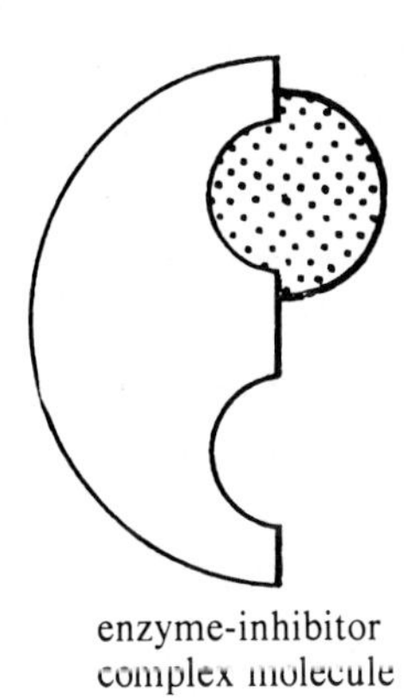

Figure 26.3 An enzyme-substrate molecule system with an inhibitor present. Do you see how the inhibitor lives up to its function, to inhibit?

26.11 Predict the effect this mercury compound will have on the liver. Use ideas about inhibitors when you write this answer.

26.12 After allowing the liver to sit overnight, describe the condition of the liver. Record any similarities and differences.

26.13 Wash the liver in distilled water. Then wash your hands. Place the piece of liver on a clean watch glass and add 20 drops of 3 percent H_2O_2 to the liver. Describe what happens. Did the mercury inhibit enzyme activity in the liver? ________

Table 26.3 Enzymes, Where They Are Found and Their Respective Function.

Enzyme	Location	Function
☐ 1. Ptyalin	Oral cavity	Changes starch to maltose
☐ 2. Pepsin	Stomach	Changes protein to peptones and proteoses
☐ 3. Renin	Stomach	Coagulates milk products
☐ 4. Trypsin	Small intestine	Changes proteins, peptones, and proteoses to peptides
☐ 5. Amylase	Small intestine	Changes starch to maltose
☐ 6. Lipase	Small intestine	Changes fats to fatty acids and glycerin
☐ 7. Erepsin	Small intestine	Changes peptides to amino acids
☐ 8. Maltase	Small intestine	Changes maltose to glucose, a simple sugar the body can use
☐ 9. Lactase	Small intestine	Changes lactose to glucose and galactose
☐ 10. Sucrase	Small intestine	Changes sucrose to glucose and fructose

26.14 The enzymes that are required to digest food types are listed in data table 26.3. While this is not a total list of enzymes present in your body, it is representative of the types needed to break down specific foods. Study the table carefully. Be able to identify any of the enzymes by their name. The suffix *-ase* identifies the substance as an enzyme. An *-ose* ending signifies the substance is a sugar. The enzyme is named in many cases by the specific material it acts on. For example, the enzyme maltase acts only on the sugar maltose. Before the body is able to use any food substance, it must be in a simple form. Notice that as the foods pass through the digestive tract, they are progressively broken down into smaller and smaller chemical units. Ultimately, they will become small enough to be absorbed into the vascular system and distributed to the cells of the body. Check (✓) the box when you are familiar with each enzyme, its location, and its function.

26.15 The phrase **alimentary canal** means a tubelike organ and includes several organs—the mouth, esophagus, stomach, and intestinal tract. The *accessory organs* that are considered a part of the digestive system include the salivary glands, liver, gallbladder, and pancreas.

After food has entered the mouth and has been masticated (chewed), the food particles move to the back of the mouth and the swallowing reflex forces the food particles into the **pharynx.** From the pharynx, food particles

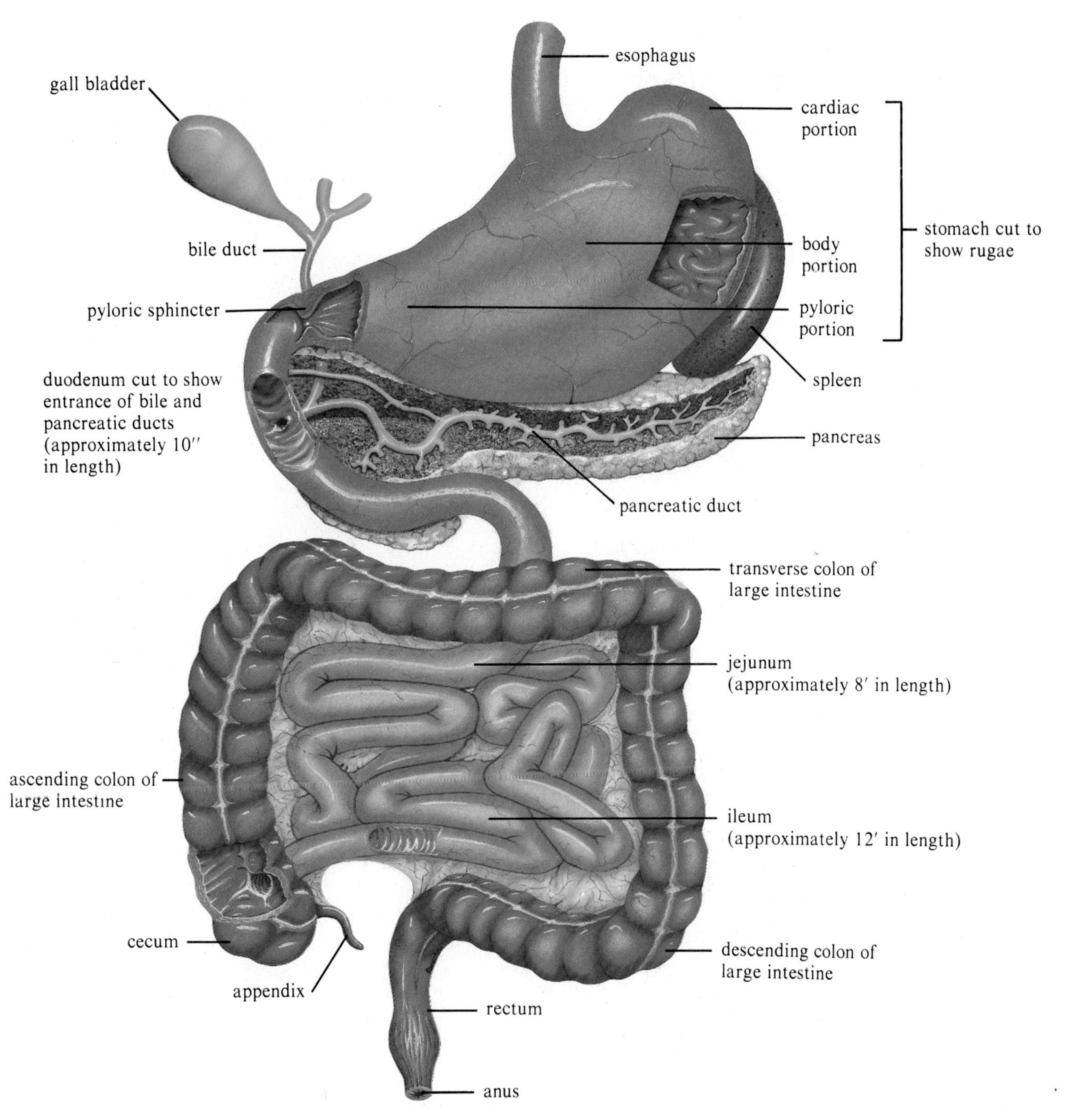

Figure 26.4 Basic anatomy of the digestive system.

enter the esophagus. A fleshy flap called the **epiglottis** prevents food from entering the trachea. If food or liquid gets by the epiglottis, a coughing reflex is activated to expel the foreign particles from the trachea.

Food continues down the esophagus by a series of rhythmic, muscular contractions called **peristalsis.** It is this process that would allow you to stand on your head and drink water. As food reaches the stomach, the muscle fibers relax as the food enters the stomach.

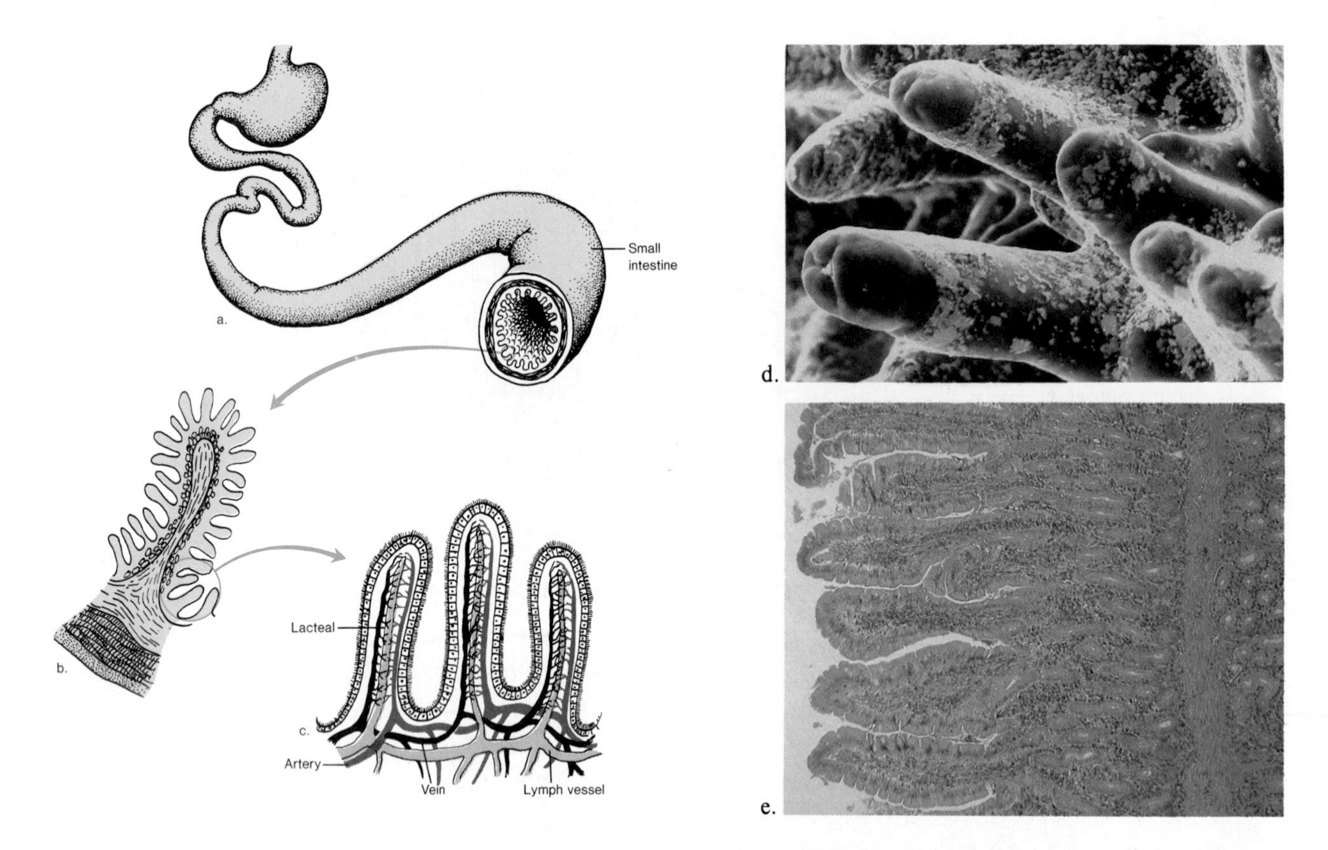

Figure 26.5 (a) The small intestine showing a cross section cut. (b) One villi system removed from the interior of small intestine. (c) Rendering of cell structure and circulatory system of villi. Note the epithelial cells are only one cell thick, thus allowing easy passage of digested micro-particles to the circulatory system. (d) SEM of villi. (e) TEM of villi.

The upper region of the stomach, near the esophagus, is called the *cardiac region.* The next region, which acts somewhat as a storage area, is called the **fundus.** The largest part of the stomach is the body, and the lower region is called the **pylorus.** The pyloric region narrows to a small canal called the *pyloric canal.* At the end of the pyloric canal is the *pyloric sphincter,* a powerful muscle that regulates the passage of food into the small intestine.

As food is broken down in the stomach, it forms a pastelike substance called **chyme.** It is chyme that enters the small intestine through the pyloric valve. What are the three regions of the small intestine? ____________ , ________________ , ________________

26.16 The lining of the small intestine consists of numerous small fingerlike projections called **villi.** The villi aid in the mixing of chyme with intestinal enzymes. They also provide a large surface area for the absorption of the products of digestion.

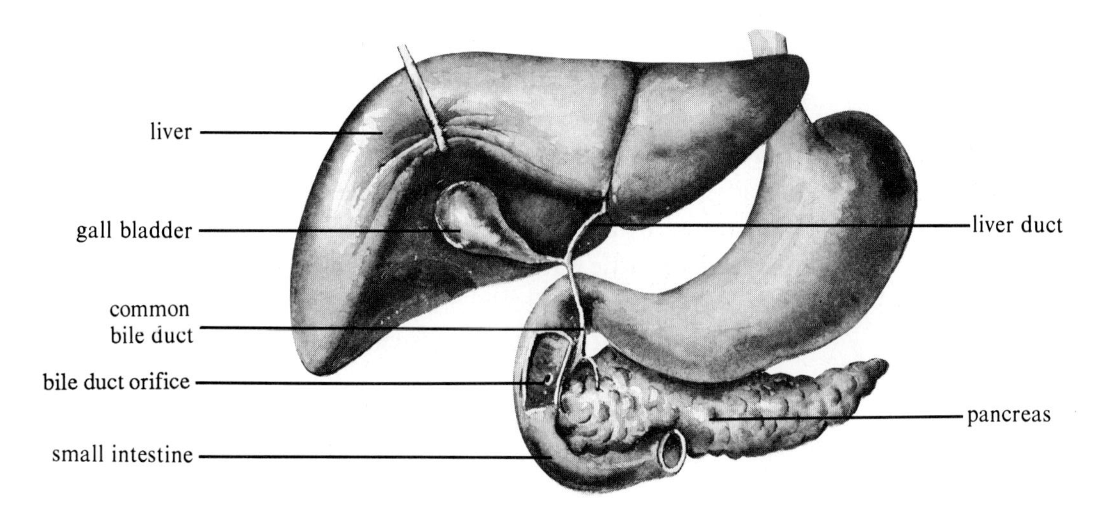

Figure 26.6 Support organs for the digestive system. (Courtesy of James Koevenig.)

26.17 The pancreas has a digestive function as well as an endocrine function. Pancreatic juices contain enzymes capable of digesting carbohydrates, fats, and proteins. The enzymes of the pancreas enter the duodenum through the pancreatic duct.

The liver, which is the largest internal organ in the body, has many functions, one of which is the secretion of **bile.** Bile is made up of bile salt, bile pigments, and cholesterol. It functions mainly to *emulsify* fats. Emulsification is a process of breaking up large fat globules into smaller particles, which allows fat-enacting enzymes to act on fat molecules more efficiently.

The bile pigments of *bilirubin* and *biliverdin* are products of red-blood-cell decomposition. If these pigments are not secreted into the bile, they accumulate in the blood and body tissues causing the skin to become yellowish in color, a condition called **jaundice.**

As bile is produced, it is collected in hepatic ducts, transported to the cystic duct, and stored in the **gallbladder.** This is a green pear-shaped sac located under the liver. When needed, bile leaves the gallbladder by way of the cystic duct and enters the duodenum through the common bile duct. Entrance into the duodenum is controlled by the *sphincter of Oddi.*

26.18 The large intestine, or **colon,** has three basic regions: the *ascending colon, transverse colon,* and the *descending colon.* The function of the large intestine is limited to the absorption of water and electrolytes that were necessary in the digestive processes. The electrolytes are absorbed by a process called **active transport,** and the water is absorbed by the mucosa by a proces called **osmosis.** As a result of these two processes, the **feces** has little water and virtually no electrolytes. Peristalic waves move the fecal material to the anal orifice, which is governed by the *anal sphincter muscle,* and the fecal material is excreted.

26.19 Before concluding this chapter, some coverage should be given to nutrition. The food we eat is the basis for all structural and functional body processes. Food is broken down to supply us with needed energy and building materials. The amount of energy available in the various foods is measured in **Calories.** The basic unit of energy is the calorie. *A calorie is defined as the amount of energy required to raise 1 cc of water 1° Celsius.* The measure used for food energy is 1,000 times as much, or a kilocalorie. Therefore, food Calories are distinguished by using a capital *C.* It requires approximately 12 Calories per pound to maintain your body weight. The actual number of Calories required to maintain body weight will vary in an individual. The number of Calories is dependent on an individual's activity, age, and physical condition.

Use the formula to determine the approximate number of Calories you should ingest on a daily basis.

12 Calories × ________ pounds = ________ Calories

26.20 List all of the food you have eaten in the past 24 hours in the space below. Using a Calorie guide, determine the number of calories for each item listed.

Breakfast:

Lunch:

Dinner:

Snacks:

26.21 Add up the approximate Calories that you consumed. Approximately how many Calories did you consume? ______
Is your consumption more, less, or the same as the required amount of Calories you need to maintain body weight? ________

26.22 To review the digestive system, simply identify the anatomical parts in figure 26.7.

a. ______________________________

b. ______________________________

c. ______________________________

d. ______________________________

e. ______________________________

f. ______________________________

g. ______________________________

h. ______________________________

i. ______________________________

j. ______________________________

k. ______________________________

l. ______________________________

m. ______________________________

n. ______________________________

o. ______________________________

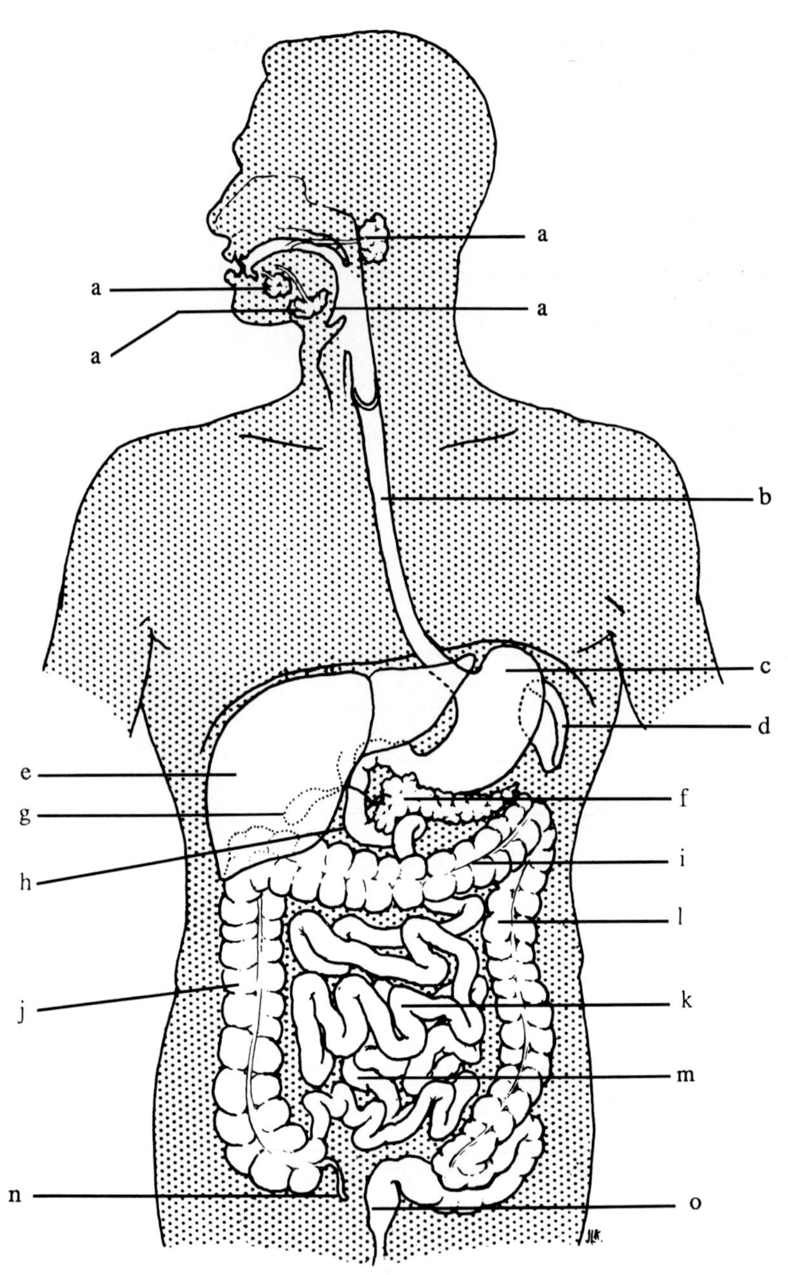

Figure 26.7 The alimentary canal. (Courtesy of James Koevenig.)

26.23 Review the study points below. Place a check (✓) in the box when you have mastered the concepts related to the given study point.

On completion of this chapter, you should be able to do the following:

☐ 1. Identify the three salivary glands.

☐ 2. Identify the role of enzymes in the body.

☐ 3. List the enzymes secreted by various organs and give the function of each.

☐ 4. Relate the function of the stomach, pancreas, gallbladder, small intestine, and large intestine as they pertain to digestion.

☐ 5. Calculate the approximate Calories needed in your personal diet.

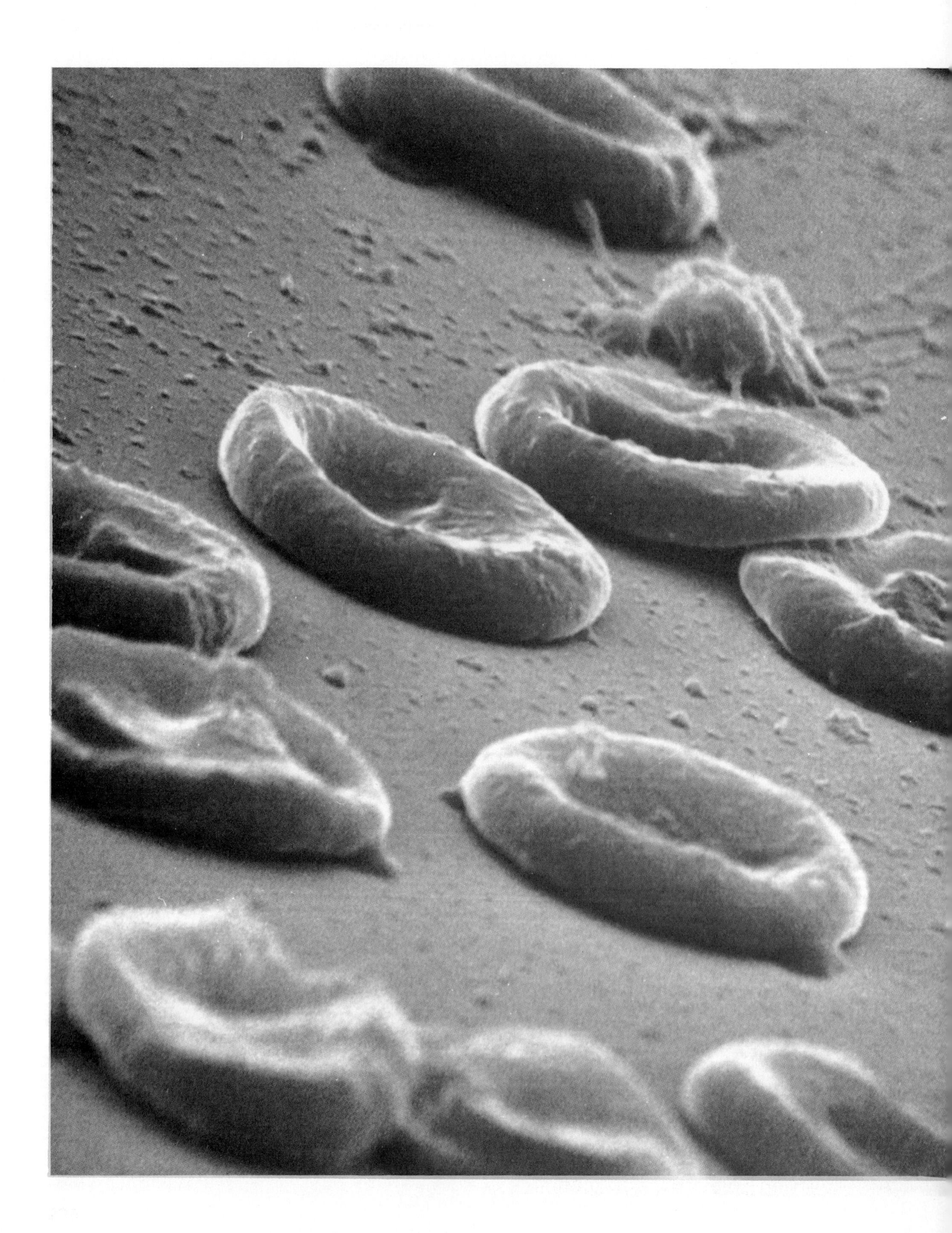

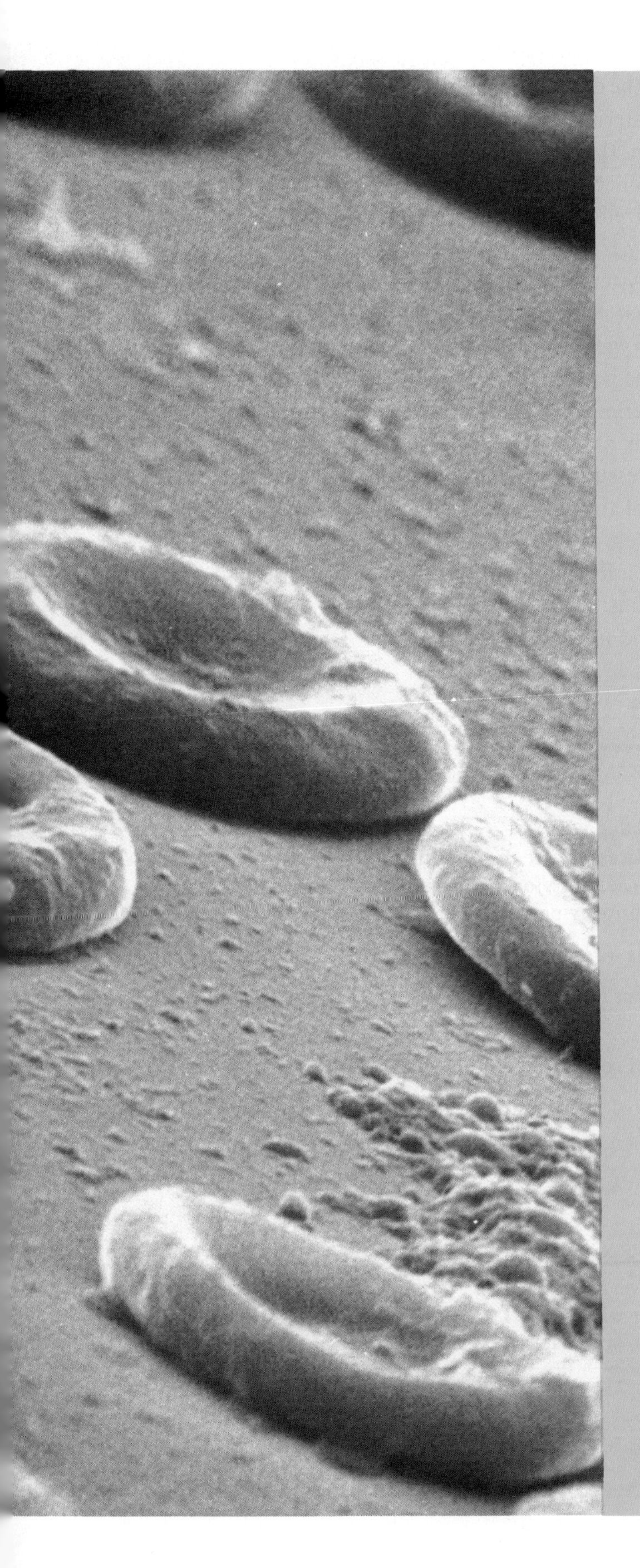

The Transport Systems: Circulatory, Lymphatic, Respiratory and Urinary

27

Transport systems?
Do you mean like in railroads?

27.1

Besides the digestive system, there are four other systems that move materials throughout your body. These four systems are the circulatory system, the lymphatic system, the respiratory system, and the urinary system. Collectively, they could be known as transport systems. In this chapter, they are being treated as such. Each system, however, will be studied individually.

Circulatory System

The circulatory system is the first transport system. Within this system, blood cells and plasma move throughout the entire body. As blood flows through the vessels, oxygen, food molecules, and waste products are transported. Each cell is the body is serviced by the circulatory system and the countless miles of arteries, veins and capillaries that make up this system. Let's take a closer look at this system. We will take a look at blood cells, blood typing, blood pressure, the heart, and several of the major veins and arteries. To know all of the parts of the circulatory system or, indeed, any system would comprise a course of its own.

Blood consists of a liquid called **plasma** in which are found many different types of materials. Data table 27.1 illustrates a few of the constituents that can be found in blood plasma.

Data Table 27.1 Constituents of Blood Plasma

Water

Plasma Proteins

- Albumins—necessary for the formation of blood clots
- Fibrinogen—necessary for the formation of blood clots
- Globulins—antibody proteins released by plasma cells

Glucose—produced as a result of sugar digestion

Cholesterol—important in metabolism and can be activated to form vitamin D

Urea—produced from cellular protein metabolism

Trace elements required for numerous cellular activities
- Sodium
- Potassium
- Calcium
- Iron

Types of cells found in blood
- Erythrocytes, or red blood cells
- Leukocytes, or white blood cells
- Thrombocytes, or blood platelets

To view blood itself, you could not see all of the items mentioned in data table 27.1. You can observe, however, several different kinds of blood cells.

The following exercise will allow you to prepare a blood slide for microscopic viewing. Obtain the following items from your laboratory supply table:

> A clean microscope slide and coverslip, a sterile blood lancet, a cotton ball, and some rubbing alcohol or an alcohol pad. Be careful to keep the lancet and your skin sterile and, likewise, be certain that the lancet is properly discarded after use. This will assure you of sanitary conditions. At no time should you ever mix your blood, even one drop, with that of another person.

The steps to follow are:

a. With a cotton ball dipped in alcohol, sponge the tip of your middle finger.
b. Prick the tip of the finger with a lancet using a quick stabbing motion.
c. Squeeze the finger until a single drop of blood is formed and place the drop on a clean microscope slide. Wipe the finger with the alcohol swab when finished.
d. Carefully place the coverslip over the drop of blood, being attentive to the possible formation of unwanted air bubbles.
e. Look at the slide with the aid of a compound microscope and describe what you see.

CAUTION: Do not mix blood from your finger with that of another student.

27.2

Red blood cells are called **erythrocytes** and white blood cells are called **leukocytes.** Figure 27.1 illustrates both. Were you able to see any leukocytes? ________ Leukocytes can be seen only with proper staining techniques. To view leukocytes, obtain a prepared slide from your supply table and view it with a compound microscope. You will see many erythrocytes, and a few leukocytes. Which of the leukocytes were you able to find? Name and draw one of each in the fields supplied in your log book.

Name ________________ Name ________________ Name ________________

Data Table 27.2 The Roles of Blood Cells

Blood Cell	Function
a. Erythrocytes	Transport oxygen to the body
b. Platelets	Involved with blood coagulation
c. Neutrophils	Phagocytic
d. Eosinophils	Phagocytic
e. Basophils	Releases histamines during inflamation
f. Monocytes	Phagocytic cells called macrophages
g. Lymphocytes	Production of antibodies

27.3

Data table 27.2 lists the blood cells and the function of each. Note that the leukocytes tend to be phagocytic, which means they fight and engulf foreign particles found in the blood stream. This is particularly important when the blood stream is invaded by viruses and bacteria.

Obtain a prepared slide that shows mononucleosis and a prepared slide that shows leukemia. Examine each slide carefully. Which white blood cell is quite prevalent with mononucleosis? ________________ Which blood cells seem to be prevalent with leukemia? ________________

27.4

The various blood cells listed in section 27.2 are illustrated in your log book. Using a Schilling or John Hopkins or other reference, label each blood cell.

Read section 27.5 entirely before you begin.

27.5

Blood type is inherited as are other hereditary traits. There are four basic blood types: A, B, AB, and O. These types result from specific **agglutinogens** found on the surfaces of the red blood cells. These agglutinogens react with **agglutinins** in the plasma of certain blood types and allow for the classification of types. The four blood types are a result of only two **antigens,** A and B. If a person's blood clumps with anti-A serum, it means that person has the A-antigen and, therefore, is type A. If the blood clumps with anti-B serum, the B-antigen is present, and the person is type B. If the blood clumps with both anti-A and anti-B serums, the individual is type AB because both antigens are present. A type O individual has neither antigen A nor antigen B, so that person's blood will not clump with either antiserum.

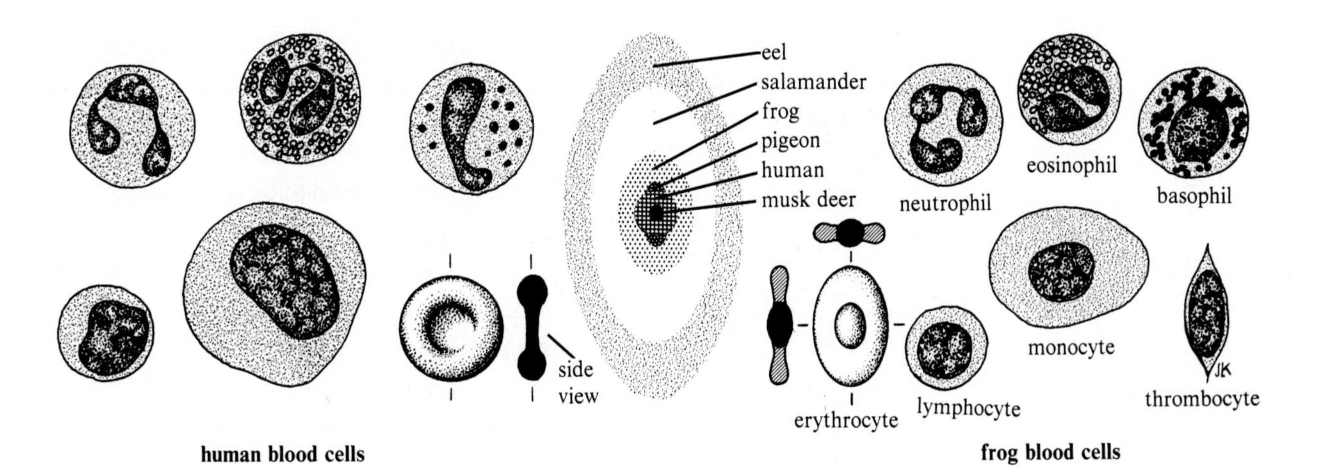

Figure 27.1 A comparison of blood cells. Note the similarity and differences between frog and human blood cells. Can you name the human blood cells? (Courtesy James Koevenig.)

Data Table 27.3 Characteristics of Blood During Blood Typing

Blood Type	Characteristics for Determination
O	No clumping of either serum
A	Clumping with anti-A serum
B	Clumping with anti-B serum
AB	Clumping with both serums

In order to type your blood, follow the steps listed below.

a. Clean and dry three microscope slides thoroughly. The slides should be depression slides, if available.
b. Mark the slides *A* and *B*.
c. Use a sterile, disposable lancet to prick your middle finger after you have wiped it off with isopropyl (rubbing) alcohol.
d. Squeeze a drop of blood out and wipe it off. Then squeeze a second drop on each slide or into the depression. One drop is enough for each slide. Do not mix your blood with anyone else.
e. Wipe the finger off with the alcohol swab.
f. Add a drop of anti-A serum to one drop of blood and stir with a toothpick. Add a drop of anti-B serum to the second drop of blood and stir with a different toothpick.
g. Examine the drops of blood carefully. Data table 27.3 will help you to determine your blood type.

CAUTION: Do not mix blood from your finger with that of another student.

Another blood factor that is inherited is the Rh factor. It is inherited independent of the ABO blood group just discussed. To determine your Rh factor, add a drop of anti-Rh serum to the third slide. If clumping occurs, you have the Rh factor and are considered Rh+ (positive). If there is no clumping, you lack the Rh factor and are considered to be Rh− (negative). What is your blood and Rh type? ____________

27.6 Figure 27.2 illustrates the circulatory system. As you read through the following discussion, label the parts listed by numbers.

The heart consists of four chambers, the right and left auricles, and the right and left ventricles. The smaller auricles are located at the superior end of the heart and are separated from each other. The larger ventricles are located at the inferior end of the heart, and these are separated by a thick muscular wall, the **interventricular septum.**

Blood enters the heart by way of the **superior** and **inferior vena cava** to the **right atrium.** It pushes through a one-way *tricuspid* **valve** to the **right ventricle.** The strong ventricular muscle tissue literally pushes the blood through the **pulmonary semilunar valve** with its pumping action. At this point, blood enters the **pulmonary arteries** and proceeds to the lungs.

While blood is in the lungs, carbon dioxide is given up and oxygen is acquired. The oxygen-rich blood is returned to the heart via **pulmonary veins.** This is the only time oxygenated blood is found in veins.

Blood from the lungs enters the **left atrium.** It moves through a bicuspid valve into the **left ventricle,** where it is once again pushed with strong muscular contractions.

Blood leaves the left ventricle through the **aortic semilunar valve** to the **aorta.** From the aorta, blood continues to all parts of the body via arteries, arterioles, and capillaries. It returns to the heart by capillaries, venioles, and veins.

27.7 After labeling figure 27.2, carefully study the flow of blood through the heart. Be able to trace the flow of blood through the heart naming each chamber, valve, vein and artery involved.

27.8 As blood flows through the heart, it moves from chamber to chamber and to the lungs and back. The movement of blood is a result of the rythmic contractions of the heart. We can listen in on the work being done by the heart. A complete cardiac cycle is related to two basic happenings called **diastole** and **systole.** As the ventricles relax (diastole), they fill with blood from the atria. As the ventricles contract (systole),

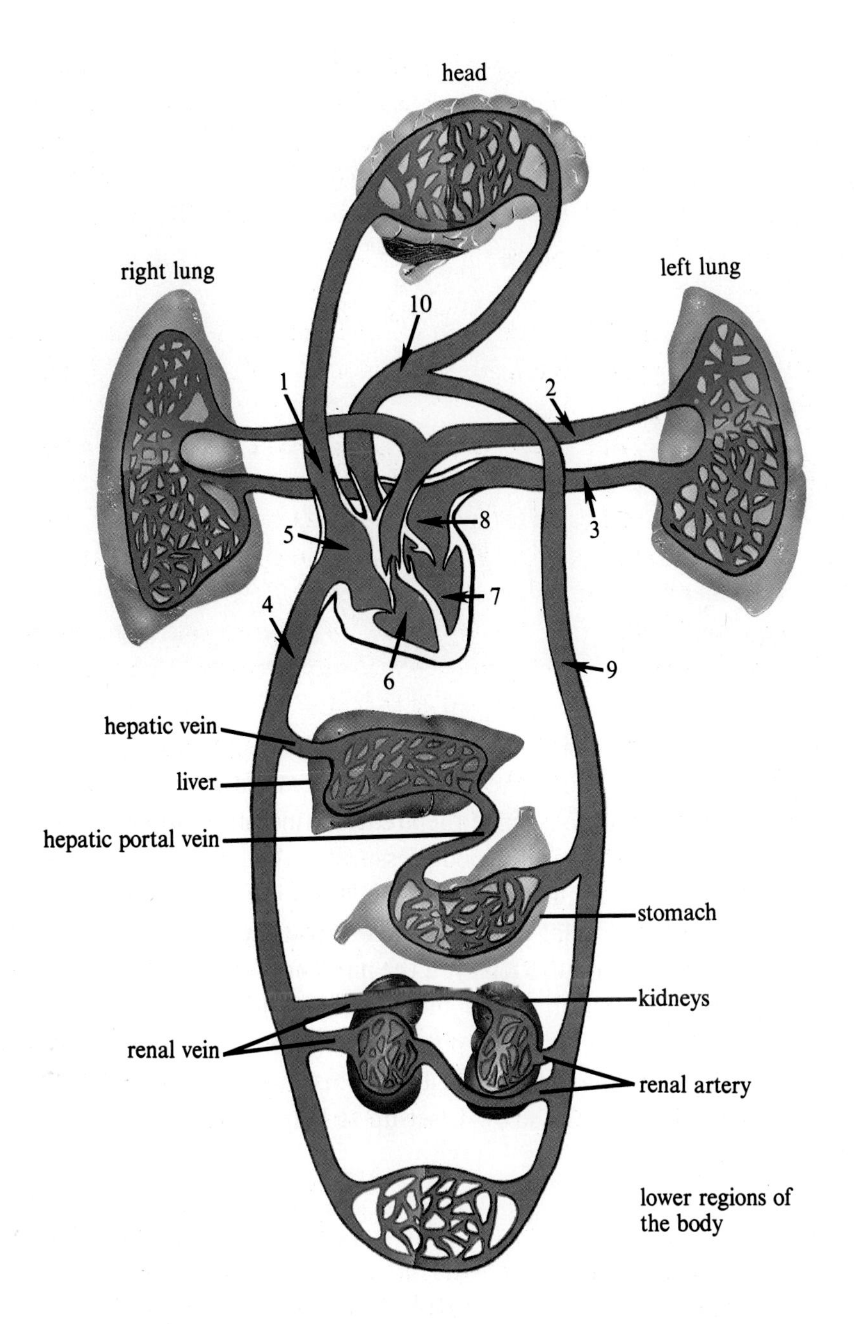

Figure 27.2 The circulatory system.

blood is forced out of the ventricles into the vessels. The sounds associated with these events are described as *lub-dub.* The lub sound is the systolic sound, and the dub sound is the diastolic sound. These sounds are very important for doctors to hear while they are giving a physical examination.

Get a stethoscope, an instrument that allows you to hear the sounds of the heart, from the supply table. Listen to your heart. If you do not hear anything, do not panic. You might be listening in the wrong location.

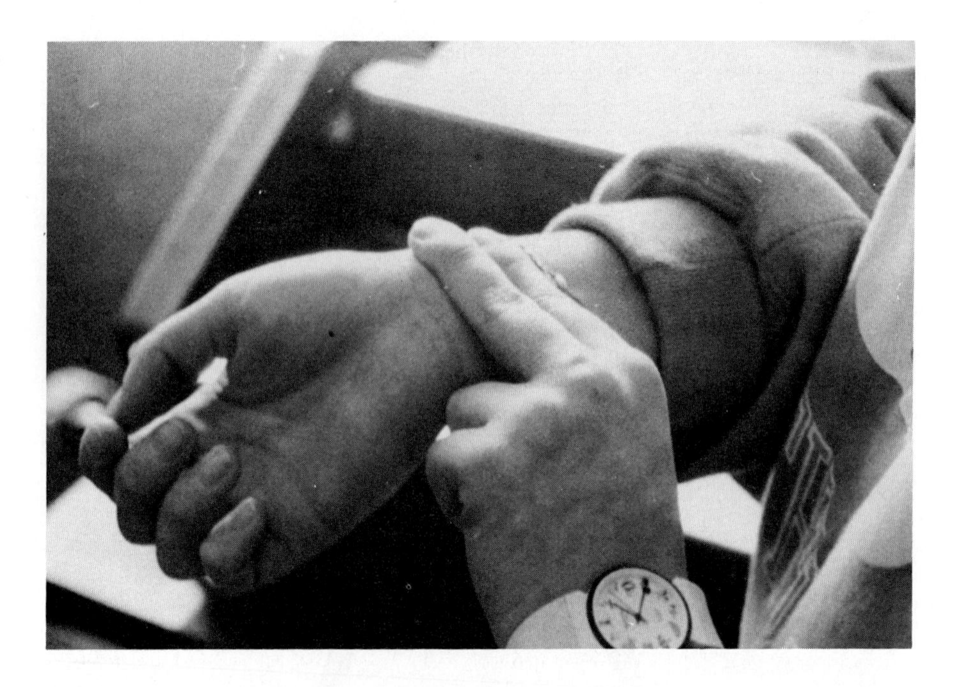

Figure 27.3 Using two fingers, find your radial pulse on the thumb side of your wrist.

27.9 Blood pressure is the pressure of blood on the walls of the arteries, veins, and the heart itself. Blood pressure is usually identified as being arteriole, since the greatest amount of pressure is found in the arteries. The changes in blood pressure can be felt as a pulse in areas where arteries are near the surface of the body. For example, figure 27.3 illustrates where the radial pulse is taken. Pulse should not be construed as blood pressure. Your pulse tells you the number of times your heart beats per minute. Before continuing with blood pressure, let us stop to investigate your pulse rate (PR). Find your pulse rate by counting the beats for 15 seconds and multiplying the number by 4. This will give you the pulse rate for 1 minute. The formula below is set up to help with your calculations.

$$\underset{\text{Beats per 15 seconds}}{\underline{\hspace{6cm}}} \times 4 = \underset{\text{Pulse rate per minute}}{\underline{\hspace{6cm}}}$$

27.10 Check your pulse three times. Record the rate in data table 27.4. Average the rate and record this figure in the table.

Now walk fast up and down the hall or around the room for 2 minutes. When you have done so, check your pulse rate again. Record this figure and two additional times in the table. Once again, determine the average of the three trials.

Finally, step up and down onto a chair for 2 minutes without stopping to rest. Check your pulse after each 2-minute segment, recording the rate in the table. Determine the average. Be certain to rest between each trial for several minutes to allow your pulse rate to return to normal.

Data Table 27.4 Determining Pulse Rate

Trial	Normal PR	Walking PR	Active PR
1			
2			
3			
Average			

How did your pulse rate react to increased activity?

____

27.11 Now we can get back to blood pressure. Determining blood pressure accurately requires use of two clinical instruments, the *stethoscope* and the *mercury sphygmomanometer.* Taking blood pressure is really quite simple once you have mastered using both pieces of equipment together. Figure 27.4 shows the proper placement of the stethoscope and the cuff of the sphygmomanometer. Proper use assures proper results.

The steps for taking blood pressure are listed below:

a. Wrap the cuff around the upper arm, as illustrated in figure 27.4. The cuff should fit snugly.
b. Place the stethoscope over the brachial artery, as indicated in the illustration. You should be able to detect a slight tapping sound.
c. Inflate the cuff to 180 mm Hg (millimeters of mercury) and allow the cuff to deflate slowly by opening the exhaust valve.
d. As the cuff deflates, listen carefully with the stethoscope for the first clear sound. At that point, read the level of the mercury. This is the systolic pressure. Continue to listen until the sound disappears. Again read the mercury level. This is the diastolic pressure.

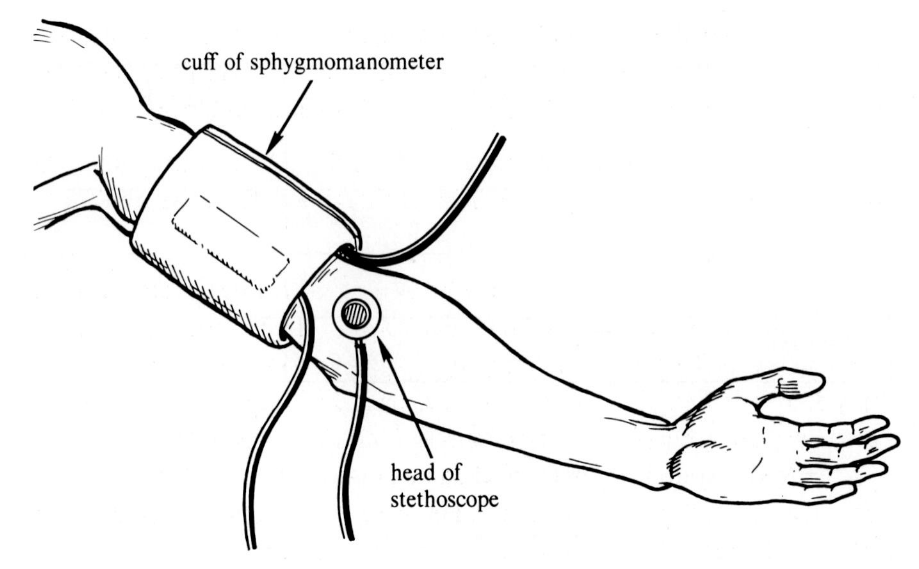

Figure 27.4 Proper placement of the sphygmomanometer cuff and stethoscope when determining blood pressure.

A normal blood pressure is in the immediate range of 120/80 mm Hg. This is read 120 over 80. Take the blood pressure of a lab partner and record his/her pressure in data table 27.5 below. Because errors occur, take your lab partner's blood pressure three times and average the three readings. Record the readings in data table 27.5. Allow at least two minutes between readings so the tension the inflated cuff puts on your vessels has a chance to wear off.

Data Table 27.5 Blood Pressure in mm/Hg

Trial	Lab Partner	You
1		
2		
3		
Average		

1. Is your blood pressure within an acceptable range? ______
2. List several reasons (for example, overweight, drugs, emotions, and so forth) why your blood pressure may not be within the usual range.

27.12 Blood pressure will also fluctuate with exercise or excitement. Smoking will also alter an individual's blood pressure. Step onto a chair as many times as you can in a 2-minute-time period. Immediately take a blood pressure reading and record ____________ mm HG How does this reading compare with the average you found at rest?

27.13 As blood passes from the heart to the systemic circuit, it travels the length of the body and back to the heart in a matter of minutes. While the arteries deliver oxygenated blood to the body, the veins carry deoxygenated blood back to the heart except for the pulmonary veins. Pulmonary veins carry oxygenated blood from the lungs to the heart.

Let's take a look at some selected arteries and veins. With the aid of references, determine the role of the major veins and arteries listed below. Figures 27.5 and 27.6 illustrate major arteries and veins.

Arteries

a. Renal
b. Radial
c. Brachial
d. Coronary
e. Common carotid
f. Pulmonary
g. Superior femoral
h. Hepatic
i. External iliac
j. Abdominal aorta

Veins

a. Common iliac
b. Basilic
c. Cephalic
d. Internal jugular
e. Femoral
f. Pulmonary
g. Hepatic portal
h. Greater saphenous
i. Renal
j. Subclavian

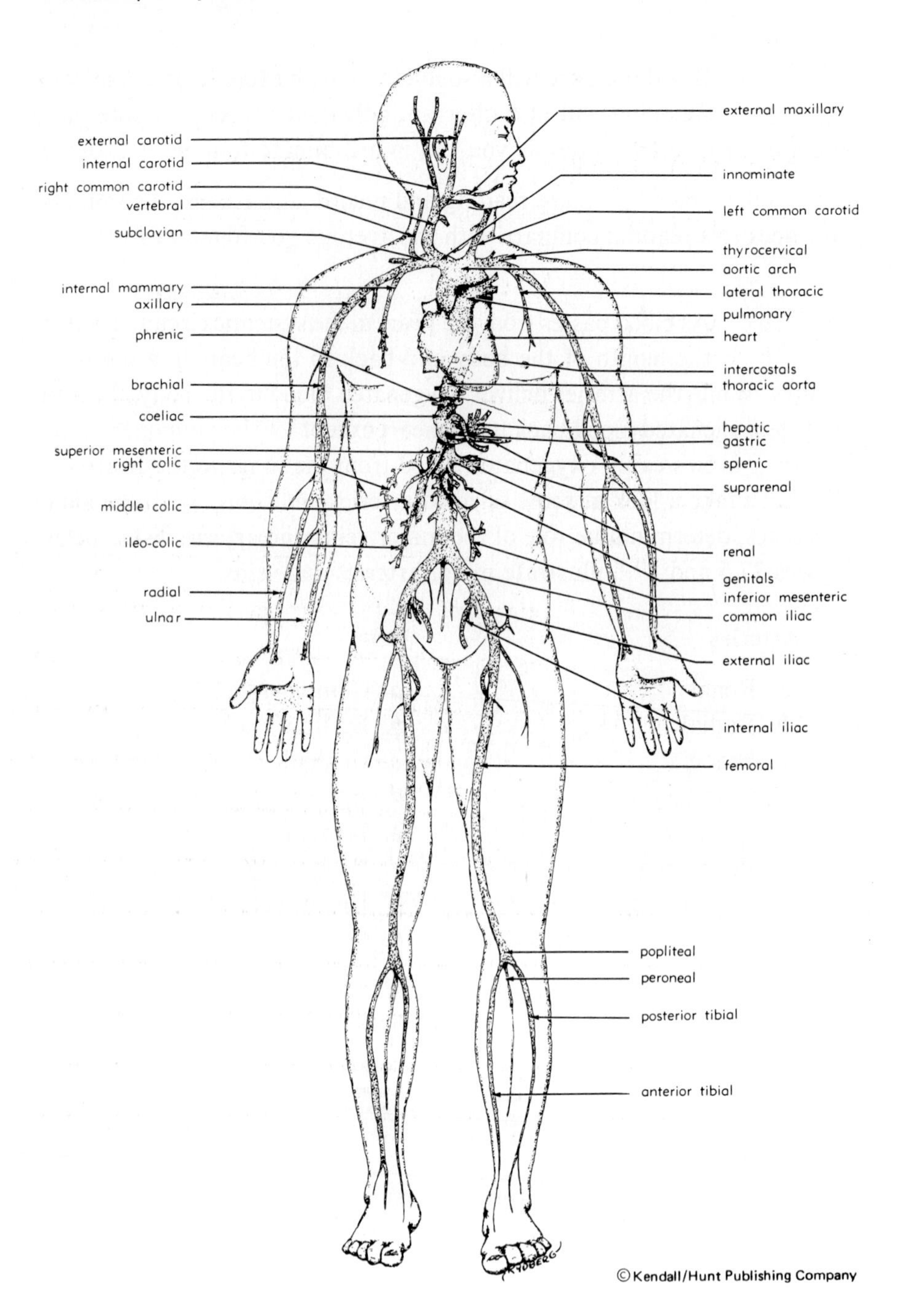

Figure 27.5 Major arteries of the human body.

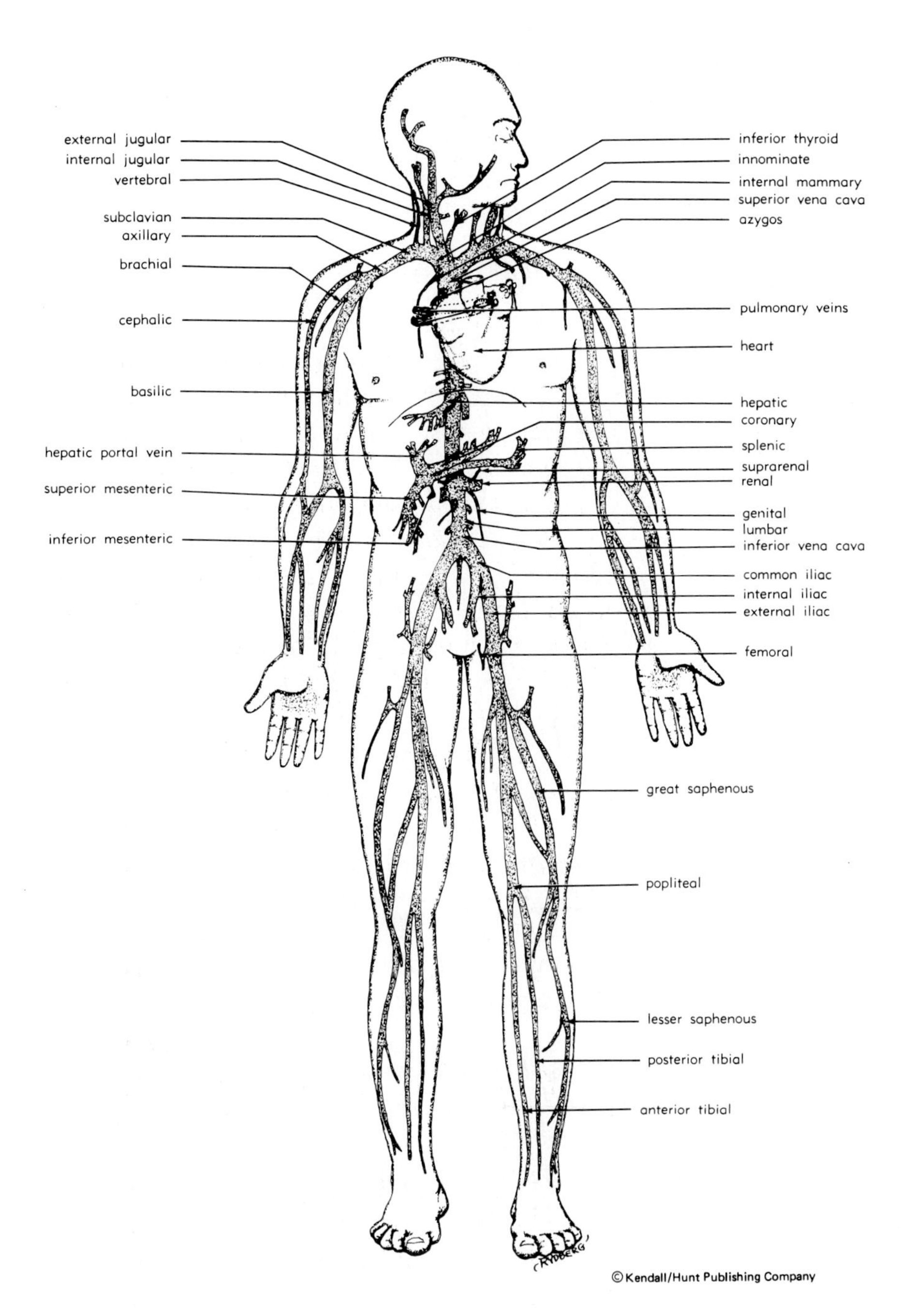

Figure 27.6 Major veins of the human body.

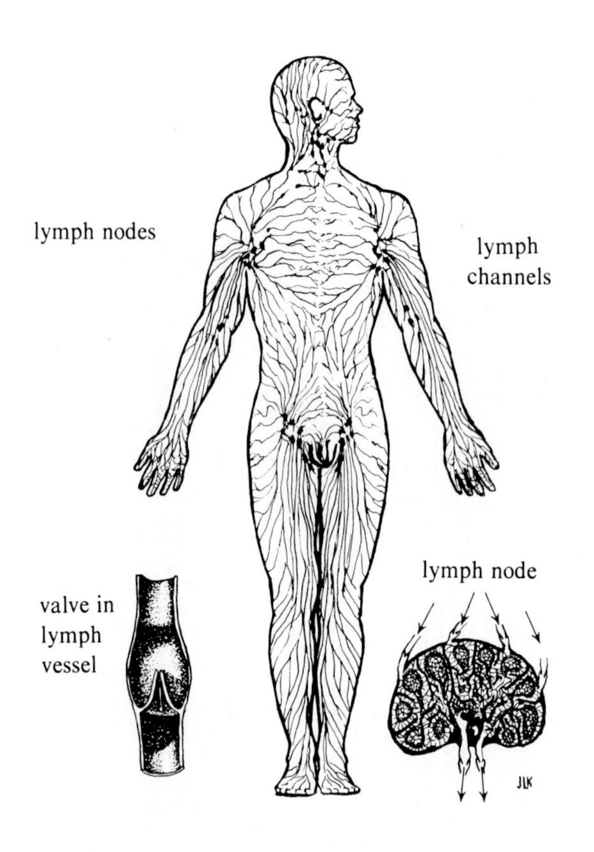

Figure 27.7 The lymphatic system noting the location of the nodes (black dots) and vessels. (Courtesy James Koevenig.)

Lymphatic System

27.14

The lymphatic system is the second transport system. The lymph ducts spread throughout the body like the veins. Figure 27.7 illustrates the lymph system. The main difference between lymph and blood plasma is that lymph has no blood corpuscles (cells).

The functions of the lymphatic system include:

a. The absorption of fat via *chyle,* a milky fluid produced by special lymphatic capillaries called *lacteals* in the small intestine.
b. The passing of lymph fluid through *lymph nodes* that act *as phagocytic organs.* Lymph nodes effectively remove foreign substances from the body, especially during a time of an infection.
c. Returning of tissue fluid back to the blood stream.

Lymphlike organs that function much the same as the lymph nodes, but are not a part of the lymphatic system, are the spleen, thymus gland, and the tonsils. All tend to serve as defense agents against the invasion of bacteria.

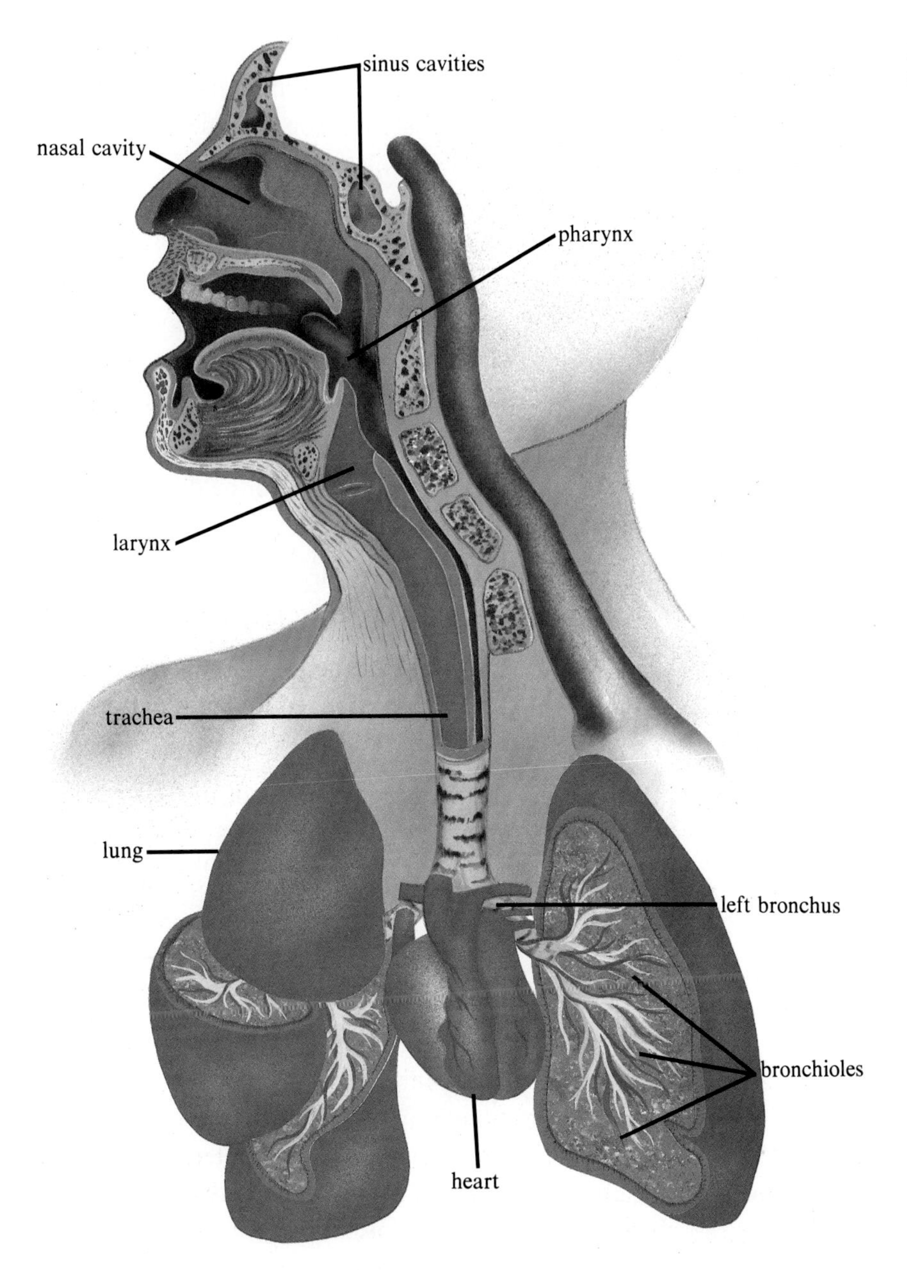

Figure 27.8 Anatomy of the respiratory system.

Respiratory System

27.15 The respiratory system is the third transport system. Figure 27.8 illustrates the respiratory system. Study it carefully.

The respiratory system functions basically to exchange gases between the external environment and the body. There is a definite route through which gases are exchanged. The organs that are involved in the respiratory system include the *nose, pharynx, larynx, trachea, bronchi,* and *lungs*. The ultimate end of the respiratory system are the microscopic **alveoli.** It is at the alveoli level where the exchange of gases is completed with the circulatory system via hemoglobin in blood.

In the exercise below, you will find six organs listed that are a part of the respiratory system. Match each function by writing its letter on the line before the number of the appropriate organ. You may need to use a reference textbook to assist you.

Organ		Function
____	1. Nose	*a.* The musculomembranous tube that is part of the respiratory tract (nasopharynx) in which the lower part is common to the digestive system as well (laryngopharynx).
____	2. Pharynx	*b.* The cartilaginous windpipe extending from the laryngopharynx to the division of the two bronchi.
____	3. Larynx	*c.* The prominent organ in the center of the face; the beginning of the respiratory tract.
____	4. Trachea	*d.* An air cell of the lung; the point at which gases are exchanged into and out of the circulatory system.
____	5. Bronchi	*e.* The primary branches of the trachea in the lungs.
____	6. Alveoli	*f.* The organ of the voice consisting of a series of cartilages: the thyroid, the cricoid, and the epiglottis. By separation of the vocal folds and by forcing air out, pitch in voice is determined.

27.16 The mechanics of breathing consist of an interaction between several sets of muscles (diaphragm and intercostals). As the **diaphragm** contracts, air is pulled into the lungs due to the decrease in pressure within the pleural cavity (*inspiration*). This occurs because of the curved shape of the diaphragm itself. When the diaphragm returns to its original position on relaxation, air is forced out of the lungs (*expiration*). Complementing the action of the diaphragm is the corresponding action of the muscles between the ribs (intercostal muscles). The actions of the diaphragm and intercostal muscles are controlled by the involuntary nervous system. Because of this, breathing is accomplished without a person being aware of the process.

As with other organ systems, the human respiratory system is extraordinarily strong, capable of resisting innumerable pressures, and endowed with ingenious safety and reserve mechanisms. For one thing, if one lung should collapse or be surgically removed, the other will continue to function at sufficient capacity to maintain the vital exchange of gases. The lungs are capable of holding about 4.5 to 6.0 liters of air. In normal breathing, only about one-eighth of total lung capacity is used, leaving plenty of reserve capacity for those occasions when the body requires a much greater flow of oxygen to generate energy. Furthermore, the lungs are never completely empty of air. For example, if a lung were collapsed, removed from the body, and placed in a tub of water, it would float.

To measure the various lung capacities, you will use the Phipps & Bird Wet Spirometer. A *spirometer* is an instrument used to measure different volumes of air involved in breathing. The Phipps and Bird Wet Spirometer is based on the simple mechanical principle that air, exhaled from the lungs, will cause a displacement of a closed chamber that is partially submerged in water.

The following experiments will measure the various lung capacities. Each experiment should be completed 3 times so an average can be determined. Several minutes rest should be allowed between each trial. The units measured are in liters and must accompany all recorded data.

27.17 In experiment 1 you will measure tidal volume. Tidal volume (TV) is the amount of air exhaled or inhaled during normal, quiet breathing.

Sit by the spirometer, breathing quietly and normally, for 1 minute. After inhaling a normal breath, place the spirometer mouthpiece between your lips and exhale in a normal unforced way into the spirometer. Record your results in data table 27.6 below. Repeat this procedure two additional times, resetting the spirometer pointer to zero after each trial.

Data Table 27.6 Tidal Volume (L)

Trial 1 = ______________________

Trial 2 = ______________________

Trial 3 = ______________________

Average = ______________________

27.18 In experiment 2, you will measure expiratory reserve volume. The expiratory reserve volume (ERV) is the amount of air that can be forcibly breathed out after normal expiration.

Stand by the spirometer breathing normally for 1 minute. After a normal exhalation, put the mouthpiece between the lips and forcibly exhale all additional air possible. Record your results in data table 27.7 below. Repeat this procedure two additional times, resetting the spirometer pointer to zero after each trial. Be sure to rest several minutes between trials.

Data Table 27.7 Expiratory Reserve Volume (L)

Trial 1 =	______________
Trial 2 =	______________
Trial 3 =	______________
Average =	______________

27.19 In experiment 3, you will measure inspiratory reserve volume. The inspiratory reserve volume (IRV) is the amount of air that can be inhaled following normal TV inhalation.

Stand next to the spirometer for 1 minute. Breath as deeply as possible and exhale through the mouthpiece normally. Do not force any air out. The IRV reading is obtained by subtracting your average TV reading from the IRV reading recorded on the spirometer. Repeat this procedure two more times, resting between each trial. Again you will be determining an average volume. Because you are using the average TV reading, your results will be slightly biased.

Data Table 27.8 Inspiratory Reserve Volume (L)

Trial 1 =	________	− TV ________	=	________
Trial 2 =	________	− TV ________	=	________
Trial 3 =	________	− TV ________	=	________
Average =	________			

27.20 In experiment 4, you will measure vital capacity. The vital capacity (VC) is the maximum amount of air that can be forcibly exhaled immediately after a maximal inhalation. Also, you should be able to calculate your VC by simply adding the figures obtained from your TV, ERV, and IRV experiments.

$$VC = TV + ERV + IRV$$

Before doing the actual experiment, add the averages of the TV, ERV, and IRV to determine your approximate VC.

$$VC = \_\_\_\_\_ + \_\_\_\_\_ + \_\_\_\_\_ = \_\_\_\_\_ L$$

Another way to determine VC is with the following formula:

$VC = \frac{W^n}{K}$ where

W = your body weight in grams
n = 0.72 and
K = 0.690

Using this formula, calculate your *VC*.

How do the two calculated vital capacities compare? Are they approximately equal?

Now measure your vital capacity. Stand next to the spirometer. Breathe slowly and deeply. Then, breathe in as deeply as possible. Place the mouthpiece to your lips and force out as much air as you possibly can. Record the results in data table 27.9. Complete this experiment two more times. Because of the force you exert here, resting several minutes between each trial is essential for good results.

Data Table 27.9 Vital Capacity (L)

Trial 1 =	____________________
Trial 2 =	____________________
Trial 3 =	____________________
Average =	____________________

How does your experimental average compare with your two calculated figures?

27.21 In addition to the previous experiment, another measurement of lung capacity can be calculated. This is the inspiratory capacity (IC). The inspiratory capacity is the amount of air that can be inhaled after normal expiration. IC is calculated by adding your IRV and TV capacities.

$$IC = IRV + TV$$

Do this with your average IRV and TV capacities.

$$IC = \_\_\_\_\_\_\_\_ + \_\_\_\_\_\_\_\_ = \_\_\_\_\_\_\_\_ L$$

27.22 Two final lung capacities need to be addressed. The residual volume (RV) of your lungs and the functional residual capacity (FRC). The residual volume is the amount of air that is always present in the lungs, even after maximal exhalation. You can calculate this amount as the value of your ERV to zero. The functional residual capacity then is your residual volume plus your expiratory reserve volume.

$$FRC = RV + ERV$$

27.23 Figure 27.9 illustrates a typical spirogram. Using your calculated averages, make your own spirogram. Include all labels and titles as given in figure 27.9.

Urinary System

27.24 The urinary system is the fourth and final transport system. The major organ of the urinary system is the kidney. Kidneys are paired organs, one on either side of the vertebral column on the posterior wall of the abdominal cavity. The medial side of the kidney is indented for accommodation of the **renal sinus.** The entrance to the sinus, called the **hilum,** supports blood vessels, lymphatic vessels, and the **ureter.**

The kidney is divided into two distinct regions: a centrally located **medulla** and an outer **cortex.** The functional part of the kidney is the **nephron.** The nephron is made up of two parts: the *renal corpuscle* and the *renal tubule.* The renal corpuscle, or *Malphighian corpuscle,* is made of a cluster of blood capillaries called the **glomerulus** and a saclike structure that surrounds the glomerulus called **Bowman's capsule.** Each capsule ultimately connects to a *collecting duct* through a continuous tube (renal tubule).

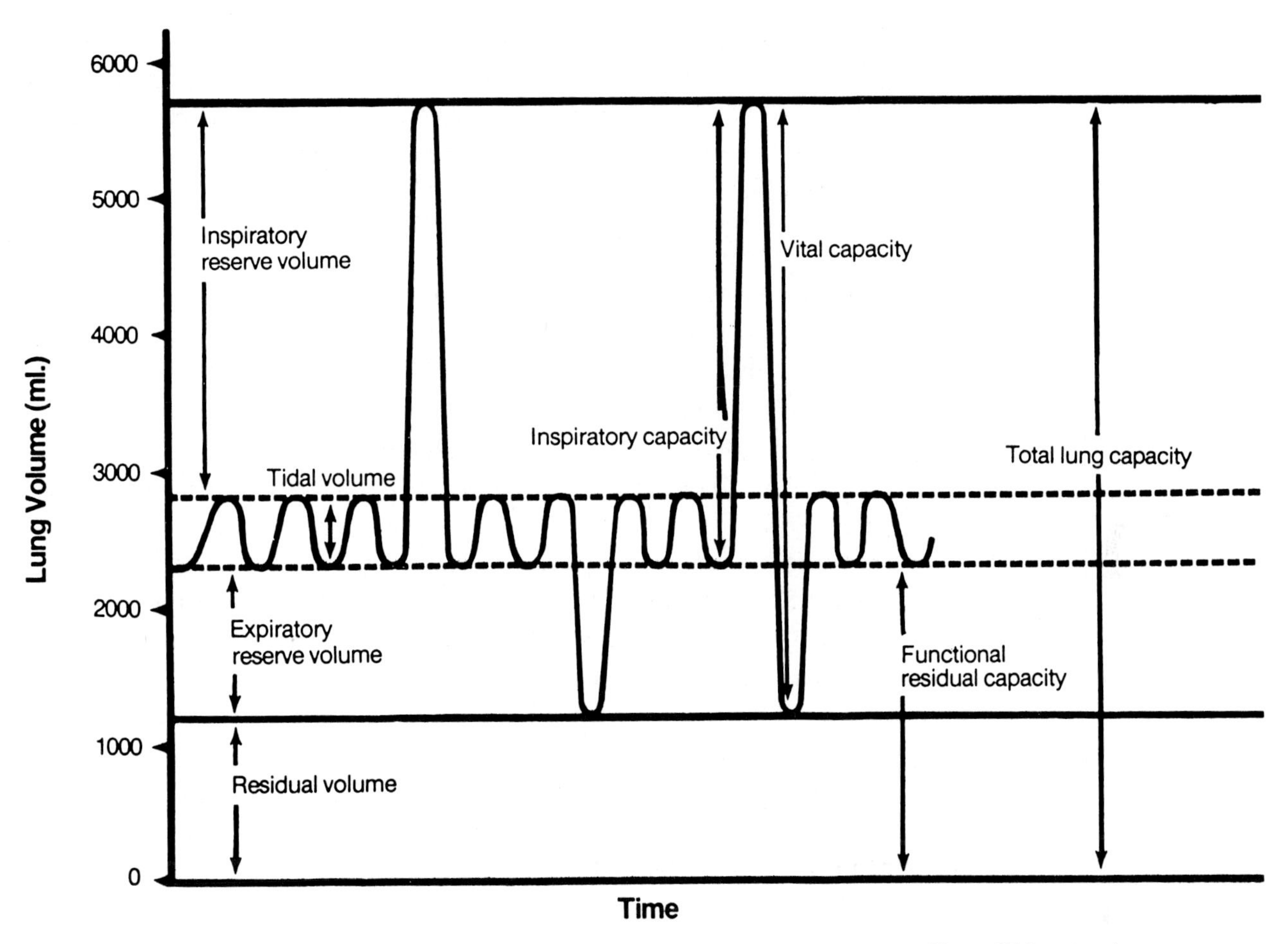

Figure 27.9 A spirogram showing the divisions of the respiratory air. (Courtesy of Phipps and Bird.)

The function of each part of a nephron is described in the table below.

Data Table 27.10 Functions of the Nephron

1. Glomerulus	Filters fluids (filtrate) containing dissolved materials from blood plasma
2. Bowman's capsule	A collecting reservoir for the filtrate
3*a*. Proximal convoluted tubule *b*. Descending and ascending loops of Henle *c*. Distal convoluted tubule	Reabsorption of glucose, sodium and potassium ions, lactic acid, water, proteins, and other usable chemicals.

Through the process of filtration by the nephron, the blood plasma is continually cleaned. Also, substances that are needed by the body are returned, thus maintaining water and electrolyte concentrations in body fluids. Substances not required by the body can be called waste and found in the **urine.** Urine is basically water containing varying amount of nitrogenous wastes, electrolytes, toxins, hormones, and various other types of foreign matter that cannot be used by the body.

After urine is formed by the nephrons, it passes to the collecting ducts that enter the *major* and *minor calyces* of the kidney (see figure 27.10). From the calyces, the urine enters the **renal pelvis** and is moved to the **urinary bladder** via the **ureter.** The urine is excreted through the **urethra.** The process of *urination,* or *micturition,* is triggered by distension of the bladder. Urine is not expelled, however, until the *external urethral sphincter muscle* is relaxed.

27.25 Because excess substances from the body are found in the urine, urine can be analyzed to determine certain abnormal conditions. For example, urine should be free of glucose. If glucose should appear in the urine, its presence could mean the body is not using the glucose that is being manufactured. Excessive glucose is one symptom of the disease diabetes mellitus.

The composition of urine is 95 percent water with varying amounts of **urea,** a by-product of amino acid metabolism, *uric acid,* a by-product of nucleic acid metabolism, *creatinine,* a by-product of creatine metabolism, and various electrolytes.

Below are several simple clinical tests that are routinely used to analyze urine samples. Fresh urine will be required to complete the next few exercises. The test will check the color, transparency, and density of urine.

Put 10 to 15 milliliters of collected urine into a *clean* tube and identify whether the urine sample is yellow or deep amber or light yellow. Record the color: ______________

The color of urine indicates the concentration of urine. If the color of the urine is dark, you have concentrated urine and probably had little fluid intake in the past hour or so. A pale color indicates dilution due to fluid intake. The pigment *urochrome* determines the color of the urine.

27.26 Holding the same tube up to a light source, determine whether the urine is clear, slightly cloudy, cloudy, or opaque. Record: ________________. Cloudy urine usually means there are excess amounts of phosphate ions present due to fat, bile, or blood. If there is a slight color to the urine, this can suggest the source of the contamination.

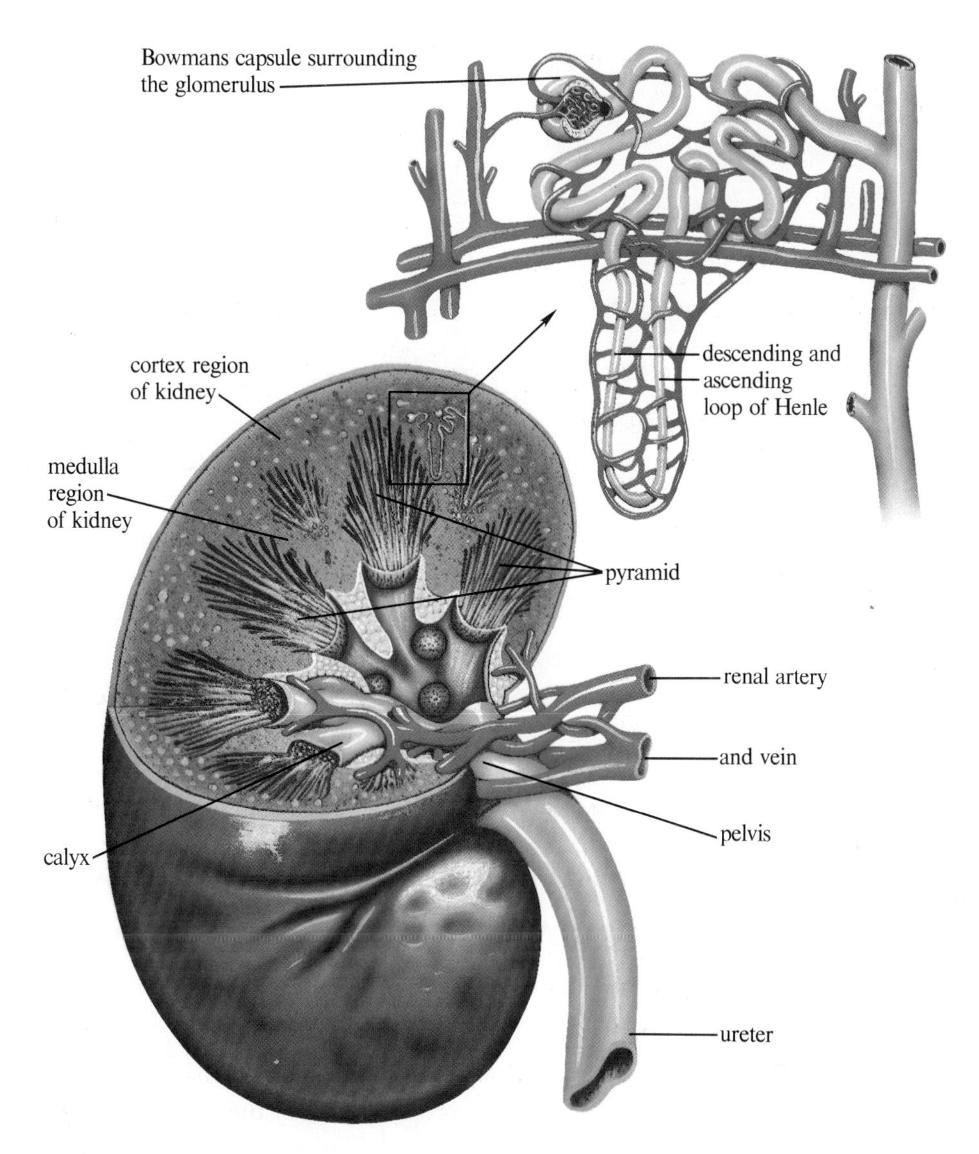

Figure 27.10 The anatomy of the kidney and nephron. As blood flows through the kidneys, waste products formed in the liver, pass into the nephrons. Needed water and salts are passed back into the blood. Surplus water and waste products become urine and is removed from the body.

If cloudy urine is reddish, then the contamination is from blood. A yellowish-green color suggests bile, and a white color indicates fat as the contaminating source.

27.27 Stopper and shake the tube vigorously. Record the color of the foam: ____________ . If the foam of the urine, after shaking, is yellowish in color, bile could be the cause.

27.28 Test the pH of the urine sample by dipping a piece of pH paper into the urine. Compare the color to the pH color chart. Record the pH of your urine sample. If the pH of the urine is checked, it will probably be acid rather than alkaline. Urine is usually slightly acidic, a pH of less than 7.

27.29 Using Clinitest paper or a comparable product, dip it into your urine sample. Wait 30 seconds. Use the accompanying color chart to determine the nature of your urine sample. Record the results of this test as negative, trace, +, ++, +++, or ++++. ___________

Clinitest paper indicates the amount of glucose in urine. Ideally, there should be none. As with the results of any of these tests, you should not read more into the results than is given here. Only a qualified medical doctor can diagnose results for certain.

27.30 The transport systems entail a lot of information. You may need to review this chapter further before looking at the study points. Or, look at the study points for this chapter. Place a check (✓) if you can assure yourself mastery of the point.

On completion of this chapter, you should be able to do the following:

☐ 1. List the four systems referred to as transport systems.

☐ 2. Identify the components of blood and the functions of each.

☐ 3. List the four blood types and characteristics for determining each.

☐ 4. Trace the flow of blood through the heart to the lungs and to the body using proper anatomy during your explanation.

☐ 5. Identify the points of determining blood pressure.

☐ 6. Determine your pulse rate.

☐ 7. Determine your blood pressure.

☐ 8. Identify specific veins and arteries and the function of each.

☐ 9. List the three major functions of the lymphatic system.

☐ 10. Identify the major components of the respiratory system.

☐ 11. Clinically complete and measure the mechanics of breathing and give the meaning for the various capacities of the lungs.

☐ 12. Identify the major components of the urinary system.

☐ 13. Give the function of the parts of a nephron.

☐ 14. Make simple tests on urine.

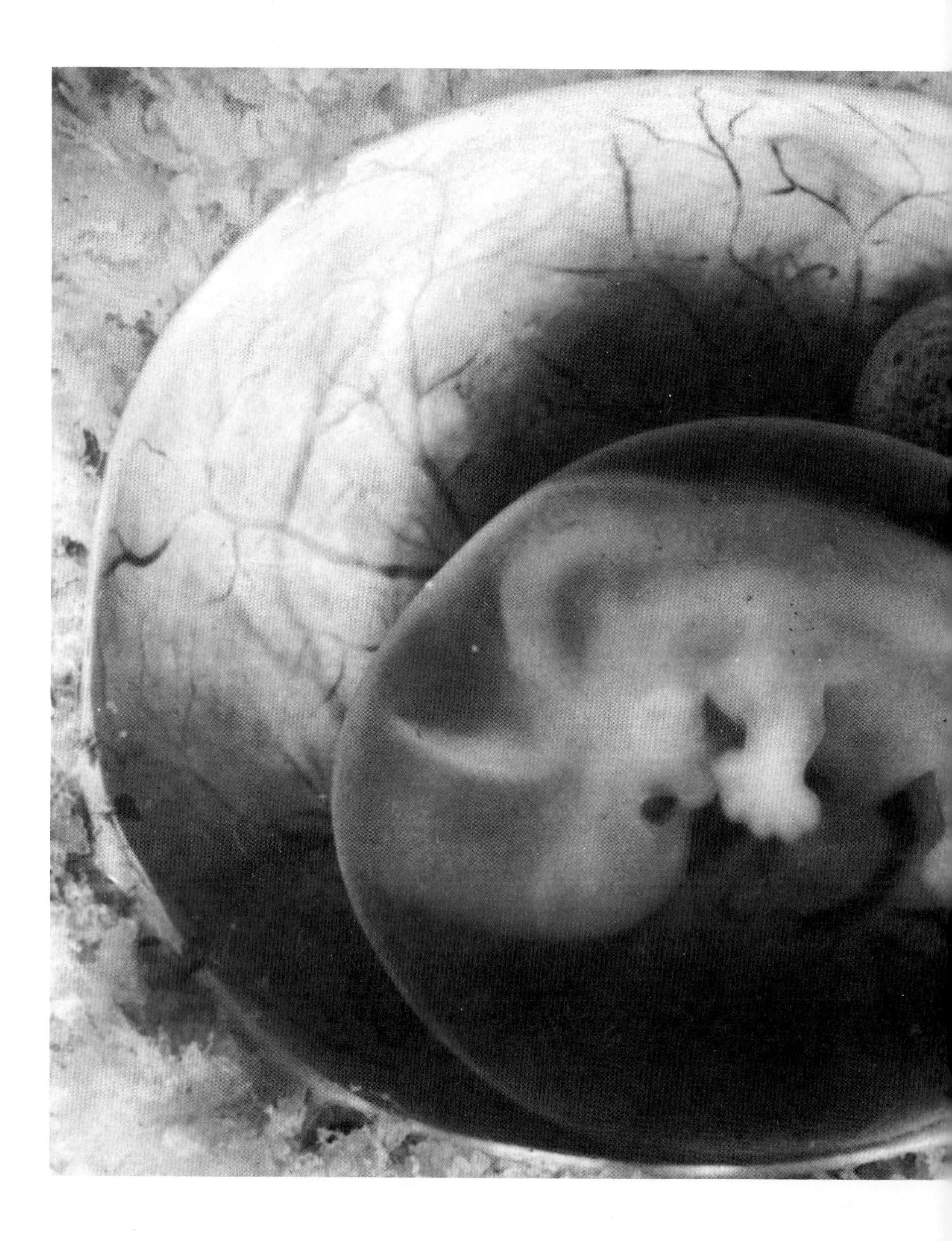

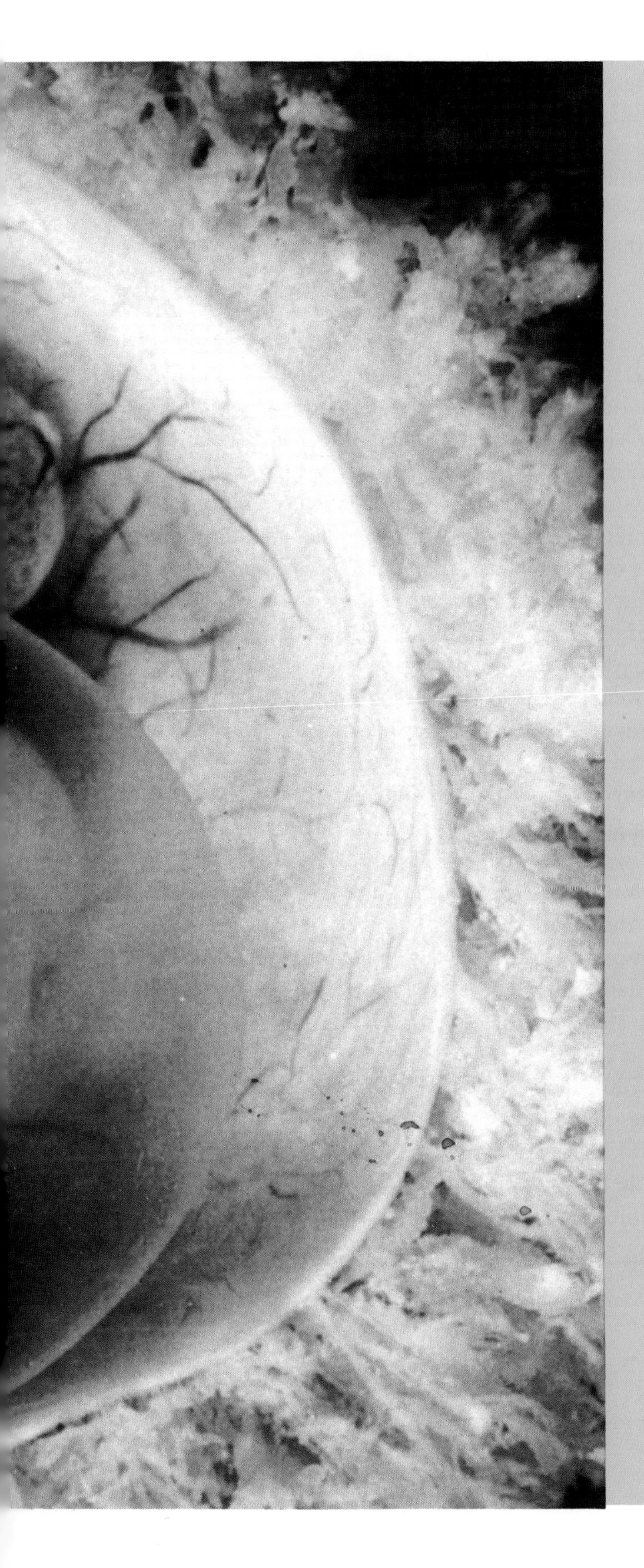

The Reproductive System 28

What is this thing called the reproductive system?

28.1

All living things share one common property. That property is the ability to reproduce. At one time, the beginning of life was believed to be one of **spontaneous generation** which presumed that living things grew out of nonliving material. The work of Spallanzani and Pasteur, however, ended such notions and showed that only life was capable of giving new life. This means that every living thing is a descendent of a living lineal past.

Our ancestors were not very knowledgeable about sexual reproduction and heredity. Aristotle, who is called the "father of biology," was unable to make the precise observations concerning reproduction that he made with many other biological observations. He believed that blood was the key to reproduction, and his view was upheld for centuries. During the Renaissance more exact information was assembled.

William Harvey studied reproduction through the use of pregnant animals. By sacrificing pregnant animals at various stages of gestation, he was able to determine the development of the young animals yet to be born. Through this work, Harvey realized that the egg was a very important part of the reproduction of offspring. He made no direct conclusion about the existence of the sperm or conception, however.

Anton van Leeuwenhoek, with his first microscope, studied the human egg. Because the sperm cell is still thousands of times smaller than the egg cell, it was not detected with the earlier microscopes. In the 1700s, when scientists studied sperm cells, they noticed a tail and a head. The head was described as having a small man in it. The term *homunculus* (Latin for "little man") was given to the sperm cell. Figure 28.1 compares a homunculus to a sperm cell. With the use of your imagination, you should be able to see homunculus in the actual sperm cell.

While the sperm cell theories were developing, the idea of conception was still not fully understood. This was the replacement of the "all egg" theory as identified by Harvey, since a sperm cell was now being considered too.

In the middle 1700s, Pierre Louis de Maupertius identified that the egg and sperm theories individually were incorrect. He claimed that individuals developed with characteristics from each parent. It was not, of course, until the middle 1800s that science was able to put the information regarding the role of the sperm cells and the egg cells into usable theories.

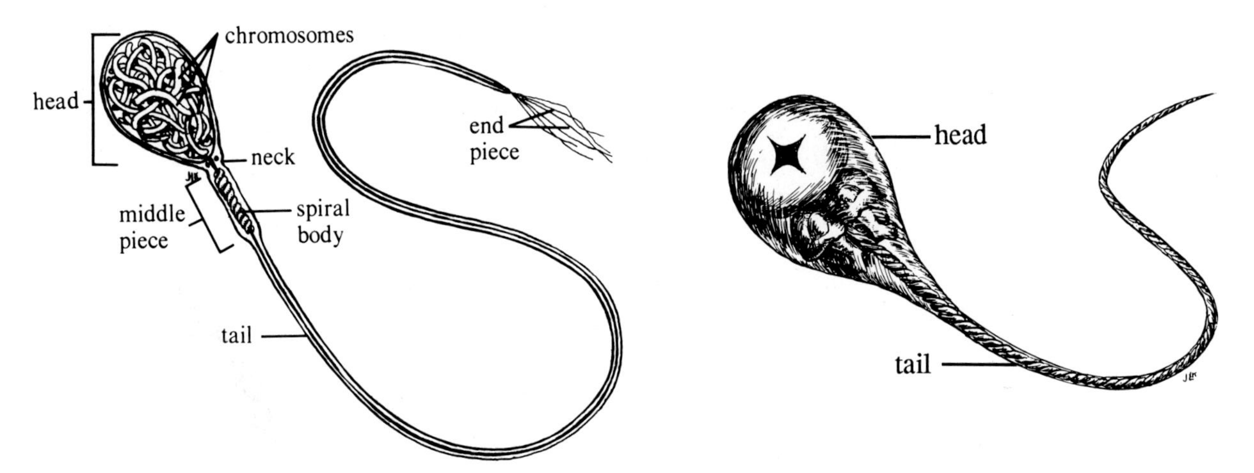

Figure 28.1 Homunculus was one of the first concepts about sperm cells. The anatomy of homunculus was quite simple compared to a current rendering of a sperm cell. Can you see how early scientists might have drawn conclusions about homunculus? (Courtesy of James Koevenig.)

For it was during this period of time that George Mendel (Augustinian monk discussed in chapter 10), Charles Darwin (a naturalist), and others began to formulate their theories of **heredity, natural selection,** and **organic evolution,** and to apply them to the living world as they knew it. The information that these scientists would gather and report would greatly alter the understanding of life science during the next one hundred years. The one thing they all agreed on was that reproduction is responsible for the continuation of a species, whether plant or animal.

Reproduction is studied by looking at the processes of **spermatogenesis** and **oogenesis** and related anatomy. The primary function of the male reproductive system is to manufacture male sex cells called **sperm cells,** or **spermatozoa.** This process, called *spermatogenesis,* occurs within the male gonad, the testis. Within the testes are found specialized cells called **spermatogonia.** During spermatogenesis, spermatogonia grow to become primary spermatocytes. Because mitotic divisions keep the supply of spermatogonia constant, primary spermatocytes are also plentiful. During the phase called meiosis I, primary spermatocytes produce secondary spermatocytes.

During meiosis II, secondary spermatocytes divide further producing haploid spermatids. Spermatids are immature sperm cells. Each spermatid will eventually mature with a distinct head and flagellum. Approximately one million sperm cells are produced daily by spermatogenesis. Figure 28.2 illustrates spermatogenesis.

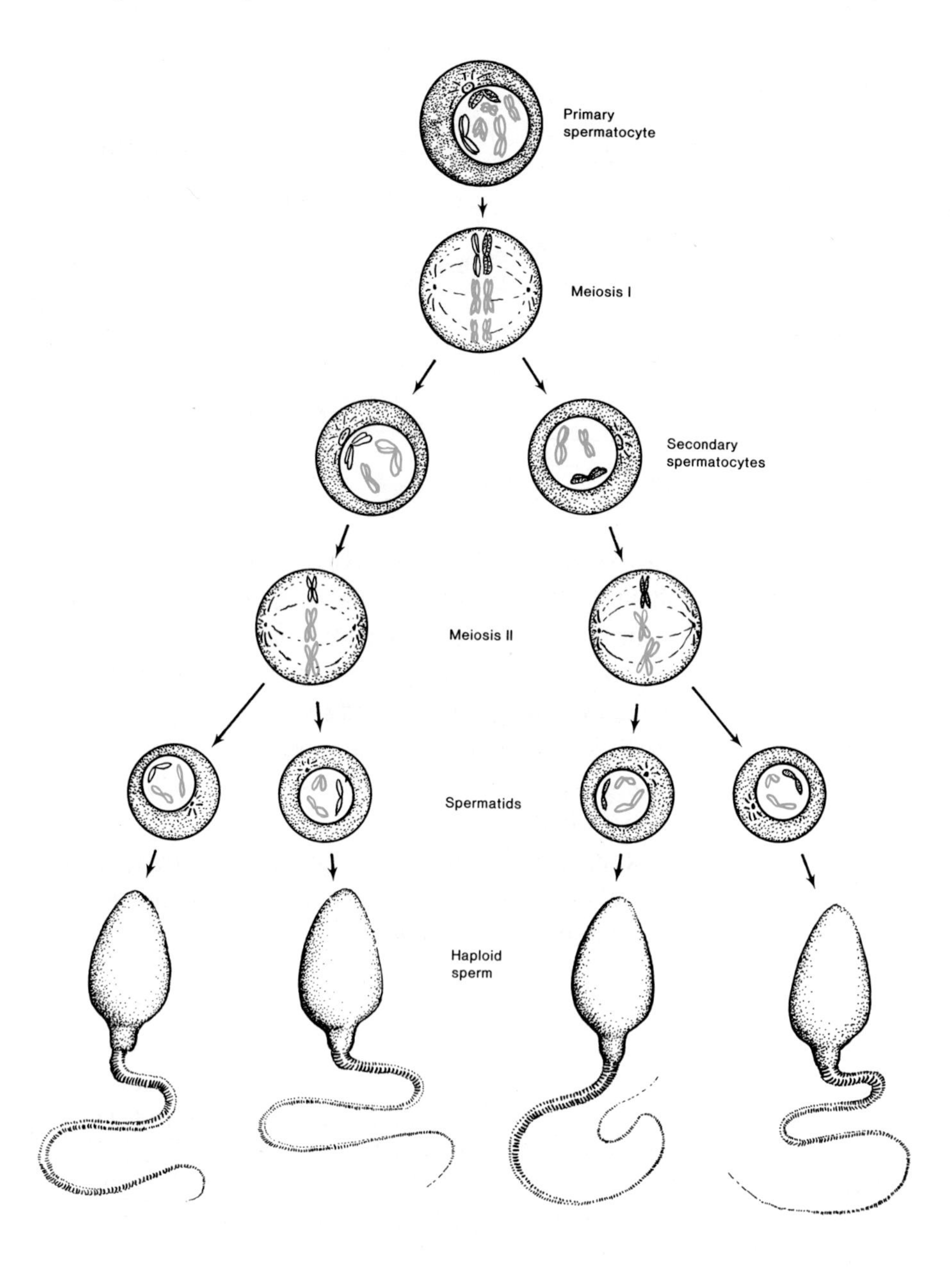

Figure 28.2 Spermatogenesis. Note each diploid spermatocyte produces four haploid sperm cells. The maturation process of spermatids to sperm cells is known as spermiogenesis.

The rest of the male reproductive tract consists of a series of ducts and accessory organs for the transport of the sperm to the outside of the body. These structures incude the **vas deferens, epididymis, ejaculatory duct, seminal vesicles, prostate gland, bulbourethral glands,** and the **urethra.** The only parts of the male reproductive system that are visible externally are the **penis** and the **scrotum.**

The testes manufacture the spermatozoa. Figure 28.3 illustrates a sagittal section of a single testicle. In this illustration, the testis is covered by a tough covering, the *tunica albuginea.* The testis is divided internally

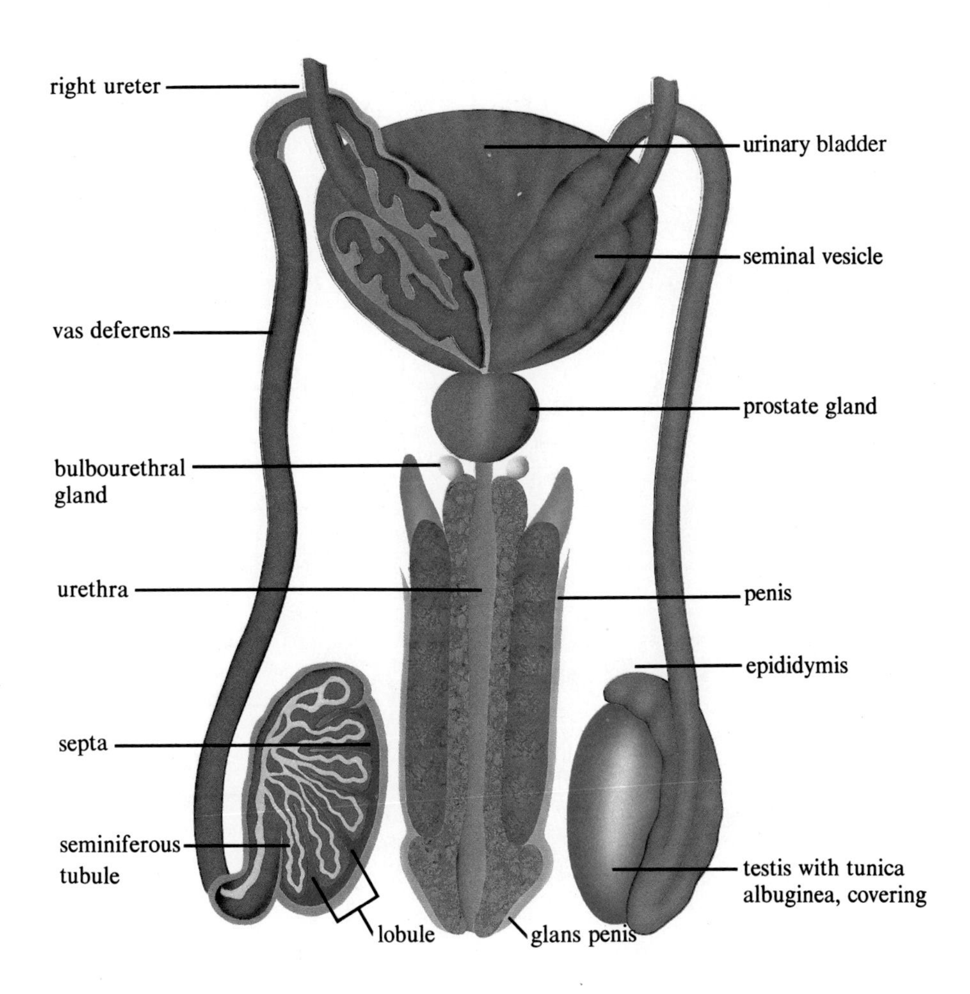

Figure 28.3 Anatomy of the male reproductive system showing a saggital section of a testicle.

into a series of compartments by **septae.** (See figure 28.3.) Each compartment contains **seminiferous tubules** that contain specialized tissue called *germinal epithelium.* This tissue differentiates into the male sex cells, ultimately the spermatozoa. *Sertoli* cells, which aid the development and manufacture of the spermatozoa, are also found in the seminiferous tubules. The space surrounding the seminiferous tubules has different specialized cells known as the *cells of Leydig,* which produce male hormones. For example, FSH causes Sertoli cells to undergo spermatogenesis.

At the time of ejaculation, the sperm cells leave the epididymis, where they are stored, into the vas deferens. Since there are two vas deferens, one for each testicle, they join at the ejaculatory duct.

At this point, the sperm cells mix with fluids from the *prostate gland* and *seminal vesicle.* Both fluids help maintain the proper pH and electrolytic balance and aid the motility of the sperm cells.

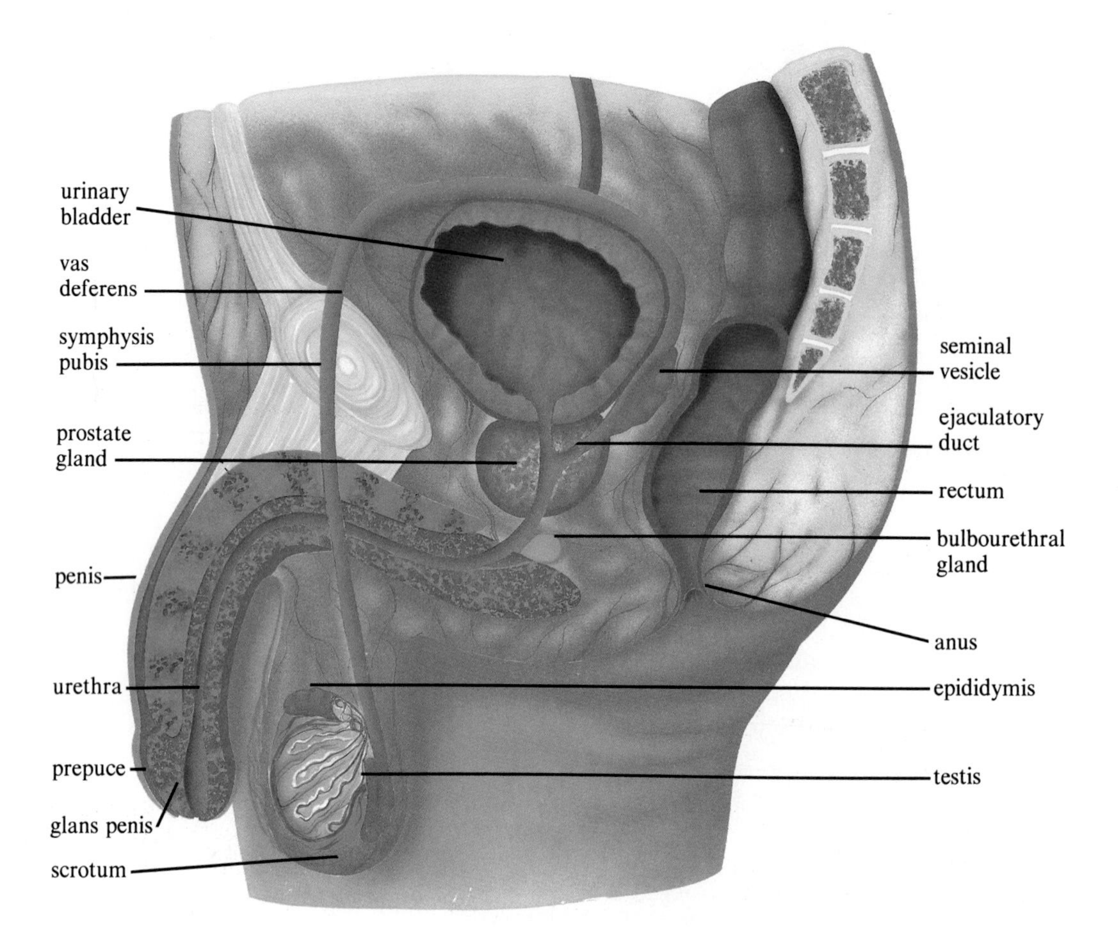

Figure 28.4 Sagittal section illustration the male reproductive system.

Another male accessory gland, the *bulbourethral gland* (Cowper's gland) secretes a fluid that lubricates the urethral lining for easier passage of seminal fluid.

28.2

The **ovaries** and the *oogenesis* process are the female counterparts to the testes and spermatogenesis in the male. During fetal development, several million *primary oocytes* (eggs) are available. But these cells continually disappear. Therefore, by birth only half remain.

The primary oocytes go into a dormant stage at birth and do not continue further development until puberty. By the time puberty is reached, approximately one-half million primary oocytes are left. Of these, only approximately one thousand will ever be released from the ovaries.

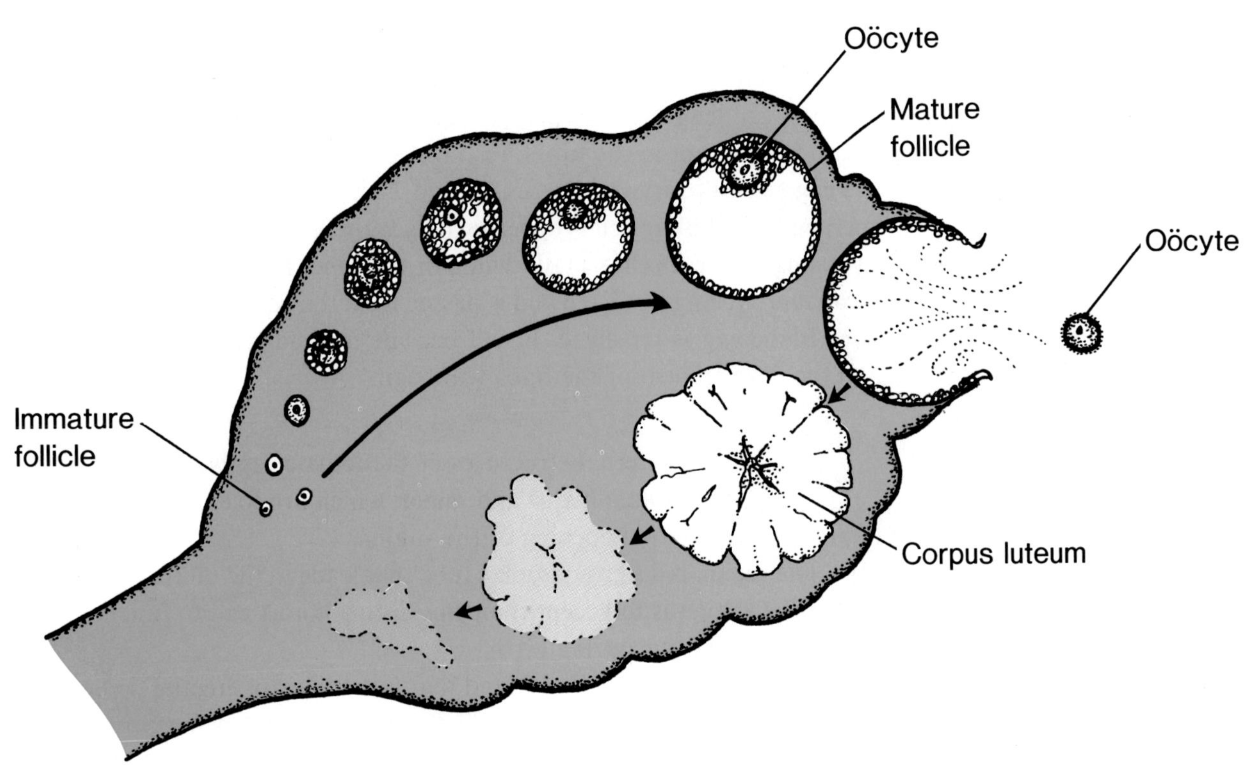

Figure 28.5 Female follicle showing the development and release of an oöcyte.

At puberty, oogenesis begins and *secondary oocytes,* or *mature eggs,* begin being released regularly each month. The release of a mature egg is called **ovulation.** The monthly release is known as the **menstruation.**

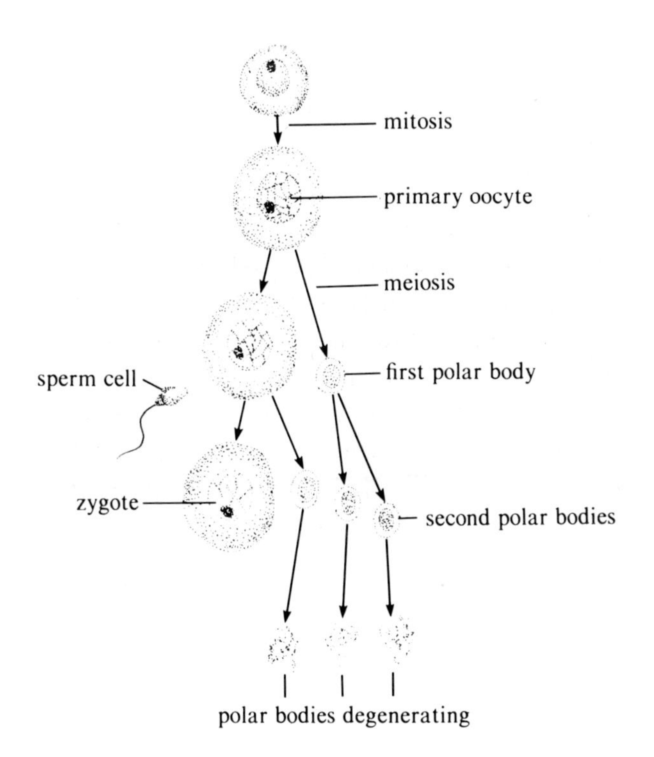

Figure 28.6 Oögenesis produces unequal sized cells. The majority of the cytoplasm, during meiosis goes into the secondary oöcyte making it a very large cell. The polar bodies are essentially nuclear waste products.

After ovulation, the egg cell enters the *uterine tube* (**fallopian tube, oviduct**) through the tube's enlarged free end, the *infundibulum* The egg will pass the length of the uterine tube and enter the uterus. However, if sperm are present in the tubes, fertilization may occur. During the travel, the egg cell will survive between 12 and 24 hours; the same is true for the sperm cell. If the egg cell is to perform its main function, it will be fertilized by a single sperm cell, usually while in the tube. If not, the egg becomes incapable of being fertilized and is passed from the body with the menstrual flow. If the egg is fertilized, it will become imbedded in the uterine wall and the development of the fetus will begin.

28.3

The external structures of the female reproductive tract are the **labium major** and **minor,** which are folds of skin that act to protect the external opening of the vagina.

The **vagina** is a fibromuscular tube that leads to the uterus. The function of the vagina is to accept the penis during coitus and to function as a passage for the fetus during birth.

The **clitoris** is a sensitive gland that is made up of erectile tissue, *corpora cavernosa,* which is similar to that found in a male penis. Stimulation triggers the secretion of lubricating fluids into the vagina.

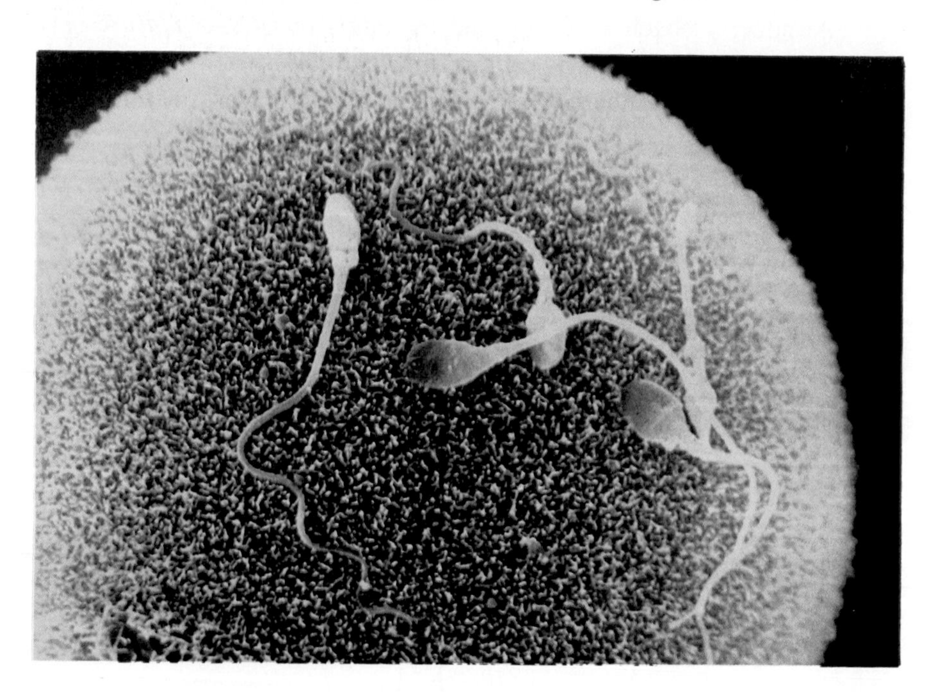

Figure 28.7 Sperm cells attempting to impregnate an egg cell.

28.4 As you conclude your studies of anatomy and physiology, keep in mind that the intent was not to give you all the anatomy or physiology that fills many textbooks. It is hoped that because of the exposure to human anatomy, you will have a greater appreciation for this marvelous machine, your body. It is also hoped that you will continue to pursue additional knowledge of the human body through further readings and discussions. Remember that though science and medicine have contributed greatly to the knowledge and understanding of your body, some things are still unknown.

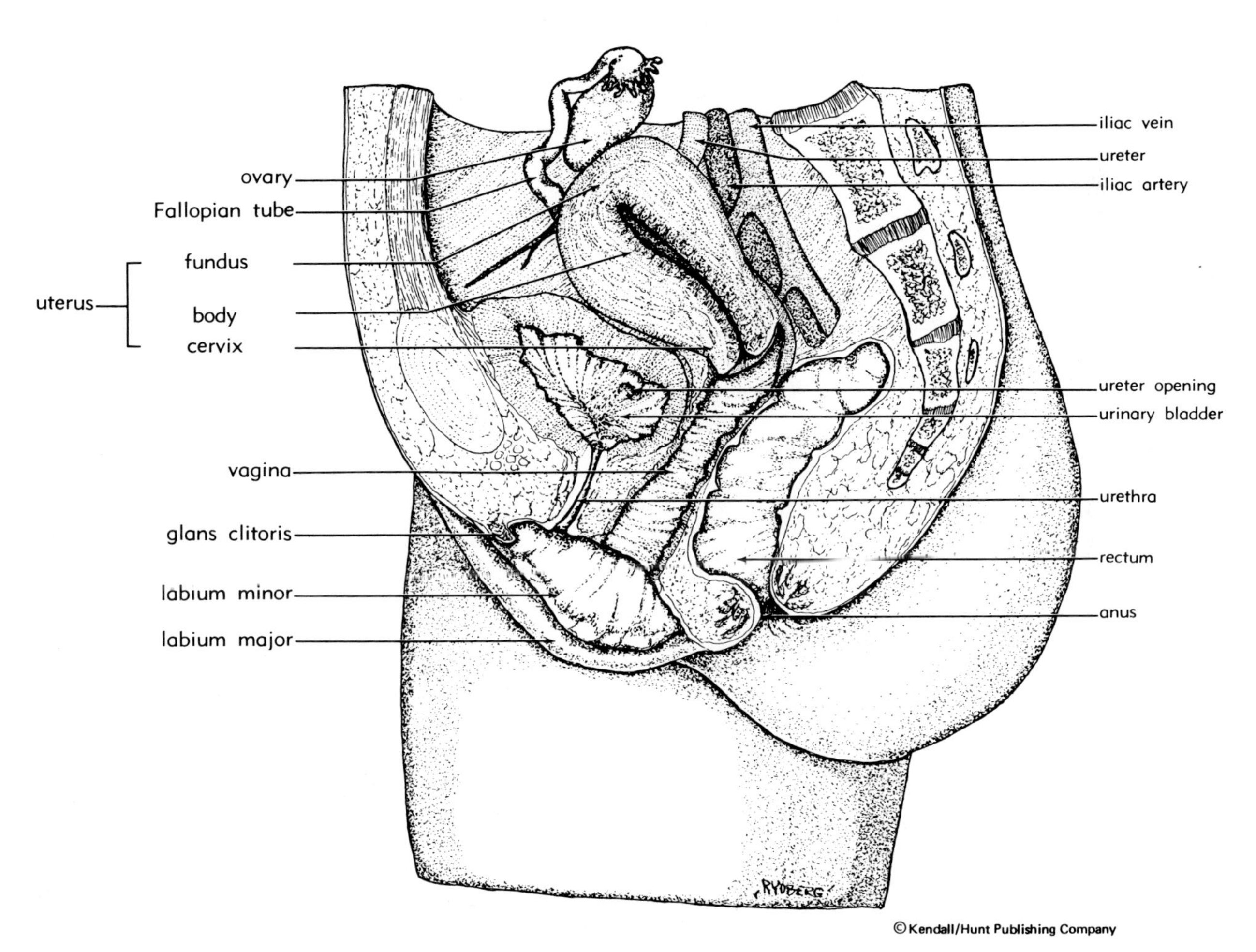

© Kendall/Hunt Publishing Company

Figure 28.8 Sagittal section illustrating the female reproductive system.

Look at the study points for this chapter. Place a check (✓) beside each study point when you are comfortable with concepts associated with it. On completion of this chapter, you should be able to do the following:

- ☐ 1. Relate some of the history behind the ideas about reproduction.
- ☐ 2. Identify the anatomical parts of the male reproductive system.
- ☐ 3. Discuss the steps of spermatogenesis.
- ☐ 4. Identify the anatomical parts of the female reproductive system.
- ☐ 5. Discuss the steps of oogenesis.

Basic Concepts in Ecology 29

Is there really a plan that ties everything together?

29.1 All living things interact with the geology, chemistry, physics, and other biological entities surrounding them. This interaction is special. The study of it is called **ecology.** The word ecology simply means "house management," from the Greek term *oikos* meaning "house."

Throughout your study of biology you have been introduced to many concepts closely tied to ecology. As a result, this chapter will touch on some of the more broad ecological principles.

Look at Figure 29.1. The artist is illustrating a concept of ecology. How would you interpret this illustration? Write a paragraph or two, using Figure 29.1 as your focus. What do you think the artist is trying to say here?

29.2 All living things can be found associated with the crust of the earth and its surrounding atmosphere. This area—the part of the earth in which life can exist—is called the **biosphere.** Other terms that often apply to the biosphere, or are used as substitutes, are "ecosphere" and "physiological atmosphere." Could the rendering in Figure 29.1 represent a part of the biosphere? ________ Is the planet earth really just one big biosphere? ________

29.3 The biosphere is so large that it is difficult to study as a whole. As a result, the biosphere is usually broken down into smaller units. For example, an **ecosystem** is a smaller unit of study. An ecosystem is a specific area having established boundaries. Within these boundaries, a biologist can study the interaction of all living things, along with their nonliving surroundings. For all practical purposes, a series of ecosystems simply overlap to cover the entire earth. In the broadest sense of the word, there are two main ecosystems, aquatic and terrestrial.

Terrestrial ecosystems can be further divided into major regions called **biomes.** Biomes are characterized by the climate that is found in the region, the major type of vegetation, and the subsequent soil type.

Not all biologists agree about how many biomes exist. Boundaries are not always clearly defined. Therefore, we will look at eight major biomes. As you research these, you might find others. That is okay. Simply be aware that breaking the biosphere into smaller units is not always accomplished with one-hundred percent agreement.

Figure 29.1 The earth's crust consists of the land, water, and surrounding air. All living things are associated with the earth's crust. If you consider all of the earth's crust, you can ask yourself, "Is the planet earth one big biosphere?"

The eight biomes we will study are (1) tropical rainforest, (2) tropical savanna, (3) temperate deciduous forest, (4) grassland, (5) deserts, (6) mixed deciduous and conifer, (7) taiga, and (8) tundra. You will need to use library references to complete this exercise.

To begin, work on the blank map in Figure 29.3. Develop a color key for each of the given biomes. Using all of the continents, draw and color the biomes where they belong on the map. Keep in mind that even though you are essentially putting in boundaries with your coloring, these boundaries are not clearly defined.

Figure 29.2 Earth, as seen from space. The crust of the earth and the surrounding atmosphere is the home of all living species.

29.4 Again using reference books, list some of the physical characteristics of each biome. Also list some of the key climatic characteristics. Identify a few of the major plants and animals that can be found in each biome.

29.5 After you have completed your map, look at the placement of the "X" on the map. This position is significant. Suppose you were to move northward to the North Pole from the "X." What biomes would you pass through? Draw an arrow to show your route. List the biomes you pass through. How does your list compare with those of other members of your class? Remember, the boundaries are not always clear. Some members of the class may have found references that showed different examples than the one you used.

29.6 While there are terrestrial biomes, similar boundaries have not been established for aquatic ecosystems, the oceans and seas. Because the vastness of the oceans and the relative lack of barriers for the organisms that live in their waters, zones have been established instead. Figure 29.4 illustrates the various horizontal and vertical marine (salt water) zones. Using library references, describe each of the horizontal and vertical zones.

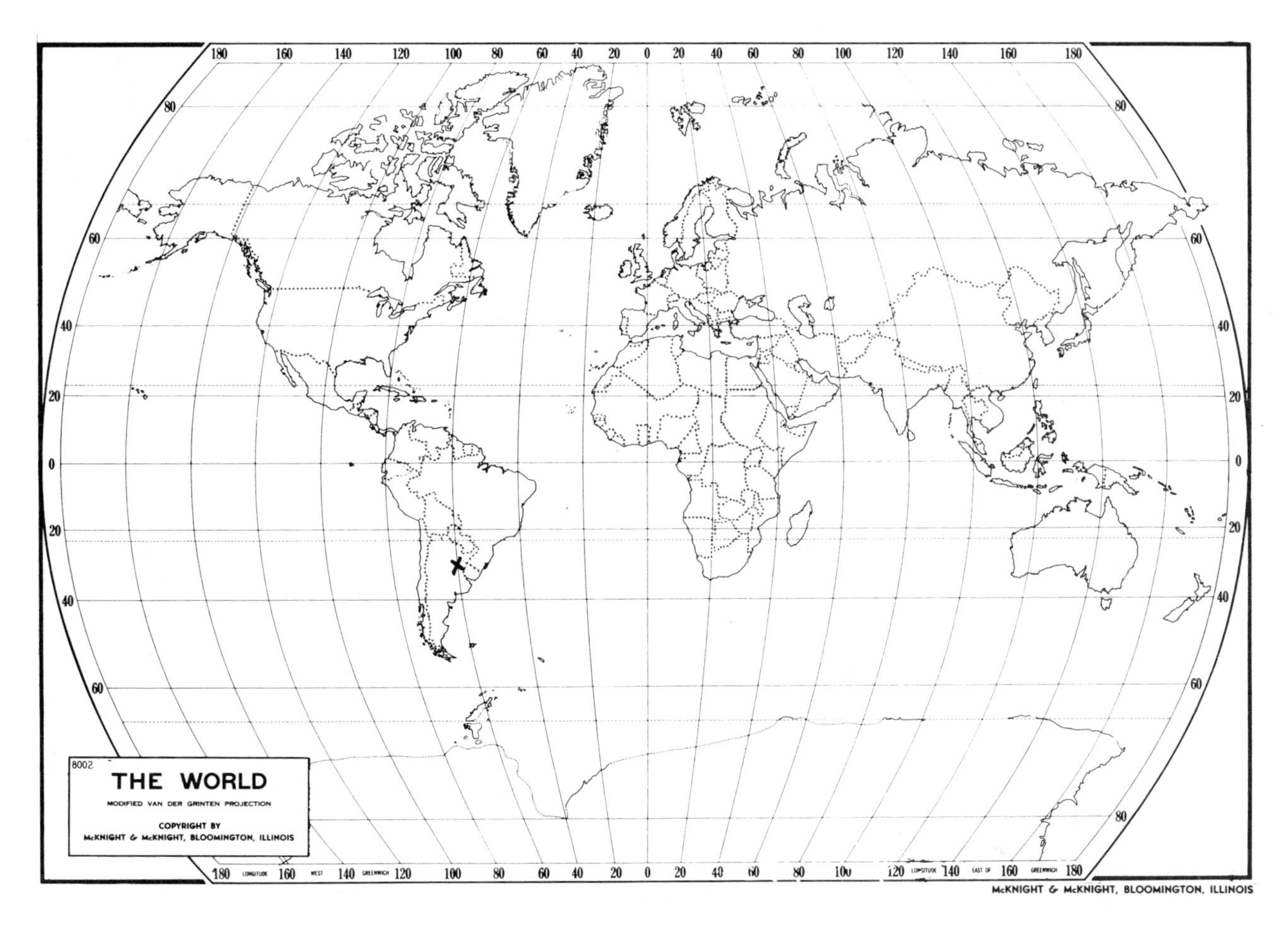

Figure 29.3 A modified Van Der Grinten projection map of the world.

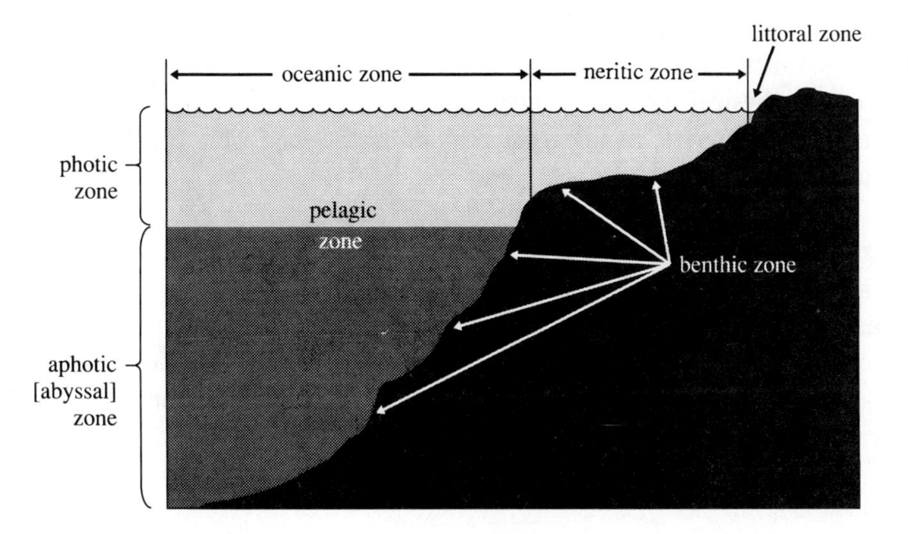

Figure 29.4 The basic vertical and horizontal marine zones. A diverse abundance of living organisms inhabit these zones.

Figure 29.5 The food web shows the interrelationships between a multitude of living organisms. The plants are the first order in the food chain for any organism. Death, and ultimately decay, are handled by bacteria and microorganisms which feeds nutrients back to the plants.

29.7

Regardless of whether you study an organism in an aquatic ecosystem or a terrestrial ecosystem, the organism has a niche.

A **niche** is an organism's functional role within an ecosystem. This is not the physical habitat. A niche sets up the organism's competition and its right to survive, its apparent position in the food web. Figure 29.5 illustrates a food web.

Competition within a food web is an active demand placed on organisms by other organisms. Competition can come from organisms vying for the same food source. It can also be the competition between a predator and its prey. One thing is for sure, the more similar an organism's niche is to that of another organism, the more likely it is that the two organisms will be in competition. If organisms avoid competition altogether, then they can coexist and not intrude on one another's niche.

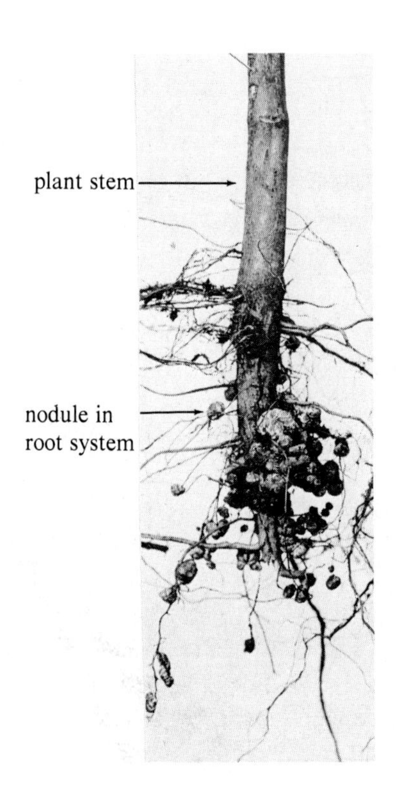

Figure 29.6 Nodules in the root systems of some plants supply a source of nitrates to plants. As a symbiotic relationship, the nodules are home for the nitrogen-fixing bacteria that inhabit them.

There are essentially three types of interspecific relationships, relationships where organisms coexist. These organisms avoid competition in their respective niches.

One interspecific relationship is called **symbiosis,** or **mutualism.** In this interspecific relationship, both species benefit from each other. One example of mutalism occurs in legume plants.

Rhizobium, a nitrogen-fixing bacteria, lives in nodules on the roots of legumes, such as alfalfa. The nodules provide a home for the bacteria. In return, the bacteria manufactures substantial amounts of nitrates that the plants use as food. You can see that both organisms benefit.

Another example of mutualism is the relationship some species of ants have with aphids. The ants essentially farm the aphids. The ants "milk" the aphids of a sugary food. In return, the ants protect the aphids from predators.

Figure 29.7 Ants and aphids illustrate a perfect symbiotic relationship.

The second interspecific relationship that exists is called **commensalism.** In a commensal relationship, one organism benefits but the other does not. Nor is it harmed. The illustration in Figure 29.8 shows a commensal relationship between a remora fish and a shark. In this relationship, the remora is transported by the shark. When the shark feeds, the remora feeds on the bits of food the shark leaves behind. The remora then reattaches to the shark and continues on with it. While the shark is not benefited by the remora, it is not harmed. Only the remora benefits in this relationship.

Finally, there exists **parasitism.** This form of interspecific relationship is definitely one-sided. Parasites live off a host. The parasite needs the host to survive, but the host receives nothing in return. Although the host may suffer, the parasite will not usually kill its host. Should a parasite destroy its host, the parasite organism would also die.

There are many parasites, many of which you probably know about. Fleas on a dog are parasites. Body lice, common to humans, is another parasite. Figures 29.9 and 29.10 show two common parasites.

Figure 29.8 A commensalistic relationship exists between sharks and remoras. While the remora does not harm the shark, the remora attains food from the fragments left in the water while the shark feeds.

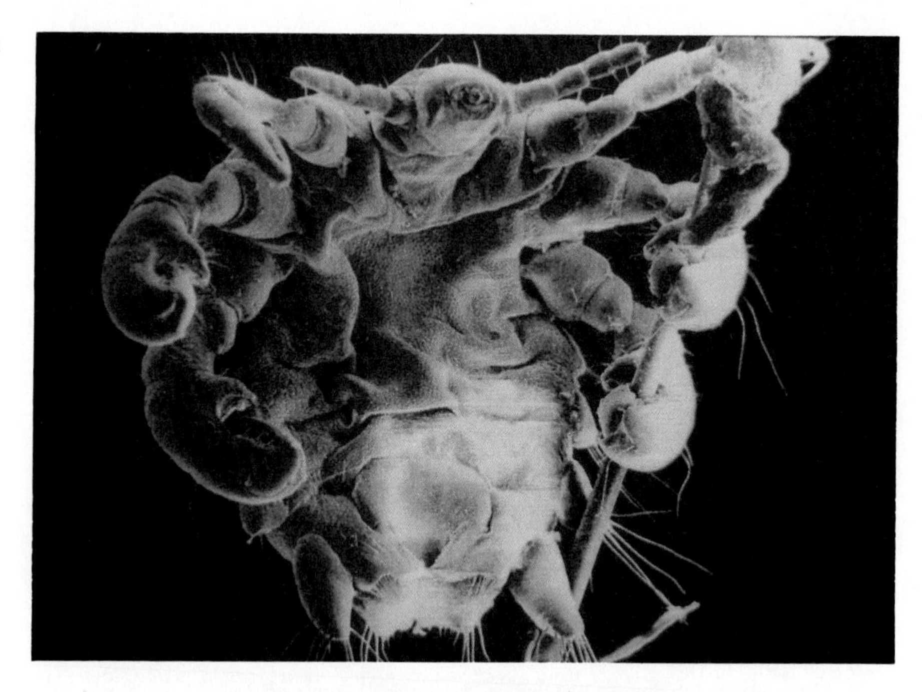

Figure 29.9 A parasite found on humans is body lice. True lice belong to the order Anoplura.

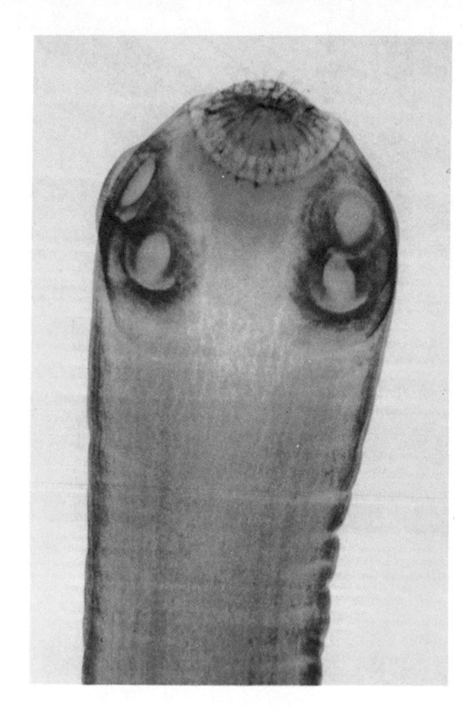

Figure 29.10 Another parasite that can invade the human digestive system is the tapeworm. Here the head (scolex) and corresponding suckers, hooks and rostellum are shown. Tapeworms belong to the class Cestoda.

29.8 Are living things tied to one another? Certainly. The entire ecology of the planet earth is tied together. There are many ways the balance of elements in earth's ecology can be upset. Certainly the delicate balance rests in the hands of everyone. The various biomes are easy to study. To understand what happens within a biome is another story.

Studying ecological principles may continue to remind you of our place in this whole picture. Look at the study points listed below. When you feel confident you know each specific point, check (✓) the point.

Upon completion of this chapter, you should be able to do the following:

- ☐ 1. Define ecology.
- ☐ 2. Define the biosphere.
- ☐ 3. Define an ecosystem.
- ☐ 4. List specific characteristics of the eight identified biomes.
- ☐ 5. Relate how individual niches are established and how three interspecific relationships (mutualism, commensalism, and parasitism) fit into their respective niches.

Environmental Science 30

What factors affect the environment or is there really such a chapter as environmental chemistry?

30.1

Historically, people have assumed that the land, water, and air around them would absorb their waste products. The ocean, the atmosphere, and even the Earth were viewed as receptacles of infinite capacity. It is clear now that we may be exceeding nature's capacity to assimilate our wastes. An American Indian adage says, "We haven't inherited this earth from our fathers, we are borrowing it from our children."

Former President Richard Nixon stated, "The recent upsurge of public concern over environmental questions reflects a belated recognition that man has been too cavalier in his relations with nature. Unless we arrest the depredations that have been inflicted so carelessly on our natural systems—which exist in an intricate set of balances—we face the prospect of ecological disaster."

Most pollutants eventually decompose and diffuse throughout the environment. When organic substances are discarded, they are attacked by bacteria and decompose through oxidation. In short, they simply rot. However, some synthetic products of our advanced technology resist natural decomposition. Plastics, some cans and bottles, and various persistent pesticides fall into this category. Many of these materials are toxic, posing a serious health danger.

The impact of the destruction of the environment cannot be measured. Today, citizens are seeking better environments, not only to escape pollution and deterioration, but to find their place in the larger community of life. Although pollution may be the most prominent and immediately pressing environmental concern, it is only one facet of the many-sided environmental problem.

A problem that has been coming to the forefront is that of acid rain, which results from atmospheric pollution. View the slide presentation called "Acid Rain Precipitation." View it once to get a feeling for this problem, and then view it a second time in order to take pertinent notes from the presentation so you can answer the following questions.

1. What is acid rain?
2. What are the forms in which acid is capable of reaching earth?
3. What are some of the sources of atmospheric pollutants that cause acid rain?
4. What effect is acid rain having on life forms in many of the regions being affected?

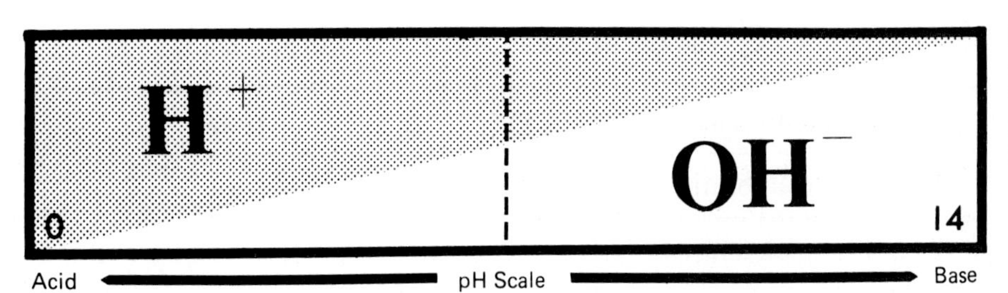

Figure 30.1 A typical rendering of a pH scale. The acid end of this scale is 7 to 0. The base end of the scale is 7 to 14. Pure water is supposed to have a pH of 7.

30.2 The key term that is a part of the study of acid rain is pH. This is a scale of 0 to 14 used to measure the concentration of hydrogen ions (H^+) available in solution. A pH of 7 is considered to be neutral or that of pure water. A pH range of 7 to 0 places the solution in an acid range. The closer to zero, the greater the concentration of hydrogen ions and, therefore, the more acidic. In figure 30.1, you can clearly see that the shaded area, which represents the amount of hydrogen ions, decreases as the pH range goes from 0 to 7.

1. What appears to be happening to the hydroxyl ion (OH^-) concentration, in the 0 to 7 range, as the hydrogen ion concentration is decreasing? ____________________

   ____

2. If the hydrogen ion concentration decreases from 0 to 14, then what happens to the concentration as it goes from 14 to 0?

   ____

3. If the hydroxyl ion concentration increases from 0 to 14, then what happens to the concentration as it goes from 14 to 0?

   ____

4. At what point is the hydrogen ion concentration equal to the hydroxyl ion concentration, causing a solution to be neither acidic or basic? __________

30.3 Let us look more closely at sulfuric acid (H_2SO_4) and its impact on the environment. Sulfuric acid causes a great deal of damage to plants, animal life, and soil. Yet when it comes in contact with carbonated materials, such as limestone, a natural scheme for combating acidic conditions occurs.

Data Table 30.1 Carbonates and Acids

Evaporating dish 1	**Data**
Weight of evaporating dish (initial)	
Weight of dish and $CaCO_3$	
Weight of dish and $CaCO_3$ after drying	
Total gain or loss of dish and contents (Circle one)	

In the reaction shown below, acid rain (H_2SO_4) is converted into harmless materials; the calcium sulfate dissolves slowly producing hard water.

H_2SO_4	+	$CaCO_3$	→	$CaSO_4$	+	H_2O	+	$CO_2(g)$
Sulfuric acid in water		Calcium carbonate (limestone)		Calcium sulfate		Water		Carbon dioxide

To illustrate this process further, complete the following laboratory. In this investigation, the acid sample will be treated with weighed calcium carbonate in the form of C.P. precipitated chalk and, finally, reweighed. The change in weight of the solid depends on its change from calcium carbonate ($CaCO_3$) to calcium sulfate ($CaSO_4$). This, along with a measure of pH, should indicate the success of using limestone to neutralize acid rain.

a. Obtain a clean, dry evaporating dish. Weigh the dish to the nearest one-hundredth of a gram and record the weight in data table 30.1.
b. Next weigh out a 0.5 gram sample of calcium carbonate ($CaCO_3$) and place it in the evaporating dish. Record this weight in the data table also.
c. Add 20 mL of 0.25 M H_2SO_4 to the dish. Add the acid very slowly or it will react too quickly and will spill over the edge of the dish.
d. Boil the contents of the evaporating dish for 3 minutes to speed up the reaction and remove any calcium bicarbonate ($Ca(HCO_3)_2$) that may be present. Do not allow your solution to boil over.
e. Place the evaporating dish under heat lamps until the contents have been evaporated to dryness (about 24 hours). Once this has occurred, reweigh the dish and complete data table 30.1.

30.4 While waiting for the contents of the evaporating dish to dry, answer these questions.

1. Determine the molecular weight for $CaSO_4$ and for $CaCO_3$.

   $CaSO_4$: ________ $CaCO_3$: ________

2. Which has the greater mole weight? ________

30.5 Before reweighing your dry precipitate, predict whether the dish and its contents will weigh more or less than it did with the original 0.5 grams of dry $CaCO_3$ in it. Be sure to explain on what basis you are making this prediction. (*Hint:* Look back at the chemical equation given in section 30.3 that shows the reaction between $CaCO_3$ and H_2SO_4. Then look at the mole weight of the calcium compounds listed above.)

Now reweigh your dry sample and complete data table 30.1. How did your prediction hold up? ____________

30.6 The experiment you just completed should have offered proof that a chemical reaction had occurred. Evidence was gathered that a new chemical compound was formed in the process. In this next experiment you will be testing the effects of calcium carbonate ($CaCO_3$) on the acidity of a solution.

The 0.25 M H_2SO_4 would represent acid rain in this experiment. Measure the pH of this solution with a pH meter. ____________ Your teacher will give you instructions on how to use your pH meter, including how to standardize the instrument.

a. Weigh out another 0.5 gram of $CaCO_3$ and place this solid in an evaporating dish. Slowly add 20 mL of 0.25 M H_2SO_4 to the dish. Heat the dish and its contents for 3 minutes.

b. Once this solution has cooled, filter the contents of the dish using a funnel and filter paper. Save the *filtrate* only. Is the filtrate the solid portion or the liquid portion? ____________

c. Check the pH of the filtrate with a pH meter. What is the pH of your filtrate? ____________ Compare this pH to the pH of your original acid solution. How do they compare?

d. Explain what has happened to the pH of the solution. Is the filtrate as acidic as before? ________ Would you expect the filtrate to be as acidic? Explain your answer.

30.7 From what you have discovered, is limestone ($CaCO_3$) a good treatment for neutralizing acid rain? ________

30.8 Knowing that H_2SO_4 has an effect on bedrock gives us a look at the natural way of handling some acid problems. Remember, since pollution is an overburden, natural systems cannot always hold up for a lengthy period of time.

How does acid rain get into water systems? Indeed, it is called rain, but sometimes SO_2 gases can invade a water system and cause similar problems even when no rain is in sight. The setup illustrated in figure 30.2 shows another principle of gas dispersion.

a. From the glassware shelf, obtain an Erlenmeyer flask that has been made to resemble the one shown in figure 30.2. Add 100 mL of tap water that has had 2 drops of phenolphthalein added to it.
b. Add several drops of 5M NH_4OH to a cotton swab and insert the stopper.
c. Allow the setup to sit for 10 to 20 minutes.

1. What has happened to the water?
2. Why has this occurred?
3. Explain how gases in the air, such as SO_2, can cause lakes and rivers to become acidic even when there is no rain.

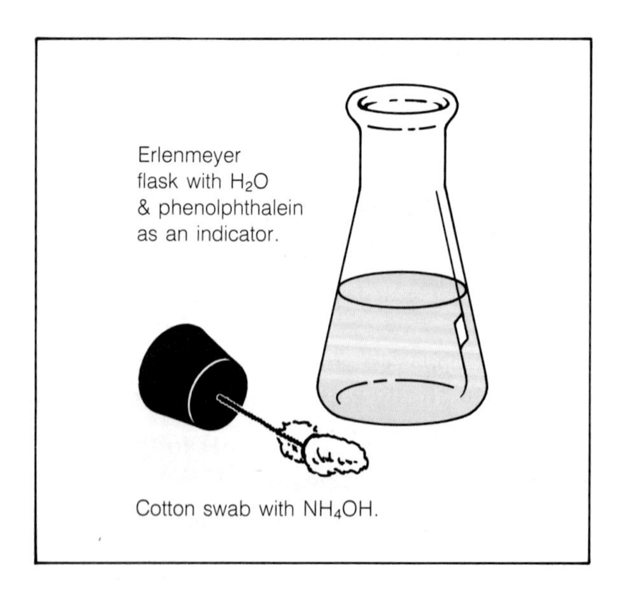

Figure 30.2 The laboratory set up for determining the means for simple diffusion of gases into water.

30.9 Acid rain affects water systems. Water certainly plays an important role in our daily lives. Recall the events of the day. How much water have you taken into your system? How much water have you come in contact with? Just how much water directly or indirectly has an effect on you? You probably have not given this much thought. Yet, your life centers around the soil at your feet, the air you breathe, and the water you drink and use.

You will recall the effect limestone had on the pH of sulfuric acid. In living systems, there are chemical solutions that exist called **buffers.** A buffer is any substance that tends to prevent a change in the pH of a solution. Your blood contains a buffer so it will not change after eating all of the varied pH foods in a balanced meal. Indeed, your stomach has a pH of about 2. If your blood pH were to change several tenths, the results could be fatal.

To illustrate how buffers work, you will need, once again, to use the pH meter.

a. Standardize your pH meter with a known buffer of 7.0.
b. Place 10 mL of buffer solution into a small beaker. Add 2 drops of 0.1M HCl to the buffer, stir the solution, and measure the pH of the buffer. Record the pH in data table 30.2.
c. Continue adding 2 drops of HCl and measuring the pH of the buffer solution until you have completed data table 30.2. This should involve making nine separate readings of the pH.

Once you have completed this procedure, empty the contents of the beaker, wash it out, and repeat the procedure with the buffered solution using 0.1 M NaOH instead of the HCl. Record your data in data table 30.2.

30.10 In order to discover the effect of a buffer on pH, it will be necessary to repeat the above experiment using water in place of the buffer. In this manner, the experiment will have a control, which will allow you to isolate the effects of the buffer on the pH of the system.

Repeat the above experiment using the same HCl and NaOH solutions. But this time, instead of using 10 mL of a buffer solution, use 10 mL of tap water. As before, standardize your pH meter before you begin the procedure. Record the results in data table 30.3.

Data Table 30.2 Acids/Bases and Buffered Solutions

Total Drops of 0.1 M HCl	pH of Buffered Solution	Total Drops of 0.1 M NaOH	pH of Buffered Solution
0		0	
2		2	
4		4	
6		6	
8		8	
10		10	
12		12	
14		14	
16		16	
18		18	

Data Table 30.3 Acids/Bases and No Buffered Solutions

Total Drops of 0.1 M HCl	pH of Water Solution	Total Drops of 0.1 M NaOH	pH of Water Solution
0		0	
2		2	
4		4	
6		6	
8		8	
10		10	
12		12	
14		14	
16		16	
18		18	

30.11 Graph the results of both data tables on a single sheet of graph paper. Include a proper title and key for this graph as well as properly labeled axes. Plot pH on the *Y* axis and total drops of acid or base on the *X* axis.

30.12 Now let us analyze the data. Which system (buffer or plain) had the more gradual pH change? ____________ The more dramatic change? ____________ How quickly did you see a significant jump of greater than one pH unit in the buffer system? ____________ In the plain system? ____________

Buffers hold pH changes to a minimum. The same kind of thing happened with the $CaCO_3$-H_2SO_4 experiment. Could $CaCO_3$ be considered a type of buffer? ____________

30.13 Water (H_20) has a pH that can change due to soil conditions, plant life, effluents, surfactants, or whatever finds its way into the water system. Your teacher has collected four soil samples and has had them soaking in water for at least 24 hours. Check the pH of each sample with a pH meter making certain you have standardized your meter against a pH buffer of 7.0. Place your data in data table 30.4.

30.14 Write a paragraph or two discussing the similarities and differences in thc pH value of the soils that you studied above. What factors do you suspect might be affecting the pH values? Why are some samples more neutral than others?

Data Table 30.4 pH of Various Soils

No.	Soil Description (Made from Bottle)	pH
1		
2		
3		
4		

Data Table 30.5 Solubility of Oxygen at Different Temperatures

Temperature (°C)	Solubility of Oxygen (ppm)
0	14.6
5	12.7
10	11.3
15	10.1
20	9.1
25	8.3
30	7.5

30.15 There are *many* factors that can be studied in relation to the environment, especially water. The factors are dissolved oxygen, alkalinity, pH, the ammonia (NH_3) level, the nitrate (NO_3) level, and the phosphate (PO_3) level. These will be studied next. Usually when water is analyzed, the sample is checked for dissolved oxygen (DO), alkalinity, pH, ammonia, nitrates, phosphates, and temperature.

What happens to the solubility of oxygen as the temperature increases? Why?

Oxygen

In lakes, oxygen can often be the limiting factor in determining the life forms present. Lakes get much of their oxygen from the atmosphere by diffusion. The remainder comes from photosynthesis of plants in the water. If oxygen is consumed by the respiration of organisms within the lake faster than it is replenished, an oxygen deficit results. This prohibits the existence of life forms that otherwise could survive there. At this time, Lake Superior and Lake Huron are saturated with oxygen at all depths. At times they become only 70 percent to 80 percent saturated in their deeper regions. In contrast, the central basin of Lake Erie is below 10 percent saturation in its deeper regions.

In the bottom of Lake Erie, for example, oxygen is consumed by the many decomposers feeding on sewage, dead algae, and the like. The oxygen is not replaced rapidly enough to prevent a deficit because this deep water is below the zone of light penetration. No photosynthesis occurs at these lower depths. Also, the movement of oxygen from the surface to the bottom requires a long time interval. Thus, although the bottom of Lake Erie may be a suitable habitat for trout with regard to food supply and temperature, the low-oxygen content is the limiting factor that prevents trout from living there.

Data Table 30.6 Thresholds of Dissolved Oxygen for Fish (Milligrams per Liter)

Fish Variety	Summer		Winter	
	Lowest Concentration at Which Fish Survived for 24 Hours	Highest Concentration at Which Fish Were Killed in 24 Hours	Lowest Concentration at Which Fish Survived for 48 Hours	Highest Concentration at Which Fish Were Killed in 48 Hours
Pike	3.1	6.0	2.3	3.1
Black bass	3.1	5.5	2.3	4.7
Black crappie	4.2	5.5	1.4	1.5
Common sunfish	3.1	4.2	0.8	1.4
Yellow perch	3.1	4.2	1.5	4.7
Sunfish	3.1	3.3	0.8	3.5
Black bulkhead	2.9	3.3	0.3	1.1
Median value	3.1	4.2	1.4	3.1

Source: Resources Agency of California, State Water Quality Control Board, "Water Quality Criteria" (Jack E. McKee and Harold Wolff), Publication No. 3–A, 1963, p. 181.

Data Table 30.7 Minimum Dissolved Oxygen Required for Survival of Fish for 84 Hours (Milligrams per Liter)

Fish Variety	Minimum Concentration Required for Survival at Indicated Temperature		
	10° C	16° C	20° C
Rainbow trout	1.89	3.00	2.64
Perch	1.05	1.34	1.25
Roach	0.65	0.71	1.42
Mirror carp	0.48	1.73	3.74
Tench	0.35	0.54	—
Dace	—	1.14	—
Bleak	—	1.50	—

Source: Resources Agency of California, State Water Quality Control Board, "Water Quality Criteria" (Jack E. McKee and Harold Wolff), Publication No. 3–A, 1963, p. 181.

Data table 30.6 shows the thresholds of dissolved oxygen requirements for several species of fish. Data table 30.7 shows the minimum amount of dissolved oxygen required for survival of fish at various temperatures. Using these two data tables as your reference, answer the following questions.

1. If a stream, lake, or river contained a DO level of 3.2 mg/L, which fish would survive during the summer months? ____________

   ____

2. If the DO level were to stay the same during the winter as during the summer, which fish would survive? ________________

   ____

   Which fish would not survive? ______________________

3. Notice that fish have minimum DO-concentration requirements at various temperatures. What do you think the reasons are for these various requirements at different temperatures?

Alkalinity

The alkalinity of a water sample refers to its capacity to neutralize acids. A solution can have a high alkalinity without necessarily having a high-pH value. In other words, a solution may have the potential to neutralize a large amount of acid without exhibiting a high pH. Alkalinity is related to the buffering systems in water. Buffers are systems that tend to moderate changes in pH. High alkalinity most commonly occurs when carbonates, bicarbonates, and hydroxides of calcium, magnesium, and sodium metals are present.

The levels and types of alkalinity depend directly on the source of water. Natural surface and well waters usually contain less alkalinity than sewage or waste water samples. High levels of alkalinity may indicate that strongly alkaline industrial wastes are present. In nutrient-rich bodies, like sewage lagoons, alkaline wastes may run as high as 500 ppm. An alkalinity value of less than 50 ppm is considered low; a value of 200 ppm is considered to be quite high.

pH

Recall that the pH of an aqueous solution represents the concentration of hydrogen ions in the solution. The pH scale runs from 0 to 14. On this scale, 7 is neutral, below 7 is acidic, and above 7 is acidic. Our major concern here is the limiting nature of pH and its relationship to other environmental factors.

Recent studies have shown that waters having a pH range from 6.7 to 8.6 will generally support a good fish population. As long as the pH is within this range, it appears to have no effect on life processes like growth and reproduction. In fact, most fish can tolerate pH values lower than 5 or greater than 9 for a short period of time. Some fish are specifically adapted to either high- or low-pH values. These species of fish cannot tolerate pH values within the normal range.

If you discover any bodies of water with a pH value beyond the 6.7 to 8.6 levels, you should be suspicious of water pollution unless you can find a natural cause. Common pollutants that can affect the pH of water are the effluents from chemical and fertilizer plants, pulp and paper plants, and steel mills.

Ammonia (NH_3)

Ammonia is a by-product of the decay of plants and animal proteins and of fecal matter. It is also formed when urea and uric acid in urine decomposes. Thus, the presence of ammonia is an indication that sewage is entering the water system. Since many fertilizers contain ammonia and ammonium compounds, runoff from farmer's fields can also contribute to the concentration of ammonia in water. A further point might interest you. Your community undoubtedly chlorinates its water supply. Chlorine kills bacteria that may be present in water. However, ammonia reacts with chlorine to destroy its germ-killing action. Thus, when ammonia is present in water, large quantities of chlorine must be added to reach a concentration sufficient to kill the bacteria.

A concentration of 0.5 ppm of ammonia is considered the maximum tolerable limit, with less than 0.01 ppm as being desirable.

Nitrates (NO_3^-)

Nitrates are present in most fertilizers, biological wastes, and as atmospheric nitrogen that has been converted to nitrates by lightning or fixed by bacteria on legume plants. A high concentration of nitrates may indicate a farm runoff problem. Other possible sources of increased concentrations of nitrates are natural decay of plants and animals, industrial wastes, and animal excrement. With respect to the last source, it has been estimated that a feedlot with ten thousand cattle can cause water pollution problems equivalent to those caused by a city of forty-five thousand people.

It is not easy to track down the exact sources of nitrates in a stream or lake. One thing is clear, nitrates are plant nutrients. The presence of too many nitrates can accelerate the growth of phytoplankton and other plants. This is followed by a lowering of the dissolved oxygen and changes in the fish populations. Foul odors, clogged filters in water purification plants, disagreeable tastes, and the lowering of recreational value of the water are some of the economic problems that occur.

Figure 30.3 There are many reasons for fish kill. One reason is an algal bloom. Algal blooms are promoted by large supplies of nutrients. As the plants die, causing large quantities of decaying organic matter, oxygen levels are decreased. The end result is death for other aquatic species such as fish.

Concentrations of inorganic nitrogen greater than 0.3 ppm can be expected to contribute to algal blooms. But wide deviations from this value do occur. With respect to human health, authorities consider that 45 ppm of nitrates in drinking water render the water hazardous to humans and to most other animals. The nitrates change to nitrites in the stomach. In infants, nitrites can cause methemoglobinemia ("blue baby" conditions). Many deaths occur every year as a result of nitrate poisoning, usually from nitrates in farm well water.

Phosphates (PO_4^{-3})

Phosphates, though essential for the growth of all living things, will upset aquatic environments if present in excess amounts. The primary source of phosphates in bodies of water is detergents. Agricultural, domestic, and industrial wastes increase the phosphate concentration resulting in algal blooms. As mentioned above, algal blooms deplete the supply of oxygen. With oxygen no longer available, the local biotic community is upset.

Phosphate content ranges from 0.01 to 0.05 ppm in most natural waters. Waste water can have a range of 5 to 30 ppm.

You may now proceed to test the water samples that your teacher has provided for you. Your teacher will show you how to use all of the water-testing equipment before you begin. Record all of your data in data table 30.8.

Data Table 30.8

	Site 1	Site 2	Site 3	Site 4	Site 5	Site 6	Site 7
Alkalinity							
Dissolved oxygen (surface)							
Dissolved oxygen (bottom)							
Ammonia							
Nitrate							
Phosphate							
pH							
Temperature							
Turbidity							
Location							
Description of sample site							

30.16 As environmental conditions are altered, the **ecology** of an affected area can also change.

By definition, ecology is the study of the interrelationships between living organisms and their environment. Ecology is a science that studies the way a living plant or animal responds to the living and nonliving world. Environmental conditions are capable of altering this relationship.

The base unit of ecology is called the **biotic community.** Various communities that interact together with their physical environment establishes an **ecosystem,** which is considered the center for most studies of ecology. Ecosystems, with similar physical and geographical characteristics, are called **biomes.** A biome is made up of specific types of vegetation and the animals that live off the vegetation. Examples of biomes are deserts, tundra, tropical rain forests, and so forth. If you put all of the biomes together with all of the nonliving elements, the result is the **biosphere.** There is only one biosphere and that is Earth.

We are all part of the biosphere. We are a part of a population, a community, an ecosystem, and a biome. As populations increase, the cause-effect relationships bring about many environmental stresses such as those produced by acid rain.

The impact that human kind has on the environment is well noted by those who care. Hopefully, you are one who cares. Let us close this chapter with the following quotes:

> We have treated our land as if it were a limitless resource. Traditionally, Americans have felt that what they do with their own land is their own business. This attitude has been a natural outgrowth of the pioneer spirit. Today, we are coming to realize that our land is finite, while our population is growing.
>
> —Richard M. Nixon

> The four laws of ecology state:
> 1. Everything is connected to everything else.
> 2. Everything must go somewhere.
> 3. Nature knows best.
> 4. There is no such thing as a free lunch.
>
> —Barry Commoner

> Climb the mountains and eat their good tidings. Nature's peace will flow into you as sunshine flows into trees.
>
> —John Muir

> We abuse land because we regard it as a commodity belonging to us. When we see land as a community to which we belong, we may begin to use it with love and respect.
>
> —Aldo Leopold

If We Sell You Our Land, Love It*

By a little church on Bainbridge Island, within sight of present-day Seattle, Washington, lies the grave of a great Indian orator, Chief Sealth (spelled Seattle by early white settlers). While fearing the intentions of white men, he welcomed them, nonetheless, and became a Christian. He died in June 1866. Twelve years earlier, in his elegant native language, Duwamish, Chief Seattle delivered the greatest, most tragic oration of his life.

> The Great Chief in Washington sends word that he wishes to buy our land. The Great Chief also sends us words of friendship and goodwill. This is kind of him, since we know he has little need of our friendship in return. But we will consider your offer. For we know that if we do not sell, the white man may come with guns and take our land.

*Reprinted from the *Lutheran Standard*.

Every part of this earth is sacred to my people. Every shining pine needle, every sandy shore, every mist in the dark woods, every clearing and humming insect is holy in the memory and experience of my people.

So, when the Great Chief in Washington sends word that he wishes to buy our land, he asks much of us.

The red man has always retreated before the advancing white man, as the mist of the mountains runs before the morning sun. But the ashes of our fathers are sacred. Their graves are holy ground, and so these hills, these trees, this portion of the earth is consecrated to us.

We know that the white man does not understand our ways. One portion of land is the same to him as the next, for he is a stranger who comes in the night and takes from the land whatever he needs. The earth is not his brother, but his enemy, and when he has conquered it, he moves on. . . . He treats his mother, the earth, and his brother, the sky, as things to be bought, plundered, sold like sheep or bright beads. His appetite will devour the earth and leave behind only a desert.

I do not know. Our ways are different from your ways. The sight of your cities pains the eyes of the red man. There is no quiet place in the white man's cities. No place to hear the unfurling of leaves in spring or the rustle of insects' wings. . . .What is there to life if a man cannot hear the lonely cry of the whippoorwill or the arguments of the frogs around a pond at night?

The air is precious to the red man, for all things share the same breath—the beast, the tree, the man, they all share the same breath. The white man does not seem to notice the air he breathes. Like a man dying for many days, he is numb to the stench.

If we sell you our land, you must remember that the air is precious to us, that the air shares its spirit with all the life it supports. The wind that gave our grandfather his first breath also receives his last sigh. And the wind must also give our children the spirit of life. . . .

We will consider your offer to buy our land. If we decide to accept, I will make one condition. The white man must treat the beasts of this land as his brothers.

I have seen a thousand rotting buffalos on the prairie, left by the white man who shot them from a passing train. [But] I do not understand how the smoking iron horse can be more important than the buffalo that we kill only to stay alive.

You must teach your children that the ground beneath your feet is the ashes of our grandfathers. So that they will respect the land, tell your children what we have taught our children, that the earth is our mother. Whatever befalls the earth befalls the sons of the earth.

This we know. The earth does not belong to man; man belongs to the earth. This we know. All things are connected like the blood which unites one family. All things are connected. Whatever befalls the earth befalls the sons of the earth. Man did not weave the web of life, he is merely a strand in it. Whatever he does to the web, he does to himself.

But we will consider your offer to go to the reservation you have for my people. We will live apart, and in peace. It matters little where we spend the rest of our days. Our children have seen their fathers humbled in defeat. Our warriors have felt shame, and after defeat they turn their days in idleness and contaminate their bodies with sweet foods and strong drink.

It matters little where we pass the rest of our days. They are not many. A few more hours, a few more winters, and none of the children of the great tribes that once lived on this earth or that roam now in small bands in the woods will be left to mourn the graves of a people once as powerful and hopeful as yours. Men come and go, like the waves of the sea.

Even the white man, whose God walks and talks with him as friend to friend, cannot be exempt from the common destiny. We may be brothers after all; we shall see.

One thing we know, which the white man may one day discover—our God is the same God. You may think now that you own him as you wish to own our land, but you cannot. He is the God of man and his compassion is equal for the red man and the white. This earth is precious to him and to harm the earth is to heap contempt on its Creator.

The whites too shall pass; perhaps sooner than all other tribes. Continue to contaminate your bed, and you will one night suffocate in your own waste. But in your perishing you will shine brightly, fired by the strength of the God who brought you to this land and for some special purpose gave you dominion over this land and over the red man. That destiny is a mystery to us, for we do not understand when the buffalo are all slaughtered, the wild horses are tamed, the secret corners of the forest are blotted. . . .

Where is the thicket? Gone. Where is the eagle? Gone. And what is it to say good-bye to the swift pony and the hunt? The end of living and the beginning of survival.

If we sell you our land, love it as we've loved it. Care for it as we've cared for it. Hold in your mind the memory of the land as it is when you take it. And with all your strength, with all your mind, with all your heart, preserve it for your children and love it . . . as God loves us all.

—Chief Seattle

30.17 A question for you to ask is, "Am I part of the solution to the environment, aiming to protect it, keep it clean, or am I part of the problem?"

On completion of this chapter, you should be able to do the following:

- ☐ 1. Describe some of the adverse effects of air pollution.
- ☐ 2. Explain what is meant by acid rain.
- ☐ 3. Explain the effect of soil type on pH.
- ☐ 4. Explain the effect of limestone on acid rain.
- ☐ 5. Define a buffer and predict the effects a buffer will have on the pH of a system.
- ☐ 6. Name the general types of tests done on water systems.
- ☐ 7. Formulate an opinion about care for our land and water resources and the use of the air as a means of permitting pollution to affect other areas away from the initial source.

Glossary

abdomen The lower part of the trunk below the level of the ninth pair of ribs; contains the major parts of the digestive and excretory systems.
abduction Withdrawal of a part of the body from the axis of the body or of an extremity.
abiogenesis Spontaneous generation; the idea that living things can arise from nonliving material.
abscission The separation of a leaf, flower, seed, or fruit from a stem.
abscission layer A thin layer of cells at the base of a petiole that forms at the end of a growing season and causes abscission to occur.
absorption The process by which substances such as nutrients and oxygen pass across a cell membrane into the cytoplasm of a cell.
acetabulum The deep socket in the pelvic girdle into which the head of the femur fits.
acid A substance that liberates hydrogen ions when mixed with water; reacts with a base to form a salt and water.
acid rain Any form of precipitation with a pH lower than 5.6; caused by the combination of sulfur oxides and nitrogen oxides with water vapor in the air.
actin One of the two proteins involved in muscle contraction. *See also* myosin.
activation energy The energy that must be supplied to a system in order to make a particular process occur.
active site That portion of an enzyme into which the substrate fits or reacts.
active transport The movement of materials across a cell membrane from a region of lower concentration to a region of higher concentration; cellular energy is required.
adaptation A change in characteristics of an organism that increases its chances of survival.
adduction Any movement resulting in one part of the body or a limb being brought toward another or toward the median line of the body.
adenine A nitrogen base found in DNA and RNA; pairs with thymine.
adenosine diphosphate, ADP A nucleotide of importance in the transfer of energy within living cells; formed when ATP loses one phosphate group.
adenosine triphosphate, ATP A nucleotide that functions in the temporary storage and transfer of chemical energy used by living cells.
adipose tissue A type of connective tissue in which each cell contains a large vacuole for the storage of fat.
adrenal gland An endocrine gland located on top of each kidney; secretes hormones and adrenaline.
adrenaline A hormone secreted by the adrenal glands that acts to increase the rate of heartbeat; dilates blood vessels of muscles, brain, and heart; constricts blood vessels of skin and viscera; dilates pupils; makes hair stand up; increases the amount of sweat produced; and speeds the change of glycogen to glucose.
adventitious root A root that grows from a part of the plant that is not the primary root; roots that grow from any node on a stem or from a bulb.
aerobic Any organism or life process that utilizes, or can only exist in, the presence of oxygen.
afferent nervous system That portion of the peripheral nervous system that acts to transmit impulses toward the central nervous system.
afferent neuron Those neurons that conduct impulses from the sense organs to the spinal cord and brain; sensory neurons.
agglutinin An antibody found in blood plasma; causes corpuscles to clump.
agglutinogen An antigen that stimulates the formation of specific agglutinins.
agriculture The process of growing plants and keeping animals to obtain food and other products.
alcohol Compounds having a hydroxyl (-OH) group attached to one or more carbon atoms in an organic molecule.
alcoholic fermentation The anaerobic decomposition of sugar to produce alcohol and carbon dioxide.
alimentary canal A long tube with an opening at each end; concerned with the digestion and absorption of food by the body; consists of mouth, pharynx, esophagus, and intestines.
alkali Substance that is basic in nature; forms hydroxyl ions in solution.
allele One of a pair of genes that control the same characteristic, at a given chromosome site.
allergen A substance that can produce a reaction in the body, such as dust, animal dander or pollen.
alternation of generations A life cycle of certain plants that alternates between an asexually reproducing generation and a sexually reproducing generation.
alveoli Microscopic saclike structure in the lung located at the end of a bronchiole; the exchange of carbon dioxide and oxygen takes place on the surface of each alveoli.

amino acid Building block of protein; an organic molecule that contains both the amino group (-NH_2) and carboxylic acid group (-COOH).
ammonia A poisonous nitrogenous waste with the formula NH_3; converted to urea by the liver in order to be excreted.
amniocentesis The removal of a small amount of amniotic fluid during pregnancy for the purpose of detecting genetic defects.
amnion The fluid-filled sac that encloses and protects the embryo as it develops.
amniotic fluid Fluid contained within the amnion.
amoeba A group of protozoa of irregular shape; continually changing shape by movement of cytoplasm.
amphetamine A class of drugs that act as stimulants on the central nervous system.
amphibious Living both on land and in water.
amylase A digestive enzyme that breaks down starch into simple sugars.
anabolism Those chemical processes occurring during the life of a cell that result in the formation of new protoplasm or the replacement of old protoplasm.
anaerobic Living or occuring in the absence of oxygen.
anal fin Single fin that is located near the tail of a bony fish.
analogous structures Structures occurring in different organisms that perform similar tasks but develop in different ways.
anal pore A structure in paramecium through which waste material is expelled from the cell.
anaphase A stage in mitosis that is characterized by the migration of chromosome halves to opposite poles of the cell; the third stage.
anatomy A branch of the biological sciences dealing with the structure of organisms.
angiosperm Any member of a class of flowering vascular plants having the seeds in a closed ovary.
ankle Joint or region between the foot and the leg.
annual A plant that completes its life cycle in one growing season.
annual ring The layer of wood produced by a single year's growth of a wooden plant.
antacids Compounds that neutralize acid, producing salt, water, and sometimes carbon dioxide.
antagonist Any muscle responsible for movement opposite to that of a prime mover; as one contracts the other relaxes.
anterior Referring to the front of the human body or located toward the head end of an animal.
anther The part of the stamen of a seed plant that contains pollen; produces the male reproductive cells in a flower.
antheridium A sperm-producing structure found in some plants.
antibiotic A substance produced by a living organism that inhibits the growth of or destroys harmful bacteria and microorganisms.
antibody Specific proteins that are produced by certain white blood cells in response to a virus or other foreign substance which has been introduced into the body.
antigen A substance that, when introduced into the body, stimulates the production of an antibody.
anus Found at the end of the digestive tract; the posterior opening of the alimentary canal through which undigested food passes from the body.
anvil The middle of the three small bones of the middle ear that conduct vibrations from the eardrum to the oval window.
aorta The main artery that carries blood from the heart.
aortic semilunar valve Valve in the heart at the base of the aorta to prevent blood from flowing back into the heart when the ventricles relax.
apical dominance The retardation of lateral bud growth caused by the presence of the terminal bud.
apical meristem The area of rapid cellular division located at the tips of stems and roots.
appendage Any structure having one end attached to the main part of the body and the other end more or less free.
appendicular skeleton The skeletal structures composing and supporting the appendages; including the shoulder and pelvic girdle as well as the arms and legs.
appendix A small fingerlike projection from the caecum of the large intestine.
aqueous humor A watery liquid that fills the space between the cornea and lens; helps to keep the shape of the eye.
arteriole The small branches of arteries that lead to capillaries.
artery A muscular blood vessel carrying blood away from the heart.
arthritis An inflammation of the joints; sometimes results in fusion of the bones.
articulation A joint; the junction of two or more bones.
asexual reproduction Production of offspring in which only one organism takes part; occurs without gametes.
association neuron A neuron in the spinal cord that, by synapses, joins a sensory nerve to a motor nerve; also passes impulses to other parts of the spinal cord and to the brain.
aster A series of cytoplasmic fibers that radiate from the centriole during mitosis.
astigmatism A defect in a lens or an eye causing improper focusing.
atom The smallest particle of an element which has the properties of that element and takes part in chemical reactions.
ATP *See* adenosine triphosphate.
atrium A chamber of the heart that receives blood.
auditory canal The outer ear canal, which leads to the tympanic membrane (eardrum).
auditory meatus Opening of the auditory canal to the exterior.
auditory nerve The cochlear branch of the eighth cranial nerve; carries impulses from the cochlea of the inner ear to the brain.
auricle The pinna of the ear; the flap of cartilage and skin of the outer ear.
autonomic nervous system Portion of the peripheral nervous system that controls involuntary actions; consists of the sympathetic and parasympathetic nervous systems.
autosome Any pair of chromosomes in humans other than the sex chromosomes.
autotroph An organism that can produce its own food by photosynthesis.

auxin A plant hormone that stimulates growth in length.
axial skeleton The central, supporting portion of the skeleton composed of the skull, vertebral column, and rib cage.
axon That portion of the neuron that carries impulses away from the cell body.
bacilli Rod-shaped bacteria.
bacteria An organism with a single cell, having no separate nucleus; it is neither plant nor animal.
balanced diet A diet that supplies enough energy for a person to live; contains a proper balance between protein, carbohydrates, and fat, and supplies the necessary vitamins; minerals and roughage.
ball and socket joint A joint in which the round end of one bone (the ball) fits into the hollow (the socket) of another bone; allows movement in all directions.
bark Tissue located to the outside of the vascular cambium in a woody stem.
basal metabolic rate The basic rate of the body's metabolism, measured at rest.
base A substance capable of neutralizing acid; combines with hydrogen ions to form a salt and water.
bicuspid valve A valve in the heart located between the left atrium and left ventricle.
biennial A plant that lives for two growing seasons, producing leaves the first year and fruiting and dying the second year.
bilateral symmetry A type of body structure in which bisection of the organism results in two halves that are mirror images of each other.
bile A green, alkaline liquid produced by the liver that aids in the emulsion and digestion of fats.
bile duct A tube from the liver to the gallbladder to the duodenum through which bile passes.
binary fission An asexual process of reproduction involving the division of one cell into two daughter cells of equal size.
binocular vision The ability to see an object with both eyes simultaneously.
binomial nomenclature The two-name system of naming organisms by both genus and species.
biogenesis The idea that all living matter arises from other similar living matter; life from life.
biological clock An inherent timing mechanism responsible for various cyclic physiological and behavioral responses of living beings.
biology A science that deals with living beings and life processes.
biome A large terrestrial community characterized by its plant and animal life.
biosphere The part of the world in which life can exist.
biotic Of or relating to life; caused by living things.
bladder (urinary bladder) A baglike structure capable of great distention that functions to store urine until it is expelled from the body.
blade Leaflike part of a nonvascular plant; also, the flattened portion of a vascular leaf.
blind spot The region of the retina that is insensitive to light because it has no rods or cones; place where optic nerve enters the eye.
blood A liquid that consists of plasma, red and white blood cells, and platelets; transports food, oxygen, and waste materials to and from cells.
bolus A round ball of food mixed with saliva ready to be passed down the esophagus.
bone A relatively hard, porous type of connective tissue containing a high percentage of calcium and phosphorous; forms the major part of the skeleton.
bone marrow The fatty tissue lying within the central cavity of many bones; functions in the formation of red and white blood cells.
botany The study of plants.
Bowman's capsule Cup-shaped structure that makes up part of a nephron in the kidney.
brain The part of the central nervous system contained within the cranium of the skull.
brain stem The lowest part of the brain; composed of the pons, midbrain (thalamus and hypothalamus), and the medulla oblongata.
breast Mammary gland; modified sweat gland for the secretion of milk, located on the front surface of the thorax anterior to the pectoralis major muscle.
breathing center A portion of the medulla oblongata that controls breathing.
bronchi A tube that branches from the trachea and leads to each lung.
bronchiole The tiny branches of each bronchi.
Brownian movement The continuous, irregular motion of particles that results from their bombardment by smaller particles.
bud A small, pointed structure on a stem that develops into either a leaf or flower.
budding A type of asexual reproduction in which a small group of cells grows from the body of an organism resulting in a miniature version of the organism.
bud scale A small, thick, modified leaf that protects a bud.
buffers A substance that tends to prevent a change in the hydrogen ion concentration of a solution.
bulb A modified, small underground stem covered with fleshy leaves that store food.
bundle sheath A layer of tightly packed cells around a leaf vein.
bursa A sac of lubricating fluid between the bones of a joint and tendons; provides protection against pressure and friction.
buccal cavity The mouth.
bulbo urethral gland (cowper's gland) Small pea-shaped gland located below the prostate gland on each side; secrete a mucouslike lubricating fluid.
caecum A small bag at the start of the large intestine in humans.
calcium An essential constituent of bones and teeth.
callus Any thickening of a tissue resulting from the normal growth of cells.
Calorie A unit of energy used to measure the energy value of different foods.
calyx A group of modified leaves called sepals that surround a developing flower bud.
cambium A layer of actively dividing cells between the xylem and the phloem of dicotyledons.

canaliculi Very small tubular channels found in bone tissue.
capillary Microscopic blood vessel through which the exchange of gases, nutrients, and waste occurs; connects an artery with a vein.
carbohydrate An organic compound composed of carbon, hydrogen, and oxygen; starch, cellulose, and all sugars are carbohydrates; major source of energy in human diet.
carbon A chemical element on which life is based.
carbon dioxide A product of the catabolism of food; necessary for photosynthesis to occur.
carbon fixation The second stage of photosynthesis in which carbon dioxide is converted into sugar molecules.
cardiac muscle A branching, lightly striated type of muscle found only in the heart; contracts rhythmically.
cardiac sphincter Circular muscle that controls the passage of food from the esophagus to the stomach.
carnivores Animals that feed on other animals.
carpel A female part of a flower consisting of a stigma, a style, and an ovary containing ovules.
carrier An individual that is heterozygous for a recessive trait.
cartilage A flexible, supporting type of connective tissue composed of a nonliving matrix within which are the living cartilage cells; supports parts of the body and lines the surface of joints.
catabolism Those processes occurring within living cells that result in the breakdown of food and the release of heat and energy.
catalyst A substance that increases the rate of a chemical reaction.
cell The smallest unit of structure and function in a living organism.
cell body Part of a neuron containing the nucleus and organelles, excluding axon and dendrites.
cell membrane A living, semipermeable membrane surrounding a cell; controls passage of materials into and out of the cell.
cell plate A structure that forms at the equator of a dividing plant cell during cytokinesis.
cell theory All living things are composed of cells and the product of cells.
cellular respiration Process by which cells convert food into a usable form of energy; this process takes place in the mitochondria.
cellulose The fibrous, woody material that makes up most of the cell wall in plants; a polysaccharide.
cell wall A nonliving, semirigid structure lying outside the cell membrane (plasma membrane) in plant cells; gives a plant the structural strength to stand upright.
central nervous system The brain and spinal cord.
centriole Two small, cylindrical structures found in animal cells near the nucleus; functions in mitosis.
centromere Point at which chromatids attach to each other.
cerebellum Second largest part of the brain; located behind and below the cerebrum; coordinates all muscular movements.
cerebral cortex Outer portion of the cerebrum.
cerebral hemispheres Two principle divisions (right and left) of the cerebrum.
cerebrospinal fluid A solution of glucose and mineral salts that cushions the brain and spinal cord.
cerebrum Largest part of the human brain; located behind the forehead.
cerumen The waxy secretion formed by glands in the external auditory meatus.
cervical Referring to the neck region.
cervix A muscular ring located at the opening of the uterus into the vagina.
chemical bond The attractive force between two atoms produced by the sharing or transfer of electrons.
chemical digestion The breakdown of food by means of digestive enzymes; involves the breaking of chemical bonds.
chemoreceptor Nervous tissue that responds to chemical stimulus.
chlorenchyma Parenchyma tissue (storage tissue) containing chloroplasts in higher plants.
chlorophyll Green pigment located in the chloroplasts of plants; functions in photosynthesis.
chloroplast Organelle found in plant cells that contains chlorophyll.
choroid layer The pigmented, highly vascular middle layer of the wall of the eye; lies between the retina and the sclera.
chromatid Name given to each strand of a chromosome pair following replication.
chromatin The material that makes up chromosomes; found within the nucleus of a cell.
chromosome Long, threadlike strands of DNA that contain genetic material; composed of genes; found in the nucleus of a cell.
chyme The fluid form of food within the stomach and intestine.
cilia Small, threadlike projections on the surface of some protozoa used for the purpose of locomotion; present in some cells of multicellular organisms.
circulation The movement of blood through the body's network of blood vessels.
circumduction Circular movement, as in a circular swinging of the arm.
class A group of related taxonomic orders.
classify Grouping of objects, ideas, or organisms on the basis of similarity.
cleavage furrow The inward folding of the cell membrane just prior to cytokinesis in animal cell division.
clitoris Erectile structure in the female, similar to the penis in the male.
cocci A spherical shaped bacteria.
coccyx The end of the vertebral column; located below the sacrum.
cochlea The spiral bony canal of the inner ear in which are located the auditory receptors.
codon A sequence of three nucleotide bases that make up the genetic code; each codon is specific for one amino acid in a protein.
coenzyme A substance that activates an enzyme.
coleoptile The first leaf of a monocot that forms a protective sheath around the growing shoot.
collagen A fibrous protein in connective tissue, cartilage, and bone.

collecting tubule A tubule into which urine drains from the nephron; leads from the nephron to the renal pelvis.

collenchyma A plant tissue specialized for strength and support.

colloid A suspension of particles dispersed in a different medium.

colon The large intestine.

commensalism A type of symbiotic relationship in which one of the organisms benefits and the other is not affected.

compact bone A type of dense bone tissue forming the outer layer of all bones; found along the shaft of long bones in the body.

companion cell A type of phloem cell found in association with sieve-tube elements.

complete flower A flower that possesses sepals, petals, stamens, and a pistal.

compound A substance that can be broken down into two or more simpler substances by normal chemical means.

compound leaf A type of leaf in which many small leaflets are attached to the same petiole.

compound microscope A magnifying device in which two lenses are used.

cone One of the two types of light-sensitive receptor cells of the retina that responds to color; seed or pollen-bearing structure in gymnosperms.

conjugation A primitive type of sexual reproduction in which two cells fuse together in order to exchange nuclear material.

conjunctiva A delicate layer of epithelium that covers the outer surface of the cornea and is continued as the lining of the eyelids.

connective tissue Tissue that supports and holds together various parts of the body.

contractile vacuole A structure found in a paramecium; maintains osmotic pressure within the cell by collecting and expelling excess water.

control A part of an experiment that is used as a standard of comparison or reference for experimental observations.

controlled experiment An experiment in which only one variable factor is tested.

convolution A fold, twist, or coil of any organ or surface.

cork The protective, water-resistant outer layer of cells in the roots and stems of woody plants.

cork cambium Layer of cells that produces cork.

corm A short, thick, fleshy underground stem.

cornea Transparent layer that covers the eye; it is continuous with the sclera.

corolla A structure found inside the calyx of a flower; made up of petals.

coronary circulation The arteries, capillaries, and veins that supply blood to the heart.

corpus callosum A band of tissue that connects the right and left cerebral hemispheres.

corpus luteum Tissue that forms a temporary endocrine gland from a ruptured ovarian follicle; produces and secretes progesterone.

cortex A general term applied to the outer region of an organ; a food-storage area in plants that extends from the epidermis to the phloem.

cotyledon An embryonic leaf of a seed plant that provides food for the plant embryo after germination.

cranial nerve A nerve that arises from the brain and passes to some region of the head or neck; there are twelve pairs of cranial nerves.

cristum A structure formed by the folding of the inner membrane of the mitochondria.

crossing-over The exchange of parts between two homologous chromatids during meiosis.

cuticle A waxy covering on the epidermis of plants and on the surface of a leaf.

cytokinesis Division of the cytoplasm in a cell during mitosis and meiosis.

cytokinin A plant hormone that stimulates cellular division.

cytoplasm The material lying within the cell membrane and outside of the nucleus.

cytosine A nitrogen base that is found in DNA and RNA; pairs with guanine.

Darwinism The idea that evolution took place by natural selection; only those organisms that are best suited to their living conditions, or environments, survive.

data The observations, both quantitative and qualitative, that are gathered during an experiment.

decibel A unit used to measure sound intensity.

deciduous A type of tree that sheds its leaves annually.

decomposer An organism that breaks down dead organisms and waste created by living organisms into chemical substances that it can use for food.

defecation The action of passing out feces, waste material, through the anus.

deglutition The act of swallowing.

dehydration synthesis A type of synthesis that takes place in living systems in which a molecule of water is removed from the site where two molecules react with each other forming a chemical bond.

deletion The loss of a part of a chromosome during meiosis; a type of mutation.

dendrite The finely divided extensions of a neuron's cell body; transmits impulses toward the cell body.

dense connective tissue A type of connective tissue with a semirigid or rigid medium; includes cartilage and bone.

deoxyribonucleic acid (DNA) A nucleic acid found in both plant and animal cells; contains genetic information and controls the cellular activities.

depressant A class of drugs that slow or inhibit the functioning of the central nervous system.

dermatology The branch of anatomy that deals with the skin (integument) and its derivatives, including the sweat glands, oil glands, mammary glands, hair, and nails.

dermis Layer of cells lying directly beneath the epidermis of the skin.

development The stages that an organism goes through from conception until it reaches maturity.

diabetes insipidus A metabolic disorder caused by a lack of secretion of ADH.

diabetes mellitus A disease resulting from the pancreas's inability to produce enough insulin.

dialysis The separation of a colloidal material from other substances through a semipermeable membrane.

diaphragm A wall of muscle separating the thoracic and abdominal cavities.
diaphysis The main part, or shaft, or a long bone.
diarthrosis A type of joint or articulation that allows varying degrees of movement; a cavity containing lubricating fluid (synovial fluid); exists between the surface of the articulating bones.
diastole The period during the beat of the heart when the muscles relax and the atria and the ventricles fill with blood.
dicotyledon A flowering plant whose seeds have two cotyledons; the vascular bundles arranged in a ring around the center of the stem.
diet The different kinds of food and the amount of each kind of food consumed by a person.
differentiation The process of cell specialization.
diffusion The movement of molecules of a substance from a region where it is in relatively high concentration to an area of lower concentration.
digestion The process of physically (mechanically) and chemically breaking food down into a form that can be absorbed by the body and used by the cells for energy and growth.
digits Referring to the fingers and toes.
dihybrid cross A genetic cross that involves two sets of traits.
dipeptide Two amino acid molecules linked together.
diploid number (2*n*) The number of chromosomes present in a somatic (body) cell; containing a full set of homologous chromosomes, or twice that of the haploid number.
disaccharide A simple sugar composed of two monosaccharides bonded together.
dislocation The forcing of a bone out of its normal position in a joint.
distal Toward the free end of an appendage.
distal tubule A thin tube in a nephron that leads from the loop of Henle to the collecting tubules.
division In plant classification, it refers to a group of related taxonomic classes.
division of labor The division or specialization of body parts into tissues and organs that perform different functions.
DNA *See* deoxyribonucleic acid.
dominant A term used to describe an inherited gene or trait that is expressed in the offspring; prevents the expression of a recessive trait.
dominant generation It is the most notable form in the life cycle of a plant that exhibits alternation of generations.
dormancy The period of inactivity that occurs during the development of a seed or spore.
dorsal Referring to the back of an animal; the side along which the backbone passes; sometimes called the posterior side.
double fertilization A type of fertilization in angiosperms in which one sperm nucleus fertilizes the egg and a second sperm fertilizes the polar nuclei.
double helix The double spiral (twisted ladder) form of the DNA molecule.
drug Any substance taken internally or used externally in the medicinal treatment, cure, or prevention of a physical or mental illness; alters the functioning of the mind or body.
duodenum The first portion of the small intestine that extends from the pyloric sphincter to the jejunum.
dura mater Strong connective tissue containing blood vessels; the outermost of the three membranes that make up the meninges.
ear drum Tympanic membrane.
ecology A study of the relations between animals, plants, and the physical conditions of the environment.
ecosystems A complete ecological unit of the biosphere in which all living and nonliving things interact and exchange materials.
ectoderm The outer layer of cells in an embryo during embryonic development.
ectoplasm In animal cells, it is the outer layer of cytoplasm.
effector A muscle, gland, or other type of tissue that responds to a definite stimulus.
efferent neuron Neurons that conduct impulses from the central nervous system to an effector organ; motor neuron.
egg Female reproductive cell.
ejaculation The expulsion of seminal fluid.
electron A negatively charged particle found in an atom.
electron microscope A microscope that uses a beam of electrons as its source of illumination rather than visible light.
electron transport system A chain of chemical reactions in which proteins and enzymes are involved and water is always formed.
element A substance which cannot be further broken down by chemical means.
embryo A young plant in a seed that is formed from an ovule in a flower; a developing zygote.
embryology The study of the early stages in the development of an organism, from the zygote to birth.
embryo sac A group of seven cells that are produced from cytoplasmic division; functions as the female gametophyte in flowers.
endocrine gland A hormone-producing gland lacking ducts and secreting directly into the blood stream.
endoderm The inner layer of cells in a developing embryo.
endodermis The innermost ring of cortex cells in a plant root and in certain plant stems.
endometrium Inner lining of the uterus.
endoplasm The inner layer of cytoplasm in an animal cell.
endoplasmic reticulum A network of fine channels passing through the cytoplasm of a cell and connecting the plasma membrane and nuclear membrane; ribosomes lie along the wall and function in protein synthesis.
endoskeleton An internal supporting framework of an animal.
endosperm Food material for the use of a developing embryonic plant within a seed during germination.
endothermic Heat energy is absorbed during a chemical reaction.
enzyme A chemical compound (protein) that affects the rate of a chemical reaction; functions as a catalyst.
epicotyl The part of a seedling (embryonic plant) above the cotyledons and below the first foliage leaves.
epidermis In both plants and animals, a specialized outer layer of cells that covers the surface of the organism and provides protection.

epididymis A coiled tube that lies on top of the testis and stores sperm.
epigeal Having growth above ground level; epigeal seedlings have their cotyledons above ground during growth; most dicotyledons are epigeal.
epiglottis A flap of cartilage that covers the opening from the pharynx to the voice box (larynx); prevents the entrance of food into the larynx.
epiphyseal line The line between the epiphysis and the diaphysis (shaft) of a long bone.
epiphysis The end portion of a long bone.
epiphyte A plant growing on another plant and using it only for support; it is not a parasite.
epithelium A type of tissue that covers the entire surface of an organism and lines all body tubes and cavities.
erythrocyte Red blood cell.
esophagus Part of the digestive tract; a tube that connects the pharynx and stomach.
estrogen One of the two major female sex hormones secreted by the ovaries. *See also* progesterone.
eukaryote An organism made of cells having membrane-bound nuclei.
eustachian tube A tube that leads from the middle ear to the pharynx; functions to maintain equal air pressure on both sides of the tympanic membrane (ear drum).
evergreen A plant that bears leaves at all times; never loses all its leaves at one time.
excretion Process in which an organism eliminates solid and liquid wastes; helps to maintain water and salt balance.
exocrine gland A gland with a duct or ducts.
exoskeleton An external supportive covering or skeleton.
exothermic Heat energy is generated and given off during a chemical reaction.
experiment A procedure or exercise to observe the behavior of objects, materials, or organisms under controlled conditions.
experimental variable A single factor in an experiment that is different from the control.
expiration Exhalation; to breathe out; to expel air from the lungs.
extension The movement of one part of the body away from another; movement that increases the angle between bones; opposite of flexion.
extensor A muscle that straightens a body part; a muscle that produces extension.
external acoustic meatus A slightly curved canal in the temporal bone that conducts sound waves from the exterior to the typmpanic membrane (ear drum).
exteroceptors Nerve endings that detect stimuli from outside the body; located in the skin, ear, and eye.
eyepiece Ocular; a lens in a compound microscope that magnifies the image produced by the objective lens.
facilitated transport The movement of molecules across a cell membrane by carrier molecules. No energy is expended in the process.
fallopian tube In females, a tube which leads from the ovary to the uterus; fertilization of an egg (ova) by a sperm takes place in the fallopian tube.
family A group of related taxonomic genera.
farsightedness Hyperopia; a defect of vision in which the focal point of the lens falls behind the retina due to a shortening of the eyeball.
fat An ester of glycerol and three fatty acids (triglyceride); a fat is solid at room temperature.
fatigue Weariness from physical exertion; the inability of nerves and tissues to respond to a stimulus following overactivity.
fatty acid A weak organic acid present in lipids (fats and oils).
feces Any solid material that is not digested; waste eliminated by the gastrointestinal tract.
fermentation A chemical change caused by yeasts and bacteria in which a carbohydrate, usually a sugar, is converted to ethanol (ethyl alcohol) and carbon dioxide.
fertilization The union of two gametes—egg and sperm—to form a zygote.
fetus Term used to describe the human embryo after the eighth week of pregnancy and before birth.
F_1 generation The first generation of offspring resulting from a genetic cross.
F_2 generation The second generation of offspring resulting from a genetic cross.
fibril Composed of filaments; makes up skeletal muscle fibers.
filament In plants, a slender stalk that supports the anther; together they make up the stamen, which is the male reproductive organ; in muscle cells, the component of fibrils.
filtration A function of the kidneys that involves the movement of plasma from the glomerulus into the Bowman's capsule.
fission A process in which one cell divides in two.
fissure A deep indentation on the surface of the cerebrum.
flagellum A long, whiplike structure that projects from a cell and is used for locomotion.
flexion The bending of one part of the body on another; movement that reduces the angle between bones.
flexor A muscle that bends a body part; contraction of a flexor muscle results in flexion.
flower The reproductive structure in angiosperms, monocotyledons, and dicotyledons.
focal point The point at which the rays of light passing through a lens are focused.
follicle Structure that contains a developing ovum in the ovary until it matures and is expelled.
follicle-stimulating hormone (FSH) A pituitary hormone that stimulates the development of ovarian follicles and the maturing of an egg.
fontanels The membrane-covered spaces between cranial bones in an infant; they allow for some skull compression during delivery.
food calorie A unit of measure that indicates the energy value of food.
foramen A small opening in a bone through which fluids, nerves, or blood vessels pass.
foramen magnum Opening in the base of the skull through which the spinal cord passes.
fossa A depression in a bone for the purpose of articulation.
fovea An area in the middle of the retina that contains only cones; site of the eye's sharpest vision.

fracture A break in a bone.
fragmentation A type of sexual reproduction in which a small piece of an organism develops into a new organism.
fraternal twins Offspring that result from two eggs being fertilized by two different sperm.
frond A highly branched leaf of a palm or fern.
frontal plane An axis that divides the body into anterior and posterior halves.
fructose A monosaccharide present in many plants.
fruit A body, formed from the ovary of a plant, containing and protecting seeds.
functional group A group of atoms that are responsible for distinctive chemical properties of organic molecules.
fundus The first portion of the stomach into which the esophagus opens.
gallbladder An organ that stores bile produced by the liver.
gamete A general term applied to either egg or sperm cells.
gametogenesis The process of producing gametes.
gametophyte The sexual stage of a plant in alternation of generations; a plant that produces gametes.
ganglion A solid mass of nerve cell bodies lying outside the brain or spinal cord.
gastric juice Fluid produced by the gastric glands in the stomach; contains enzymes and hydrochloric acid that aid in the digestion of food.
gemma A sexual reproductive structure in some liverworts and mosses.
gene A portion of a DNA molecule that carries the code for a particular inherited trait.
generative nucleus One of two haploid nuclei in a pollen grain.
genetics The study of inherited characteristics.
genotype The genetic makeup of an individual.
genus A group of similar species.
geotropism The growth of a plant in response to gravity.
germination The sprouting or development of a seed.
gestation The period of development between fertilization and birth.
gibberellin A plant hormone that promotes cell elongation.
girdling The process of removing a ring of bark from a tree.
glenoid cavity Depression (socket) in the scapula with which the head of the humerus (ball) articulates.
glial cell Specialized cell that surrounds a neuron; nourishes and protects the neuron.
gliding joint Diarthroses type of joint; allows limited movement between bones; the surface of one bone moves over the surface of another.
glomerulus The network of capillaries imbedded in a Bowman's capsule of a nephron; functions to filter the blood.
glucagon A hormone that converts glycogen to glucose; secreted by the Islets of Langerhans in the pancreas.
glucose A monosaccharide (simple sugar) present in all plants and animals; the product of photosynthesis stored as starch.
glycerol An alcohol with three hydroxyl groups.
glycogen A polysaccharide (starch); used as a food-storage molecule in the liver.
glycolysis The process by which glucose is changed chemically, and energy is released in the form of ATP.
glycosuria The excretion of sugar in the urine.
goiter Hypertrophy of the thyroid gland.
Golgi apparatus An organelle in the cytoplasm of a cell; concerned with the production of secretions.
gonads A general term applied to male and female reproductive organs, ovaries and testes; produces gametes.
grafting A type of vegetative propagation in which a stem (scion) from one plant is attached to another, rooted, growing plant (stock).
growth ring A band of xylem cells formed in one growing season.
guanine A nitrogen base found in DNA and RNA; pairs with cytosine.
guard cells Pair of specialized epidermal cells on either side of the stoma of vascular plants; regulate the size of the stoma.
gymnosperm A type of seed plant in which unprotected seeds develop on the scales of cones.
gynoecium The entire female reproductive organ of a flower, consisting of one or more carpels.
hair follicle A down-growth of the epidermis that surrounds the root of a hair.
hammer Malleus; first of the three bones of the middle ear; conducts sound vibrations from the ear drum (tympanic membrane) to the oval window.
haploid number (***n***) A nucleus possessing unpaired chromosomes; half of the diploid number ($2n$).
Haversian canal Small tubes in bone that carry blood vessels and nerves.
heart A hollow organ with muscular walls responsible for pumping blood through the circulatory system.
heartwood The central portion (xylem) of a tree stem that no longer conducts water; becomes filled with waste products.
hemoglobin A protein found in blood that combines readily with oxygen and transports it from the lungs to the tissues.
hemophilia An inherited defect in which the blood does not clot normally.
hemopoiesis The process of forming blood.
hemorrhage The loss of blood from the circulatory system.
hepatic Concerned with the liver.
herbaceous Having no woody stem tissue; soft, green stem tissue that lives for one growing season.
herbivore A plant-eating animal.
heredity The passing of traits or characteristics from one generation to another.
heterotroph An organism that cannot produce its own food and must obtain nourishment from other living things.
heterozygous An individual possessing two different alleles for the same characteristic; hybrid.
hibernation A prolonged period of inactivity and reduced metabolism in certain animals; occurs during winter.
hilum A scar or mark on the hard, outer covering of a seed; it shows where the seed was fixed to the parent plant (ovary).
hind-brain Two divisions of the brain, the cerebellum and medulla, make up the hind-brain; it joins the spinal cord.
hinge joint A joint in which the round end of one bone turns on the flat surface of another bone, allowing the joint to bend in one plane only; permits flexion and extension.

homeostasis The maintenance of a constant internal environment.
homoiothermic A term used to describe warm-blooded animals in which the internal temperature remains constant and is not affected by the surroundings.
homologous chromosomes A pair of chromosomes.
homologous structures Parts of an organism that are similar in structure and origin.
homozygous An individual that possesses two identical alleles for a particular characteristic; purebred.
honey guide The pattern of stripes on a flower that directs bees to nectar.
hormone A chemical messenger produced by an endocrine gland that regulates body functions; also regulates plant functions.
host The organism that a parasite lives on or in.
hybrid The offspring that result from crossing two different purebreds for the same trait.
hybrid vigor A hybrid offspring with increased size and strength from that of the purebred parents.
hydrolysis A biological reaction in which a macromolecule is split into two smaller molecules by the addition of a water molecule; digestion.
hydrosphere The part of the earth covered by water.
hydrotropism A response to the stimulus of water.
hyper- A prefix meaning in excess; above.
hyperopia Farsightedness.
hypertonic A solution having a higher concentration of solutes than the solution to which it is compared.
hypo- A prefix meaning below or beneath.
hypocalcemia A deficiency of calcium in serum.
hypocotyl The part of a seedling below the cotyledons and above the root; it is the stem of the embryo seed plant.
hypogeal Having growth below the ground; hypogeal seedlings have their cotyledons below ground level during growth; most moncotyledons are hypogeal.
hypothalamus A part of the brain that controls many of the body's internal activities; functions to maintain homeostasis.
hypothesis A possible explanation for a series of observations (natural phenomenon).
hypotonic A solution having a lower concentration of solutes than the solution to which it is compared.
identical twins Offspring that result from a fertilized egg splitting into two separate embryos.
ileum The third and last part of the small intestine; connects the jejunum and cecum of the large intestine.
imbibition The process of absorbing water and swelling.
immovable joint Synarthroses type of joint that allows no movement between the articulating bones.
immune system Collection of cells and tissues that enables a plant or animal to resist the attack of antigens and pathogens.
imperfect flower A flower that is missing either stamens or a pistil.
inbreeding The crossing of two closely related individuals.
inbreeding depression Poor health and decreased fertility that may result after many generations of inbreeding.
incomplete dominance A type of inheritance in which neither of a pair of alleles for contrasting characteristics is completely dominant or recessive; blending.
incomplete flower A flower lacking sepals, petals, pistils, or stamens.
incus Anvil; the middle of the three small bones of the middle ear that conduct sound vibrations from the ear drum (tympanic membrane) to the oval window.
infloresence A group of flowers occurring together.
infrared radiation The invisible wavelengths of light lying just beyond the red end of the visible spectrum.
ingestion Intake of solid or liquid food materials into the body.
inheritance Those characteristics that are determined by the chromosomes passed from generation to generation.
inhibit An action that slows down an activity or prevents it from occurring.
insertion Point of attachment of a muscle to a bone that moves on contraction of the muscle.
inspiration To take air into the lungs; inhalation.
insulin A hormone secreted by the Islets of Langerhans of the pancreas; controls the utilization of blood sugar by the body.
integration The power to coordinate, speed up, or inhibit activities so that they work together effectively; cells in the brain and other organs of the nervous system are extremely active in integration.
integument The skin, hair, and fingernails.
internal respiration The exchange of oxygen and carbon dioxide between the blood and cells of the body.
interneuron A nerve cell that transmits information between a sensory neuron and motor neuron.
interoceptors Receptors located in the viscera that receive stimuli connected with digestion, excretion, circulation, and so forth.
interphase The period in the life of a cell when it is not dividing; period of cell growth.
intestine The part of the alimentary canal between the stomach and the rectum.
involuntary muscle Muscle tissue that is not under conscious control; smooth muscle and cardiac muscle.
iris The colored portion of the eye; regulates the amount of light entering the eye by adjusting the size of the pupil.
Islets of Langerhans Small groups of endocrine cells embedded in the pancreas; produce and secrete insulin and glucagon.
isotonic A solution having the same concentration of solutes as the solution to which it is being compared.
jaundice A yellow coloration of the skin due to hyper amounts of bilirubin in the blood.
jejunum Second portion of the small intestine that connects the duodenum and the ileum.
joint Articulation; the point of union between two bones.
keratin Fibrous protein found in skin, hair, and fingernails.
kidney A major organ of the urinary system in the body of most animals; removes waste from the blood and produces urine.
kinesthetic sensations Information received from nerve receptors located in muscles and tendons (proprioceptors) as to the position of the body.
kinetic energy The energy of motion.
kingdom The broadest division in taxomic classification.

Klinefelter's syndrome Abnormal sexual development resulting from the presence of an extra *X* chromosome in males.
kneecap Patella.
labor A period of uterine contractions necessary for human childbirth.
large intestine Portion of the alimentary canal that connects the small intestine and rectum.
larynx The voice box, or Adam's apple; located between the pharynx and the trachea.
lateral Pertaining to the side of the body; situated on either side of the median vertical line.
lateral bud A small side bud above each leaf scar; develops into new growth on a twig.
lateral meristem An area of rapid cell division that accomplishes secondary growth in plants.
layering A type of vegetative propagation in which roots are induced to form from a stem.
leaf A flat structure, usually green, growing on a stalk from the node of a stem or branch of a plant; functions in photosynthesis and transpiration.
leaflet One of many small blades attached to the same petiole.
leaf scar A mark left on a twig indicating the previous attachment of a leaf.
lens An optical device in the eye that focuses light; made of transparent, semicrystalline protein.
lenticel A group of loosely packed cells formed in woody stems; functions in gas exchange.
leucocyte White blood cell.
leucoplast A colorless plastid that stores starch.
ligament A tough band of connective tissue that connects bones to each other.
light reactions First stage of photosynthesis; converts the sun's light energy into chemical energy.
lipase A digestive enzyme that functions to break down fats.
lipid Fatty substances including fats, oils, and waxes.
lithosphere The solid part of the earth; upper crust.
liver The largest gland in the body; functions in the secretion of bile, storage of glycogen, and removal of toxic materials from the blood.
long-day plant A plant that flowers only when the photoperiod exceeds a certain length of time.
longitudinal plane A plane running lengthwise through the body.
lordosis An abnormal curvature of the lumbar vertebrae in an anterior direction resulting in a swayback condition.
lumbar Referring to the region of the back extending from the thorax to the sacrum.
lumen The hollow space inside the alimentary canal.
lung Major organ of the respiratory system; functions in the exchange of oxygen and carbon dioxide.
luteinizing hormone (LH) A hormone secreted by the pituitary gland; causes a follicle to rupture, releasing an egg, and then to become the corpus luteum.
lymph A colorless liquid consisting of tissue fluid and white blood cells that circulates in the lymphatic vessels.
lymphatic system A network of very small tubes, lymph vessels and lymph capillaries that transport tissue fluid.
lymph node An area of swelling in a lymph vessel that produces lymphocytes, removes bacteria from lymph, and filters out foreign bodies.
lymphocyte A white blood cell that produces antibodies.
lyse- A prefix that means to digest or breakdown.
lysosome A membrane-bound cell organelle that contains digestive enzymes.
macromolecule A large organic molecule made up of a chain of repeating smaller molecules.
macronucleus An organelle in paramecia and other ciliated protists that manufactures RNA.
malleus Hammer; the first of the three small bones of the middle ear; attached to the tympanic membrane.
malnutrition A condition in which the body does not get enough of the required nutrients.
Malpighian layer The bottommost layer of the skin's epidermis; consists of actively dividing cells.
maltase A digestive enzyme that breaks down maltose.
maltose Disaccharide; a simple sugar made of two molecules of glucose.
mammary glands Milk-producing glands in females.
mandible Lower jaw; aids in the process of chewing.
marrow The soft tissue within the medullary cavities of bones that produces blood cells and stores fat.
mastication The act of chewing.
matrix The nonliving, solid material that is produced and secreted by osteocytes.
maxilla Upper jaw; aids in the process of chewing.
mechanical digestion The physical breakdown of food by chewing and the muscular churning of the stomach.
mechanoreceptor A specialized nerve ending that detects touch, pressure, or vibrations.
medial Toward the midline of the body, as opposed to lateral.
medulla A general term applied to the internal region of a solid organ.
medulla oblongata A part of the brain stem; controls involuntary body processes.
megasporangium A female reproductive structure of the conifers that produces haploid megaspores.
megaspore A haploid cell giving rise to the female gametophyte in gymnosperms and angiosperms.
meiosis The type of cell division that occurs during maturation of gametes; results in the formation of sperm and ova containing the haploid number of chromosomes.
meiospore A haploid spore resulting from meiosis.
melanin A dark pigment found in the epidermal cells of the skin.
melatonin The only hormone secreted by the pineal gland. It decreases skin pigmentation.
meninges Three layers of protective tissue that form the covering of the brain and spinal cord.
menstruation The breakdown and expulsion of the lining of the unfertilized egg.
menstrual cycle Female reproductive cycle.
meristem Area of rapid cell division in plants.
mesophyll Specialized tissue that lies between the epidermal layers of the leaf.

messenger RNA (mRNA) One of the three types of RNA found in cells; carries information required for protein synthesis from the nucleus to the ribosomes.
metabolic rate The rate at which an organism consumes energy.
metabolism The sum of all chemical reactions of an organism; anabolism and catabolism.
metaphase The third stage of mitosis, in which the chromosome halves align at the cell's equator.
microfilament A solid threadlike structure made of protein; able to contract and thereby create movement.
micronucleus An organelle in a paramecium containing the chromosomes.
microorganism Single-celled, microscopic living thing.
micropyle A very small hole in the testa of a seed; water enters a seed through its micropyle, and then the seed begins to germinate.
microsporangium A male reproductive structure of the conifers that produces haploid microspores.
microspores A haploid cell giving rise to the male gametophyte, or pollen grain, in gymnosperms and angiosperms.
microtubule A long, thin, hollow tube found in cilia and flagella; creates movement by contracting.
microvilli The microscopic folds of the surface of villi of the intestines.
midsaggital plane A plane that passes through the skull and spinal cord, thus dividing the body into right and left halves.
mineral Inorganic essential nutrient; elements classified as metals.
mitochondria An organelle found within the cytoplasm of a cell; functions in energy production; known as the powerhouse of the cell.
mitosis The process of cell division in which the chromosomes are duplicated and separated prior to cytoplasmic division.
mixture Particles of two or more substances intermingled with each other but keeping their own properties.
molecule The smallest particle of a substance that has all the properties of that substance.
monocotyledon A flowering plant with one cotyledon in its seed; has parallel venation and irregular arrangement of vascular bundles in stem.
monohybrid cross A genetic cross involving only one pair of alleles for a particular characteristic.
monosaccharide The building block unit of all carbohydrates; a simple sugar.
monosomy Lacking a chromosome.
motor neuron A nerve cell that transmits impulses to an effector (muscle or gland) from the central nervous system.
mucosa Layer of specialized epithelial cells lining the alimentary canal, reproductive tract, and urinary tract; secretes mucus.
mucus A thick, lubricating fluid secreted by the mucous membrane; lubricates internal body surfaces.
multicellular Composed of more than one cell.
multinucleated Cell containing more than one nucleus.
multiple alleles A single trait or characteristic that is determined by three or more alleles.
multiple fruit A type of fruit that develops from a single ovary of each flower in a cluster.
muscle fiber One of the fibers that composes skeletal muscle tissue; consists of a single multinucleated cell.
muscle tissue A type of tissue that has the ability to contract.
muscle tone The constant state of partial contraction of a muscle.
mutagen Any substance that causes a mutation.
mutation An error in the replication of DNA that results in a change in the genetic material and causes the sudden appearance of a new characteristic.
myelin sheath A protective covering of fatty material surrounding a nerve fiber.
myofibril A subdivision of a striated muscle fiber.
myopia Nearsightedness; a defect of vision resulting from an elongation of the eyeball and the failure of the lens to focus the image on the retina.
myosin One of the two proteins involved in muscle contraction; component of myofibrils.
nares Nostrils.
nastic movement A plant response that results from a change in turgor pressures.
natural selection The process by which organisms less adapted to their environment tend to perish, while those that are well adapted tend to survive.
nearsightedness *See* myopia.
nectar A sugary solution produced by flowers to attract pollinators.
needle A modified leaf of a conifer.
nephron The working unit of the kidney; a system of tubes that filter the blood.
nerve A cablelike structure composed of axons, dendrites, or both.
nerve impulse A wavelike impulse passing over a nerve cell process that is characterized by both electrical and chemical effects.
nerve tissue A specialized tissue that is adapted to respond to a stimulus.
neuron The unit of structure of the nervous system; a nerve cell.
neurotransmitter A chemical that transmits impulses across a synapse.
neutron A neutral particle found in the nucleus of an atom.
night blindness A deficiency disease (caused by a lack of vitamin A) in which a person sees well by day but badly by night or in poor light.
node The part of a stem from which leaves grow; adventitious roots may also grow from nodes.
nodule A swelling on the roots of some plants.
nonvascular plant A plant that lacks xylem and phloem.
nuclear membrane A semipermeable membrane surrounding the nucleus of a cell and separating it from the cytoplasm; present only in eukaryotic cells.
nucleic acid A complex type of organic acid found in chromatin.

nucleolus An organelle within the nucleus of a cell; associated with the formation of proteins.
nucleotide The basic structural unit of nucleic acids; the combination of a purine or pyrimidine base with a sugar and a phosphoric acid.
nucleus A membrane-bound organelle containing the chromosomes and acting as the controlling center of a cell's activities.
nutrient A substance that can be used in the nutrition of an organism; supplies energy and materials to the body.
nutrition The process of ingesting food, digesting food, and, through absorption and assimilation, using it to provide energy for living and materials for growth.
obesity A condition characterized by excessive body fat.
objective The lens in a microscope nearest to the object being viewed.
ocular A lens in a compound microscope that magnifies the image produced by the objective lens.
oil A type of relatively simple lipid that is liquid at room temperature.
olfaction The process of smelling.
olfactory nerve The first cranial nerve; transmits information concerning the sense of smell.
omnivore An animal that eats both plant and animal material.
oogamy Type of sexual reproduction in which the male gamete is flagellated and motile (sperm), and the female gamete is larger and nonmotile (egg).
oogenesis The process of forming the female gamete (egg).
oogonium An egg-producing structure found in some plants.
optic nerve The second cranial nerve; passes from the retina of the eye to the occipital lobe of the brain; responsible for the sense of vision.
oral groove A channel located on the side of a paramecium into which cilia sweep food to be ingested.
orbital cavity The eye cavity.
order A group of related taxonomic families.
organ A group of tissues that function together to perform a definite body function.
organelle A structure lying within the protoplasm of a cell and performing a specific function.
orifice An opening; an entrance to a cavity or tube.
origin The point of attachment of a muscle to a bone that does not move when the muscle contracts.
osmosis Diffusion through a semipermeable membrane.
osmotic pressure The pressure that is established on one side of a semipermeable membrane as a result of the passage of a material through it.
ossicles Small bones in the ear: malleus, incus, and stapes.
ossification The process of bone formation.
osteoblast A cell concerned with the formation of bone.
osteoclast A cell that destroys bone to form channels through the bone for the passage of blood vessels and nerves.
osteocyte A bone-forming cell.
oval window A membrane that separates the middle ear and inner ear; it is connected to the stapes (stirrup).
ovary The female reproductive organ; produces the female gamete (egg).
oviduct Fallopian tube; a tube through which an egg passes from an ovary to the uterus.
ovulation The release of an egg from the ovary.
ovule In seed plants, a structure that contains a female gamete that, after fertilization by a male gamete, forms an embryo.
ovum The female gamete; egg.
palisade layer A layer of cells located beneath the upper epidermis of a leaf; contains chlorplasts.
palmate venation A branching or net type of venation in which several veins radiate from a single point.
palmately compound A type of leaf in whch the leaflets join together before attaching to the petiole.
pancreas A gland that secretes digestive enzymes and produces insulin and glucagon.
pancreatic juice Digestive fluid produced by the pancreas and secreted into the duodenum.
paramecium A group of protozoa of a shoelike shape whose surface is covered with cilia, which aid the organism in movement.
parasite An organism living on or in another organism and obtaining all its food from its host.
parasympathetic nervous system A division of the autonomic nervous system; maintains homeostasis.
parathyroid gland Four small endocrine glands located just behind the thyroid gland; secretes hormones which regulate the levels of calcium and phosphate ions in the blood.
parenchyma A plant tissue specialized for storage of food.
passive transport The movement of materials across a membrane by diffusion and osmosis; requires no cellular energy.
pathogen A bacteria, virus, fungus, or protozoan that causes disease.
pectin A group of carbohydrates found in the walls of some plant cells, such as apples, grapes and plums.
pectoral girdle Shoulder girdle; the scapula and clavicle together form the attachment for the upper limbs.
pelvic girdle The hips; the ilium, ischium, pelvis, and sacrum together form the attachment for the lower limbs.
penis Part of the male reproductive system through which seminal fluid and urine are passed out of the body.
pepsin A digestive enzyme secreted by glands in the wall of the stomach; helps to breakdown proteins.
peptide bond A bond that forms between two amino acid molecules.
perennial A plant that continues to live from year to year.
perfect flower A flower having both male and female reproductive parts (stamens and pistil).
pericardium A double-walled sac of endothelial tissue surrounding the heart.
pericycle A layer of cells around the vascular tissue in the primary root from which the secondary roots grow.
periosteum The tissue that covers the outside of a bone.
peripheral nervous system The part of the nervous system lying outside of the central nervous system.
peristalsis A rhythmic, wavelike series of contractions moving over the walls of a tube and resulting in the passage of materials through the tube.
peritoneum A layer of tissue that lines the abdominal cavity and surrounds many of the abdominal organs.

permeable Allowing most substances to pass through freely.
petal A modified leaf that surrounds the reproductive organs of a flower; is often brightly colored and many produce special odors and nectar.
petiole The stalk of a leaf; it grows from a node.
P_1 generation The first parental generation in a genetic cross.
pH A measure of how acidic or basic a solution is; also denotes hydrogen ion concentration.
phagocyte A type of white blood cell that devours bacteria and other microorganisms that invade the body.
phagocytosis The surrounding of a particle by the cell membrane in order to engulf and ingest the particle into the cytoplasm.
pharynx The cavity in back of the mouth that is a common passageway to both the respiratory and digestive systems.
phenotype The visible characteristics of an individual as determined by the genes.
phloem The part of the vascular system containing tubelike conducting vessels in which dissolved food material passes from the leaves to all parts of the plant.
phonation The ability to produce sounds by means of vocal cords.
photoperiodism The effect on flowering plants that varying periods of light and dark have.
photoreceptor Rods and cones; a receptor that detects light.
photosynthesis Process by which chlorophyll-containing organisms in the presence of sunlight convert carbon dioxide and water into carbohydrates.
phototropism The response of a plant to grow in the direction of a light source.
phylum In animal classification, a large group of related taxonomic classes.
physiology The study of the functions or activities of the various parts of the body.
pigment A substance that imparts color by reflecting light.
pinnately compound A type of leaf in which the leaflets are separately attached to the petiole.
pinnate venation Parallel venation; a type of venation in which smaller veins branch off one main vein.
pinocytosis A process by which extracellular fluid is taken into a cell.
pistil The female reproductive organ in a flower.
pith The central spongelike part of the stem of a plant; it helps in the storage of plant food.
pituitary gland A small endocrine gland at the base of the brain; composed of two lobes; secretes hormones that regulate all other endocrine glands.
placenta The organ on the wall of the uterus to which the embryo is attached by the umbilical cord and through which it receives its nourishment.
plasma The extracellular fluid in blood vessels.
plasmolysis The loss of turgor pressure in a plant cell; the shrinking of the cytoplasm away from the cell wall.
plastid An organelle in the cytoplasm of plant cells that functions to produce starch or protein, or to store these materials.
pleura A two-layered sac of endothelial tissue surrounding each lung.
polar body A small haploid cell that is formed during the maturation of the ovum; cannot be fertilized by a sperm.
pollen A small grain that contains two male gametes in a seed-producing plant; produced in the stamens of flowers.
pollination The process by which pollen grains are carried from anther to stigma; accomplished either by wind or by insects.
polypeptide A series of three or more amino acids bonded together; protein.
polyploid Having one or more extra sets of chromosomes.
polysaccharide Complex carbohydrate; a series of three or more monosaccharides bonded together.
polyunsaturated A term indicating the presence of one or more double bonds in an organic molecule; having less than the maximum number of hydrogen atoms.
positive tropism The growth response of a plant toward a stimulus.
posterior Referring to the back of the human body or toward the tail end of an animal.
potential energy The energy of position; stored energy.
primary growth The cellular division of meristematic tissue that results in the lengthening of the stem and roots.
primary tissue Plant tissue that arises from the meristem.
prime mover Any muscle that is responsible for a definite motion.
progesterone One of the two major female sex hormones.
prokaryote A cell lacking a membrane-bound nucleus.
prone Lying face downward; the opposite of supine.
prophase The second stage of mitosis, in which individual chromosomes become visible.
proprioceptor Receptors located in muscles, tendons, joints, and the semicircular canals of the inner ear; detect body movements and position.
prop root A root growing from a node on a stem down into the earth; it supports the stem.
protease A group of digestive enzymes that break down protein.
protein A polypeptide chain of amino acids; one of the major nutrients required by the human body.
proton A positively charged particle found in the nucleus of an atom.
proximal The part of an appendage that is toward the main part of the body.
pseudopod False foot; temporary extension of the body of a protozoan for the purpose of locomotion or obtaining food.
ptyalin A starch digesting enzyme found in saliva.
puberty A stage in human development when the sex hormones cause the production of gametes to begin and the appearance of secondary sex characteristics.
pulmonary circulation The path of the blood through the lungs.
pulse A wave of blood flowing along the arteries.
Punnet square A method of determining the results of a genetic cross.
pupil The opening in the iris through which light enters the eye.
pyloric sphincter A strong circular muscle that controls the passage of chyme between the pylorus and the duodenum.
pylorus The last section of the stomach; lies between the body of the stomach and the duodenum of the small intestine.

radicle The part of an embryo that grows to form the main root of a new plant.
ray Conducting tissue that transports materials laterally in woody stems.
reabsorption The second stage in the formation of urine in which materials pass from the tubule of a nephron back into the circulatory system.
receptacle The enlarged tip of a stem that forms the base of a flower.
receptor The ending of a nerve fiber that is specialized to receive one particular type of stimulus.
recessive Refers to an allele that has no genetic effect in a heterozygous person; a trait that is masked by a dominant trait.
rectum The end of the large intestine; stores feces before passing it out of the body.
red blood cell A nonnucleated, biconcave disk containing hemoglobin.
red bone marrow The principal site of the formation of red blood cells.
referred pain A feeling of pain in one part of the body resulting from the stimulation of nerves in another region.
reflex An automatic, immediate response to a stimulus.
reflex arc The path followed by an impulse; a response to a particular stimulus with no delay; the arc consists of a receptor, sensory neuron, association neuron, motor neuron, and effector.
renal circulation The pathway of blood through the kidneys.
renal cortex The outer layer of the kidney.
renal medulla The inner layer of the kidney.
renal pelvis A funnel-shaped structure in the kidney in which urine collects.
replication The process by which a molecule of DNA makes an exact duplicate of itself.
reproduction The ability of a living organism to produce similar living things.
resin A group of substances found in trees, either as a sticky solid or as a solution.
resolving power The ability of a microscope to separate and distinguish between two magnified objects.
respiration The exchange of oxygen and carbon dioxide between cells and their surroundings; also, the process of converting food energy into a form of energy that cells can utilize.
respiratory center An area of the brain that responds to the concentration of carbon dioxide in the blood.
response An organism's reaction to a stimulus.
retina The innermost layer of the eye composed of nerve endings that are sensitive to light.
rhizome A horizontal underground stem.
rib cage Protective structure formed by the ribs, sternum, and thoracic vertebrae; houses vital internal organs.
ribonucleic acid (RNA) A nucleic acid found principally in the cytoplasm of a cell; functions in protein synthesis.
ribosome Tiny organelles located on the endoplasmic reticulum; site of protein synthesis.
RNA *See* ribonucleic acid.
rod A light receptor in the retina; sensitive to low intensities of light but not color; responsible for black and white vision.
root The part of a plant that grows downward into the earth in order to anchor the plant and absorb water and nutrients.
root cap Group of cells that protect the tip of the root.
root hair Small, hairlike growth from a root; has a very thin cell wall; functions in the absorption of water and nutrients.
root pressure The pressure that forces water up from a root into the stem of a plant.
runner A type of stem that grows along the ground and sends out roots at the end of the stem.
sacrum The region of the vertebral column, composed of fused vertebrae, that forms part of the pelvic girdle.
saliva A liquid secretion produced by glands in the mouth; contains mucus and the digestive enzyme ptyalin.
sap The liquid in a plant; it can be seen when a stem or a root is cut.
sapwood The younger portion of a woody stem that contains xylem vessels.
sarcolemma The delicate sheath surrounding a muscle fiber.
sarcomere The contractile unit of a muscle cell.
saturated Refers to an organic molecule with no double bonds; has the maximum number of hydrogen atoms bonded to each carbon atom.
Schwann cell A cell that forms a protective covering around the axon of a neuron.
scion A stem cut from one plant that is to be grafted onto the rooted stock of another plant.
sclera The tough, outer layer of the eye; the white of the eye.
sclerenchyma A plant tissue specialized for strength, support, and protection.
scrotum External sac that contains the testes.
sebaceous gland Gland that secretes oil; located in the epidermis.
secondary growth Growth of cambium tissue that causes the stems, branches, and roots of a plant to increase in width.
secondary sex characteristic The physical changes that are a result of the increased level of hormones at puberty.
secretion The act of releasing a product of a gland from the cells that manufactured the product; secreted material specialized to perform certain functions.
seed The plant embryo produced by flowering plants from which a new plant grows; contains a food supply and is wrapped in a protective seed coat.
seed coat The hard, outer covering of a seed that develops from the walls of the ovule.
seed cone A female reproductive structure of the conifers containing spore-producing megasporangia.
seedling A newly formed organism growing from a seed.
self-pollination Pollen is carried from anther to stigma of the same flower, or to a stigma of another flower on the same plant.
semen Seminal fluid; sperm together with secretions from the accesory glands of the male reproductive system.
semicircular canals Three bony canals in the inner ear that contain fluid; concerned with maintenance of balance.

seminal vesicles Small elongated glands in most male mammals; opens to the vas deferens.
seminiferous tubule A coiled tube in the testes; produces sperm.
semipermeable membrane A type of membrane that will permit the passage of certain materials but prevent the movement of others through it.
sensory neuron A nerve cell that transmits impulses from a receptor to the central nervous system.
sepals A green, leaflike structure that surrounds and protects a developing flower.
septum A wall of tissue dividing a cavity.
sex chromosomes A pair of chromosomes, designated *X* and *Y*, that determine the sex of an offspring.
sex-influenced trait A characteristic or trait that is dominant in one sex and recessive in the other.
sex-linked A gene carried on the *X* chromosome but not on the *Y* chromosome.
sexual reproduction Reproduction in which two organisms take part; two haploid gametes join to form a diploid zygote.
shoot The stem of a young plant.
short-day plant A plant that flowers only when the period of daylight is shorter than a critical period.
simple fruit A type of fruit that develops from a single ovary in a single flower.
simple leaf A leaf having only one blade attached to the petiole.
simple sugar A monosaccharide, the basic structural unit of carbohydrates.
sinoatrial node A bundle of nerve fibers in the wall of the right atrium that serves as the pacemaker for the heart beat.
sinus A normal cavity within a bone or other organ.
skeletal muscle Striated, voluntary muscle that allows for movement of the bones of the skeleton.
skeleton The bony and cartilaginous framework of the body.
skull A group of bones fused together that protect the brain.
small intestine Portion of the alimentary canal between the stomach and large intestine.
smooth muscle Elongated, thin, spindle-shaped muscle fibers that are involuntary in their action; found in many internal organs.
solute Any solid or gas that is dissolved in a liquid (solvent); the component of a solution present in lesser concentration.
solution The result of dissolving a solute in a solvent.
solvent The liquid in which a solute is dissolved; the component of a solution present in greater concentration.
somatic nervous system Part of the peripheral nervous system that connects the central nervous system to all voluntary functions; controls skeletal muscles.
species A taxonomic group of closely related organisms capable of breeding and producing fertile offspring.
sperm A male reproductive cell produced in the testes.
spermatogenesis The process of producing sperm.
spermatophyte A seed-producing vascular plant.
sphincter A layer of smooth muscle arranged in a circular manner around a tube or opening; controls the movement of material from one area to another.
spinal cord A cylindrical mass of nervous tissue that runs through and is protected by the vertebrae.
spindle fibers Microtubules that extend throughout the cytoplasm of a cell and connect the centrioles during mitosis.
spine A modified leaf that is very sharp and functions in protection of the plant.
spirilli Spiral-shaped bacteria.
spleen An organ of the lymph system that manufactures lymphocytes.
spongy bone Bone tissue found in the epiphyses of long bones and the core of short bones.
spongy layer A loosely packed layer of cells in a leaf beneath the palisade layer.
spontaneous generation The idea that life can arise from nonliving matter; abiogenesis.
sporangium Reproductive organ found on the underside of fern leaves that produces spores.
spore A reproductive cell that originates from asexual division and usually has a hard, protective covering.
sporophyte Refers to a plant in alternation of generation that is in the diploid stage and produces spores.
sprain The result of tearing or straining of tendons and ligaments at a joint.
stamen The male reproductive structure in a flower.
stapes Stirrup; the innermost of the three small bones of the inner ear, articulating with the incus at one end and the membrane of the oval window at the other.
starch Carbohydrate; a polysaccaride made up of glucose units; used as a food storage molecule by plants.
stem The part of a plant that bears leaves and buds; conducts water and minerals up from the roots and food down from the leaves.
steroids A specific hormone category characterized by a complex basic ring structure.
stigma The tip of the pistil that is usually sticky; pollen grains stick to it and germinate.
stimulus Any change in the environment that brings about a response.
stirrup *See* stapes.
stipe The stalk of a nonvascular plant.
stock The rooted, growing plant to which a scion is grafted.
stolon A stem that grows along the ground and roots at each node, developing a new plant.
stoma A very small opening in the surface of a leaf through which water, oxygen, and carbon dioxide pass.
stomach The part of the alimentary canal that connects the esophagus and small intestine; its muscular wall churns the food into a liquid mass called chyme.
striated muscle Voluntary muscle tissue.
style The stalk of a pistil bearing the stigma.
substrate The particular substance on which an enzyme acts.
sulcus A shallow groove, often only associated with the brain.
sweat gland A gland located in the dermis; removes water and salts from the blood and releases them onto the surface of the skin in the process of controlling body temperature.
sympathetic nervous system A division of the autonomic nervous system that prepares the body for an emergency.

synapse The gap between the axon of one neuron and dendrite of another.
synergist Any muscle that aids the prime mover by keeping one joint steady while the prime mover applies force to a neighboring joint.
synovial fluid A clear, lubricating fluid present in joint cavities.
systemic circulation The circulation of blood from the heart to all parts of the body except the lungs and tissues of the heart.
systole The period of active contraction of the atria and ventricles.
tactile Concerned with touch.
taproot The main root; often functions in food storage.
taste bud A chemoreceptor located on the tongue.
taxis Movement directly toward or away from a stimulus.
taxonomy The science of classifying organisms.
telophase The final stage in mitosis, in which the cytoplasm of the parent cell divides into two halves and a new cell membrane appears.
tendon A tough band of connective tissue that attaches a muscle to a bone.
tendril A long, slender modified leaf that wraps around nearby structures for support as the plant climbs.
terminal bud An area of undeveloped tissue at the tip of a woody stem.
testes The male reproductive organs; produce sperm and male sex hormones.
testosterone A male sex hormone produced in the testes; responsible for male secondary sex characteristics.
thalamus A small region of the brain located at the bottom of the cerebrum; relays impulses between the cerebrum and brainstem.
thallus The body of a nonvascular plant.
thermoreceptor A nerve ending that detects warmth and cold.
thoracic Referring to the region of the thorax.
thorax The upper part of the trunk.
threshold The minimum stimulus required to bring about a response.
thymine A nitrogen base found only in DNA and pairs with adenine.
thyroid An endocrine gland located on the trachea that secretes the hormone thyroxine.
thyroxine Hormone secreted by the thyroid gland; controls the rate of metabolism.
tissue A group of similar cells performing a specific function.
toxin Any poison produced by a plant or animal (particularly by bacteria); interferes with the normal functioning of body cells.
trachea The wind pipe; a tube leading from the larynx and branching to form the right and left bronchi.
tracheid An elongated hylem-conducting element. Symbolic to tracheids are their oblique end walls.
transpiration The loss of water through the stomates of leaves in a plant.
transverse plane A cross section.
tropism The tendency of a plant to have curved growth under the effect of its environment; growth toward or away from a stimulus.
trunk The main part of the body, excluding the appendages and head.
turgor The state of a plant cell when fully expanded because of water absorbed by its cytoplasm and vacuole.
turgor pressure Water pressure that acts to keep the cell wall rigid.
twitch The brief contraction followed by relaxation of a muscle fiber.
tympanic membrane Eardrum; thin membrane that separates the external and middle ear; vibrates at the same frequency as the sound waves that strike it.
ultraviolet radiation That region of the light spectrum lying just beyond the shortest visible violet rays.
umbilical cord A tube that connects the embryo to the placenta.
unicellular Single-celled organism; made of only one cell.
unsaturated Refers to an organic molecule containing double bonds; normally refers to fats.
uracil A nitrogen base found only in RNA; pairs with adenine in DNA.
urea A by-product of the digestion of protein; excreted in the urine.
ureter A tube through which urine passes from each kidney to the bladder.
urethra Tube through which urine passes from the bladder to the exterior.
urinary bladder A collapsible structure capable of great distension that stores urine until it is expelled from the body.
urine A liquid produced by the kidneys; consists of water and waste materials that have been filtered from the blood.
uterus A muscular structure in the female reproductive system that receives the fertilized egg and in which the fetus develops.
vacuole A space in cytoplasm enclosed by a membrane and filled with liquid.
vagina A tube leading from the cervix to the exterior in the female reproductive system; receives sperm for fertilization and serves as the birth canal.
variable Any condition that can be changed or varied in an experiment.
variation Inherited differences that exist between the members of a species.
vascular bundle A bundle of xylem and phloem vessels passing from the end of a root to the end of a stem.
vascular cambium A thin layer of cells responsible for secondary growth of xylem and phloem in plants.
vascular plant Plant containing vessels that conduct fluid.
vascular tissue Xylem and phloem; specialized tissue that transports water, minerals, and food.

vas deferens A tube leading from the epididymis through which sperm travels.
vaso- A prefix referring to a blood vessel.
vegetative propagation Asexual reproduction in plants in which a new plant is produced from a living portion of another plant.
vein A blood vessel carrying blood toward the heart.
vena cava Either of two main veins that carry deoxygenated blood from the body to the right atrium.
venation The arrangement of small tubes (veins) seen in a leaf; there are two kinds of venation, parallel and branching.
ventral Referring to the side of an animal opposite to that along which the backbone passes; the anterior side in humans.
ventricle A chamber of the heart from which blood flows; pumps the blood to various parts of the body.
venule A small branch of a vein.
vertebral column The backbone; the arrangement in a line of vertebrae joined by ligaments and separated by elastic cartilage.
vertebrate An animal with a backbone.
villi Small, finger-like projections of the mucosa of the small intestine; function to increase surface area.
viscosity The apparent force that prevents fluids from flowing easily.
vitamin An essential nutrient; an organic substance that functions as an accessory food substance.
vitreous humor A jelly-like material that fills the eyeball from the crystalline lens to the retina.
vocal cords Specialized structures within the larynx that vibrate to produce sound when air rushes over them.
voluntary muscle Striated muscle; skeletal muscle under conscious control.
white blood cells A relatively large, nucleated blood cell that functions to protect an organism against infection; leukocyte.
wilting The loss of turgor of a plant; results in the inability to stand upright.
***X* chromosome** A sex chromosome in humans; a pair of *X* chromosomes produces a female.
xiphoid process Small, cartilaginous portion of the sternum.
xylem The part of the vascular system of a plant; conducts water and minerals from the roots to all parts of the plant.
***Y* chromosome** A sex chromosome in humans; an *X* and *Y* chromosome produces a male.
zone of elongation Region located behind the meristem where elongation of plant cells takes place; growth in length.
zone of maturation Region located behind the zone of elongation where plant cells begin to differentiate.
zygomatic arch Bridge between the zygomatic and temporal; cheekbone.
zygote The diploid cell that results from the joining of two haploid gametes during sexual reproduction; a fertilized egg.

Photo Credits

Chapter 1

Opener: Carl W. Bollwinkel

Chapter 2

Opener: Kenneth L. Weik

Chapter 3

Opener: Harold R. Hungerford; **3.1,** 3.2, **3.3:** Cambridge Instruments, Inc., Optical Systems Division; **3.6:** VU/L. Maziarski; **3.7a:** VU/K.G. Murti; **3.7b** :VU/David M. Phillips; **3.7c:** John D. Cunningham; **3.11:** VU/G. Musil.

Chapter 4

Opener: Kenneth L. Weik; **4.1, 4.2, 4.3:** John D. Cunningham; **4.4:** VU/Martha Powell; **4.9a:** *Lysosome,* VU/K.G. Murti—*Centrioles,* VU/David M. Phillips—*Golgi Apparatus,* VU/Martha Powell—*Plasma Membrane,* VU/Harold H. Edwards—*Nucleus,* VU/K.G. Murti—*Mitochondrion,* VU/K.G. Murti; **4.9b:** *Chloroplast,* L.K. Shumway—*Rough ER,* VU/David M. Phillips—*Mitochondrion,* VU/David M. Phillips; **4.10a:** VU/David M. Phillips; **4.11:** VU/Martha Powell; **4.12:** Science Dept., Price Laboratory School, University of Northern Iowa; **4.13a, 4.16, 4.17, 4.20:** VU/David M. Phillips; **4.22:** John D. Cunningham; **4.23b:** VU/D.W. Fawcett; **4.23c:** John D. Cunningham; **4.24a:** Richard Sjoland; **4.25a:** VU/Martha J. Powell; **4.25b:** John D. Cunningham; **4.29a:** VU/K.G. Murti; **4.29b:** VU/R.J. Post; **4.35:** VU/Sidney Fox.

Chapter 5

Opener: VU/Stanley Flegler; **5.1:** Photosynthesis/Biology Media © 1976; **5.13:** John D. Cunningham; **5.15a:** VU/David M. Phillips; **5.15b:** VU/Stanley Flegler.

Chapter 6

Opener: VU/K. G. Murti; **6.8a,b:** VU/David M. Phillips; **6.8c:** John D. Cunningham.

Chapter 7

Opener: Carl W. Bollwinkel; **7.6, 7.7, 7.10, 7.11:** Energy and Life/© Biology Media 1978; **7.14:** Photosynthesis/© Biology Media; **7.18:** Science Dept., Price Laboratory School, University of Northern Iowa; **7.20:** Photosynthesis/© Biology Media; **7.25b:** John D. Cunningham; **7.26:** Richard Sjoland; **7.27:** L.K. Shumway.

Chapter 8

Opener: VU/David M. Phillips; **8.4:** VU/K.G. Murti.

Chapter 9

Opener: VU/K.G. Murti; **9.17a, b:** VU/David M. Phillips.

Chapter 10

Opener: Harold R. Hungerford.

Chapter 11

Opener: John D. Cunningham.

Chapter 12

Opener: Carl W. Bollwinkel.

Chapter 13

Opener: Carl W. Bollwinkel; **13.3:** From *Botany: Principles and Applications,* Roy H. Saigo and Barbara W. Saigo. Copyright © 1983 by Prentice-Hall, Inc. Reprinted by permission of the authors; **13.4, 13.5:** Bill Witt; **13.7:** Carl W. Bollwinkel.

Chapter 14

Opener: Harold R. Hungerford; **14.1:** From *Botany: Principles and Applications,* Roy H. Saigo and Barbara W. Saigo. Copyright © 1983 by Prentice-Hall, Inc. Reprinted by permission of the authors; **14.2:** John D. Cunningham; **14.4:** Carolina Biological Supply Company.

Chapter 15

Opener: Carl W. Bollwinkel.

Chapter 16

Opener: Carl W. Bollwinkel; **16.3:** VU/R.A. Gregory; **16.4:** VU/Randy Moore; **16.8, 16.11, 16.12:** John D. Cunningham; **16.13, 16.14:** Bill Witt.

Chapter 17

Opener: Carl W. Bollwinkel; **17.2:** Carl W. Bollwinkel; **17.4, 17.5:** Bill Witt; **17.8, 17.10:** John D. Cunningham; **17.13:** Carl Bollwinkel; **17.14:** John D. Cunningham.

Chapter 18

Opener: Harold R. Hungerford; **18.1:** VU/Gerald VanDyke; **18.4:** Carolina Biological Supply Company; **18.6:** Ray F. Evert; **18.7, 18.8:** Carolina Biological Supply Company; **18.14:** Jon M. Holy; **18.15:** From *Botany: Principles and Applications,* Roy H. Saigo and Barbara W. Saigo. Copyright © 1983 by Prentice-Hall, Inc. Reprinted by permission of the authors; **18.16:** Carolina Biological Supply Company.

Chapter 19

Opener: Harold R. Hungerford; **19.3:** Eberhard Fritz; **19.4:** C.E.I. Botha; **19.5:** From *Botany: Principles and Applications,* Roy H. Saigo and Barbara W. Saigo. Copyright © 1983 by Prentice-Hall, Inc. Reprinted by permission of the authors.

Chapter 20

Opener: Harold R. Hungerford; **20.1:** From *Botany: Principles and Applications,* Roy H. Saigo and Barbara W. Saigo. Copyright © 1983 by Prentice-Hall, Inc. Reprinted by permission of the authors; **20.3:** D. R. Farrar; **20.6:** VU/H. Oscar; **20.10:** From *Botany: Principles and Applications,* Roy H. Saigo and Barbara W. Saigo. Copyright © 1983 by Prentice-Hall, Inc. Reprinted by permission of the authors; **20.11:** Bill Witt.

Chapter 21

Opener: Carl W. Bollwinkel; **21.1:** David Dilcher; **21.2:** VU/H. Oscar; **21.3:** John D. Cunningham; **21.4:** D.R. Farrar; **21.5:** Carolina Biological Supply Company; **21.7:** Field Museum of Natural History; **21.8, 21.9:** Bill Witt; **21.10:** Carolina Biological Supply Company; **21.11:** From *Botany: Principles and Applications,* Roy H. Saigo and Barbara W. Saigo. Copyright © 1983 by Prentice-Hall, Inc. Reprinted by permission of the authors; **21.12:** Bill Witt; **21.13:** Carolina Biological Supply Company; **21.14:** Bill Witt; **21.16:** VU/D. Newman; **21.19:** VU/David M. Phillips; **21.21, 21.22, 21.23:** Carolina Biological Supply Company.

Chapter 22

Opener: General Biological Supply House, Inc., Chicago.

Chapter 23

Opener: VU/D. W. Fawcett

Chapter 24

Opener: Ward's Natural Science Establishment, Inc.

Chapter 25

Opener: Science Dept., Price Laboratory School, University of Northern Iowa.

Chapter 26

Opener: Merck & Co., Inc.; **26.5d,e:** VU/David M. Phillips.

Chapter 27

Opener: VU/Michael C. Webb; **27.3:** Don Findlay, Price Laboratory School, University of Northern Iowa.

Chapter 28

Opener: Chester F. Reuther, Director of Photographic Services, John Hopkins Hospital, Baltimore, MD; **28.7:** VU/David M. Phillips.

Chapter 29

Opener: VU/Max Hunn; **29.2:** VU/C. Zeiss; **29.6:** Science Dept., Price Laboratory School, University of Northern Iowa; **29.7:** VU/W. Ormerod; **29.9:** VU/Veronica Burmeister; **29.10:** VU/Triarch.

Chapter 30

Opener: U.S. Soil Conservation Service; **30.3:** John D. Cunningham.

Index

H

I

J

K

L